HYDRAULIC AND PNEUMATIC POWER AND CONTROL

HYDRAULIC AND PNEUMATIC POWER AND CONTROL

Design, Performance, Application

Edited by

FRANKLIN D. YEAPLE

***Fluid Power and Control Editor,* PRODUCT ENGINEERING**

McGRAW-HILL BOOK COMPANY

New York St. Louis San Francisco Düsseldorf Johannesburg
Kuala Lumpur London Mexico Montreal New Delhi
Panama Rio de Janeiro Singapore Sydney Toronto

HYDRAULIC AND PNEUMATIC POWER AND CONTROL

 Library of Congress Catalog Card Number 65–28832

ISBN 07-072257-9

567890 HDBP 765

PREFACE

For the first time you can work from practical step-by-step design procedures in every area of hydraulic and pneumatic design on machines and in processes. Included here are graphic shortcuts in solving problems of flow, high temperature and pressure, pressure loss, fluid hammer, cushioning, actuation, control, adjustable-speed drives, filtration, sealing, safety, and system design.

The material is exclusive and was sought, researched, and written to establish positive design procedures where none previously existed. *Product Engineering* editors participated heavily in developing the actual techniques. In some instances original engineering research was sponsored in industry by the editorial staff to settle controversy. Every major area of fluid power and control, including the new *fluidics* logic (pages 166–168), is covered.

There is no comparable book. The material is original in concept. Most equations, graphs, illustrations, and techniques will be found only here, and will not be available elsewhere.

Franklin D. Yeaple

CONTENTS

1

FLUID PROPERTIES

Oil, synthetics, air, and gases

a guide to choosing and using . . .

MODERN FLUIDS for power and control

Hydraulic power-and-control fluids are available in abundance—one for almost every need—but how do you find the right one? Even the fluid manufacturers admit their catalogs cannot keep up wth the labs. One answer—given here—summarizes the basics underlying all such fluids, and supplies charts of typical properties.

FRANK YEAPLE, associate editor

Today there are power-and-control fluids engineered for almost any purpose. But the fluid types are so numerous, their properties so dependent on how the fluids are used, that manufacturers' catalogs can't give all the answers. This is why making a correct choice takes a team: design engineer; fluids expert; and the ultimate user. But only you—the design engineer—can finally decide what performance is essential to proper equipment operation, and what compromises are acceptable.

To help you, here is a guide summarizing the hydraulic basics, pointing out typical characteristics of general fluid types, but avoiding extensive detail. It helps you make a first guess before calling in the fluids expert for final help.

Matching Fluid to Job

Stating the selection problem in consistent language comes first. In simplest cases this is done by expressing needed system performance in terms of measurable fluid properties. For example, system flow and response are dependent on viscosity, density, vapor pressure, air solubility and bulk modulus. Additional properties not directly related to performance, but sometimes essential for safety, convenience and long life, are: fire resistance (if necessary), compatibility, fluid stability (temperature, pressure, radiation, etc.), lubricity, leakage, dirt-purging ability and cost.

But it is not always possible to express performance in simple numbers. In almost any system that has a pump or relief valve, localized flow is complex and it is difficult to predict peak pressure loadings, maximum temperatures and other conditions peculiar to a particular installation. The answer, although not covered in this

ACKNOWLEDGMENT: The material in this article is based on papers and talks given at two recent fluid-power conventions: National Conference of Industrial Hydraulics (NCIH); and National Fluid Power Association (NFPA). We thank R. L. Leslie and W. W. LeRoy of Vickers Inc for their graphs and detailed data on fluid properties, and Thomas Smith, Celanese Corp, for his concise evaluation of fire-resistant fluids, including the table of compatibility.

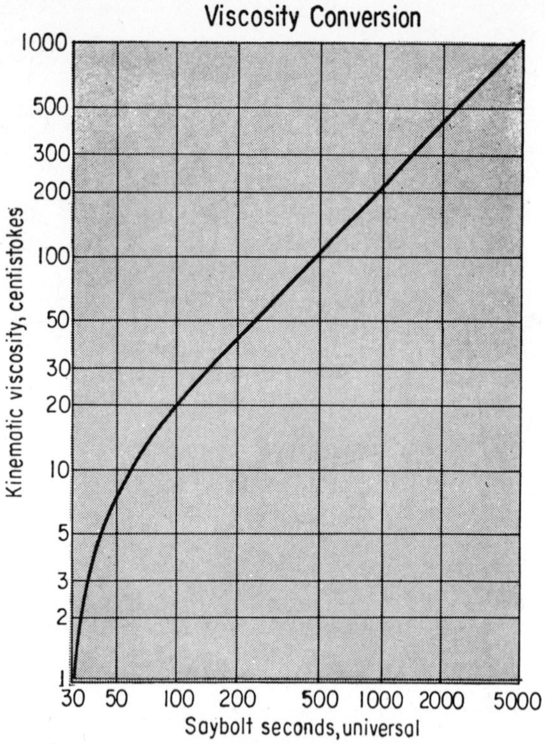

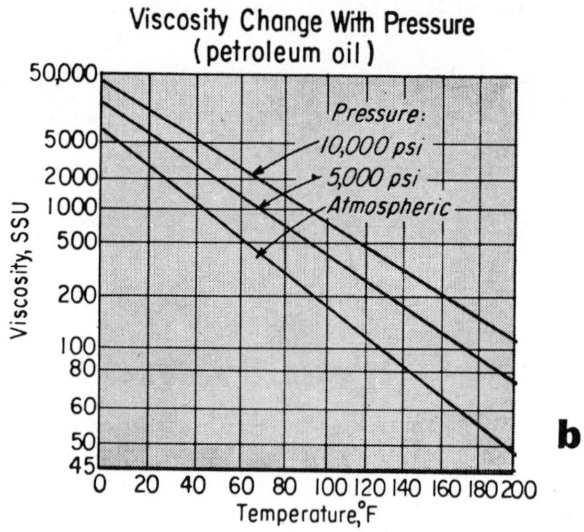

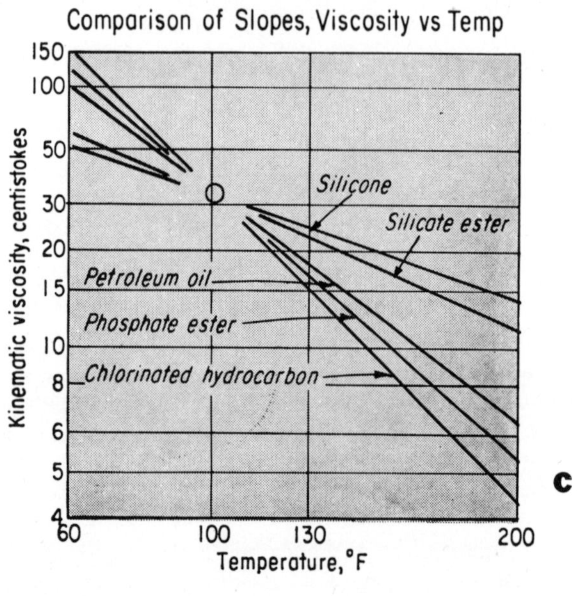

a VISCOSITY IN SSU is an accepted way to express kinematic viscosity of fluids for values of 100 SSU and above. Below that value, the relationship between true kinematic viscosity (centistokes) and SSU units is not linear on this scale.

b PRESSURE AFFECTS VISCOSITY of petroleum oils—particularly at elevated pressures (values are approximate). For most hydraulic fluids, viscosity will increase by 10 to 25% if pressure is increased from atmospheric to 1000 psi. Increase will be 30 to 70% from atmospheric to 3000 psi.

c TEMPERATURE AFFECTS VISCOSITY in widely different ways, even though the fluids selected for this example have identical viscosities at 100 F. None of the fluids are limited to the levels shown, however. Silicone fluids, for example, are available with viscosities ranging from less than 1 to over 1,000,000 centistokes.

article, is to test the fluid in a setup that duplicates or simulates expected field conditions.

The fluid finally chosen will most likely be one or a special combination of some of these common types: petroleum (naphthenic, aromatic, paraffinic, inhibited, uninhibited); water; glycol (glycerin is a glycol); phosphate or silicate ester (that is, an ester is formed by replacing hydrogen of an acid with a hydrocarbon radical); and various hydrocarbons (chemical compounds containing only carbon and hydrogen). Silicone fluids are not common in ordinary industrial hydraulic systems, but find many applications in special circuits and devices, such as fluid springs, damping pistons, fluid couplings, and some power transmission units. These fluids have exceptional thermal stability (see *"Liquid Spring"*).

Many variations of the above fluids are possible—chlorinated hydrocarbons, for example. However, knowledge of chemistry is not essential, because the manufacturer will help select the exact type and grade to meet the need.

These Properties Affect Performance

Viscosity and density together determine pressure loss for a given flow (see *"Pressure-drop Calculations"*). Viscosity is frequently expressed in SSU (Saybolt Universal Seconds) units, a measure of the time it takes for 60 cc of oil to flow through an orifice 0.176 cm dia and 1.225 cm long. In the accompanying graphs, SSU is related to kinematic viscosity, in centistokes. Values of viscosity as well as specific weight are given for several fluids. Kinematic viscosity in centistokes equals absolute viscosity, in centipoises, divided by mass density.

The slope as well as position of each viscosity-vs-tem-

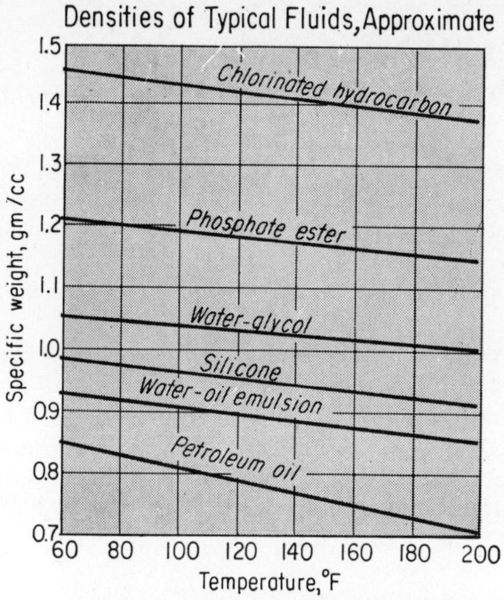

a

a DENSITIES VARY not only with temperature (shown) but also with differences between fluids of the same type (not depicted—only one representative example of each fluid type is given).

b VAPOR PRESSURES of water-based fluids are relatively high, magnifying the tendency to cause cavitation at pump inlets, particularly at high temperatures.

c & d AIR SOLUBILITY at any given pressure is the ratio of dissolved air volume (converted to 32 F and atmospheric pressure) and volume of fluid. Air released is actual volume of air that would come out of solution if petroleum oil, saturated with air (at atmospheric pressure, for two different temperatures) was then subjected to reduced pressures. All data are for equilibrium conditions.

e AVERAGE BULK MODULUS of each fluid during a given pressure increase is the pressure change in psi, divided by the percent compression expressed as a decimal. The average modulus for a typical petroleum oil from zero to 20,000 psi is 20,000/0.061—about 328,000 psi. For the silicone fluid shown (100 centistokes at 77 F) the average modulus is 20,000/0.11—about 182,000 psi.

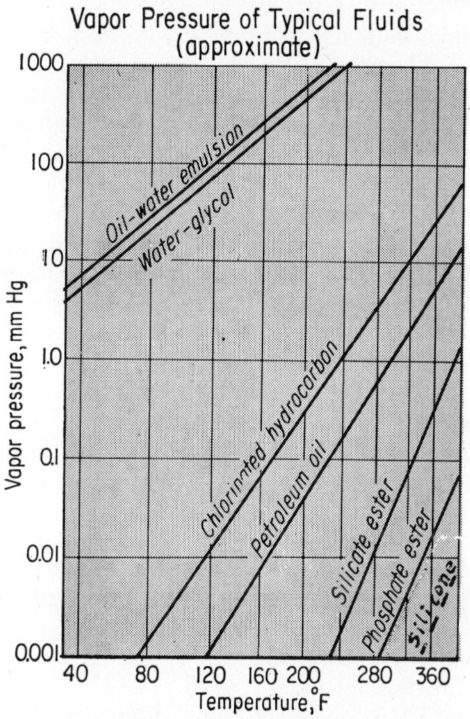

b

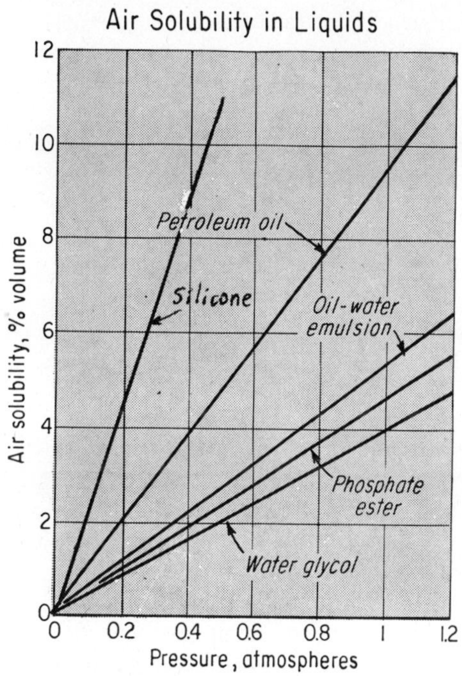

c

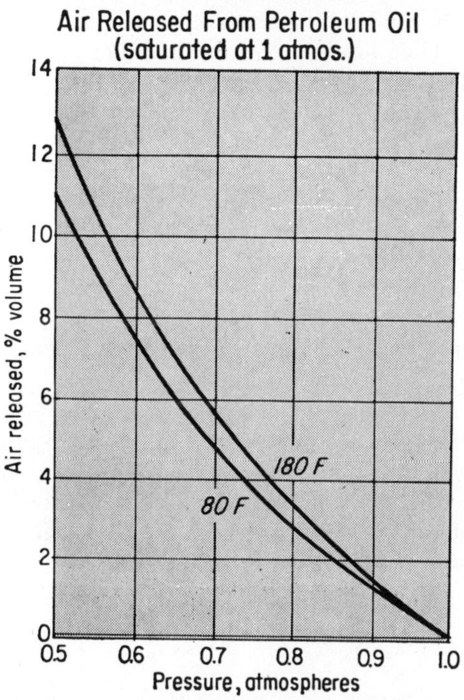

d

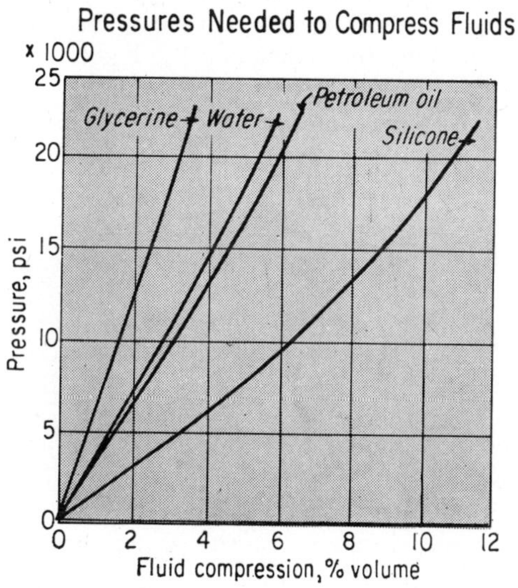

e

perature curve is important—a flat slope indicates a large allowable temperature change before exceeding minimum and maximum values of viscosity. The term Viscosity Index (VI) was invented to compare those slopes. It is based on two standard oils that have been given arbitrary values of zero and 100 VI, respectively, not plotted here. The zero VI is for the steeper slope (a naphthenic-petroleum oil, sensitive to temperature changes); the 100 VI shows less sensitivity to temperature (a paraffinic-base oil). Today's fluids range in VI from negative to over 200.

Although viscosity is usually considered constant for a given temperature and pressure regardless of the flow, it is sometimes greatly influenced by the rate of shear. In fluids that have large-polymer molecules, rapid shear rates temporarily lower the viscosity—suggesting that the very-long-polymer molecules align in direction of flow. Permanent damage can also occur if there is high flow through small clearances. This tends to break down large molecules into smaller ones.

Density not only influences flow rate and pressure drop, but also affects the natural frequency of a system. This frequency is proportional to the square root of the moving mass.

Vapor pressure of the fluid, if too high, will cause gas pockets to form in areas of low pressure—such as in the inlet of a pump (see "*Cavitation in Pumps*"). The graph shows that water-based fluids have fairly high vapor pressure . . . the system temperature must therefore be kept low to prevent vaporization and loss of water. Non-water-based hydraulic fluids, including silicones, have reasonably low vapor pressures.

Air solubility in many liquids is directly proportional to absolute pressure of the system. High solubility is undesirable because the air (or other gas) dissolved at reservoir conditions may come out of solution at a lower pressure (pump inlet, for example). Pump cavitation can result. If the reservoir is pressurized, then the problem is even greater because more air can be dissolved, coming out of solution later.

Bulk modulus—the reciprocal of compressibility—has units of psi. Bulk modulus is similar in concept to a mechanical spring rate, and measures the pressure change needed to cause a given percent volume change. The higher this modulus, the stiffer the system and the higher the resonant frequency. A fast-responding servosystem requires a fluid with a high bulk modulus.

Fire Resistance

Those fluids that meet stringent safety requirements usually sacrifice one or more performance qualities. Best known of the fire-resistant fluids are water, water-glycol, water-oil emulsions, phosphate esters, and chlorinated hydrocarbons. Each type falls into one of two broad categories: water-based or non-water-based.

Water-based types rely mainly on their water for fire resistance. Care must be taken to prevent loss of water from the mixture. Such loss in the glycol type increases the viscosity; but for the oil-emulsion type, it decreases the viscosity. In either case, change in viscosity (if noticed in time) can warn of a dangerous condition.

Water-based fluids are not usually recommended for high bearing loads found in ballbearings and axial-piston pumps, or for fluid pressures over 1000 psi. Also, rusting may occur if water-based fluids only partly fill a system—the water vapor increases oxidation rate on the exposed areas.

The non-water-based types such as phosphate esters and chlorinated hydrocarbons are not volatile, and can stand extremely high pressures and temperatures to 250 F. However, they are heavier than water-based fluids (and petroleum), and have the disadvantage of high viscosity at low temperatures—auxiliary heating is often required for outdoor startups.

Compatibility: Fluid and System

Hydraulic fluids must not give trouble with seals, filters, diaphragms and other materials. The accompanying compatibility table summarizes the conversion problems faced when petroleum oils are replaced with fire-resistant types. In no instance should a fire-resistant fluid be dumped into a system after merely draining the petroleum—although this has been "gotten away with" occasionally.

Before changing over to a fire-resistant fluid, be sure to completely clean the system. Residual amounts of petroleum and sludge can cause emulsion fluids to "break," and water-based materials to emulsify. Replace conventional 100-mesh strainers with 50- or 60-mesh units rated at 4 times' pump flow, instead of the normal 1½ times' pump flow. Limit pump-inlet vacuum to about 5 psi. Do not use fuller's earth or activated clay filters when additives are in the fluid . . . although they work well for chemically pure phosphate esters. Change all seals and packings where possible to materials that are compatible. At least change the dynamic seals, saving the static-seal change-over until a regular maintenance period. Don't grease O-rings with a noncompatible substance. Do not test the system with petroleum oil—use the intended fire-resistant fluid.

Ordinary joint compounds will be unsatisfactory. Stick to products recommended by the fluid manufacturer. One compound acceptable for many fire-resistant fluids, called "Loctite," is made by American Sealants Co. It can be applied in drops to the made-up joint; capillary attraction draws it into the thread, where it "sets" to become a vibrationproof seal. Another acceptable joint sealer is Teflon-type tape that can be wrapped around the male thread. Upon assembly, the threads cut the tape, forming individual gaskets for each thread. These tapes are available from Minnesota Mining and Manufacturing Co, Fluoro-Plastics Inc, and Friesland Plastics Co.

Elastomer-type materials suitable for various fluids are listed in the compatibility table:

Butyl is the best all-around elastomer for phosphate esters, but shows increased swell with chlorinated esters and other chlorinated fluids—care should be taken in close-tolerance dynamic applications. It is unsuitable where it may be exposed to petroleum oil in any quantity. Usable temperature range is −65 to 225 F.

Silicone material is recommended for chlorinated fluids. It has good temperature resistance, but is poorer than

Materials Compatible with Typical Fluids

	PETROLEUM OIL	WATER-BASE		NON-WATER-BASE		
		Oil-water Emulsion	Water-glycol Mixture	Chlorinated Fluids	Mixtures of Both	Phosphate Esters
Seals and Packings	Neoprene, Buna N	Neoprene, Buna N, (no cork)		Silicone, Viton, Vyram, Teflon, FBA Butyl (with caution)		Butyl, Viton, Vyram, Silicone, Teflon
Paints	Conventional		Consult Fluid Mfr.	Air-cured epoxy per mfr. recommendation		
Pipe-sealing Compound	Conventional		Special "dopes" such as "Loctite"			
Suction strainers	100-mesh wire, 1½ times pump capacity	40-mesh wire, 4 times pump capacity	50 mesh wire, 4 times pump capacity			
Filters	Cellulose fiber, 200–300 mesh wire, knife edge, or plate	Glass fiber, 200–300 mesh wire, knife edge or plate	Cellulose fiber, 200–300 mesh wire, knife edge, or plate	Cellulose fiber, 200–300 mesh wire, knife edge or plate type (fuller's earth or micronic type may be used on non-additive fluids.)		
Metal parts	Conventional		Avoid galv., zinc and cad. plating	Conventional		

butyl in tensile strength, tear and abrasion resistance—so needs care for dynamic sealing applications. Available temperatures range from −100 to 500 F.

Vyram is a trade name for an acrylic elastomer. It is suitable for all non-water fluids including petroleum, permitting fluid change without seal change. It is not suitable where exposed to water in any quantity. It can be used from −40 to 275 F.

Viton A is a fluorocarbon elastomer that seems to have considerable potential. It is suitable for all types of fire-resistant fluids as well as petroleum.

Teflon is well known for its chemical and thermal stability; however, its rigidity and tendency to cold-flow limit its application for dynamic seals.

Other Requirements

Fluid stability depends on the fluid's tolerance to temperature, pressure, atmospheric oxygen, water, radiation and catalysts. Petroleum oils have good oxidation resistance, except at high temperature—and oxidation is speeded up by the presence of certain metals. In long-chain organic compounds, high temperatures will degrade the large molecules into smaller ones, reducing viscosity. There are fluids already designed to be stable in almost any atmosphere, but usually a sacrifice must be made in other properties.

Lubricity of the phosphate esters is better than that for any other fire-resistant fluid, which helps justify their higher price. Water-glycol mixtures, unlike the fixed-composition esters, have varying proportions and ingredients, depending on the manufacturer. Therefore, the lubricity and other properties will vary. All types of fire-resistant fluids, however, will give satisfactory performance if applied correctly.

Maintenance and other costs cannot be predicted by rules, because high costs in one area—such as initial fluid cost—may reduce costs drastically in another. However, here are two important considerations. Leakage of fluid past seals is influenced by factors such as surface tension. Fluids which hold dirt particles in suspension are more troublesome than those which let the particles drop harmlessly to the bottom of the tank. And, of course, the compatibility of a fluid with the system and with other fluids it may contact is also a guide to costs.

What is Viscosity?

Only absolute viscosity (centipose; lb-sec/ft²; etc) has physical units that can be visualized. Kinematic viscosity is more or less a defined term, being absolute viscosity divided by mass density.

A simple way to visualize absolute viscosity—let's select lb-sec/ft² as the example—is to reproduce the definition pictorially. So, imagine the sketched cube to be 1 ft on a side. Apply a sufficient force at the upper surface of the fluid to move that upper surface at a velocity of 1 ft/sec relative to the fixed lower surface.

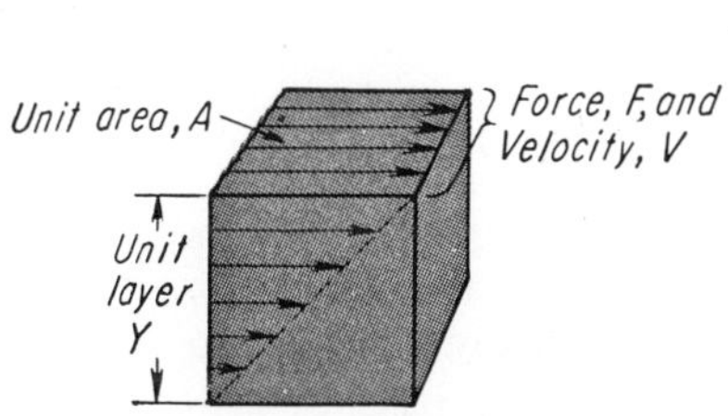

Absolute viscosity is a measure of the value of that shearing force, expressed as follows:

$$\mu = \frac{(F/A,\ \text{lb/ft}^2)}{(V/Y,\ \text{ft/sec per ft})} = FY/VA,\ \text{lb-sec/ft}^2$$

For hydraulic systems, drives, dashpots

What's hot in fluids?

It's not the heat—it's the versatility!
400 to 850-F fluids are available, but you've got to sacrifice one characteristic to gain another

H. M. SCHIEFER, Development Supervisor, Fluids and Lubricants Section, Dow Corning

Blowtorch tests a 6-gpm, 3000-psi gear pump made by Pesco Products, Div of Borg-Warner Corp. Fluid is MIL-L-7808, a jet engine lube.

NO one in the world can name every synthetic high-temperature fluid on the shelves today. If you compiled such a list it would be obsolete tomorrow, because new fluids are introduced every day.

How then can you make a decision? Suppose your design demands a hydraulic cylinder fluid for 4000-psi service, 500-F average temperature, 650-F hotspots, two hours between peak temperatures, total minimum life 100 hours, plus assorted special requirements. Must you phone every chemical and oil company in America to pick a fluid?

The answer is no, because the new fluids are chiefly variants of the old. Only the chemistry changes, not the critical performance limitations—except, perhaps, to improve.

The silicones (hundreds of them) still have a top temperature of around 700 F, the chlorinated hydrocarbons still degrade under certain severe conditions, and the polyphenyl ethers still withstand the highest temperatures of the organics. All this is explained in the accompanying charts and tables, based on 9 of the chief types of fluids for high temperature.

text continues, next page

Short history of fluids. The history of hydraulic fluids is like a pile of toy blocks. The highest ones are the exotic synthetic liquids, and are the most difficult to balance. But all the bottom ones are still being used.

Water, castor oil, and petroleum were the first, and they are still around. Petroleum, super-refined, handles some of the hot jobs, being adequate in special sealed systems up to 400 or 500 F.

But new aircraft and missiles demand even higher temperatures. When it became apparent in 1945 that only synthetics could withstand the higher temperatures projected for future designs, the diesters (see Fluids Primer p. 9) first were tried out. Others followed. From 1945 until today, the approximate order of appearance of the high-temperature fluids has been: diesters, polyglycols, halogenated hydrocarbons, silicones, phosphates, silicates, neopentyl glycol esters, super-refined mineral oils, polyphenyl ethers, polyaromatics. And the work goes on.

Table I is an abbreviated list of the current hydraulic fluids for the range from 300 to 900 F. Arbitrary code letters A, B, C, and D are for normal-temperature fluids. Codes A-1 to L are for 9 high-temperature groups, and these are covered in more detail here.

FLUIDS PRIMER

Chemistry and hydraulics are bedfellows, not necessarily by choice. It's just because there are so many kinds of hydraulic fluids that it takes a chemist to give them all names. Table 2 lists 9. Table 15 tells where to get them.

Only a chemist can love a hexa (2-ethyl butoxy) disiloxane, and even he is unable to describe it—except to another chemist. But we cannot proceed without some kind of explanation, at least for the nine typical fluids in this article. Let's give it a try.

First, it is nearly impossible to list every chemical structure because the chemical names can be stated in many ways. For instance, a halogenated hydrocarbon, a chlorofluorocarbon, and a chlorotrifluoroethylene polymer might be the same fluid depending on how specific the given structure is.

But, there are certain terms (radicals, endings, prefixes, etc) that may crop up in any of the nine fluids. Let's examine a few of the common ones.

Mono = one as in monoacid. *Di* = two as in diesters and disiloxanes. *Poly* = repeating units such as the repeating silicon oxygen units in polysiloxanes. The disiloxane is the lowest member of a silicone in a series, whereas all the higher molecular weights are polysiloxanes. The silicates also come as disiloxanes or polysiloxanes, but these are not their common form.

Alkyl (also aliphatic, paraffin or alkane) = straight chain and branched chain hydrocarbons where all the available bonds are taken up by hydrogen (saturated). *Alkene* (alkylene, ie ethylene, propylene) = hydrocarbons with some hydrogens missing (unsaturated).

Alkoxy = alkyl groups with an oxygen at the end. *Naphthenes* = five to six carbon ring structure hydrocarbons which are saturated with hydrogen. *Aryl* (aromatic, phenyl, benzene) = six carbon unsaturated ring structure hydrocarbons. A silphenylene contains the aromatic group between the silicon atoms.

Fluoro, chloro, and *bromo* = fluorine, chlorine and bromine, which are also called halogens. These can replace hydrogen in the hydrocarbon alkyl, alkene, alkoxy, naphthene, and aryl structures described above.

Hydroxy = oxygen hydrogen as in alcohols, glycols, and water. *Oxy* = only oxygen as in oxyalkylenes. *MLO* = Materials Laboratory Oil designation of the Air Force. It is a prefix to an experimental oil developed by the Air Force or their contractors. *Organo metallic* = a class of organic materials which also contain a metal atom such as silicon, aluminum, tin, and germanium.

Esters = hydrocarbon types of oils which contain what is termed an ester or carbonyl linkage (–COO–) in the molecule. More specifically, if two of these occur in one molecule, they are referred to as diesters (most common ones). These materials are prepared from an alcohol (–COH) and a hydrocarbon acid (–COOH). Common acids are sebacic (ten carbon atoms), azelaic (nine carbon atoms) and adipic (six carbon atoms). Common alcohols are 2-ethyl hexanol (branched C_8), C_8, C_9, and C_{10} alcohols. The esters are named after both the alcohol and acid that they contain (ie, di (2-ethyl hexyl) sebacate). If the alcohol contains fluorine, the final product is called a fluoroester.

Neopentyl-polyol esters are very similar to diesters in having the same groups in the molecule. However the alcohol is specific and has a unique five-carbon structure (one carbon atom contains four others around it). This structure adds to the thermal stability of the molecule. The alcohols possessing this unique structure are neopentyl glycol, trimethylolethane, trimethylolpropane, and pentaerythritol. The esters are usually named after the alcohol portion. Organic acids used for these esters are generally of the mono-acid type containing six to ten carbon atoms.

Silicate esters contain a silicon to oxygen to carbon linkage. Hydrocarbon alcohols (similar to ones in diesters) are used to produce these oils. If the oil contains only one silicon atom, it is named as a tetra alkyl silicate (silicate implying the silicon oxygen linkage to the alkyl group) or tetra alkoxy silane (here the alkoxy group implies a carbon oxygen linkage to the silicon). Oils containing more than one silicon atom are called alkoxy siloxanes (implying also a silicon to oxygen linkage as occurs in the silicones).

But don't forget the variants *below* the top limitations. Any given fluid type can be tailored to almost any selected characteristic as long as it is safely within the ultimate boundaries of performance. Silicones, for example, are readily available in viscosities from 1 to 1,000,000 centistokes—figuratively from water to tar. But top temperature of the most stable is 700 F, and minimum temperature of one of the thinnest is −150 F.

And degradation with time and temperature is often a function of refinement in chemistry and manufacture. So be careful which source you call upon.

Do it yourself. A selection procedure is offered here that starts with the problem (Selection Table, next page), not with a hypothetical list of fluid properties. You immediately eliminate all but the applicable fluids in one quick pass through the table, and then go on to the latest conventional performance tables and explanations following. And of course let the fluid manufacturer make the final recommendation: Table 1 is approximate.

WHAT'S YOUR PROBLEM?

How many separate fluid-selection problems can there be? Only a handful, but they can be beauties. First, what do we mean by high temperature?

Let's define high temperature as anything from 400 to 850 F. Above that, extreme temperature might be a more apt description (*PE*—Dec 11 '61). Moderate temperature goes from 250 to 400 F. Normal temperature is from −40 to 250 F. Low temperature starts at −40 and drops down to cryogenic temperature at −200 F.

No single fluid covers the whole range, although the silicones come as close as any. In fact, no fluid meets all the performance requirements in even a single range of temperature. The unvarnished truth is that you must sacrifice important characteristics to achieve essential characteristics. And remember that any special fluid is expensive, at least compared with long-run refinery products.

Normal temperatures are no problem. Mineral oils are usually selected for temperatures to 300 F. Water-in-mineral oil emulsions are fire-resistant and suitable to 200 F. Water-glycols, also fire-resistant, are useful to 200 F. Alkyl phosphates can withstand 225 F, Between 250 and 400 F any of the synthetics listed under high-temperature fluids are applicable.

Super-refined oils, synthetic hydrocarbons, and alkyl silanes also are being developed for temperatures to

1 Preliminary selection

°F	Performance Requirements (see code in table 2 below): Stability (Closed), Bulk oil: Poor to fair	Bulk oil: Good to exc	Hot spots: Poor to fair	Hot spots: Good to exc	Resistance to oxidation: Poor to fair	Resistance to oxidation: Good to exc	Volatility or vapor-pressure: High	Volatility or vapor-pressure: Low	Lubricity: Poor to fair	Lubricity: Good to exc	Viscosity (Starting @ 100cs @ 77°F): Below 1.5cs	Viscosity: Above 1.5cs	Bulk modulus (0-4000 psi): Below 80,000 psi	Bulk modulus: Above 80,000 psi	AIT °F: Above	AIT °F: Below
400	F (see code, below)	A-1, E G, H I, J, K L		A-1, E F, G H, I, J, K, L	A-1, F G, E	F, I J, K L	A-1, I J	E, F, G H, K L	K	A-1, E F, G, H I, J, K L	A-1, I J, L	E, F G, H K	K	A-1, E F, G, H I, J, K L		A-1, E F, G, H I, J, K L
450	E, F G	A-1 H, I J, K L	F	A-1, E G, H, I J, K, L	A-1, E F, G H	I, J K, L	A-1, F G, I J	E, H K, L	H, J K	A-1, E F, G I, K L	A-1, F G. I, J L	E, H K	K	A-1, E F, G, H I, J, K L		A-1, E F, G, H I, J, K L
500	E, F G, H	A-1, I J, K L	E, F G	A-1, H I, J, K L	A-1, E F, G H, I	J, K L	A-1, E F, G, I J	H, K L	H, J, K, L	A-1, E F, G I, K	A-1, E, F, G, I J, L	H K	H, K	A-1, E F, G, I J, K		A-1, E F, G, H I, J, K L
550	E, F G, H, I	A-1, J K, L	E, F G	A-1, H I, J, K L	A-1, E F, G, H I, K	J, L	A-1, E F, G, I J	H, K L	E, F G, H J, K L	A-1, I	A-1, E F, G H, I J, L	K	A-1, E F, G H, K	I, J, L		A-1, E F, G, H I, J, K L
600	A-1, E F, G H, I, J	K, L	E, F G, H I	A-1, J K, L	A-1, E F, G H, I J, K	L	A-1, E F, G H, I J, L	K, L	E, F G, H J, K L	A-1, I	A-1, E F, G H, I J, L	K	A-1, E F, G H, I K	J, L	A-1, E F	G, H, I J, K, L
650	A-1, E F, G H, I, J K	L	E, F G, H I, J	A-1, K L	A-1, E F, G H, I J, K	L	A-1, E F, G H, I J, L	K	E, F G, H I, J K, L	A-1	A-1, E F, G H, I, J K, L	K	A-1, E F, G H, I J, K	L	A-1, E F, G	H, I J, K, L
700	A-1, E F, G, H I, J, K	L	E, F G, H I, J	A-1, K L	A-1, E F, G H, I, J K, L		A-1, E F, G H, I J, L	K	A-1, E F, G H, I J, K L		A-1, E F, G H, I, J K, L	K	A-1, E F, G H, I J, K	L	A-1, E F, G H	I, J K, L
750	A-1, E F, G, H I, J, K	L	A-1, E F, G H, I J, K	L	A-1, E F, G H, I, J K, L		A-1, E F, G H, I, J K, L		A-1, E F, G H, I J, K, L		A-1, E F, G H, I, J K, L	K	A-1, E F, G H, I, J K, L		A-1, E F, G H	I, J K, L
800	A-1, E F, G, H I, J, K L		A-1, E F, G H, I J, K	L	A-1, E F, G H, I, J K, L		A-1, E F, G H, I, J K, L		A-1, E F, G H, I, J K, L		A-1, E F, G H, I, J K, L		A-1, E F, G H, I, J K, L		A-1, E F, G H	I, J K, L

2 Operating temperature

Code (for article)	Chemical type	Approximate upper limit, °F: Long duration (500-1000 hrs)	Moderate duration (10-100 hrs)	Hot spot short duration <1 hr
A, B, C, D	Mineral oils, emulsions, water-glycol, alkyl phosphates	see text	see text	see text
A-1	Super refined mineral oils, synthetic hydrocarbons, and silanes	*300-450	550	700
E	Polyglycols	325-350	375-400	450-500
F	Dibasic acid esters	350	400	425
G	Neopentyl-polyol esters	375	425	450
H	Silicate esters	375-400	475-525	575
I	Aromatic phosphate esters	400	525	575
J	Halogenated aromatics	450	575	600
K	Silicones	‡425-550	550-625	700
L	Polyphenyl ethers	600	800	900

*Depending on oxidation resistance. ‡Range due to many viscosities available

3 Special qualities

Fluid type	Outstanding characteristics	Main limitations
A-1	Very good lubricity, good thermal stability and additive response	Flammability, poor viscosity vs temperature, and oxidation stability
E	Good lubricity and additive response, low sludging	Flammability, limited stability
F & G	Very good lubricity and good additive response	Flammability, corrosiveness, limited stability (acid test and sludging)
H	Good viscosity vs temperature properties, and low sludging	Limited hydrolytic and oxidation stability, limited lubricity
I	Good flame resistance and lubricity	Poor viscosity vs temperature, and poor compatibility with elastomers and paints
J	Good flame resistance	Very poor viscosity vs temperature, and corrosiveness
K	Flat viscosity vs temperature slope, good flow characteristics at high viscosity levels, low volatility, and good overall stability	Limited lubricity for steel against steel; low oxygen tolerance at high temperature
L	Stability	Limited lubricity, and poor viscosity vs temperature

4 Viscosity vs temperature (ASTM slopes, 210 to 100 F)

Code	Fluid type	Typical commercial brands	ASTM slope
A-1	Super refined mineral oils	Penn State MLO 7557 or MLO 60-294 Humble Oil 3160	0.76 0.78
E	Polyglycols	Dow Ambiflo C-44 UCC Ucon LB-170X	0.55 0.64
F	Dibasic acid esters	Rohm & Haas Plexol 201 (di 2-ethyl hexyl sebacate)	0.70
G	Neopentyl-polyol esters	Celanese 704	0.70
H	Silicate esters	Monsanto tetra 2-ethyl hexyl silicate UCC hexa (2-ethyl butoxy) disiloxane	0.72 0.55
I	Aromatic phosphate esters	Monsanto tri-cresyl phosphate & Celanese Cellulube 220	0.89
J	Halogenated aromatics	Monsanto Aroclor 1248	1.15
K	Silicones	Dow Corning 200 Fluid Dow Corning 510 Fluid Dow Corning 560 Fluid	0.24 0.27 0.39
L	Polyphenyl ethers	Dow ET 492	0.86

500 F. Viscosities are low at the higher temperatures, and oxidation will deteriorate the oil in open systems. However, closed systems and proper bearing and seal design help solve both problems.

High temperatures

Here, in the region from 400 to 850 F, the fluids are asked to withstand as much as they can bear. It is easy to say *don't* design systems for over 400 F. But have you any choice? Assuming not, take this as your guide:

General requirements. A fluid for a high-temperature hydraulic device or system of any sort usually needs a high level of thermal stability, a low level of flammability, reasonable noncorrosiveness, compatibility with the many materials encountered in the flow path, low vapor pressure so that the liquid doesn't flash into vapor at every pump suction, and a low order of toxicity.

Even normal-temperature systems benefit from those stipulations, but in high-temperature systems the problem is magnified. Mild fluids become corrosive when hot, stable fluids break down into useless new compounds, and pumps lose suction when the liquid disappears into thin air. Additional requirements are:

Hydraulic systems. Viscosity ought to exceed 1.0 centistoke (eg, water at 0°C) yet not be more than 5000 cs (maple syrup at room ambient) at any operating temperature. Lubricity should be at least equal to that of an oil-water emulsion, even though the fluid isn't used in an actual bearing. It still has to be pumped, and the piston needs some lubrication.

Sludging and varnishing are taboo, particularly in servosystems where clearances are small. Bulk modulus ought to be high, for fast response and good stiffness. Viscosity must not degrade even under shearing action. Hydrolytic stability (unaffected by water) must be good, and foaming must not be excessive. The fluid should be a good conductor of heat, and have a low coefficient of thermal expansion.

Fluid couplings and dashpots. Here, the fluid gets hot because of the shearing action. Viscosity ought to stay as constant as possible with temperature, and be stable. Viscosity level depends on the application, and all values should be readily available from the fluid manufacturer. No deterioration is permitted, particularly in precision dashpots where timing is important.

Liquid springs. High compressibility (low bulk modulus) is the prime requirement. Viscosity should be as flat with temperature as possible.

Thermal control devices. A high thermal coefficient of expansion, linear

5 Viscosity-temperature curves

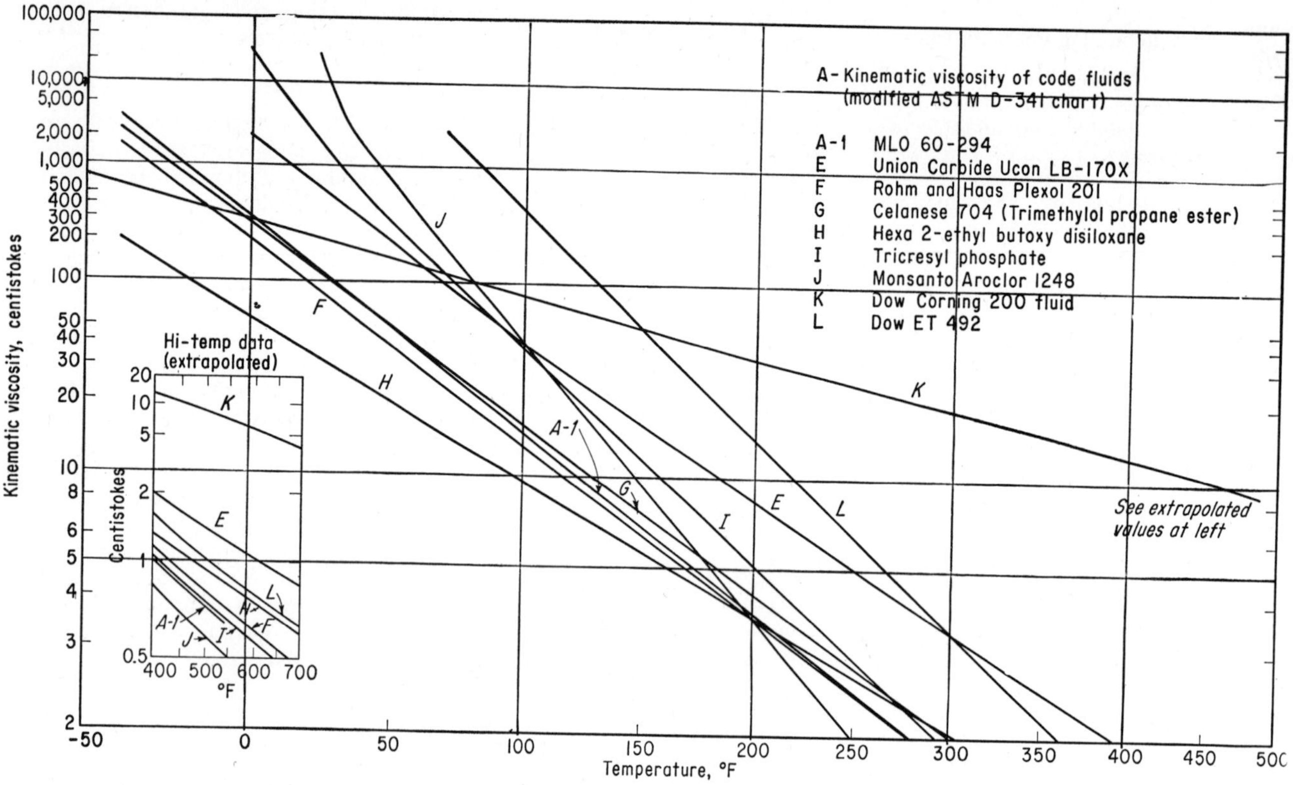

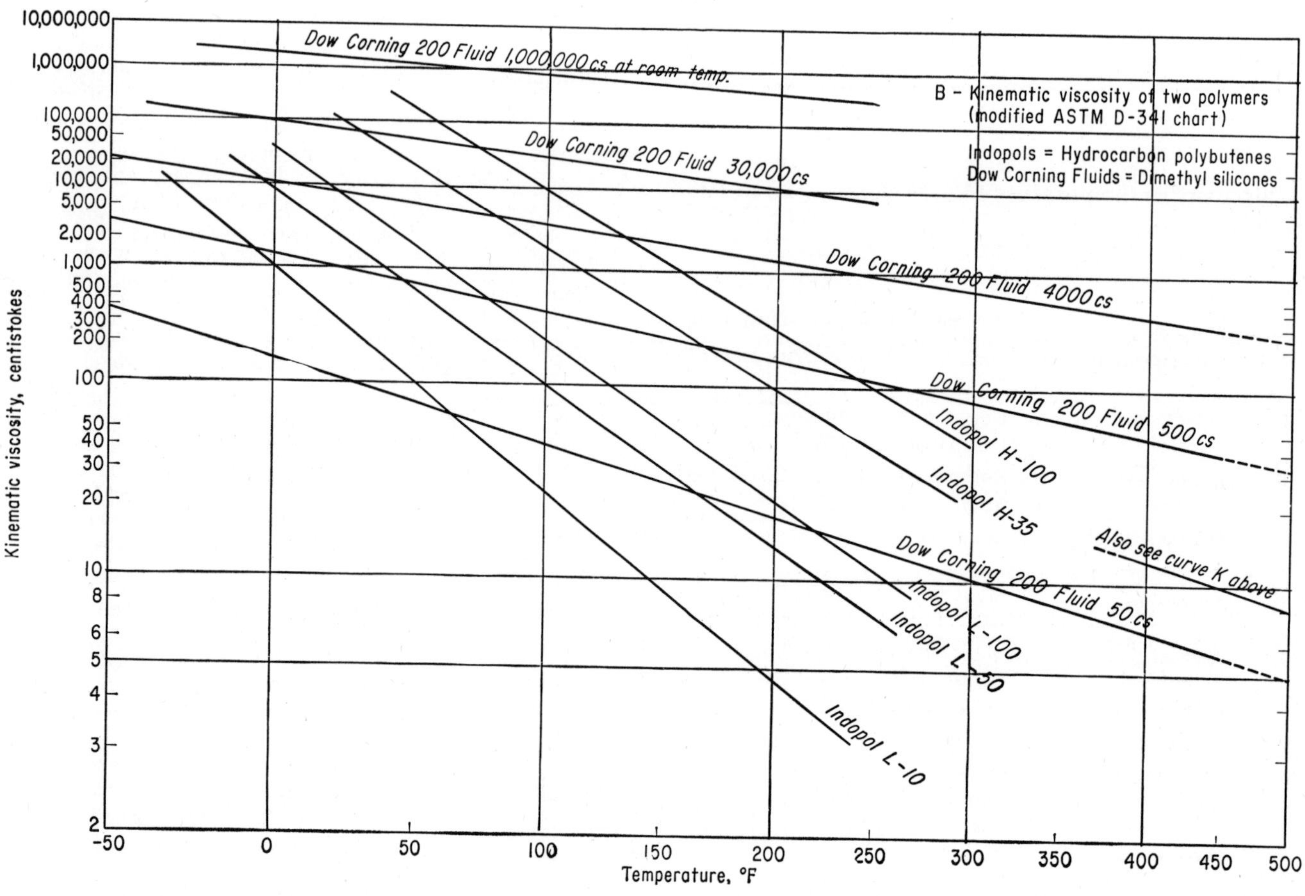

6 Deterioration temperature

Code	Incipient thermal degradation temperature, °F
A-1	600 — 700
E	500 — 550
F	525
G	575
H	550 — 600
I	600 — 700
J	600 — 650
K	625 — 800
L	750 — 850

7 Lubricity

Code	Steel against steel	Steel against bronze	Steel against rubber and plastics
A-1	Excellent	Fair	Fair-good
E	Good	—	Good
F	Excellent	Good	Poor to fair
G	Excellent	Good	Poor to fair
H	Fair	Fair to good	Fair
I	Good	—	Poor
J	Fair to good	—	—
K	Poor to good	Good-excellent	Excellent
L	Fair to good	—	—

8 Oxidation and varnishing

Code	Resistance to oxidation	Tendency to form solids or varnish
A-1	Fair to good	Large
E	Fair to good	Small to medium
F	Good	Large
G	Good	Large
H	Fair to good	Small to medium
I	Good	Medium
J	Excellent to 500 F	Small
K	Excellent to 400 F Good above 400 F	Small Small to medium*
L	Excellent to 500 F	Small

*Excessive oxidation can produce a gel

9 Compressibility

Code	Average bulk modulus (0 to 4000 psi) at 77 F, psi
A-1	250,000
E	270,000
F	260,000
G	260,000
H	240,000
I	320,000
J	350,000
K	150,000 to 240,000
L	430,000

10 Hydrolytic stability

Code	Comparative rating
A-1	Excellent
E	Good
F	Fair
G	Fair
H	Poor to fair
I	Fair to good
J	Excellent
K	Excellent
L	Excellent

11 Volatility

Code	Comparative rating
A-1	High
E	Medium
F	Medium-high
G	Medium-high
H	Medium
I	Medium
J	Medium
K	Low
L	Low to medium

12 Specific gravity

Code	Specific gravity at 77 F
A-1	0.87
E	1.00 — 1.20
F	0.85
G	0.85
H	0.87 — 0.92
I	1.10 — 1.25
J	1.20 — 1.50
K	0.97 — 1.30
L	1.18

with temperature, is required. Viscosity and vapor pressure should be low. All these characteristics must remain the same even after continued temperature cycling.

FLUID PROPERTIES

The tables here define and list advantages and limitations, but you must temper the values with judgment. Remember:

- Large variations exist within each class.
- Mechanical design of the system, particularly relating to shear-inducing flow restrictions or elements, greatly affects fluid life. High-viscosity complex polymers tend to degrade quickly if exposed to a combination of maximum temperature and extreme shear, such as in pumping or throttling.
- A continuous temperature slightly below the incipient thermal degradation level (Table 6) is more damaging than occasional hotspot temperatures above the degradation level.
- Transient or hotspot temperatures of 25 to 100 F over the incipient value will degrade only that portion of the fluid in contact. If the byproducts are not objectionable, useful fluid life might not be reduced appreciably.
- For long duration operation, the oxidation, thermal stability, and viscosity stability will determine the limiting temperature.
- Viscosities of certain fluids drop severely with an increase in temperature. For instance, aromatic phosphates usually will be of too low viscosity if operated at 500 F, even though they might withstand the temperature.
- Paints and elastomers might limit use of a fluid otherwise suitable for high temperatures. Aromatic phosphates and esters, for example, damage paints and elastomers at 500 F.

Twelve properties and characteristics define the performance capabilities of high-temperature fluids: viscosity, viscosity stability, thermal stability, lubricity, noncorrosiveness, oxidation stability, compatibility with other materials, compressibility or stiffness (bulk modulus), volatility, hydrolytic stability, coefficient of expansion, and specific gravity.

Viscosity vs temperature. This characteristic varies with each chemical class and is best compared by plotting slopes on ASTM paper (Table 5A).

Halogenated hydrocarbons and aryl phosphates are poorest (steepest slope), silicones are best. Only silicones and some synthetic hydrocarbons are available in high-viscosity grades (Table 5B). Viscosity is very important in fluid couplings and damping units, but of minor importance in thermal control devices.

13 Flash points

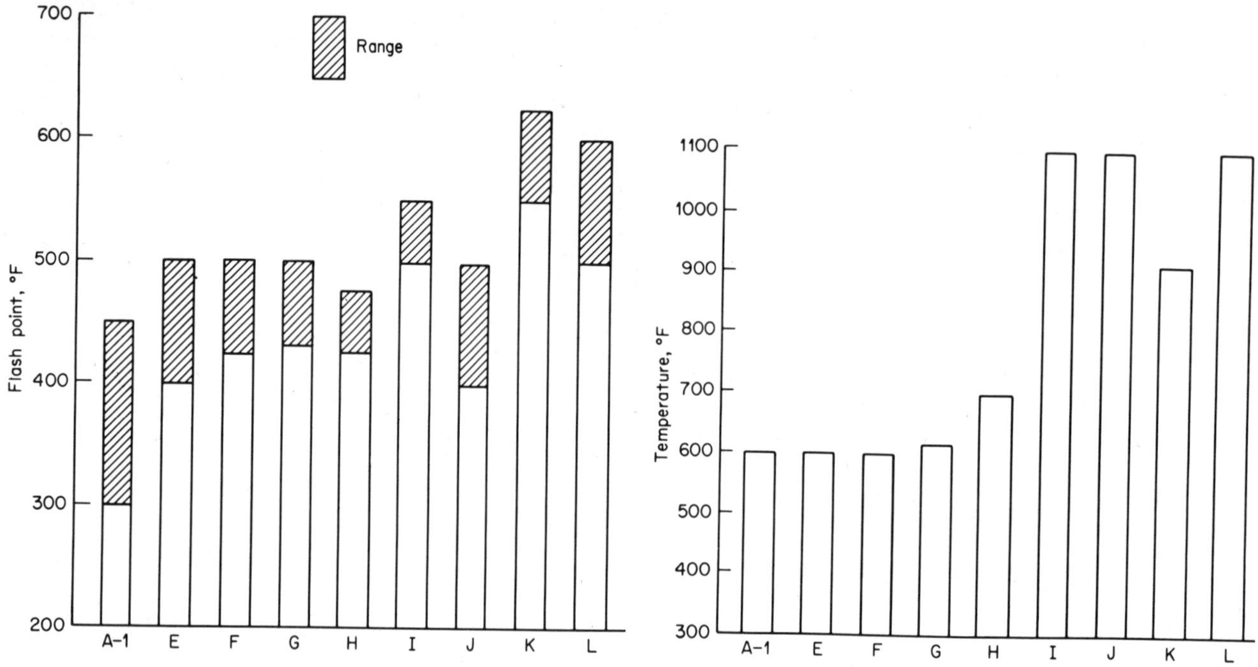

Viscosity stability. For fluids under continuous shear action, high stability is required. Couplings and damping applications are the best examples, but any hydraulic system can benefit. Of the high-viscosity fluids (1000 cs at 77 F), the silicones work best. All fluids of lower viscosity generally have good stability if they do not contain a polymeric viscosity index improver. Many fluids that have shortcomings in viscosity versus temperature characteristics do have a viscosity index improver. It is impossible to rate such fluids because of the large variety of improvers available.

Thermal stability. The polyphenyl ethers have the best stability, followed by the aromatic silicones and other aromatic fluids. The incipient thermal degradation temperatures are given in Table 6.

Several general rules of thermal stability can be stated: Aromatic groups add to stability. Examples are silphenylenes and phenyl silicones vs methyl silicones, aromatic phosphates vs aliphatic phosphates, polyaromatics vs straight-chain hydrocarbons, and aromatic esters vs aliphatic esters. Neopentyl esters have better stability than dibasic acid esters; chlorinated and brominated groups can harm stability; and stability decreases in polymers as viscosity is increased.

Viscosity stability at high temperature is particularly important in coupling applications. High-viscosity fluids are limited to lower continuous operating temperatures than low-viscosity fluids.

Flammability. No high-temperature fluid can be considered nonflammable.

The aromatic phosphates, halogenated aromatic silicones, and polyphenyl ethers are least flammable by the two test methods in Tables 13 and 14. Variation in flash point in each class varies with molecular weight.

Lubricity. In hydrodynamic lubrication, the viscosity under temperature, pressure, and shear is most important: Silicones normally have the advantage. For boundary conditions (some rubbing contact), the hydrocarbons, esters, fluorosilicones and polyglycols normally are best. Some adjustments within a chemical class, such as halogenation, can change the boundary lubricating properties. Some halogenated fluids are fluorinated and chlorinated silicones, fluorinated esters, etc. Bearing materials also affect lubricity. Ratings for three combinations are given in Table 7.

Noncorrosiveness. Below the recommended long-duration temperature, the fluids will not corrode or deteriorate common construction materials. Above this temperature, the diesters, neopentyl esters, phosphates, and halogenated aromatics can liberate acidic products corrosive to metals. Silicate fluids will also form acidic products, but these are not as detrimental to metals. Dibasic and neopentyl esters can be corrosive to lead, copper, magnesium, brass, cadmium, and zinc plating. Phosphates and chlorinated aromatics can attack copper. Hydrocarbons, polyglycols, silicones, and polyphenyl ethers are relatively noncorrosive even at their decomposition temperatures.

All of the fluids can be compounded with rust inhibitors to provide protection to metals for long-term storage. Also, corrosion additives can provide high temperature protection in most fluids.

Some metals such as copper and magnesium are not recommended for any of the fluids at temperatures much above 450 F.

Oxidation stability. Fluids can be rated in three ways: (a) resistance to reaction with oxygen; (b) oxygen tolerance, which is the amount of oxygen that can be reacted with the fluid before significant property changes occur; and (c) resistance to the formation of sludge, solids, and varnish.

Table 8 shows that halogenated aromatics, silicones (phenyl silicones in particular), and polyphenyl ethers, are excellent in oxidation resistance. The other fluids generally have some reaction with oxygen at temperatures of 300 F and higher.

Fluids have to be excellent in either category (a) or (b), and also in (c), for use in couplings, open hydraulic systems, and shock absorbers, where oxy-

text continues, page 16

15. THE NINE HI-TEMP FLUIDS

(See Fluids Primer, p. 9

Code A-1

Super-refined mineral oils = natural petroleum products consisting of complex mixtures of hydrocarbons (carbon and hydrogen in the form of paraffins, naphthenes, and aromatics). They differ from normal mineral oils in that additional refining steps are taken to eliminate weak spots in the molecules, and contaminants.

Synthetic hydrocarbons are similar to mineral oils in containing only carbon and hydrogen, except that they are built up to specific molecules by chemists. Silanes are similar to the synthetic hydrocarbons, and are named silane because they contain one silicon atom per molecule.

Super-refined mineral oils were developed by Penn State University (under Air Force contract), which supplies experimental quantities of MLO 60-294, MLO 7557, etc. Commercial suppliers include Humble Oil (3160, 3158, 3152 and 3153), and California Research.

Polymeric synthetic hydrocarbons are supplied by Amoco Chemicals Corporation (Indopol polybutenes), and others. Lower-viscosity synthetic hydrocarbons and silanes are mainly experimental. Sun Oil Co and others supply the synthetic hydrocarbons, whereas chemical companies such as Dow Corning (XF6-7009) and Olin Mathieson (MLO-59-92) supply the silanes. Typical silanes are tetra alkyl or mixed aryl alkyl silanes such as diphenyl (aryl) didodecyl (alkyl) silane. A military specification is MIL-H-27601 (USAF, −40 to 500 F).

Code E

Polyglycols (poly alkylene glycols, polymers of oxy alkylenes) contain a large portion of carbon and hydrogen in repeating alkylene groups (polymers) but also have an oxygen between each repeating group. The alkylene groups normally used are ethylene (C_2), propylene (C_3), butylene (C_4), or mixtures of them. Dow Chemical Company (Ambiflo and other polyglycol lubricants) and Union Carbide Chemicals Co (UCON fluids) are chief producers. Others are Jefferson Chemical, Olin Mathieson, and Wyandotte Chemical.

Code F

Ester lubricants are supplied as base oils by several chemical companies such as Rohm & Haas (Plexol type), Emery Industries (Emolein lubricant ester), E. F. Drew (Drew Esters), Tennessee-Eastman (Eastman Esters), Union Carbide, and Grace Chemical Company.

Some oil companies produce their own esters but most only formulate them into finished products. Companies include Esso, Humble, American, Shell, Sinclair, Sunoco, Socony Mobil, Texaco, Bray, Lehigh Chemical, and Royal Lubricants.

Military specifications for the diesters are mainly for lubricant applications.

Code G

Commercial producers of **neopentyl-polyol-ester** base oils (see Primer) are Celanese Chemical Co (Celanese 700, 701, 702, 703, 704), Heyden Newport (Pentalubes), and Hercules Powder Co (Hercuflex 600, etc). Oil and chemical companies are using this type of base oil to formulate products to meet MIL-L-23699 (Wep).

Code H

Silicate esters are defined in the Fluids Primer. Alkyl silicate based oils are produced by Monsanto and Union Carbide. Dow Corning sells two aromatic silicates as XF-1-0184 and XF-1-0185. Formulated oils are produced by Oronite Chemical and Royal Lubricants. MIL-H-8446B covers a hydraulic fluid for the temperature range of −65 to 400 F.

Code I and D

The **phosphate ester** structure is an organic alcohol (as in silicates) which is attached to a phosphorus atom. The phosphate differs from the silicate in that it contains only three carbon-oxygen groupings on phosphorus (four for silicates) and also has an additional oxygen attached to phosphorus.

As in silicates, the phosphate can be alkyl (Code D) but for high temperature use the organic group has to be aromatic. Commercial producers are Monsanto Chemical Co (Pydraul Fluids F-9, 60, 150 and 625), Celanese Chemical Co (Cellulube Fluids 90 to 1000) and E. F. Houghton Co (Houghton-Safe Fluids). MIL-H-19457 (Ships), covers aromatic phosphates.

Code J

The **halogenated aromatic** structures are exactly as the name implies—aromatic (phenyls) compounds that have hydrogens replaced with halogen groups, mainly chlorine (ie, chlorinated polyphenyls).

Halogenated hydrocarbons are a bigger group classification which also includes the aliphatic (paraffins) compounds that have some or all of their hydrogen atoms replaced with chlorine, fluorine, or bromine atoms.

The main producer of halogenated aromatics is Monsanto (Aroclor 1248 and Pydraul A-2000). A number of companies produce halocarbons: Hooker Chemical (Fluorolubes), Minnesota Mining & Manufacturing (Kel K oils), and Halocarbon Manufacturing.

Code K

The **silicones** (organo polysiloxanes) have similarities to all hydrocarbon type oils previously mentioned as well as to the structure of glass. They are polymers with the repeating unit being the stable silicon oxygen linkage (as present in glass) and containing side groups which are hydrocarbons (silicon carbon linkage). The hydrocarbon side groups vary: aliphatic (ie methyl), aromatic (ie phenyl), halogenated aromatic, and fluorocarbons.

There are three producers of methyl silicones: Dow Corning 200, 330 and 331 fluids, General Electric SF 96 and 81 fluids, Union Carbide L-45 fluids. Two producers of phenyl methyl silicones are: Dow Corning 510, 550, 710, F-6-7024, and General Electric 81705. Chlorinated phenyl silicones are made by Dow Corning (560) and General Electric (F-50). One producer of fluorosilicones is Dow Corning (Fs-1265).

The military specification for methyl silicones is MIL-S-21568A. A specification for the halogenated type of silicones is MIL-L-27694.

Code L

Polyphenyl ethers contain repeating units of phenyl with an oxygen between them. **Polyaromatics** (polyphenyls) are similar but without the oxygen.

Commercial producers of polyphenyl ethers are Dow Chemical (ET378, 492 and 540), Monsanto Chemical Co (OS-124 and 138), and Shell Chemical Co (VRT Fluid E). Polyphenyls by Monsanto are Santowax R.

gen is mixed with the fluid. However, in closed high-temperature hydraulic systems, category (c), antisludging, etc, is the most important. Although solids and varnish are generally a result of oxidation, in some cases they are formed from hydrolytic and thermal breakdown, or from evaporation of the fluid.

The esters and mineral oils are limited in high-temperature use for mechanical applications because they tend to sludge with the amounts of oxygen present in many systems. The polyglycols will oxidize but generally leave little residue. The silicate and phosphates cause some solid formation, whereas the remaining three classes of fluids form relatively small amounts of solids except at very high temperatures.

Compressibility. Because this characteristic can be either an advantage or a disadvantage, the fluids are compared on the basis of average bulk modulus at 77 F and 4000 psi (Table 9). The low bulk modulus (high compressibility) of some special dimethyl silicones is desirable in liquid springs. The high bulk modulus (stiffness) of the phosphates, aromatics, and polyphenyl ethers is desirable for hydraulic cylinders.

Volatility. Table 11 compares classes of fluids. All high-temperature fluids should have low vapor pressure to prevent pump cavitation and evaporation of fluid. Excessive vapor pressure can also cause high pressure in a closed system, in a thermometric device, for instance.

Mineral oils and esters have the highest volatility; silicones have the most favorable (lowest) volatility versus viscosity relationship. Most of the fluids can be synthesized to have low volatility, usually at a significant increase in viscosity.

Hydrolytic stability. Most of the high-temperature fluids are stable enough in closed systems. However, if water contamination occurs, precautions should be taken for the four ester types (Table 10).

Coefficient of expansion. All organic fluids expand approximately 4 to 5% per 100 F. Low-viscosity silicones have the highest coefficients, useful in thermal-expansion control devices.

Specific gravity. Mineral oils, esters, and silicates have the lowest specific gravities, whereas the aromatic and halogenated fluids have the highest (Table 12). Low-gravity fluids will reduce system weight. High-gravity fluids transfer heat better.

Additives and compatibility

In many cases, a weak property can be improved by the use of special additives such as viscosity-index improvers, lubricity agents, corrosion and rust inhibitors, oxidation inhibitors, and metal deactivators.

These additives improve one property of a fluid, but in many cases harm another high-temperature property. Viscosity-index improvers usually reduce the shear and thermal viscosity stability of fluids. Lubricity and corrosion inhibitors can cause solids formation. Oxidation inhibitors reduce sludging of the base oils, but in turn can turn into solids upon oxidation.

Corrosion and metal deactivators at times can reduce the lubricity of the fluid. All of the additives in most cases will harm the thermal stability of the base fluid. All of the fluids can be formulated with rust inhibitors to act as preservative oils in storage of equipment. If this oil is drained or removed before operation, there should be no detrimental effect from this inhibitor.

Super-refined mineral oils, esters, and polyglycols have very good additive response to lubricity agents and antioxidants. The silicates and silicones have poor additive response to lubricity additives.

Compatibility. Fluids, and materials of construction (metals and elastomers) have to be compatible to make a successful system.

Metals must not act as thermal degradation catalysts and should have good bearing surfaces to exploit lubricity. Fluids that produce acidic degradation products can be corrosive. Examples are esters and halogenated fluids. Silicates, silicones, and polyphenyl ethers in many cases need special bearing combinations in hydraulic systems to overcome lubricity deficiencies of sliding steel surfaces.

The two ester classes and phosphate fluids cause many elastomers to swell. Mineral oils, chlorinated aromatics, and polyphenyl ethers cause moderate swelling. The other fluids usually cause only small amounts of swelling.

Sometimes the swelling effect is desired. For silicate and silicone fluids, a high-swelling ester can be added in specific amounts to provide a controlled amount of swelling.

Selection of elastomers for use with phosphate fluids is the biggest problem. Phosphates are not compatible with Buna N, but are compatible with certain formulations of butyl, silicone, and ethylene-propylene elastomers. Ester oils have poor compatibility with neoprene and should be used with the nitrile, silicone, or fluorinated elastomers.

Fluorinated elastomers, such as Dow Corning LS-63 elastomer, DuPont Viton A and Teflon, and 3M Kel F, are compatible with all of the fluids listed. They can withstand higher temperatures better with low-swell fluids than with high-swell fluids. A combination suitable for many fluids is a silicone elastomer backed up with Teflon.

Compatibility with paints also has to be considered in hydraulic fluids. Phosphate fluids are strong solvents and need special paint formulations. Dibasic and neopentyl esters, although not as severe on paints, still need special formulations.

Toxicity

Toxicity becomes more of a problem as temperature is increased because decomposition products are introduced. Toxicity varies widely within each class of fluid, and you had better contact the individual suppliers.

Volatile fluids, and fluids containing phosphorous and halogen, require more precautions than the less volatile fluids such as silicones, silicates, and high-molecular weight esters and polyphenyl ethers.

MACHINE VS FLUID

Choose the fluid before you design the machine. A system designed for one fluid will not necessarily perform as well with another. In some instances, only one class of fluid can satisfy the requirements for a specific application. An example is the high-viscosity dimethyl silicone selected for automotive fan couplings. In other instances, no fluid can meet the need, and compromise is the only answer.

new data for...

HIGH-PRESSURE AIR

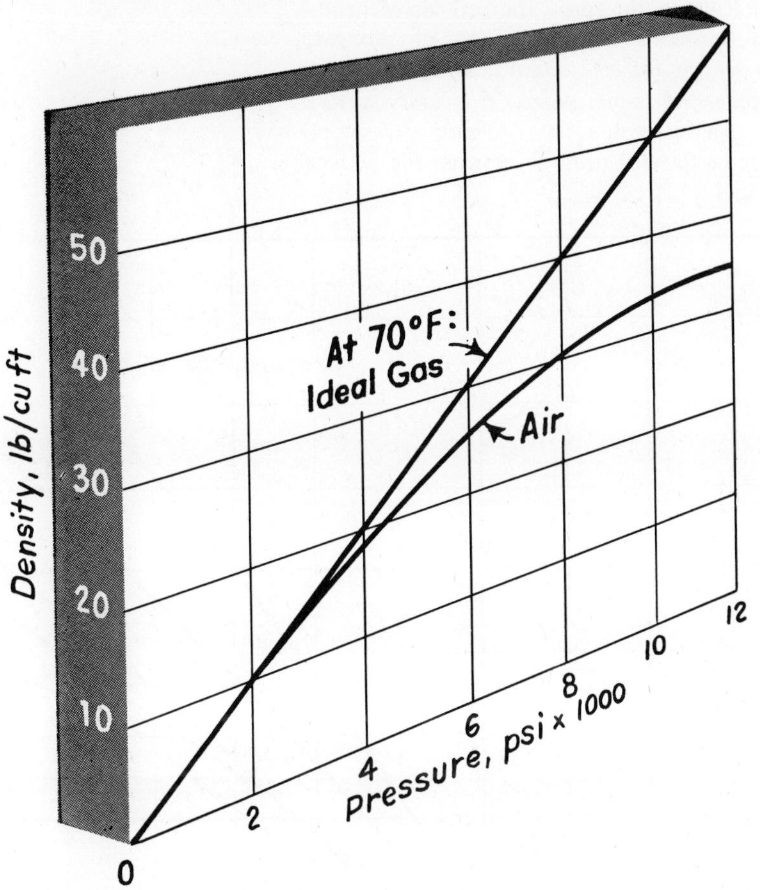

Tests by Amsterdam Institute have extended the range of air properties far above the 100 atmospheres covered by the Bureau of Standards. Although there is still disagreement on many high-pressure properties among various groups, the author has correlated diverse data to give these useful curves from 0— 12,000 psi and —150 to 250 F.

T H CHADWICK, *engineering staff specialist*
P L BRADY, *pneumatics group engineer*
Convair, div of General Dynamics Corp
San Diego, Calif

It is predicted that air at relatively high pressures—to 6000 psi and even higher—will be commonplace in design in several years. The reason? Air is plentiful, fast acting, non-toxic and it doesn't burn up or wear out when the going gets tough. And with high pressures plenty of power can be packed into little space.

However, a major obstacle in the development of high pressure air as a power medium is lack of consistent data on its properties—therefore its behavior in systems—at pressures over 1500 psi.

Part of the answer is found in the accompanying curves. They are based on a selective compilation of test data from all known sources. Most of the useful information for pressures above 1500 psi, however, was from Amsterdam Institute in Holland.

Values given in all curves are reasonably accurate, with these limitations:

- Velocity-of-sound data above 2000 psi was obtained partly by extrapolation.
- Sources disagree concerning effect of pressure on viscosity, particularly at the higher temperatures and pressures.

(CONTINUED ON NEXT PAGE)

Compressibility Factor

A compressibility correction factor, Z, inserted in the ideal equation of state, $Pv = RT$, will make the equation valid for a real gas. Thus, $Pv = ZRT$, where P is pressure in lb/sq ft, v is specific volume in cu ft/lb, R is a perfect-gas constant (53.3 for air), and T is temperature in deg Rankine (absolute).

Factor Z, plotted on this page for real air, depends on temperature, pressure and the nature of the gas. If the value is known for each condition, then perfect gas relations apply. At points where Z is unity, a real gas behaves as a perfect gas.

A way to express the deviation of air from the perfect

Compressibility factor

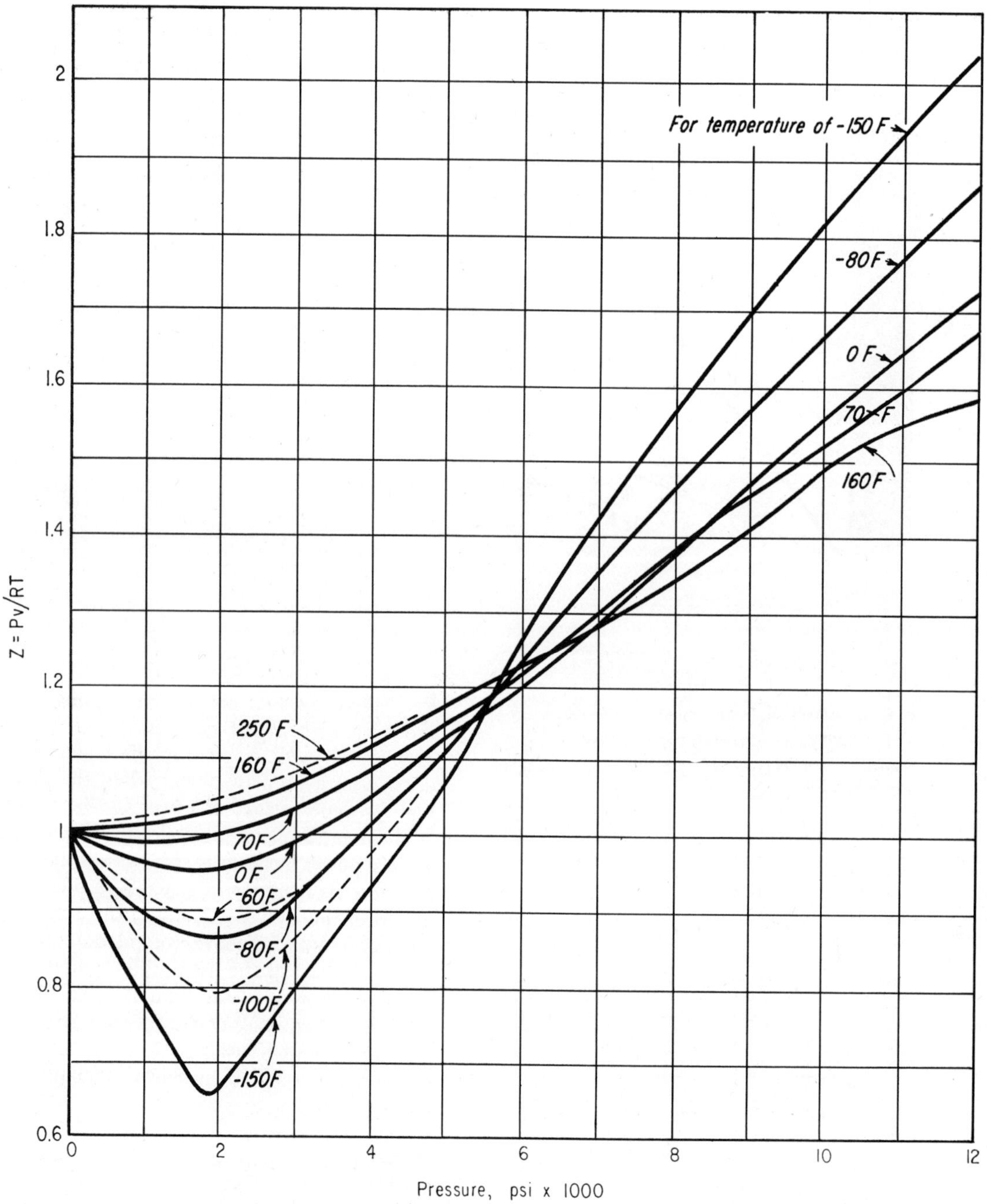

state is as a percentage of ideal compressibility: 100 (Z − 1)/1. The next plot, obtained from separate source data than those for factor Z, gives this deviation in the range from 0-5000 psia, −100 to +250 F. A positive value indicates greater specific volume than that for a perfect gas at the same temperature and pressure.

Specific-Heat Coefficients

Individual values for coefficients C_v (at constant volume) and C_p (at constant pressure) are plotted below. The ratio, $k = C_v/C_p$, is important to thermodynamic calculations and is affected considerably by changes in temperature and pressure. In general, k

Compressibility deviation from perfect gas

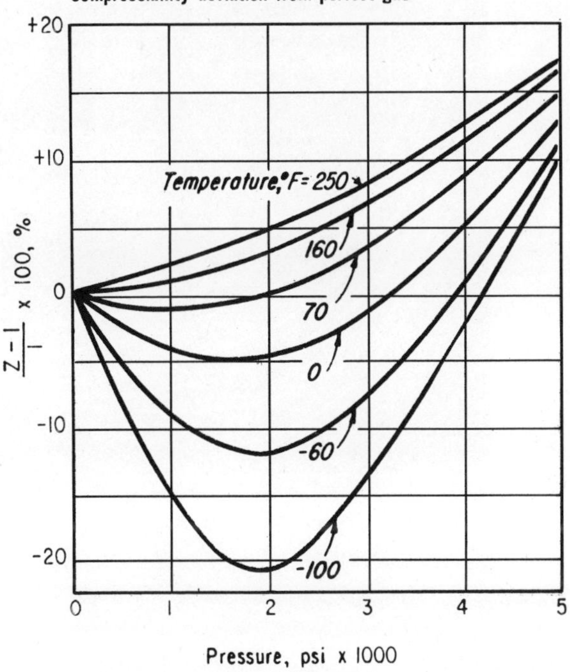

Specific heat at constant volume

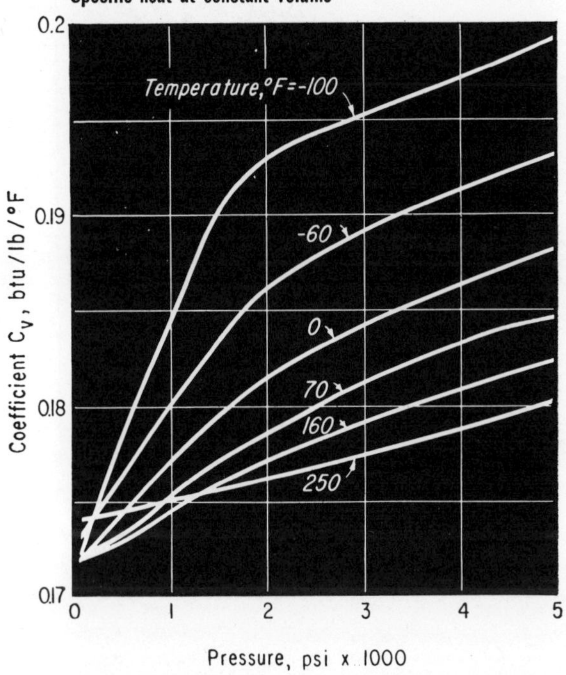

Specific heat at constant pressure

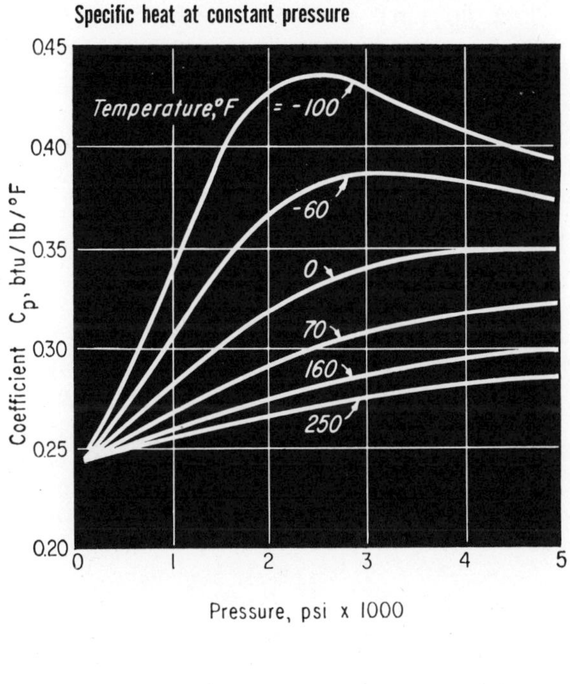

Specific-heat ratio

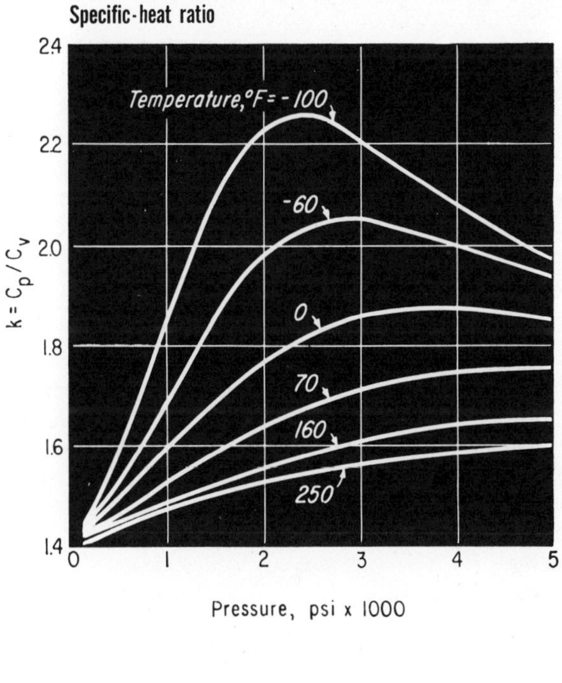

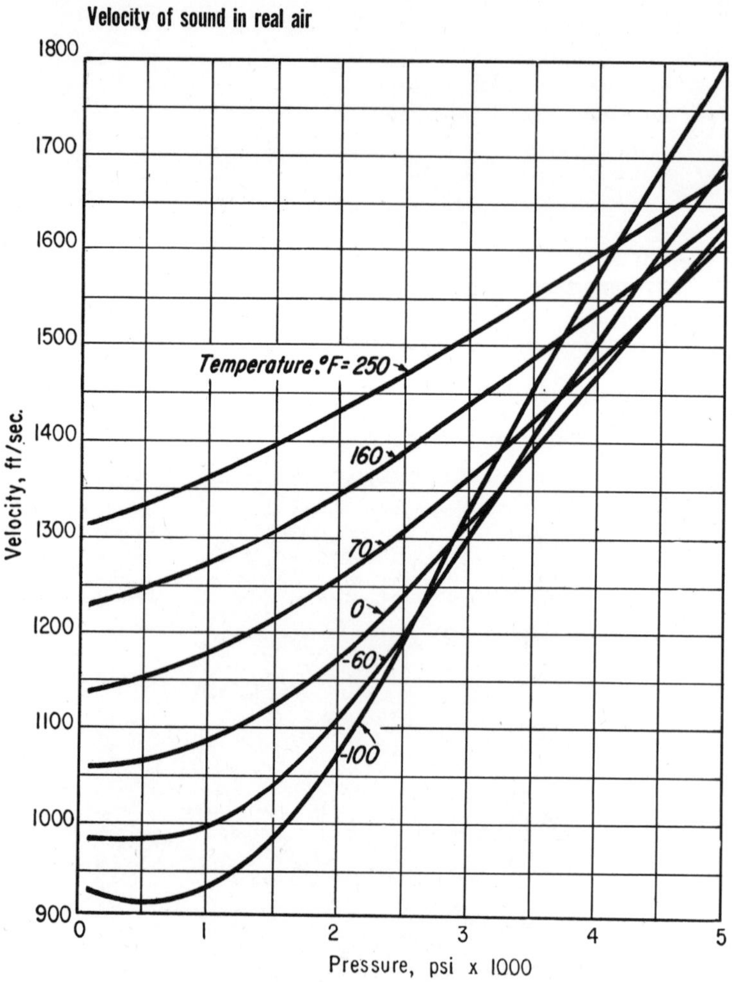

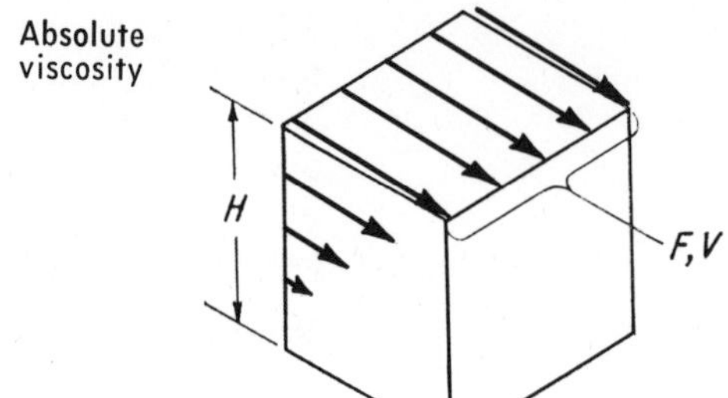

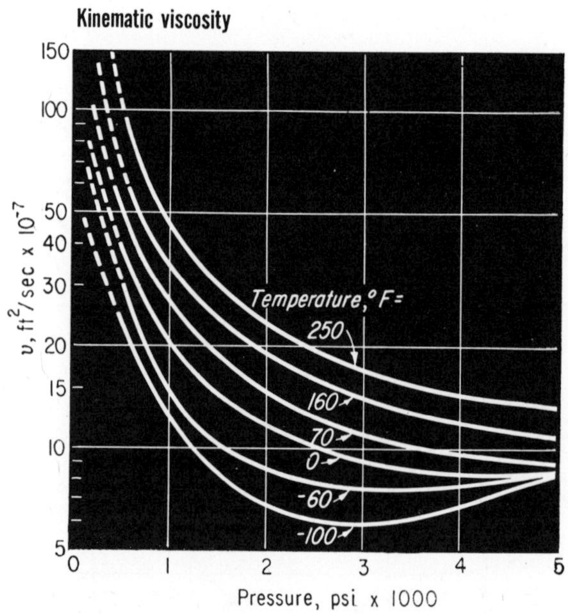

increases directly with pressure, but inversely with temperature.

Velocity of Sound

For a perfect gas, the velocity of sound $a_o = \sqrt{kgRT}$, where g is 32.2 ft/sec^2; other units are already defined. For air, velocity of sound is 1126 ft/sec at 68 *F*, 14.7 psia. For low pressures (less than 100 psia) this velocity is relatively independent of pressure (changes in density are compensated by changes in elasticity) and is roughly proportional to the square root of temperature. In round figures $a_o = 49\sqrt{T}$. Example: At 70 *F*, 14.7 psia, $a_o = 49\sqrt{530} = 1128$ ft/sec.

At higher air pressures, next page, the approximate equation does not apply because the velocity is not constant at all pressures for each temperature. The greatest variation is for -100 F air: sound velocity at 5000 psia is 1801 ft/sec, which is almost twice the value of 928 ft/sec based on $49\sqrt{T}$.

Viscosity

Absolute viscosity, μ, is a measure of the force, *F*, to shear a 1-ft cube of fluid at a specified velocity, V. $\mu = (F/A, \text{lb/sq ft}) \div (V/H, \text{ft/sec-ft}) = FH/VA$, lb-sec/ft^2. Alternate units, slug/ft, are obtained by replacing force *F* with its equivalent in mass and acceleration units.

Kinematic viscosity, υ, plotted above, is absolute viscosity, slug/ft-sec, divided by mass density, slug/cu ft, giving units of ft^2/sec.

An Example

Find the properties of air at 3000 psi, 160 *F*. Compressibility factor Z is approximately 1.064, therefore deviation is plus 6.4%. Mass density $= P/ZRT = (144 \times 3000)/(1.064 \times 53.3 \times 620) =$ about 12.3 lb/cu ft or 0.381 slug/cu ft. Specific heat $C_p = 0.287$ btu/lb/F; $C_v = 0.179$. Ratio $k = 0.287/0.179 = 1.61$. Velocity of sound = 1433 ft/sec. Kinematic viscosity is 14×10^{-7} or 0.0000014 ft^2/sec. Absolute viscosity $= (14 \times 10^{-7}) \times (0.381) = 5.33 \times 10^{-7}$ lb-sec/ft^2. In metric units, absolute viscosity is $5.33 \times 10^{-7} \times 47{,}880 = 0.0255$ centipoise.

2
RESERVOIRS AND CONTAINERS

Tanks, cooling, bubbles, gas containers

Designing the . . .

Fluid-system reservoir

Often taken for granted, the reservoir is vital to performance and reliability of the hydraulic system. Here's a guide to practices and pitfalls.

LOUIS DODGE, Consultant, New Richmond, Ohio

THERE are several dozen separate requirements for an oil reservoir —and just that many ways to stub your toe on the design. But each requirement is simple and you can use this article for a checkoff list.

Before getting to the reservoir you will have settled on flows, pressures, and temperatures in your overall hydraulic system. The following are basic parameters around which the reservoir is designed.

Tank volume

Plan for a reservoir capacity that can feed the system pump for two or three minutes, neglecting any return flow. With a tank that big, several related requirements usually are met automatically: There will be enough reserve fluid to fill the hydraulic system at startup without exposing filter and strainer (make sure of this); a fairly stable oil level despite normal fluctuations in flow; enough fluid to sustain the system while the rotating parts coast to a stop during emergency shutdown if a return line breaks; thermal capacity to absorb unexpected heat for short periods or store heat during idle periods in a cold environment; and enough surface area (tank walls) for natural cooling during normal operation.

If tank volume in gallons is less than twice pump flow in gallons per minute—that is, if the tank can be pumped dry in less than 2 min—add a heat exchanger to avoid excessive temperature fluctuation (a future article will show how). For any size of tank, specify an oil-level indicator or sight glass, in addition to whatever automatic level controls are provided.

Piping and accessories

Study the sketch of a typical tank below and the in-the-tank piping schematic on the next page. You should include each element or have a good reason for leaving it out.

The suction-line strainer (or filter) should be ½ to ¾ in. above the tank bottom. Strainer flow capacity should be two to four times pump capacity. A vacuum gage on the pump suction will indicate if the strainer is clogged. A permanent-magnet filter can be specified as a drain plug or mounted on the baffle plate in a region of concentrated return flow.

The major return flow should discharge below oil level about one inch above the tank bottom; back pressure will be 5 to 10 psi and more. Atmospheric return lines, including seal-leakage lines, are at zero pressure and should be discharged above oil level.

If atmospheric lines have high flow and high content of air, they should be discharged above oil level into a chute sloping gradually (5 to 10 deg) into the fluid. The chute slows and

Important elements in a typical

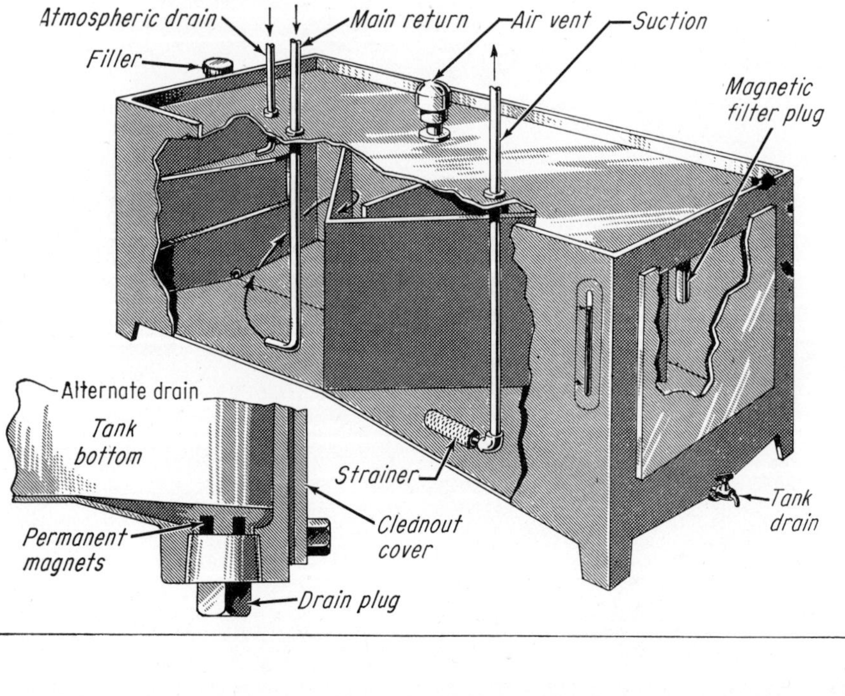

fans out the flow, enabling the oil to free itself of air. This is important, because oil saturated with air and operating at high pressures will run 25% hotter than air-free oil. This is caused partly by heat of compression of the air and partly by its low thermal conductivity.

Leakage lines have low flow but must discharge above oil level to release entrained or dissolved air and prevent airlock in these vital lines.

Internal baffles between the return pipe and pump inlet will slow down fluid circulation, help settle out dirt particles, give air a chance to escape, and allow dissipation of heat. The top of the baffle should be submerged about 30% below the surface of the fluid.

A breather opening with an air filter should be installed on top of the tank. Its air capacity should be twice pump capacity to assure atmospheric pressure inside the tank even if the liquid level drops rapidly during peak demands of hydraulic system. Remember that it is atmospheric pressure that forces fluid into the suction pipe of a pump.

A filler pipe for new oil should be on or near top of the tank and easy to reach. Include a strainer to keep out foreign particles. All top-mounted pipes and accessories should be protected against damage from dropped objects and workers' feet. Tank-top leakage can be kept from spilling on the floor if the tank cover is recessed one half to one inch inside the tank walls.

Temperature and viscosity

Keep oil temperature between 120 and 150 F — preferably the lower value for oil viscosities from 100 to 300 SSU (Saybolt Seconds Universal) based on 100 F. Temperatures up to 160 F are permissible if viscosity is from 300 to 750 SSU based on 100 F. Higher operating temperatures require special design.

Without a heat exchanger, natural cooling (convection and radiation) must be depended on to dissipate generated heat. You should utilize all external and internal surface areas. Put baffles inside the tank to direct hot incoming oil against tank walls. Design legs under the tank to make the undersurface available for natural cooling and to help level the tank during assembly.

Tank walls should be thin for good thermal conductivity. Make them approximately 1/16 in. thick for tank capacity up to 25 gal; 1/8 in. for capacity up to 100 gal; and 1/4 in. for 100 gal or more. Use slightly heavier plate for the bottom. Give the top plate four times wall thickness to assure vibration-free operation and to hold alignment of pump and motor shafts.

In addition to whatever automatic temperature-limiting controls you provide, specify a thermometer to be mounted on the tank top where the operator can see it.

Ease of maintenance

Avoid designing machines with integral tanks. It is better to have a separate unit, accessible from all sides. Small tanks, in fact, are sometimes mounted on castors. Tanks within the machine frame are troublesome to maintain; be sure to work out maintenance details of such a tank before committing yourself to the design.

Equip the tank with cleanout doors and slope the bottom toward the doors. Provide a drain cock or discharge valve at the low point of the bottom and at other low points if this is necessary for complete drainage. Put a manhole cover on the tank top for removing filters and strainer. Provide a connection for hooking up to a portable filtration unit.

Write the assembly, painting, testing, and maintenance instructions before settling on final details of the design. Here are the trouble spots:

If the tank is made of cast iron, don't paint the interior surfaces, but be sure to remove all grit and core sand. Surfaces must be sandblasted.

If the tank is made of sheet steel, clean it thoroughly to remove all dirt, chips, and burrs. Use solvent and rags to remove any grease or oil and then dry with clean rags. If the tank is slightly rusted, clean by wire brushing or grinding. Heavily rusted and scaled surfaces should be shotblasted.

Painting technique depends on which hydraulic fluid will be used in your system. If the fluid has a petroleum base, coat all interior surfaces with an oil-resistant paint recommended by a reputable supplier. For fire-resistant fluids with a phosphate-ester base, use no paint. Special paints resistant to phosphate esters are available, but the fluid itself is noncorrosive and provides the necessary protective coating. Where painting is required specify air-dry paints, such as epoxy primer and enamel, and polyurethane coating. If the fluid is an oil-water or water-glycol emulsion, don't apply paint to the surfaces because the emulsion will probably peel it off. But if you must paint, obtain detailed information from the manufacturer of the emulsion.

fluid reservoir

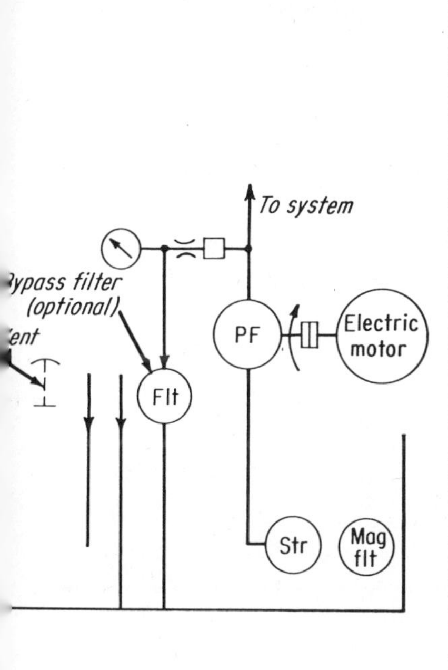

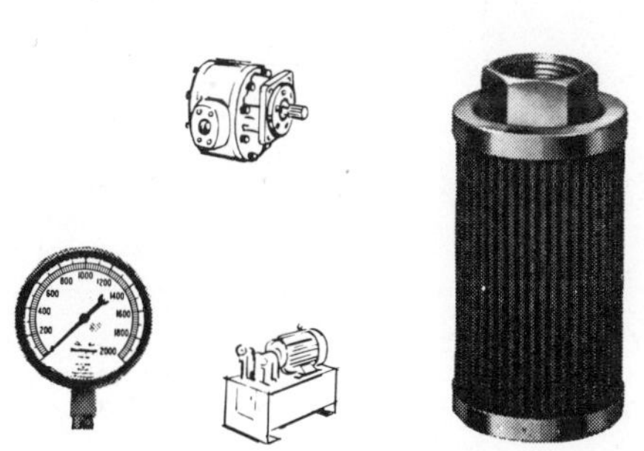

Simple math estimates . . .

Oil-system cooling

Avoid overheating three ways: pump only the oil you need; provide plenty of cooling surface; add a heat exchanger if necessary.

LOUIS DODGE, Consultant, New Richmond, Ohio

THE hottest recommended temperature for oil in a reservoir normally is 120 F. This article shows ways to keep it from going above that level. There are exceptions, of course. Some conventional systems operate at 150 F, and there are super-systems with special fluids and seals that are good up to 500 F and higher. For any level of temperature, the same heat-transfer principles apply.

First step, assuming you've already established the basic system and reservoir design, is to make a simple heat balance as described in the accompanying text and tables. If the reservoir peak temperature calculated this way is less than 120 F (or some other desired temperature) no further work is necessary. If the calculated temperature is higher, you have two alternatives: improve heat dissipation by modifying the tank, components or piping, or add a heat exchanger, using the rating method in this article.

Source of heat

Most of the heat comes from in-the-system components. Exceptions are systems in hot environments or adjacent to heat-producing equipment, but the same heat-balance principles can be applied here.

Heat is generated wherever oil is throttled or otherwise restricted. Examples are pressure regulators, relief valves, undersize piping, dirty or undersize filters, leakage points, and areas of turbulence anywhere in the system.

A good measure of internally generated heat loss is the difference between pump input power, Btu/hr, and system useful work, Btu/hr.

$$\text{Energy loss } E_L = E_{in} - E_{used} = E_{in}(1 - \mu) \quad (1)$$

Where μ is system efficiency. The relationship may also be written:

$$E_L = 1.48 \times Q \times P(1 - \mu) \quad (2)$$

Where Q is flow, gpm, and P is pump gage pressure, psi.

The first graph (this page) shows a hypothetical duty cycle, pressure vs time. The utilization pressure, measured at the inlet to the working device (fluid motor or cylinder), will always be lower than the source pump pressure, depending on the amount of throttling or other regulation required.

The difference in energies, pumped vs utilized, must be absorbed during transients and eventually dissipated by the system. An approximate measure of overall efficiency, μ, is the ratio of mean effective pressure P_{mean} to pump pressure P_{pump} where P_{mean} is calculated from the area under the

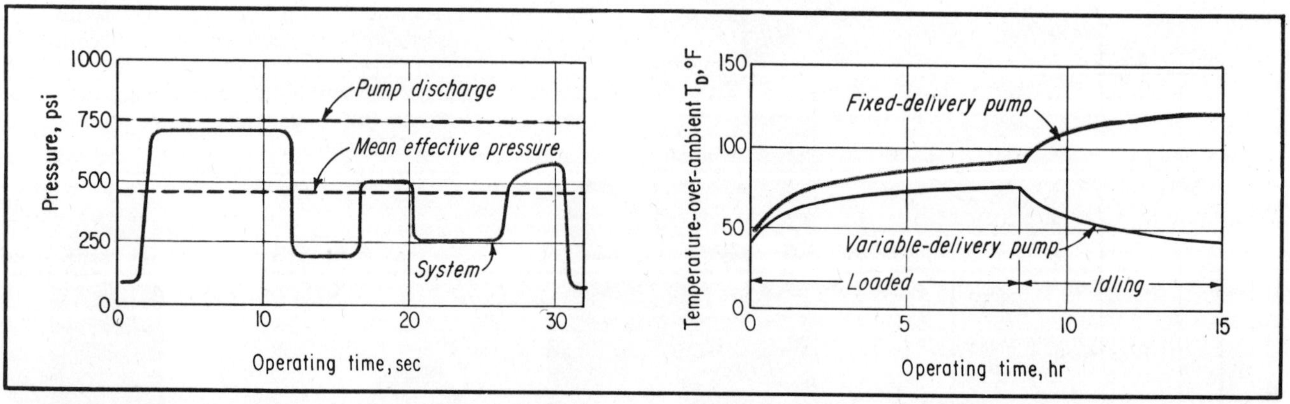

TYPICAL DUTY CYCLE . . . in graph on left demonstrates mep (mean effective pressure) calculation. area under curve is 14,000 psi-sec; time base is 32 sec; mep is 14,000/32 = 440 psi. Graph on right shows how proper choice of pump can conserve energy and reduce system temperature: the variable-delivery pump is more efficient, has negligible loss at idle.

curve divided by the time base. Measure the area with a planimeter.

The second graph compares tank temperature for two different pumps: a fixed-delivery pump, with constant flow and pressure and a full-flow relief valve for bypass flow during idling of the workload; and a variable-delivery pump, with pressure and flow automatically varied to match load requirements.

Note that in the constant-pressure system the greatest rate of heat generation is during idling of the workload; all the flow is throttled back to the reservoir and does no useful work. On the other hand, a variable-delivery pump does not waste energy at idle, because the flow is automatically reduced to nearly zero.

Both graphs indicate that savings in energy are possible if only the needed oil is pumped. Excess capacity is forced back to the reservoir through the relief valve, and the energy is converted to waste heat. Auxiliary pumps are great offenders if they are operated when not needed.

An alternate to the fixed-delivery and variable-delivery pump is the high-pressure low-pressure pump, consisting of two fixed-delivery pumps with intervalving and controls to permit automatic selection of the most efficient of two pressure levels. Any pump chosen, of course, must be able to handle the maximum transient flow of the system regardless of how infrequently it occurs, or there must be an accumulator or similar storage device to handle such transients.

Heat balance

Heat balance is the mathematical summation of heat entering and leaving a system: what goes in must be accounted for either as a temperature increase or a transfer of heat out of the system.

$$E_L = E_A + E_D \qquad (3)$$

where E_L is internal heat loss generated by the pump and other components in the system; E_A is heat absorbed by the oil, tank, and components; and E_D is heat dissipated to atmosphere or to a coolant.

The balance is usually a transient one, except where long-term steady-state operation is reached. The general relationship is:

$$E_L dt = (\Sigma cW) dT_D + (\Sigma kA) T_D dt \qquad (4)$$

where t is operating time, hr; c is mean specific heat, Btu/lb-°F; W is combined weight, lb, of oil and system components; T_D is temperature difference, °F, between oil system and

Heat-transfer terminology

HEAT LOSS AND EFFICIENCY

μ = System efficiency, E_{used}/E_{in}, %

E_{used} = Energy utilized in system, Btu/hr

E_{in} = Pump input power, Btu/hr

E_L = Heat loss generated in system, Btu/hr

E_A = Heat absorbed by oil, tank and components, Btu/hr

E_D = Heat dissipated to atmosphere or coolant, Btu/hr

E_{exch} = Heat exchanger load, Btu/hr

FLUID CONDITIONS AND FLOW

t = Operating time, hr

Q = Flow, gpm

P = Pump gage pressure, psi

T_D = Temperature-over-ambient for oil, °F: $T_D = T_{oil} - T_{air}$ (values are mean)

ΔT = Heat exchanger only: $\Delta T_{water} = T_{out} - T_{in}$; $\Delta T_{oil} = T_{in} - T_{out}$; ΔT_{mean} = log-mean ΔT, oil-to-water

EQUATION CONSTANTS

c = Specific heat (mean value), Btu/lb-°F

k = Overall heat-transfer coefficient, Btu/sq ft-hr-°F

W = Combined weight of oil and system components, lb

A = Surface area for dissipating heat, sq ft

e = Base for natural logs = 2.718

Σ = Summation sign. ΣcW = effective cW for all of system components

TYPICAL VALUES FOR c AND k

c, Btu/lb-°F:

Oil, 0.40; aluminum, 0.18; iron, 0.11; copper, 0.09

k, Btu/sq ft-hr-°F:

2 to 5—Tank inside machine or with inhibited air circulation

5 to 10—Steel tank in normal air

10 to 13—Tank with good air circulation (guided air current)

25 to 60—Forced air cooling or oil-to-air heat exchanger

80 to 100—Oil-to-water heat exchanger (k values increase slightly with temperature)

surrounding air; k is overall heat-transfer coefficient, Btu/sq ft-hr-°F; and A is heat-dissipating area, sq ft. Typical values of c and k are given in the tables.

The wanted quantity is temperature-over-ambient T_D. The differential equation first must be solved. Briefly:

$$dt = \frac{(\Sigma cW)dT_D}{E_L - (\Sigma kA)T_D} \qquad (4a)$$

Call the initial system-to-air temperature difference (init T_D). Integrate the equation between limits initial T_D and T_D and solve for T_D:

$$T_D = \frac{E_L}{\Sigma kA}\left(1 - e^{\frac{-\Sigma kA}{\Sigma cW}t}\right) + ({}_{\text{init}}T_D)e^{\frac{-\Sigma kA}{\Sigma cW}t} \qquad (5)$$

where $e = 2.718$; t = time, hr. Maximum T_D prevails when t approaches infinity: a steady-state temperature balance is reached, and oil temperature remains constant.

$${}_{\max}T_D = E_L/\Sigma kA \qquad (6)$$

If the calculated maximum temperature difference exceeds the specified limit, and if adding surface area or increasing the oil capacity does not lower it sufficiently (see the worked-out examples at the end of the article), then add a heat exchanger.

As a general rule, no heat exchanger is required when oil supply is adequate and circuit pressure is below 1000 psi. An exception is in precision servo circuits: a heat exchanger will prevent undesirable viscosity changes by holding oil temperature fairly constant.

Heat exchangers

Water-cooled shell-and-tube exchangers, shown in sketch, are most common, although air coolers are used where water is not readily available and where the air is at least 10 F cooler than the oil. Water coolers are compact, reliable, and efficient and can have simple temperature controls. Exchangers usually are built of copper-base alloys, sometimes of aluminum or stainless steel. Stainless is recommended if the oil is corrosive.

The water flows through the tubes, and the oil across the tubes. This facilitates maintenance—water is more likely than oil to contaminate a metal surface, and the inside of a tube is easier to clean than the outside. Baffles direct the oil across the tubes and also support them. Pressure drop of oil should not exceed 10 or 15 psi; typical flow velocity is 3 fps.

Single-pass and double-pass coolers are available. In single-pass coolers the water flows straight through once; in two-pass coolers the water makes one 180-deg. turn. Choice depends on calculated temperature difference between outlet water and outlet oil: if the outlet oil will be considerably cooler than the outlet water, a single-pass cooler is sufficient. If the outlet oil temperature is likely to approach the outlet water temperature, multiple-pass exchangers can be justified, but you should increase the total calculated surface area 20% to compensate for lower efficiency of a multiple exchanger.

Heat balance in an exchanger is:

$$E_{\text{exch}} = \Delta T_{\text{oil}} \times Q_{\text{oil}} \times 210 = \Delta T_{\text{water}} \times Q_{\text{water}} \times 500 \qquad (7$$

where the specific heat of oil is assumed to be 0.4 Btu/lb-°F, and ΔT_{oi} and ΔT_{water} are input-to-output temperature changes in oil and water respectively.

The balance depends on the overal heat-transfer coefficient k between the water and the oil.

$$E_{\text{exch}} = kA\ \Delta T_{\text{mean}}\ \text{Btu/hr} \qquad (8$$

Value of k for most oil-to-water exchangers is 80 to 100 Btu/sq ft-hr-°F A is the surface area, sq ft. ΔT_{mean} is the logarithmic mean temperature difference computed as follows:

Heat-exchanger performance

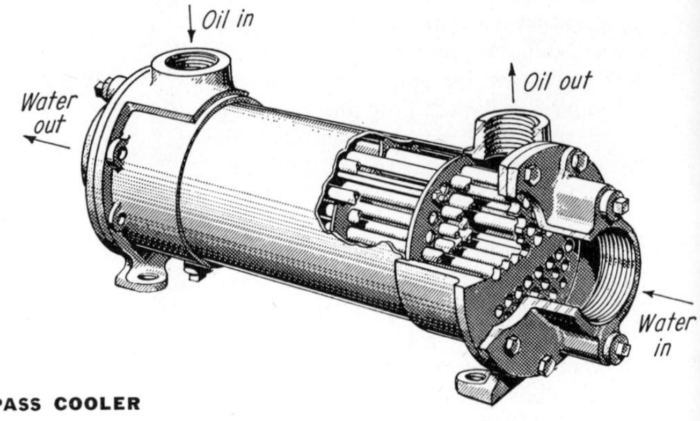

SINGLE-PASS COOLER

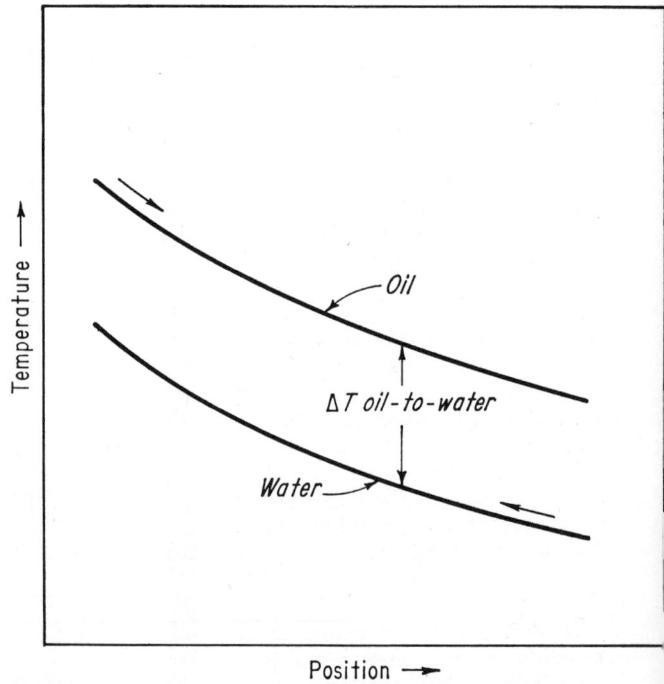

FLUID TEMPERATURES

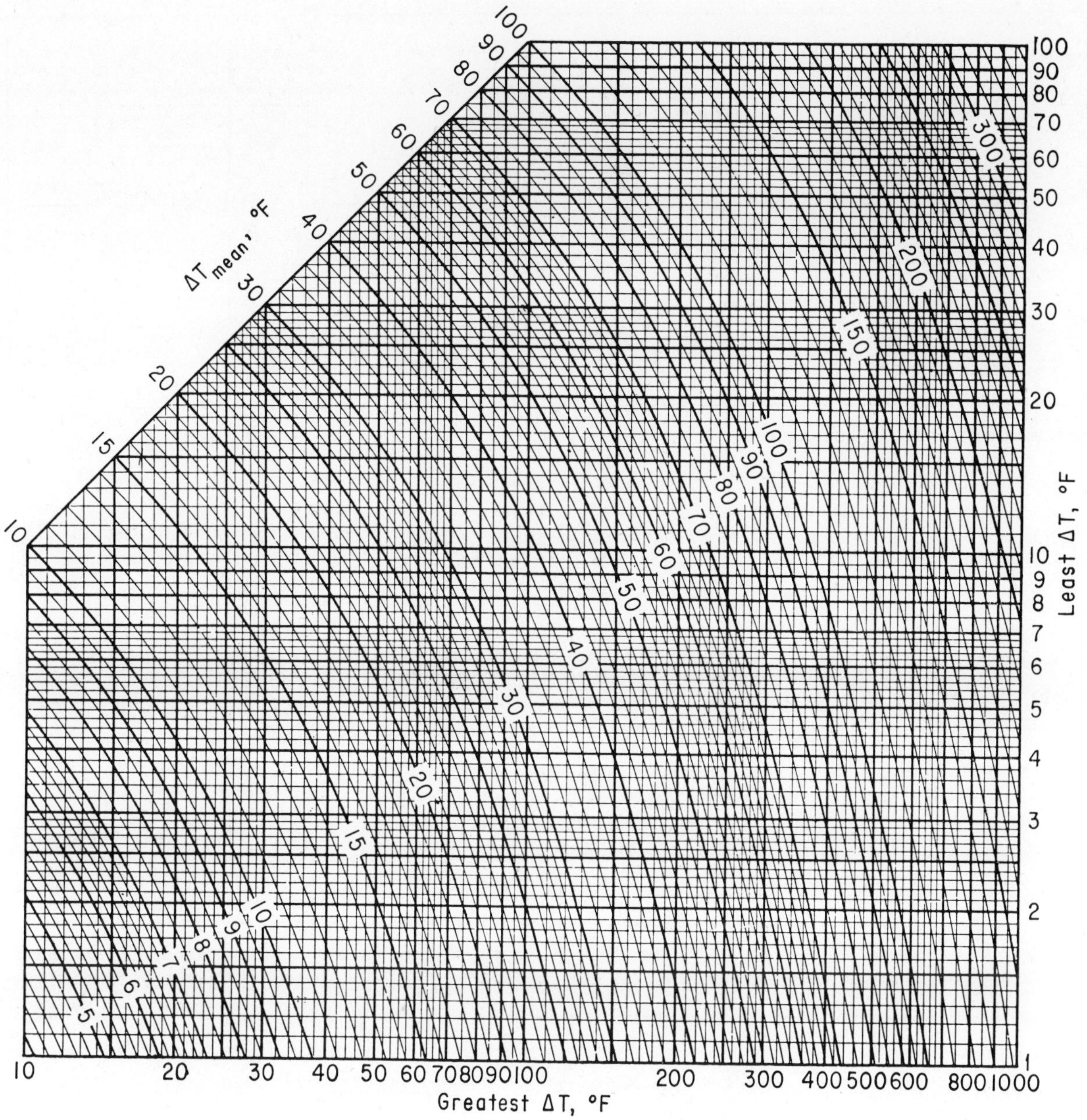

LOG-MEAN TEMPERATURE DIFFERENCE FOR ANY PARALLEL OR COUNTERFLOW HEAT EXCHANGER

$$\Delta T_{\text{mean}} = \frac{\Delta T_{\max} - \Delta T_{\min}}{\ln \Delta T_{\max}/\Delta T_{\min}} \quad (9)$$

where $\Delta T_{\max}$ is the greatest and $\Delta T_{\min}$ the least temperature difference of any measured between oil and water throughout the exchanger. Convenient values of ΔT_{mean} are plotted on this page. The method is valid for any parallel or counterflow exchanger using any materials and any fluids.

EXAMPLES

1. To determine whether additional cooling is needed for a steady-state system without an exchanger, apply Eq. 2 and Eq. 6 as follows:

Required: Determine if reservoir can dissipate enough heat to keep oil temperature below 120 F in a 70 F ambient (50°F difference T_D). The source of heat is a 20-gpm constant-delivery pump operating in a 60%-efficient overall system. Assume first that the working unit, including piping, is not helping to dissipate heat. Tank size is small: reservoir cooling surface is 21 sq ft, based on a tank volume of twice pump flow—40 gal. System pressure = 750 psi; overall heat-transfer coefficient k = 5 Btu/sq ft-hr-°F (conservative). Total generated heat is calculated with the heat-loss equation, Eq. 2:

$$E_L = 1.48 \times 20 \text{ gpm} \times 750 \text{ psi} \times (1 - 0.60) = \text{about } 9000 \text{ Btu/hr}$$

Required surface area from maximum-temperature equation, Eq. 6:

$$A = 9000/(50 \times 5) = 36 \text{ sq ft.}$$

Conclusion: Actual tank area is not great enough and some means of cooling should be provided.

2. In choosing a heat exchanger, calculate the necessary heat-transfer area according to Eq. 7, as shown here:

Required: Oil system return flow in a hypothetical system must be

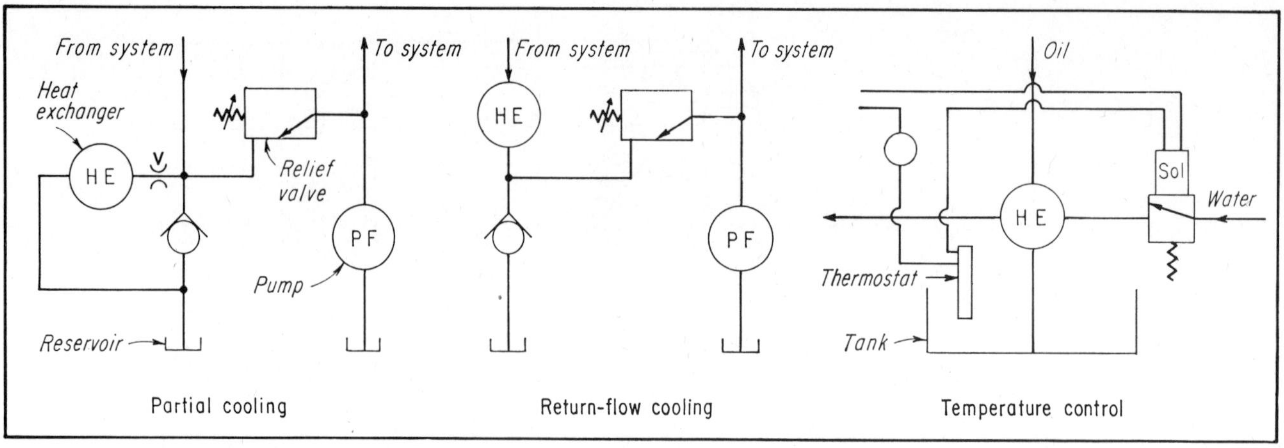

THREE APPLICATIONS OF HEAT EXCHANGERS

cooled continuously to 125 F. Hottest drain temperature, uncooled, is 140 F. Flow is 12 gpm, water inlet = 65F, and outlet = 85 F; k = 90 Btu/hr-sq ft-°F. Assume a counterflow, single-pass exchanger.

Conclusion: Heat-transfer area of exchanger must be at least 7.32 sq ft and water flow 3.78 gpm, calculated with Eq. 7 and 8 as follows:

$$E_{exch} = (140 - 125) \times 12 \times 210$$

$$= 37{,}800 \text{ Btu/hr}$$

$$\Delta T_{max} = 125 - 65 = 60 \text{ F}$$

$$\Delta T_{min} = 140 - 85 = 55 \text{ F}$$

$$\Delta T_{mean} = 57.5 \text{ F (from chart)}$$

$$A = 37{,}800/(90 \times 57.5)$$

$$= 7.32 \text{ sq ft}$$

$$Q_{water} = 37{,}800/(500 \times 20)$$

$$= 3.78 \text{ gpm}$$

See manufacturers' catalogs for size of exchanger with proper surface area. Check allowable exchanger pressure if your system return exceeds 150 psi.

3. If oil temperature must be predicted after a given period of time in a system without a heat exchanger, Eq. 5 applies:

Required: Calculate temperature of a standard 60-gal reservoir after 5 hr of operation. Pump discharge is 20 gpm at 750 psi. Cooling surface is 28 sq ft. An attached heat-dissipating working unit weighs W = 800 lb and its effective surface area A = 5.5 sq ft. Assuming an initial system temperature T_{oil} = 70 F and an ambient temperature T_{air} = 50 F, the initial temperature-over-ambient initial T_p = 20 F. Estimated median value of k = 4 Btu/sq ft-hr-°F. Sixty gallons of oil weigh 7.4 x 60 = 444 lb.

$$E_L = 9000 \text{ Btu/hr}$$
(same as first example)

$$\Sigma cW = 0.4 \times 444 + 0.1 \times 800$$

$$= 257 \text{ Btu/°F.}$$

ΣkA for tank and working unit

$$= 4(28 + 5.5)$$

$$= 134 \text{ Btu/hr-°F}$$

The temperature-above-ambient T_D at t = 5 from the general heat-balance equation, Eq. 5:

$$T_D = \frac{9000}{134}\left(1 - e^{\frac{-134}{257}\times 5}\right) + 20e^{\frac{-134}{257}\times 5}$$

$$= 67(1 - 0.0745) + 20 \times 0.0745$$

$$= 63.5 \text{ F}$$

The maximum operating temperature over ambient, from Eq. 6: max T_D = 9000/134 = 67 F. Oil temperature = 50 + 67 = 117 F.

Conclusion: No heat exchanger required.

Note: The results mainly depend on the correct evaluation of k, which depends on air circulation. Influence of initial temperature difference (init T_p) is minor. Practical experience is most important.

Combating a problem

Air bubbles in oil

Here is a three-pronged attack: avoid bubbles where possible; remove them if they form; allow for their effect on oil properties if they persist.

A. T. J. HAYWARD, National Engineering Laboratory, East Kilbride, Glasgow, Scotland

Bubbly oil is ordinary hydraulic fluid in which there are discrete bubbles of air separated by relatively thick films of oil. It differs from foam, in which air bubbles are packed so closely that they interact. Bubbly oil is still a liquid; foam is not. Because it is spongy—compressible—bubbly oil produces undesirable side effects:

- Cavitation in pumps. As pressure drops at the impeller inlet, the bubbles expand, interfering with the pump suction.
- Noise and local overheating. Bubbles collapse suddenly under pressure while passing through the pump.
- Increased effective volume, requiring larger lines to handle the flow.
- Reduced system stiffness. The large pockets of air formed when bubbles coalesce make cylinders sluggish and servos erratic.

How bubbles are formed

The primary cause of bubble formation is a process of mechanical entrainment. Air may be entrained by impinging of jets, vortex formation at suction intakes, churning when rotating parts run partially submerged in oil, oil leakage, and drainage from pressure-fed bearings. Mechanical air entrainment is reduced when inlet piping, flow paths through the reservoir, oil-to-air seals, leakage lines, and other components that handle or contact air are carefully designed.

A second source of bubbles is air already dissolved in the oil. This is more difficult to control, particularly if the oil is saturated with air at high pressure. Any reduction in pressure, at

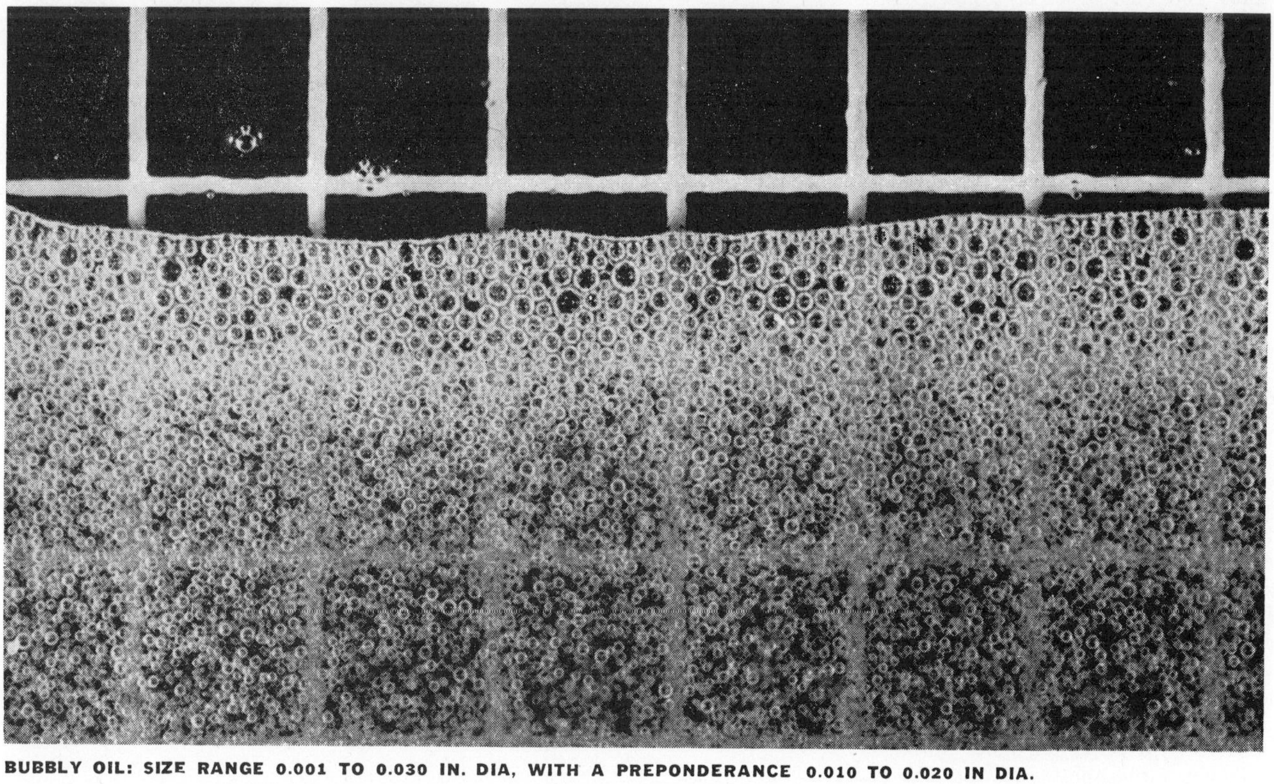

BUBBLY OIL: SIZE RANGE 0.001 TO 0.030 IN. DIA, WITH A PREPONDERANCE 0.010 TO 0.020 IN DIA.

a throttling valve or hydraulic motor, for example, releases the air in the form of bubbles. Fortunately oil is normally saturated at atmospheric pressures, and the air can be kept in solution if the system reservoir pressure is held slightly above atmospheric.

How to get rid of them

If bubbles are unavoidable, try to take them out before the oil is recirculated. One way is to put air vents in the return lines and reservoir. A more positive method is to separate the bubbles from the oil with a screen (60-mesh, 36-SWG wire gauze). Mount the screen in the reservoir below the oil surface at an angle of 30° from the horizontal. Let the bubbly drain oil enter the reservoir above the screen; locate the suction line below the screen (photograph at right).

Air trapped in hidden pockets of the system piping cannot be easily removed. It dissolves very slowly because there is little oil-air interface in relation to the air volume that is trapped. Modify the system to avoid such pockets.

How to live with them

If bubbles persist, learn the properties and characteristics of bubbly oil and include them in your computations. You will be concerned primarily with air solubility (absorption and release); bubble population (number, size, distribution, and life); viscosity; and compressibility (bulk modulus and heat of compression).

In most calculations it may be assumed that undissolved air acts independently of the oil; overall performance is a composite of the individual performances. If the air is entirely dissolved, the effects are nil, and the air may be neglected.

Air solubility—Tests show that a commercial petroleum fluid of normal viscosity dissolves about 8.5 ± 0.5% by volume of air at atmospheric pressure and normal ambient temperature. The tests were made, with and without additives, on more than a dozen oils from all parts of the world. Viscosity of these oils ranged from 24 to 69 centistokes (115 to 320 SSU), and solubility of each oil tested was never less than 7.9% or more than 9.1% at room conditions. It was found that most additives and inhibitors have no appreciable effect on solubility.

For pressures other than atmospheric, absorption follows Henry's law, that is, solubility is directly proportional to absolute pressure. This means that oil will not dissolve air in a vacuum. Oil holds ten times more air at ten atmospheres (about 150 psia) than it does at atmospheric pressure.

High-viscosity oil dissolves somewhat less air than low-viscosity oil. The relationship is shown in the graph on this page; every test point was within 3% of the plotted line, and most were within 2%.

Rates of absorption and release of dissolved air in actual systems are not well established, but the principles are understood. The following conclusions are reasonably valid:

- Absorption of bubbles in still oil at atmospheric pressure is negligible, and the bubbles rise and escape quickly. If pressure is increased to more than 100 psi, the bubbles dissolve rapidly at first and then very slowly as the skin of oil around each bubble becomes saturated with air. About 25% of the air volume dissolves immediately; the remainder takes a minute or more.
- Agitation increases the dissolving rate of pressurized oil manyfold. Low-viscosity oil absorbs air more quickly than high-viscosity oil, in some instances so rapidly that absorption seems instantaneous. Not all bubbles are dissolved in every case, however. It has been observed that some remain undissolved after passing through a pump and 5 ft of ¾-in. pipe at 300 psi.
- For still oil at atmospheric pressure, release of dissolved air as bubbles is very slow. But in a vacuum (deaeration chamber) the release is rapid. For example, during a violent foam-

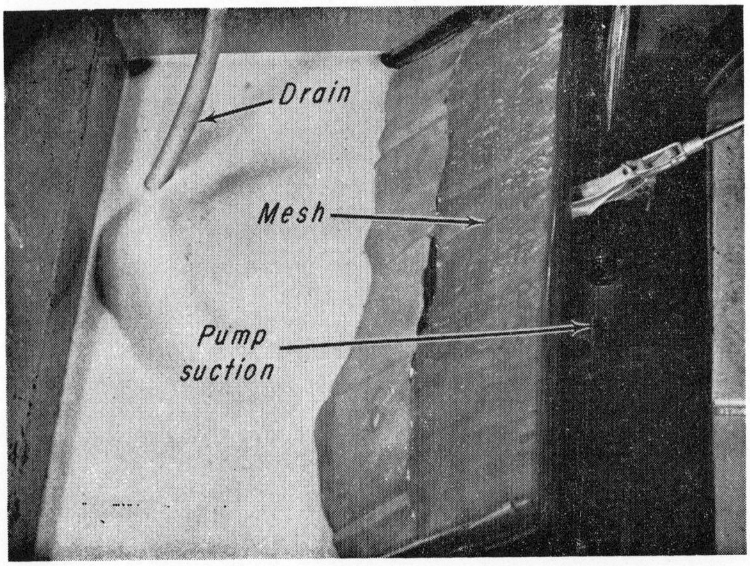

SCREEN TO REMOVE BUBBLES.

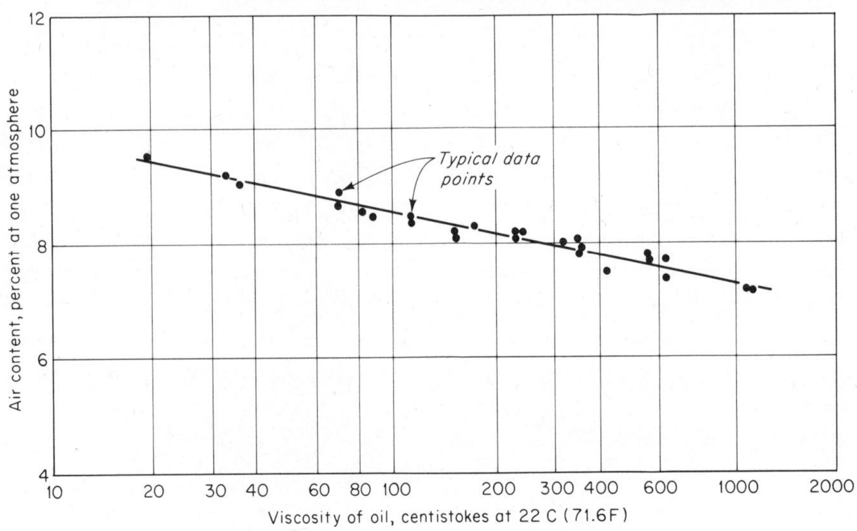

ing action lasting about 2 min, 75% of the dissolved air is released from a film of oil 0.2 in. thick. The remaining air is released by diffusion. For low-viscosity oil, this takes half an hour; for high-viscosity oil, more than an hour.

Measuring bubbles—Bubbly oil at atmospheric pressure can contain up to 30% by volume of air. With more air than this, the oil becomes foam. By definition, the percent of bubbles is the total volume of the bubbles divided by the total volume of bubbles plus oil, assuming that the bubbles are intact and inside the oil. The air inside each bubble is slightly compressed by oil-film surface tension and if released will expand and occupy slightly more volume.

Bubble size is predominantly 0.01 to 0.02 in. dia and it is difficult, for purposes of experimentation, to produce oil with a preponderance of bubbles of any other size. Bubble diameters cannot be measured individually because the bubbles disperse too quickly. However, if the oil is poured into a transparent container (photo on page 32) that has ruled grids, the bubbles can be photographed immediately for later measurement.

Measurement of cumulative bubble volume is fairly easy and can be done in one of four ways, depending on oil flow conditions (see table for limitations). The crudest method is to compare the weights of a given volume of bubbly oil and an equal volume of non-bubbly oil; the difference in weight indicates the bubble content. Another method is to pour bubbly oil into a calibrated glass measuring tube and wait for the bubbles to disappear; the reduction in volume depends on initial bubble content. For more rapid measurement a deadweight tester, of a type designed by the author, compresses the bubbly oil and dissolves the bubbles immediately; here too, the reduction in volume depends on initial bubble content. The simplest and most accurate method is to float a hydrostatic balance (a ball) on the surface of the oil; the degree of submersion depends on oil density, therefore on bubble content.

Temperature corrections should be

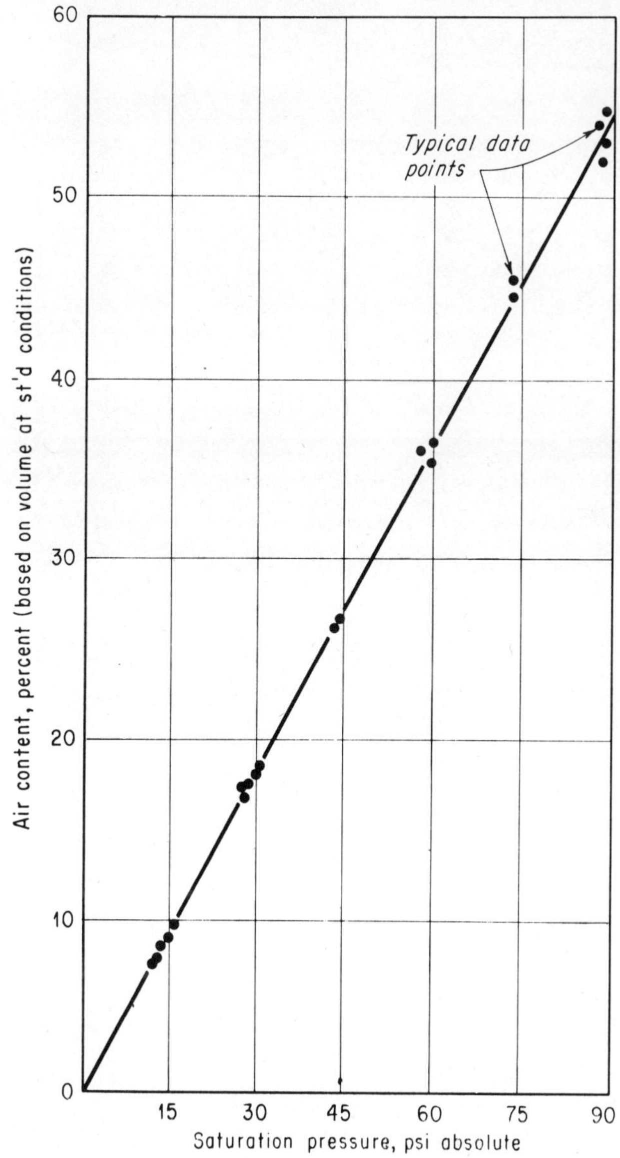

EFFECT OF PRESSURE ON SATURATION.

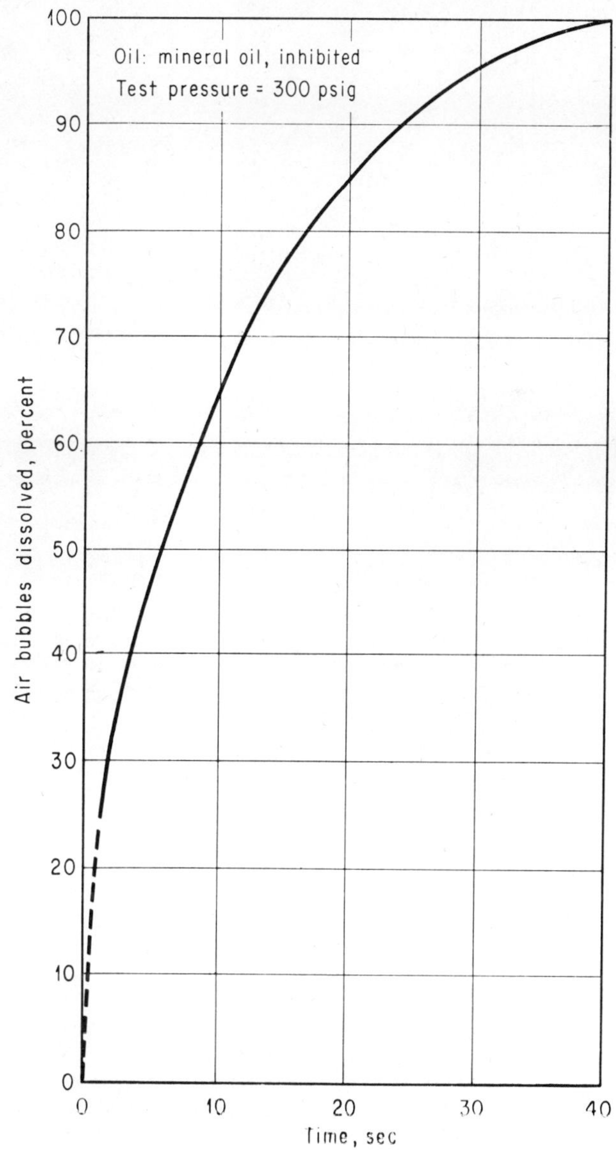

RATE OF ABSORPTION OF AIR IN STILL OIL.

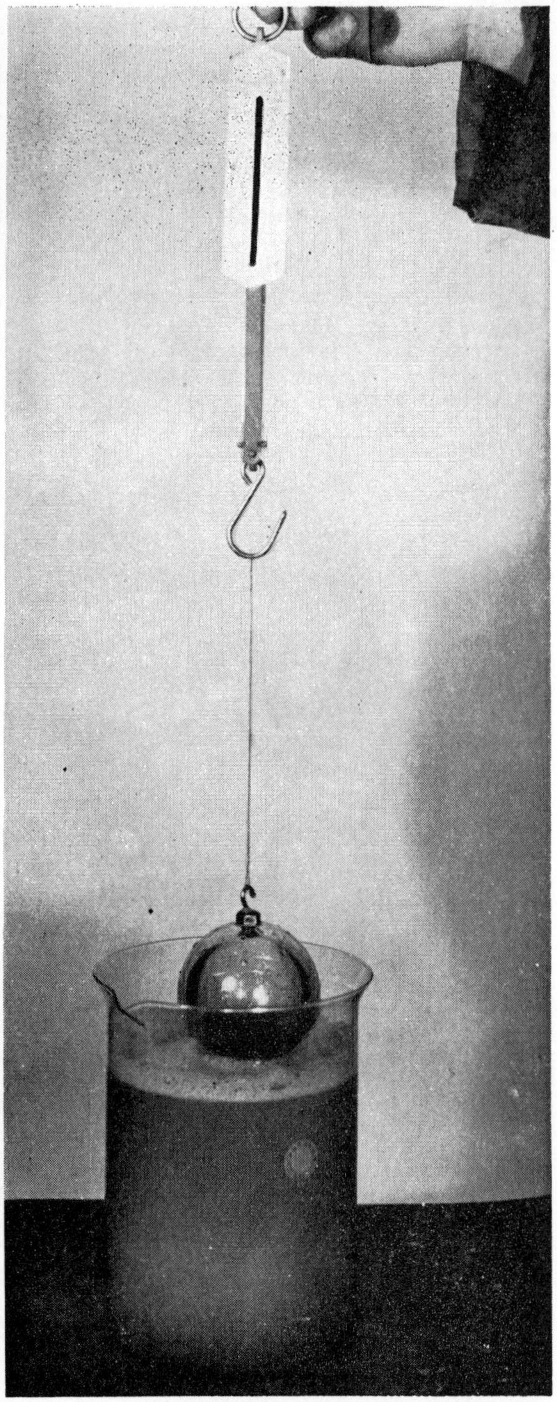

HYDROSTATIC BALANCE TO MEASURE DENSITY OF BUBBLY OIL.

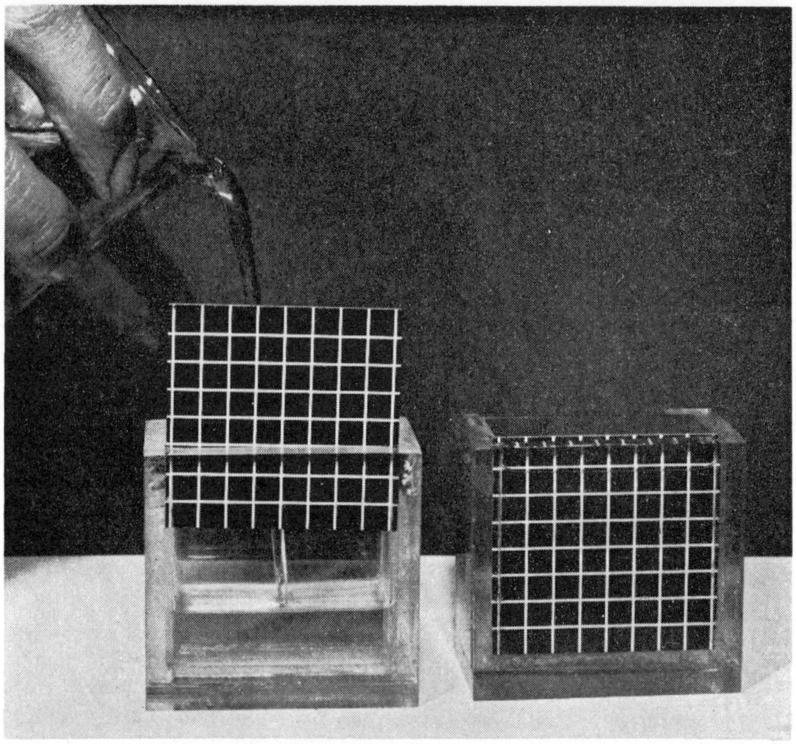

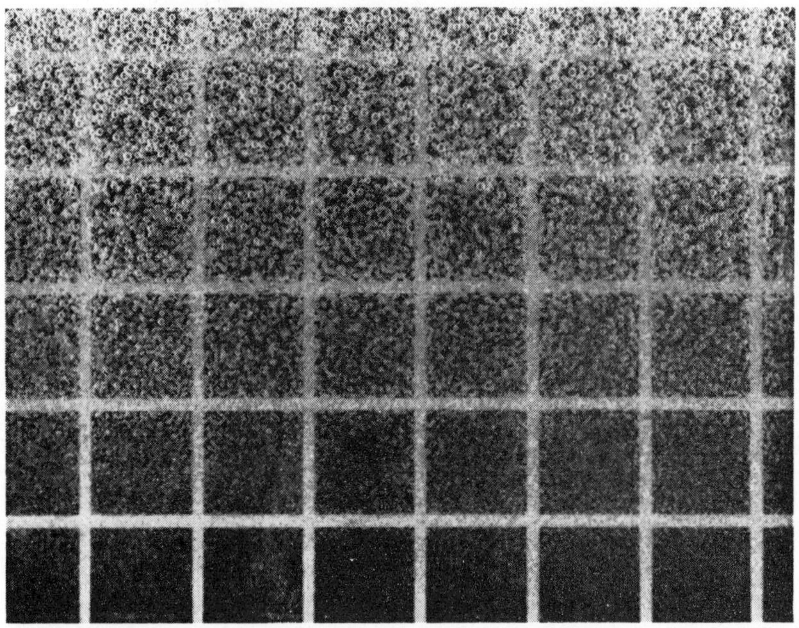

RULED TRANSPARENT PLASTIC BOX TO TRAP INDIVIDUAL BUBBLES FOR MEASURING AND COUNTING.

made if oil temperature differs from container temperature. Oil expands 1% in volume with an increase of 14° C (25.2° F) in temperature.

Viscosity—Bubbles change the viscosity of an oil slightly, and in a consistent manner. The empirical relationship, derived from dozens of tests, is this: $\mu_b/\mu_o = 1 + 0.015\ (\%\ B)$ where μ_b is viscosity of bubbly oil, μ_o is viscosity of non-bubbly oil, and $\%\ B$ is percent of bubbles. For example, if B is 10%, $\mu_b/\mu_o = 1 + 0.015 \times 10 = 1.15$. This equation holds true for bubble contents from zero to 25%, pressures near atmospheric, temperatures near ambient and probably far beyond, bubbles evenly dispersed, and oil not pumped.

Average bubble diameter has no effect on viscosity. Additives don't affect viscosity except for silicone anti-foaming additives; they increase data "scatter" but not average value. Value of μ_b/μ_o is not greatly different whether oil has high or low viscosity.

Direct measurement of viscosity of bubbly oil is difficult, because bubbles are unstable. It takes only 15 sec for a bucket of bubbly oil at room temperature and 20 centistokes (100 SSU) to release half its bubbles. Under agitation, they persist slightly longer, but not long enough to permit accurate readings of viscosity with any conventional viscometer.

But an indirect technique works quite well. Bubbly oil and non-bubbly oil are tested with the same simple

WAYS TO MEASURE OVERALL BUBBLE CONTENT

OIL FLOW CONDITION	PREFERRED APPARATUS
Open sump: approximate method	Pycnometer (weight of oil)
Open sump: accurate method	Hydrostatic balance
Slow-flowing effluent (later reading)	Measuring tube (low-cost apparatus)
Slow-flowing effluent (immediate reading)	Deadweight (pressure dissolves bubbles)
Fast-flowing effluent	Hydrostatic balance

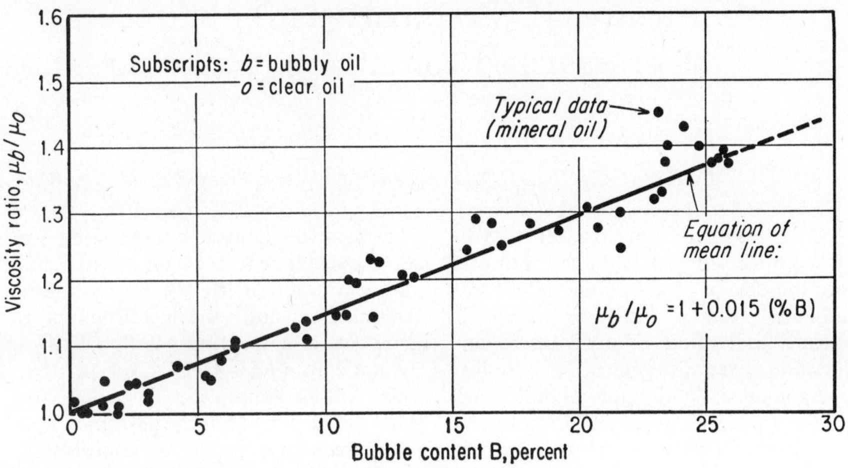

EFFECT OF BUBBLE CONTENT ON VISCOSITY RATIO.

flow tube (a short straight nozzle). Viscosity for each oil is calculated from flow, pressure, and tube dimensions. Absolute errors caused by turbulence are large—up to ±25% of the true viscosity—but are nearly alike for bubbly and non-bubbly oil. The ratio μ_b/μ_o can be established fairly accurately by test and μ_b can be calculated if you know μ_o from more conventional sources of data.

Dimensions of the viscometer tube greatly affect accuracy. In a short tube, high-velocity oil is turbulent; in a very long tube, low-velocity oil is subject to segregation of the bubbles, and this invalidates the readings; in a capillary tube, bubbly-oil flow is severely restricted because the bubbles are about the same size as the tube and the apparent viscosity is greatly magnified. A 0.25-in. dia tube 6 in. long is best. Most errors then are from turbulence, and although the errors are large they are fairly consistent and predictable.

Compressibility—Compressibility is usually expressed in the reciprocal form called bulk modulus, which has units of psi/cu in./cu in. This is a measure of the external pressure that must be exerted on the six faces of a hypothetical cube to compress it a given percent by volume. For example, if it takes 10 psi to reduce a 1-in. cube of oil to 0.99996 cu in. (a reduction of 0.00004 cu in.), the bulk modulus is nominally 10/0.00004 = 250,000 psi/cu in./cu in., or more simply 250,000 psi.

When the oil contains bubbles, the analysis is more difficult, because compression of the air may be adiabatic, isothermal, or something in between. However, tests on bubbly oil indicate that the air-bubble compression under normal conditions is nearly isothermal (the oil quickly absorbs the heat of compression) and the so-called isothermal bulk modulus may be used.

There are several basic variations in the definition of bulk modulus, including the one for isothermal bulk modulus. If the fluid is already under operating pressure, and you want effective bulk modulus for rapid slight deviations from that pressure, use tangent isentropic bulk modulus. For slow changes near the operating pressure use tangent isothermal bulk modulus. For overall change from atmospheric to a given pressure, the proper term is average or secant bulk modulus—isentropic for sudden pressure change, isothermal for slow. The applicable ASTM (American Society for Testing Materials) standards are explained in the *Proceedings of the 16th National Conference on Industrial Hydraulics,* Vol. 14, 1960, available from the Illinois Institute of Technology.

An approximate relationship for overall bulk modulus of bubbly oil is:

$$\beta_b = \frac{\beta_{oil}\,(V_{oil}/V_{air} + 14.7/P)}{V_{oil}/V_{air} + 14.7\,\beta_{oil}/P^2}$$

where: β_b is isothermal bulk modulus of bubbly oil, psi; β_{oil} is isothermal bulk modulus of oil alone, psi; V is volume and P is pressure of bubbly oil. It is assumed that the air and oil compress independently, which is a reasonable assumption for nearly atmospheric pressure.

Example: Assume bulk modulus β_{oil} is 300,000 psi, operating pressure 20 psia, and bubble content 10% by volume based on atmospheric conditions. Then:

$$\beta_b = \frac{300{,}000\left(\dfrac{100}{10} + \dfrac{14.7}{20}\right)}{\dfrac{100}{10} + \dfrac{14.7 \times 300{,}000}{20 \times 20}} = 292 \text{ psi}$$

Note that the combined modulus is drastically below that of oil alone. Even for a low air content of 0.1%, combined modulus is only 25,000 psi.

One limitation: the equation doesn't hold for high pressures because the bubbles dissolve in the oil and the bulk modulus changes as the bubbles dissolve. At zero time (before the bubbles start to dissolve) the equation is correct; at infinite time (all the bubbles dissolved) the modulus will be that of the oil alone. To predict the modulus at times in between you must know the rate of solution.

How to design, select, and apply

Pressurized containers

With good reason, the Bureau of Explosives prepares ICC regulations for shipment of pressurized containers. Make no mistake!

CARTER C. HIGGINS, President, Worcester Pressed Steel Co, Worcester, Mass

JUST filling a tank or bottle with compressed gas and integrating it into the design of a machine is not too difficult. What's hard is doing it cheaply and safely. You've got to understand three disciplines: gas physics, stress analysis, and safety engineering. Of the three, safety engineering exerts the greatest influence. If you cannot satisfy state, national, and industrial safety regulations, you may end up with a machine that is un-insurable and which nobody will buy.

Iron bottles of laughing gas, imported from London in 1870, were the first gas cylinders in the US. By 1911, Pittsburghers were bottling oxygen and acetylene in tanks of "tough, pliable steel . . . tested to at least twice the working pressure."

Today, a surprising variety of machines are bottle-fed from one type of pressurized gas container or another. For instance, helium compressors for some nuclear reactors have hydrostatic gas bearings that lift the heavy rotors friction-free until the rotating speed generates a supporting gas film hydrodynamically (*PE*—Oct 14 '63, p 116). Gas bottles supply the pressure for that initial lift. Diesel, gasoline, and jet engines have bottles of compressed gas for powering the pneumatic cranking motor. Metalworking shops have tanks of compressed nitrogen to accelerate the high-velocity pistons of certain impact machines. Airbrakes on trains and buses always need pressurized tanks. Refrigeration systems are pressurized. Even industrial hydraulic systems have compressed-gas chambers if accumulators are used or if the reservoir is pressurized with inert gas.

This article concentrates on containers small enough to mount on a machine—up to about 10 gallons. Site-mounted tanks are discussed briefly; the principles are just the same. The chief problem always is: how to get as much usable pressurized liquid or gas (or both) as possible into the container without creating a hazard.

Thar she blows

What constitutes a hazard is carefully, perhaps arbitrarily, spelled out by various regulatory agencies. The Interstate Commerce Commission (ICC), Coast Guard, and Federal Aviation Agency (FAA), take great interest, and any container shipped loaded is fair game to them. On-site pressurized tanks are scrutinized by state agencies, backed up by ASME boiler and unfired pressure vessel codes.

Even the aerosol cans that squirt your shaving cream are designed in

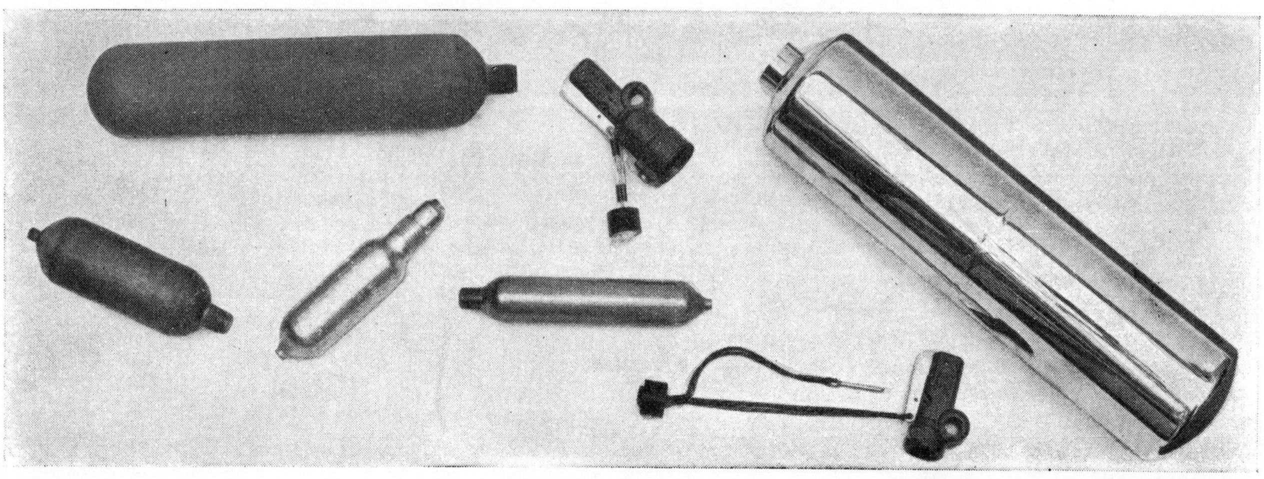

TABLE I..**Safety rules**

1..Gas containers are marked to note contents in accordance with Booklet Z48.1-1954 of ASA (American Standards Assn). Do not substitute other gases.

2..Purge before filling any small cylinder from a large one, and then only with the help of an expert on gas characteristics, filling densities, and safety.

3..Detect leaks with soapy water, never with a flame.

4..Empty any leaking or damaged cylinders, and those that have been exposed to fire. Close the valve, replace any protective caps or outlet plugs, and consult the supplier. DO NOT SHIP.

5..Keep below 120 F. Shield against direct sun's rays, never use a flame to melt accumulated ice or snow, never weld to cylinder even if it is empty.

6..Avoid extreme cold on high-pressure cylinders because the alloys may lose shock resistance.

7..Store cylinders in dry, cool, well-ventilated area. Avoid damp places, salt air, corrosive atmosphere.

8..Keep volatile gases away from oil, gasoline, and waste. Oxygen should be stored at least 20 ft from gasoline or fuel gas. LPG gases are not allowed in certain tunnels.

9..Open pressurized containers slowly. Do not tamper with the safety devices—your life may depend on them.

10..Tighten connections between tanks firmly but do not force; they are designed to be made up easily.

11..Store cylinders of argon, nitrogen, and oxygen in an upright position to make vapor venting possible.

12..Provide gas masks or self-contained breathing apparatus for handlers of poisonous gases, with appropriate and prominent instructions.

13..Use old cylinders first, because of outdating, possible corrosion, or aging of steel, affecting flexibility.

14..Avoid electrical currents through container, such as accidental grounding near arc welders.

15..Avoid mechanical damage, because small scratches and dents create high-stress points that later might fail. In other words, use tanks as containers, not as rollers or supports. Protect them from loose or falling objects by suitable covers.

16..Move cylinders with fork truck or other proper lifts, never by dragging. Do not grasp by caps or valve handles.

17..Build safe stands for topheavy cylinders.

18..Retest all pressure containers periodically in accordance with applicable regulations, some of which are mandatory.

19..Avoid horse play, and jet-dusting of clothing. High-pressure gases are as lethal as a bullet.

strictest accordance with carefully written safety standards. The Aerosol Guide, published by the Chemical Specialty Manufacturers Assn, summarizes applicable tests and regulations covering steel, glass, and aluminum containers of up to 32 oz of water capacity. And the Compressed Gas Assn in New York is a good source of information on working pressures of various gases.

The dangers are greater than you might realize. Some special bottles have test pressures up to 43,000 psi! Even an aerosol can of shaving cream (less than 100 psi) can make an awful mess in your suitcase if punctured. A cartridge of carbon dioxide at 838 psi and 70 F will reach 2265 psi if the temperature is increased to only 130 F. A 1-in. dia Hi-Pac CO_2 cylinder for life belts (1000 psi) can soar 200 feet in the air when punctured. Oxygen cylinders in military aircraft are wrapped with steel wire to prevent shrapnel effects should the cylinder be punctured by a bullet.

Obviously this is a field to be conservative in. For one thing, learn and accept recommended safety factors. And secondly, keep the tanks away from heat. Don't leave aerosols lying around in the hot sun. Never scratch or dent a high-pressure container. Rules of thumb are given in Table I. The rest of the text will help you design, select, and apply small pressurized containers with a minimum of trial, error, and danger.

TABLE II..**Vapor pressures, psi**

	At these temperatures, deg F						
	—35	**0**	**35**	**70**	**105**	**140**	**175**
Freon 12	8 in. Hg vacuum	9	33	70	126	206	320
nitrous oxide	165	295	475	745	1500	2450	3450
propane	3	24	57	109	185	290	435
carbon dioxide	145	294	512	835	1580	2530	3520
ammonia	5 in. Hg vacuum	16	52	114	214	365	579

Above cylinder pressures assume cylinders charged to recommended filling densities.

*Filling density=percent by weight of liquified gas actually in container at 70 F, related to weight of water the container will hold.

TABLE III..**Volume-pressure relationships**

	Perfect gas
Chemical symbol	none
ICC regulation:	none
Liquid* filling density (typical %)	none
Volume, in.^{3}, to hold one ft^3 of free gas at 70 F and at various psi	221.5 at 100 psi; 14 at 1800 psi
Expansion ratio, %: volume free/ volume compressed	7.8; 123.5 (for conditions above)
Comments	none

(table continues, p. 36)

Gas or liquid

A typical small pressurized vessel contains gas, or gas and liquid. Ideally, the liquid is a puddle of liquefied gas, chosen to have the desired vapor pressure at the expected operating temperature (see Tables II and III). As the vessel is depleted, the liquefied gas vaporizes and keeps the pressure up until nothing but gas remains. Then pressure drops rapidly.

Aerosols, which are the commonest and cheapest pressurized containers, are all based on that principle. Thus they maintain constant pressure throughout their life, and are relatively safe at all ordinary temperatures. At elevated temperatures, however, vapor pressure increases dangerously, as shown in Table II.

Many gases, including nitrogen and oxygen, cannot be liquefied at room temperature. These are stored in one of two ways: as a cryogenic liquid; or as a gas at high pressure but at room temperature. Only the latter is within the context of this article. Cryogenic liquids such as liquid oxygen (LOX) are a subject in themselves.

Vessels for plain liquids such as hydraulic oil and water may also operate under pressure. Examples are fuel filters built to withstand fuel pump pressure, cylinders for piston-type accumulators (*PE*—Aug 6 '62, p 60), heat-exchanger drums, and even beer kegs. However, the dangers are less severe because the likelihood of explosive expansion is less—except in the case of water expanding into steam.

Working pressures

Aerosol cans for paint and insecticides rarely exceed 100 psi. Shop air goes up to 130 psi, as does city water pressure. Propane and butane cylinders usually are charged up to 240 psi. Low-pressure vessels according to ICC definitions have less than 1000 psi fill pressure, and most are less than 500 psi. Nitrous-oxide pressurization in food aerosols goes as high as 1087 psi at 98 F. Sunken ocean buoys must withstand external pressures of over 7000 psi, which means you must either make the chamber strong or prefill it with high-pressure gas. If there were a practical way to equalize external water pressure with internal gas pressure as the chamber was lowered, its wall could be very thin.

Test or proof pressure under ICC procedures, for conventional high-pressure cylinders from about 1000 psi and above, is 5/3 working pressure. Low-pressure cylinders are tested at twice working pressure. The ICC performs some tests in water jackets, though not all tests are run that way. A jacketed test usually is applied to one sample out of 200 low-pressure vessels, and measures the total or plastic expansion and permanent expansion that occurs when test pressure is applied internally. Permanent expansion allowed is 10% of total expansion.

The ICC demands that pressurized cylinders be stress relieved and flexible, meaning that yield strength must not be more than 67% of tensile strength. This lowers the absolute yield strength but eliminates variations due to work hardening.

ASME tests differ from those of ICC. Test pressures are 1.5 times working pressure, and the vessels are struck with hammer blows, making the tests more severe and thus more conservative. Burst strength generally is specified over 3 times working pressure.

Comparing ICC and ASME: Wall thickness in accordance with ICC regulations sometimes can be about 86% of that under ASME requirements. The UL (Underwriters' testing laboratories) specifies ICC standards where applicable.

On low-pressure cylinders with pressures below 200 psi, such as ordinary fire extinguishers, UL specifies burst strength 6 times working pressure, and for cartridges, 5 times. Actual testing is at 3 times working pressure. Stress relieving afterwards sometimes is required.

Container styles

There is nothing proprietary about good container design, and the most economical method of manufacturing becomes the key consideration. Quantities naturally play their part. Here we shall primarily discuss containers under 10 gallons, which is 2300 in.,

continued

TABLE III.. **Volume-pressure relationships (gases and vapors)**

Air	Acetylene	Ammonia	Argon	Carbon dioxide	Freon 12	Helium	Nitrogen	Nitrous oxide	Oxygen	Propane
none	C_2H_2	NH_3	A	CO_2	none	He	N	N_2O	O_2	C_3H_8
varies	8; 8AL	4A480	varies	3A1800 3E	9; 41; 4B225	varies	varies	3A1800; 3E	varies	4A150; 41; 9; 4B225
none	acetone used as filler	54	none	68	119	none	none	68	none	42
221 at 100 psi; 13.97 at 1800 psi	15 at 250 psi	2.22 at 114 psi	219.7 at 100 psi; 13.22 at 1800 psi	4.65 at 835 psi	7.74 at 70 psi	222.3 at 100 psi; 14.87 at 1800 psi	221.3 at 100 psi; 14.13 at 1800 psi	4.66 at 745 psi;	219.5 at 100 psi; 13.19 at 1800 psi	7.68 at 109 psi
7.82; 123.7	115.2	778.3	7.86; 130.7	371.6	223.2	7.77; 116.2	7.81; 122.3	370.8	7.87; 131	225
none	dangerous to load	unpleasant odor	inert	high expansion; cools	good mixer	leak detector	inert	harmless to foods	dangerous with oil	gaseous fuel

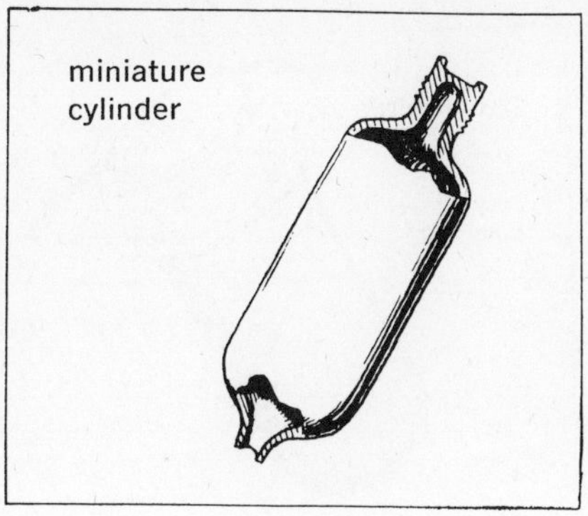

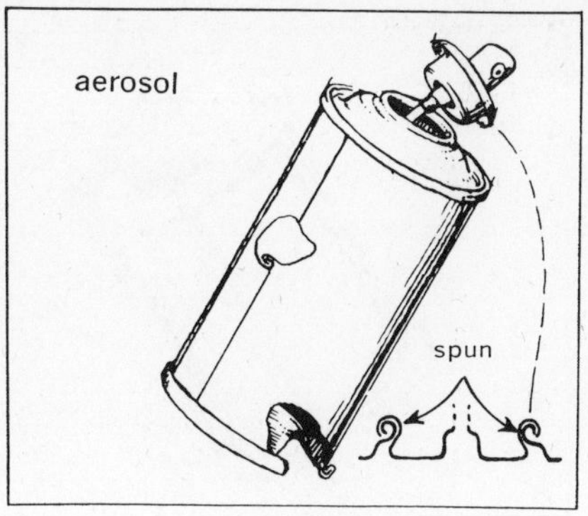

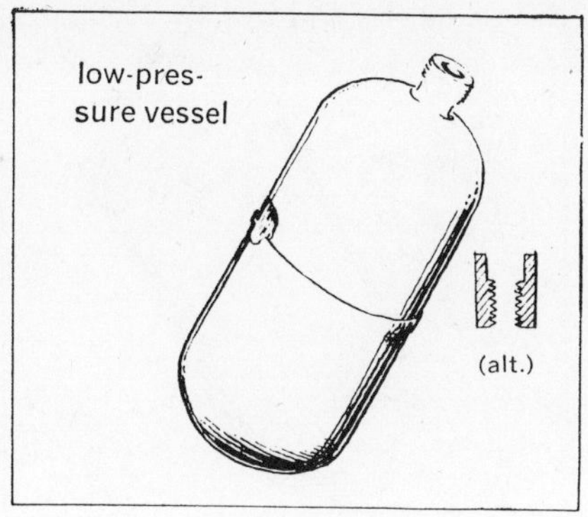

83½ lbs of water, or about 12 in. dia by 24 in. high.

Choose a standard container, preferably one suited for a higher pressure and rather larger than your need. It often will prove less costly than a precise fit, and will be safer.

Most containers are cylindrical because that shape is easiest to make. However, a sphere can withstand higher pressures and contains more volume for a given total wall area.

An aluminum sphere 13 in. OD with 0.5 in. wall can withstand 7500 psi. Some are applied as deep-sea submerged buoys. A compromise here could be a sphere of thinner wall pressurized internally with inert gas at half the external test pressure. The wall stresses would be halved, but would go from tension to compression during descent.

For practical purposes, three classifications cover almost any type of cylindrical pressurized container: miniature cylinders, low-pressure vessels, and high-pressure vessels. The terms container, cylinder, vessel, bottle, flask, tank, and tube mean essentially the same thing: a pressurized reservoir of some sort.

Miniature cylinders: Not actually cylinders but steel bottles, they come in assorted sizes from 8 gm (about ½ in.^{3} content) to 35 gm (about 6¼ in. long and 1 in. OD). They are mass-produced with highly automated tooling, and usually filled with CO_2. A 12-gm miniature cylinder of CO_2 contains about ¾ in.^{3} and costs less than 20¢. The larger sizes range up to about $1. For sizes greater than standard, you can manifold the cylinders together.

Miniature cylinders are punctured to release the contents and are not reusable. They are manufactured by deep drawing and are seamless. The filling end is sealed in any of several ways, including a welded-in or a spun-in pierce disc, or a staked end. Manufacture to other than standard designs would be uneconomical.

Low-pressure vessels: In sizes from 1/3 to 2 lb of water (1 lb of water occupies about 27.7 in.^{3}), lock-seam construction is chosen. Aerosol cans are made this way. The amount of charging gas in each is small, keeping pressures below 100 psi.

For fairly large sizes, consider using beer kegs or carbide drums. Or try wrap-arounds. A number of air tanks for compressors, reservoirs for truck brakes, etc, are made by wrapping a sheet around (see sketches), welding them longitudinally, and adding two elliptical heads. In standard sizes, these are made with elaborate jigging, and welded with special submerged-arc fixtures. The spuds (attached threaded bosses for the con-

TABLE IV. . Typical causes of failure

1. . Design errors (see text for recommended practices)
2. . Failure to inspect or retest properly
3. . Overpressurizing (or excessive vacuum)
4. . Corrosion or erosion of the metal
5. . Malfunction of automatic devices permitting overpressure
6. . Excessive heat or excessive cold
7. . Water hammer or carry-over of system material
8. . Ventline not functioning, or not used

Stress and wall thickness

The stress on the wall of a sphere or an infinitely long tube can be figured, but most cylinders are of irregular shape. The stress we seek to measure is at working pressure (or at required pressure for temperature-sensitive gases) in temperature ranges from −20 to 130 F, without special allowances for bumping, corrosion, and aging of steel. Hence retesting is called for on a regular basis.

Sidewall strength is the criterion. The minimum yield strength of materials used in a stress-relieved and flexible condition determines the minimum wall thickness of a cylindrical shape that contains fluid under pressure. Failure rarely will occur at the ends if they are hemispherical or at least elliptical.

S = working stress, psi
t = minimum wall thickness, in.
P = proof-test pressure, psi
D = outside diameter, in.
d = inside diameter, in.
R = inside radius, d/2, in.

(continues on p. 38)

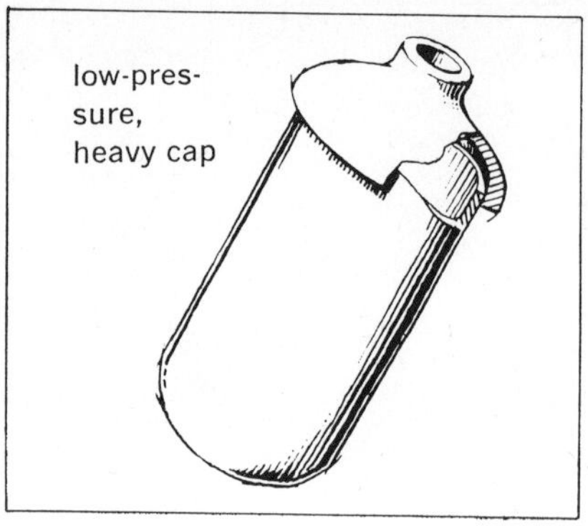

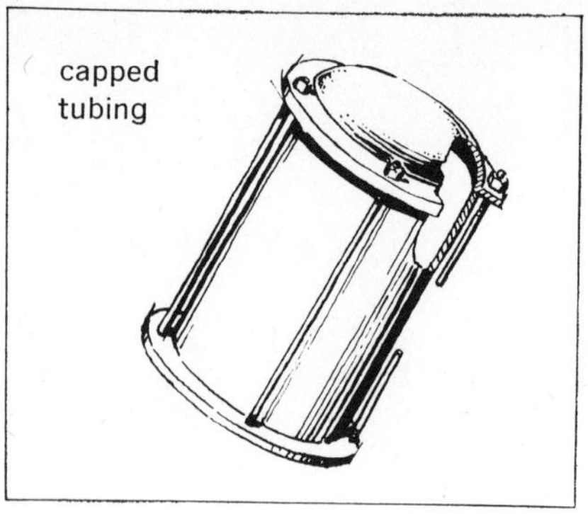

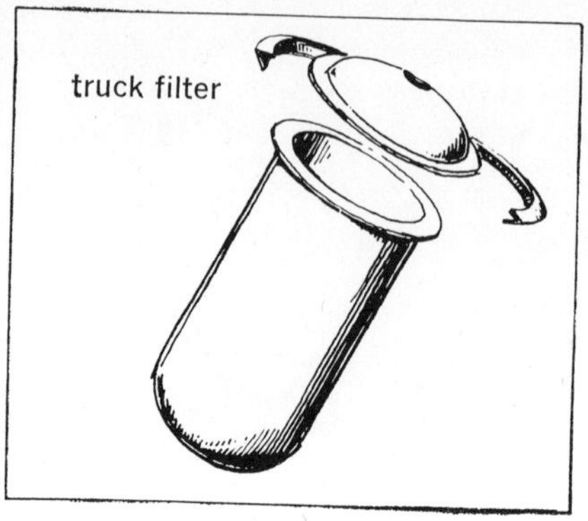

tainer valves) are projection welded, or welded metal projections tapped.

We now see more low-pressure containers made of two drawn halves overlapped and brazed or girth welded, and with the spuds welded on. Because long-stroke drawing presses needed are comparatively slow, and tooling is costly, most containers over 6 in. long are centrally joined to avoid using a deep shell.

Some capped containers are made of tubing with two end caps. These are cheap to tool, but in quantity cost more than joining two halves. For example, a 3-in. dia, 12-in. long dry-powder fire extinguisher shell (ICC-4B240) costs about \$1.45 each vs about \$0.80 each to join two halves and add a 9¢ spud. Sometimes, for a shorter container, no spud is needed. A heavier cap is used, which can be threaded at the extruded outlet, and welded or brazed to a lighter shell.

Tubing is useful in refrigerant lines, where screens holding decontaminants are placed in lengths of copper or stainless tubing, and the ends spun.

Aluminum float cylinders are sometimes impact-extruded. Specification ICC-4E for drawn welded halves with 225-500 psi is the only aluminum specification. The finished container, good for holding LPG (propane) for welding or camp stoves, is shipped under special permit and has a rupture disc safety release. Aluminum is used for high pressures in England, and is being considered here for CO_2 cartridges to charge beer kegs at home.

High-pressure vessels: If a container is not too large, the ICC favors seamless construction. Spun tubing is limited by the diameter of tubing available (about 16 in.) but can be quite long when used for gas storage. ASME-designed vessels such as accumulators, vulcanizers, and carbon-dioxide sublimating chambers are welded and may have caps held by bolts. Larger high-pressure vessels, 7 to 54 in. dia., may call for forgings, or perhaps shell castings with heavy walls. Investment castings are being studied by the missile people primarily for hollow vessels with integral lugs for holding purposes. These vessels are costly but do get away from welded lugs.

Spun tubing is used in varying quantities and outlet configurations. In large quantities, such as for life-raft cylinders, drawn shells with the outlet end spun in are used. The bottom is closed; the spun wall of the outlet end will be considerably thicker than the side wall, handy for threads and O-ring grooves. Half-pound carbon-dioxide cylinders (21 in.^{3}) and some larger fire extinguishers are standard.

If one end is to be completely open, or at least have a large opening,

Stress and wall thickness *continued*

Applicable equations.

- ICC (based on the Bach Clavarino formula)

$$S = \frac{P\,(1.4D^2 + 0.4d^2)}{D^2 + d^2}$$

$$S/P = \frac{1.3\,(D/d)^2 + 0.4}{D/d - 1.0}$$

Poisson's ratio is assumed = 0.3.

- ASME code (Barlow formula modified by Boardman)

$$S = \frac{P\,(R + 0.6t)}{t}$$

$$S/P = \frac{0.6\,(D/d) + 0.4}{D/d - 1.0}$$

This is not used for very heavy small vessels, where t is greater than $R/2$. From that formula is derived the following, where inside diameter is known:

$$t = \frac{PR}{S - 0.6P}$$

Thus for a 2.8-in. ID cylinder at 3000 psi proof pressure, and 70,000 psi allowed for S, $t = (3000 \times 1.4)/(70{,}000 - 1800) = 0.0615$-in. minimum wall.

In the event outside diameter is governing, the formula is:

$$t = \frac{PD}{2S + 0.8P}$$

For a 2.87 OD cylinder this is $(3000 \times 2.87)/(141{,}400) = 0.0605$ minimum t. Drawing strip would be ordered 0.072 plus or minus 0.003 in., to allow for material tolerance, a degree of thinning in drawing, and a slight margin for decarburization.

drawn cylinders may be preferred over tubing. One company supplies drawn tanks to 100-gal capacity. Necking-in a drawn cylinder in a press is not easy or cheap due to possible bulging. To reduce a 4-in. dia shell to 2.4 in. on the end (40% reduction) calls for three necking operations and probably two anneals with appropriate tooling.

Both spun and forged containers may have heavy walls and substantially heavier ends, adequate to receive male threaded valves, drain plugs, etc. Drawn cylinders have extruded ends for the same purpose, and necking down adds some thickness to the end. Bottoms remain at the original stock thickness, of course.

Where pistons are to be installed, there may be problems in sealing. The ID of seamless tubing is rather inaccurate, and hot spun tubing, though more accurate, is not smooth. Welded tubing, resized, comes closer. Tolerances on drawn shells are often tighter: a 0.006- to 0.010-in. spread on 3- and 5-in. dia shells, not counting out-of-roundness. However, a slight draft is needed—in the order of 0.002 in., closed end smaller, for each 3 in. of length, to help in stripping parts off punches without scratching. It might be necessary to grind or hone. Heat treating, of course, increases out-of-roundness.

For small-lot production, such as special valves or machine hydraulic cylinders, we still find uses for tubing lengths and caps that are gasketed and held by through bolts. An advantage is that they can be completely disassembled.

Stand-up cylinders are a separate problem. To get a drawn or spun cylinder with a rounded end to stand upright, rings and collars can be welded on if the tank is intended for low pressures, like those for propane. Where welding is not permitted and adhesives are not sufficiently reliable, the bottom can be pushed back into the body. On a drawn cylinder, a rather generous internal radius is needed so the die post will be substantial. Some spun and forged cylinders have almost flat bottoms.

Correct wall thickness

The one place where error can spell disaster is in the design of the container wall. You must consider all parameters, and allow for every possible abuse. Table IV lists typical causes of failures. Fortunately there are equations for about every situation, and most of them appear in the ASME Section VIII Code for Unfired Pressure Vessels, available from the American Society of Mechanical Engineers, New York. Special equations used by the pressurized-container industry are summarized on page 72.

Proper specifications for construction should consider: working pressure ranges and possible surges; working temperature ranges; description of contents including recommendations on whether the vessel should be lined with rubber, glass, alloy metal, or painted to prevent corrosion; gas volumes; information affecting metal properties, including welding; possible explosion of lubricating oil vapors in air tanks; and vibration, particularly if a compressor is mounted on the tank.

Stress on the ends of the container varies with shape. A flat end is under the greatest stress. The closer the end comes to being a hemisphere, the less it will expand under pressure. Frequently, cylinder ends are shaped with internal or external elliptical shapes, usually external. In testing at twice working pressure, such ends tend to expand permanently, and may lead to rejections. If a shallow elliptical head is designed on a cylinder that is only two or three times as long as its diameter, the problem is greater than on a longer cylinder. In a long cylinder, sidewall expansion helps take up out-of-roundness. The ICC does permit prestressing to within 10% of proof pressure, which helps reduce the degree and percent of permanent expansion.

This stress on cylindrical walls must be contained. Steel seems to be the best answer. Aside from aerosols, there are few aluminum cylinders in this country. Reinforced fiberglass containers, although very dependable, are not widely used where lowest cost is a requirement.

Low carbon steels are figured to a maximum stress of 24,000 psi, multiplied by joint efficiency if less than unity. Certain steels usually sold as having 50,000-psi minimum yield (GLX-50 is an example) are figured at 35,000 psi. These steels often are preferred on a strength-to-cost ratio, and give lighter cylinders. An *A* is added to the specification (eg, ICC-4B240 and ICC-4BA240).

High-pressure cylinders usually are made of alloy steels such as AISI-4130 (chrome molybdenum), heat treated and drawn, and are figured at 70,000 psi. This would be covered by ICC-3E (always 1800 psi) or ICC-3A followed by the working pressure marking. ASME watches steel analyses very closely.

In some cases, such as 3-in. dia brazed dry chemical extinguishers, a minimum wall of 0.041 in. is called for because of possible mishandling. Cylinders of low alloy steel over 6-in. diameter holding up to 1000 lb of water (2774 in.^{3}) must have a 0.078-in. minimum thickness wall.

Raw material for drawn cylinders is usually ordered about 10% over figured gage with sheet, strip, or plate tolerance above that. This permits some thinning in fabrication, and slight decarburization in alloy steels.

Inspection: Cylinder inspection is complicated by the wide variety of designs. Manufacturer or user inspection by "competent" inspectors is less costly than inspection by outside "disinterested" inspectors, who must follow complicated regulations. Manufacturer inspection is permitted for low-pressure cylinders such as ICC-3B (seamless), ICC-4B, ICC-41, ICC-9 and ICC-3E, and also for high-pres-

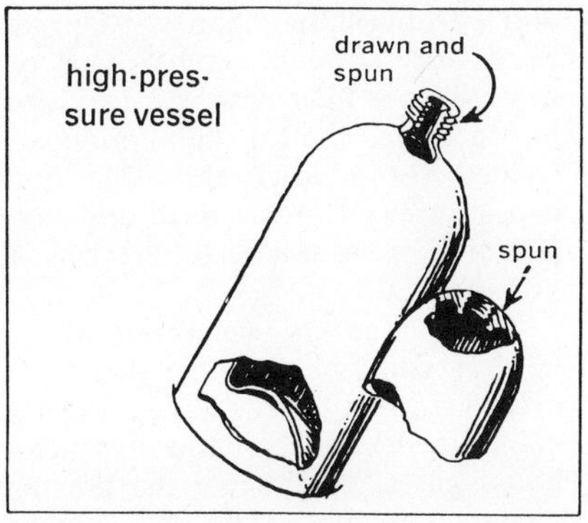

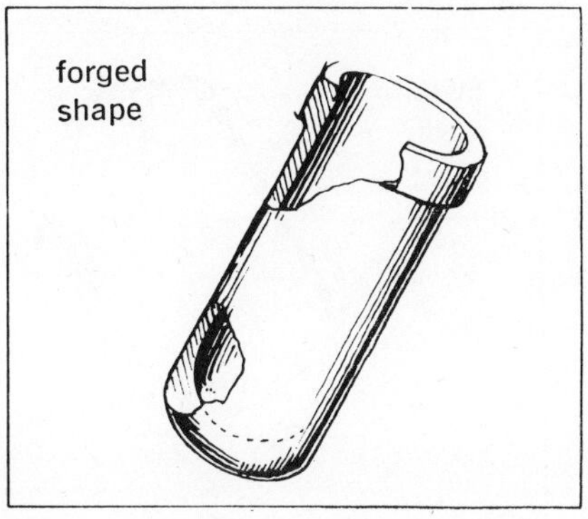

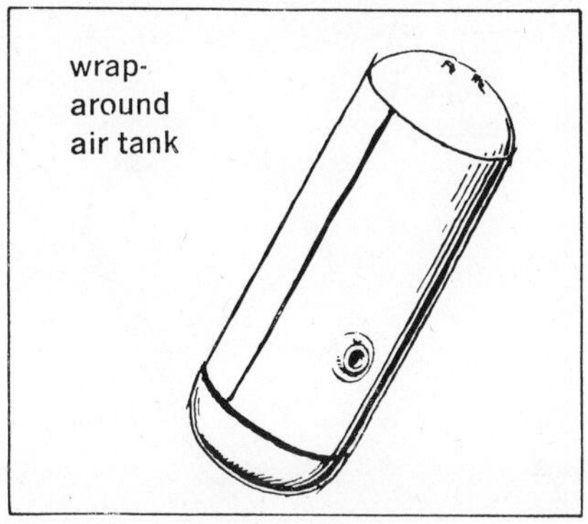

sure cylinders containing less than 4½ oz of water (8.9 in.³). ASME vessels up to 5 ft³ and 250 psi, or 1.5 ft³ at any pressure, are manufacturer- or user-inspected and stamped. Inspection of course is subject to audit by regulating agencies such as ICC.

Records of pressure tests or retests, after proper periods of time, are filed. They are useful for future testing after the factors of metal aging, corrosion, and knocks or local thinning have been at work. It is generally illegal to ship an ICC-3A, 3AA, 3B, 4A, 4AA, 4B, or 9 cylinder which has had no retest and stamp within 5 years. The month and year of manufacture are stamped with the marks of the specification, pressure, inspector, and manufacturer's marks near the neck. It may be necessary to re-heat-treat such cylinders before they can be restamped for use.

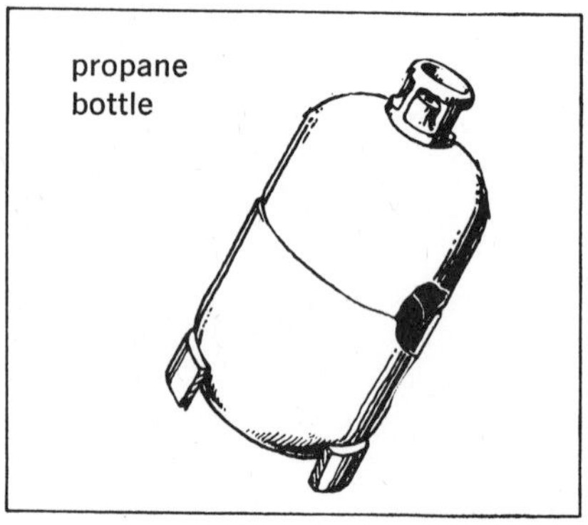

Non-critical pressure vessels below ASME minimums and shipped with no pressure are exempt by their nature, though ASME testing principles will help in good design. Where weight saving is vital, some military cylinders at high pressures are permitted to have harder materials than ICC specifications, and are shipped on special permits after testing.

Valves and caps

Safety valves may be spring-loaded, frangible-disc, or pin-type (for medical gases). An ICC-approved safety valve must be used on cylinders over 4½ in. dia and over 12 in. long exclusive of neck. For CO_2, N_2O, and gases at 1800 psi or over, relief valves are required. They may be located within the release valves. Safety releases in lower-pressure cylinders permit 10% more fill.

Aerosol valves often are spun over a curl on the can outlet. Releases for powder fire extinguishers are male projections screwed into spuds with an O-ring above the threads. Propane and refrigerant tanks have similar valves. Miniature cylinders may have threads below the pierce end, and are screwed into releases with central piercing pins. On sophisticated military bottles, the pins may be actuated by an electric current. Larger high-pressure bottles may be filled through a valve screwed into threads cut in an extruded bottle neck. Generally, O-ring construction is simple and effective. It is also used on drain plugs.

There are so many types of valves, spring valves, valves supported by gas pressures, cylinder packings, valves with replaceable puncture discs for release or safety, and special explosion-protected designs for missiles, that much ingenuity with little standardization goes into containers and controls. The desirable standard sizes for fittings suggested by the Compressed Gas Assn are covered by pamphlet ASA B57.1.

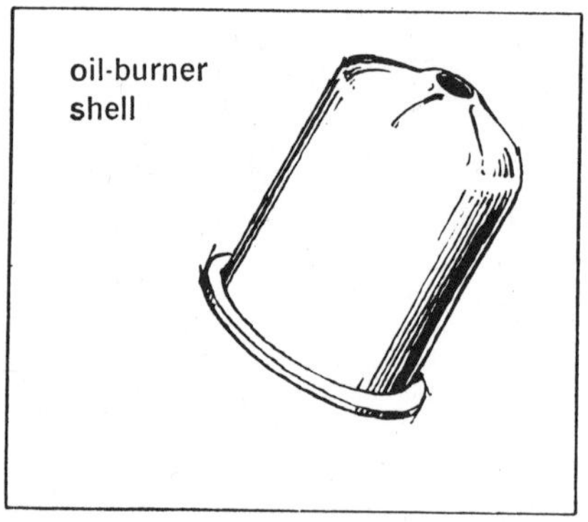

Caps: On a part like a pressure-reduction valve or an autoclave, one must be able to remove the cap. Top covers made in various ways may be gasketed and clamped over lugs on the end of the container. Some low-pressure containers are held by latch mechanisms. Truck oil filters have mating flanges on shell and cap held by circular hinged V-section clamps.

If the container is drawn, it is not good construction to try to get holding threads for screws in the flange, because the metal is almost always too thin, even if extruded. Using nuts and bolts, or adding weld nuts under the flange, are two answers. If a drawn container wall is to be threaded to hold the closure, thickness will be reduced. One may start with heavier metal, iron it about 20% most of the way up the wall, then use a rather fine thread. (Ironing means drawing a cup between a punch and a die with clearance less than the original stock thickness.) A flange or head, screwed on, supports the threads, but not for their full length. Thus there is a weak point.

Cast or forged open-end shells can have lugs or heavier flanges, but this may call for more costly fabrication. On the low-pressure side of a piston-cylinder pressurized container, the head merely holds and centers the piston rod; sheet-metal screws may serve to hold the cap.

Gaskets on a flanged container should be thick enough to take up out-of-flatness, unless the flange is ground at extra cost. To prevent a heavy gasket from squeezing out, some designs call for wide flanges press-trimmed and formed-up around a small radius. The small radius will require a stress-producing operation and perhaps another anneal cycle.

The heads of many filter shells are ribbed in a petal design to add rigidity and permit flatter shapes. Up to a third more stiffness results. One type of oil-burner filter case, not very deep, has a long central bolt projection-welded to the closed end. The outlet cap is screwed to the head and might flatten it, thus the stiffening ribs are very helpful.

The numerous specifications and lack of common practice designating cylinder sizes (viz: water capacity, cubic inches or cc, and weights of liquid gases) complicate the pressure-container picture. The field is nevertheless growing rapidly; more and more designers will need reliable and consistent information.

EDITOR'S NOTE: ICC, ASME, and other regulations set the pattern of design. Here is where you can get them:

ICC Regulations, listed in the Federal Register, Title 49, parts 70-78, are available from the Government Printing Office, Washington, DC. They are also incorporated in T. C. George Tariff 15, published by the Assn of American Railroads, Bureau of Explosives, New York.

ASME Boiler and Pressure Vessel Code (Sections II, VIII, and IX) are available from the American Society of Mechanical Engineers, New York.

Marine Engineering Requirements and Material Specifications (CG 115, part 54) can be obtained from the Marine Inspection Office of the Coast Guard, New York.

Aerosol Guide is available from the Chemical Specialties Manufacturers Assn, New York.

Standard 58 (LGP) and **Standard 51** (Welding and Cutting Systems) are published by National Board of Fire Underwriters, New York. **Code #58** (LPG) and **Code 51** (also welding and Cutting Systems) are obtainable from the National Fire Protection Assn, Batterymarch St, Boston.

—Franklin D. Yeaple

3

FILTERS AND CONTAMINANTS

Particle size, filter selection

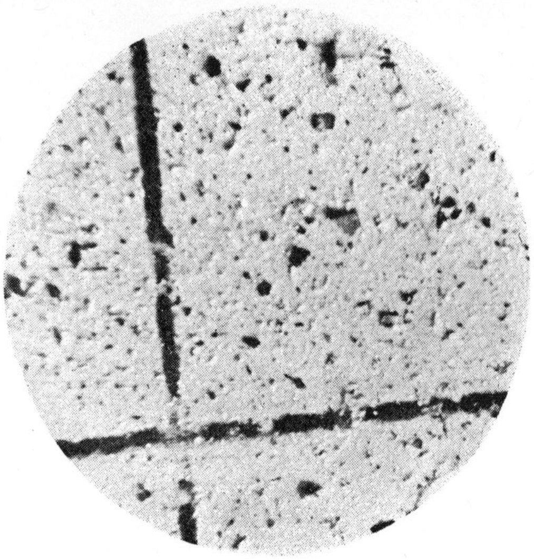

new criteria for MEASURING FLUID CONTAMINATION

It's no snap to measure and count dirt particles smaller than 40 microns in hydraulic fluids. Yet such data are often essential. Here are latest available measurement techniques.

FRANK YEAPLE
Associate editor

Lord Kelvin said that if you can't measure something and express it in numbers, then you don't have full knowledge of it. But this isn't always easy. The measurement of contaminant particles in a fluid is a good example: counting and measuring them seems more an art than a science, particularly in the microscopic sizes.

Scientific work on contaminant analysis is being done, however, because many modern fluid-control devices cannot tolerate dirt. And where system specifications call for extremely clean systems, ways are being found to monitor them before and during service. The need for such care was pointed out in the next chapter. It showed that passing fluid through an absolute 10-micron filter doesn't necessarily ensure freedom from 10-micron-and-over particles in an over-all system—there are other sources of dirt that filters can't reach until the system has recirculated awhile.

Dirt particles larger than 40 microns (1 micron = approx. 1/25,000 in.) are relatively easy to measure and remove. For one thing, they are visible to the naked eye. But some systems must be cleaner; some servovalves have less than 0.0002-in. (about 5 microns) clearance between operating parts, and require super-methods of cleaning and monitoring. Today's state of the art is at about the 5-micron level in absolute removal rating: particles 5-25 microns in size can be removed and measured with some consistency, but below 5 microns, only extremely rigorous measures will work.

HOW CLEAN IS CLEAN?

Clean fluid from an oil drum is too dirty for a missile servo system, yet is clean enough for an ordinary industrial control. Cleanliness therefore is relative.

Ordinary contamination by solids differs from "film" and "bacterial" types. Clean liquid-oxygen systems must be free of hydrocarbon films, but can tolerate some inorganic dirt. Hydraulic systems must be free of solids, but are not troubled by films. Surgica"y clean equipment can contain some dirt, so long as the dirt is sterilized!

The dirt particles

Particulate dirt—the type discussed in this article—is rarely uniform in shape. (The full-color photomicrograph on the cover of this issue is an example of actual filtered dirt.) Most particles are jagged and irregularly shaped, varying from submicron metal dust (worn from moving parts), to organic fibers several inches long. For comparision, the diameter of a red blood corpuscle is 5 micron; of bacteria, 1 micron; of human hair, 75 micron.

The method of dimensioning a particle differs with the purpose. Smallest dimension of a fiber should be given if its tendency to pass through a surface-type filter is being evaluated. However, its longest dimension is more important when evaluating fluid contamination downstream of the filter, because the long dimension may easily bridge and block an orifice or clearance space.

Sources of contamination are virtually impossible to list because there are thousands. Common ones are: lubricants used on pipe fittings; atmospheric dust of almost any material; metal scraped off one or both surfaces of mating or moving parts; metal scraped off during making-up of fittings; particles from pumps and accumulators; elastomer seals; sand in castings; oxides; lint; oil additives that decompose or become gelatinous; bacterial growth in fluid or system; chemical action; paint; and residue from overheated oil.

It is not enough to say "clean" or even "superclean." You've got to say exactly how clean, of what, and where. Many users of extremely clean hydraulic fluids—the aircraft and missile industries are leaders in this area—have defined the limits in terms of their own specific needs.

Company standards

Douglas Aircraft's missiles division requires that a 100-milliliter sample of system fluid must not produce "background color" when filtered through an 0.8-micron, 47-

millimeter-dia cellulose membrane (made by Millipore Filter Corp, Bedford, Mass). Fewer than 10 particles over 40 microns are allowed. A fluid so pure is termed "super-clean" by Douglas; extreme precautions are taken during manufacture, assembly, test, and operation at frequent intervals.

Government-specification fluids such as petroleum MIL-H-5606A already have strict quality control during manufacture, including a required passage through 10-micron filters before packaging in containers. For special applications some users (government agencies) get even closer control by specifying a random sampling of the oil. Here is one example, for quart and gallon cans:

- Take as many 100-milliliter test samples as the cube root of number of containers filled in one day.
- Total of particles over 100 microns in size to be less than number of samples.
- Filtration time for each 100-ml sample to be less than 50 minutes.
- Weight of sediment from any 100-ml sample not to exceed 3.0 mg.

Some users have suggested including a detailed count of particles in the sampling of MIL-H-5606A petroleum fluids. The suggested limits (not yet a standard) are:

Particle-size range, microns	5-15	16-25	26-50	51-100
Max number allowed in 100-ml sample	20,000	9000	2500	150

These counts are actually higher (in 25-100 micron range) than some users already allow. For example, in-the-factory standards set by The Martin Co for airborne and ground operational equipment require the following contamination limits to be met for MIL-H-5606A and 6083B fluids:

Particle-size range, microns	0-25	25-100	over 100
Max number allowed in a 100-ml sample	unlimited	250	10

STANDARDS OF MEASUREMENT

It is not enough to specify cleanliness; you must also specify the way to measure it. All standards and methods of contaminant-particle measurement are based on particle count and size in a given volume of fluid.

Sampling-type methods require a sample of fluid to be taken, then passed through a special filter to collect all particles for examination.

Onstream-type methods count particles in a given size range as they flow through the actual system.

No method is completely satisfactory, but the most accepted one is sampling-type technique ARP 598 developed by SAE (Society of Automotive Engineers).

Ruled-membrane sampling (ARP 598)

Aeronautical Recommended Practice ARP 598 issued by SAE specifies filtering a 100-ml sample of fluid through a membrane of cellulose nitrate. The membrane is ruled as a grid, with 100 equal squares covering the entire effective area. All particles can be counted, or just those particles in a representative (random) number of squares. The random count can be extrapolated to represent all

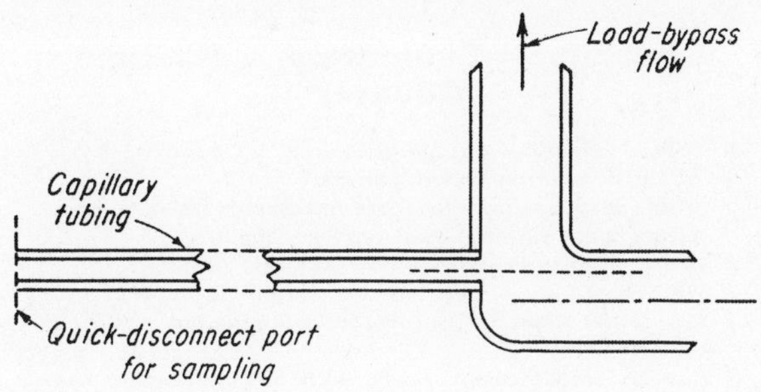

SAMPLING ELBOW designed by Douglas Aircraft has offset capillary outlet and gives free path from oil stream to sample bottle via quick-disconnect valve. Offset outlet extracts fluid at point halfway from centerline to pipewall.

the squares. (A microscope is needed to make the count.)

The general acceptance of ARP 598 as a standard is based partly on its carefully detailed instructions regarding cleaning apparatus, transferring of samples, filtration, examination, microscopic equipment, lighting techniques, and resolution. Tests can be duplicated reasonably well by different operators. And it is a self-checking procedure: the slide is viewed by more than one lab. Standard calibrated slides are available, and both operator and equipment can be checked with standard method.

Accuracy in particle count of plus or minus 20% may be expected when using ARP 598, if the operator is highly qualified. This estimate is based on an interesting series of "round-robin" tests sponsored by Marshall Space Flight Center (formerly Redstone Arsenal) in Huntsville, Ala. The tests were made this way:

- Three Millipore Corp membrane disks containing varying degrees of particulate contamination were sealed between glass plates. Each of these slides was examined independently by 10 companies: 4 filter manufacturers, 3 missile manufacturers, 2 servovalve manufacturers, and a research institute.
- Each company followed counting procedure in ARP 598 as rigorously as possible. Some already had experience in using the technique.
- The average count was calculated, and deviation plotted from this average.

Results showed that although some inexperienced operators disagreed considerably in total particle count (as much as 50% based on an established average), the qualified operators were able to count consistently close, within 20% deviation from the average. However, some system designers do not think the 20% figure is realistic, and would choose a much higher one.

Other membrane-type tests

Filter-disk tests are as old as chemical knowledge. The fluid passes through, and the residue is weighed (gravimetric method); counted; or just looked at (patch background color). Any of these tests may be used in evaluating fluid contamination.

Gravimetric tests simply weigh a membrane that starts out clean and dry (heated and dessicated) and then weigh it again after it has filtered a given volume of contaminated

Checklist—Evaluating a Particle Counter

√ Will it distinguish dirt particles from air bubbles?
√ Will it detect transparent particles?
√ What particle dimensions are measured (volume, cross-section area, surface area, random length)?
√ Can it distinguish between two adjacent particles?
√ What about small particles near large particles?
√ Are results reproducible throughout expected range of contaminant sizes?
√ How do results compare with ARP 598 over entire concentration range?
√ What is saturation point (where there too many particles to count accurately in each size range)?

test fluid. If desired, the weight of noncombustible residue can be found by burning filter in special oven.

Counting-type tests are usually similar to ARP 598, and differ only in ways the sample is gotten, the volume of the sample, and in how it is forced or allowed to pass through the membrane. In the end, you simply count the particles retained by the membrane.

Color-background tests use visual experience of the operator to compare a discolored test filter with a non-contaminated new filter. With practice, over-all particle count can be estimated closely.

Getting the sample

Perhaps the most difficult part of sample-testing is getting the sample. It can be obtained as simply (and sometimes futilely) as dipping a cup of oil from an open tank; or by way of elaborate bypass piping with special fittings. Remember: a sampling valve might generate more dirt than you are trying to measure. (One suitable valve to use is ball-type with teflon seat.)

It is desirable to pick off the sample from a representative portion of the stream. But a simple pipe tee is not always the answer, because larger particles may not make the turn successfully. Inline takeoff (the sampling pipe concentric with the main pipe) is complicated, and might interfere with normal flow. Compromise is the sampling elbow-fitting designed by Douglas Aircaft (previous page).

AUTOMATIC PARTICLE-COUNTERS

Automatic devices are available that count particles in a sample (on a filter slide, or in a fixed chamber), or onstream. Of the six best-known types, three count from samples, and three onstream.

Sample-counter types

One of these is the Coulter counter, made by Coulter Electronics Inc, Chicago. It draws a sample of contaminated fluid (fluid must be made electrolytic) through a very narrow passage. Each particle that goes by displaces some of the liquid. This changes effective electrical resistance between two electrodes in the passage. Measurement of resistance gives indication of particle size, and a counter keeps track of total number passing through.

For counting particles on slides, there is the Cintel Flying-spot Particle Resolver, made by Rank Cintel Ltd, London. It is an optical system combined with a cathode tube and related circuitry. A moving spot of light scans the specimen (particles on the slide) and registers on the cathode tube via a microscope. Visual resolution is about 0.1 micron, and the device will count or measure contaminant particles as small as 1.0 micron.

Another counter for slides is the Casella Automatic Particle Counter and Sizer, made by Casella Electronics Ltd, London. It is an optical system combined with a photocell and related circuitry. The illuminated specimen slide is made to oscillate back and forth under the objective of a microscope, and contaminant particles on the slide interrupt a narrow beam of light created by an adjustable slit within the optical system. Photocell senses number and size of particles.

Onstream-type counters

These are similar to counters for fixed chambers, but can be connected directly into actual system.

One available device is the HIAC Automatic Particle Counter and Sizer (for liquids) made by High Accuracy Products Corporation, Claremont, Calif. It is connected into the line, and flow is forced to reverse within the device, creating turbulence. A small amount of liquid is continuously bled from the most turbulent point and is channeled through a narrow passage between a light source and a photocell. Particles (passing only one at a time under all normal conditions) are counted at the photocell. Total bled volume is captured in a graduated container, so that particles per unit volume can be calculated. Sensitivity can be adjusted to count particles above any given size within a 5 to 250-micron range.

Another available onstream-type counter (for air) is made by Royco Instruments Inc, Mountain View, Calif. It works on the same general principles as the HIAC —particles pass between a light source and a photocell.

An ultrasonic counter (for liquids) has been designed and built by Sperry Products Company under contract to Marshall Space Flight Center and now is undergoing evaluation tests at the Marshall laboratories. A transducer produces sound waves which pass through the liquid in a restricted portion of the flow. Sound waves reflected by particles back to the transducer are electronically detected, sized, and counted. Sensitivity can be adjusted for each of the four size ranges, detecting particles as small as 10 microns. Flow rate is small—about 0.1 to 1 gpm—through cell of present design, but bypass flow (in parallel with main stream) can be used. Instrument will be capable of continuous indication in fully closed and operating hydraulic system and will be adaptable to automatic operation.

ACKNOWLEDGMENT

Much of the material in this article is based on papers and discussions at the recent NCIH (National Conference of Industrial Hydraulics) meeting in Chicago. Our special thanks go to G. L. Wenner and C. B. Davis (of Martin—Denver), W. A. Riehl (Marshall Space Flight Center, Huntsville, Ala), D. R. Walker (Douglas Aircraft), and R. Q. Sharpe (Socony Mobil Oil Co) for permission to excerpt certain tables and data from their papers. Thanks also to John Farris, Aircraft Porous Media Inc, subsidiary of Pall Corp, for sharing his knowledge of automatic particle-counters. Cover photo was supplied by The Martin Co, Denver Div.

A 10-micron filter may or may not remove every 10-micron (or even every 100-micron) particle from the fluid. It all depends on which rating method was used. Such areas of misunderstanding can handicap the choice of filters for hydraulic and pneumatic systems. But this article, with its comprehensive filter-media comparison tables, strips away the mystery.

Examine these basics before EVALUATING FLUID-FILTERING MEDIA

CHARLES H HACKER
Chief engineer
Micro Metallic Div
Pall Corporation

A nominal "10-micron" filter for fluid earns that rating because it removes most of the dirt particles down to 10 microns in size. The fact is, however, this filter might actually allow some (or even many) much larger particles to pass through. Yet technically, such a filter can still be considered flawless and meet every appropriate commercial and military specification.

Is this a contradiction? No—the standard filter ratings are usually specific enough, but reading between the lines is very important. Best approach—the one given here—is to evaluate the basic filter "media" (materials that filter),

continued, next page

definitions

Micron – In metric measurement, one millionth of a meter; in the English system, approximately 1/25,000 of an inch. Particles as small as 40 microns are visible to the naked eye

Absolute Rating – Removal of 100% of all incident particles larger than a given size

Clean Pressure Drop – The differential pressure across a new filter, based on using a perfectly clean fluid

Maximum Allowable Pressure Drop – Usually the highest pressure differential which filter element can sustain without collapse – can also mean maximum drop before system is affected–and filter must be cleaned

Onstream Time – Elapsed time between cleaning or replacement of filter element

Effective Filter Area – Actual area in a filter medium which is open to flow of the fluid being filtered

Filter Medium – Any material capable of removing solids from a fluid (liquid or gas) stream

Filter Cake – An accumulation or "bed" of solids removed from a fluid and deposited on a filter medium; is often depended on to give a finer degree of filtration than that possible with the medium itself

Precoat – A special filter cake, usually diatomaceous earth, purposely laid down on a filter medium to provide either extremely fine filtration unfeasible with the filter medium itself; or to prevent clogging of the filter medium and make cleaning a simple, mechanical operation

Body Feed – The addition of diatomaceous earth or other filter aid to the slurry or fluid being filtered, in order to render a slimy or compressible filter cake more permeable, thus increasing the onstream time

Slurry – Term sometimes used for fluid to be filtered; implies high solids concentration

Filtrate – Clarified effluent from filter

Clarification – Final absolute filtering of an almost clean fluid

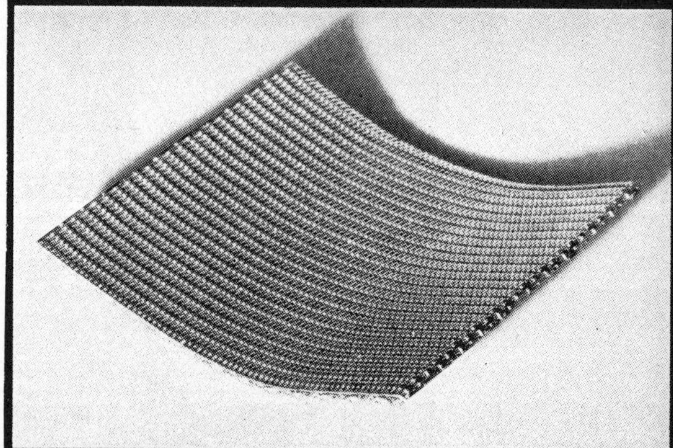

Composite filter, made of fine mesh sintered to coarse, has fine filtration characteristics; is essentially rigid.

then consult filter manufacturers before selecting or specifying an actual filter. These are the points to consider:

- Is the fluid a liquid or gas? (A filter of given rating can generally remove much smaller particles from the latter.)
- Is it "nominal" or "absolute" filtration you need–will an occasional oversize particle harm the system?
- Can the system tolerate coarse particles during initial start-up? (Some filters improve with time onstream–a porous cake of contaminant particles builds up on surface of filter medium.)
- What type of dirt will be filtered, and how does it compare with the arbitrary "typical" particles specified by test standards? Remember: some gelatinous particles might deform enough to squeeze through tight openings.
- Is fluid's dirt concentration large enough to build a filter cake? And will this cake formation pass liquid or block it?
- Will built-in dirt and other loose particles found in some off-the-shelf filters be damaging to system? How about those filter media that occasionally "flake" or break off particles, adding them to the stream?
- Will any single filter have large enough surface or volume to remove solids at rate required?
- What about total system pressure–and allowable pressure differential across the filter? Compressibility of media? Flow velocity?
- Will fluid and filter be compatible at all operating temperatures? If some corrosion is allowable, how much will this shorten onstream time of filter?

CHOOSING THE FILTER

Follow these three general steps in choosing a filter.

First, know your fluid system. Not just the fluid's tradename and expected flow, but something of its chemical and physical nature and behavior under various temperatures and pressures.

Second, evaluate available filtering-media types that might perform well with the fluid under question.

Finally, consult filter manufacturers to decide on specific filter design. (The last step is not discussed here–it will be wiser to benefit directly from the specialized experience of each manufacturer.)

Gas, Liquid, Slurry?

First step–knowing the fluid–is not difficult if fluid has been analyzed by you or the manufacturer. This article is limited to engineering fluids carrying dirt particles smaller than 200 microns average diameter. Here are basic facts concerning such fluids:

Gases are usually easier to filter than liquids. From gas, under certain conditions, some filter media will remove particles ½ to 1/10 the size of particles removable from liquid. One explanation: lack of lubricity of dry gases plus their generation of electrostatic charges helps hold particles that would otherwise slip through.

Dirt carried by fluid (gas or liquid) can be a mixture of fine and coarse, compressible and non-compressible, slimy and crystalline solids. Filtering is easiest when most particles are coarse, noncompressible and crystalline. The fine, compressible and slimy solids tend to pack together and restrict fluid flow drastically. It sometimes pays to add "body feed"—a coarser, more permeable material—to the fluid; this helps prevent compaction of the slimy solids.

Concentration of dirt (volume of dirt vs volume of fluid) must be known, because any filter has a dirt-capacity limit. Concentrations over 2% may be called slurries, although the definition varies with each industry. Remember this: if flow is 1 gpm and concentration is 1%, then dirt-flow alone is 2.3 cu in. per minute. If the dirt is compressible, then only a thin filter cake is allowable—a fraction of an inch. Coarse, noncompressible dirt can build up to many inches without excessive pressure drop across the filter medium. The performance-comparison table on the next page shows effects of various dirt concentrations.

Construction of Filtering Media

Second step is evaluating the filtering media. Only the common ones are discussed here; not covered are developmental types such as porous titanium, aluminum and copper. And tables only apply to basic media—not to any finished commercial product such as a filter cartridge or bed. For instance: wire cloth is a filtering medium, but a convoluted or pleated wire-cloth cartridge with housing and fittings is a filter product and is not discussed in detail here. Avoided, too, is the complexity and controversy of a wide variety of configurations and housing materials available from dozens of manufacturers. Given instead are limitations applying to any filter.

Broadly, there are two types of media: surface and depth. In surface media (wire mesh, membrane and simple edge-type are three examples) dirt is removed by a plane surface with fairly uniform orifices. Depth media (paper, felt, glass fiber, sintered powders, matted wool, and wound spools are examples) depend on tortuous paths through the material to remove particles.

With surface-type media, the number of openings available for clogging determines the time between cleanings for a given, allowable pressure drop. This cleaning is especially important where dirt particles are small and compressible: relatively small accumulation of cake gives disproportionate increase in pressure drop. It is possible to convolute the surface (as done with wire mesh) to give up to nine times the surface area within a given envelope volume compared with simple cylinder.

Depth-type media have a high dirt capacity for a given filter volume, particularly if concentration of solids is small; the particles work themselves into interstices between fibers. However, for large concentrations of solids, cake buildup may be quite rapid, and the filter becomes essentially surface-type.

Each type—surface and depth—has some characteristics of the other, and simple categorizing is not always possible. The following examples do not attempt to show direct distinction. Also, most comments apply to both metals and nonmetals.

Granular (porous) media have adjacent grains and these particles join to form a porous, rigid structure. (Sintered powder is a good example of a joined-particle medium; sand is an exception, it is often used in loose form.) The grains can be joined with a compatible adhesive or by sintering at temperatures slightly below melting point of the particle material. Sintering may be done with or without the simultaneous application of pressure. Degree of filtration depends on compression and on smallness of the grain chosen. Dirt particles are trapped in the interstices.

Media made of bonded granular materials are normally thicker in cross-section (for manufacturing reasons) than woven-wire media, for the same fineness of filtration. Pore sizes are more fixed in granular media than in ordinary unsintered square-weave mesh, but special interlocking weaves (dutch twill, for example) overcome this disadvantage.

Smaller contaminant particles can be filtered out by granular media than by woven types; but there is wider variation in particle size stopped, and a greater tendency towards contaminant migration. Extreme fineness is possible: a recently developed leached-glass porous medium has a removal rating better than 0.01 micron—it is not an industrial-type filter, however. Granular media are most suitable where solids concentrations are very low.

Porous depth-type filters can be cleaned to a limited extent by backwashing. Progressive fouling will occur after many backwash cycles, but element can be restored to original state with suitable cleaning solution.

Edge-type filter media may be either metal or nonmetal. The fluid filters between edges of flat disks or wires of circular or other cross-section. Spacing determines the degree of filtration. In the disk type, spacing is with alternate disks that are smaller in diameter, or with a central ring structure that has radial projections. The wire type is spaced by winding the wire around a mandrel or by using projections on the wire itself.

Long particles can get through an edge filter if one dimension (diameter or thickness) of the particle is smaller than the edge opening. (Round or cubical particles are stopped.) If downstream devices have small orifices or small-clearance moving parts, just a few fibers passing downstream can readily bridge the orifice or clearance, restricting flow or movement. Composite filters (described on page 46) will solve this problem.

Woven media are cloths of wire or yarn (natural or synthetic). Both are produced on looms and use loom terminology: *warp* is the wire or yarn that runs the length of the cloth; *shute* (sometimes called *woof*) runs perpendicular to and is woven into the warp by means of a shuttle.

Many types of weaves are available. The most common are: square (plain), dutch, twill, and twilled-dutch. Plain and twilled weaves present straight-through openings to

PERFORMANCE COMPARISON OF

	METALS							
	Sintered Particles		Edge Type			Mesh and Cloth		
						Sintered		Nonsintered
	Fiber	Powder	Disk	Ribbon	Wire	Wound on mandrel	Woven	Woven
General appearance of medium	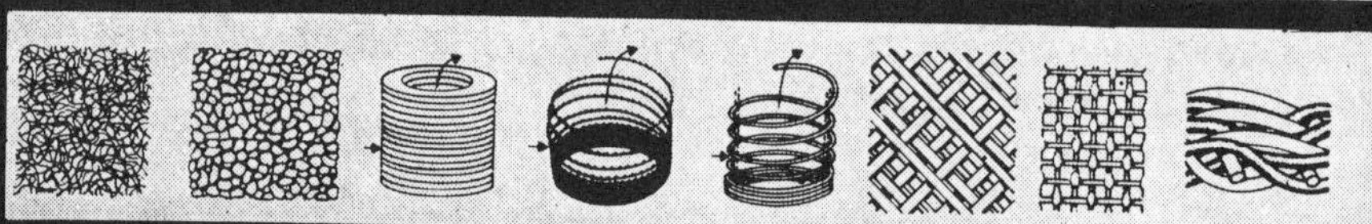							
Filtering action	Primarily surface		Surface					
Removal ratings, microns (based on smallest particle size removed). 98% removal (nominal)	No data	2-60	40-125	25-500	75-700	2 and over		
100% removal (estimated absolute rating)	No data	4-60	No data	50-500	125-700	12 and over		
Applicability—assuming reasonable flow conditions, is medium suitable for these dirt concentrations? >5% of liquid volume	Limited application only					Recommended		
0.1 to 5%	Recommended							
<0.1% (clarification)	Recommended			Not recommended		Recommended		
Cleanability	Cleanable (re-usable) based on nondestructive methods such as: chemical, ultrasonic, backwash, blowback							

*Estimated data based on media only, not on manufactured cartridges or systems; removal ratings based on liquid—with gases, filter performance is usually better, particularly in low-micron ratings.

fluid flow; twilled-dutch and dutch weaves present a somewhat tortuous path, hence finer filtration.

Wire cloths may be sintered or unsintered. Sintered wire cloth has warp and shute wires sintered together where they cross, fixing the pore size permanently. In addition, sintering allows integral bonding of coarser meshes to finer facings to produce a stronger composite.

Woven cloths are most often used where concentration of solids is high enough to permit rapid cake buildup for finer filtration. They also work without a cake—typical are those wire-cloth media having nominal (98%) removal ratings about 70 microns and smaller. If cake is used to improve filtration, recirculation (until cake builds up) will be necessary before best rating is achieved.

Woven yarn media are basically flexible and changes in pressure (including instantaneous reversal) can temporarily displace filter cake away from the filtering surface. Fine solids will then pass through the filter—bypassing the cake—until the displaced cake has settled again.

Wound wire mesh is a form of cloth made by helically cross-winding a fine flattened wire on a mandrel. Wires are then sintered where they cross, and finished mesh can be pulled from the mandrel to be slit into sheet stock or kept as a cylinder.

Felted media are masses of fibers matted together to produce a porous structure. Good felts are carefully designed to have interlocking fibers. They may or may not be resin-bonded. The degree of filtration obtained, regardless of the fiber type, depends on fiber diameter and length. Smaller fibers give finer filtration. Paper filters are similar in structure to felted media.

Felted media are usually of the depth type, are thrown away after use, and are limited to relatively low pressure differentials. Media migration will occur under some circumstances, particularly if fibers are loosely interlocked.

Unfelted fibers are initially loose, but are packed into a woven or knitted bag. Cotton or other cellulosic fibers are the types usually chosen, and major application is for filtering oil. Application is limited by temperature and pressure, and media migration is likely to occur. Filtering ability is improved with fillers such as diatomaceous earth.

Wound yarn media are commonly made in cylindrical form, helically wound around a support core that is part of the filter. The weave gives a diamond-shape pattern externally, without pores in the strict sense of the term. Filtration here is the depth type; it takes place through (and in) tight channels formed by nap (furry surface) of the yarn.

TYPICAL FILTER-MEDIA CONSTRUCTIONS*

NONMETALS												
Fibers (natural and synthetic)				Granules and Powder						Edge Type		Mono-lithic
Compressed or packed		Woven Cloth	Wound Yarn	Sand	Sintered Plastic	Fired Porcelain	Fritted Glass	Bonded Carbon	Bonded Stone	Ribbon	Disk	Membrane
Felted	Nonfelted											
Depth and surface	Depth	Primarily surface	Depth	Depth	Surface	Primarily surface				Surface		
5-100	10 and over	25 and over	1-100	30 and over	3-15	0.2-25	0.1-125	10-70	5-100	Available: 40 only	0.5-10	0.1-12
No data			10-150	No data	13-20	0.2-25	0.1-125	No data			5-10	0.1-12
Not recommended		Recom-mended	Limited applic.	Not recom-mended	Limited application only							Not recom-mended
Recommended		Not Recom-mended	Recom-mended	Recommended								Limited applic.
Throw-away type		Cleanable (re-usable) by nondestructive methods								Throw-away type		

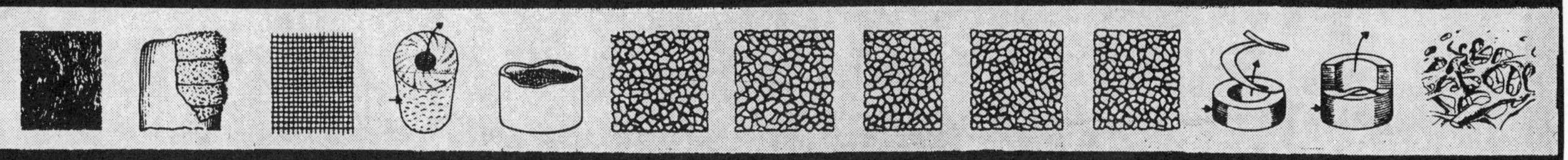

Membrane media essentially are thin porous sheets produced by the evaporation of solvents from a solution of cellulose acetate or cellulose nitrate. The resulting structure is monolithic (a single piece). Because it is not a fused or bonded mass of fibers or granules, it is not subject to media migration.

Composites of various media improve over-all characteristics for some applications. Without going into details of manufacturers' special products, here are several useful combinations: edge-type media with metal powder (powder is bonded to itself and to the edge-type medium by sintering); sintered stainless steel mesh with metal powder; and glass fiber with membrane. In first two the coarser medium is downstream, the finer is upstream. The two media are bonded together to prevent bypass leakage around one or the other.

For glass fiber and membrane, the fiber is upstream, and is not bonded to membrane.

There are also magnetic composite filters which combine mechanical separation of nonmagnetic solids with magnetic removal of iron particles. Two common constructions are: Alnico magnets with ordinary mesh; or mesh with magnetic wires. Extremely fine filtration is possible. Efficiency of the magnetic portion is inversely proportional to velocity of fluid past the magnet. One drawback is that as the filter cake increases in thickness during onstream time, particles are more likely to break loose.

USING THE PERFORMANCE TABLE

Just the filtering media, not the finished filter assemblies, are put under scrutiny here. Values given in the table on this page are practical for the media, but might look ridiculous if applied indiscriminately to any given product. For instance, the 5% solids-concentration level given in performance table will pile up 320 cu ft of filter cake in one workday (8 hours) if flow is 100 gpm and filter is 100% effective. Such an application is not normally recommended. However, the low flow through cloth media used in rotary vacuum-operated filters or in plate-and-frame units are certainly suitable for high concentrations of solids. And for many filter assemblies there are ways (such as mechanical scrapers) to automatically and continuously clean off the filter cake.

Another example: take the 0.1%-and-below solids-concentration level. Here it might seem that individual size and not concentration of particles is the only logical basis for comparing filters. This is not necessarily so. Some

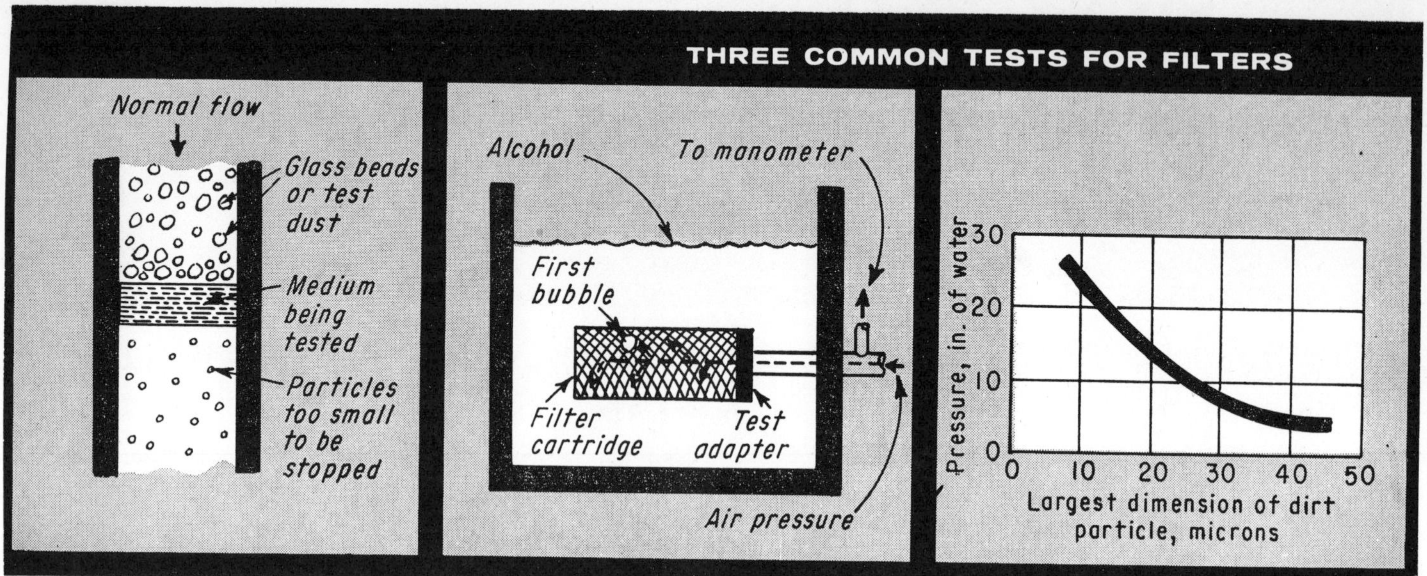

Known test particles such as glass beads (graded) or AC test dust are weighed before and after test to establish nominal or absolute filtration ratings.

Bubble-test simply measures the air pressure necessary to force the first bubble through any pore in submerged filter.

Bubble-test data correlate consistently with measured pore sizes and with actual particle sizes that pass through the pores. Graph is for sintered stainless steel mesh filter cartridge; particles are assumed to be hard (glass beads used for this data); alcohol is Solox 190.

coarse filtering elements such as edge type and cloth depend on cake buildup to achieve manufacturer's rating. If concentrations are too low, bridging of the pores by the particles might not occur. In finer media, dirt-particle size is critical.

In-between concentrations such as 1.0% can be handled by almost any filter, but the table points out certain practical limitations. Sand filters, for example, are not recommended here. They have a relatively small apparent filter surface, which cannot be conveniently increased to any extent. Therefore sand filters are used primarily for filtering water, where solids concentrations are low—measured in parts per million.

Some of the examples on page 40 are practical ones and they take filter housing and configuration into consideration. It would be impractical to cover every situation—a book is needed. The tables are guides. They will help you understand basic limitations, and determine what types of filters can do the job. But individual manufacturers must be consulted for the special details.

Nominal Rating for Filters

Military specifications MIL-F-5504A and MIL-F-5504B are the only established standards for any filter media, but are not considered industry standards. Version A defines a 10-micron filter as being capable of removing 98% by weight of all particles of the selected contaminant (AC fine test dust, supplied by AC Spark Plug Div of GM) larger than 10 microns at a certain high concentration. Version B defines a 10-micron filter as being capable of removing 95% by weight of 10 to 20-micron glass beads (supplied by Minnesota Mining and Manufacturing and others) at a certain high concentration. For many conventional applications, this nominal filtration is adequate.

Two loopholes exist, however. They make the use of filters meeting only these requirements unsatisfactory for the protection of certain vital equipment such as servovalves. First loophole is that the military specifications don't limit maximum size of the 2% AC fine test dust or 5% glass beads allowed to pass through the filter element. Tests by many laboratories have found that filters meeting these requirements readily pass particles in the range of 100 to 200 microns. Second loophole: contaminant concentrations used in the test simulate the high concentrations of a total pump failure and are not at all indicative of performance under normal system conditions. With the lower normal concentrations, efficiency proves to be much less. And be sure to find out if manufacturer's rating means "initial" removal rate, or the rate after a filter cake has built up.

Absolute Ratings

Absolute ratings differ from nominal in one important way: Rating is based on largest particle that can get through at all, rather than on a percentage of such particles. But there are differences of opinion even about absolute ratings. Here is one. If an edge-type or square-weave filter will absolutely stop a 10-micron glass bead, but could conceivably allow a 9-micron-dia fiber 50 microns long to slip through endwise, should it be rated at 10 microns or 50 microns absolute?

Despite some disagreement in approach, the need for absolute ratings is apparent. True, some systems can tolerate a little dirt, and the conventional nominal ratings will suffice here. But for more critical applications, the true effectiveness of a filter is not realized until ALL contaminants which might have a detrimental effect on the system are removed. Nominal ratings based on the military specifications and many commercial ones are not always enough. Just one fragment of dirt, 100 microns dia, can effectively shut down a complex, expensive automatic machine tool if it lodges in a critical location. You've

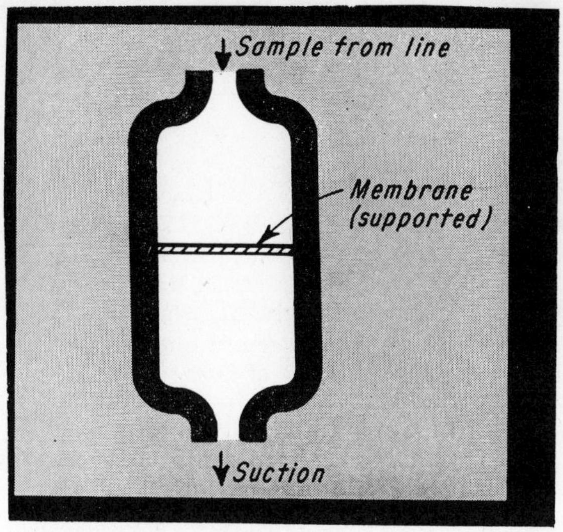

Bottle test is a simple lab operation with one sample of fluid taken downstream of an operating filter. Fluid is sucked through a finely ruled membrane. Microscope is needed to count particles.

got to match potential loss in time and productivity against demanding a more suitable, perhaps costlier, filter that can meet an absolute rating.

Absolute ratings are a fairly recent innovation—aircraft and missile industries began to use them about 1955. First application was for sintered woven-wire mesh, but now ratings can be determined for a wide variety of media, including porous metals, ceramic and cellulose acetate. Such ratings are still not in general use in the product design field.

Absolute filtration ratings are defined by specification MIL-F-8815 and proposed specification MIL-F-25682 (USAF). They are determined by measuring the diameter of the largest glass bead to pass through the filter. It has been established that this is approximately the same size as the maximum dimension of normal system-contaminant passed. An exception is a long narrow fiber that can pass through a square-weave wire mesh or edge-type media. Changes in contaminant type or concentration do not affect absolute filter rating, so this rating method is very useful.

Absolute ratings of rigid filter media are not normally affected by pressure. Some flexible media, such as paper, sand and fiberglass, are shifted or compressed by changes in differential pressure. Someday there may be absolute ratings qualified by pressure to cover all media; but at present, no.

Test Methods to Rate Filters

Four rating tests are common: special test dust; glass beads; bubble-point test; and bottle test. The first two—test dust and glass beads, both adapted as MIL specs—were described earlier.

Bubble-point test measures pressure needed to force an air bubble through pores of a submerged filter. Pressure reading is inversely proportional to size of largest pore opening, and once the setup is calibrated for a given type of medium and bubble-point fluid, the reading gives absolute rating of filter directly. Here is how the test works:

The filter cartridge or medium is immersed in a fluid with known constant surface tension (Solox-190 alcohol is commonly used; hydraulic oil will work too, but gives less reliable results. Water is often used with porous glass and porcelain—its greater surface tension here gives higher pressure readings). Low air pressure is applied inside the cartridge, but surface tension of the fluid covering each pore prevents breakthrough of the air. The cartridge is rotated slowly and air pressure is gradually increased until surface tension is overcome; a stream of bubbles will appear

TEMPERATURE LIMITATIONS OF COMMON FILTER MEDIA

Metals	Normal Temp. Range, °F
Carbon steel	−100 to 1000
Bronze	−400 to 400
Nickel	−400 to 1000
Alloys:	
Stainless 304	−400 to 1000
Stainless 316	−400 to 1000
Monel	−400 to 1000
Inconel	−400 to 1200
Hastelloy C	−400 to 1000
Noble:	
Silver	−400 to 1000
Platinum	−400 to 1000

Nonmetals	Normal Temp. Limit, °F
Natural Fibers:	
Cotton	250
Wool	180
Asbestos	1000
Synthetic Fibers:	
Dacron	200
Dynel	200
Glass	750
Nylon	300
Orlon	300
Paper	275
Polyethylene	150
Saran	160
Teflon	450
Vinyon	130
Natural Particles:	
Carbon	650
Ceramic	2000
Glass	1000
Sand	Amb.
Stone	2500
Synthetic Particles:	
Teflon (−400 to:)	350
Kel—F	400
Polyvinylchloride (pvc)	120
Membrane:	
Cellulose acetate	260

at the largest pore opening. If pressure is increased further, bubbles will appear at other pore openings that are smaller than the first, thus giving a useful pattern of hole size and distribution. Pressure is inversely proportional to hole size (expressed as a circle), and results match the more conventional glass-bead test very closely.

Bottle test (so-called) is simply a technique that measures particle size of effluent (filtered fluid) directly. A sample of filtered fluid is put in a bottle, and a vacuum is used to draw fluid through a special filtering membrane that is ruled with extremely fine grid lines. Examination under a microscope makes possible an accurate count and measurement of all particles over 0.1 micron in the sample.

Of the four tests, only the bubble test does not require that the filter be put onstream and contaminated with solids.

ONSTREAM PERFORMANCE

Onstream conditions almost always differ from test conditions, and none of the standard rating tests cover all eventualities. These basic principles prevail, however; if they are recognized, proper allowance can be made.

Migration in filter—The filter itself sometimes causes or allows dirt to get into the effluent. Called migration, it can happen in one or more of these four ways: (1) Media migration—a part of the medium itself breaks away and gets downstream of the filter. This is common with fiber media but might also occur in many other types. (2) Contaminant migration can occur in some flexible media; contaminants on the upstream side work through to downstream side as pressure drop increases. Or flexing of the upstream surface might loosen filter cake, thereby breaking the particle-bridges across filter pores, and releasing fine solids through the filter into the effluent. (3) Built-in migration is dirt that gets into the filter during manufacture. (4) Abrasion migration is dirt generated by poorly designed parts that rub together and wear during vibration or shock induced by flow or other stimuli.

Migration is not necessarily harmful if it is anticipated. For example, a depth-type fiber filter with known migration tendencies can be easily followed up with a low-capacity surface-type filter that removes the bits of migrated fiber.

Flow—The flow pattern through filter media is usually of the laminar type, and pressure drop is directly proportional to viscosity and velocity. By comparison, flow through the housing is usually turbulent and pressure drop is proportional to the square of the velocity—normal viscosities have no effect.

Flow rates depend on open area, total thickness of filter medium, pore size. It is possible to have two wire cloths of same size with substantially the same removal rating but with widely different effective open areas—giving widely differing pressure drops for the same flow rate. For example: each of two cloths might be rated at 43 microns. The first could be a calendered 200 x 200 mesh with 0.0026-in.-dia wires; the second could be a 325 x 325 mesh with 0.0014-in.-dia wires. The first has about 40,000 holes per sq in. and 11.4% open area; the second has about 105,000 holes per sq in. and 30% open area. It is fairly obvious that the second has the lower pressure drop and the greater dirt capacity at a given rate of flow.

Pressure differential—Most media can be designed to support almost any pressure drop encountered in service, but a lot depends on size of element and design of support structure. Absolute pressure has very little effect, so long as differential pressure across filter element is low.

High differential pressure has this effect: compressible media might deform or even flake or break, thus releasing contaminants, or in extreme cases be compressed enough to restrict flow. Small filter sizes are the least affected, because filtering area is small compared with other dimensions—flexing is less likely. However, even in large sizes, support structures can be built to help media withstand high differentials.

In practice, filter media can usually be categorized as high-pressure or low pressure—though almost any medium can be adapted for almost any differential pressure.

Fiberglass, paper, felts, random-packed fibers and membranes are usually considered 75-125-psi media. Wire and sintered-metal media are normally considered applicable to 10,000-psi differential. In-between these 125 and 10,000 psi categories are found media such as edge-type, ceramic, stone, wound yarn and sintered plastic.

Temperature—Limit is usually a direct function of the lowest-temperature material in the filter assembly. Glass fibers can serve at temperatures over 1000 F, but if resin binders are used, they might have lower limits. Temperature values in the table of common media materials are significantly lower than known melting points—deterioration of properties starts slightly above values given.

Secondary effect of temperature is a different rate of cooling and heating between a filter medium and its solid supports. This effect is difficult to predict because published data are lacking. Design can allow the medium to expand or contract with relative freedom. Gradual temperature changes within range will not harm the medium itself. Avoid sudden changes of over 100 deg F.

CHEMICAL AND PHYSICAL COMPATIBILITY

The compatibility tables on next pages summarize data obtained from several dozen chemical companies, plus data obtained from filter manufacturers. Some are based on tests, but others are estimated from chemical characteristics (a practical way to judge the new synthetic media and fluids). Although the tables are on the conservative side, the filter manufacturer should be allowed to adjust the data where his own product is concerned. This is because he may use special-alloy supports and bonding resins that are more (or less) compatible than the basic media themselves.

(text continued, p. 55)

COMPATIBILITY OF FLUIDS AND FILTERING MEDIA

HYDRAULIC AND TRANSMISSION LIQUIDS

LIQUID			ESTIMATED COMPATIBILITY with typical media (table)
Chemical Base Mil. (or Federal) Spec.	**Tradenames**	**Normal Temp. Range, °F**	C — Fully compatible at normal temp. LC — Limited compatibility if water present NC — Not compatible NA — Not applicable (for nonchemical reasons)
Petroleum Mil-H-5606A	Aeroshell Fluid 4, Brayco 756, Codes 566, 3587, RL-102-A, Rayco 756, Texaco 641, Univis J-43	−65 to 275	LC with carbon steel C with all others
Petroleum Mil-H-6083B	Brayco 783, Rayco 783, Code 3591, RL-94A-2, Univis PJ-44-45	−65 to 175	LC with carbon steel C with all others
Petroleum Mil-F-17111	Chevron PTF, Kendex 7054, Petrofluid 17111, Univis J-58	−65 to 225	LC with bronze NC with PVC, cellulose acetate C with all others
Phosphate Ester Mil-H-19457	Cellulube 220, Houghto-Safe 1120, Ohio-Apex Grade MP-280, OS-70	+50 to 200	NC with Saran, Vinyon, PVC, cellulose acetate C with all others
Phosphate Ester (No Mil. Spec.)	Skydrol 500A	−65 to 225	LC with bronze C with all others
Phosphate Ester (No Mil. Spec.)	Skydrol 7000	−65 to 225	NC with PVC, cellulose acetate C with all others
Phosphate Ester (No Mil. Spec.)	Pydraul F-9, 625 Pydraul 60, 150	+50 to 300 −10 to 200	LC with bronze NC with PVC, cellulose acetate C with all others
Fluorinated Hydrocarbon (Chlorotrifluoro-ethylene) (No Mil. Spec.)	Kel-F oils	−70 to 500	C with all
Silicate Ester Mil-H-8446B	Dow Corning 1-0054, Oronite 8515	−65 to 400	C with all
Silicate Ester (No Mil. Spec.)	Oronite 8200 Coolanol 45, 35, 59 OS 45 type 4	−65 to 500 −65 to 400 To 500	C with all
Diester	Ucon JL-1, 2, 4, 6	−65 to 300	C with all except cellulose acetate
Trifluoro-vinyl Chloride	Fluorolubes	To 575	C with all except cellulose acetate
Silicone Oils	Silicone Fluid 81644 Versilube F-50 Dow Corning 200	−100 to 600 −100 to 450 −75 to 350	C with all
Chlorinated Hydrocarbon (No Mil. Spec.)	Pydraul A-200	+50 to 300	NC with PVC, cellulose acetate C with all others
Polyglycol Ether	Ucon Lubricants	−50 to 500	C with all except cellulose acetate
Glycol-Water Mil-H-7083A	Ucon Hydrolubes	−65 to 150	C with all (except carbon steel: LC)
Glycol-Water (No Mil. Spec.)	Houghto-Safe 620	0 to 130	C with all (except carbon steel: LC)
Glycol-Water Mil-H-22072	Houghto-Safe 271	−20 to 130	C with all (except carbon steel: LC)

LUBRICATION LIQUIDS

LIQUID			ESTIMATED COMPATIBILITY with typical media (table, p. 51)
Chemical Base Mil. (or Federal) Spec.	**Tradenames**	**Normal Temp. Range, °F**	C — Fully compatible at normal temp. NC — Not compatible
Petroleum Mil-O-6081 Grade 1005	K-80, K-877, Turbine Oil 1, Turbo Oil 5, 7A, 10A	−40 to 250	C with all
Petroleum Mil-O-6081 Grade 1010	Commavio J-77, Jet Oil 1010, Kendex 7042, NY3955, Petrolube 1010, Rayco 460, TJB Oil, Turbine Oil 2, Turbo Oil 10, Turbojet Oil 1010, Windsor Lube L-110, 3B, 10B, 31100, 460	−65 to 250	C with all
Petroleum Mil-L-9000E	D-X Motor-Oil, Delomil, Nator, Talona, Naturalube, Neptune, NST, Super Tenol	To 225	C with all
Fluoroalkyl Camphorates (No Mil. Spec.)	Freon	0 to 600	C with all (no data on nonmetals)
Silicone Oils (No Mil. Spec.)	Silicone Fluid 81644 Versilube F-50 Dow Corning 200	−100 to 600 −100 to 450 −75 to 350	C with all
Oil-Water Emulsion (No Mil. Spec.)	No tradenames	No data	C with all except steel
Fluorinated Hydrocarbon BB-F-671A	Freon, Genetron, Ucon	−25 to 180	C with all
Fluorinated Hydrocarbon (No Mil. Spec.)	Kel-F Oil	−40 to 500	C with all
Petroleum Mil-L-2105	Nator 2105	−10 to 225	C with all
Diester	Ucon JL-1, 2, 4, 6	−65 to 175	C with all except cellulose acetate
Diester Mil-H-6085A	A-107B, Aeroshell Fluid 12, Anderol L-401-D, Brayco 885, Cosmolubric No. 270, KC-85A, L-245X, No Rust 600, PQ Rust Preventative No. 160, Product 80, Univis P-38	−65 to 175	C with all except cellulose acetate
Diester Mil-L-7808C	Aircraft Turbo S Oil, Anderol L-807A, Cellutherm 2712, CISAF, Conojet, Eastman Synthetic Lube No. 3A, Hatco 111, Turbine Oil 15, Turbo Oil 15, WS-3650, WS-4040	−65 to 300	C with all except cellulose acetate
Diester Mil-L-9236B	Eastman Synthetic Lubricant 9	−40 to 370	C with all except cellulose acetate
Diester Mil-L-9236A	Anderol L-825	−40 to 370	C with all except cellulose acetate
Tetraester	Fluid 307	−25 to 400	C with all
Tetraester Fluoroalcohol	Fluid 350	+45 to 600	C with all
Neopentyl Ester Mil-L-9236B	Cellutherm 2505A, Pentalube TP 653-0	−65 to 450	C with all except cellulose acetate
Polyglycol Ether (No Mil. Spec.)	Ucon Lubricants	−50 to 500	C with all except cellulose acetate
Mixed Base (No Mil. Spec.)	OS81, Pydraul AC	+20 to 300	C with all except cellulose acetate

AND FILTERING MEDIA *(continued)*

COOLING, HEAT-TRANSFER, AND MISCELLANEOUS FLUIDS

FLUID			ESTIMATED COMPATIBILITY with typical media (table, p. 51)
Chemical Base (No Mil. or Fed. Specs. except where indicated)	**Tradenames**	**Normal Temp. Range, °F**	C — Fully compatible at normal temp. NC — Not compatible NA — Not applicable (for nonchemical reasons —mainly temperature limitations)
Silicate Ester	Coolanol 45	−65 to 400	C With all
Fluorinated Hydrocarbon (BB-F-671A)	Genetron, Freon, UCON	−250 to 180	C With all
Fluorinated Hydrocarbon	Fluorochemical FC-75	−80 to 225	C With all except cellulose acetate
Fluorinated Hydrocarbon	Fluorochemical FC-43	−40 to 355	C With all
Chlorinated Biphenyl	Aroclor, Pyranol, Inerteen, Noflamol	500 to 600	NA Bronze, cotton, wool, Dacron, Dynel, Orlon, paper, polyethylene, Saran, Teflon, Vinyon, nylon, sand, Kel-F, PVC, cellulose acetate
Biphenyl Bi-phenyl Oxide	Dowtherm A	500 to 750	
Terphenyl	Santowax R, Santowax OM, Santowax OMP	500 to 750	
Biphenyl	No tradenames	500 to 750	C With all others
Polyphenyl Ether	OS-124	500 to 825	
Polyglycol Ether	UCON heat-transfer fluids	−50 to 500	C With all except cellulose acetate
Liquid Sodium	No tradenames	220 to 800	C Only with carbon steel, nickel, stainless alloys, asbestos, glass, carbon, ceramic, stone NA Sand
Liquified Gas (Several Mil. Specs.)	No tradenames	Cryogenic	C Only with bronze, nickel, stainless alloys NA Silver, platinum
Water	No tradenames	32 and up	C With all except carbon steel
Air	No tradenames	Any temp.	C With all

Theoretically, corrosion resistance of a porous-metal filtering medium is the same as that of solid metal. However, the porous structure consists of particles joined by small bridges that are sometimes only 0.0003 in. dia; this and the large exposed surface area are the controlling factors when measuring filter life under corrosive conditions. Filter cloth with a wire diameter 0.0015 in. or larger will, under corrosive conditions, remain onstream longer than a porous medium having 0.0003-in.-dia bridges. But remember: a porous structure with the same nominal removal rating has greater effectiveness than mesh, and in the final choice you must weigh higher effectiveness against longer onstream time.

Adsorption and absorption—Adsorption is a surface phenomenon, absorption occurs in depth. Certain media, including uncoated, unimpregnated cellulose fibers and various filter aids (granular material added to fluid to act as or help build filter cake) tend to remove chemical additives from fluid. Such additives are necessary to improve fluid viscosity and inhibit corrosion in machine parts, and must not be removed.

EXAMPLES: CHOOSING FILTER MEDIA

1—High temperature and viscosity

The fluid to be filtered here is hot (600 F), highly viscous, and will be allowed to reach 5000-psi differential pressure. Solids concentration is low, but absolute filtration must be in microns. Cleaning will usually be with solvents, but hot nitric acid is sometimes used.

Only porous stainless steel and stainless steel wire mesh can meet the requirements of temperature, pressure differential, and cleaning. The mesh type will allow fewer oversize particles to pass (based on nominal rating) and should be selected if this is a critical requirement. Its first cost will be lower, its size smaller, and its clean pressure drop less than that for an equivalent porous type. Mesh filters can be small because they can be convoluted to increase effective flow area. And the smaller over-all volume reduces cost of the containing vessel—important at high operating pressures.

2—Gelatinous and compressible solids

The fluid here is an aqueous solution containing gelatinous or compressible solids—no corrosion, pressure, or temperature problems. Degree of filtration required is 10 microns. Purpose: clarification of fluid.

A depth-type filter medium of almost any variety is recommended: surface will remain relatively unclogged throughout most of the filtration cycle.

Surface-type filter media would tend to clog rapidly with these solids and would be unsatisfactory unless a diatomaceous-earth precoat and body feed are used—adding cost.

3—Clarification (absolute final filtration)

The filter medium needed will be downstream of a diatomaceous-earth filter that has wire cloth or fabric as its main filtering medium. Purpose of the downstream filter: final clarification of fluid.

Depth type or granular-type media will work for this downstream filter. While initial cost of a throwaway type will be lower, maintenance and replacement costs of a permanent filter medium are less, making it more attractive. This is particularly true when flows are 50 gpm or more, and operation is continuous, 24 hours per day.

Any of the granular types will satisfactorily filter out the diatomite. However, if the porous medium does not have homogeneous permeability, penetration of the pores by the diatomite may necessitate chemical cleaning. The cleaner—a hot alkaline solution—is so strong that it will attack ceramic, glass and stone. Therefore only porous-carbon and porous-stainless-steel media will be suitable. Area for area, porous carbon costs less initially, although it has somewhat higher maintenance costs. Porous carbon is the choice unless floor space is limited—a porous-stainless-steel element can be smaller in envelope volume.

4—Slurry (2 to 5% solids)

Any surface-type filter medium can handle slurries; but yarn cloth, paper, or compressed fiber-pad media are preferred if slurry has a reasonably large solids content. Yarn cloth would very likely require some recirculation to build up a cake, but would be an excellent choice. Paper and compressed fiber can give clear slurries on the first pass, and should be considered if recirculation is not feasible.

Initial costs are almost the same for all three types, but replacement costs for paper and compressed fiber will run considerably higher than the fabric cloth medium. Other media are too expensive for this application, or are not available in a suitable configuration. For any filtering medium, a plate type of unit is more economical than a cylindrical type when fluid is a slurry.

5—Miniature filter for oil

Lubricating oil (or hydraulic oil) in this example is in a closed system, with severe limitations in envelope size and weight. Filtration of 10 microns nominal (per military specification) is required.

Mesh- or paper-type media are available in convoluted form, and can be packed with a large filter area in a very small volume. But final choice depends on operating conditions and fluid type. For low temperatures and pressures, and for standard fluids, paper is the most economical. When pressures and temperatures are high and fluid is corrosive, stainless steel wire mesh, woven or nonwoven, is recommended.

What went into this report

PRODUCT ENGINEERING'S editors have for some time recognized the need for a comprehensive report comparing all filtering media on some consistent basis. But there was a problem. Each filter manufacturer has his own techniques for making and applying filters, and there are so many catalogs available that it requires an extremely intensive literature search to digest them.

Then Pall Corporation gave a paper on servo-system filters at the 1959 NCIH (National Conference on Industrial Hydraulics) meeting in Chicago. It had the desired down-to-earth approach. We asked Pall to tackle a more comprehensive story covering ALL filter types. Result is this article by Micro Metallic's chief engineer.

Before it was finished, over 500 man-hours of writing and research had gone into the article, with direct help—checking for accuracy—from a half-dozen filter companies and three dozen chemical companies. Each filter company—they are all competitors—agrees on this at least: use the article as a preliminary guide to the pinpointing of several suitable types of filtering media; but then consult the manufacturers for highly specialized advice in choosing the exact filter package.

Our thanks specifically to these filter companies who looked over and commented on the performance table and text: R P Adams Co Inc, Filter Div of Bendix Corp, Commercial Filters Corp, Corning Glass Works, Cuno Engineering Corp, Permutit Co, Purolator Products Inc, and Selas Corporation of America.

4

FLEXIBLE PIPING AND FITTINGS

Hose, couplings, joints, and manifolds

BRAIDED, CORRUGATED TYPE gets its axial burst strength from the braid. This strength can be calculated from tensile strength, size and angle Δ of wire, number of wires used, and average hose diameter.

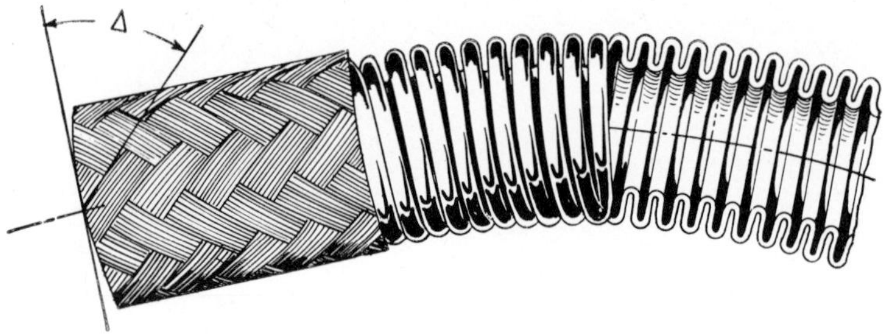

for the first time . . .

STANDARDS FOR METAL HOSE

Two new methods—now being studied by ASME—introduce uniformity into the choice of corrugated or interlocking hose. An equation rates pressure-type hose; a 5-digit code indexes nonpressure types on bases of material and construction.

L B DONKLE, *engineering manager*
Industial Hose Division
Flexonics Corp, Rock Falls, Ill

Choosing a flexible metal hose for a design frequently involves a survey of manufacturers' catalogs to find such specific information as size, pressure ratings, and temperature limits. Now, for the first time, here are uniform standards for many of the most controversial ratings, simplifying initial choice of a hose type. The final choice, of course, should be made only after consultation with the manufacturer.

These standards have been submitted to Committee on Code for Pressure Piping of ASME (American Society of Mechanical Engineers) and the Flexible Metal Hose and Tubing Manufacturers Association. If accepted, they will become an industry-wide guide.

A hose's primary feature is its ability to move—whether to follow motion, facilitate assembly, correct misalignment, or to absorb and isolate vibration. But that's not the only consideration. Pressure and temperature are important factors, too. You must know how long the hose can transmit fluid without leakage or failure at the temperatures and pressures to be encountered, in the imposed configuration, and with the necessary movement.

But a pressure rating at any temperature is meaningful only if related to some measurable point of failure or deformation. And this failure-pressure reference point will be different for each type of metal hose—whether corrugated or interlocked—depending on its construction.

CORRUGATED HOSE

Braided type

Maximum working pressure rating of the hose assembly under given service conditions depends on: tensile strength of the braided-wire jacket; ability of the corrugated tube (internal-pressure carrier) to resist permanent deformation under pressure while restrained by the braid; operating temperature; pulse or shock loading of the hose; and the method used to attach the end fittings. How these factors influence final pressure rating of hose assembly is expressed in this empirical relationship:

$$P_{\text{rated}} = F_T \times F_S \times F_E \times P_{\text{nominal}}$$

where F_T, F_S and F_E are factors for temperature, shock or pulse loading, and end-fitting effect, respectively.

Nominal hose pressure P_{nominal} in the equation is, by definition, the lesser of these two quantities: one-third of fluid pressure P_B (calculated) that will burst the external braid in an axial direction; or four-fifths of pressure P_D (determined by test) that will start permanent deformation of the inner corrugated tube.

For a single-layer braided hose, the axial burst pressure obeys this relationship:

$$P_B = \frac{1.57N\, d^2 T \sin \Delta}{A}$$

where P_B = fluid pressure, psi, that will burst the braid in an axial direction; N = number of braid wires spiraling about hose in one direction; d = diameter of wire, in.; T = effective tensile strength of braid wire, psi; A = total (not annular) cross-sectional area of hose, sq in., based on mean diameter; and Δ = braid angle, deg, measured from a perpendicular to the hose centerline.

This axial-burst-pressure equation is mathematically exact, and makes two major assumptions: (1) all of the axial pressure force is sustained by braid; (2) all the radial pressure force is sustained by the corrugated tube. Typical values for wire tensile strength T are: 40,000 psi for commercial and phosphor bronze; 50,000 for mild steel; and 115,000 for AISI (American Iron and Steel Institute) 300-series stainless steels. Published maximum hose pressure ratings must show values of T assumed.

For a double-braided hose, the same axial-burst equation applies, but the equivalent axial tensile strength of the combined jacket is assumed to be 1.8 times the value for a single-braided hose. This assumption is valid (and conservative) if braid wire can elongate at least 15% (based on a 10-in. length) when tested prior to braiding.

Summarizing: to find $P_{nominal}$ in the hose-assembly rating equation, compare one-third of the final value of hose axial-burst pressure P_B with four-fifths of the radial-deformation pressure P_D determined by test. The lesser of the two will be $P_{nominal}$.

Temperature factor F_T is defined as the ratio of the 0.2% offset yield strength for the hose metal at the elevated (or reduced) temperature, to its yield strength at 70 F. See graph on next page for typical values at elevated temperatures. The factor will increase above 1.0 at lowered temperatures for some metals, but to be conservative use values no greater than 1.0.

Pulse or shock factor F_S cannot be determined by simple calculation, but is an empirical value. Typical values for F_S range from 0.5 (moderate pulses) to 0.2 (severe shock waves). For a manufacturer to estimate probable derating required, he must know peak pressure, rate of pressure rise, number of pulses per second, and expected total operating cycles per year. He will probably recommend a metal hose specially designed for shock or impulse service.

End-fitting factor F_E is 1.0 for a standard brazed or silver-brazed connection, and varies from 0.6 to 2.0 for other attaching methods. For heliarc-welding, the factor is 0.8 to 1.0; atomic-hydrogen welding, 0.6 to 0.8; soft solder, 1.0 to 2.0; and a mechanical coupling, 1.0 to 1.8. The design of the end fitting itself can influence the factor's value, but to a lesser extent.

INTERLOCKED TYPE OF METAL HOSE gets flexibility from sliding metal components in the interlock. Below are common interlock designs, with code numbers based on table, p. 61.

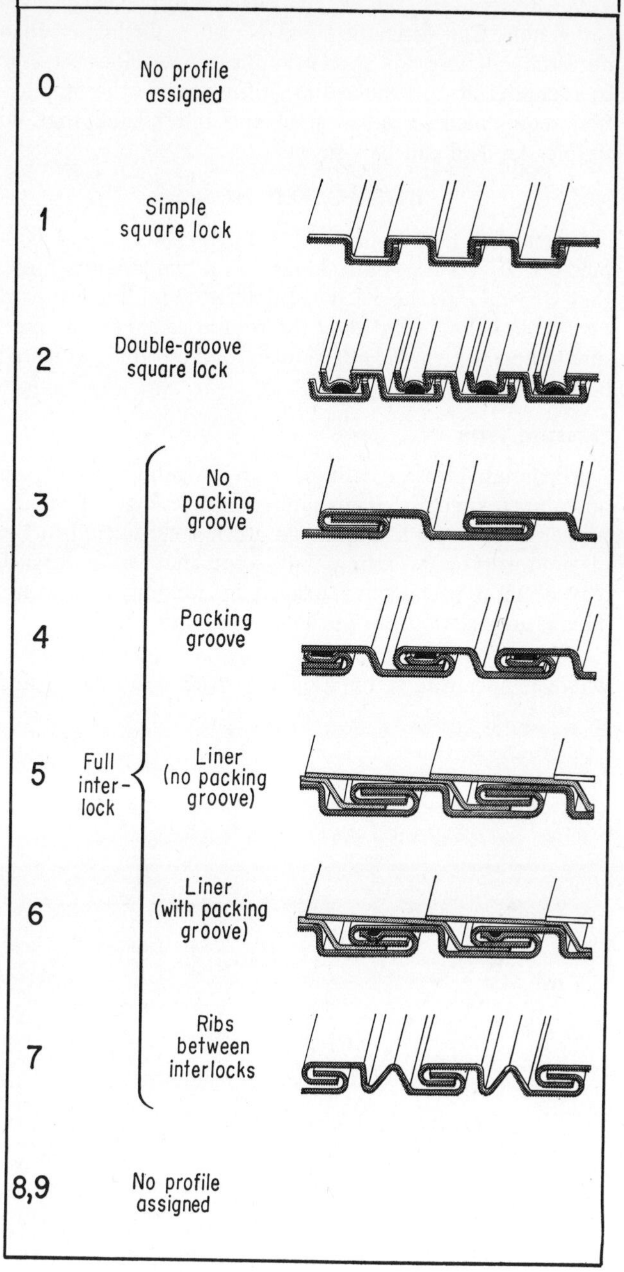

Unbraided, unrestrained type

The maximum pressure rating for unbraided corrugated hose when free to elongate, depends only on permanent axial elongation and temperature. Hose-deformation pressure P_D and end-fitting design are unimportant, because the pressure levels permissable in unbraided corrugated tubing are relatively low—axial deformation will occur before the other factors come into play. Shock factor F_S is left out because pulsing flows are not normally recommended for unbraided hose. The relationship between rated and nominal pressure in a hose assembly is:

$$P_{rated} = F_T \times P_{nominal}$$

The temperature factor F_T is the same as for braided hose. $P_{nominal}$ is that pressure at which permanent elongation (set) equals 0.25% at 70 F, based on this equation:

$$\text{Permanent elongation, } \% = 100 \times (L_F - L_O)/L_O$$

where L_F is final free length of a hose subjected to, and released from, the test pressure; and L_o is the original free length. The test is run at room temperature. The hose rests on a smooth flat surface with one end free to move. Both hose ends are blanked, but the fixed end has a pressure connection attached. Free length includes active corrugations only—soldered, brazed or clamped corrugations are not included.

Unbraided, restrained type

With this hose, ends are fixed axially to prevent over-all elongation. Can often be rated considerably higher than unrestrained hose for special applications. However, the free length, offset, bend radius, stiffness and other application requirements are so varied and interrelated that no simple standard can be written.

INTERLOCKED HOSE

Interlocked hose is made both for pressurized and nonpressurized service. Both kinds have temperature limits that depend on the packing materials and metals used (see table below). But only the pressurized type has pressure-temperature standards similar to those for corrugated hose.

Pressure type

Maximum pressure rating for an assembly depends on hose burst strength, temperature, and leakage. Axial deformation and end-fitting design are not considered in the maximum pressure rating, but some end-fitting leakage may occur if packing is damaged by welding or brazing. This rating relationship for hose assemblies applies:

$$P_{\text{rated}} = F_T \times P_{\text{nominal}}$$

where temperature factor F_T is the same as for corrugated hose, but P_{nominal} is defined as the lesser of these two values: one-third the burst pressure at 70 F; or one-third the minimum weep (leakage) pressure.

Burst pressure will vary from test to test; the value selected can be an average value but must be at least 15-20% higher than the minimum burst pressure. Minimum weep pressure is the lowest at which there is no continuous leakage of water (or SAE 10 oil) at 70 F through the hose convolutions. (Continuous leakage means that drops of fluid form on the outside of a hose bent to its minimum radius within one minute). However, most interlocking hose will burst before leaking.

Nonpressure type

This type is not classified by pressure but by major construction features. A five-digit number code is sufficient to cover the important details:

first digit—type of interlock
second digit—thickness range of metal strip
third digit—metal used
fourth digit—packing
fifth digit—jacket or covering

To designate an interlocked hose with the five-digit method, look up the key number for each digit in the classification table on this page. Here are some examples, the first of which is worked out in detail:

• Fully interlocked lightweight galvanized steel ventilating ducting will be classified 31100. The "3" means fully interlocked without a packing groove; the first "1" is light-gage metal—0.008 to 0.015 in. thick; the second "1" is galvanized steel; the first "0" means no packing; and

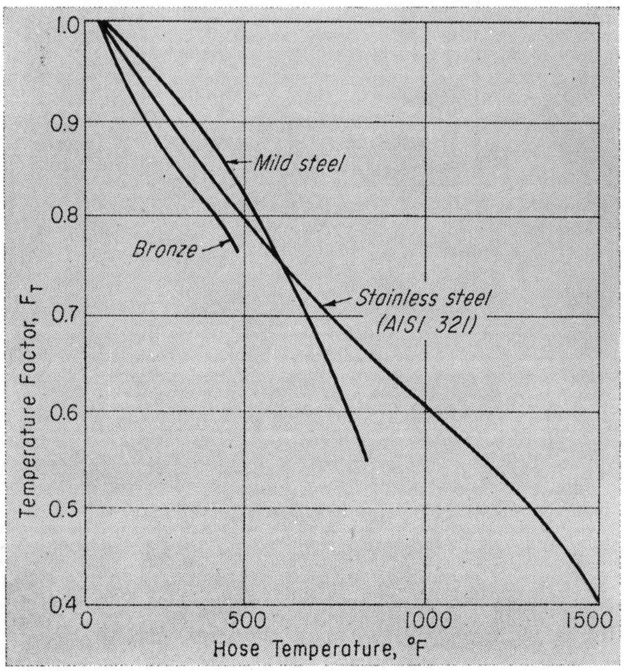

TEMPERATURE FACTORS for three hose metals (max rated pressure at elevated temp = $F_T \times$ max rated pressure at 70 F)

TEMPERATURE LIMITS OF HOSE MATERIALS
(recommended)

Material	Max temp, °F
Aluminum	600
Asbestos packing, grade:	
commercial	400
Underwriters	450
A	550
AA	600
AAA	750
AAAA	900
Brazing alloys:	
copper base	850
silver (AMS 4770)	750
silver (AMS 4772)	850
Brass	450
Bronze	450
Cotton-cord packing	200
Galvanizing	450
Monel	800
Neoprene packing	250
Solder, soft	
60-40% lead-tin	250
95-5% lead-tin	350
Steel, mild	850
Steel, stainless AISI type:	
302, 304	850
316	850
316 ELC	1500
321, 347	1500

5-DIGIT CLASSIFICATION OF INTERLOCKED HOSE

Category→	1st digit INTERLOCK	2nd digit THICKNESS, in.	3rd digit METAL	4th digit PACKING	5th digit COVERING
Code:					
0	....	<0.008	bright steel	none	none
1	simple square lock	0.008–0.015	galvanized steel	asbestos	wire braid
2	double-groove square lock	0.016–0.025	stainless: AISI-400	cotton	thermoplastic (PVC, etc.)
3	full; no packing groove	0.026–0.035	AISI-300	neoprene or Buna N	synthetic rubber
4	full; packing groove	0.036–0.045	combined (code 1 and 3)	natural rubber	asbestos or fiberglass
5	full; no packing groove; liner	0.046–0.055	brass	copper	reinforced synthetic rubber
6	full; packing groove; liner	0.056–0.065	bronze	....	....
7	full; ribbed	0.066–0.075	aluminum	....	....
8	....	0.076–0.085	copper-nickel alloys	....	....
9	other	>0.085	other	other	other

the last "0" means no covering. If a synthetic rubber cover were desired, last digit would become 3.

• Packed, galvanized-steel, abrasion-resistant, heavy-weight tubing with stainless steel liner—62410.

• Fully interlocked bronze capillary casing—31600.

To distinguish between corrugated and interlocked hose, type numbers are included in the classification. Type I is corrugated, braided; Type II is corrugated, unbraided, unrestrained; Type III is interlocked, pressurized; and Type IV is interlocked, nonpressurized. The first example of the preceding three is completely defined as: Type IV, class 31100, ventilating ducting.

DEVIATIONS FROM THE STANDARD

Higher-than-standard pressures and temperatures are allowed in special circumstances if the manufacturer knows that life and reliability will be adequate. Here are typical circumstances.

Intermittent temperatures—If the hose is at peak temperature only part of the time, then hose-deterioration rate will be slower. This is important when failure is related to excessive oxidation, carburization, graphitization, diffusion of injurious gas into the hose metal, metallurgical phase changes, or other time-dependent changes. A somewhat higher peak temperature can therefore be recommended by the hose manufacturer.

Favorable environment—If the atmosphere (or fluid) is essentially noncorrosive, even at high temperatures, then temperature-induced metallurgical phase changes in the hose metal are sometimes tolerable. For example, an AISI type-304 stainless steel hose suffers from carbide precipitation at grain boundaries at 1200 F—usually undesirable because the metal loses its resistance to corrosion. But in a noncorrosive environment, the hose will not deteriorate, and can be rated at the high temperature.

Temperature gradient—The limiting temperature is not always the highest measured temperature, because cooler portions of the hose cross-section can add strength to the whole. An uninsulated hose will have a slight temperature gradient from inside to outside, and the fluid temperature can safely be somewhat above the maximum service temperature of the hose. An insulated hose, however, has negligible temperature gradient, and fluid must be at rated temperature.

Tube-wall thickness—A thick-walled hose exposed to a corrosive environment will corrode at the same rate as one with a thin wall. But for the same depth of penetration, the thick wall retains more of its original mechanical strength. Therefore, in corrosive environments, thick-walled hose can be rated for slightly higher temperatures, or for longer service at the same temperatures.

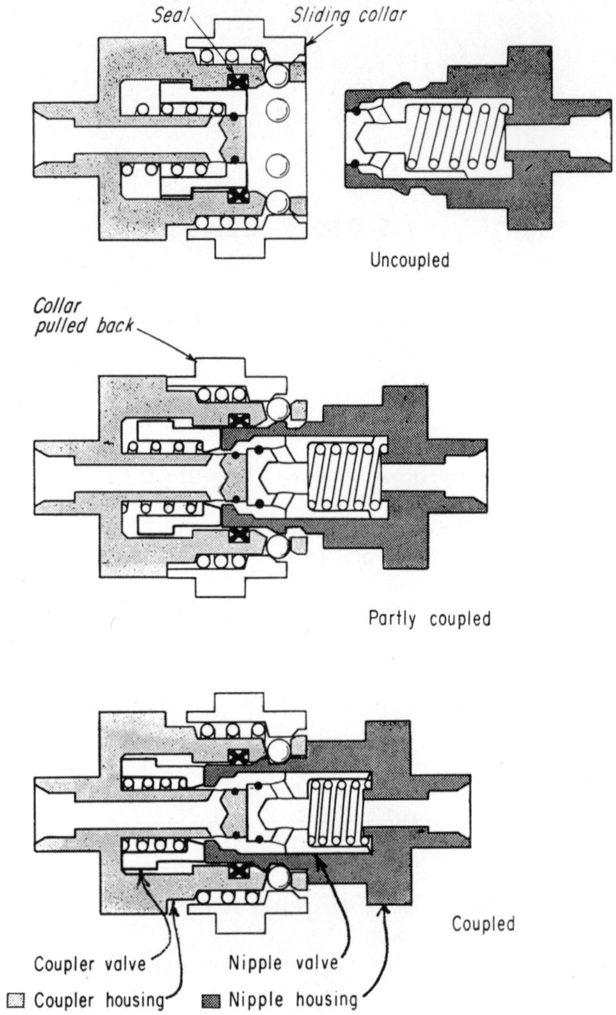

LAP SEAL BALL LOCK (double valve)

FOR HYDRAULIC AND PNEUMATIC SYSTEMS

QUICK-CONNECT COUPLINGS CUT COSTS

And these plug-ins have a tight joint you can make or break in a hurry. Some are valved types that act fast when the coupling is parted, holding spillage to a minimum.

GEORGE HAGGER

Vice-president and chief engineer
Snap-Tite Inc, Union City, Penna

The double-valve arrangement in quick-connect couplings lets you join together two parts of a hydraulic or pneumatic system about as easily as plugging an appliance into an electric outlet. When the two parts mate, a valve in each automatically opens and the liquid or gas flows through the union immediately. Disconnecting is just as simple—when the halves of the couplings are disengaged the two valves close, stopping the flow.

Two other kinds of quick-connect couplings are available. One is plain (straight-through); it has no valves. This type has been used for years at your local gas station to join the rubber air hose to the pipe from the compressor. The other has a valve in its coupler; its simpler half, the nipple, is plain. This combination can be economical where many units containing nipples, such as a

REPRESENTATIVE TYPES AND HOW THEY WORK

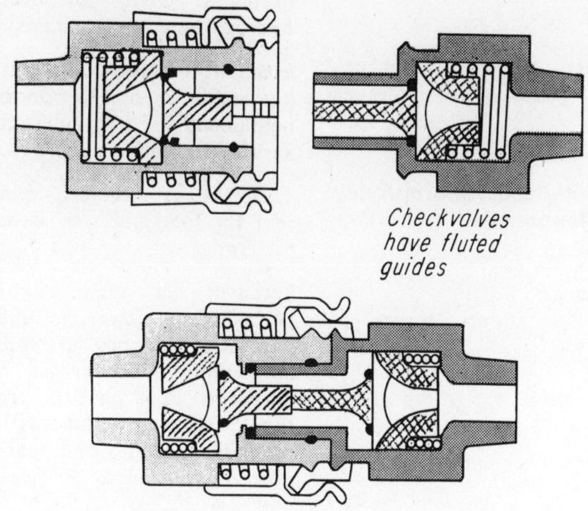

Lap-Butt Seal, Dog Lock (double valve)

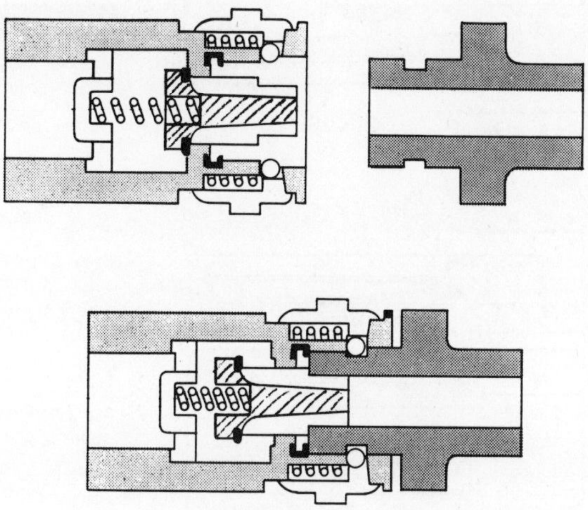

Butt Seal, Ball Lock (single valve)

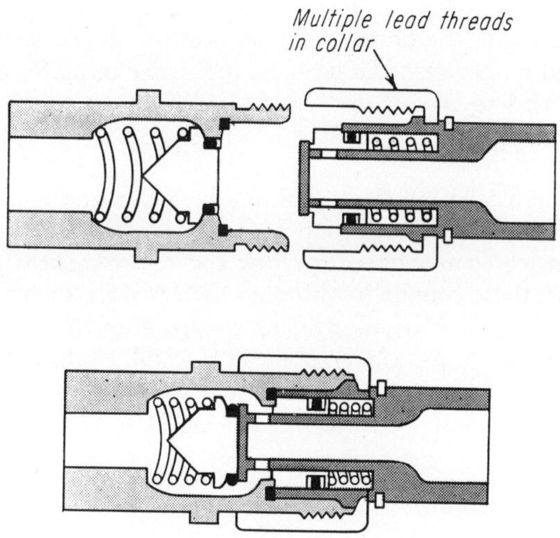

Butt Seal, Twist Lock (double valve)

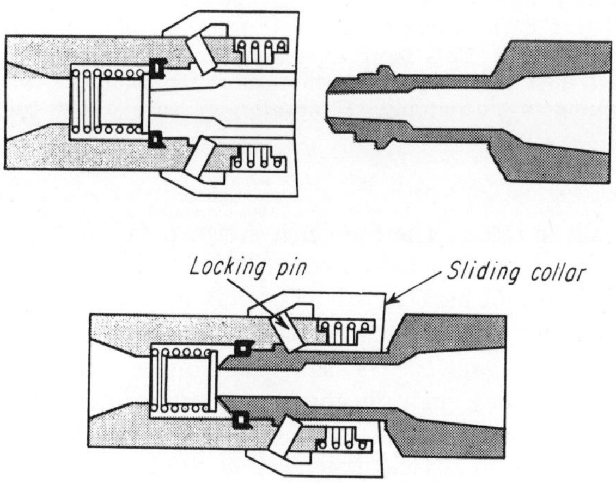

Lap Seal, Pin Lock (single valve)

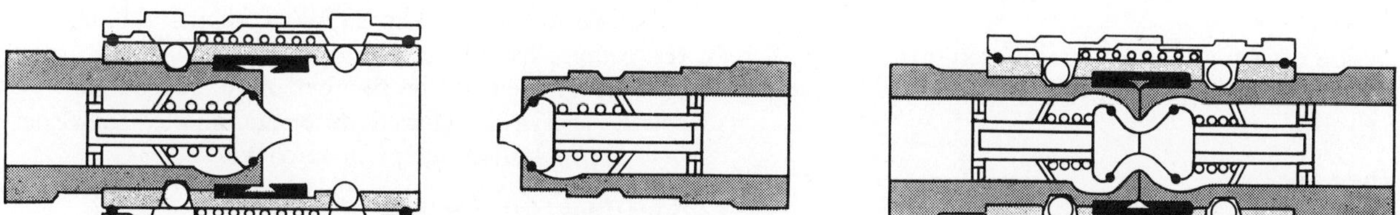

Double-Lip Seal, Ball Lock
(Double valve; either nipple can be uncoupled)

Seals for Quick-connects

Buna N	Excellent for petroleum oils, gasolines. Good for aromatic hydrocarbons, acids, bases.
Butyl	Excellent for vegetable oils, dilute acids or alkalis. Good for nonflammable hydraulic fluids. Resists gas permeation.
Polymerized chloroprene (Neoprene)	Excellent as tasteless, odorless seal for food industry. Excellent for Freon.
Silicones	Excellent for temp. extremes (—100 F, or lower, to 500 F). Excellent resistance to weathering, aging. Good for oils, many chemicals. Poor physical properties (compared to Buna N) such as tensile strength and tear and abrasion-resistance.
Polysulfide (Thiokol)	Excellent for solvents, ozone, low-temperature flexibility. Poor physical properties, heat resistance.
Fluorocarbon rubber (Teflon, Kel-F, Viton A)	Excellent for fuming acids, strong bases, petroleum oils, synthetic lubricants, fuels, silicate esters, some solvents, especially up to 500 F.

set of pneumatic power tools, connect to a single coupler. Several types are shown on these pages. See p. 62 for color code used to label parts.

When to Use Them

Where downtime is critical in a hydraulic or pneumatic system, quick-connect couplings can simplify removal of the major components for repair and replacement.

And these couplings will more than pay for themselves in such applications as window air conditioners, portable dishwashers, car-wash equipment, fuel-oil carriers, brake systems, and some production operations where sealed assemblies must be initially charged (such as refrigerant systems).

What's Available

You can get quick-connect couplings for various temperatures from −300 to 500 F. Some are designed to withstand vibration from 100 to 2400 cps over extended periods of time, pressures to 3000 to 4000 psi, and as many as 1000 connects and disconnects. Most commercial applications have far less stringent requirements.

Sizes generally range from ¼ to 1½ in. (pipe-thread size of the fittings); for special applications they go up to 3 or even 4 in.

There are couplings for practically every gas or liquid—corrosive and noncorrosive; organic (fuels) and nonorganic (acids, alkalis).

How They Lock

Different manufacturers use different methods to lock the coupling halves. These four types (illustrated on p. 62) are the most popular:

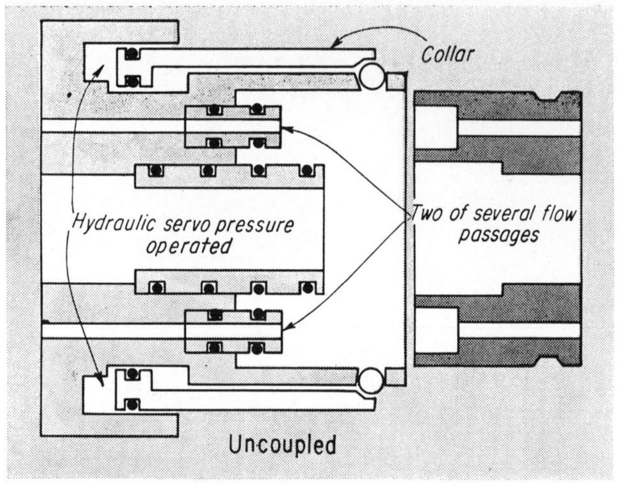

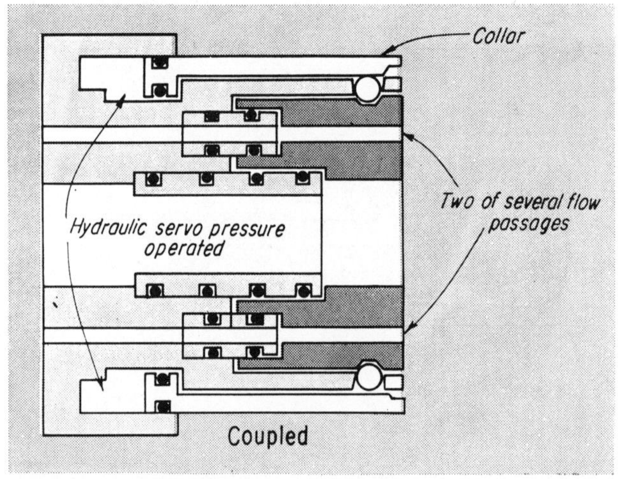

SPECIAL CONNECTOR has multiple passages for handling several fluids. This is an unvalved, straight-through type, which means internal pressure must be removed when coupling or uncoupling. Hydraulic pressure keeps collar in locked position.

Balls in races. The coupler is designed to hold a number of circumferentially spaced balls. The nipple is machined and heat-treated to provide an inner race that retains the balls when the assembly is coupled.

Pins and grooves. A radial cam drives locking pins into a retaining groove in the nipple; spring-loaded detent secures the assembly. Sometimes cams are mounted around the internal circumference of the nipple. A rotation of about 90° draws the two parts together; they are then locked by a spring detent.

Dogs in the coupler body. The dogs pivot into a locking groove in the nipple as the locking sleeve is moved into position.

Simple threads. The nipple contains the male and the coupler has the female threads (usually they are multiple-lead to reduce the number of turns required). The coupler sleeve is screwed onto the nipple, then locked by a spring-loaded detent or ratchet.

How They Are Sealed

Lap seal uses a flat, U-shape packing seal. The coupler and the nipple overlap to form the seal. This design provides tighter sealing as pressure increases. However, it is subject to seal extrusion at higher pressures. The manufacturer can minimize this tendency by holding close tolerances. The lap seal is not suited to low-pressure or vacuum applications.

Butt seal. Here the packing seal is at the end of the nipple, forming a butt with the coupler. This method is suitable for all pressures, including vacuum.

In some seals, an O-ring substitutes for the U-shape packer. One variation of the lap seal is the double-lip seal; another is a combination lap-butt seal.

Materials and Finishes

Couplings are usually made of brass plated with nickel and chrome; anodized aluminum; steel plated with cadmium, cadmium chromate, or nickel; or stainless steel, Types 303 and 316.

Seal materials are generally natural or synthetic rubber, depending on type, temperature, and pressure of the fluid. The table on preceding page lists some typical compounds and appropriate conditions. High pressures add problems of permanent set and cold-flow extrusion.

How They Perform

Speed of disconnect is generally a fraction of a second, and is accelerated by the internal line pressure. The check-valves are usually spring-loaded to ensure rapid action. One special application specifies a time of 2/5 sec at 3000-psi internal pressure.

Couplings can be joined or separated by hand; but for safety, larger sizes are limited to lower internal pressures. For instance, a man can safely connect a ¼-in. coupling at 200-psi internal pressure, or a 1-in. unit at about 65 psi. Systems handling higher pressures are usually manually connected and disconnected while internal pressure is reduced. They can be remotely disconnected under high pressure by using solenoid valves, lanyards, or other means. One special high-pressure connector (*PE—Sep 14 '59, p 97*) can be disconnected with one hand while internal pressure is 4000 psi. It is used with compressed air to start jet engines.

How Much Liquid Spills?

When a coupling is disconnected under pressure, it is only natural that some of the liquid will spill. But not much. For instance, ¼-in. couplings, depending on design, will spill anywhere from 6 drops to about 1.5 cc during the disconnect cycle, with an internal pressure of 60 psi. Stepping up to 1½ in., with an internal pressure of about 10 psi, spillage will be about ten times as great.

those pipefittings with bearings . . .

SWIVEL JOINTS

MAY WORK WHERE NOTHING ELSE CAN

Here's your guide for choosing between swivels and other flexible connections such as rubber or metal hose. Swivels often supply the best answer where the joint must be strong as the pipe.

G. M. BAGNARD, *Chief Engineer, Chiksan Co., Brea, Calif.*

A flexible connection is a nuisance needed in any piping system that requires excessive movement. A nuisance because it adds to cost and may be a weak spot —and necessary because it can be manipulated where rigid piping cannot. Flexible rubber or metal hoses have the greatest degree of bending freedom, weigh least, and are usually lowest in initial price. But swivel joints often handle the pipe-joining job better because they are stronger and can negotiate tighter turns.

The conditions that call for flexible connections include: vibration, relative motion, thermal expansion, twisting and bending, need for frequent disassembly, multiple connections demanding a pipe or hose, and coupling that can be moved to several positions.

The choice between the two types of flexible connections depends on knowing the extremes of movement required, and matching them with cheapest components that meet service specifications (Table I). Other considerations are: fluid type being handled, pressure, temperature, geometry of installation, service (intermittent or continuous), pressure drop, and leakage. Bolting flanges—pipe- or straight-thread types—and welding flanges are available for either swivel joints or hose and seldom are problems.

One successful trick in many applications is to combine swivel joints with flexible hose. This retains hose versatility and—permitting easy rotation and reduced stresses—overcomes drawbacks of hose. End result is a saving in total pipe because tighter bends are possible.

For use by themselves, the rubber or metal hoses can be made very limber or stiff—depending on wall thickness, materials used, and construction. They can be absolutely leakproof when specified. And they usually have no sliding parts—so wear of seals by grit or other contamination presents no problem.

But hoses are limited by strength and endurance. The greater the flexibility, the more delicate the construction, both for mechanical strength and for wear of hose by abrasion or corrosion (metal hose). A hose may be the weakest link in the piping system.

Twisting is a problem. For example, when one end of a hose is moved laterally, the hose twists. If excessive, twisting will eventually cause fatigue failure. Sometimes simple metal fittings are used at end connections to allow some rotation. However, under high pressure, the torque required to rotate one of these fittings may be greater than shear strength of the hose itself, resulting in a ruptured line. Another limitation: hose is elastic and kinks easily when bent in a short radius, Table II.

Rubber hose is not usually recommended at working pressures above 1000 psi. Flexible metal hoses are generally limited to pressures of about 750 psi, although reinforcement and several-ply construction can increase this to over 3000 psi. Even at relatively low pressures, 200 psi or more, a section of hose can become nearly as rigid as a piece of metal pipe.

TABLE II—Min Radius of Bend, in. (approx.)

OD, in.	Hose		Swivel fitting
	Metal corrug.	Rubber	
1/2	14	5	3
1	18	10	4
2	25	20	6
4	51	40	8
	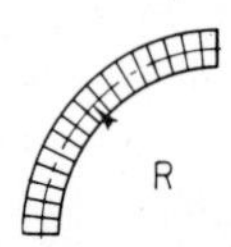	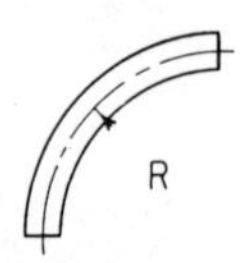	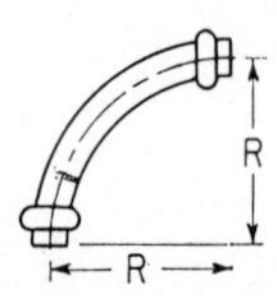

O

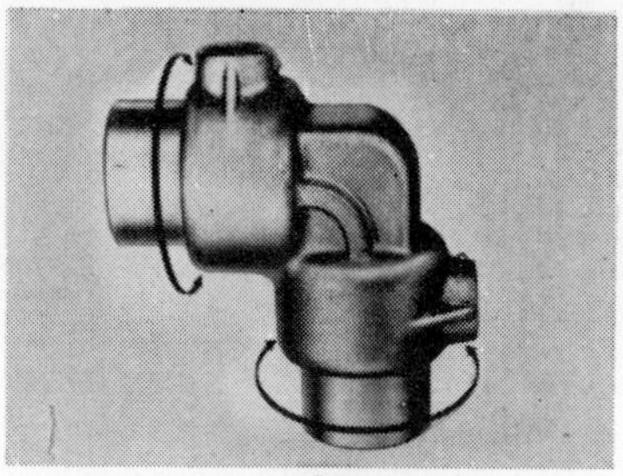

A

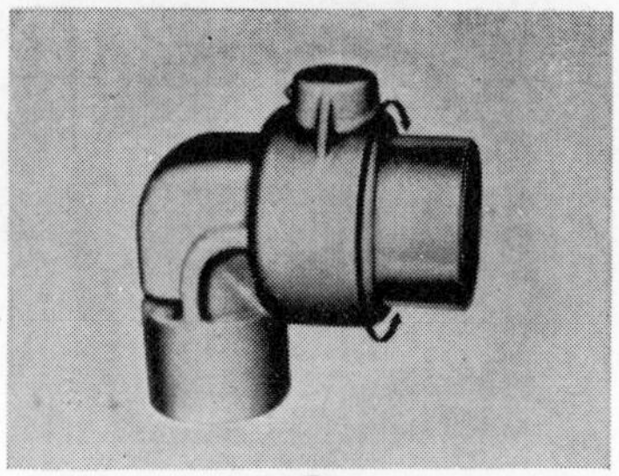

B

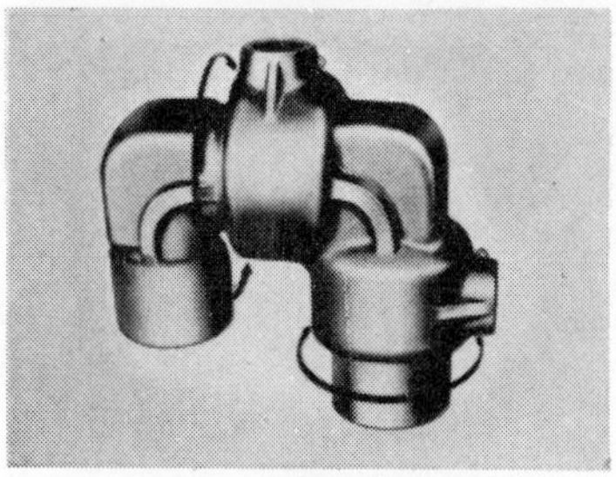

C

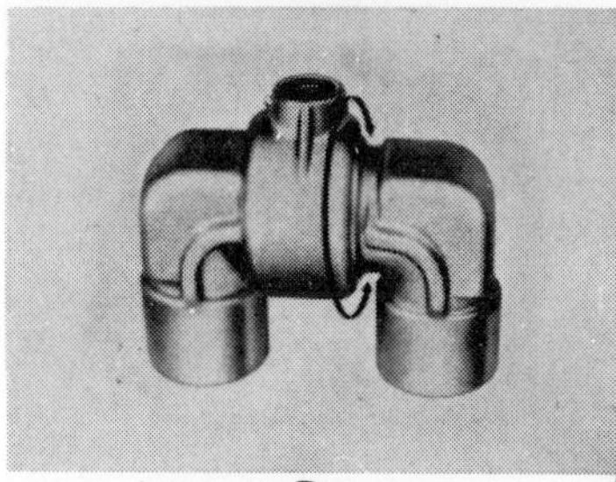

D

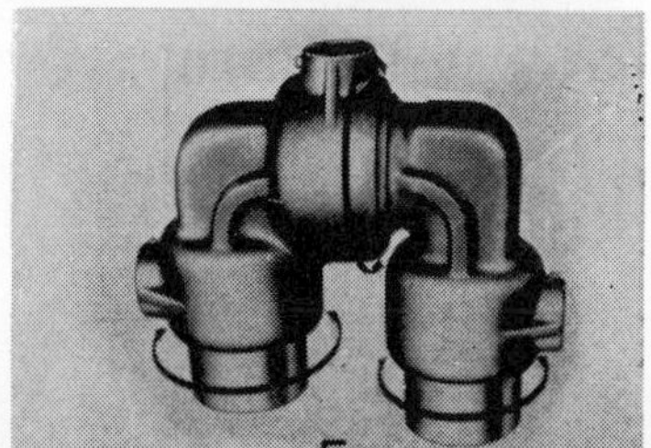

E

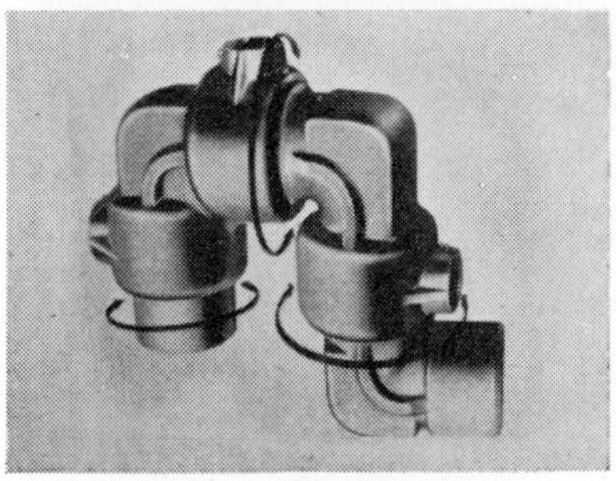

F

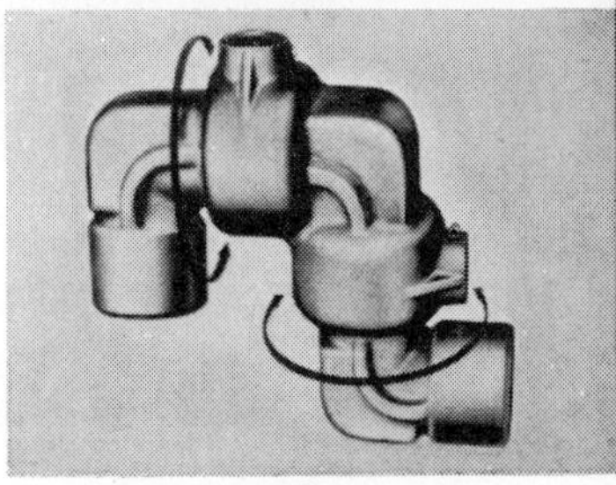

G

Table III – Styles of swivel joints

	No. of elbows	Planes of rotation
O	0	1
A	1	2
B	1	1
C	2	2
D	2	1
E	2	3
F	3	3
G	3	2

TABLE I—Over-all performance of hoses and swivels

E—excellent G—good F—fair P—poor	Hose		Swivel Joints
	Metal	Rubber	
Max pressure	G	F	E
Flexibility:			
Force required	F	F	E
Planes of motion	G	E	F
Min bend	P	P	E
Safety	G	F	E
Fluids handled	F	F	E
Min pressure drop	F	G	G
Life expectancy	F	F	E
Weight	G	G	F
Leakage	G	G	F
Cost-(initial)	G	G	F

Rubber and most metal hoses using special materials can handle fluid temperatures up to 500 F without adverse effects. However, conditions of max temperature and pressure cannot be met simultaneously because the safe working-pressure level for hose decreases with increasing temperature.

SWIVEL JOINTS

Rival to flexible hose is piping with joints that allow rotation or other relative movement, yet seal the contained fluid. They are constructed of steel, cast iron, bronze, aluminum or other rigid materials, and can usually withstand the same environment, pressures, and types of service as the rest of the piping system. Low-pressure joints operate from 28-in. vacuum to 1000 psi, and high-pressure joints to 15,000 psi. Most joints will work at 250 F, and special designs permit as low as −300 F or high as 1000 F. Corrosion is not too great a problem because there is a lot of material to corrode away. Small-radius bends are easily accommodated because the rotating joint is not part of the bend and therefore does not limit the bend radius significantly (Table II). A secondary advantage of a single-plane swivel is its movement in a predetermined arc and plane.

Almost any degree of freedom of movement can be

TABLE IV—Allowable cantilevered moments for high-pressure swivel joints with flame-hardened ball races..

Pipe size, in.	Moment, in.-lb	
	standard	special
2	6,525	......
3	14,400	46,000
4	28,200	90,000

obtained by combining two or more joints. Single-plane swivels are common, but multiplane, spherical, or combinations of swivels and hose are available. Seven swivel joints will give unlimited flexibility. Sizes from ¼ to 36 in. dia can be built.

Two limitations to swivel joints are: difficulty in maintaining a leaktight rotating or sliding seal, particularly when installation and maintenance personnel forget that these are precision devices and require special care; relatively few planes of rotation available to any type except the spherical.

BEARINGS FOR SWIVELS

The joint itself may swivel on ball bearings, a spherical joint, thrust plates, or a bushing. The seal packing may also serve as the bearing in some applications. Full 360° rotation in the plane of the swivel is usually available, and the spherical (ball-joint) type may also pivot a total of 40° about its ball center.

Ball bearings are used to reduce torque without hurting smoothness. They have less wear than sliding bearings, and can carry thrust and radial loads simultaneously. Thrust loads come from thermal expansion and other forces exerted by the piping on the coupling, and from the internal fluid pressure. Radial loads are caused by weight of the piping, which also contributes a bending moment at the joint. With proper spacing between bearing races and careful choice of heavy-duty rolling elements, up to 90,000 in.-lb can be handled by a nominal 4-in.-dia swivel joint.

Thrust bearings (shoulders and washers) cannot take radial loads, and will not rotate freely if the axial load is heavy or cantilevered. Applications are for small-size hydraulic lines.

Sleeve bearings are satisfactory for light radial loads.

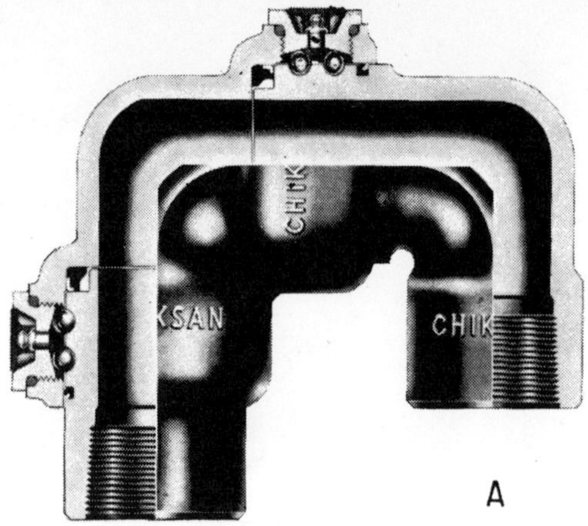

A

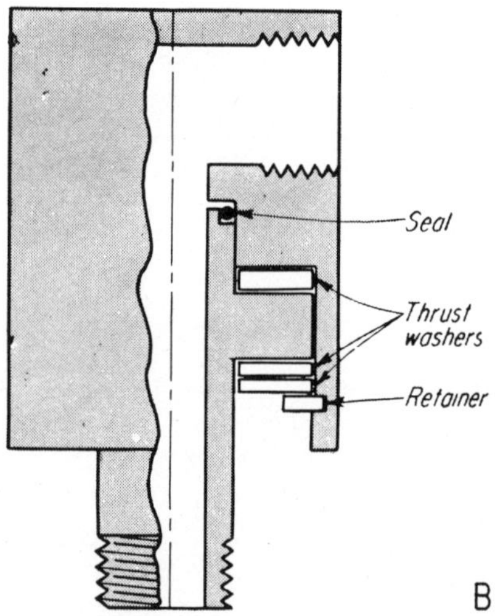

B

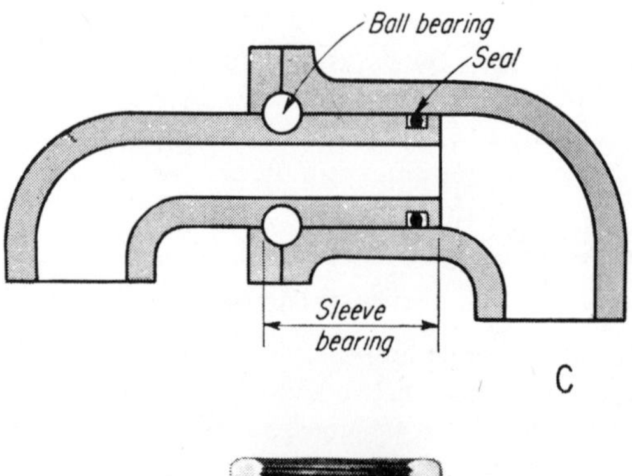

C

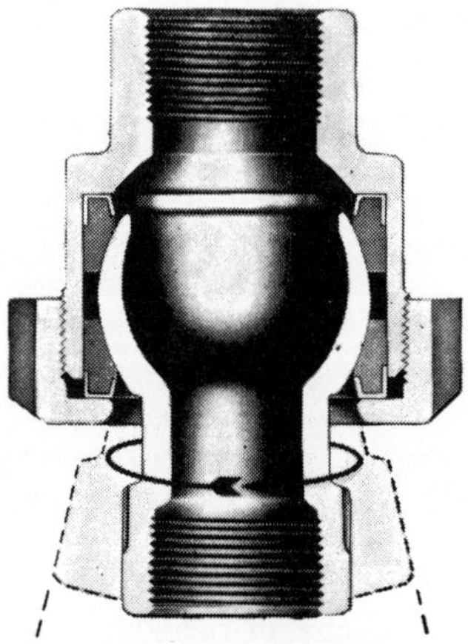
D

Bearing and seal . . .
design help determine application of swivel fittings. Style "C" swivel (photo A) features sealed bearings arranged to take heavy axial and thrust loads. Thrust-bearing type (B) and radial-bearing type (C) are for light duty: each can handle limited amounts of both thrust and radial loads. Spherical joint (D) has no metal-to-metal contact (ball seats itself in gaskets for either pressure or vacuum service) and flexing angle (dotted lines) can be 40° total.

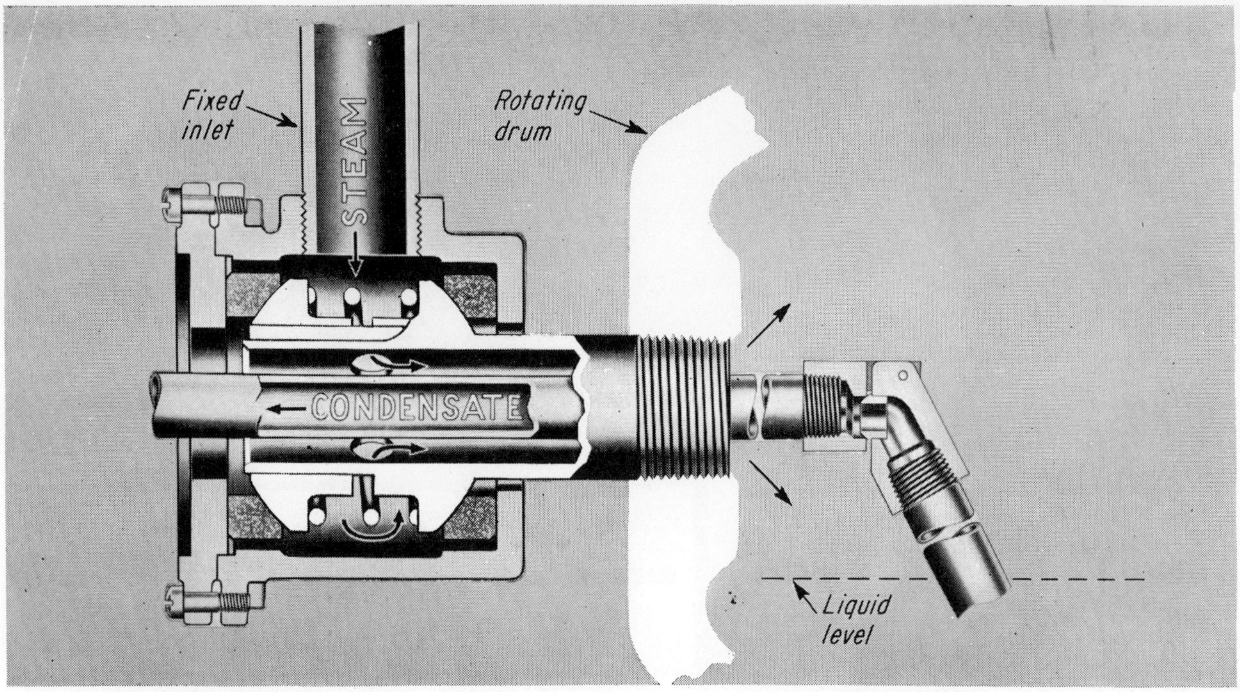

Revolving joint . . .
allows steam to enter and condensate to leave revolving drum. Gasket-type spherical bearing also acts as seal, and loading is light enough to give long service. Drum bearings not shown.

Heavier loads increase torque considerably and accelerate wear. Axial loads cannot be taken.

Spherical-type fittings have circular gaskets that seal the joint and also act as spherical bearings. Each joint is self-aligning, but a single one should not be depended on to solve misalignment in other parts of the piping system. However, two will often correct misalignment better than three or more single-plane swivels.

CONTINUOUS ROTATION

Among swivel joints one type permits continuous rotation like that required when fluid from a stationary source is fed into a revolving drum. Special swivel fittings may be required for fast, continuous rotation, and knowledge of peripheral speed is important. The bearings and packings might be expected to meet certain velocity, low-torque, and low-wear requirements, in addition to those involving temperature, pressure, flow and mechanical stress. The larger the joint dia, the lower max rpm is allowed—this is to limit surface speed and help packing life. There are small joints designed for speeds to 10,000 rpm, pressures to 250 psi, and temperatures to 400 F.

For high temperatures, choose materials, packings and lubricants carefully, and borrow from the experience of manufacturers and users of standard high-temperature bearings as well as from swivel makers. The relationship between high-temperature and load-carrying capabilities of lubricants is complex, and expert advice is essential.

MATCH THE FLUID AND JOINT

Fluid type influences material selection for swivel joints. Any metal is satisfactory for ordinary oils, alcohols, aromatics (benzene, napthalene, toluene), esters (acetates), natural gas, kerosene and so forth. Stainless steel, monel metal, or nickel-bronze should be specified for acids, mercury, molasses, and sewage. Asbestos chevron-type packing is suitable for the aromatics and esters, but Teflon or synthetic rubbers will be required for most of the other fluids mentioned above. The subtle effect of temperature, plus compatability problems, suggest that experts be consulted to assure proper material selection.

fewer wrong guesses when...

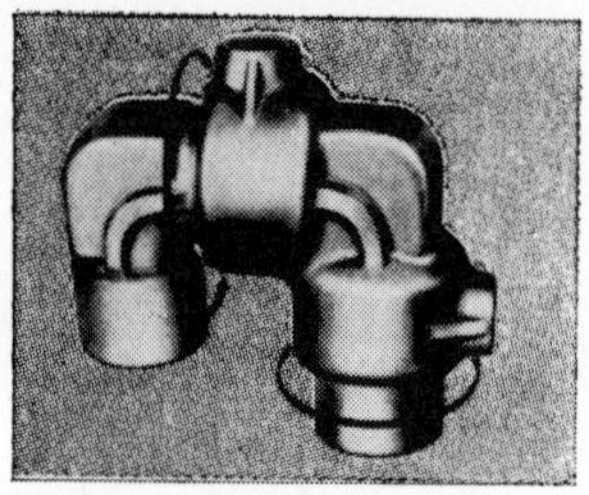

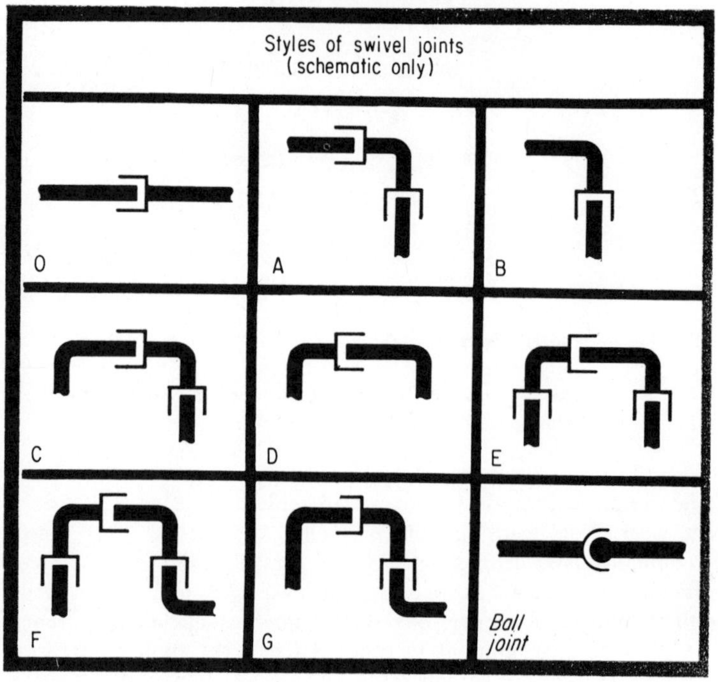

When the pipe connection isn't hose but has to turn in different directions—especially when it's a 3-D affair—final design becomes answer to a puzzle. Here's a straightforward basic approach that will eliminate much of the trial and error.

G M Bagnard, *Chief Engineer*
Chiksan Company, Brea, Calif.

The problem of choosing swivel joints for a fairly complicated piping system need not be as difficult as it looks at first. True, thinking in three dimensions is usually required before you can hit on the combination of swivels and pipe that will do the job at lowest cost–and without the built-in errors of judgment that result in mismated flanges, jammed assemblies or more joints than actually needed. But three basic techniques are given below to determine the best configuration of piping–and they work for 3-dimensional as well as 1- and 2-dimensional problems.

Of course, some compromise will be necessary. For example, it may not be possible to have simultaneously the minimum number of joints and the shortest total length of piping. And there may be specific problems and limitations that will affect design. It will be necessary to determine in advance whether space will be tight or loose, maintenance easy or difficult, obstructions present or absent, and loading on supports excessive or negligible.

Each of these problems will have to be considered separately. This explains why successive revisions of the original design can not only be expected but will be a necessary part of the design procedure.

Three Ways to Plan Joints

Despite a certain amount of trial-and-error needed to arrive at the final design, work all along the way will be reduced by following the three basics.

Method 1–Models. A scale model of the proposed installation will usually save design time. With this way, also, points at which the assembly will lock can be readily determined in advance–thereby simplifying the necessary changes. But models can be expensive–or not available.

Method 2–Plan View. A simplified procedure without need for models is made possible by determining, in a series of views, what motion is required about each fixed point. Starting with the plan view, examine each critical point in a separate view that shows true lengths of the sections of the assembly required to move to that point. Then connect sections of pipe and swivel joints to provide the flexibility required in each view.

DESIGNING WITH SWIVEL JOINTS

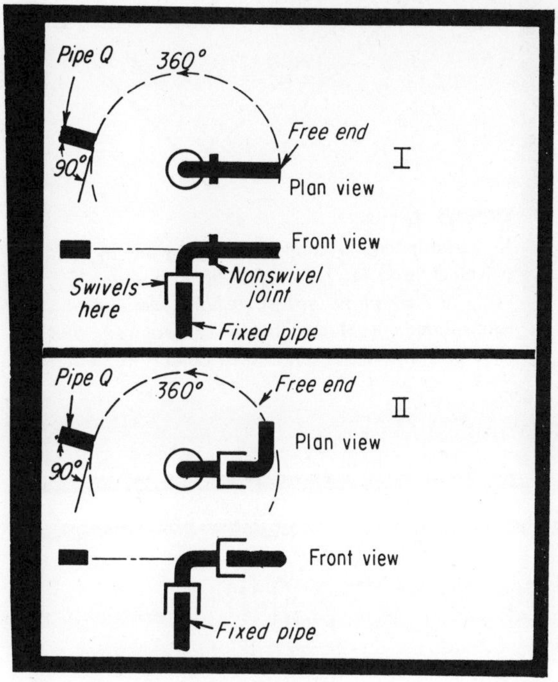

. . . AND MORE ABOUT THE SWIVELS

The flexible connection you choose need not necessarily be swivel joints—pipefittings with bearings. As the author pointed out in his previous article (*PE—Aug 4, p 40*), metal hose or rubber hose also have advantages, and his discussion of the swivels weighed what the other two types can offer when you consider such factors as temperature, pressure, flexibility, type of fluid, fluid friction, leakage, moments and forces. That article also cited nine previous ones that covered such subjects as rotating seals and joints, flexible hose, quick-disconnects, calculation procedures to compute pressure drop and hp in valves, piping and fittings.

Mr. Bagnard's present article assumes that the reader is choosing swivels—for such reasons as greater mechanical strength or pressure capability. But his advice will also help design with any other flexible connections.

Plan views help avoid . . . impractical connections. Scheme I using style "B" joint does allow simple connection at free end to pipe Q; but scheme II does not—an additional style "C" joint, at the end, would be needed to overcome angularity.

Method 3—Three Planes. A more rigorous approach—usually best for very difficult 3-D connections—is to select three mutually perpendicular planes, and examine all motions in each. An assembly of joints can then be established to give desired movement in each plane. The designs may be modified afterwards if corrections or special conditions are required. In this method, it is often necessary to lay out several auxiliary views to get the exact dimensions of the various sections of the assembly.

Direction of Movement

Most of the normal flexibility problems can be solved by Method 2. Inasmuch as the same reasoning applies to all three methods, let us examine the second method's simplified procedure:

At each swivelling connection start with the one fixed point about which the mating lines must move. If both ends move, one line should be considered stationary, and the other line mobile in relation to the stationary line.

Next step is to determine the path that the moving end of the line is to follow. The maximum and minimum distances it must travel are very important. Line length should allow parts of the assembly to move to all required locations without mechanical interference. If the line must move in at least two planes, the critical points in each plane must be determined.

Radial position of the line is important. The pair of sketches on this page show two different fittings that give the same rotation (a full 360°) cannot be given the same hookup at the free end because of difference in offset—therefore radial position—in each fitting.

The Torque Requirement

At increased fluid pressure, the joint packing makes a tighter fit and bearings also get increased loading. The torque required to turn the swivel becomes greater, and must be evaluated in advance: to compute force needed at end of the pipe to swing the joint, and to compute forces imposed on the rigid lines.

These factors must be considered in determining the allowable torque: Will the line be moved by hand, by equipment motion, or by some special actuator designed for the purpose? How much effect will weight of the completed assembly have on the total torque developed?

Torque and related forces must sometimes be controlled or limited by modifying the piping design. For example, the force needed to overcome swivel friction

DESIGNING WITH SWIVELS—an example

This simple 3-dimensional design is a typical problem that can be solved with Method 2—the plan-view approach. Pipe *P* is a feed line that must be able to pass fluid to any one of three receiving pipes, X, Y or Z.

Choice is a single, flexible-line connection to serve the three pipes. It is less expensive than the alternative—valves and manifolds. Steel pipe with swivel joints will suit this hypothetical application best.

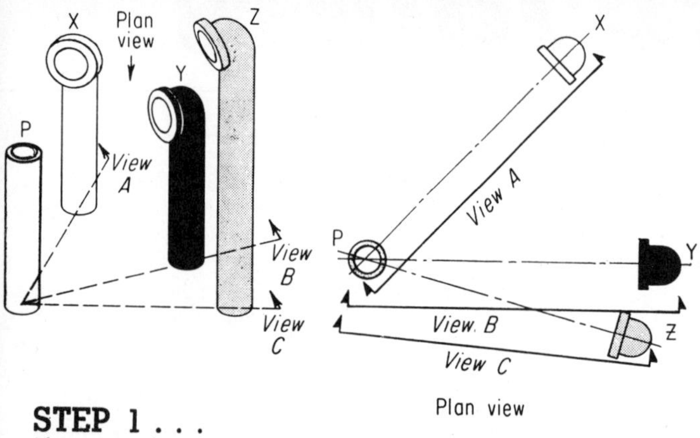

STEP 1 . . .
Plan view shows that rotational movement about point at top end of pipe *P* in a plane perpendicular to its centerline is required. This calls for attaching a style "A", "B", "C" or "F" swivel joint. The style chosen depends on motion required in other planes.

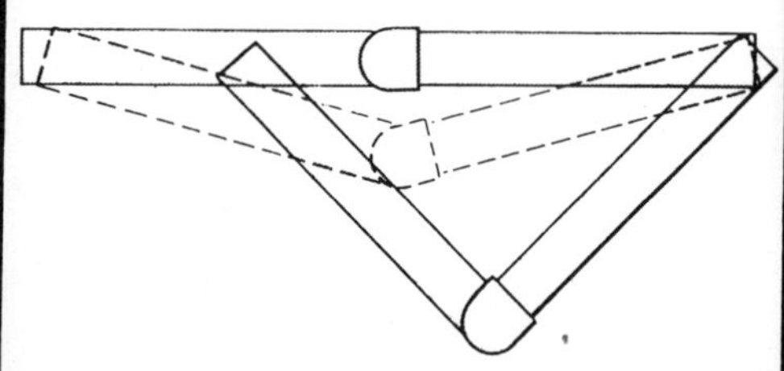

STEP 3 . . .
A variable-length connection is given by this swiveled "dog leg" that can have a style "C", "D", or "E" plane swivel. Choice depends on two related problems: flexibility required, and style of plane swivel selected at point *P* in Step 1.

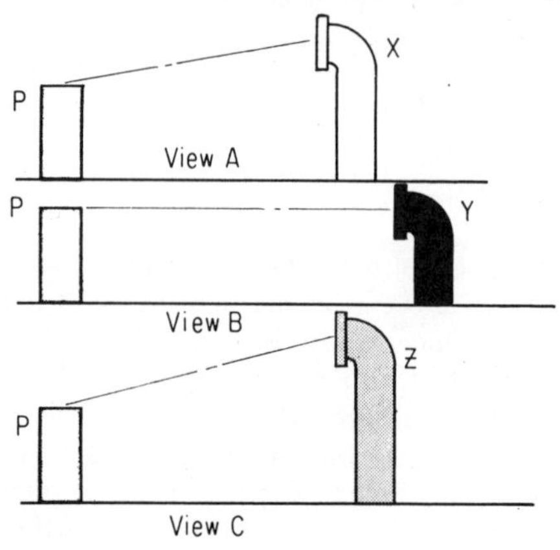

STEP 2 . . .
Three auxiliary views are then developed in planes perpendicular to the flanges on pipes *X, Y,* and *Z*. Distances between top of *P* and each of the three pipes are different, and a variable-length connection is needed.

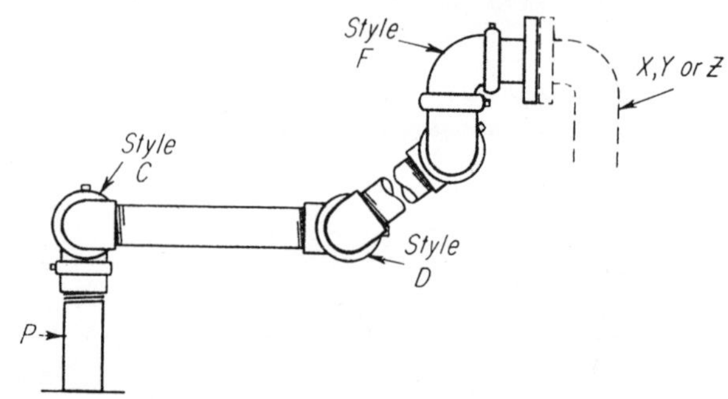

STEP 4 . . .
Selection of end fittings requires care because choice of center joint will be affected. Style "C" swivel is the best selection at *P*. A style "E" would have more flexibility than needed and be too expensive for the job (it costs $30 more than style "C" in 4-in. size); and a style "D" has too few degrees of freedom. Style "D" is suggested for center of dog leg, and style "F" can be attached to opposite end of the dog leg to allow rotation of the flange to match bolt holes.

can be reduced by lengthening a section of pipe—therefore improving the moment arm. Furthermore, design of an assembly must sometimes be altered to avoid torques that tend to unscrew pipefittings. Counterweights can solve still another problem—that of unbalance caused by overhanging of piping.

The Cost

Costs can be evaluated for any given application, but it is not possible to establish a general rule. In some designs a single joint with two or more swivels will be cheaper than separate single-swivel joints for the same job. In general, the design with fewest interconnections will cost the least and have the lowest pressure losses. But flexibility and safety also frequently decide the issue.

No tubing needed—passages in manifold connect hydraulic components

when valves belong together keep them together with . . .

SANDWICH MANIFOLDING

If the prototype of a hydraulic circuit design turns out to be a forest of tubing and fittings, almost hiding the hydraulic components, then look to this method for an economical answer.

ALLEN A. DICKE
Monarch Machine Tool Co.
Sidney, Ohio

The curse of automatic hydraulics is the complex of piping and controls it brings into assembly. Here is a proven way to eliminate half the problem—the piping It's done with a sandwich manifold containing a base plate that not only mounts the hydraulic components but also contains the interconnecting passages.

Simplest version of the manifold is an assembly of two plates: the thicker has machined or cored-out passages on one side (resembling a printed-circuit board). The second plate is brazed on top as a cover, and porting holes are drilled into the passages. Tapped holes in the cover permit mounting of valves and other devices.

Or both sides of the thick plate can be machined out and two covers brazed on. This 3-ply method permits mounting components on both sides of the sandwich. Sandwiches beyond 3-ply are also possible.

SAVES SPACE AND MONEY

The old way, with tubing, requires large-bend radii, bulky connections, and wrench-room. Sandwich manifolding eliminates tubing fittings, shortens passages and keeps them in one (or two) horizontal planes. Simultaneously it provides a strong mounting surface. Aluminum plates will help keep weight down.

Cost can be less for sandwich manifolding because quantity production is possible. Here the plates can be die-cast rather than machined, giving more accurate and consistent results. Piped components, by contrast, always require considerable labor no matter how many identical assemblies are made; and errors, leaks, unsightly connections, vibration failures can result.

BEST FOR GROUPED COMPONENTS

Manifolding works best with components that are grouped functionally and physically close. Sequencing and safety-interlock circuits are two good examples. It is worthwhile to draw design schematics with this in mind, showing functional boundaries by dotted lines. Interconnections between groups can be designed with tubing, or with porting through the frame or body of the main machine. It is easy to catch errors made in drawings by first checking each complete manifold

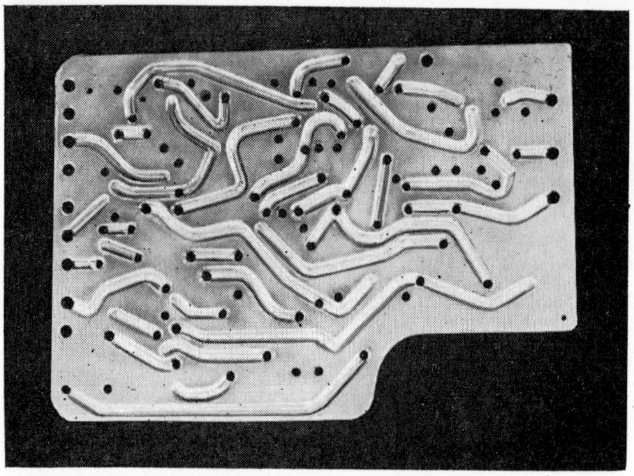

Machined grooves . . .
in top and bottom face of center plate of 3-ply manifold serve as hydraulic passages between preselected points. Top face is connected with bottom face by drilled holes where inter-connections are necessary.

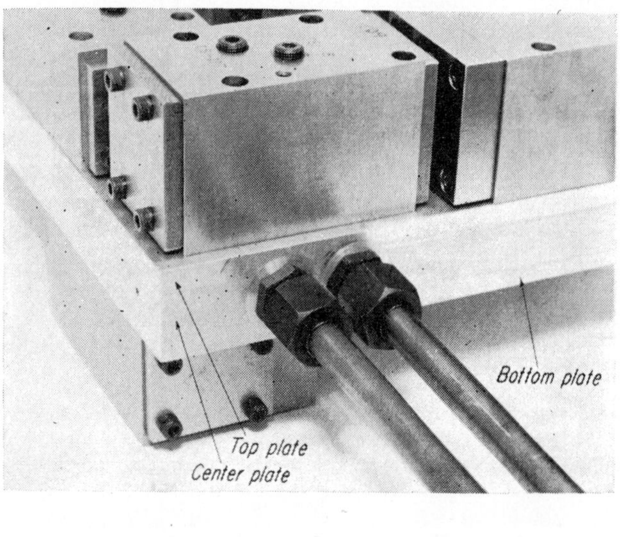

grouping, then the interconnections between manifolds.

Components mechanically connected to the machine (such as hydraulic pump driven by a shaft on an accessory gearbox) must be located near edge of the manifold. Passages machined in the center plate should avoid interference and thin walls. Where passage crossing is necessary, both sides of the plate can be grooved, and cross-over accomplished by drilling through. This becomes a 3-ply sandwich.

With 3-ply construction, it is wise to orient the general passage direction on one side north-to-south and on the other east-to-west to assure crossing of passages that will be drilled through. Friction loss in operation will compare favorably with tubing and fittings if total passage length and turns are kept to a minimum.

Manifold shape depends on contour of the machine frame to which it must be mounted. Installation is easy if frame-to-manifold ports are incorporated, because the sandwich assembly can be dropped into position without pipe connections. Sealing is accomplished with O rings.

The draftsman will usually be able to convert from tubing to manifolding without an engineering redesign. There is one important exception: All devices must have inlet and outlet connections on one face, capable of leakproof attachment (using O rings, for example) with this face against the sandwich. To prevent mismounting, different bolting dimensions are selected for each different component.

QUALITY CONTROL REQUIRED

Aluminum plates are favored because of inertness to oil and moisture, ease of manufacture, durability, and general availability. Brazing of aluminum, however, requires excellent equipment and a skillful operator.

Whatever the plate, it must be of high quality to avoid cracks, porosity, warping, and poor brazing characteristics. Brazing should be done carefully to avoid drips into the channels or leaks between channels. Complete operational testing is especially important, at least on a sampling basis, because there is no way to check inside joints visually.

Oil supply . . .
and return come into edge of plate. Other porting methods are: at manifold bottom, using O rings, to match similar parts in machine frame; or tube connection brazed on top or bottom of manifold.

5

FLOW AND PRESSURE DROP

Pressure-drop equations for flow of liquids and gases through piping, fittings, and devices; survey of flow meters

A new look at fluid friction

Flow resistance in pipe

There's no lack of data on pressure losses in pipe—just difficulty in keeping units straight in calculation. Here are easy conversions, for liquids

LOUIS DODGE, Hydraulic Consultant, New Richmond, Ohio

NOMENCLATURE

d = pipe inside dia, in.
D = pipe inside dia, ft
f = friction factor, dimensionless
g = gravity constant, 32.2 ft/sec²
H_L = head loss, ft
L = pipe length, ft
$L_{in.}$ = pipe length, in.
Δp = pressure loss, psi
ΔP = pressure loss, psf
q = flow, gpm
Q = flow, ft³/sec
$Q_{in.^3}$ = flow, in.³/sec
R_e = Reynold's Number $\rho DV/\mu$. Typical units are:
$3162\ q/d\nu_{cst}$ gpm, in., centistoke
$50.6\ q\gamma/d\mu_{cp}$ gpm, lb/ft³, in., cp
$\gamma DV/g\mu$ lb/ft³, ft, ft/sec, lb-sec/ft²
v = oil velocity, in./sec
V = oil velocity, ft/sec
μ = absolute viscosity, lb-sec/ft²
ν = $\frac{\mu}{\rho}$, kinematic viscosity, ft²/sec
ρ = mass density, lb-sec²/ft⁴ = slugs/ft³
γ = weight density, lb/ft³
ϵ/D = relative roughness of pipewall. Values of ϵ, ft: drawn tubing, 5×10^{-6}; steel or wrought iron, 150×10^{-6}. D = inside dia, ft.

REVIEW OF PIPE LOSSES

Variables (μ, ν)						
L	V	D	ΔP	Q	f	μ
ALL FLOWS						
ft	$\frac{ft}{sec}$	in.	psi	$\frac{ft^3}{sec}$	f	–
ft	$\frac{ft}{sec}$	in.	psi	gpm (q)	f	–
LAMINAR ONLY						
ft	$\frac{ft}{sec}$	in.	psi	$\frac{ft^3}{sec}$	–	cst
ft	$\frac{ft}{sec}$	in.	psi	gpm	–	cst
in.	$\frac{in.}{sec}$	in.	psi	$\frac{in.^3}{sec}$	–	$\frac{lb\text{-}sec}{in.^2}$
ft	$\frac{ft}{sec}$	ft	psf	$\frac{ft^3}{sec}$	–	$\frac{lb\text{-}sec}{ft^2}$
ft	$\frac{ft}{sec}$	ft	psf	gpm	–	$\frac{lb\text{-}sec}{ft^2}$

HOW do you convert a hydraulic-pressure loss relationship from units of gpm, psi, and inches into ft³/sec, psf, and ft? Or from anything to anything? You probably head right for Crane Company's Technical Paper 410, "Flow of Fluids," or some other fluid power reference such as Crocker's *Piping Handbook* (McGraw-Hill, New York), McClain's *Fluid Flow in Pipe* (Industrial Press, New York), *Marks' Handbook* (McGraw-Hill), and the *Pipe Friction Manual* (Hydraulic Institute, New York).

Then, if you are lucky, you might find the conversion you need, or one you can adapt after a half hour of misplacing decimal points, mistaking absolute for kinematic viscosity, and miscalculating feet head of fluid. It never seems to work out that the data you've taken on the test stand have the right units for the handbook you pull from the shelf.

What's needed is a master conver-

What is Viscosity?

Only absolute viscosity (centipose; lb-sec/ft²; etc) has physical units that can be visualized. Kinematic viscosity is more or less a defined term, being absolute viscosity divided by mass density.

A simple way to visualize absolute viscosity—let's select lb-sec/ft² as the example—is to reproduce the definition pictorially. So, imagine the sketched cube to be 1 ft on a side. Apply a sufficient force at the upper surface of the fluid to move that upper surface at a velocity of 1 ft/sec relative to the fixed lower surface.

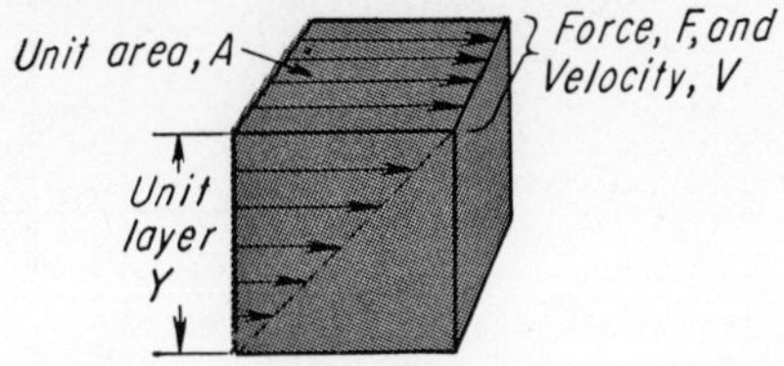

Absolute viscosity is a measure of the value of that shearing force, expressed as follows:

$$\mu = \frac{(F/A,\ \text{lb/ft}^2)}{(V/Y,\ \text{ft/sec per ft})} = FY/VA,\ \text{lb-sec/ft}^2$$

A poise has units of dyne-sec/cm², but is otherwise a carbon copy of the principle described above.

Viscosity of oil will change with changes in temperature, pressure, rate of shearing, and purity. Some synthetics, particularly those with long-chain molecules, are adversely and permanently affected by high rates of shear, such as would be experienced in throttling the oil at high pressure. The shearing breaks down the chains, reducing viscosity.

Temperature change has the greatest effect on viscosity. The best way to simplify calculation of pressure drops in fluid systems is to keep the temperature stabilized between 70 and 160 F. Even then, oil viscosity changes about 2:1.

General empirical relationships:

All flows, $H_L = f\frac{L}{D}\frac{V^2}{2g}$; laminar only, $H_L = 32\frac{\mu LV}{\varrho g d^2}$

Useful conversions, for oil only (= 55 lb/ft³)

Equations

pressure loss	velocity	flow
$\Delta p = f\frac{L}{d}V^2 \times 0.072$	$V = \sqrt{\frac{d\Delta p}{fL}} \times 3.73$	$Q = \sqrt{\frac{d^5\Delta p}{fL}} \times 0.0203$
$\Delta p = f\frac{L}{d^5}q^2 \times 0.0123$	$V = \sqrt{\frac{d\Delta p}{fL}} \times 3.73$	$q = \sqrt{\frac{d^5\Delta p}{fL}} \times 9.1$

pressure loss	velocity	flow
$\Delta p = \nu\frac{L}{d^2}V \times 0.0006$	$V = \frac{d^2\Delta p}{\nu L} \times 1670$	$Q = \frac{d^4\Delta p}{\nu L} \times 9.1$
$\Delta p = \nu\frac{L}{d^4}q \times 2.45 \times 10^{-4}$	$V = \frac{d^2\Delta p}{\nu L} \times 1670$	$q = \frac{d^4\Delta p}{\nu L} \times 4080$
$\Delta p = \mu\frac{L_{in.}}{d^4}Q_{in.^3} \times 40.75$	$v = \frac{d^2\Delta p}{\mu L_{in.}} \times 0.0312$	$Q_{in.^3} = \frac{d^4\Delta p}{\mu L_{in.}} \times 0.0245$
$\Delta P = \mu\frac{L}{D^2}V \times 32$	$V = \frac{D^2\Delta P}{\mu L} \times 0.0312$	$Q = \frac{D^4\Delta P}{\mu L} \times 0.0245$
$\Delta P = \mu\frac{L}{D^4}q \times 0.091$	$V = \frac{D^2\Delta P}{\mu L} \times 0.0312$	$q = \frac{D^4\Delta P}{\mu L} \times 11$

sion table, similar to the one on page 77. This one has only seven variations, but you can add a few of your own and eventually take some of the mistakes (and maybe the thinking) out of your computations for pressure losses in piping. (Editor's note: Send in any handy ones you dream up—we'll be pleased to have them.)

Bernouilli, Poiseuille, Darcy, Weisbach, Moody, Pigott, Reynolds, and other pioneers set down the rules for flow of hydraulic fluids years ago, and the equations haven't changed much since. However, only the simplest of the relationships are used at all extensively by fluid power and control engineers. Bernouilli's balance of energy equation, which involves velocity change, elevation, and static pressure, is far less convenient than this accurate-enough empirical relationship for flow losses in pipes:

$$H_L = f\,\frac{L}{D}\,\frac{V^2}{2g}$$

where head loss in feed depends on pipewall friction factor f, length over diameter ratio, and velocity in ft/sec.

That empirical equation has limitations, of course. Pipe diameter is assumed constant for the length examined, to assure constant velocity: And the friction factor is assumed constant at some average value even though it is known to change with temperature, pressure, and variations in pipewall conditions.

Friction factor f is a function of the Reynold's Number and the roughness of the pipewall. When flow is laminar ($R_e < 2000$) and isothermal then

$$f = \frac{64}{R_e}$$

When oil system temperature is variable, as when system components are subject to heating and cooling, the friction factor tends to vary as follows:

$$f = \frac{75}{R_e}$$

For turbulent flow based on smooth steel or brass tubing, an average relationship is:

$$f = \frac{0.3164}{(R_e)^{1/4}}$$

Reynolds Number $R_e = \dfrac{\rho D V}{\mu}$ is dimensionless and thus independent of the particular set of units chosen. The nomenclature on page 77 suggests several alternatives. However, it's tough enough even staying within a given set of units.

Viscosity μ usually is the culprit in most wrong calculations, so let's examine that. Absolute viscosity (the one in R_e before) is a direct measure of the fluid shearing forces per unit area, velocity, and layer thickness. Kinematic viscosity includes density, and is useful mainly because most viscosity-measuring instruments give readings in kinematic viscosity. You still have to convert to absolute units for many equations. Fortunately, viscosity conversion tables are readily available in the literature.

Velocity V is important because losses vary as V^2 does. In fact, a popular rule of thumb is to size piping to give some specified velocity. Typical suitable maximum values are indicated in the table at the right, converted in the graph to gpm for several pipe sizes. That is a good place to start figuring your system losses.

Friction factor for laminar flow

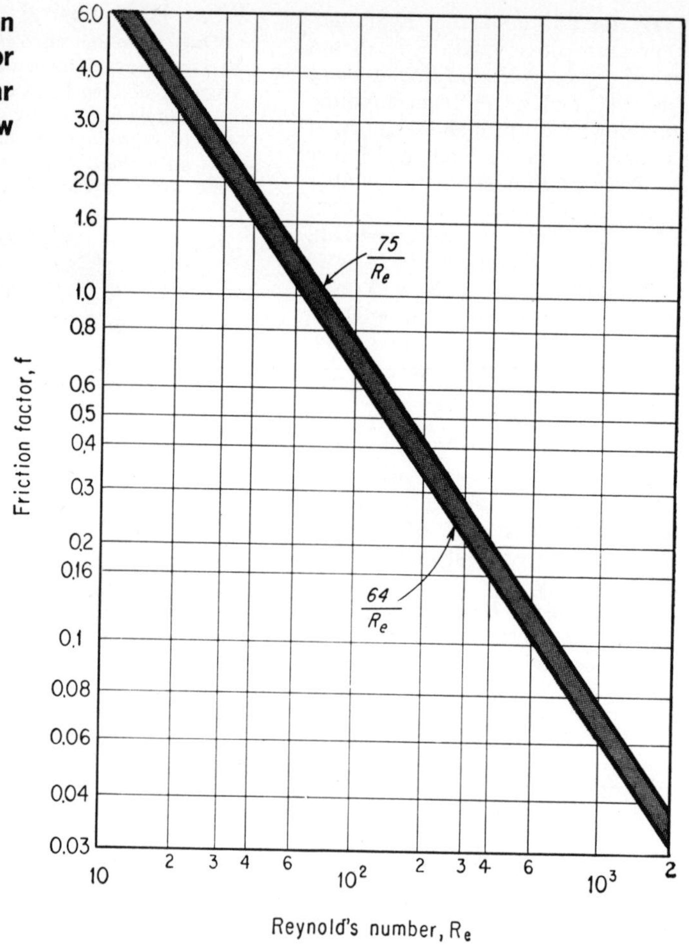

Recommended maximum velocities

Piping	Velocity ft/sec	Flow, gpm
Pressure lines,		curve:
800 psi and up	16	A
400 to 800	13	B
up to 400 psi	10	C
Return lines	7	D
Suction lines	5	E

Example: Required flow is 50 gpm, system pressure is 900 psi. What pipe sizes are needed to assure recommended maximum velocities?

On graph at right, draw a horizontal line from 50 gpm, intersecting all curves. Draw vertical lines to establish that:

Pressure lines need 1¼-in. pipe
Return lines need 2-in. pipe
Suction lines need 2-in. pipe

Friction factor for turbulent flow

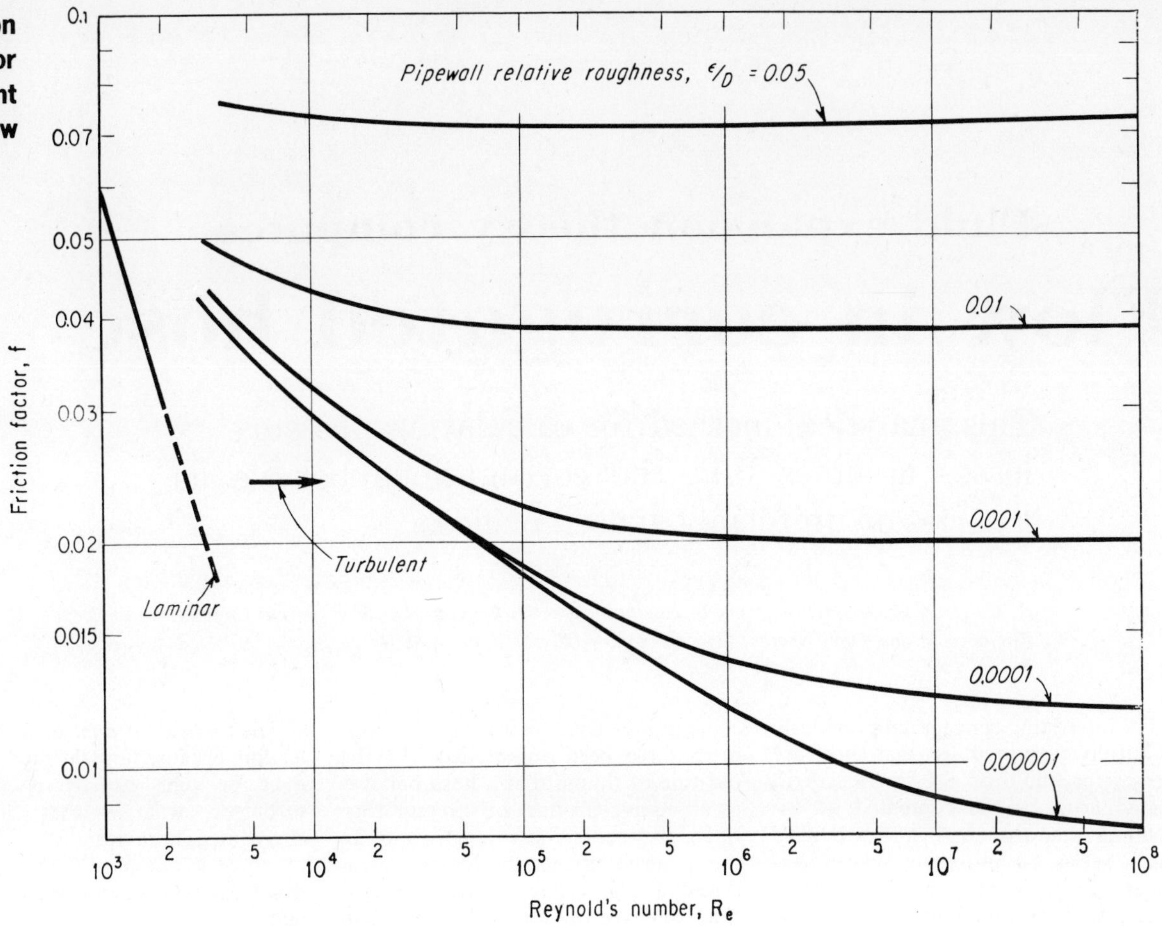

Recommended maximum flow

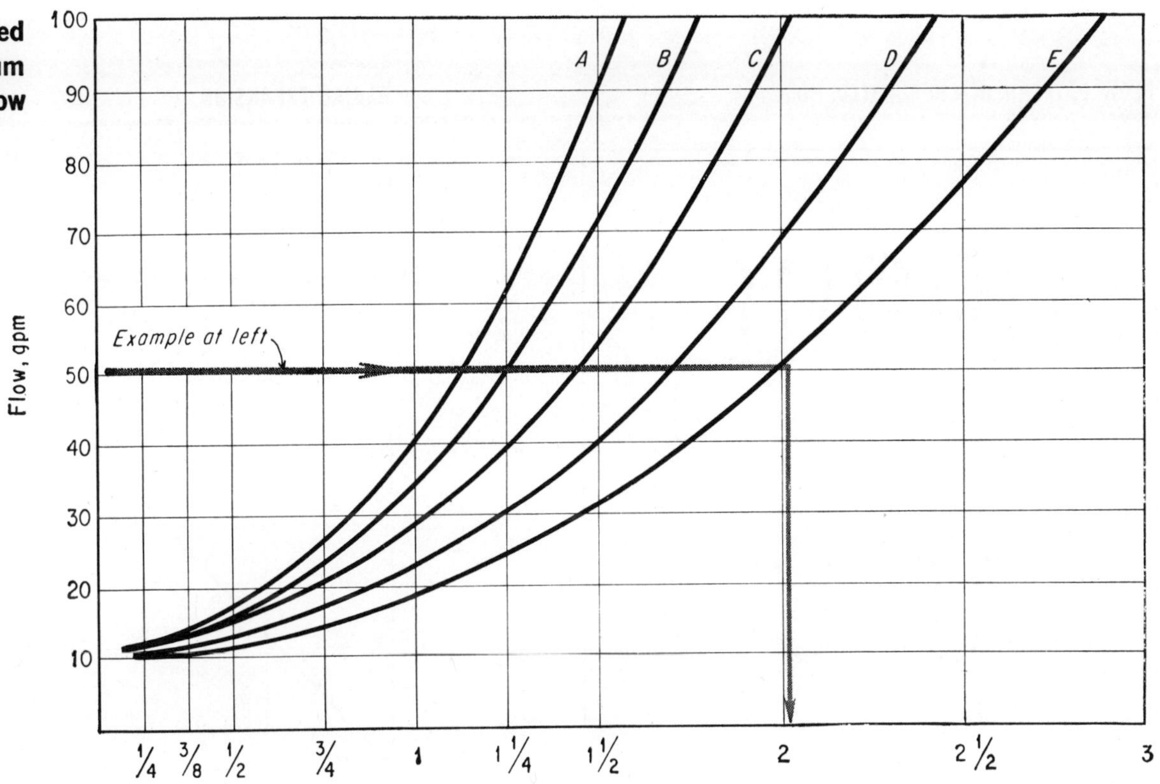

Fluid expansion theory computes

Flow in corrugated hose

This analytical method for calculating pressure losses assumes that the corrugations behave as a series of uniformly spaced orifices.

R. C. HAWTHORNE, Supervising Engineer, and H. C. von HELMS, Senior Engineer
Engineering and Development Div, Flexonics Div, Calumet and Hecla Inc

AN interesting concept not previously fully explored is that pressure losses for liquid or gas flow in corrugated hose can be computed by assuming that the corrugations behave as a series of uniformly spaced orifices. Pressure drop is caused by a succession of individual flow expansions.

That it's a valid assumption has been recently proved in an extensive test program at Flexonics and some cooperating labs. Two earlier concepts have been superseded: 1) the old rule of thumb that a hose behaves as an equivalent duct of three or four times the actual hose length and 2) the assumption that the losses are induced in the valleys of the corrugations and are in some way related to corrugation height — specifically, roughness ratio e/D where e is corrugation height and D is the equivalent diameter of the duct.

The *equivalent duct* concept is misleading because flow through a smooth duct or pipe usually is not fully turbulent, whereas that through a corrugated hose usually is. It is not practical to compare the two flows on this basis unless both are fully turbulent.

Flow through a corrugated hose or bellows becomes turbulent almost immediately upon reaching a Reynolds Number of $R = DV/n = 2000$ (see

FLOW PATTERN IN CORRUGATED HOSE

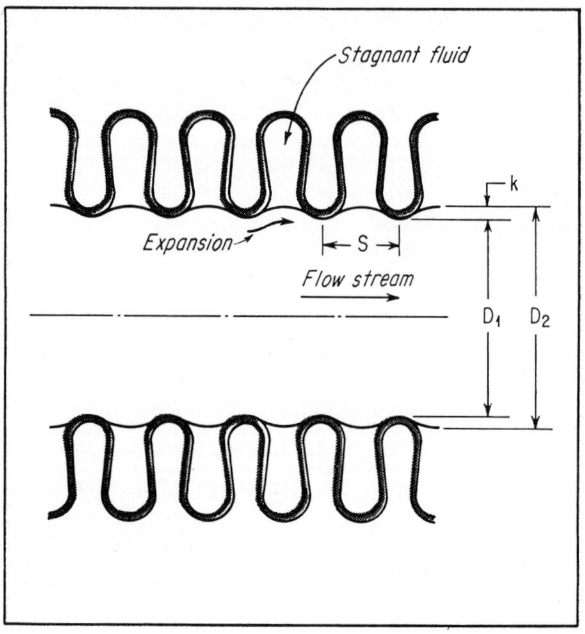

RADIAL EXPANSION, *k*

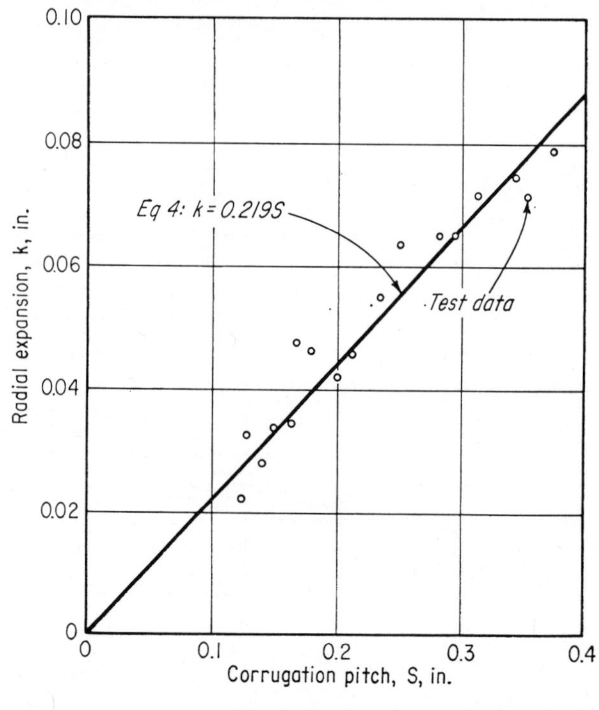

Symbols at right), and 2000 is exceeded in most applications. Flow through a duct may not become fully turbulent until R = millions, depending on surface finish and duct diameter.

For example, take a stainless steel pipe of 4-in. diameter having a 100-microin. finish. This surface is relatively smooth: the roughness ratio $e/D = 100 \times 10^{-6}/4 = 0.000025$ in./in., where e is the height of the surface roughness and D is the equivalent duct or pipe diameter. Complete turbulence does not occur until R = millions, a magnitude not usually found in commercial systems or even in aircraft and missile piping systems.

The *corrugation height* concept is also wrong. In no cases examined during the test program was flow affected by the full corrugation height. The sketch at left shows what probably does happen—simple expansion from corrugation to corrugation. The empirical flow equations and charts in this article have been developed on that basis.

Liquid and gas

Liquid and gaseous flow are treated alike. The only essential difference is the compressibility of a gas, and this usually can be ignored at subsonic velocities. For example, in uninsulated ducting or piping having a 5-psi pressure drop in a 100-psi line, the calculated pressure drop ignoring compressibility is only 2½ % less than that including compressibility.

It is assumed here, in calculating liquid flow, that cavitation does not occur. Cavitation could greatly increase pressure loss, but this did not occur in any of the tests, and it is not likely to occur in any pressurized system. In open discharge at high flow rates, however, there might be cavitation.

Pressure loss in the context of this article is always a loss of static head. Velocity head does not vary unless some of the fluid is lost or altered or the duct diameter is changed.

Laminar flow is excluded from the graphs because the pressure drop in pounds per square inch is easily solved with this general-relationship Darcy-Weisbach resistance equation:

$$\Delta P = f \frac{L}{D} \times \frac{V^2 \rho}{9266} \quad (1)$$

where f = friction factor = $64/R$ for laminar flow, L = pipe length, in.; D = pipe or hose inside diameter, in.; V = flow velocity, ft/sec; ρ = weight density, lb/ft³; and 9266 is a conversion constant.

Straight lengths

The new method of calculating pressure drop of turbulent flow in corrugated hose assumes that each convolution is an orifice plate. Loss in contraction of the fluid while passing through the orifice is ignored because all losses eventually appear in the expansion and are evaluated there.

Head loss for a sudden expansion in a straight pipe is expressed as

$$h_L = C \frac{V_1^2}{2g} \quad (2)$$

where h_L = head loss, ft; C = loss coefficient = $(1 - D_1^2/D_2^2)^2$, plotted below: and g = 32.2 ft/sec². Terms

SYMBOLS

C=loss coefficient for straight run
C_B=loss coefficient for bend
$D_{1,2}$=stream diameters, in. (see sketch)
e=surface roughness, in.
f=friction factor=$64/R$ for laminar flow
h_L=head loss, ft
k=expansion of the stream radius, in.
L=conduit length, in.
N=total number of corrugations
n=kinematic viscosity, ft²/sec
ΔP=pressure loss, psi
r=radius of bend centerline
R=Reynold's number
S=corrugation spacing (pitch), in.
V_1=flow velocity, ft/sec at D_1
W=corrugation height, in.
ρ=weight density, lb/ft³
θ=angle of bend (zero to 180°)

CALCULATED LOSS COEFFICIENT, C

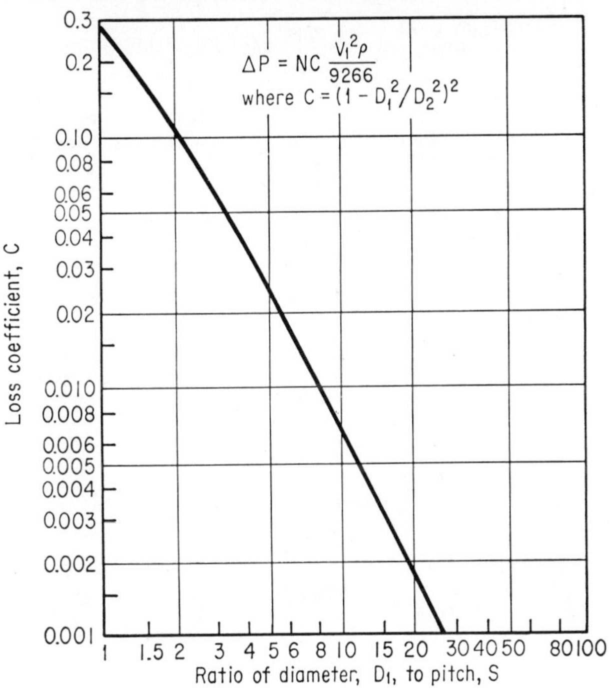

CALCULATED 90-DEG-BEND LOSS COEFFICIENT, C_B

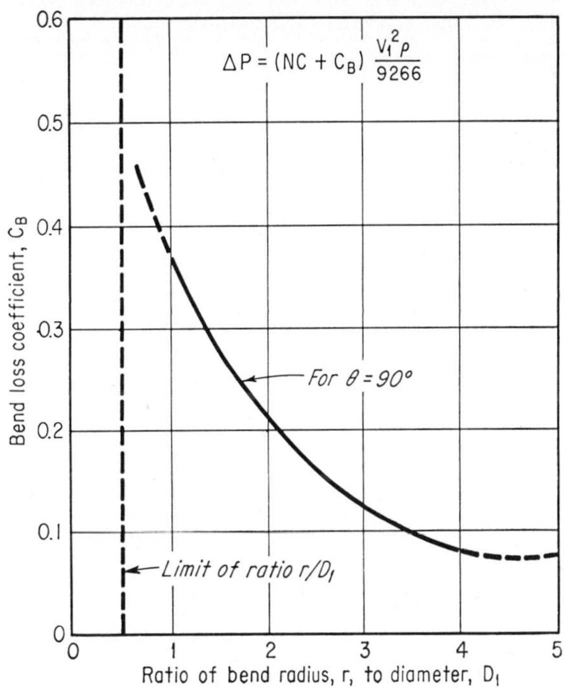

D_1 and D_2 are effective stream diameters, in., defined in the first illustration.

Eq 2 can be rewritten in terms of the actual radial expansion k, in., if it is noted that $D_2 = D_1 + 2k$:

$$h_L = \left[1 - \left(\frac{D_1}{D_1 + 2k}\right)^2\right]^2 \times \frac{V_1^2}{2g} \qquad (3)$$

The empirical part of this development was to find a relationship for radial expansion k. This was done for enough different hose configurations and fluids to establish the graph on the first page. Note the linearity, which makes this general expression valid:

$$k = 0.219S \qquad (4)$$

where S is the corrugation spacing or pitch, in.

Eq 2, converted to ΔP and with the new expression for C, becomes

$$\Delta P = N\left[1 - \left(\frac{D_1}{D_1 + 0.438S}\right)^2\right]^2 \times \frac{V_1^2 \rho}{9266} \qquad (5)$$

where N = the total number of corrugations or expansions.

Bends and elbows

Once a suitable friction factor or loss coefficient is established for straight lengths of corrugated hose or bellows, pressure loss in bends and elbows can be handled in the same manner as pressure losses in piping and ducting.

The method is to calculate the pressure loss in the developed length assuming that the bend has been straightened. Then add to this value a loss based on the kinetic energy of the fluid in the bend. The combined expression is

$$\Delta P = N\left[1 - \left(\frac{D_1}{D_1 + 0.438S}\right)^2\right]^2 \times \frac{V_1^2 \rho}{9266} + C_B \frac{V_1^2 \rho}{9266} \qquad (6)$$

Values of C_B for 90-deg bends are plotted in the graph on page 81. Values of pressure drop for other angles have been developed by many experimenters and can be determined by using these approximate relationships:

Zero to 90°: $\Delta P_\theta = \Delta P_{90}\left(\frac{\theta}{90}\right)^{1/2}$

90° to 180°: $\Delta P_\theta = \Delta P_{90}\left(\frac{\theta}{90}\right)$

where ΔP_{90} is computed with the help of the bend-loss graph.

The tests

Tests were run on straight sections of corrugated hose only. The parameters varied were length, pitch and height of corrugations, lay of corrugations (annular or helical), Reynold's number (24,000 to 1,300,000), fluid (water and air), and fluid flow. The table on this page lists some of the results.

The water-flow tests were closed-loop, whereas the air-flow tests were open-discharge. The change in corrugation spacing (pitch) was accomplished for the 4-in. size by simply extending conventional 4-in. ID bellows to many different lengths (four are shown in the table below). None of the parameter changes greatly affected the basic relationship of Eq 4, namely, $k = 0.219S$.

TEST DATA WITH WATER AND AIR (based on Flexonics, Boeing, and Martin tests)

METAL-HOSE DIMENSIONS, IN.						WATER-FLOW DATA (closed circuit)		PRESSURE-LOSS CALCULATIONS		
Dia. D_1 Nominal	Actual	Length L	Pitch S	Height W	Number of corrug.	gpm	Reynold's Number R	Loss coeff. C for one corrug. (note)	Test Friction Factor f	Radial flow expansion k, in.
¾	0.765	12	0.125	0.226	96	5.5	2.44×10^4	0.01225	0.0750	0.0225
1	1.000	20.9	0.128	0.159	164	72.0	2.04×10^5	0.01350	0.1050	0.0318
	1.016	12	0.138	0.239	87	10.0	3.44×10^4	0.01015	0.0745	0.0275
1¼	1.284	12	0.150	0.273	80	12.0	3.18×10^4	0.00927	0.0793	0.0330
1½	1.500	20.3	0.163	0.207	124	216.0	4.02×10^5	0.00724	0.0666	0.0340
	1.517	12	0.200	0.327	60	17.0	3.82×10^4	0.01050	0.0795	0.0420
4	3.990	16	0.252	0.234	64	871.0	7.45×10^5	0.00376	0.0598	0.0640
	3.990	24	0.375	0.213	64	898.0	7.68×10^5	0.00548	0.0584	0.0785
	4.016	3	0.178	0.337	17	942.0	8.00×10^5	0.00198	0.0448	0.0460
	4.016	6	0.353	0.306	17	935.0	7.93×10^5	0.00454	0.0515	0.0710
						AIR-FLOW DATA (Discharge to atmos.)				
2	2.000	36	0.212	....	170	0.2 lb/sec	1.18×10^5	0.00683	0.0647	0.0455
4½	4.460	2.5	0.167	0.375	15	5 lb/sec	1.32×10^6	0.00168	0.0448	0.0470

Note: $C = f\,S/D$ **for one corrugation based on** $h_L = C\frac{V^2}{2g} = f\frac{L}{D}\frac{V^2}{2g}$ **when** $L = S$

Calculate pressure drop the easy way

Local resistance to flow

Here is a compilation of every type of loss-inducer from elbows to filters, with easy-to-use flow coefficients for each

LOUIS DODGE, Hydraulic Consultant, New Richmond, Ohio

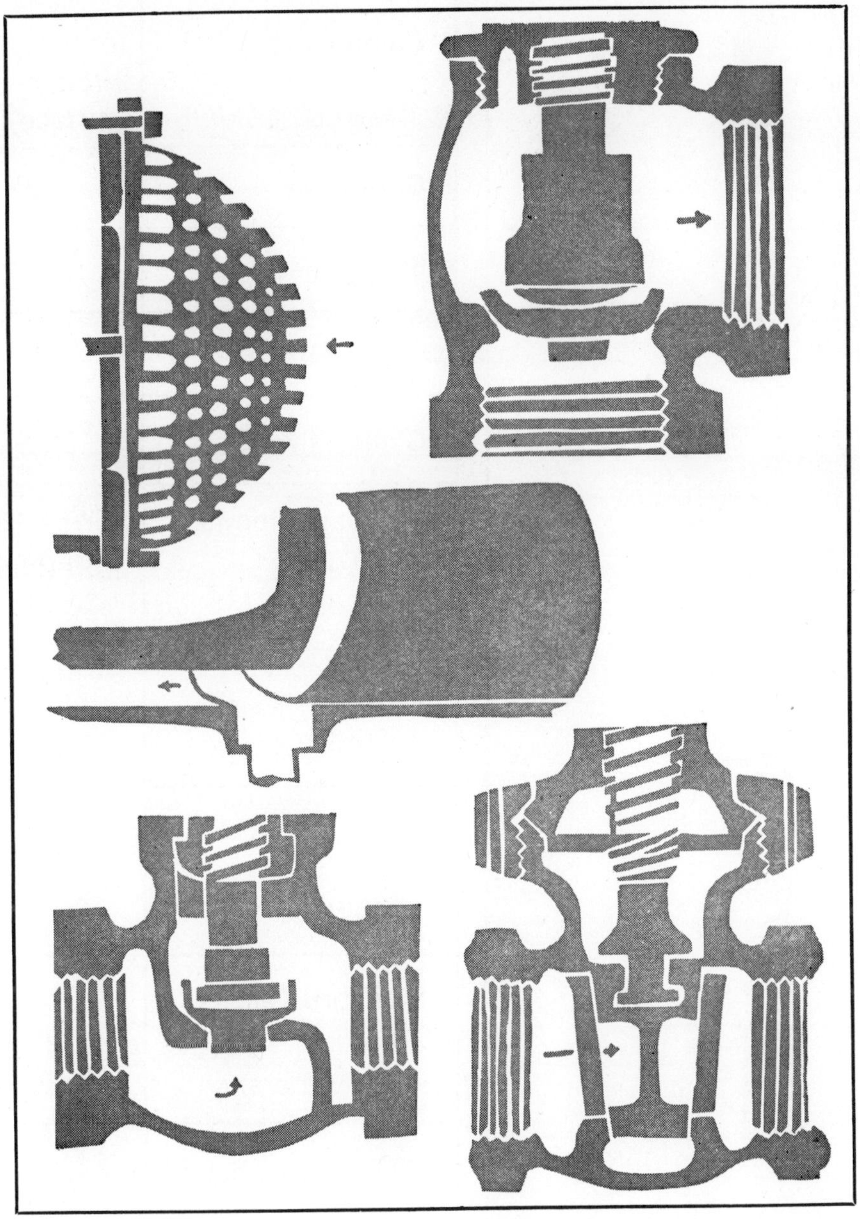

CALL it what you will—flow friction, head loss, pressure drop, ΔP—it has always been the problem at the heart of almost any hydraulic system design.

What's new is that modern investigators have developed empirical data for practically every type of restriction, assuming you can find the specific reference when you need it. (Straight pipe losses were covered in the Feb 17 issue of *PE.*)

Here we will summarize in one place what's been published in a dozen sources on local flow resistances such as pipe fittings, and add a few data that are new.

"Local resistances" means unintentional, though unavoidable, loss-inducers such as bends, filters, and wide-open stop valves. Many are irregularly shaped obstructions to flow, and the resistance is impossible to calculate.

Basic equation

Fortunately a simple empirical relationship exists for rough computation of pressure loss for almost any restriction:

$$H_L = KV^2/2g \qquad (1)$$

$$\Delta p = KV^2\,(1.08\times10^{-4})\,\gamma \qquad (2)$$

$$\Delta p = K\,(q^2/d^4)\,(1.8\times10^{-5})\,\gamma \qquad (3)$$

where H_L = head loss in ft, Δp = head loss in psi, γ = weight density, lb/ft^3, V = average velocity, ft/sec, with the exception of section changes like those in the graphs on page 67, q = flow in gpm, d = diameter in in., and K is an empirical constant, different for each type of restriction. See page 85 for a capsule history of K.

These empirical relationships fall far short of perfection, and a better answer is always possible if you want to break down the flow resistances into entrance losses, friction losses, turbulence losses, velocity head changes, and exit losses. But the K-factor method is much simpler and is accurate enough.

If you want to improve accuracy somewhat, assume that wall friction losses will be higher when flow is laminar. Increase the indicated K values 30% for R_e near 2000, and 50% for those near 100.

You can convert K to equivalent pipe length measured in diameters if you divide by friction factor f:

$$L_E/D = K/f \tag{4}$$

where f is a function of Reynold's Number, R_e, for laminar flow, and of both R_e and wall roughness for turbulent flow. The $L_E/D = K/f$ relationship is derived from these equations:

$$H_L = f\frac{L}{D}\frac{V^2}{2g} \text{ for pipe}$$

$$H_L = K\frac{V^2}{2g} \text{ for restrictions} \quad \text{(Eq 1)}$$

Where test values of flow resistance are unobtainable, or where values vary with time, such as in filters, rough estimates of the K-factor can be made.

Reynold's Number R_e is the key. Coefficient K depends on R_e, and R_e can be calculated if you know the perimeter or some other critical dimension of the restriction.

For round holes:

Nomenclature

A = cross-sectional flow area, ft^2
A_O = open area in coarse mesh or screen, ft^2
A_2 = total effective area of filter, ft^2; $A_2 = A_O$ + solid area
C = coefficient of pressure loss for filter
d = flow diameter, in.
D = flow diameter, ft
f = friction factor, dimensionless
g = acceleration of gravity, 32.2 ft/sec^2
H_L = head loss, ft
K = K-factor, dimensionless
L = length, ft
L_E = equivalent length, ft
L_W = wetted perimeter, ft
Δp = pressure drop, psi
ΔP = pressure drop, psf
q = flow, gpm
Q = flow, ft^3/sec
$Q_{in.^3}$ = flow, $in.^3/sec$
R_e = Reynold's Number, $\rho DV/\mu$.
R_H = hydraulic radius, ft; $R_H = A/L_W$
V = velocity, ft/sec
W = wall thickness of filter element, in.
ρ = mass density, $lb\text{-}sec^2/ft^4$ or $slugs/ft^3$
γ = weight density, lb/ft^3
μ = absolute viscosity, $lb\text{-}sec/ft^2$
ν = $\frac{\mu}{\rho}$, kinematic viscosity, ft^2/sec

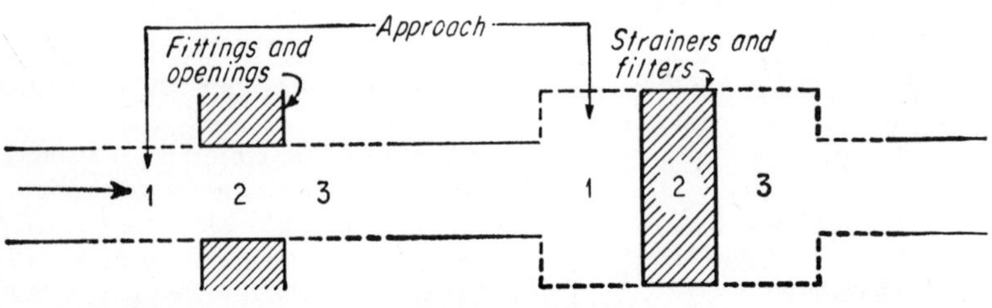

K-factors for standard valves

Component	L_E/D (assumes rough surface)
Gate valve	13
Swing check	135
Angle valve	145
Globe valve	340
Standard elbows:	Note: Smooth
90 deg	30 (high)
45 deg	16
Long elbow:	
90 deg	20
Street elbow:	
90 deg	50
45 deg	26
Standard Tee:	
through run	20
through branch	60
Close return bend	50

$$R_e = \frac{\rho D V}{\mu} = \frac{DV}{\nu} \quad (5)$$

For holes of noncircular cross-section:

$$R_e = \frac{4VR_H}{\nu} \quad (6)$$

where R_H = hydraulic radius = A/L_W = (cross-sectional area)/(wetted perimeter of opening).

Coarse-mesh suction filters:

$$R_e = 34 \frac{V_1 d}{(A_O/A_2)\,\nu} \quad (7)$$

where R_e = Reynold's Number, dimensionless, μ = absolute viscosity, lb-sec/ft^2, ν = kinematic viscosity, ft^2/sec, V_1 = approach velocity, ft/sec, based on effective cross-sectional area of filter element, d (or D) = mean diameter of flow passages, in. (or ft), and A_O/A_2 = ratio of open flow area to total cross-sectional area of filter element exposed to flow.

Deriving K-factors

Most are already derived, from tests by manufacturers over a period of many years. The results, an average of data from many sources, are listed with the sketches of the components and fittings on the pages following.

K-factors for screwed fittings are 25% higher than those for flanged fittings. Any value given will increase if the fitting is other than clean and conventional.

Valve and fitting *K*-factors in the

and fittings (Crane Co)

Internal diameters			
0.5 in.	1 in.	2 in.	4 in.
K = 0.36	0.3	0.25	0.21
3.6	3.0	2.5	2.1
3.8	3.3	2.7	2.3
9.5	7.9	6.6	5.7
fittings have much lower values			
0.82 0.43	0.68 0.36	0.58 0.3	0.5 0.26
0.55	0.45	0.38	0.33
1.4 0.7	1.2 0.59	0.96 0.5	0.82 0.43
0.55 1.7	0.45 1.4	0.38 1.2	0.33 1.0
1.4	1.2	0.96	0.82

The K factor

Invented to simplify pressure-loss calculations in ordinary fittings and valves, the *K*-factor takes some liberties with true energy-balance laws. It neglects changes in velocity, elevation, viscosity, and wall friction. It also discounts the effects of piping roughness preceding and following the fitting.

In fact, the main basis for using *K* is that observation of numerous test data on many fluids over the years has indicated that pressure losses vary with V^2 of the fluid, give or take 10% on the exponent 2.0. It was logical and useful to express the relationship in this way:

$$\text{head loss } H_L = K\frac{V^2}{2g}$$

$$\text{or:} \quad \Delta p = KV^2\,(1.08 \times 10^{-4})\,\gamma$$

thus tying in nicely with the velocity-head concept in Bernoulli's conservation of energy theorem. Factor *K* is dimensionless and has a special value for each fitting or restriction. The value tends to decrease somewhat with diameter. For example, a large elbow has a smaller *K* than a small elbow. Exceptions are sudden enlargements and contractions: The *K*-factors are less dependent on nominal diameter and thus remain fairly constant, regardless of size.

If the *K*-factor for a given component is based on actual tests, good accuracy can be expected—assuming your operating conditions are similar to the test.

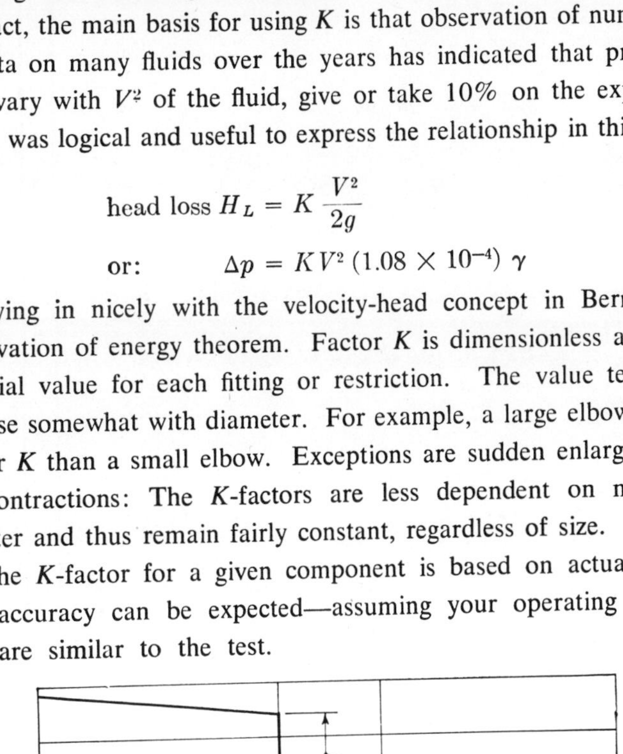

table are taken directly from Crane Bulletin 409. Valves are assumed fully open. When other than fully open, then the valve function is controlled throttling, the subject of the next article.

Orifices other than ordinary inlet and outlet holes are not covered here, because those are control-type restrictions, also to be covered in the next article.

Straining and filters are a special breed of restriction and cannot be calculated by analytical means. However, K-factors for strainers and coarse-mesh filters can be simply though crudely estimated by two simple relationships, one for R_e above 400, and one for R_e below 400:

$R_e > 400$:

$$K = 1.3\left(1 - \frac{A_O}{A_2}\right) + \left(\frac{A_2}{A_O} - 1\right)^2$$

$R_e < 400$:

$$K = Z \times (K \text{ for } R_e > 400)$$

where K is the value for one thickness of wire mesh or perforated screen, and Z is the correction for R_e <400. To calculate flow resistance of several sheets, just add the K-factors. Typical values are plotted in the graphs on the last page.

For fine filters (fiber or sintered metal), Reynold's Number is meaningless because the passages are too small to be considered channels. Pressure drop in psi can be estimated with this empirical relationship:

Entrance losses

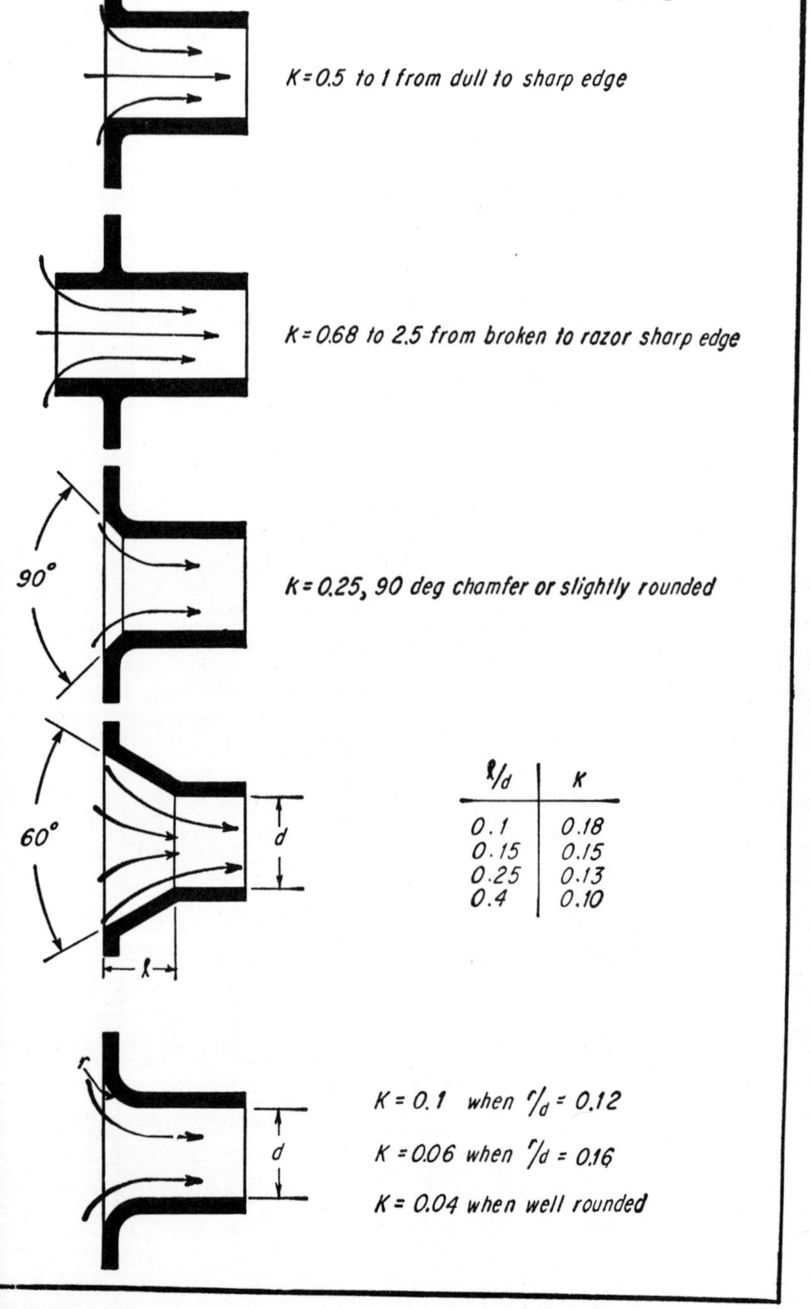

ℓ/d	K
0.1	0.18
0.15	0.15
0.25	0.13
0.4	0.10

Conduit branches in tubing

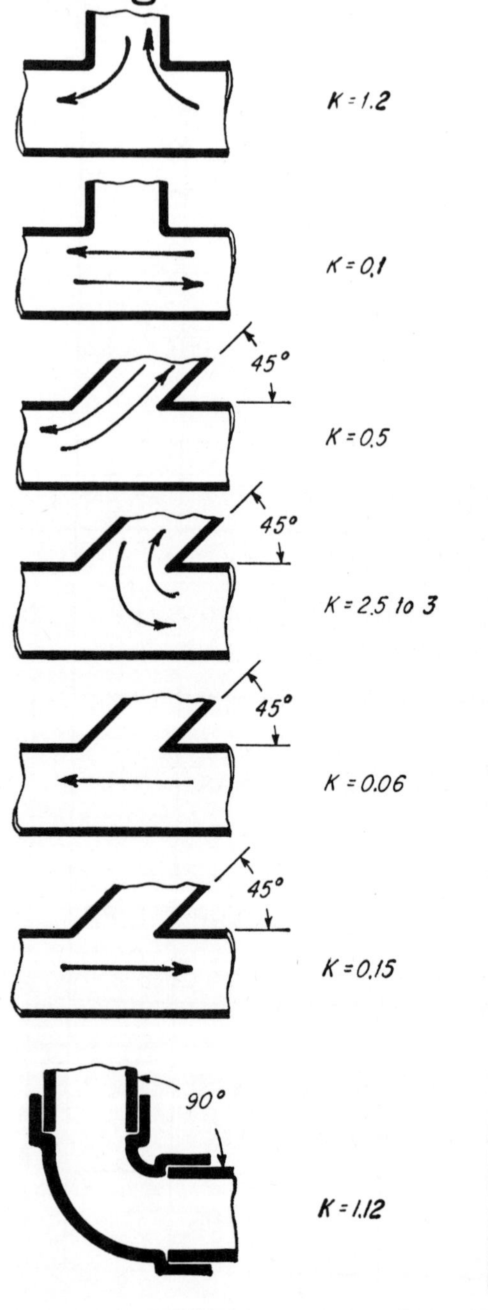

$$\Delta p = \frac{Q_{in.^3} C \mu W}{A_2}$$

where W = wall thickness in direction of flow, in., and C is a constant peculiar to the filter element. The filter manufacturer runs tests to establish C. Typical values for a 40-micron (nominal) element of impregnated cellulose fiber are:

For SAE 10 oil at 100F: $C = 1.22 \times 10^6$
" 20 " $C = 1.28 \times 10^6$
" 30 " $C = 1.36 \times 10^6$
For SAE 40 oil at 100F: $C = 1.48 \times 10^6$

Add 25% to the pressure loss to include the housing, because the C value pertains to the element only. And remember that Δp increases as the filter clogs, and eventually the entire pressure drop occurs across the element.

Validity of tests

All the K values in the article are based on actual tests. Some of these tests (for mating holes in valve blocks) were run by the author.

The danger is not that the K values are wrong, but that they are often misapplied. For one thing, velocity V usually is average velocity, ½ $(V_1 + V_2)$, but sometimes is inlet or outlet velocity, as in the graphs of changes in section, below. Furthermore, slight differences in inlet contour or finish can alter K threefold (see Entrance Losses graphs).

text continues next page

Mating holes in valve blocks

Gradual changes in section

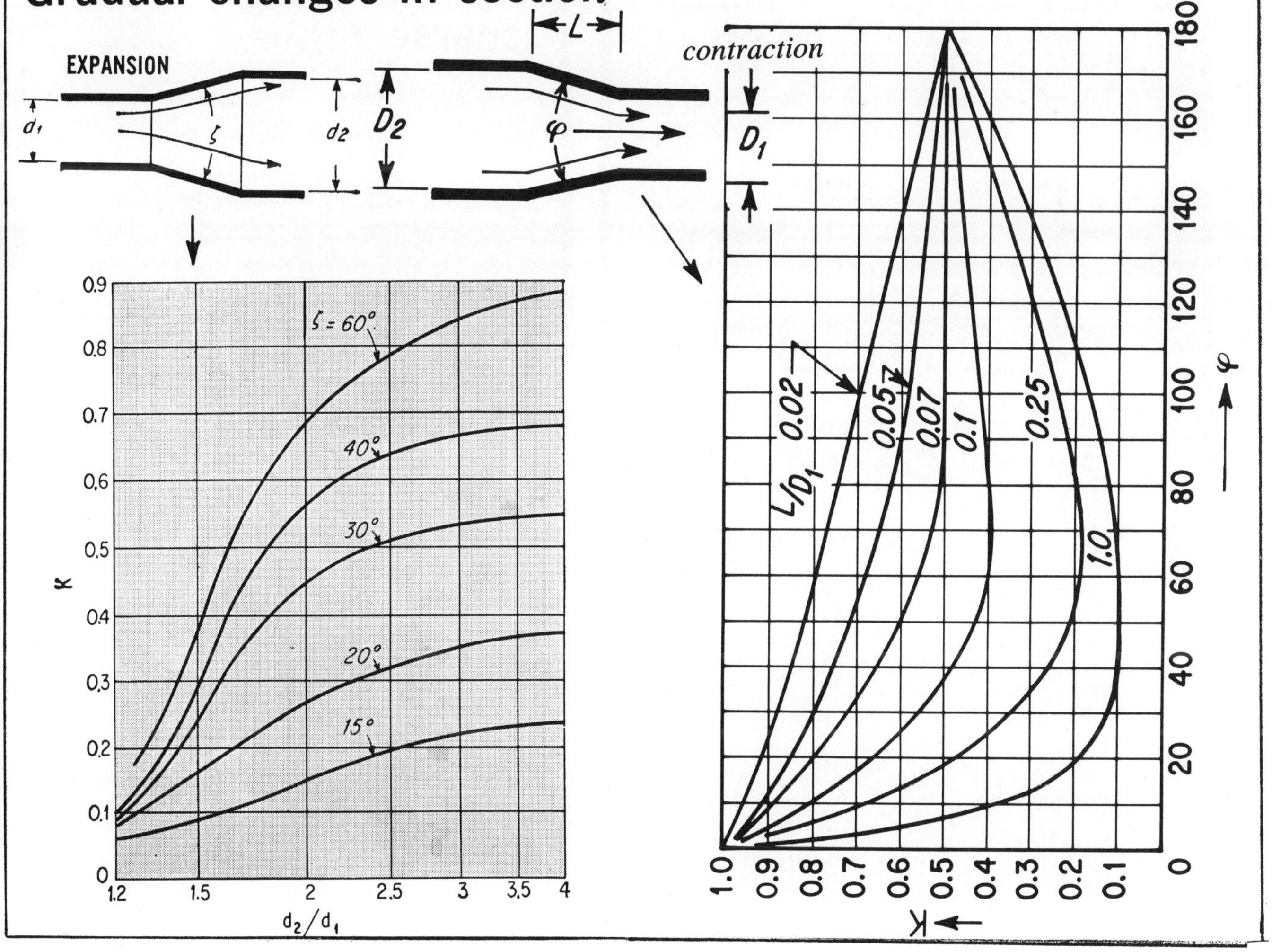

Bends of tubing and hose

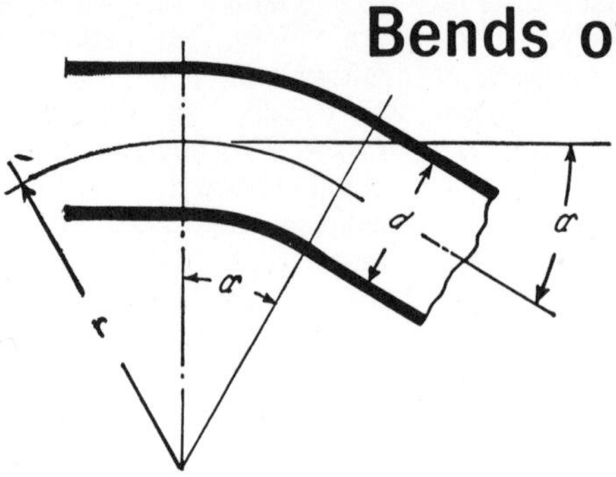

If bend is angular (or smooth with $r/d < 1$ or > 6) use this table for K: (conservative)

α	10	20	30	40	50	60	70	80	90
K	0.04	0.1	0.17	0.27	0.4	0.55	0.70	0.90	1.12

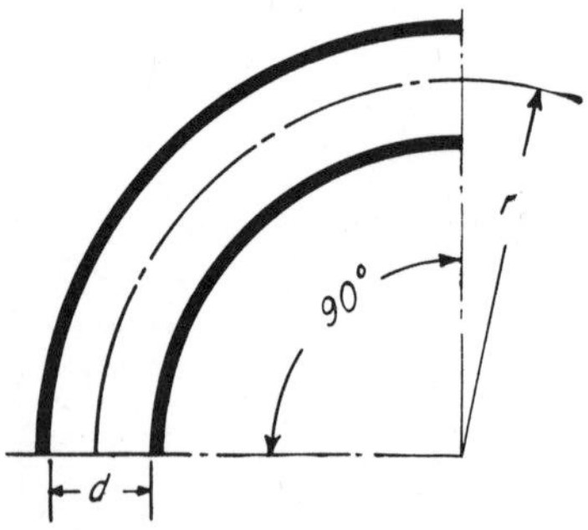

Best $K = 0.1$ to 0.15 for smooth 90°-bend, and r/d at ideal 2 to 4

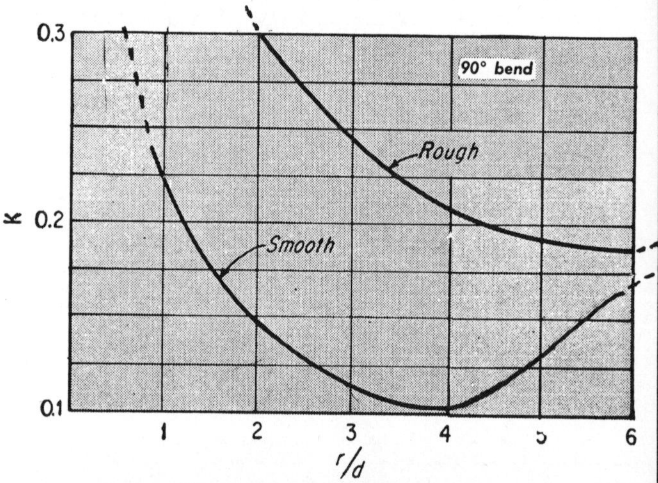

Sudden changes in section

K	0.81	0.64	0.49	0.36	0.25	0.16	0.09	0.04	0.01
a/A	0.1	0.2	0.3	0.4	0.5	0.6	0.7	0.8	0.9
K	0.4	0.38	0.34	0.30	0.24	0.18	0.1	0.05	0.015

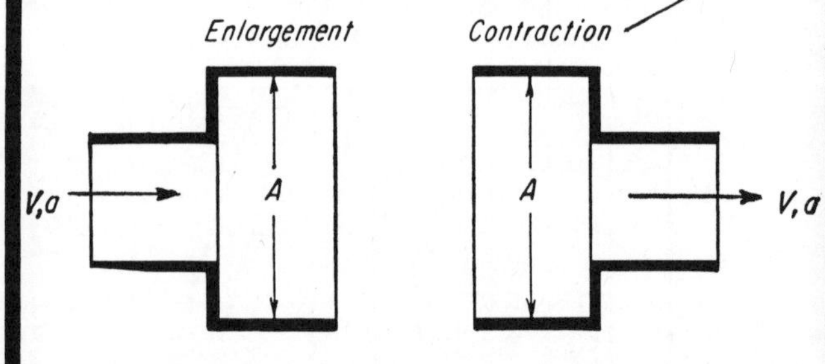

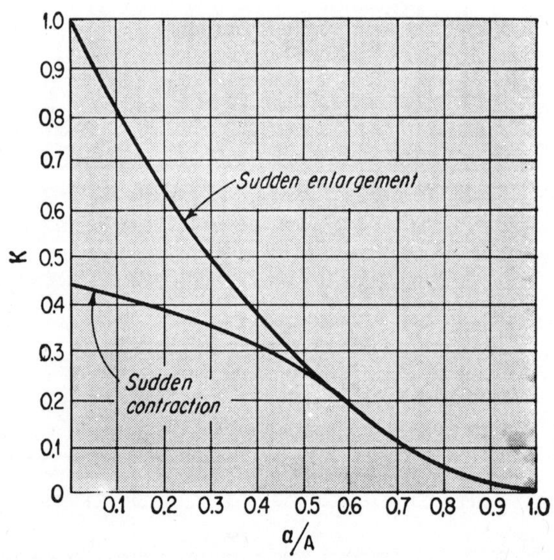

Strainers and coarse filters

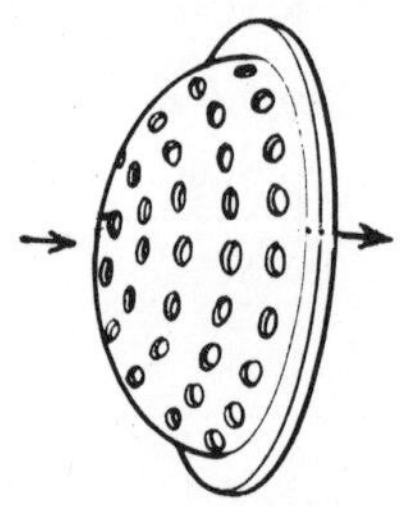

For $R_e > 400$: $K = 1.3\,(1 - A_0/A_2) + (A_2/A_0 - 1)^2$

For $R_e < 400$: $K = Z \times (K \text{ for } R_e > 400)$.

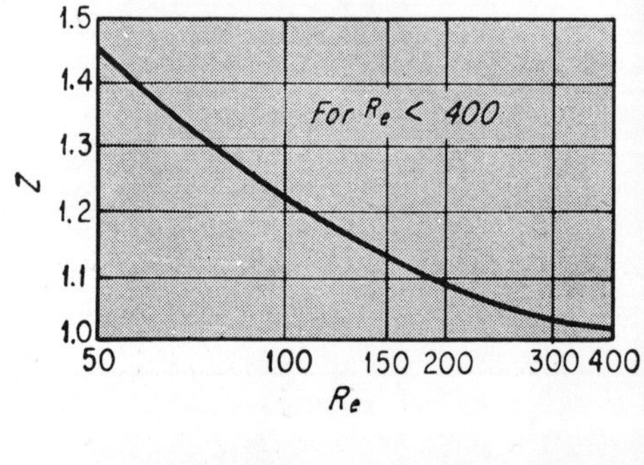

For controlled flow resistance

Fluid throttling devices

When you *want* a pressure drop, this is the article to use. Two prior articles predict unintentional ΔP

LOUIS DODGE, Hydraulic Consultant, New Richmond, Ohio

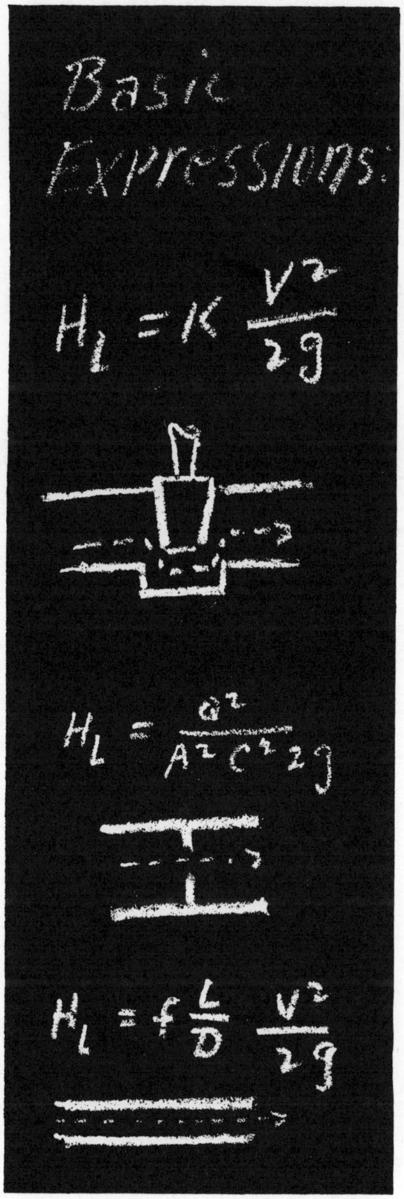

PLANNED (and, we assume, desirable) pressure drops include those through orifices, nozzles, control valves, capillaries, clearances, and porous plugs (not filters). Many of the equations are the same, whether the pressure drops are wanted or not. The coefficients are variable when the restriction is a manipulated one.

Orifices and nozzles

More is known about orifices and nozzles—perfect ones, that is—than about any other planned restriction in a fluid-power system. ASME specifications describe every dimension, and the flow coefficients are accurate within a few percent. The basic relationship is

$$Q = AC\sqrt{2gH_L} \qquad (1)$$

where Q = flow, ft^3/sec, A = orifice or nozzle area, ft^2, H_L = head loss, ft, and C = the flow coefficient, **plotted in the graphs on page 90.** If you want a precise ΔP for a given flow, or vice versa, that's the way to get it. Note that oil temperature and viscosity are not factors.

Not all orifices and nozzles are perfect, however, and many have nonstandard dimensions to solve special problems. Here are the principal types, perfect and otherwise:

- Square-edged orifice (ASME standard)
- Drilled hole in thin plate
- Contoured nozzle (sketched on page 90)
- Drilled hole in plug or thick plate
- Orifice plus change in pipe diameter
- Punched hole or other crude opening

The square-edged orifice flow coefficient is fairly well defined. The dimensions must be sharp and true, however. A square-edged hole in a thin plate has nearly the same flow characteristics as a square-edged orifice.

Contoured nozzles smooth out the flow at high Reynold's Numbers, and therefore have higher flow coefficients than orifices. They are as accurate as square-edged orifices, but more expensive.

Long orifices such as drilled plugs or plates introduce wall friction losses, and these must be accounted for in any expression for pressure drop. The equations on the next page summarize losses for several types.

Punched, cast, or rough-drilled orifices and other crude openings are useful flow restrictions, but flow might easily differ 20% from what you calculate based on drill or punch size.

Take a 0.25-in. drilled orifice, for example. A 0.01-in. error in the assumed diameter gives an 8% difference in area; a 0.1 error in an assumed 0.8 flow coefficient is equivalent to a 12% change in flow; and any rounding of the hole entrance might boost flow another 10 or 20% because of the nozzle effect.

Where you are not sure about the edge sharpness of a drilled hole, compromise by using a flow coefficient midway between that plotted for orifices and nozzles.

Special coefficient K**:** Although flow coefficient C for orifices and nozzles (Eq 1) is the conventional one to use, expressions involving coefficient K (the K-factor) are available too. These are convenient when other calculations in your hydraulic system also

Simple orifices and nozzles (from Crane Co.)			
Units		Equations based on flow coefficient, C	
Flow	Pressure drop	Flow	Pressure drop (measured across flange taps)
$\frac{ft^3}{sec}$	ft	$Q = d_0^2 C\sqrt{H_L}\, 4.38\times10^{-2}$	$H_L = \frac{Q^2}{d_0^4 C^2} 5.21\times10^2$
$\frac{lbs}{sec}$	psi	$W = d_0^2 C\sqrt{\gamma\Delta p}\, 5.25\times10^{-1}$	$\Delta p = \frac{W^2}{d_0^4 C^2 \gamma} 3.63$
gpm	psi	$q = d_0^2 C\sqrt{\frac{\Delta p}{\gamma}}\, 2.36\times10^2$	$\Delta p = \frac{q^2\gamma}{d_0^4 C^2} 1.8\times10^{-5}$

involve K (*PE*—Mar 2 '64, p 63). The general expression is

$$H_L = K\frac{V^2}{2g} \qquad (2)$$

where V is the average velocity, ft/sec, of the fluid just before and just after the restriction, and H_L is head loss in ft.

The K-factor is awkward to use if the inlet and outlet have greatly different flow areas, or if they are irregular in shape. Assuming a value of K, just which velocity do you select for calculating pressure loss?

Velocity for K

In this article, which follows general practice, K is based on the average velocity measured through the restriction. Specifically:

Orifices and Nozzles, Eq 1

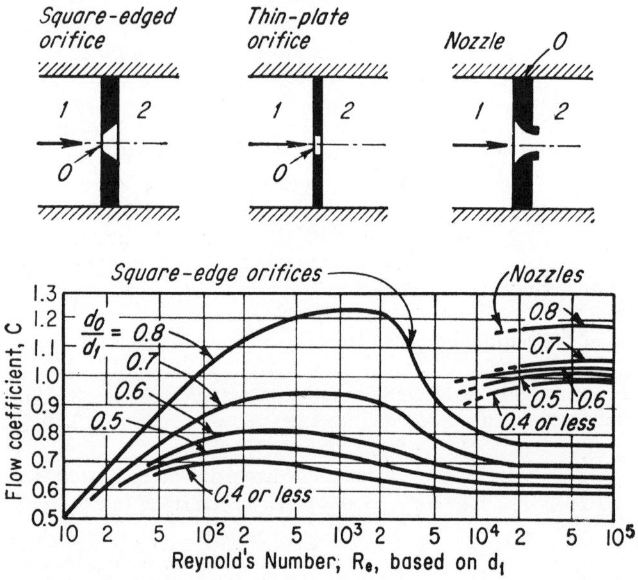

Long Holes, Eq 3

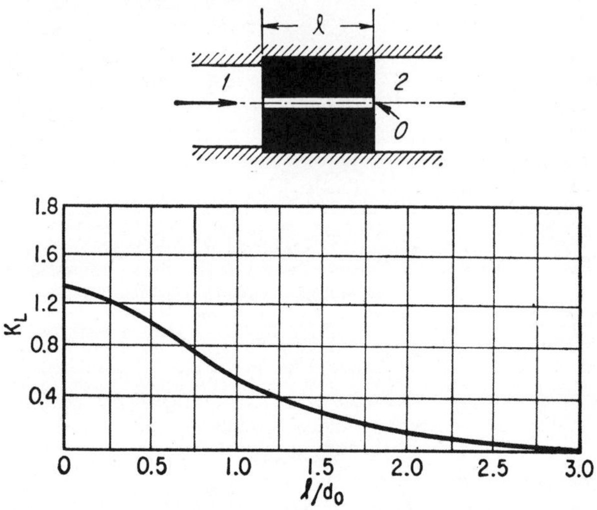

Pipe Diameter Change (long hole)

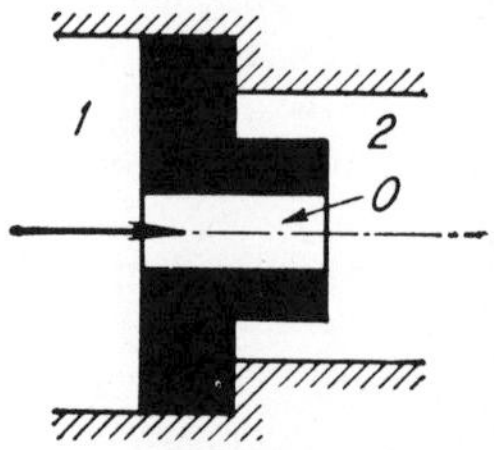

$K = K_1 + K_2 + K_3 + K_4$
(See text, Eq 4)

Pipe Diameter Change (orifice)

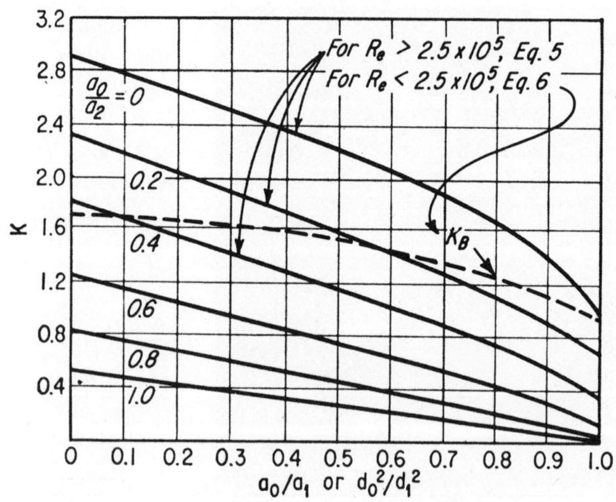

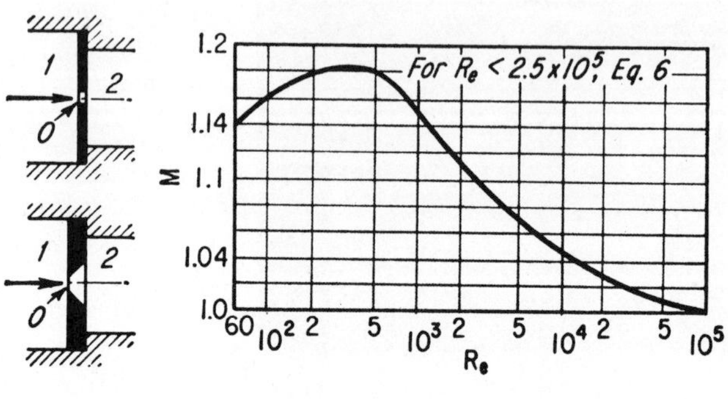

For orifices and holes, except the nozzles and sharp-edged orifices covered by flow coefficient C, the velocity V is measured through the orifice area A_o.

For valves, and for slots and notches, velocity V is approach velocity, based on inlet port area.

Equation (2) applies in every case. Here are the special derivations of graphs and constants used in this article, starting with orifices:

Long-hole orifices: Friction coefficient f creeps into the calculations when the orifices have significant length (high L/D). If D_o is small in relation to D_1 and D_2, this handy expression for K will suffice:

$$K = 1.5 + K_L + f\frac{L}{D_o} \qquad (3)$$

Nomenclature

a, A = cross-sectional flow area, in.^{2}, ft^2
C = flow coefficient
C_{dia} = diametral clearance
C_{rad} = radial clearance
d, D = diameter: in., ft
f = friction factor, dimensionless
g = acceleration of gravity, 32.2 ft/sec^2
H_L = head loss, ft
K = K-factor, dimensionless
l, L = length, in., ft
L_W = wetted perimeter, ft
$\Delta p, P$ = pressure drop: psi, psf
q = flow, gpm
Q, Q_{in} = flow: ft^3/sec, in.^{3}/sec
R_e = Reynold's Number, $\rho DV/\mu$ for round holes and $\frac{4\rho VR_H}{\mu}$ for non-circular openings.
R_H = hydraulic radius, ft $R_H = A/L_W$
V = velocity, ft/sec
W = weight flow, lb/sec
w = width of slot
ρ = mass density, lb-sec^2/ft^4
γ = weight density, lb/ft^3
μ = absolute viscosity, lb-sec/ft^2
ν = $\frac{\mu}{\rho}$, kinematic viscosity, ft^2/sec

Relief Valves

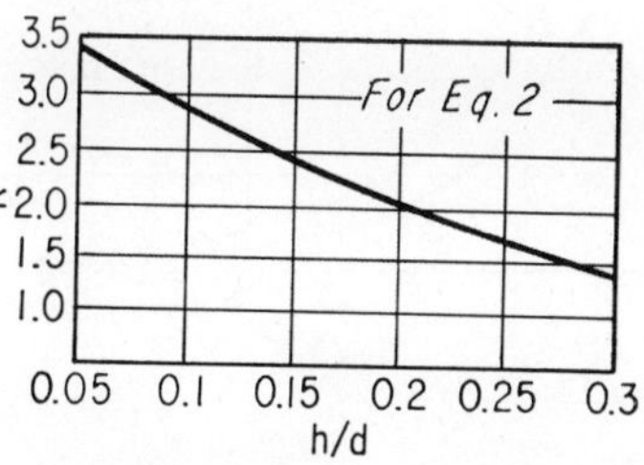

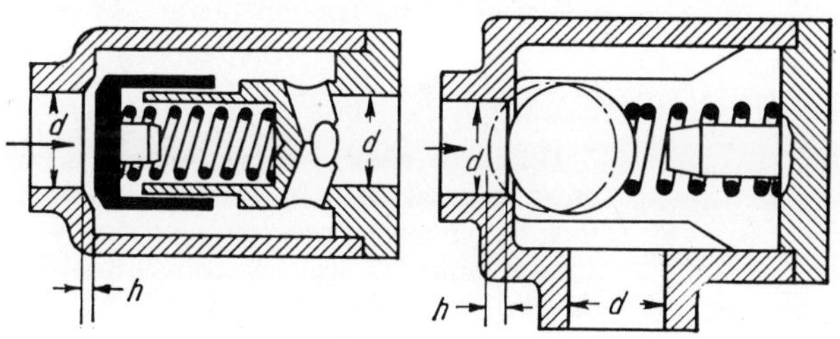

Butterfly Valve

T/D	K
0.5	2
0.4	1
0.3	0.5
0.2	0.2
0.1	0.1

Various wafer thicknesses, assuming fully open (ξ = zero) in closed system. For free discharge, add 1.0 to K-values.

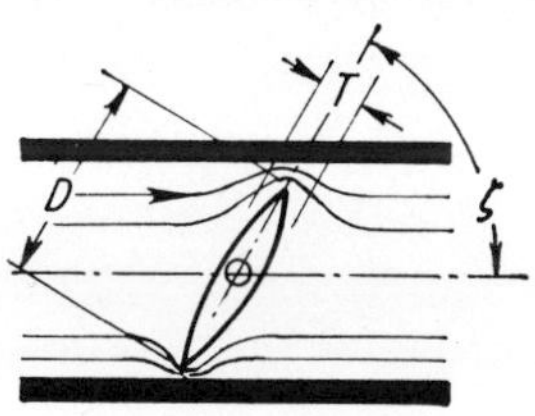

ξ	K
0	1
10	1.2
20	1.8
30	4
40	12
50	35
60	145
70	600
80	closed

All for T/D = 0.4

Gate Valve

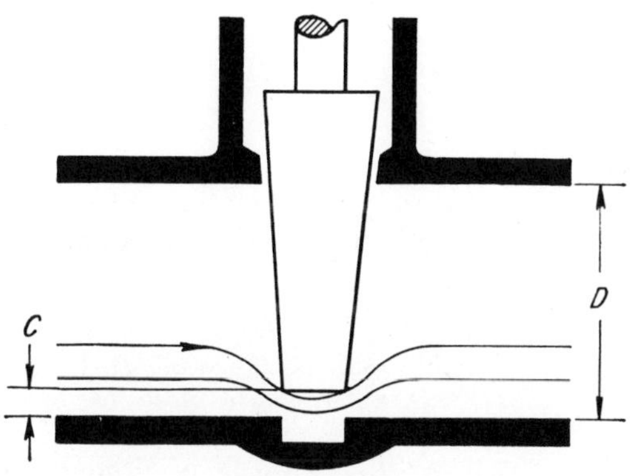

C/D	1	0.9	0.8	0.7	0.6	0.5	0.4	0.3	0.2	0.1
K	1.3	1.6	2.0	3.0	4.5	6.2	10	20	50	200

Ball Valve

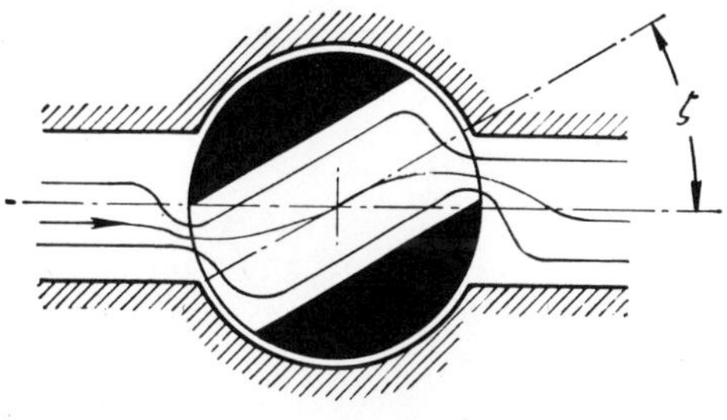

ξ	0	10	20	30	40	50	60	70	75
K	0	0.3	1.6	5.5	18	54	210	1000	closed

Disk Valve

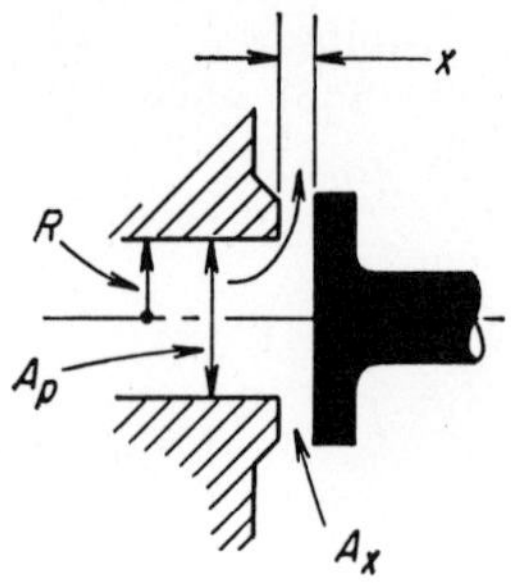

Orifice area

$A_x = 2R\pi X$

$A_x = A_p$ when $X = R/2$

Wetted perimeter (for R_e)

$$L_W = 4R\pi$$

Resistance coefficient

$$K = 1.3 + 0.2\left(\frac{A_p}{A_x}\right)^2$$

Needle Valve

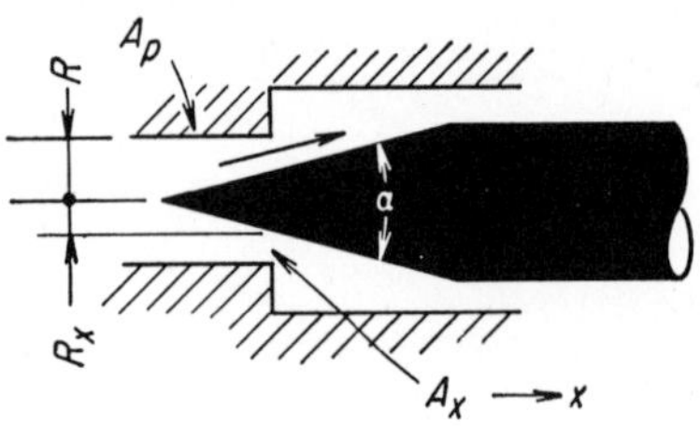

$$A_x = \pi\left(2RX\tan\frac{\alpha}{2} - X^2\tan^2\frac{\alpha}{2}\right)$$

$A_x = 0$ when $X = 0$

$$L_W = 2\pi\left(2R - X\tan\frac{\alpha}{2}\right)$$

L_W = optimum if X = nearly zero

$$K = 0.5 + 0.15\left(\frac{A_p}{A_x}\right)^2$$

K is valid from $A_x = 10$ to $A_x = 100\%$ of entry area, A_p

Simple Ball Valve

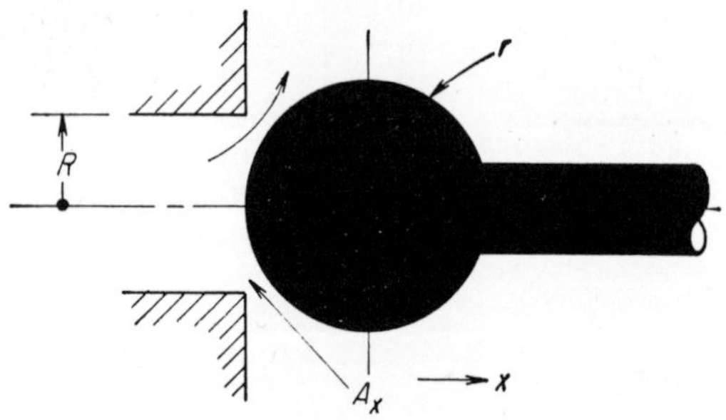

A_x = approx. $1.5R\pi X$

A_x is valid when r = approx. $1.3R$

L_W = approx. $4\pi R$

$$K = 0.5 + 0.15\left(\frac{A_p}{A_x}\right)^2$$

Four-Way Valve

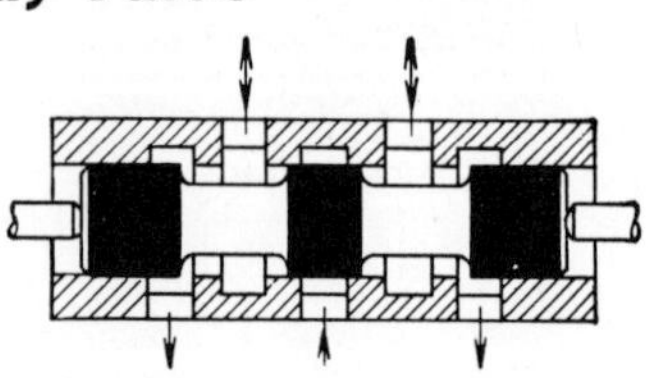

K = 3 to 5.5 at fully open position

Typical Ports

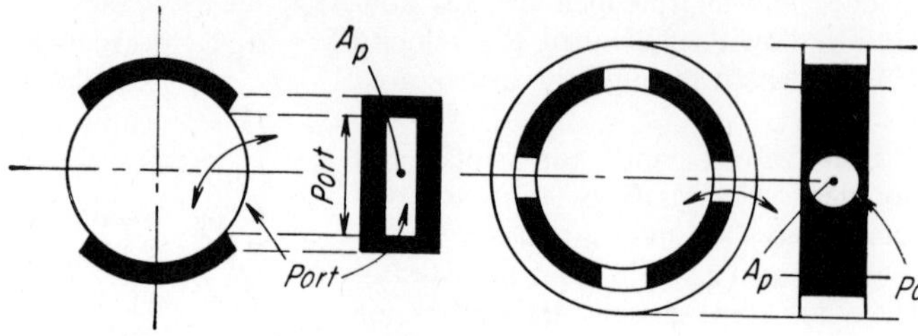

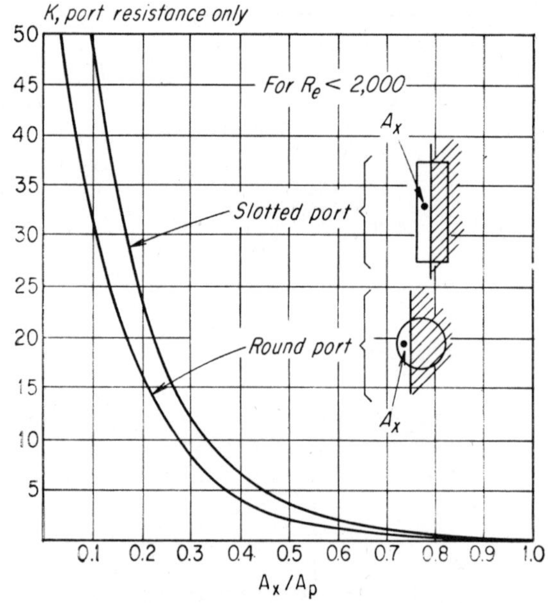

where f = the friction factor (*PE*—Feb 17 '64, p 94), and K_L is a correction factor based on the L/D_o ratio, plotted on page 90.

Diameter change (long hole): This more general case for the long hole includes effects of upstream and downstream pipes as well as conventional losses:

Loss at entrance:

$K_1 = 0.5(1 - D_o^2/D_1^2)$

Contraction:

$K_2 = \sqrt{1 - D_o^2/D_1^2} \times K_L(1 - D_o^2/D_2^2)$

Enlargement:

$K_3 = (1 - D_o^2/D_2^2)^2$

Length:

$K_4 = fL/D_o$

The ratio (D_o^2/D_1^2) is of course identical with A_o/A_1. The combined terms are

$$K = K_1 + K_2 + K_3 + K_4 \quad (4)$$

Note that if the (D_o^2/D_1^2) and (D_o^2/D_2^2) terms are close to zero, the composite $K = 0.5 + K_L + 1.0 + fL/D_o$, which is identical to Eq 3.

Diameter change (thin orifice): Overall K-factors are plotted on page

Notches in Plunger

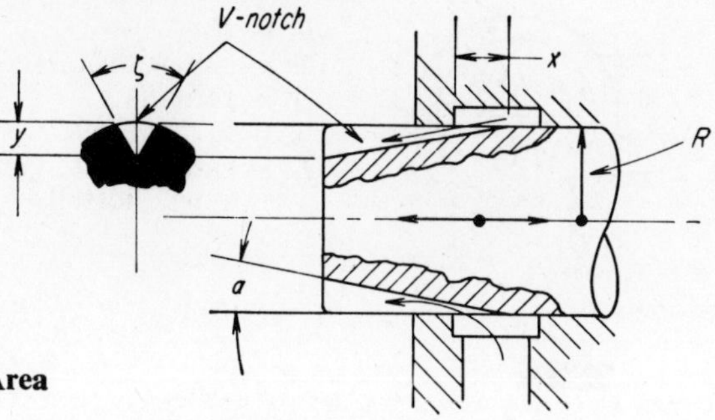

Area

$$A_x = \text{approx.}\ \frac{\zeta}{2} y^2 \quad \text{where} \quad \zeta = \text{radians}$$

$$y = X \tan \alpha \quad \text{when} \quad \zeta = \frac{\pi}{3} = 60°$$

$$A_x = N \frac{\pi}{6} X^2 \tan^2 \alpha$$

N = number of notches

Wetted perimeter (for R_e)

$$L_W = \pi y \frac{\zeta°}{180°} + 2y$$

$$L_W = (3.05 \tan \alpha)\, X \quad \text{if} \quad \zeta = 60°$$

Resistance coefficient

K = approx. $400/R_e$ for $R_e < 150$

K = approx. $10/R_e^{0.25}$ for R_e = 150 to 2000

Partial Orifice

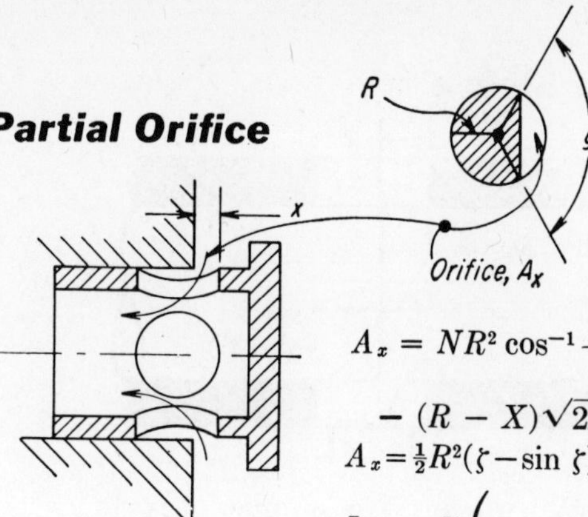

$$A_x = NR^2 \cos^{-1} \frac{R - X}{R} - (R - X)\sqrt{2RX - X^2}$$

$$A_x = \tfrac{1}{2} R^2 (\zeta - \sin \zeta); \zeta = \text{rad}$$

$$L_W = R\left(\zeta + 2 \sin \frac{\zeta}{2}\right)$$

K same as typical port, round, previous page

Ramp Slot

$$A_x = Xw \sin \zeta$$

$$L_W = 2(w + X \sin \zeta)$$

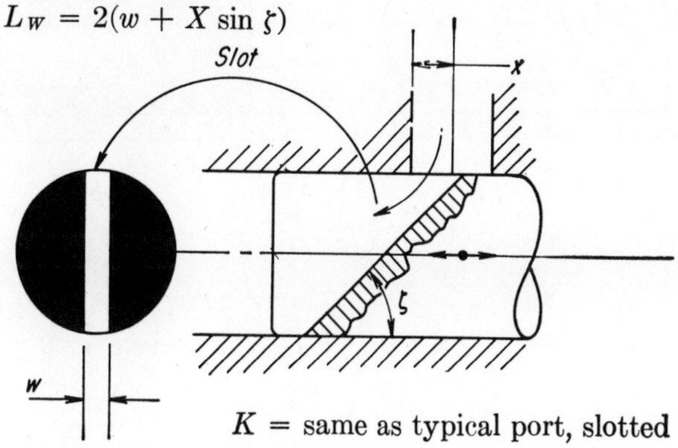

K = same as typical port, slotted

90 for various values of (D_O^2/D_2^2). The equations are

When $R_e > 2.5 \times 10^5$:

$$K = (0.707\sqrt{1 - D_O^2/D_1^2} + 1 - D_O^2/D_2^2)^2 \quad (5)$$

When $R_e < 2.5 \times 10^5$:

$$K = (K_B/M - D_O^2/D_2^2)^2 \quad (6)$$

where K_B and M are indicated on the graphs.

Special cases of Eq 5 and 6 are when (D_O^2/D_1^2) and (D_O^2/D_2^2) are less than 0.1. Then $K = 2.9$ for $R_e > 2.5 \times 10^5$ and $K = 2.9/M^2$ for $R_e < 2.5 \times 10^5$.

Valves and ports

The purpose of any flow-control valve is to introduce a flow restriction. The sketched examples cover almost every type of moving element and port.

The pressure-loss equations are based on the *K*-factor, Eq 2. (See the bottom of page 95 for a description of a similar coefficient $-C_v$ The values for K are average ones based on valve sizes from ⅜ to 1½ in. Pertinent test conditions such as range of R_e are indicated on the graphs. Where no conditions are marked, flow can be laminar or turbulent.

For each type of restriction, the pressure loss is for the restriction alone, and does not include flow resistances of the valve passages preceding and following the restriction.

Conventional valves: Relief valves, ball-checks, butterfly valves, four-way spools, ball valves, gate valves, and needle valves, are typical valve restrictions. These are grouped, accompanied by *K*-factor graphs for a wide range of openings.

The poppets of spring-loaded relief valves vary their position with flow, and the *K*-factor varies inversely with the amount of the opening (graph on page 91). Just what the final position is for any given flow depends on jet pressures, spring gradient, back pressure, and velocity conversions.

Four-way spools have two ports in series. Throttled flow is in accordance with the graph on the previous page. Wide-open flow is restricted only by the ports, and the *K*-factor typically is between 3 and 5.5. Closed flow, meaning leakage through clearances, is covered in the section on clearances.

Slots and segments: Even round holes have complicated expressions for flow if they are intersected by a plunger edge or another round port.

The *K*-factor for any opening is easily estimated. First calculate the cross-section area of the opening, using the equations accompanying each sketch.

Then read the graph of *K*-factors for slotted and round ports (page 92). If the opening is hard to define, take the average K between the slotted and round ports.

Clearances, capillaries, and pores

Laminar flow through an annular clearance varies as the first power of pressure drop, but as the third power of the clearance. Thus a 0.001-in. diametral clearance (a normal allowance) passes eight times as much oil as a 0.0005-in. clearance (a close fit), and 1000 times as much oil as a 0.0001-in clearance (a lapped fit). The latter must be carefully tested under actual operating conditions if you are to be sure that thermal differential expansion doesn't close up the clearance and hang up the plunger. Or you can exploit the thermal expansion to control flow (*PE*—Nov 11 '63, p 72).

Slotted Sleeve

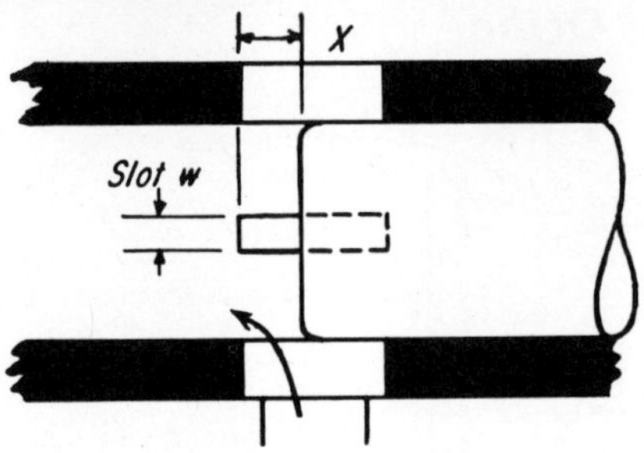

$A_x = NwX$ where N = number of ports
$L_W = 2N\,(w + X)$
K = same as for typical port, slotted

Intersecting Holes

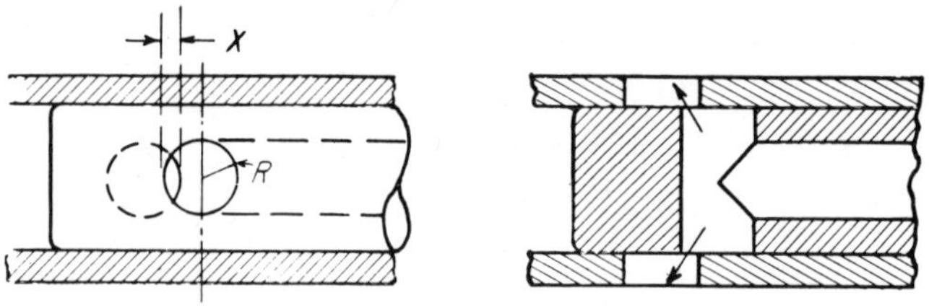

$$A_x = 2\left[R^2 \cos^{-1}\left(\frac{R - \frac{X}{2}}{R}\right) - \left(R - \frac{X}{2}\right)\sqrt{2R\,\frac{X}{2} - \left(\frac{X}{2}\right)^2}\,\right]$$

$A_x = R^2(\zeta - \sin\zeta)$ if ζ = radians
$L_W = 2R\zeta$
K = same as for typical port, round

Rotary Notch

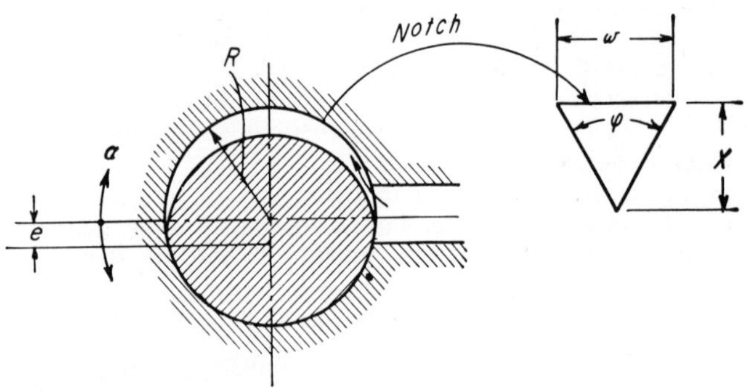

Notch section height

$$X = \sqrt{e^2 + R^2 - 2eR\cos\alpha} - R$$

$X = 0$ when $\cos\alpha = e/2R$

Area

$A_x = wX/2$

$$w = 2\left(\tan\frac{\zeta}{2}\right)X$$

$A_x = 0.577X^2$ if $\zeta = 60°$

Wetted perimeter (for R_e)

$$L_W = 6\left(\tan\frac{\zeta}{2}\right)X = 3.46X$$

K = same as for notched plunger

Rotary Slot

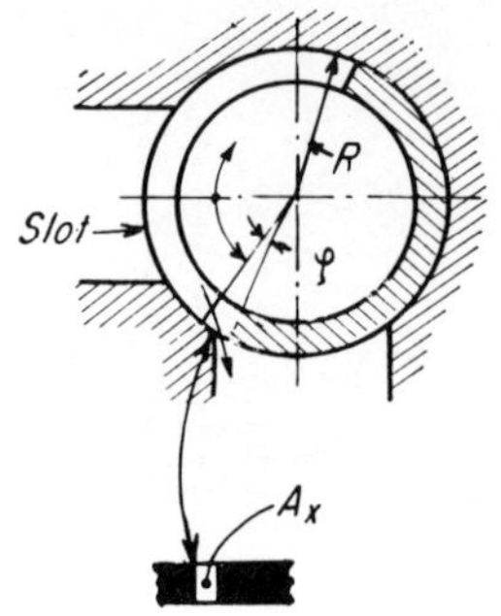

$A_x = Rw\zeta$
w = slot width,
ζ = radians
$L_W = 2(R\zeta + w)$
K = same as for typical port, slotted

Rotary Wedge

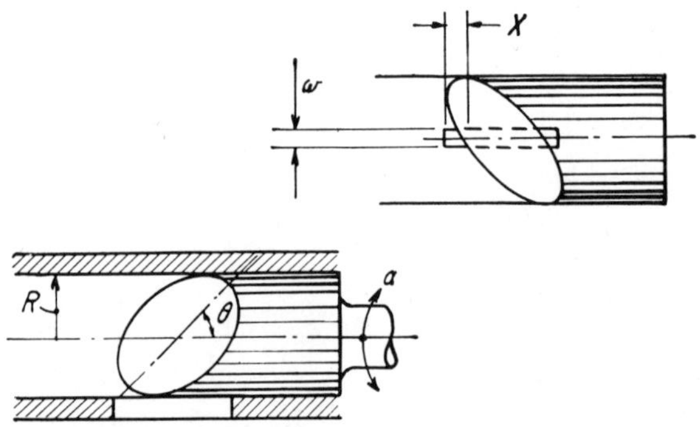

$A_x = w(1 - \cos\alpha)R\cot\theta$
$L_W = 2(X + w)$
$X = R\cot\theta(1 - \cos\alpha)$
K = same as for typical port, slotted

K-factors don't apply to clearances (or to any other laminar-type restriction) because Δp is not a function of V^2. The expression covering flow is

$$Q_{in} = \frac{(C_{rad})^3\,\Delta p d_M \pi}{12\mu_M l} \quad (7)$$

where Q_{in} = flow, in.^{3}/sec, C_{rad} = radial clearance, in., Δp = psi, d_M/l = ratio of clearance's mean diameter to land length, and μ_M = mean absolute viscosity, lb-sec/in.^{2} The companion expression for Δp is:

$$\Delta p = \frac{12Q_{in}\,\mu_M l}{(C_{rad})^3 d_M \pi}$$

Capillaries and small-diameter tubing have similar expressions for laminar flow:

$$Q_{in} = \frac{d^4\Delta p}{\mu l}\,2.45 \times 10^{-2}$$

Annular Clearance

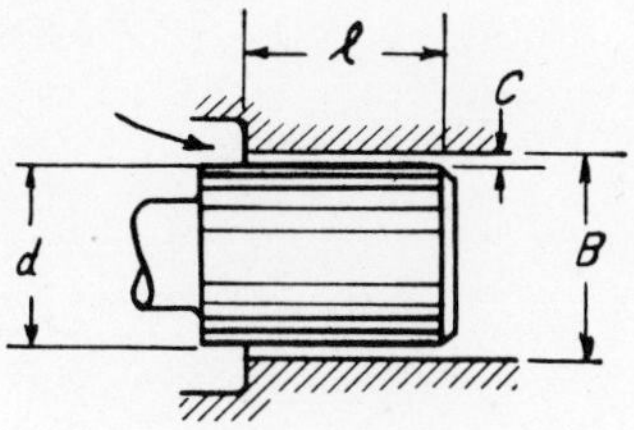

$$\Delta p = \frac{12Q_{in}\,\mu_M l}{(C_{rad})^3 d_M \pi}$$

Capillary Tubing

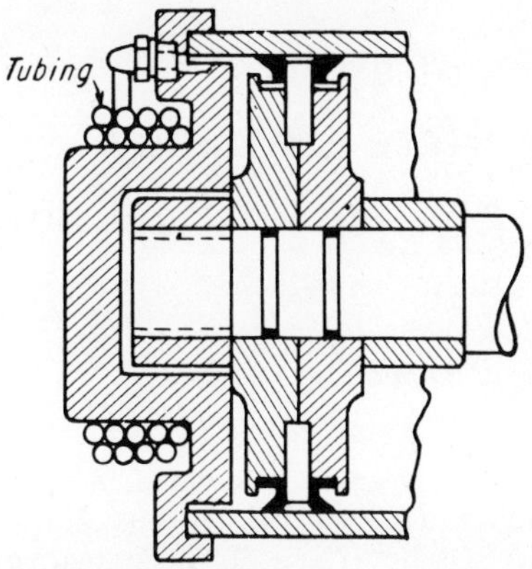

$$\Delta p = \frac{Q_{in}\,\mu l}{d^4}\,40.75$$

Screw Path

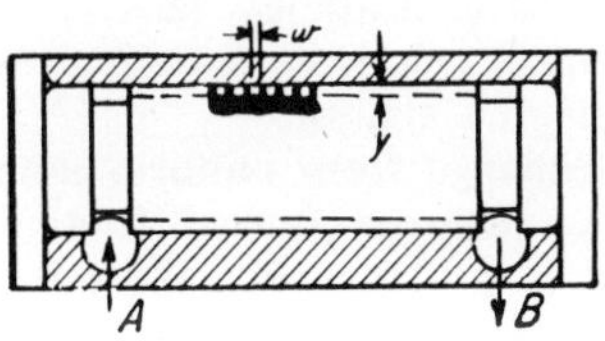

$$\Delta p = \frac{Q_{in}\,\mu l}{a^2}\,32$$

where $a = wy$.

Porous Plug

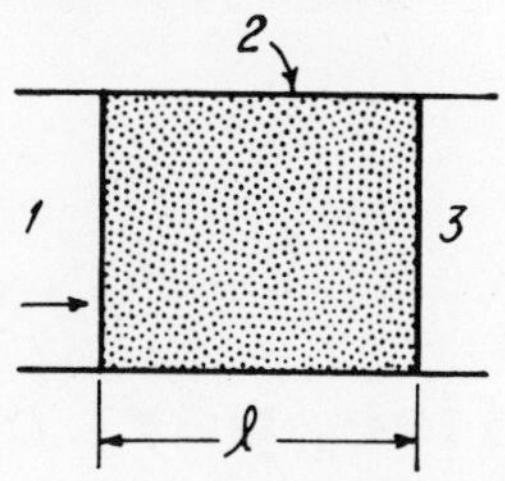

$$\Delta p = CQ_{in}\,\mu\,\frac{l}{a_2}$$

VISCOUS FLOWMETER FOR GAS

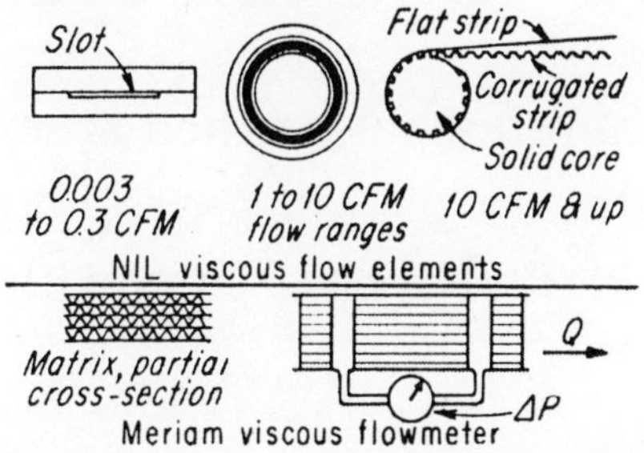

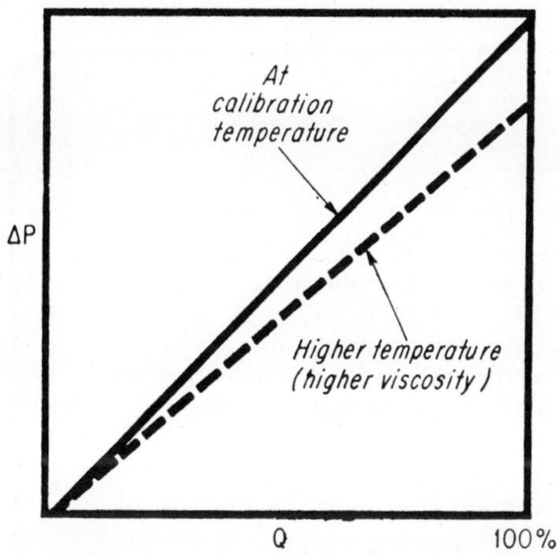

$$\Delta p = \frac{Q_{in}\,\mu l}{d^4}\,40.75$$

where d is tube inside diameter, in. A variant is a male screw thread in a smooth bore, which forces the fluid to follow a helix.

One advantage of using capillaries and annular clearances for flow control is that for any given oil temperature, flow is directly proportional to pressure drop. And oil temperature tends to stay low and constant because the surface-to-volume ratio is favorable for dissipating the heat to atmosphere. Some flow meters are based on capillaries (*PE*—June 5 '61, p 50) and have bundles of small-diameter tubes in parallel to carry the necessary flow.

Porous plugs of sintered metal, and metal tubes filled with tiny round beads, are also used as flow throttlers. The object in each instance is to provide many flow passages and obstructions, assuring laminar flow. Convenient linear relationship between flow and pressure drop is thus achieved. A general expression for pressure drop through a porous plug is:

$$\Delta p = CQ_{in}\,\mu\,\frac{l}{a}$$

where C is a constant, different for each material, l = plug length, in., and a = plug cross-section area, in.^{2} Typical value: $C = 1.3 \times 10^6$ for a 40-micron cellulose filter, passing SAE 20 oil at 100 F.

Flow coefficient C_v

A very common flow coefficient, not covered in this article, is C_V. Valve manufacturers in particular favor C_V because it is an easy coefficient to assign by test, and is readily convertible for use in air-flow equations.

Based on water flow, the relationship is gpm $= \sqrt{C_V\,\Delta p}$, where p = psi. A restriction has a flow coefficient $C_v = 1.0$ if it will pass one gpm of water with a pressure drop of one psi.

Flow coefficient C_V can be converted readily to the K factor with this expression:

$$K = \frac{1460\,a^2}{C_V{}^2}$$

where a = flow area, in.^{2}

A more detailed discussion of the interesting relationships among C_V, K, and other flow coefficients may be found on p 112.

Based on tests

The author's tests are the basis for the graph of Typical Ports. The remaining data were taken from many sources.

Part orifice, part capillary

Close-clearance orifices

When flow clearance is only a few thousandths of an inch, ordinary pressure-drop equations don't apply. Here are handy empirical charts to do the job—for liquid, or gas (sub-critical flow)

ANDREW LENKEI, Project Engineer, Research Department, Worthington Corp, Buffalo

SYMBOLS

A = Flow area, ft^2
C = Radial clearance, ft
c_p = Coefficient of specific heat at constant pressure, btu/lb°F
d = Minor diameter of annular orifice, ft
D = Major diameter of annular orifice, ft
g = gravitational constant, 32.2 ft/sec^2
H = Head, ft
K = Orifice coefficient (dimensionless)
J = Mechanical equivalent of heat, 778 ft lb/btu
L = Flow length; also labyrinth groove axial distance, ft
m = seal configuration factor
n = Polytropic exponent ($n=k=1.4$ for air in isentropic process)
N = Number of sealing points
L_p = Pitch of sealing points, ft
P = Pressure, psf
R = Gas constant (53.3 for air)
R_H = Hydraulic radius (R_H=flow area/wetted perimeter)
R_e = Reynold's number, $W(D-d)/A\mu g$
T = Temperature, °R
V = Fluid velocity, ft/sec
μ = Absolute viscosity, lb sec/ft^2
W = Mass rate of flow lb/sec
ρ = Density, lb/ft^3
ln = Log_e
Z = Slot width, ft

Subscripts

1 First term
2 Second term
n nth term

ORDINARY orifice flow is easily calculated, and so is flow in long tubes. (Recent *PE* reprints "Flow Resistance" (p 76) and "Dashpots" (p 263) summarize all important relationships.)

But when the restriction is too long for pure orifice flow, and too short for line flow, as is the case for many short capillary orifices and close-clearance labyrinths (Figs 1-6), a good mathematical solution seems impossible. Several facts complicate the analysis:

- The velocity profile will change from uniform flow at the entrance to parabolic flow later, as sidewall friction retards the outer layers of fluid progressively along the passage.
- Turbulent flow differs from viscous flow, and this further complicates the flow pattern when Reynold's number is between 1000 and 4000.
- The flow passages in most close-clearance restrictions are rarely symmetrical or even uniform, and different entrance configurations are numerous.
- Boundary layer flow occurs when clearances are extremely small—measured in microns.

Empirical solutions are the only way. Fortunately there have been many test results published, and the data agree surprising well (within ±15%). The accompanying charts portray the mean values in readily usable fashion.

KEY EMPIRICAL EQUATIONS

The basic equation selected (it could have been another) is:

$$V = K\sqrt{2gH}$$

or

$$V = K\sqrt{2g(P_1 - P_2)/\rho} \qquad (1)$$

where V is velocity, K is an empirical coefficient (charts I and II), and H is head loss, ft. $H = (P_1 - P_2)/\rho$, where P_1 is inlet pressure, psf, P_2 is outlet pressure, psf, and ρ is density, lb/ft^3. As you can see, it is simply the equation for incompressible flow through an orifice.

Let's qualify it. Obviously the restrictions are not pure orifices, but neither are they long tubes. The correction charts, based on actual tests, take care of the discrepan-

1.. Circular capillary orifice (Use chart I)

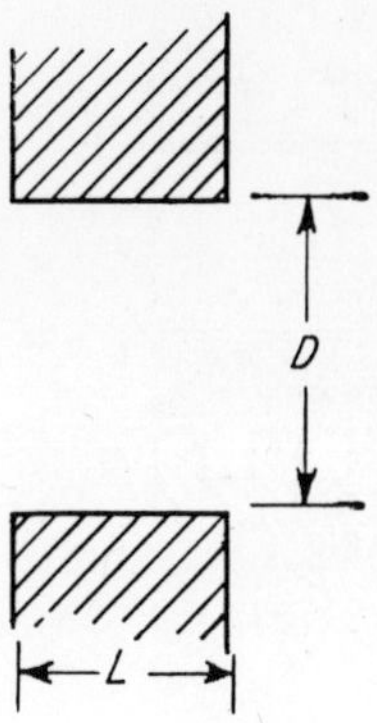

$$R_e = \frac{WD}{A\mu g}$$

$$V = K\sqrt{\frac{2g(P_1 - P_2)}{\rho}}$$

$$W = \rho \frac{V\pi D^2}{4}$$

2.. Annular orifice (Use chart I)

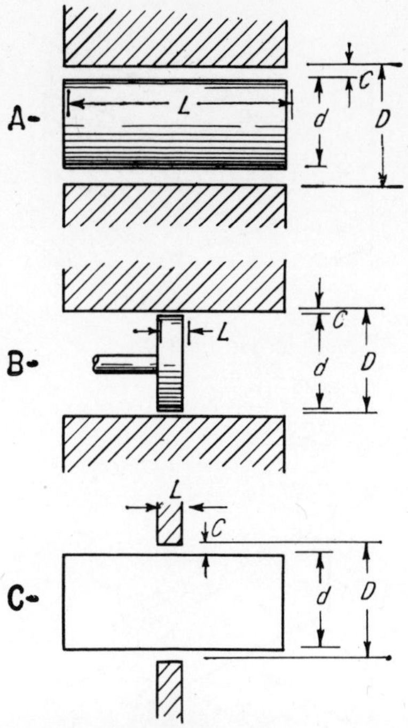

$$V = K\sqrt{\frac{2g(P_1 - P_2)}{\rho}}$$

$$R_e = \frac{W(D-d)}{A\mu g} = \frac{W(2C)}{A\mu g}$$

$$W = \rho V \pi \frac{(D^2 - d^2)}{4}$$

$$A = \pi \frac{(D^2 - d^2)}{4}$$

3.. Off-center orifice, tangent at 1 point (Use chart II)

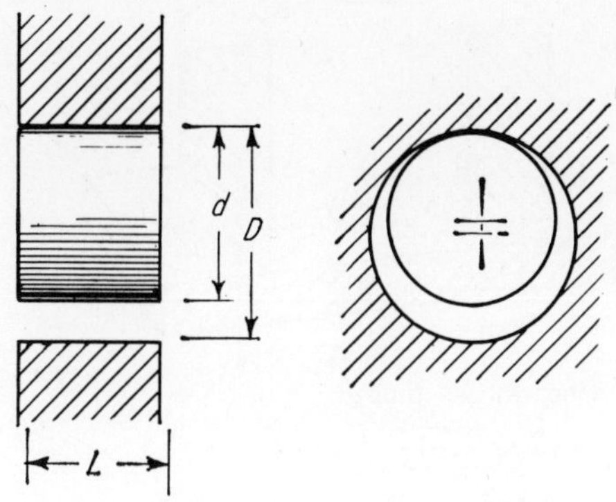

$$\text{avg } C = \frac{D-d}{2}$$

$$R_e = \frac{W(D-d)}{A\mu g}$$

$$W = \rho V \frac{\pi}{4}(D^2 - d^2)$$

$$V = K\sqrt{\frac{2g(P_1 - P_2)}{\rho}}$$

cies. We could have chosen an equation for tubes, and the method would still work, except that a different set of correction charts would be needed.

Flow of compressible fluids also can be calculated with the method. If the density change from inlet to outlet is slight, then assume the gas is incompressible and continue to use Eq 1. If there is considerable expansion, then substitute the following conventional equation for polytropic compressible flow:

$$W = \frac{KA_2P_1}{R} \times \sqrt{2gJ\frac{c_p}{T_1}\left[\left(\frac{P_2}{P_1}\right)^{2/n} - \left(\frac{P_2}{P_1}\right)^{(n+1)/n}\right]} \qquad (2)$$

The units for Eq 2 are given in the symbols box. Use Chart I for the value of coefficient K. In the case of a labyrinth seal, special versions of Eq 2 have been developed. These accompany the sketches of labyrinths, Fig 6. The seal configuration coefficient m is plotted in Chart III for all common values of the length-clearance ratio L/C.

One problem with compressible flow is in determining exactly what kind of expansion is taking place. Is it isentropic? Polytropic? In most applications you can guess fairly close. In example 3, an error of 10% in n changes the answer approximately 11%. For accuracy, build and test a model.

Reynold's number

Reynold's number, R_e, is at the heart of this method. Here is a convenient form for round orifices and capillaries:

$$R_e = WD/A\mu g \qquad (3A)$$

where W is mass flow rate, lb/sec, D is diameter, ft, A is flow area, ft^2, μ is viscosity, lb sec/ft^2, and g is the gravitational constant, 32.2 ft/sec^2.

For annular orifices, a different version is needed:

$$R_e = W(D - d)/A\mu g \qquad (3B)$$

And for slots:

$$R_e = W(2C)/A\mu g \qquad (3C)$$

Eq 3B and 3C were derived using the concept of hydraulic radius, R_H, where $4R_H$ is known to be roughly equivalent to diameter in figuring Reynold's number. Hydraulic radius is

$$R_H = \frac{\text{area of flowing fluid}}{\text{wetted perimeter}} \qquad (4)$$

For annular orifices, $A = \pi(D^2 - d^2)/4$ and perimeter $= \pi(D + d)$. Thus radius $R_H = (D - d)/4$,

4..Tube-end gaps (Use chart I)

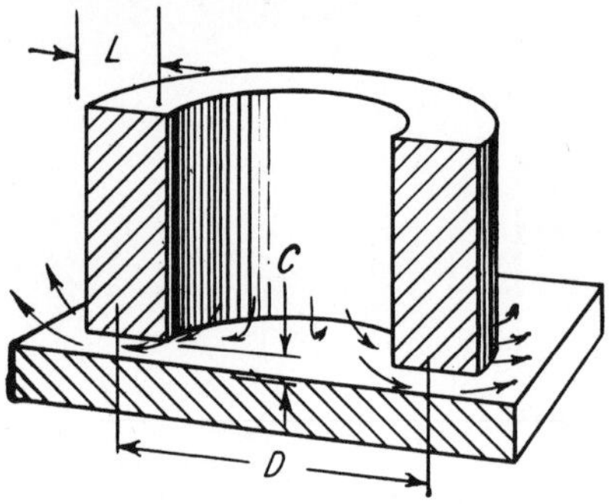

D much larger than L

$$V = K\sqrt{2g(P_1 - P_2)/\rho}$$

$$R_e = \frac{W(2C)}{A\mu g}$$

$$W = \rho V \pi DC$$

$$A \cong \pi DC$$

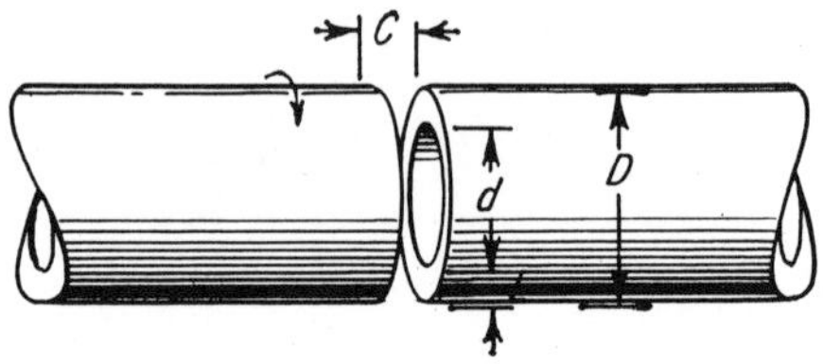

d much larger than L

$$R_e = \frac{W(2C)}{A\mu g}$$

$$V = K\sqrt{2g(P_1 - P_2)/\rho}$$

$$W = \rho V \pi dC \qquad A = \pi dC$$

5..Long flat channels, slits (Use chart I)

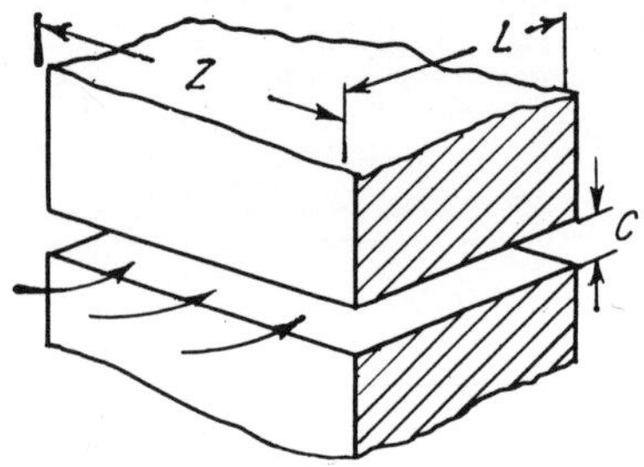

Z much larger than C

$$V = K\sqrt{2g(P_1 - P_2)/\rho}$$

$$R_e = \frac{2W}{\mu g(Z + C)}$$

$$W = \rho VZC$$

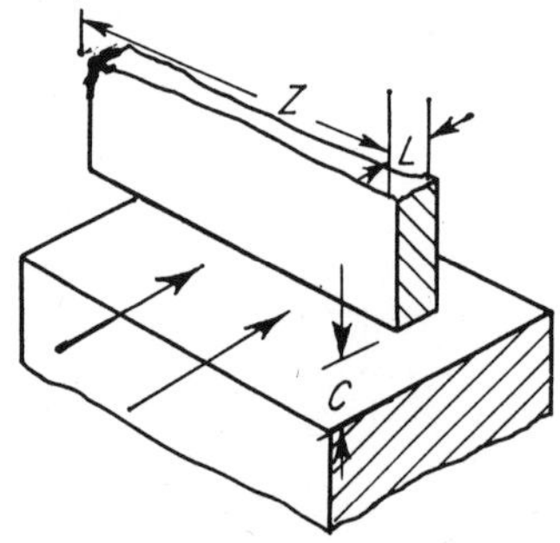

Z much larger than C

$$V = K\sqrt{2g(P_1 - P_2)/\rho}$$

$$R_e = \frac{2W}{\mu g(Z + C)}$$

$$W = \rho VZC$$

6..Annular, non-uniform orifices, for gases (Use charts I and III)

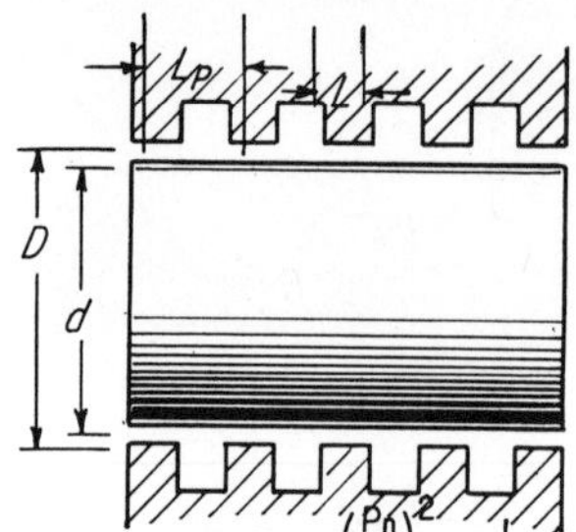

$$W = 5.7\,KA\frac{P_1}{\sqrt{RT_1}}\left[\frac{1-\left(\frac{P_n}{P_1}\right)^2}{n - \ln\left(\frac{P_n}{P_1}\right)}\right]^{\frac{1}{2}} y$$

$$y \cong \frac{1}{\left[1 - \frac{8.5}{\frac{L_p - L}{C} + 7.2}\right]^{\frac{1}{2}}}$$

$$R_e = \frac{W(2C)}{A\mu g} \qquad A = \frac{\pi}{4}(D^2 - d^2)$$

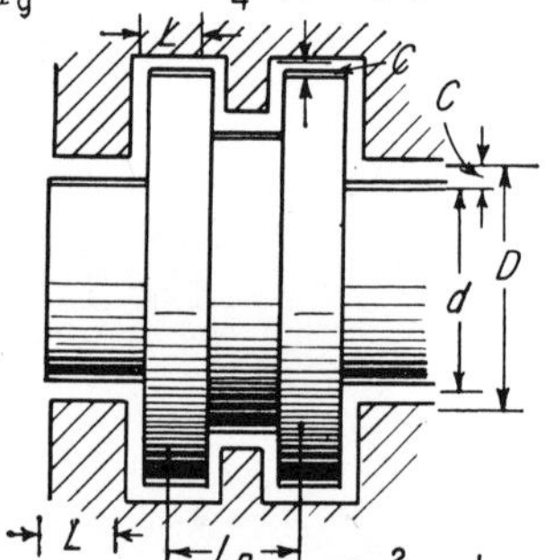

$$W \cong 5.7\,KA\frac{P_1}{\sqrt{RT_1}}\left[\frac{1-\left(\frac{P_n}{P_1}\right)^2}{\frac{nm}{2} - \ln\left(\frac{P_n}{P_1}\right)}\right]^{\frac{1}{2}}$$

m is obtained from chart III

$$R_e = \frac{W(2C)}{A\mu g} \qquad A = \frac{\pi}{4}(D^2 - d^2)$$

and $4R_H = D - d$ or $2C$, where C is radial clearance.

For slots, $A = CZ$, perimeter $= 2(C + Z) =$ approximately $2Z$. Thus $R_H = 2C$, where C is the clearance.

Note: There is some disagreement among experts here. For instance, Crane Technical Paper 410 (Crane Co, Chicago), pp 1-4, states that the hydraulic radius for a very thin slot is equal to the narrower dimension, C. Thus the equivalent diameter would turn out to be $4C$ instead of $2C$. Eq 3B and 3C would be affected.

Using the equations

Solve Eq 1 or 2 first, yielding velocity V in terms of the unknown coefficient K. Convert to mass flow W with equations provided. Then solve for Reynold's number using Eq 3. It will still contain the unknown coefficient K.

Solve for coefficient K by trial and error. First assume a value for K, then calculate Reynold's number. Check the value against that in Chart I or II for the given value of $(D - d)/L$. If it's wrong, try another value for K. Three or four guesses ought to do it. Knowing K, you can calculate the desired velocity and mass flow.

Accuracy of this method is as good as that for simple orifices or pipes, but the same chances for inaccuracy exist. Remember that poor surface finish, slight rounding of the edges, inaccurate dimensions, and many other physical variations will affect the results greatly. It is better to build and test a model where possible.

Example 1: Water through a concentric orifice

Problem: Find mass rate of flow for these conditions: Diameter = 5 in., radial clearance $C = (D - d)/2 = 0.001$ in., water temperature = 140 F, pressure drop = 20 psi, flow length = 0.01 in., viscosity = 0.00001 lb sec/ft².

$$V = K\sqrt{2g(P_1 - P_2)/\rho}$$
$$= K\sqrt{64.4 \times 20 \times 144/62.4}$$
$$= 54.5K \text{ ft/sec}$$
$$W = \rho VA$$
$$= 62.4 \times 54.5K \times \pi 5 \times 0.001/(12 \times 12)$$
$$= 0.371K$$
$$R_e = W(D - d)/A\mu g,$$

or

$$\text{or } W(2C)/A\mu g$$
$$= \frac{0.371K(2 \times 0.001)12}{\pi 5 \times 0.001 \times 0.00001 \times 32.2}$$
$$= 1760K$$

Ratio $(D - d)/2L = (2 \times 0.001)/(2 \times 0.01) = 0.1$

Trial and error: Assume $K = 0.5$. Then $R_e = 1750 \times 5 = 875$. But Chart I indicates $R_e = 200$ at $(D - d)/2L = 0.1$, so try again. Try $K = 0.6$, giving $R_e = 1750$

I . . . K values for symmetrical restrictions

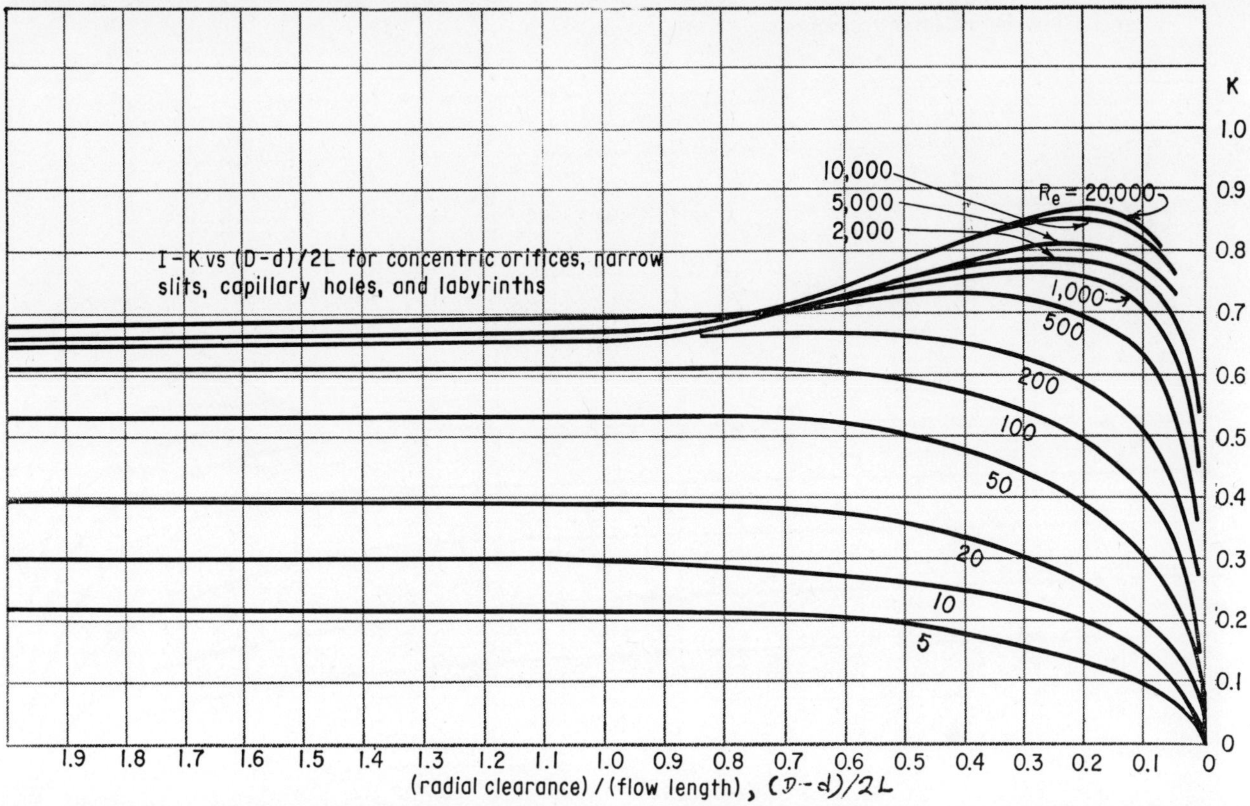

$\times$ 0.6 = 1050, and Chart I shows R_e = 450—still wrong.

Answer: K = 0.71 proves to be a close enough guess, with R_e = 1750 $\times$ 0.71 = 1240, and Chart I shows R_e = 1250. Therefore mass rate W = 0.371 $\times$ 0.71 = 0.263 lb/sec.

Example 2: Air through a capillary (small ΔP)

Problem: Find mass rate of flow for these conditions: Length = 0.27 in., diameter = $\frac{1}{16}$ in., air density = 0.0743 lb/ft^3, viscosity = 4.79 $\times$ 10^{-7} lb sec/ft^2, and pressure drop = 0.32 in. H_2O (1.67 psf).

$$\begin{aligned}
V &= K\sqrt{2g(P_1 - P_2)/\rho} \\
&= K\sqrt{64.4 \times 1.67/0.0743} \\
&= 37.9K \text{ ft/sec} \\
W &= \rho V A \\
R_e &= WD/A\mu g \\
&\quad \rho VAD/A\mu g \\
&= \rho VD/\mu g \\
&= (0.0743 \times 37.9K \times 0.0625)/ \\
&\quad (12 \times 4.79 \times 10^{-7} \times 32.2) \\
&= 950K
\end{aligned}$$

First trial: Let K = 0.5, then R_e = 950 $\times$ 0.5 = 475. Chart I shows that for $D/2L$ = 0.0625/0.54 = 0.116, and K = 0.5, R_e = 180. Wrong guess. Try K = 0.7; then R_e = 950 $\times$ 0.7 = 665. Chart I shows R_e = 900. Try K = 0.65, then R_e = 950 $\times$ 0.65 = 617. Chart I shows R_e = 600, which is close enough.

Answer: Velocity = 37.9K = 37.9 $\times$ 0.65 = 24.6 ft/sec, and mass rate W = 0.0743 $\times$ 24.6 $\times$ π (0.0625)^2/(4 $\times$ 144) = 3.89 $\times$ 10^{-4} lb/sec.

Example 3: Air flow through a capillary (large ΔP)

Problem: Find mass rate of flow for these conditions: Diameter = 0.0156 in., length = 0.135 in., air temperature = 200 F, air viscosity = 4.79 $\times$ 10^{-7} lb sec/ft^2, P_1 = 21 psia, P_2 = 14.7 psia, flow is isentropic (n = 1.4), and coefficient of specific heat at constant pressure c_p = 0.24 btu/lb −°F. Use Eq 2 and 3:

$$W = \frac{223.8AP_1}{R} \times \left(\sqrt{\frac{c_p}{T_1}\left[\left(\frac{P_2}{P_1}\right)^{2/n} - \left(\frac{P_2}{P_1}\right)^{(n+)/n}\right]}\right)K \qquad (2)$$

$$R_e = \frac{W(2C)}{A\mu g} = \frac{223.8P_1}{R} \times \left(\sqrt{\frac{c_p}{T_1}\left[\left(\frac{P_2}{P_1}\right)^{2/n} - \left(\frac{P_2}{P_1}\right)^{(n+1)/n}\right]}\right)K\,\frac{2C}{\mu g} \quad (2)\ (3)$$

$$= \frac{223.8 \times 21 \times 144}{53.3 \times 4.79 \times 10^{-7}} \times \sqrt{\frac{.24}{660}\left[\left(\frac{14.7}{21}\right)^{2/1.4} - \left(\frac{14.7}{21}\right)^{(1.4+1)/1.4}\right]} \times K\,\frac{0.0156}{12 \times 32.2} = 4900K$$

Ratio $D/2L$ = 0.0156/0.27 = 0.058. Use Chart I and assume values for K. First time: K = 0.5, and R_e = 4900 $\times$ 0.5 = 2450. From Chart I, R_e = 350, which is wrong. Try K = 0.6, then R_e = 4900 $\times$ 0.6 = 2940. Chart I shows R_e = 700. Wrong again. Try K = 0.715, then R_e = 4900 $\times$ 0.715 = 3500. Chart I shows R_e = 3550, which is close enough.

Answer: From Eq 2, W = 0.55 $\times$ 10^{-4} lb/sec.

(text continues, p. 100)

II . . . K values for eccentric restrictions

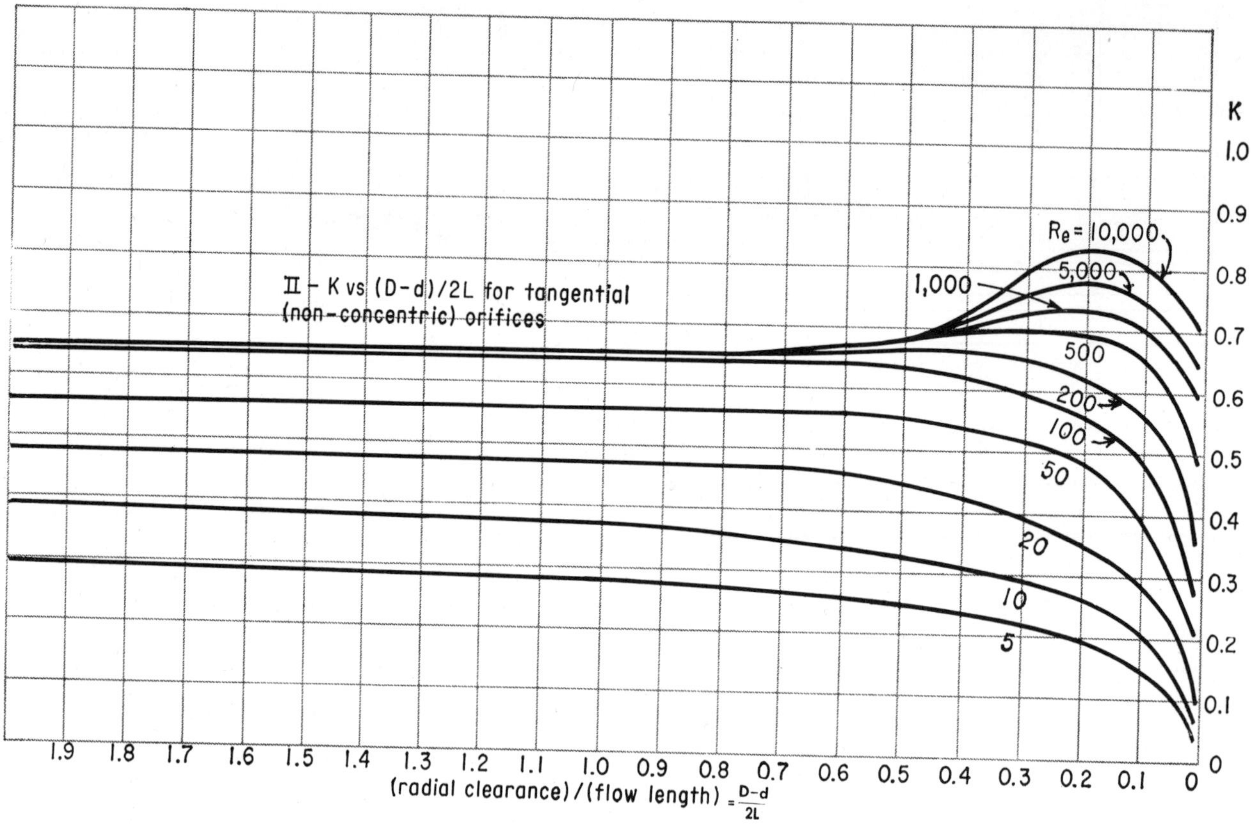

III . . . Seal configuration coefficient m vs L/c, for labyrinth restrictions

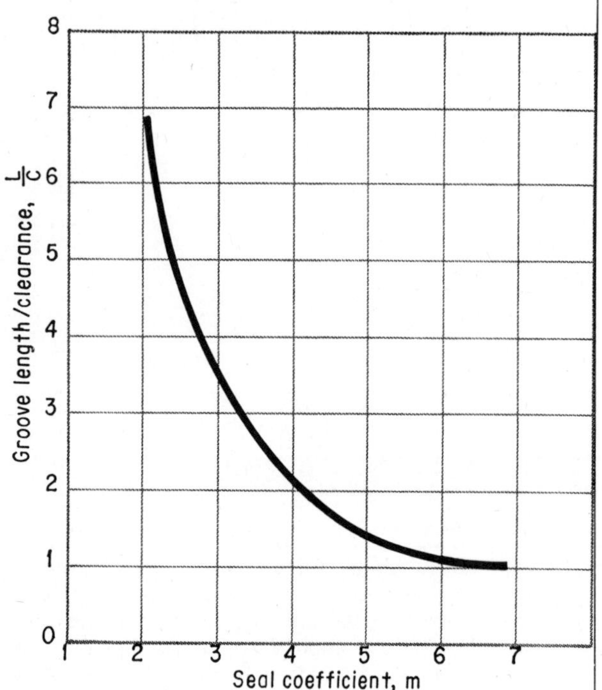

For more data on the special orifices that the author describes, see these scientific papers:

Pressure drop and flow characteristics of short capillary tubes at low Reynolds numbers, F. Kreith and R. Eisenstadt, ASME Transactions, 1956

Flow through annular orifices, K. J. Bell and O. P Bergelin, ASME Transactions, 1956

A fluid mechanics approach to the labyrinth seal leakage problem, G. Vermes, ASME Transactions, 1960

A new semi-empirical equation for gas flow through capillaries, E. S. Creutz and L. R. Zumwalt, Journal of Applied Physics, Vol 33, No. 9, Sept 1962

Leakage of steam through labyrinth seals, A. Egli, ASME Transactions, 1935

Labyrinth seal leakage analysis, W. Zabriskie and B. Sternlicht, ASME Transactions, 1959

Flow through capillary tubing with eccentric annular sections, C. H. Bachman and H. G. Hottenrott, Review of Scientific Instruments, Vol 30, No. 2, 1959.

They work even at low flow

Errorless orifices

When Reynold's number drops below 10,000, no single standard orifice has a constant discharge coefficient. Here's a way to achieve it with two

GERALD BLOOM, Project Engineer, Hoke Inc, Cresskill, NJ

IT would be a lot easier to control flow in an orifice or nozzle if the discharge coefficient would hold still. **Even with handy charts (see pages 76 and 83 for discussions) it's annoying to have to adjust a constant every time the flow or viscosity changes.**

What's the problem? Basically, the discharge coefficient of a typical sharp-edge orifice (in a relatively large conduit) drops from about 0.7 to about 0.6 when Reynold's Number increases from 100 to 10,000. Conversely, the discharge coefficient for a typical rounded orifice or nozzle rises from about 0.7 to about 0.9 in the same range. No standard orifice stays at one value.

Answer: Two heads are better than one. Simply by combining a sharp-edge orifice in series or parallel with a rounded orifice (or cone, or nozzle) you practically can cancel out the variation. Fig 1 shows examples of what improvements can be made.

Just putting any sharp-edge orifice in series or parallel with any rounded orifice is not exactly the answer—although it is bound to help. For really constant discharge coefficients you must find the right combination. Fortunately, it can be discovered by calculation using the techniques that follow.

Sizing the orifices

For best results, limit the study to a few known types. Beveled-inlet orifices are fairly easy to make and have the right characteristics. The values of discharge coefficient C for most sharp-edge orifices are fairly consistent for all sizes, as long as the orifice diameter is small relative to the inlet channel diameter. Above an orifice-channel diameter ratio of 1:2, C will tend to be higher for the same Reynold's number. So to be safe, stay below a diameter ratio of 1:3 where almost all size effects, including those of inlet and discharge velocity changes caused by crowding, will be eliminated —to within 1% or so.

The first step mathematically is to either: set the flows equal if the assumed orifices are in series; or set the sum of the flows equal to the main flow if the orifices are in parallel.

Testing your results

If you aim for constant discharge coefficients (within a few percent) then to prove your results you must be able to *measure* the flow with even better accuracy.

In series

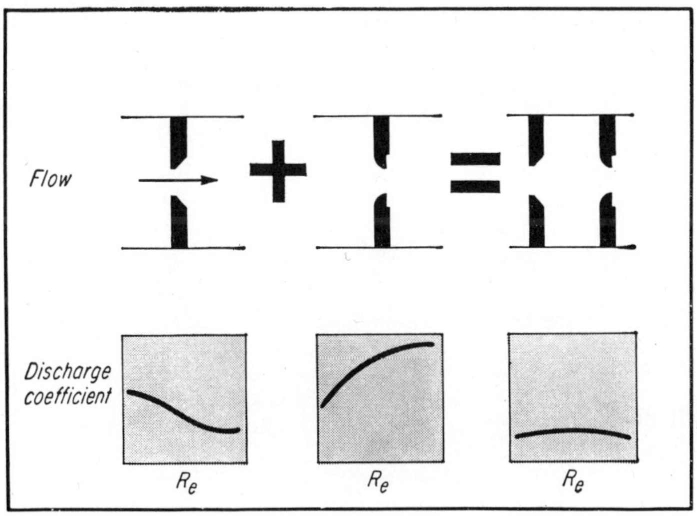

In parallel

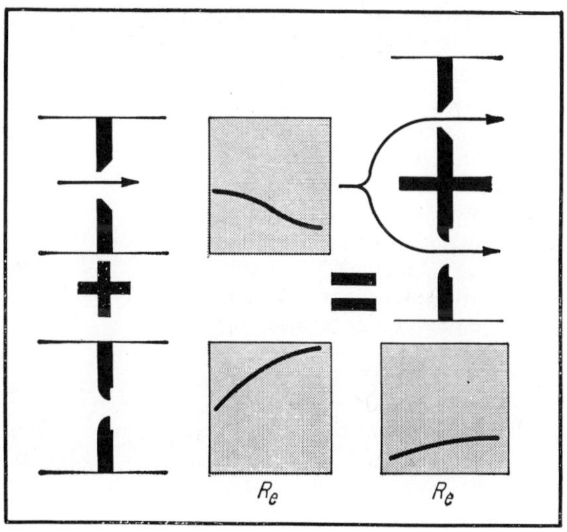

Charts simplify calculation of

Airflow in components

Flow, flow factor and pressure loss in valves and fittings can be determined easily with these nomographs and conversion charts.

DOMINIC J LAPERA, design engineer, Marotta Valve Corp, Boonton, NJ

What is the flow factor?

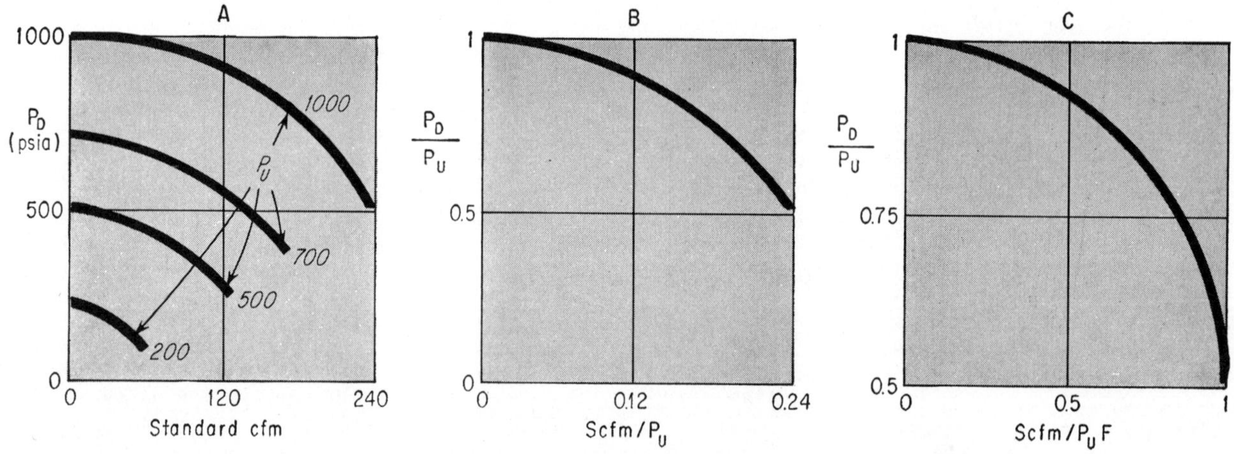

Gas or airflow through a pneumatic component may be fairly approximated with this relationship:

$$\text{scfm} = \text{constant} \times \sqrt{(P_U^2 - P_D^2)}$$

where the constant depends on passage configuration, friction, and inlet temperature; P_U and P_D are upstream and downstream pressures, psia. Scfm is actual flow in cfm converted to "standard" conditions—14.7 psia, 68 F. Although the equation does not represent isentropic flow, and assumes considerable loss in stagnation pressure, its convenient mathematical form simplifies manipulation and the error will probably not exceed 5 to 10%. Test results can be plotted in three ways. Curves in *A* are flow data in scfm for a typical component. Curve *B* is from the same data converted to single curve by dividing by upstream pressure. Curve *C* is from the same data, further converted to "normalized" form by dividing by a defined flow factor, *F*, that is a measurable quantity.

Flow factor, *F*, is the ratio

$$\frac{\text{flow (scfm)}}{P_U\ \text{(psia)}} \quad \text{when} \quad \frac{P_D}{P_U} = 0.5$$

The constant in the flow equation equals $F \times \sqrt{4/3}$, therefore the normalized equation for curve *C* becomes

$$\frac{\text{scfm}}{P_U \times F} = \sqrt{(4/3)} \times \sqrt{(1 - (P_D/P_U)^2)}.$$

or

$$Q = P_U \times F \sqrt{(4/3)} \times \sqrt{(1 - r^2)}$$

This variation of the NBS equation has proven to be a better approximation of actual flow:

$$Q = P_U F \sqrt{(8/5)} \sqrt{[r(1-r)(3-r)]}$$

It is the mean between the $\sqrt{(1-r^2)}$ relationship and adiabatic expansion. Author Lapera used the mean equation to develop his flow nomograph, Fig 5.

1a..Typical test (water)

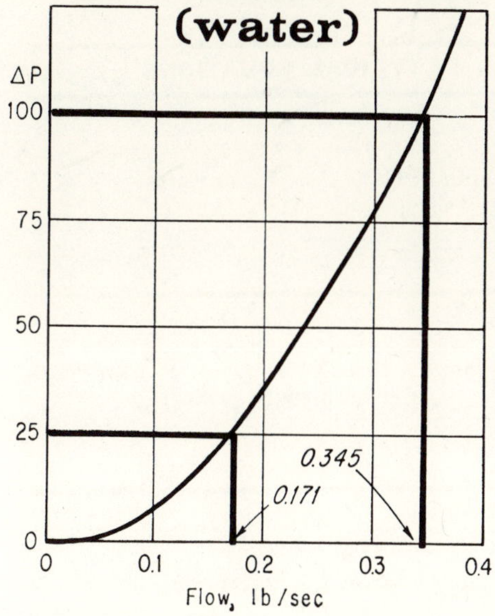

1b..Effective orifice, D_o

Effective orifice diameter is derived from performance curve of water flow through a component as follows: On the water-flow-test locate any convenient pressure drop and find the corresponding water flow rate *W*; locate this *W* value on the nomograph and extend a line from that point through pivot *B* to intersect the pivot *A* scale: locate the ΔP value on the nomograph and extend a line from that point through the point on the pivot *A* scale to intersect the D_o scale, giving the answer. Two examples will show the uniqueness of the capacity constants: 1) $\Delta P = 25$ psi; $W = 0.171$ lb/sec; $D_o = 0.119$ in.; 2) $\Delta P = 100$ psi; $W = 0.345$ lb/sec; $D_o = 0.119$ in.

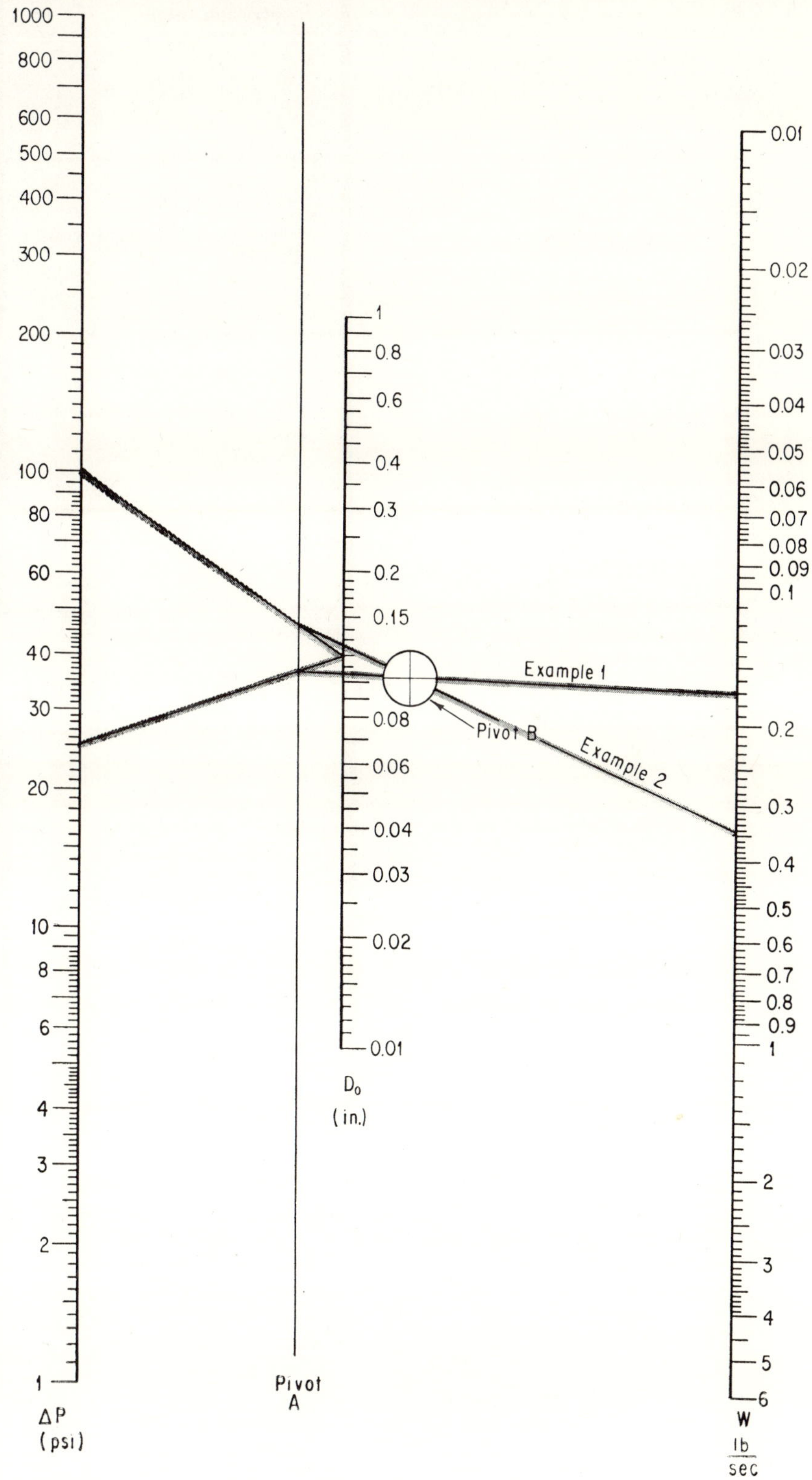

ESTIMATING airflow and pressure loss through a component in a pneumatic system is tricky for two reasons: pressure-loss data are expressed differently in different catalogs (*Fig 1a above is just one of the ways*) and subsequent conversion calculations are likely to produce errors; and flow is affected by gas properties.

This article helps solve both problems. The first, differences in catalogs, is handled by providing simple charts for direct conversion from catalog flow constants to the National Bureau of Standards flow factor, *F*. The NBS flow factor is widely accepted as a standard, and the method in this article is based on it. A brief summary of the NBS method is given on the facing page.

The second problem, effects of gas properties on flow, is handled by first converting the gas flow to an equivalent flow of air at standard atmospheric conditions (here chosen to be 14.7 psia, 70 F). Charts and tables based on standard air simplify conversion and eliminate that source of error.

The over-all procedure for calculating airflow, starting with the catalog or data sheet and producing a numerical answer, is given in the accompanying series of charts and instructions. They are in proper sequence to take care of *all* problems; just skip the charts that don't apply to yours. Answers are given in units of flow, pressure-loss, or flow-factors; starting with any two you can easily read off the third.

Only a single pneumatic valve or component at a time is considered here. However, piping and system calculations based on the same NBS

TABLE I.. Comparison of flow constants

BASIS FOR CONSTANT	COMMENTS	TYPICAL EQUATIONS
$C_v \propto Q/(P_U - P_D)^{\frac{1}{2}}$	Although based on liquid, C_v and D_o are applicable to gas. It depends on the expression used.	Q, gpm $= C_v \sqrt{P_U - P_D}$ for water
$D_o \propto Q^{\frac{1}{2}}/(P_U - P_D)^{\frac{1}{4}}$		Q, gpm $= 18.1\, D_o^2 \sqrt{(P_U - P_D)}$ for water (sharp-edged orifice)
$F \propto Q/(P_U^2 - P_D^2)^{\frac{1}{2}}$	For air at standard conditions.	Q, scfm $= F \sqrt{\frac{4}{3}} \sqrt{P_U^2 - P_D^2}$
$K \propto \frac{A^2}{Q^2}/(P_U - P_D)$	$\rho V^2/2g$ is velocity head in psi.	$\Delta P = K\left(\frac{\rho V^2}{2g}\right) = K\left(\frac{\rho Q^2}{2gA^2}\right)$

TABLE II.. Flow factor, F, for tubing (NBS equations)

For short tubes ($L \leqslant 10$ dia)	A = cross-section area, sq in.; d = ID, in.; f = friction factor; L = length, in.	$F = 16.5A$
For long tubes ($L > 10$ dia)		$F = 23.5A\sqrt{(d/fL)}$

method are easily made and are among the topics covered in a previous article (*PE—Oct* 26 '59).

Catalog data

Actual tabulated or plotted performance is the most basic flow information in a catalog—though not often provided. Conversion to flow factor F is accomplished in three steps with the help of Fig 1. Plot flow (lb/sec of water) vs pressure drop (psi) using whatever catalog information is provided; graphically determine effective orifice diameter D_o using the nomograph; and convert D_o to F with Fig 2.

More common than plotted performance are the various flow constants used in flow-vs-pressure-drop equations. Two types are easily converted to flow factor F: they are flow coefficient C_v and effective orifice D_o. Other methods such as the K-factor technique (*PE—Apr.* 14 '58) are not mathematically similar to flow factor F and require special conversion techniques (one is to plot calculated test points and convert to flow factor F graphically). The constants are sum-

2..C_V-D_o-F conversion

Flow constants C_V, D_o and F are directly related mathematically as follows: $D_o = 0.236\sqrt{C_V}$; $C_V = 1.77\,F$; $F = 9.9\,D_o^2$. These three examples show how to read the values directly: if $D_o = 0.03$ in., $C_V = 0.016$; if $C_V = 0.02$, $F = 0.01125$; and if $F = 62$, $D_o = 2.5$.

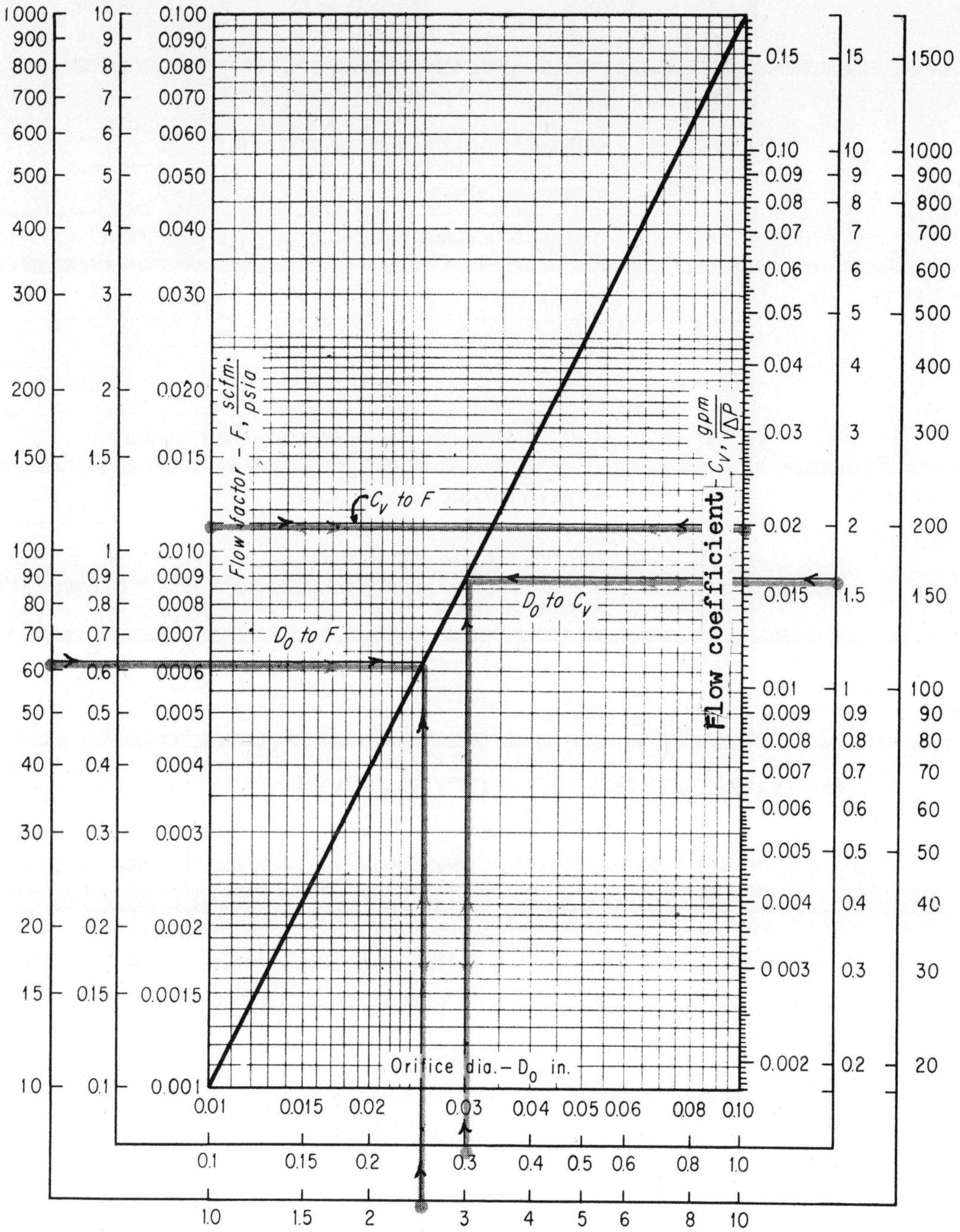

marized in table I. C_V, D_o and F may be converted in Fig 2.

In brief, here is how each of the flow constants works: Flow coefficient C_V by definition equals 1.0 if the component will pass 1 gpm of water with a pressure drop of 1 psi. Intermediate values of water flow, flow coefficient and pressure drop are determined with the relationship gpm $= C_V\sqrt{\Delta P}$. Compressible flow does not have the same equation, but the value for C_V obtained with water adequately defines the component flow characteristics and may be used in gas equations or it may be converted to the NBS flow factor.

Be careful when applying C_V values obtained from different sources: some companies combine C_V with other constants to develop special flow coefficients unique to their own products. However, they will usually give you standard flow coefficients on request. Or you can plot several calculated points in the recommended system of units, and convert to flow factor F starting with Fig 1.

Effective orifice D_o simply defines the diameter of a hypothetical sharp-

TABLE III .. Nomenclature

FLOW	W	weight flow, lb/sec
	Q	volume flow, cfm
	Q_A	airflow, scfm (14.7 psia, 70 F)
PRESSURE	P_U	inlet (upstream) pressure, psia or psig
	P_D	outlet (downstream) pressure, psia or psig
	P_D/P_U	pressure ratio, psia/psia
	Z	function of pressure ratio, plotted on next page
GAS CONSTANTS (see table of values)	ρ	density, lb/cu ft
	R	gas constant (53.3 for air)
	γ	ratio of specific heats, C_P/C_V
	$(\rho_G/\rho_A)_{std}$	relative density of a gas and air at standard conditions
FLOW CONSTANTS and passage dimensions	C_D	coefficient of discharge for orifice
	C_V, K, F, D_O	flow constants, described in text
	A	cross-section area
SUBSCRIPTS	U,D	upstream, downstream
	A,G	air, gas

TABLE IV .. Thermodynamic constants

	Air	H_2	N_2	O_2	He	A	CO	CO_2	NH_3	CH_4
ρ_G/ρ_A	1.00	0.0695	0.968	1.105	0.1381	1.379	0.967	1.530	0.596	0.554
R	53.36	766.6	55.16	48.29	386.2	38.69	55.19	35.12	90.74	96.40
γ	1.402	1.409	1.400	1.399	1.66	1.66	1.400	1.301	1.319	1.319

edged orifice that will pass the required flow (assumed to be water) with the given pressure drop. The basic equation is gpm $= 18.1\ D_o^2 \sqrt{(P_U - P_D)}$ and includes a discharge coefficient of 0.61—an average value for such orifices.

K-factor method assumes that pressure drop in piping and components is a function of the stream velocity: $\Delta P = K\rho V^2/2g$. For a given valve configuration the pressure drop will quadruple if stream velocity doubles.

Equivalent pipe length is another measure of pressure loss in a component or piping system, and it too can be converted to flow factor F. This conversion has been covered in a previous article (*PE—Apr* 14 '58). The basic relationships—dependent on friction factor—are briefly summarized in table II.

Convert airflow to scfm

Before you can insert values into the final nomograph of the series you must first convert volume flow of whatever gas you are using into volume flow of air at standard conditions (14.7 psia, 70 F). The final nomograph is based on scfm—standard cu ft/min. The values of P_U, P_D and their ratio P_D/P_U remain the same

3..Expansion factor Z

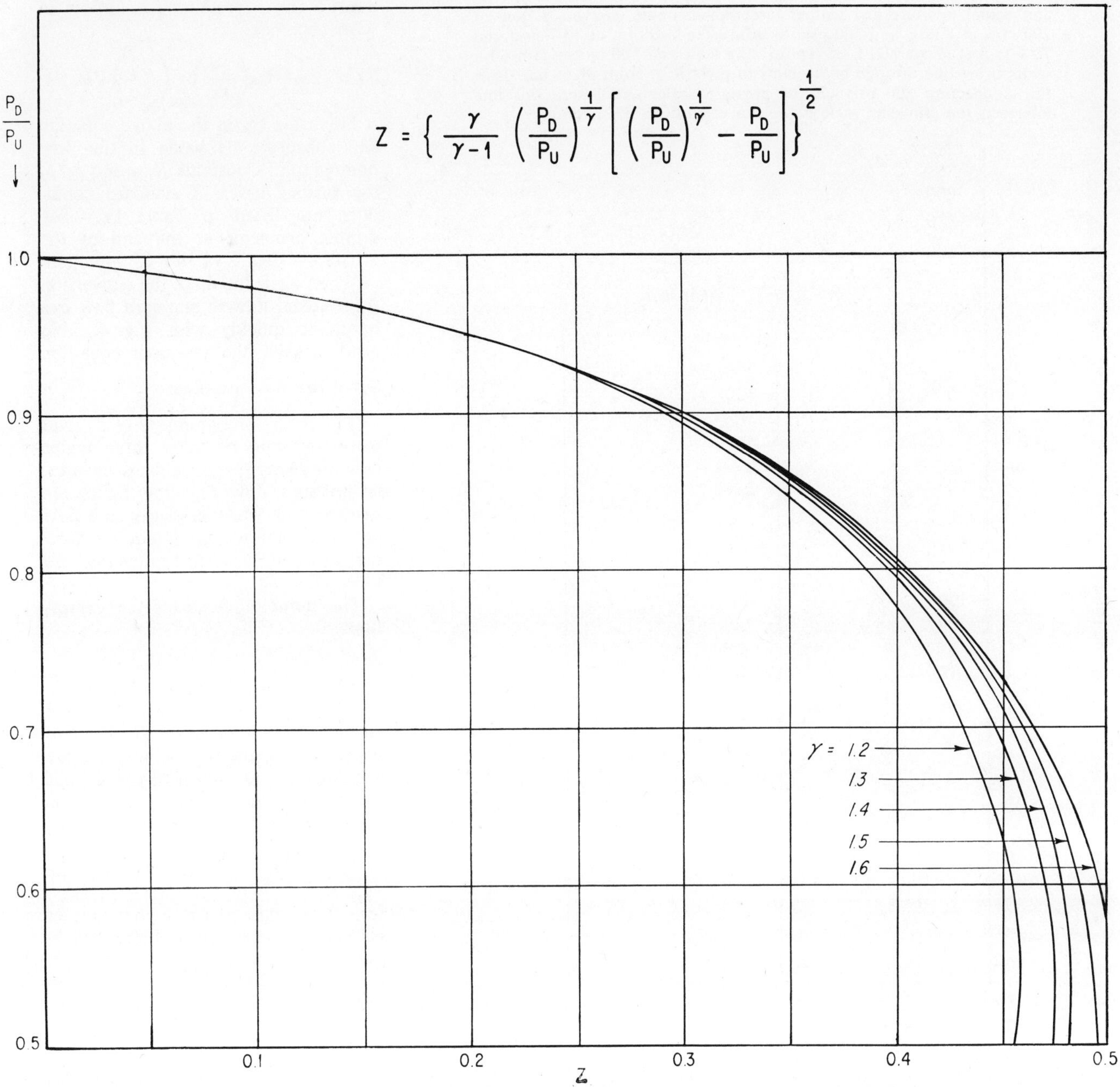

This graph is needed only when the gas has a different γ than air because Z_A/Z_G (see airflow equations below) will be unity if $\gamma_G = \gamma_A = 1.402$. Nitrogen, oxygen and hydrogen, for example, have γ's of about 1.4 and conversion to standard airflow units is simple—see Case 5 on the last page of the article for an example using nitrogen.

before and after the conversion from actual flow to scfm. Here is why.

Flow is a function of passage geometry and the thermodynamic properties of the gas. The thermodynamic expression for weight flow in lbs/sec is:

$$W = 12AC_D\rho_U \sqrt{2g\frac{\gamma}{\gamma-1}\frac{P_U}{\rho_U}\left[1-\left(\frac{P_D}{P_U}\right)^{\frac{\gamma-1}{\gamma}}\right]}$$

If two flow passages are identical and each carries a different gas at the same pressure conditions, the ratio of flow between the two gases depends only on the thermodynamic properties. For air (subscript A) and any gas (subscript G) the ratio of weight flows is expressed as follows:

$$W_A = W_G \frac{Z_A}{Z_G}\sqrt{\frac{R_G T_G}{R_A T_A}}$$

where:

$$Z = \left\{\frac{\gamma}{\gamma-1}\left(\frac{P_D}{P_U}\right)^{\frac{1}{\gamma}}\left[\left(\frac{P_D}{P_U}\right)^{\frac{1}{\gamma}} - \frac{P_D}{P_U}\right]\right\}^{1/2}$$

The coefficient Z combines the specific-heat constants and pressure ratio; values of Z vs P_D/P_U are plotted in Fig 3.

Weight flow must be converted to volume flow in order to use the flow nomograph, Fig 5; this is done by in-

4..Airflow in scfm

Airflow in scfm can be read directly if you know actual airflow in cfm and actual pressure and temperature. For example, convert a flow of 200 cfm at 800 F and 500 psia to scfm (standard cfm at 14.7 psia and 70 F): first locate 800 F on temperature scale and 200 on the cfm-scfm scale; draw line through both points to pivot line; from pivot line draw line connecting 500 psia on the pressure scale; and where this line intersects the cfm-scfm scale is the converted flow, 2700 scfm.

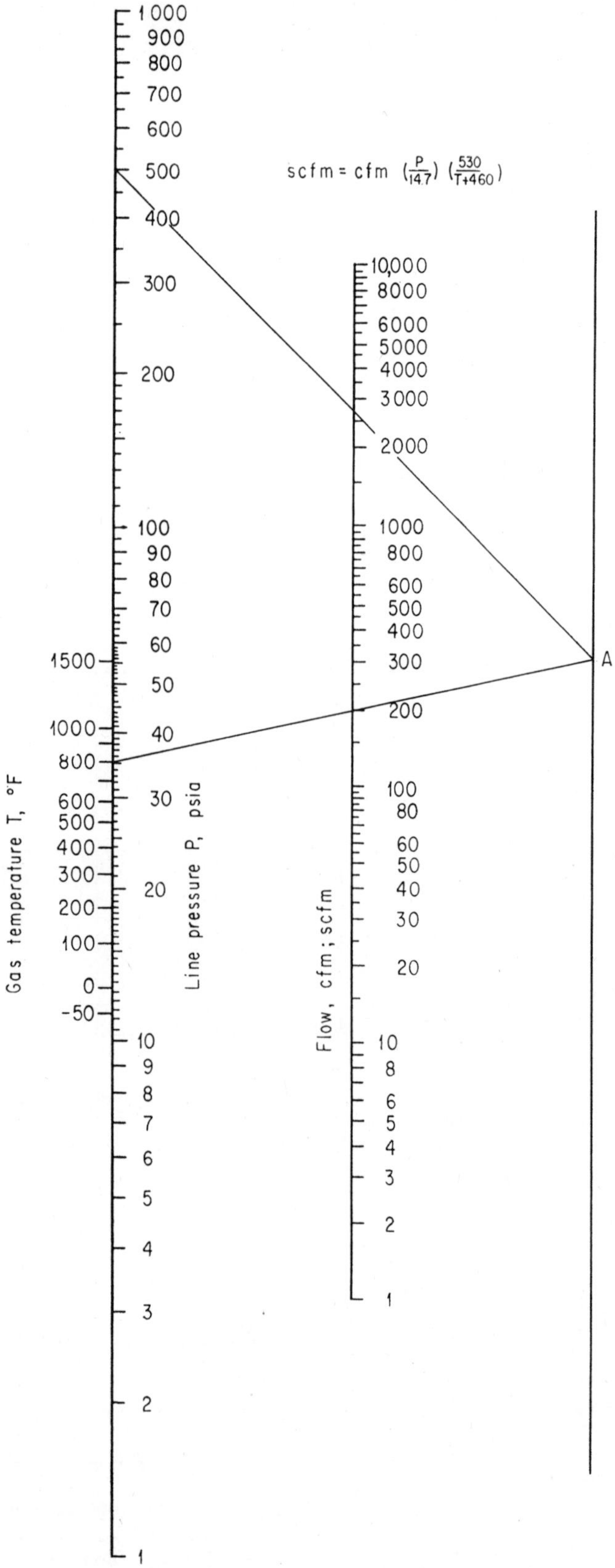

troducing gas-to-air density ratio (ρ_G/ρ_A), measured at standard conditions. The expression for airflow in scfm becomes:

$$Q_A = \frac{1}{168.1} Q_G \left(\frac{\rho_G}{\rho_A}\right)_{Std} \left(\frac{Z_A}{Z_G}\right) \sqrt{R_G T_G}$$

Solve for Q_A in the above equation and substitute its value in the flow nomograph. Constants R, γ and ρ_G/ρ_A for various gases at standard conditions are listed in Table IV. See sample problems at the end of the article for details of this computation.

When air is the fluid, the conversion from actual flow to standard flow can be made quickly with Fig 4. No need to solve the gas-to-air equation.

Solve the flow problem

The final nomograph (Fig 5) will solve any one of these three major flow problems: pressure drop when inlet pressure, flow, and flow factor are known; flow when pressures and flow factor are known; and required flow factor when flow and pressures are specified.

The nomograph solves the expressions:

$$Q_U/F = \sqrt{(8/5)}\sqrt{[r(1-r)(3-r)]}$$

where $r = P_D/P_U$ in absolute units; and $Q_A = F\,P_U(Q_U/F)$ which is based on the NBS definition for F. The two scales for P_U on the nomograph are necessary because the two expressions have unlike nomographic characteristics. In addition, one of the two scales is in psig and the other is psia to help avoid errors in choosing the right scale. Note that this expression for Q_U/F differs from the simpler NBS relationship given before: it is the mean between ideal isentropic expansion and the basic NBS equation which assumes high stagnation loss.

The best way to explain the nomograph is to work out examples. The

(text continued, p. 110)

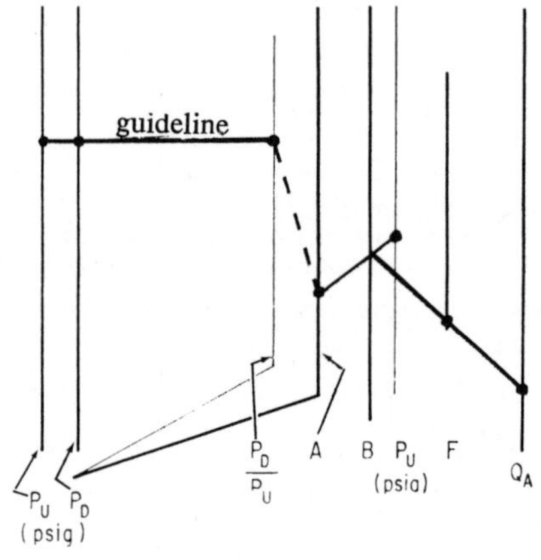

5..Flow, flow factor, pressure drop

Flow, flow factor, and pressure drop can be found on this nomograph by following the examples on the following page. Note that flow must be in units of scfm, flow factor in NBS *F*-units, and pressure drop in psi.

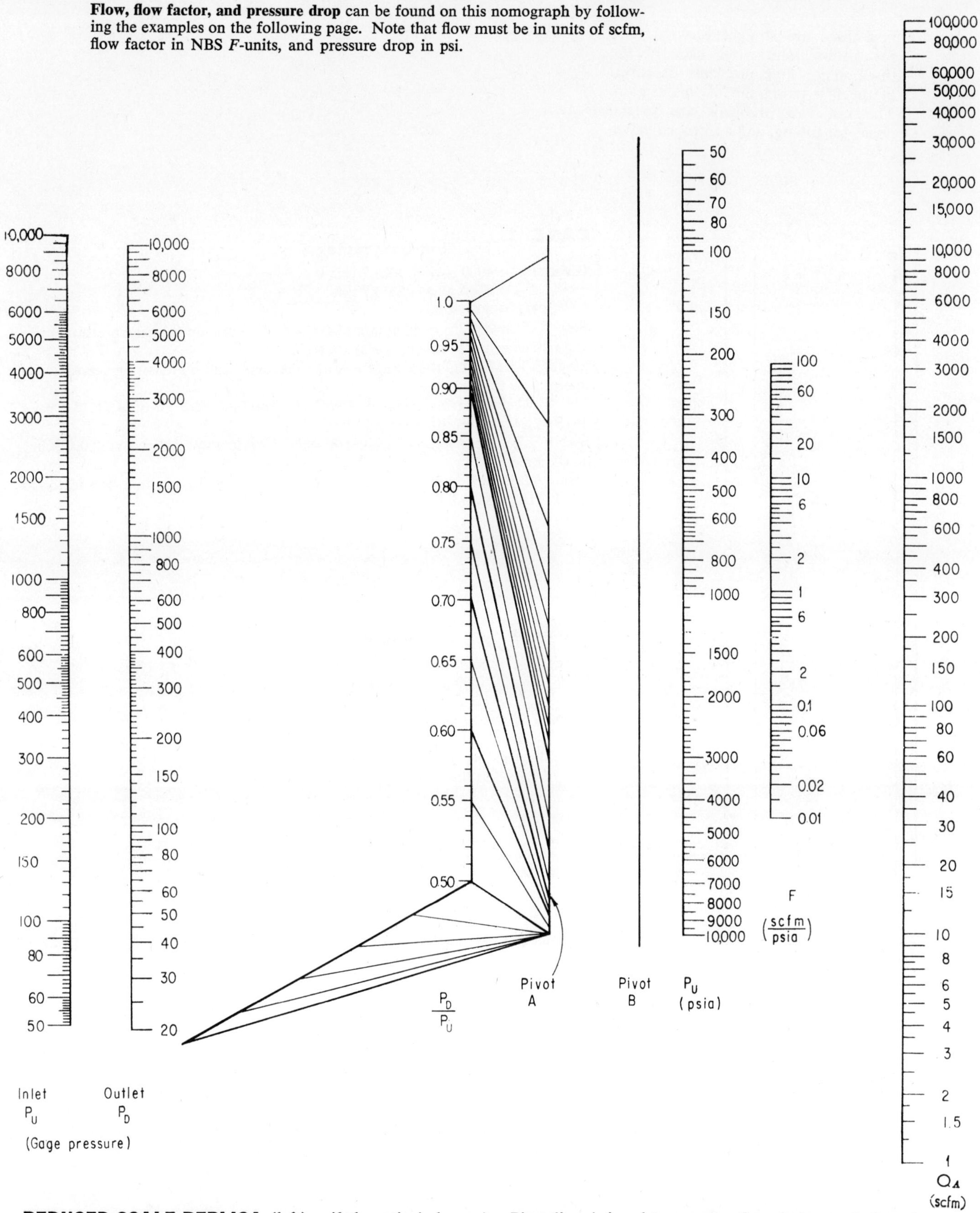

REDUCED-SCALE REPLICA (left) of the flow nomograph above shows how it works. The three leftmost scales (guideline) simply solve for pressure ratio (or for the unknown pressure if the ratio is known). Pivot line A is there to transfer the value for pressure ratio to a different logarithmic scale for the right half of the nomograph. The five right scales relate pressure ratio, inlet pressure, flow factor, and flow in standard cfm: if any three are known, the fourth can be found. See examples of the application of this nomograph to five realistic problems on next page.

first three are straight forward ones for airflow, and cover each of the three major flow problems described in the first paragraph of this section. The last two problems are special ones: for tubing, and a nitrogen valve.

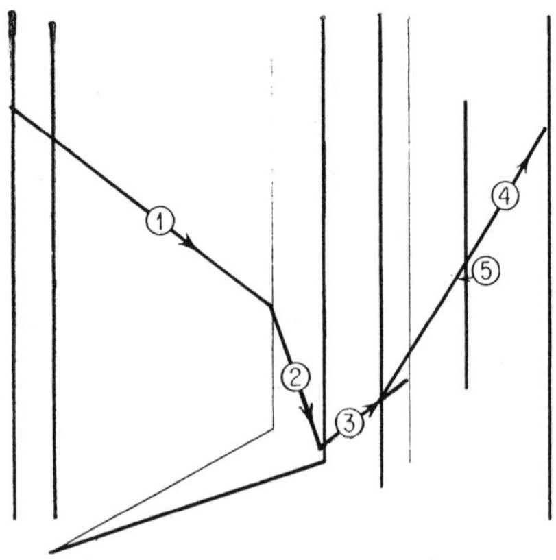

CASE 1 . . find flow factor F

Given: P_U = 4000 psig = 4014.7 psia
P_D = 2500 psig = 2514.7 psia
Q_A = 2000 scfm

Step 1. . Connect P_U = 4000 psig and P_D = 2500 psig on the gage pressure scales and extend a line to intersect P_D/P_U scale.

Step 2. . Transfer point on P_D/P_U scale to pivot A scale by following nearest guide line.

Step 3. . Connect point on pivot A scale to absolute inlet pressure (4014.7 psia) on P_U (psia) scale with a straight line.

Step 4. . From intersection on pivot B scale, draw line to connect flow (2000 scfm) on Q_A scale.

Step 5. . Intersection of line at 0.5 on F scale is then the required flow factor to satisfy the given conditions.

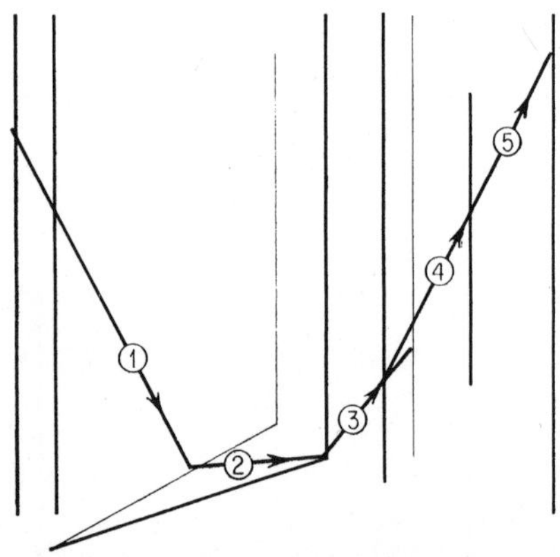

CASE 2 . . find flow Q_A

Given: P_U = 3000 psig = 3014.7 psia
P_D = 1000 psig = 1014.7 psia
F = 2.0 scfm/psia

Steps 1, 2, and 3: identical to those of Case I solution.

Step 4. . From the intersection on pivot B scale, draw a line to connect flow factor (2.0) on the F scale.

Step 5. . Extend line to cross Q_A scale at 6250 scfm.

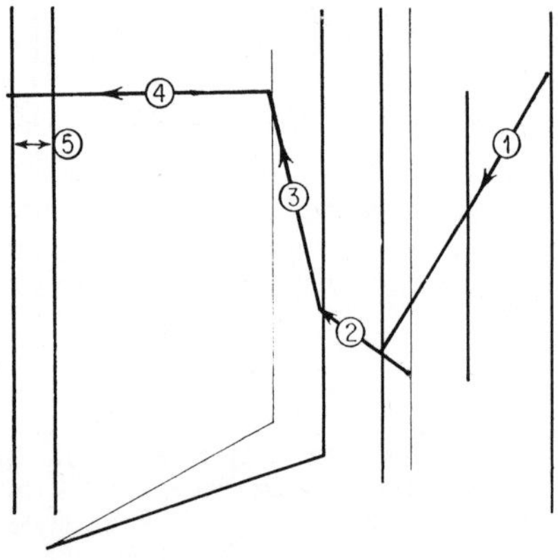

CASE 3 . . find pressure drop $P_U - P_D$

Given: P_U = 4000 psig = 4014.7 psia
Q = 4000 scfm
F = 2.0 scfm/psia

Step 1. . Connect 4000 on Q_A scale and 2.0 on the F scale. Extend line to intersect pivot B scale.

Step 2. . Connect point on pivot B scale to 4014.7 psia on P_U (psia) scale.

Step 3. . Transfer point to P_D/P_U scale in a manner similar to that outlined in **Step 2**. . Case I solution,

Step 4. . Connect this point on the P_D/P_U scale to 4000 psig on the P_U (psig) scale.

Step 5. . The pressure drop is the difference $P_U - P_D$ or 300 psi.

CASE 4 . . Find pressure drop in tubing

Given: $L = 10$ ft (120 in.)

$d = 0.391$ in. ($\frac{1}{2}$ in. nominal).

$A = \pi d^2/4 = 0.12$ sq in.

$f = 0.02$ (typical for AN-type tubing)

$P_U = 200$ psig (214.7 psia)

$Q = 5.0$ cfm at 120 F and 200 psig

Step 1. . Calculate equivalent flow factor F from table II.

$F = 23.5A\sqrt{(d/fL)} = 23.5 \times 0.12\sqrt{[0.391/(0.02 \times 120)]} = 1.14$ scfm/psia

Step 2. . Solve for scfm with fig 4. $Q_A = 70.0$ scfm.

Step 3. . On flow nomograph (fig 5) connect 70 scfm to 1.14 on F scale to intersect pivot B scale. Then join 214.7 psia on P_U absolute scale with point on pivot B scale to intersect pivot A scale. Transfer this point up to P_D/P_U scale and connect to P_U gage scale. Read P_D scale intersection at 195 psig. Pressure drop is 200 − 195 = 5 psi.

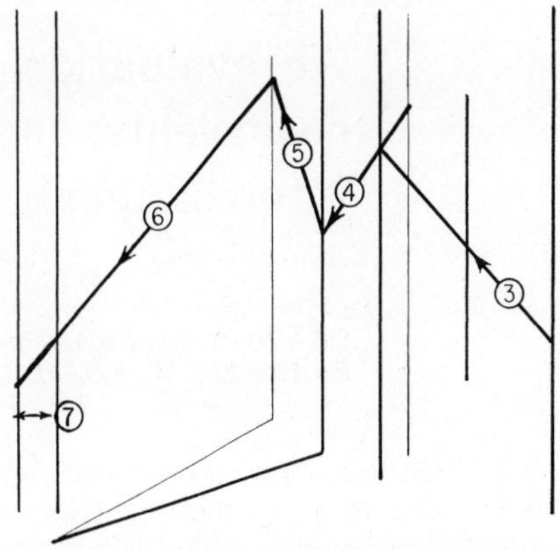

CASE 5 . . nitrogen, special flow constant

Given: $C_V = 0.02$

$T_G = 500\ F = 960\ R$

$P_U = 1000$ psig $= 1014.7$ psia

$W_G = 0.01$ lb/sec nitrogen flow rate

Step 1. . From the C_V-D_O-F conversion chart, a valve with a C_V of 0.02 has an equivalent flow factor of 0.01125 scfm/psia.

Step 2. . Nitrogen weight flow must be converted into air flow (weight units) at standard temperature (530 R) by using the formula:

$$W_A = W_G \frac{Z_A}{Z_G} \sqrt{\frac{R_G T_G}{R_A T_A}}$$

The ratio Z_A/Z_G is practically unity throughout the whole pressure range because γ_G is nearly equal to γ_A. This is fortunate, because for gases having unlike γ's the problem becomes complex and unless more information is known a solution must be found by trial and error. You would use the airflow chart and work from sonic velocity (maximum pressure drop) until a minimum pressure drop was found.

For nitrogen, it's easy: the conversion formula reduces to:

$$W_A = W_G \sqrt{\frac{R_G T_G}{R_A T_A}}$$

where

$R_G = 55.16$ for nitrogen

$R_A = 53.36$ for air

$T_A = 530$ R for air at standard temperature

then

$$W_A = 0.010 \sqrt{\frac{55.16 \times 960.0}{53.36 \times 530.0}}$$

$$W_A = 0.01369 \text{ lb/sec equivalent air flow}$$

Step 3. . Convert airflow W into standard air volume flow (scfm units) with the equation $Q_A = 800W_A$ where the constant 800 is $60/\rho_o$ and ρ_o is air density at standard conditions. $Q_A = 800 \times 0.01369 = 10.95$ scfm (airflo

Step 4. . Follow Case III to solve the airflow problem. $\Delta P = 240$ psi.

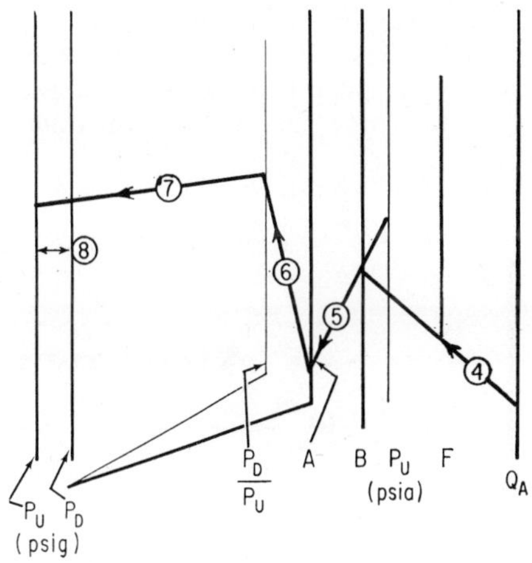

Experiments verify

Which air-flow equation

An evaluation of air-flow equations and their coefficients, and a simple way to make them interchangeable in system calculations.

DOMINIC LAPERA, Chief Engineer, Kemp Aero Products, West Milford, NJ;
FRANKLIN D. YEAPLE, Associate Editor

HAVE you ever picked one air valve in preference to another simply because the catalog flow equation for calculating pressure drop was easier to apply? If so, you've got lots of company—particularly if you are designing systems with components from many manufacturers.

There are at least ten different flow equations published today and when you are forced to use more than one of them in a single set of flow calculations it's tricky.

Now there is an easier way. Using the tables in this article you can convert quickly any equation or flow coefficient into a single consistent expression. You can apply the results immediately and repeatedly in a pneumatic circuit of any complexity.

But remember that no coefficient can be better than its defining equation, which is an approximation at best. And converting a poorly defined coefficient to one that is more acceptable does not necessarily correct inaccuracies. Relative accuracies are discussed in detail.

Converting to one coefficient

For this article the National Bureau of Standards flow factor F is the

(text continued, p. 114)

Air-flow tests

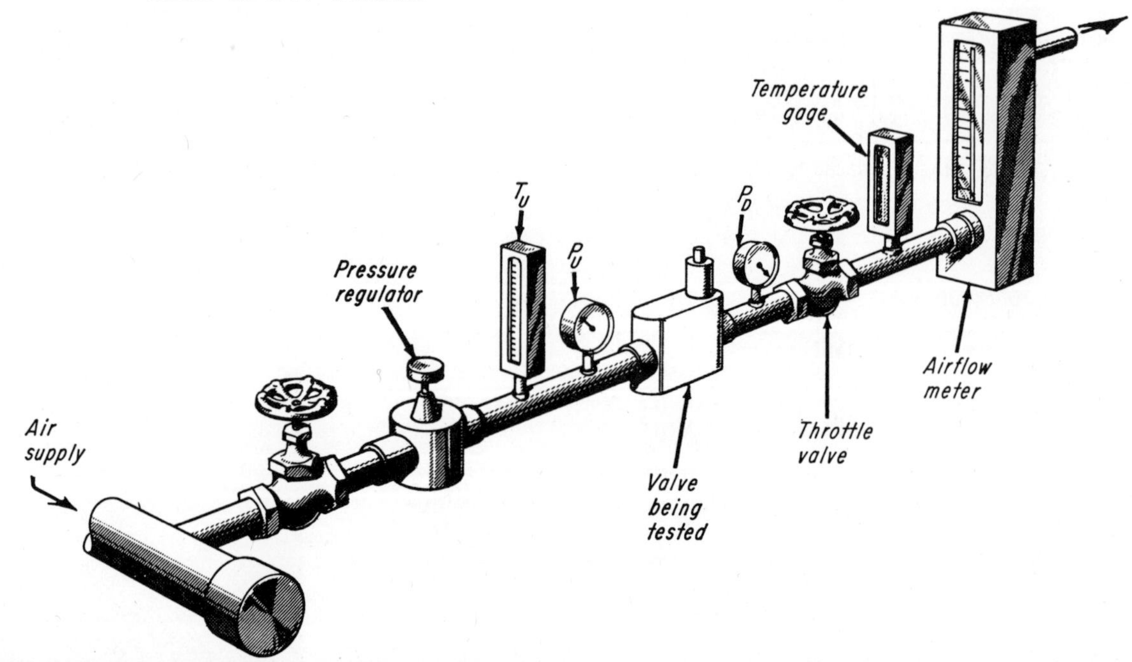

System with several flow coefficients

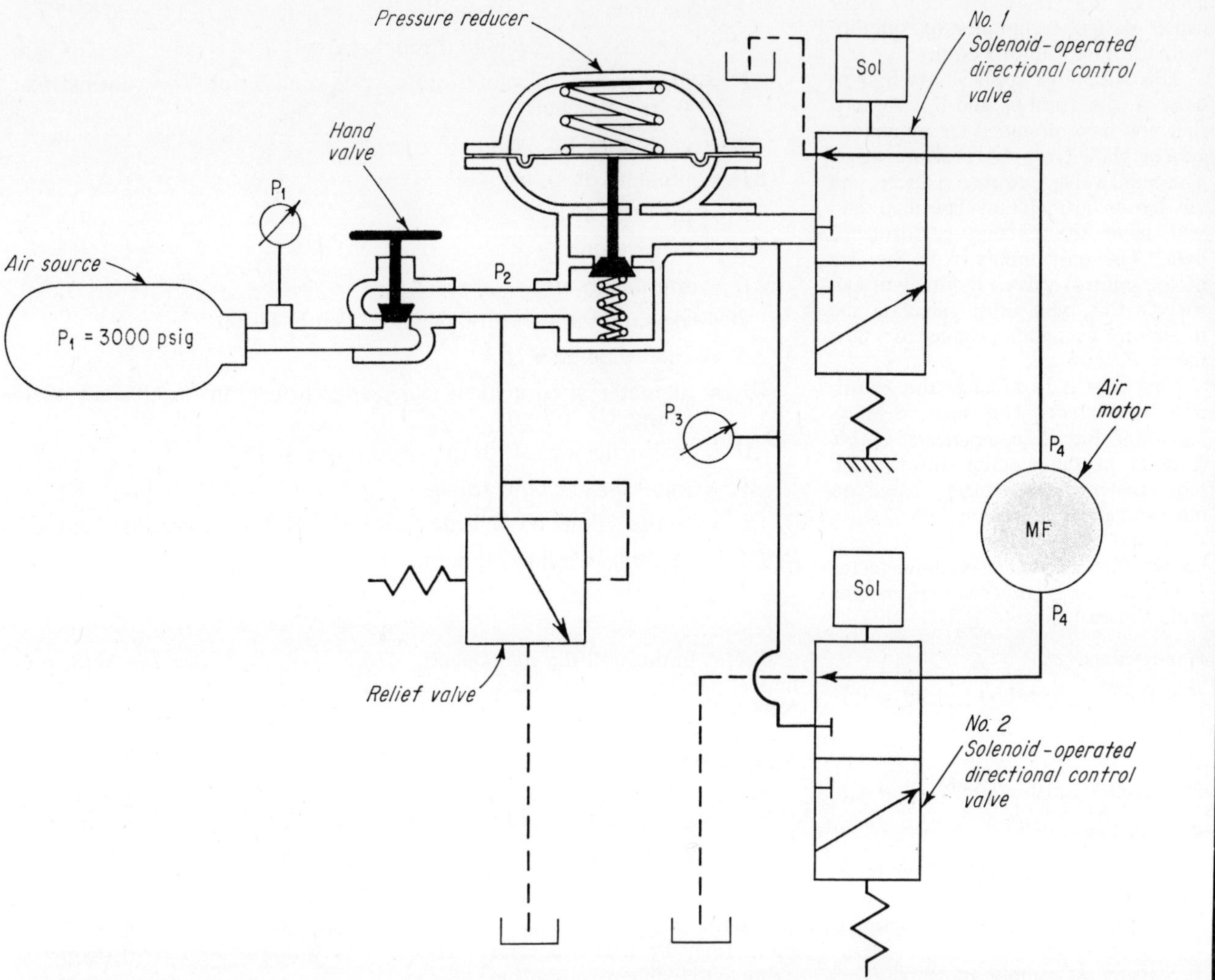

Catalog information

Component	Flow coefficient	Defining equation
Hand valve	$C_V = 1.26$	$C_V = \frac{Q \times 60}{1360} \sqrt{\frac{GT_U}{\Delta P \times P_U}}$
Pressure reducer	$D_O = 0.25$	$D_O = \sqrt{\frac{Q}{33P_U}} \times \frac{1}{\sqrt{r(r^{0.43} - r^{0.71})}}$
Control valve	$F =$ unknown	$F = \frac{Q}{P_U \sqrt{8/5}} \sqrt{\frac{1}{r(1-r)(3-r)}}$
Air motor	$Q = 250$ scfm; $P = 600$ psia (given)	

standard flow coefficient — although any other will work. The equations based on flow factor F are the most accurate, and there are many published design techniques to simplify their use (see Editor's Note).

This simple example shows how to convert to F from C_V and D_O: Assume that you have designed the pneumatic system shown on the previous page. The hand valve, pressure reducer, and air motor are already specified and you have the catalog performance data. The only unknown is the size of the control valve. Before you can specify this, you must calculate the minimum acceptable value for flow factor F.

First step is to tabulate the known flow coefficients and their defining equations for all components. (This is done in the Catalog Information table under the schematic, based on the typical air-flow equations.)

Second step is to convert each known flow coefficient to flow factor F, using the Conversion Equations table shown below.

Symbols

Q = air flow in standard units, scfm (14.7 psia, 68 F)

q = air flow at actual conditions, cfm. $Q = q(P/14.7)(528/T)$

V = velocity, fps (average through valve)

P = pressure in absolute units, psia (subscript D = downstream, U = upstream)

p = gage pressure, psi

ΔP = pressure drop, psi

r = pressure ratio P_D/P_U

ρ = density, lb/ft³

G = specific gravity, ρ_{gas}/ρ_{air}

T = absolute temperature, deg R = deg F + 460

A = Inlet pipe area, in.²

D_o = diameter of equivalent sharp-edge orifice, in. (coefficient of discharge $C_D = 0.6$)

M = molecular weight, lb (M = 29 lb for air)

W = water weight flow, lb/sec

C_v, K, F, D_o = typical flow coefficients (also called flow constants and flow factors) in a flow equation

Hand valve:

$$F = 0.556C_V = 0.556 \times 1.26 = 0.7$$

Reducer:

$$F = 10D_O^2 = 10 \times (0.25)^2 = 0.625$$

Control valve:

Calculate control-valve flow factor F from the known flow and pressure at P_4 and the calculated pressure at P_3. Here is how.

Starting at supply pressure P_1 = 3014.7 psia, and knowing that flow is 250 scfm through all the components, find P_2 and P_3 by substitution in the NBS flow equation (Eq 9 in table):

Hand valve:

$$F = \frac{Q}{P_U\sqrt{8/5}}\sqrt{\frac{1}{r(1-r)(3-r)}}$$

where $P_U = P_1$ = 3014.7 psia, Q = 250 scfm, F = 0.7, and $r = P_2/P_1$. Solve for r by trial and error. This abbreviated table of values for r (1-r) (3-r) will help:

r	$r(1-r)(3-r)$
0.5	0.625
0.7	0.483
0.8	0.352
0.9	0.189
0.92	0.153
0.94	0.116
0.96	0.0784
0.98	0.0396
0.99	0.0199
0.995	0.0100
1.000	0.0000

(text continued, p. 116)

Conversion equations

		D_O	C_V	F	K: $r = 1.0$	K: $r = 0.75$	K: $r = 0.5$
D_O			$= 0.236\sqrt{C_V}$	$= 0.316\sqrt{F}$	$= 1.456\frac{\sqrt{A}}{K^{1/4}}$	$= 1.521\frac{\sqrt{A}}{K^{1/4}}$	$= 1.641\frac{\sqrt{A}}{K^{1/4}}$
C_V		$= 18.0D_O^2$		$= 1.8F$	$= 38.2\frac{A}{\sqrt{K}}$	$= 41.5\frac{A}{\sqrt{K}}$	$= 48.3\frac{A}{\sqrt{K}}$
F		$=10D_O^2$	$=0.556C_V$		$= 21.2\frac{A}{\sqrt{K}}$	$= 23.1\frac{A}{\sqrt{K}}$	$= 26.9\frac{A}{\sqrt{K}}$
K	$r =$ 1.0	$= 4.5\frac{A^2}{D_O^4}$	$= 1460\frac{A^2}{C_V^2}$	$= 450\frac{A^2}{F^2}$	NOTE: The K factor varies with r and A and you must know which values the manufacturer used to derive his published K. For example, if K was derived at $r = 0.75$ and valve inlet pipe area $A = 0.2$, then $F = 23.1 \times 0.2/\sqrt{K} = 4.62/\sqrt{K}$.		
	$r =$ 0.75	$= 5.36\frac{A^2}{D_O^4}$	$= 1725\frac{A^2}{C_V^2}$	$= 534\frac{A^2}{F^2}$			
	$r =$ 0.5	$= 7.29\frac{A^2}{D_O^4}$	$= 2330\frac{A^2}{C_V^2}$	$= 724\frac{A^2}{F^2}$			

Typical air-flow equations

(taken from catalogs)

Eq No	Equation for flow (sub-critical) (Q = *scfm-standard ft³/min*)	Flow coefficient defined	
1	$Q = 33\,D_o^2 P_U \sqrt{r\,(r^{0.43} - r^{0.71})}$	D_o = equivalent sharp-edged orifice (coeff of discharge $C_D = 0.6$)	$= \sqrt{\dfrac{Q}{33\,P_U} \times \dfrac{1}{\sqrt{r(r^{0.43} - r^{0.71})}}}$
2	$Q = \dfrac{963}{60}\,C_V \sqrt{\dfrac{\Delta P\,(P_U + P_D)}{GT_U}}$	C_V = valve coefficient	$= \dfrac{Q \times 60}{963} \sqrt{\dfrac{GT_U}{\Delta P\,(P_U + P_D)}}$
3	$Q = \dfrac{1360}{60}\,C_V \sqrt{\dfrac{\Delta P \times P_U}{GT_U}}$	C_V = flow coefficient	$= \dfrac{Q \times 60}{1360} \sqrt{\dfrac{GT_U}{\Delta P \times P_U}}$
4	$Q = \dfrac{1390}{60}\,C_V \sqrt{\dfrac{\Delta P \times P_D}{GT_U}}$	C_V = capacity factor	$= \dfrac{Q \times 60}{1390} \sqrt{\dfrac{GT_U}{\Delta P \times P_D}}$
5	$Q = \dfrac{5180}{60}\,C_V \sqrt{\dfrac{P_U^2 - P_D^2}{MT_U}}$	C_V = valve-flow coefficient	$= \dfrac{Q \times 60}{5180} \sqrt{\dfrac{MT_U}{P_U^2 - P_D^2}}$
6	$Q = \dfrac{963}{60}\,C_V \sqrt{\dfrac{P_U^2 - P_D^2}{GT_U}}$	C_V = flow coefficient (Eq 2 = Eq 6)	$= \dfrac{Q \times 60}{963} \sqrt{\dfrac{GT_U}{P_U^2 - P_D^2}}$
7	$Q = \dfrac{2.32^{0.443}}{60}\,C_G \dfrac{\Delta P^{0.443} \times P_U^{0.6}}{\sqrt{GT_U / 520}}$	C_G = gas-flow coefficient	$= \dfrac{Q \times 60}{(2.32)^{0.443}} \times \dfrac{\sqrt{GT_U / 520}}{\Delta P^{0.443} \times P_U^{0.6}}$
8	$Q = FP_U \sqrt{4/3}\,\sqrt{1 - r^2}$	F = NBS flow factor (#1)	$= \dfrac{Q}{P_U \sqrt{4/3}} \sqrt{\dfrac{1}{1 - r^2}}$
9	$Q = FP_U \sqrt{8/5}\,\sqrt{r(1 - r)(3 - r)}$	F = NBS flow factor (#2—better)	$= \dfrac{Q}{P_U \sqrt{8/5}} \sqrt{\dfrac{1}{r(1 - r)(3 - r)}}$
10	$Q = 38.1\,P_U A \sqrt{\dfrac{1 - r}{K}}$	K = K factor	$= \dfrac{2g}{\rho V^2}\,\Delta P$

In the example, r turns out to be 0.995. Therefore $P_2 = 0.995 \times 3014.7 = 3000$ psia, and ΔP = about 15 psi.

Reducer:

Determine P_3 in the same way. Answer: $r = 0.994$, $P_3 = 0.994 \times 3000 = 2982$ psia, and ΔP = about 18 psi.

Control valve:

Compute minimum flow factor F for the control valve:

$$F = \frac{Q}{P_U\sqrt{8/5}}\sqrt{\frac{1}{r(1-r)(3-r)}}$$

where Q = 250 scfm, $P_U = P_3$ = 2982 psia, and $r = P_4/P_3 = 600/2982 = 0.201$.

Note: $r = 0.201$ is less than critical flow ratio $r = 0.5$, therefore F by definition is $Q/P_U = 250/2982 = 0.084$. (With a flow-factor nomograph you can read F directly without calculation.)

A valve manufacturer will accept a flow coefficient such as F as a specification for minimum flow because it completely defines flow and pressure drop.

The relief valve ($A/\sqrt{K} = 0.021$) is not part of the flow path, but you can convert its flow coefficient to F and easily compute its relieving capacity. Pressure ratio r is less than 0.5 because it discharges to atmosphere. Therefore the conversion equation is

$$F = 26.9\frac{A}{\sqrt{K}} = 0.565$$

Maximum flow is then known to be $Q = FP_U = 0.565 \times 3000 = 1695$ scfm—more than adequate.

Errors in flow coefficients

The major source of potential error is in the technique used to develop the original flow coefficients—C_V, D_O, F, or K. If each valve could be tested with air at the normal operating conditions, any type of flow coefficient based on these tests would accurately depict performance. (Accurate in this context means ±10 or 20%, depending on the constant, because even the gages cannot be read consistently closer than 5%.)

But most manufacturers find it uneconomical to air-test each valve, and they fall back on one of these two simpler techniques:

- No test at all. If the valve or component is very similar to another valve for which flow data are available, some manufacturers will assign a flow coefficient by inspection or calculation, based on port size or flow area.
- Water test. A dependable way used by most manufacturers to determine effective flow area of an irregular passage is to run a water flow test and convert the results to a hypothetical diameter of an equivalent sharp-edged orifices, D_o. (Oil is also used, but water is less viscous and behaves more as an ideal fluid). Once you know D_O you can apply all air-flow formulas based on sharp-edge orifices. However, no valve restriction is exactly equivalent to a sharp-edge orifice and errors can result.

Coefficients C_V and D_O

Coefficients C_V and D_O can be derived directly from the water flow test. A valve has a $C_V = 1.0$ if it will pass one gpm of water with a pressure drop of 1 psi. The relationship is gpm $= C_V\,(\Delta P)^{1/2}$ and is valid at all pressures and flows—for water.

Equivalent orifice D_O is simply part of the statement of Bernoulli's conservation of energy theorem for water: lb/sec $= 2.5D^2_o\,(\Delta P)^{1/2}$ with the coefficient of discharge = 0.6.

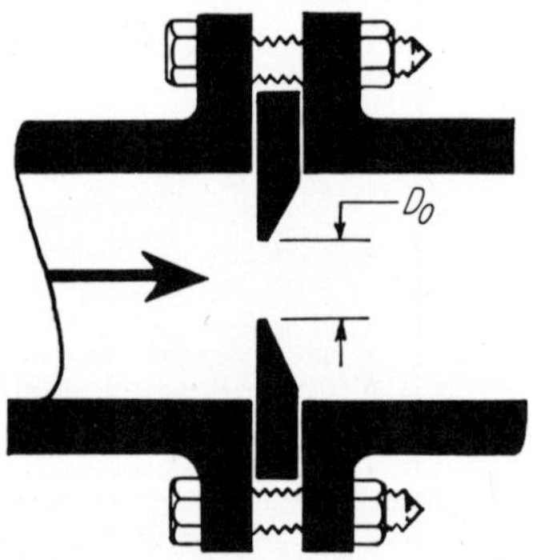

Once the value of C_V or D_O has been calculated, the water flow equation is not used again because you need compressible-flow equations for air valves. This is true even though the coefficients are derived with water or oil. The commercial air-flow equations given in the typical air-flow equations tables were developed by test and theory. Assuming that the coefficients C_V or D_o represent a physical opening or restriction (as they already do by definition), the performance of such openings in air is predictable.

Flow factor F

Flow factor F normally is not calculated from a water-flow test; it is based on air-test data. However, a surprisingly accurate value for F can be derived if the equivalent orifice diameter D_O is known. The converting is done by combining Eq 9 (for F) and Eq 1 (for D_o) from Typical Air-Flow Equations table on page 115, and expressing F in terms of D_O:

$$F = 26.1D_o^2\sqrt{\frac{r^{0.43} - r^{0.71}}{(1-r)(3-r)}}$$

This complex form has very little practical value, but fortunately the total value of the terms under the square root is fairly constant for a wide range of r. (For $r = 1$, $F = 9.99D_o^2$; for $r = 0.9$, $F = 9.5D_o^2$; for $r = 0.5$, $F = 8.45\,D_o^2$.) Tests show that a good workable equation is $F = 10\,D_o^2$.

K factor

The K factor normally is based on air tests. In theory K can be converted mathematically to any other coefficient, but the variables don't cancel out as nicely as they do for D_o or C_V, and the conversion "constant" doesn't stay constant.

Take for example the conversion from K to F: if you combine Eq 10 (for K factor) and 9 (flow factor F), here is the resulting expression for F in terms of K:

$$F = \frac{A}{0.0332\sqrt{K[r(3-r)]}}$$

The relationship depends on inlet pipe area A and pressure ratio r and is therefore impractical for rapid conversions. However, if you know what the range of r will be, you can assign each range an approximate conversion constant.

Conversions from K to C_V and D_o are similarly dependent on pressure ratio and the same approximations can be applied. The Conversion Equations table lists the calculated conversion constants for r = 0.5, 0.75 and 1.0.

Actual lab tests

Each air-flow coefficient and its defining equations were checked out with air-flow tests on several valves: ¼-in. two-way solenoid-operated shut-off valve; ⅜-in. two-way pressure-operated shut-off valve; and ¾-in. hand valve. Graphs on next page are for the ⅜-in. (tube size) valve but very similar data were obtained on the other two valves.

To be fair, two methods were used to calculate and plot the equations:

- Purely mathematical, based on a value of D_O derived from water tests. (The coefficients were computed from the Conversion Equations table without reference to actual air-flow test data.)
- Purely empirical, based on the best average plot of actual airflow. (The coefficients were computed from ac-

Actual vs calculated air flow

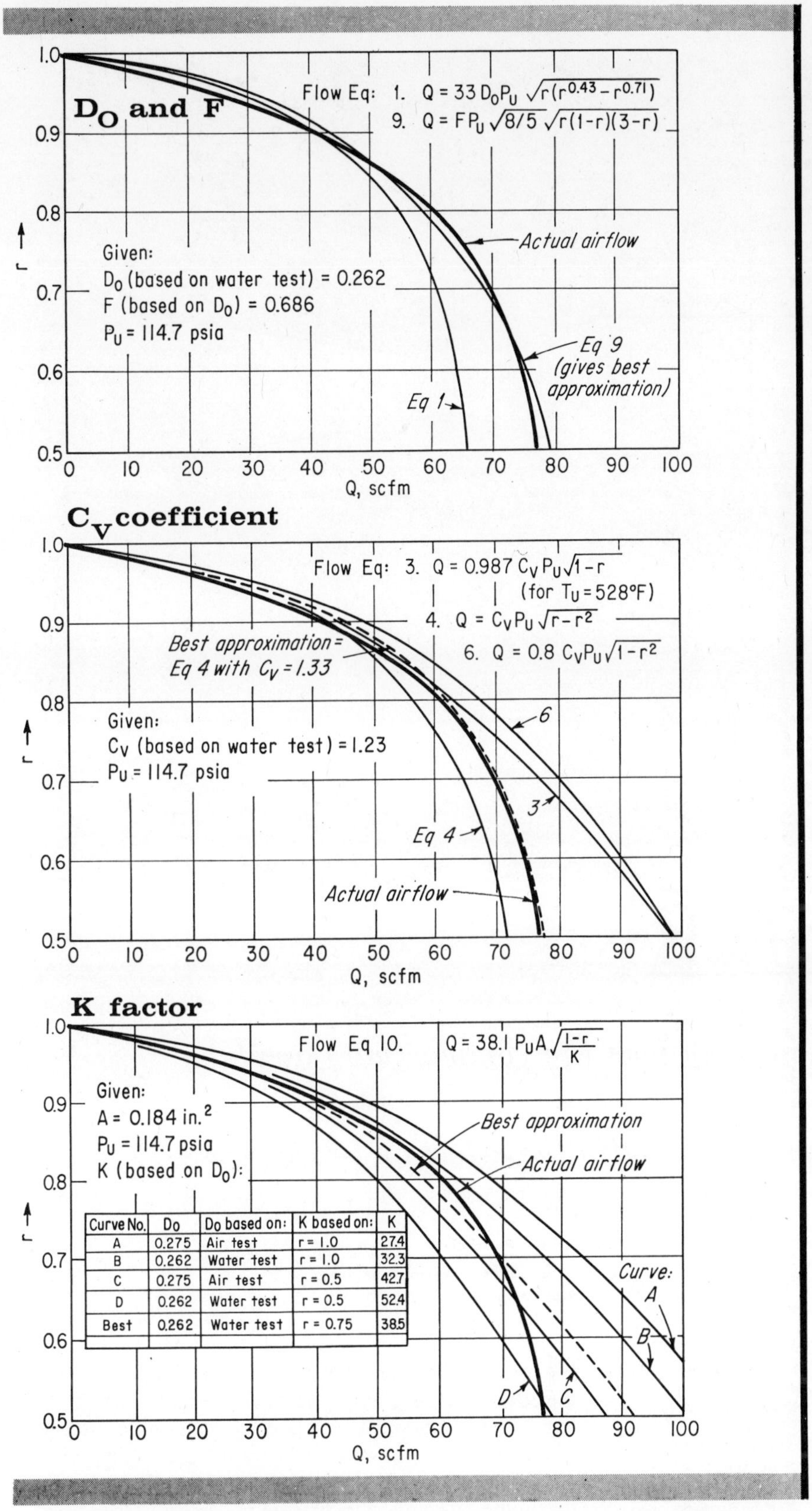

Curve No.	Do	Do based on:	K based on:	K
A	0.275	Air test	r = 1.0	27.4
B	0.262	Water test	r = 1.0	32.3
C	0.275	Air test	r = 0.5	42.7
D	0.262	Water test	r = 0.5	52.4
Best	0.262	Water test	r = 0.75	38.5

ACTUAL AIR-FLOW TESTS on a typical ⅜-in. valve were run with the test setup shown on first page of article. The colored curve (same for each graph) shows actual performance. The black curves are calculated: solid where the air-flow computation was purely mathematical (based on a flow coefficient derived from a water test alone), dashed where the coefficient was derived empirically to match actual air-flow data. (Tests on ¼-in. and ¾-in. valves gave similar results.)

tual test points, referring to the Typical Air-Flow Equations table. No conversion equation was needed.)

Relative error is measured by the divergence of calculated data from actual test data. It can be seen that the NBS flow factor F gives consistent results even if D_o from a water test is the only quantity known. The other coefficients can also be made to work but must be more carefully applied to avoid conversion errors. Any coefficient can be made accurate if the air-flow test is run first and the coefficient calculated afterwards.

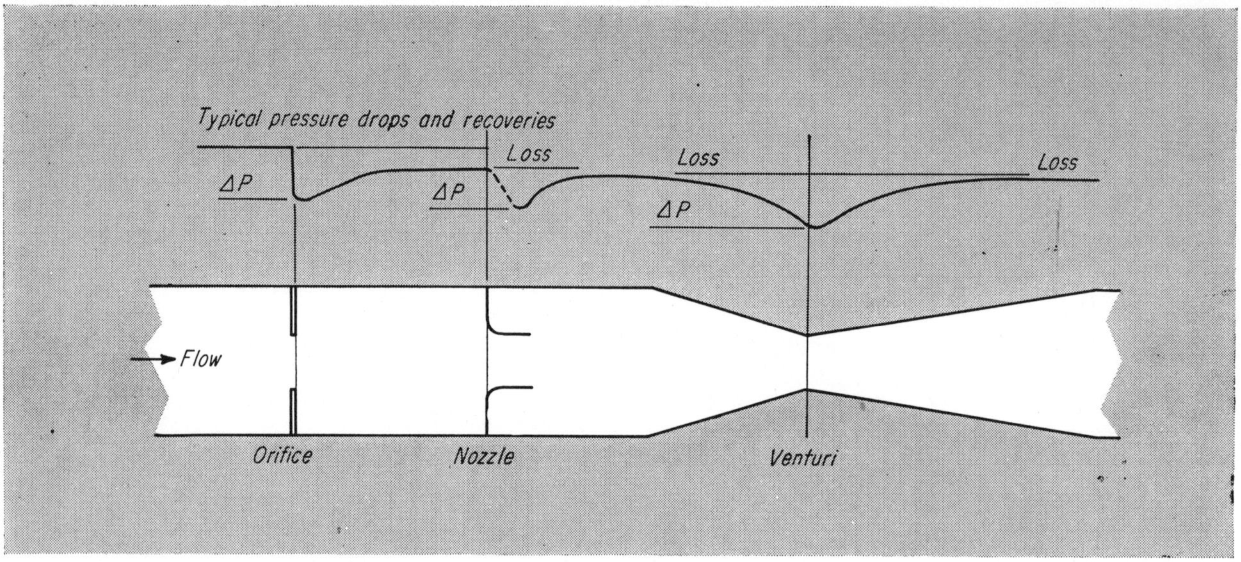

HOW TO SELECT A FLOWMETER

This up-to-date roundup not only lists the commercially available flowmeters, plus distinctive special types—it also shows how two steps will simplify the selection procedure

DR C C MIESSE, *senior scientist*
O E CURTH, *research engineer*
Armour Research Foundation

ABOUT THE AUTHORS

As supervisor of the Fluid Mechanics and Combustion Research Section, Armour Research Foundation, since 1956, Dr Miesse has directed research in flow measurement, seal development, cycle analysis, liquid-propellant rockets, jet engines, marine boilers, and automobile afterburners. Previously, he was research engineer at Battelle Memorial Institute and later a physicist for Aerojet-General Corp—where he analyzed unstable burning, atomization and combustion of rocket fuels, and vibration of missile components.

Mr Curth researches problems of fluid flow and controls, and has investigated flowmeters, servo systems, high-temperature high-pressure seals, refrigeration, compressors, valves, 3-dimensional flow sensors, and engines. Before joining Armour, he spent five years under Office of Naval Research sponsorship, investigating vibration and sound-transmission problems for Illinois Institute of Technology.

Here is a familiar situation—you need a flowmeter to measure fluid flow, and it must be designed into the system. One important factor that will determine the choice of meter is—do you want it to measure gas or liquid? The other factors are flow rate, accuracy, temperature range, maximum pressure, and maximum viscosity.

Special considerations also enter the picture: for instance, if you must measure laminar flow in a relatively large area such as a wind tunnel (where you must traverse the area and calculate an average flow) then only a pitot tube will do. Or if you are measuring fuel and need mass-

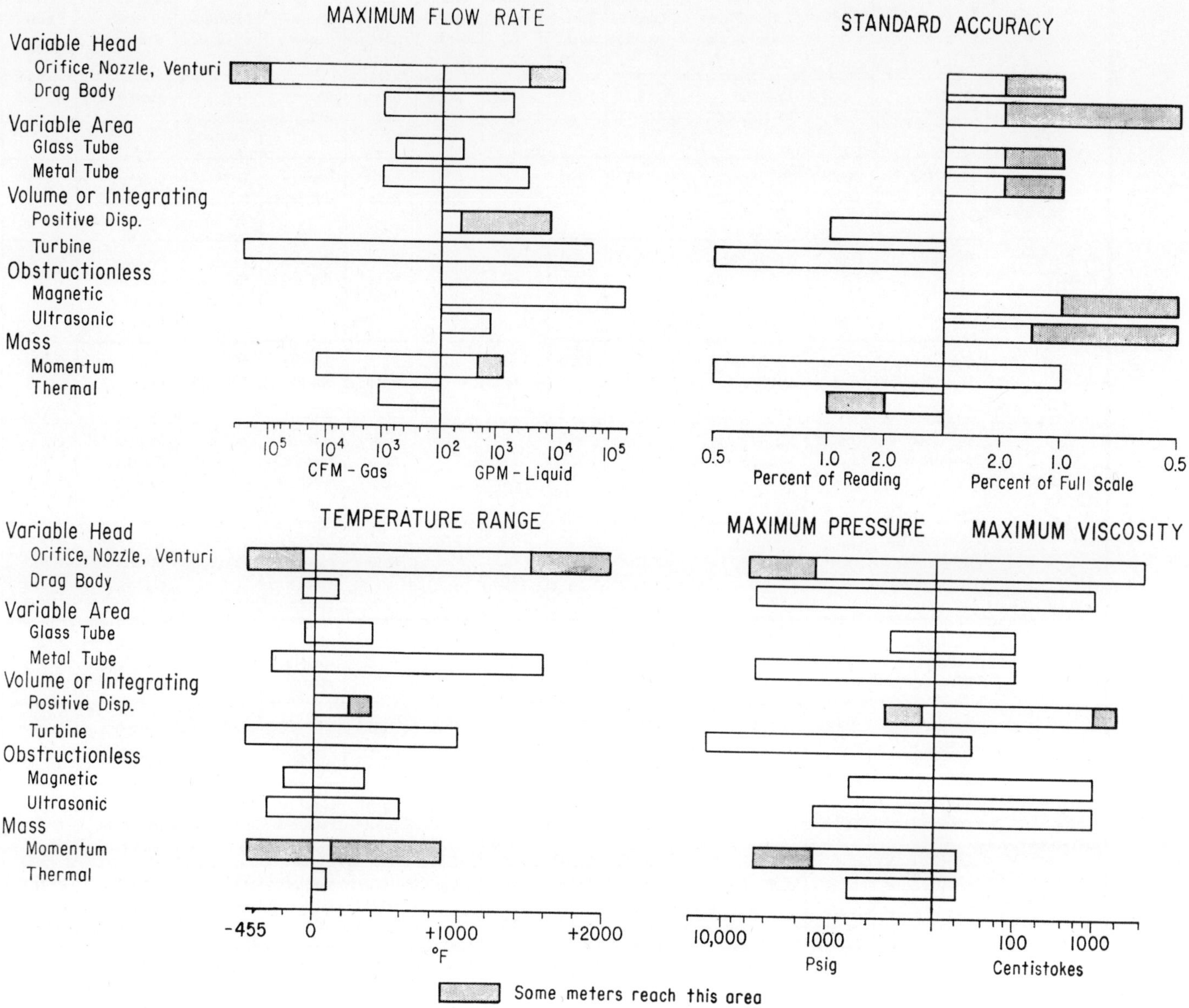

flow or weight-flow information for horsepower or thrust computations, mass flowmeters may be the answer—although many inferential or volume meters can be adapted for mass readings with the right auxiliary equipment.

This article shows how two steps will simplify the selection procedure. If one or more of the other factors are critical, a glance at the bar graphs will quickly eliminate inappropriate flowmeter types. Then take a closer look by turning to the performance table on page 120, where all basic types are compared in detail. Now you are ready to make a final choice, or to query your supplier.

Preliminary selection

When only one of the factors in the bar graph is critical, selection of a flowmeter type is relatively simple. For example, if you need high flow rate for a liquid, the magnetic type stands out, along with the turbine and variable head. Next you turn to the performance table and compare other criteria such as cost and condition of fluid.

If two or more factors are critical, choosing a flowmeter may be more complicated and possibly require some compromise. For instance, suppose you must measure a gas and you need the high flow rate provided by the orifice meter, yet you need the accuracy of the drag body. If the application demands both factors, you may have to turn to the turbine meter as a compromise, but cost will be higher.

PERFORMANCE CHARACTERISTICS OF FLOWMETERS BY

METER TYPE		FLUID FACTORS						APPLICATION FACTORS				
		Fluid	Range of Max Flow: Liquid, gpm Gas, cfm	Useful Range-ability	Max Pres-sure, psig	Temp. Range, °F	Max Vis-cosity, centi-stokes	Scale	Indi-cation	Standard Accuracy	Construc-tion Materials	Pres-sure Loss, psi
	Orifice	Liquid Gas	0.2–3500 50–100,000	5:1	6000	−455 to 2000	4000	Sq rt. Rate	Remote diff. press.	1–2% full scale diff. press.	Most metals	0.1–30
	Flow Nozzle	Liquid Gas	0.5–15,000 100–500,000	5:1	1500	−60 to 1500	4000	Sq rt. Rate	Remote diff. press.	1–2% full scale diff. press.	Bronze, iron, steel, stainless stl.	0.1–20
VARIABLE HEAD	Venturi	Liquid Gas	0.5–15,000 100–500,000	5:1	1500	−60 to 1500	4000	Sq rt. Rate	Remote diff. press.	1–2% full scale diff. press.	Bronze, iron, steel, stn. stl., plastic	0.1–15
	Drag Body (Var. Force)	Liquid Gas	0.5–2000 5–1000	5:1	5000	−60 to 180	1000	Sq rt. Rate	Remote electric	½–2% full scale	Steel, stn. stl.	2–50
VARIABLE AREA	Glass Tube Rotameter	Liquid Gas	5cc/min–250 400cc/min 700	10:1	250	−50 to 400	100	Linear Rate	Local or remote electric or pneumatic	1–2% full scale	Most metals, plastics & ceramics	0.2″ H_2O to 1.0 psi
	Metal Tube Rotameter	Liquid Gas	0.5–4000 0.6–1000	10:1	5000	−300 to 1600	100	Linear Rate	Local or remote electric or pneumatic	1–2% full scale	Most metals, plastics & ceramics	1.0″ H_2O to 10 psi
	Recipro-cating Piston	Clean liquids	0.002–200,000	10:1 to 20:1	250	0 to 250	1000	Linear Total	Local or remote electric	1% reading	Bronze, iron steel, stn. stl.	1–20
	Rotary Piston	Clean liquids	5–8750	10:1 to 20:1	3000	0 to 300	2000	Linear Total	Local or remote electric	1% reading	Bronze, iron, steel, stn. stl.	1–20
VOLUME OR INTEGRATING	Nutating Disk	Clean liquids	0.25–1000	10:1	150	0 to 300	2000	Linear Total	Local or remote electric	1% reading	Bronze, iron, steel, stn. stl.	1–20
	Diaphragm	Gas	0.25–500	20:1	100	−30 to 180	all gas	Linear Total	Local	1% reading	Steel, aluminum, syn rubber	1″ H_2O to 1.5 psi
	Turbine or Propeller	Clean liquid or gas	0.004–50,000 0.5–300,000	15:1 10:1	15,000	−455 to 1000	30	Linear Total	Remote electric	½% reading	Aluminum, stn. stl.	2–10
	Magnetic	Con- ductive liquids only	0.002–50,000	20:1	600	−200 to 360	1000	Linear Rate	Remote electric	½–1% full scale	Plastic, stn. stl. with non-conduc- tive liner	same as line
OBSTRUCTIONLESS	Sonic or Ultrasonic	Liquid	0.1–800	20:1	1500	−300 to 600	1000	Linear Rate	Remote electric	1–1½% full scale	Bronze, steel, stn. stl.	same as line
	Momentum Gyroscopic	Liquid Gas	100–4000 ppm 20–600 ppm	10:1	1500	0 to 165	20	Linear Total Rate	Local or remote electric	1% full scale	Bronze, steel, stn. stl.	1–10
MASS	Momentum Axial	Liquid Gas	40–100,000 ppm 10–15,000 ppm	10:1	5000	−430 to 850	20	Linear Total Rate	Remote electric	½% reading	Aluminum, stn. stl.	5–20
	Thermal	Gas	1.0 gm/min to 80 ppm	20:1	600	0 to 100	all gas	Loga- rithmic Rate	Remote electric	1–2% reading	Bronze, steel, stn. stl	low

BASIC GROUPS

INSTALLATION FACTORS			
Line Size, in.	Special Piping Considerations	Typical Power Requirement	Relative Cost
1½–12 & larger	Straight pipe: 10 diameters upstream; 3 downstream	None	1.0
1–24 & larger	Straight pipe: 10 diameters upstream; 3 downstream	None	1.4
1–24 & larger	Straight pipe: 10 diameters upstream; 3 downstream	None	1.5
½–12	Vertical or horizontal	115 v 60 cps 20 w	1.0–4.0
¼–4	Vertical only	None	1.0
½–12	Vertical only	None	1.2
¾–3	Horizontal recommended	None	1.0–2.0
⅝–16	Horizontal recommended	None	1.0–2.0
½–6	Horizontal recommended	None	1.0
¼–4	Horizontal recommended	None	1.0
⅛–36	Any position	115 v 60 cps 50 w	5.0
1/10–78	Any position	115 v 60 cps 200 w	6.0
¼–2	Any position	115 v 60 cps 100 w	10.0
1–3	Horizontal recommended	115 v 60 cps 50 w	4.0–5.0
1–12	Horizontal recommended	115 v 60 cps 50 w	6.0–8.0
⅛–6	Any position	115 v 60 cps 10 w	1.0–4.0

HOW TO SELECT A FLOWMETER

The performance table

Flowmeters are listed in the performance table by basic group and by subgroup within each group. Several types that are described later on are not listed in the table—either because they are not commercially available or are beyond the requirements of this article.

Fluid factors. These six columns compare fluids to be measured by quality, quantity, pressure, temperature, and viscosity. Some meters can only handle especially clean fluids and one type, the magnetic, needs a conductive liquid. Most types of meters can measure flow of either gas or liquid, but some are good only for gas (diaphragm, thermal) and some only good for liquid (volume types except diaphragm, and obstructionless types).

Range or maximum flow does not show range of any one meter, but the flow range for each type within which a meter may be procured. The rangeability column gives an indication of how broad each meter's range will be within each type.

Pressure and viscosity are listed by maximum rather than by range because the minimums are unimportant—if you have enough of each to have flow, you can measure it. Temperature is listed in range rather than maximum because of the growing importance of low-temperature measurement.

Application factors. These columns cover the characteristics of the meters themselves—scale, indication, accuracy, materials, and pressure loss. Some types are rate meters, giving instantaneous flow rate. Others are totalizers, giving total quantity of flow, like the water meter in your house. Equipment can be obtained that will adapt a rate meter to a totalizer or vice versa, but this adds to cost. If possible, it's better to decide which function you need, then pick the type that does it naturally.

Almost all flowmeter types provide remote indication (readout), but some such as the rotameter are direct-reading, or local.

Accuracy must be carefully noted because some types are accurate within a certain percentage full scale, and others within a percentage of the reading. Construction materials become important when you're faced with special conditions—high temperatures, corrosive fluids or atmospheres, light weight, etc. Pressure losses are hard to specify exactly by basic type because the flow range and operating pressure of the particular meter have a large influence on this factor. The exception here is the obstructionless types—they handle high rates of flow with no more pressure loss than a straight pipe.

Installation factors. These are factors of line size, special piping, power required, and relative cost. For purposes of this article, only minimum line sizes are significant. For other applications, some types listed will handle 2-, 3- or even 6-foot pipelines.

Special piping is required for the variable-head meters

because they measure differential pressure which must be picked off at certain distances upstream and downstream from the restriction. Other types must be installed in certain positions—for example, the rotameters must be vertical in order to operate.

Some of the more complex and expensive types require external power to operate. Mass flowmeters even need a special motor to drive them at a known constant speed.

Cost is listed on a relative basis only, because it will vary by manufacturer, flow range, accuracy, extra equipment, and so on, depending on your particular requirements within a basic type.

STANDARD FLOWMETER TYPES

Variable-head meters

Variable-head meters measure pressure differential across a restriction placed in a pipeline. Volume flow for liquid is calculated using the equation:

$$Q = CA\sqrt{2gh}$$

where Q is flow, cu in. per sec; C is discharge coefficient; A is orifice area, sq in.; g is acceleration of gravity, in. per sec^2; and h is pressure drop across the restriction, in inches of column height. For gas, a more complex equation is needed, but often the liquid equation is used with suitable correction factor.

Head meters consist of a primary element, which is a restriction containing an aperture, and a secondary element, such as a manometer, bell gage, or bellows or diaphragm. Primary elements can be an orifice plate, nozzle, venturi, drag body, capillary tube, or pitot tube. Because these meters have long been the most popular and versatile type, much empirical data has been published on discharge coefficients, and sizes and combinations of primary and secondary elements. Head meters are available in a wide range of flow quantities, line sizes, and materials.

When used with other equipment, such as a densitometer which compensates for density changes, head meters can measure mass flow. The equation becomes:

$$W = CA\sqrt{2\rho\Delta p}$$

where W is flow rate, lb; C is discharge coefficient (not necessarily same as C used for volume flow); A is cross-sectional area, sq in.; ρ is density; and Δp is pressure drop across the restriction, ft.

ADVANTAGES. Accuracy, economy, availability. Also, versatility. You can buy any of the primary devices and supply your own secondary, or you can buy a combination of the two. Adaptable to remote indicating and recording.

DISADVANTAGES. Pulsing flow, as well as changes in density, temperature, pressure and viscosity, affect accuracy. Because relationship between differential pressure and flow is nonlinear, flows less than 20% of the instrument's maximum capacity are impractical to measure. Rangeability is about 3:1.

Pressure recovery is never 100% (pressure downstream of the restriction never recovers from the pressure drop caused by the restriction). Illustrations on p 35 show effect of three primary head meter elements on recovery.

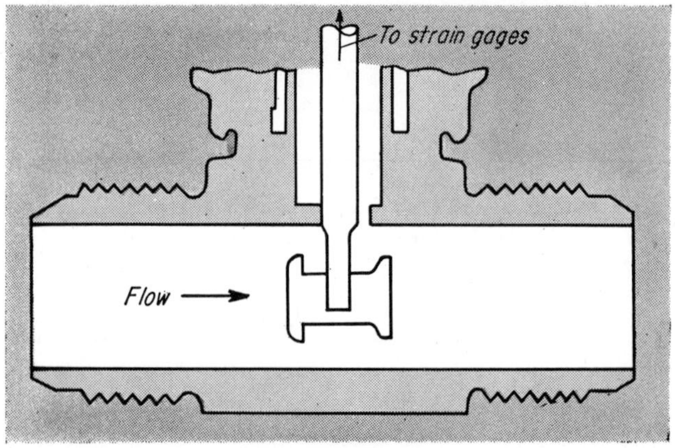

DRAG BODY, Ramapo Instrument Co

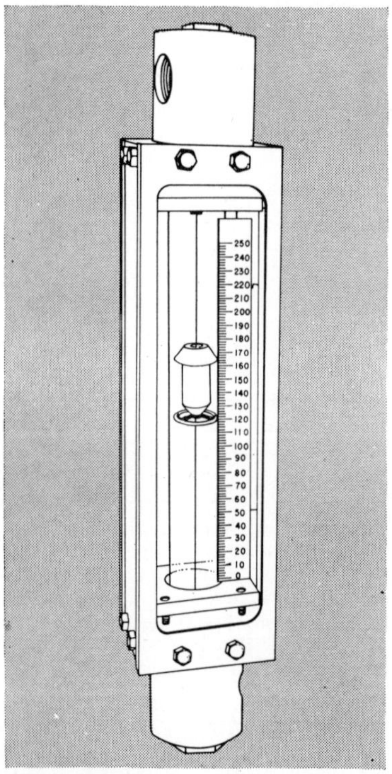

ROTAMETER, Brooks Rotameter Co

Orifice type consists of a thin plate containing a sharp-edged orifice, usually installed between pipe flanges in a line. Orifice diameter is ordinarily fixed at some proportion to line diameter, from 0.2 to 0.8. Instruments with smaller or larger diameters (0.1 or 0.9) require individual calibration. Exact orifice diameter can be calculated from application conditions of line size, flow quantity and condition, and differential pressure. Flow coefficients have been empirically established for all practical conditions, and are available in flow-calculation manuals or from your supplier.

A method for obtaining mass flow rate by recirculating the flow through two identical orifices will be covered in a forthcoming article now in preparation.

Pressure taps for secondary elements must be located

HOW TO SELECT A

FLOWMETER

DIAPHRAGM GAS METER, Rockwell Mfg Co

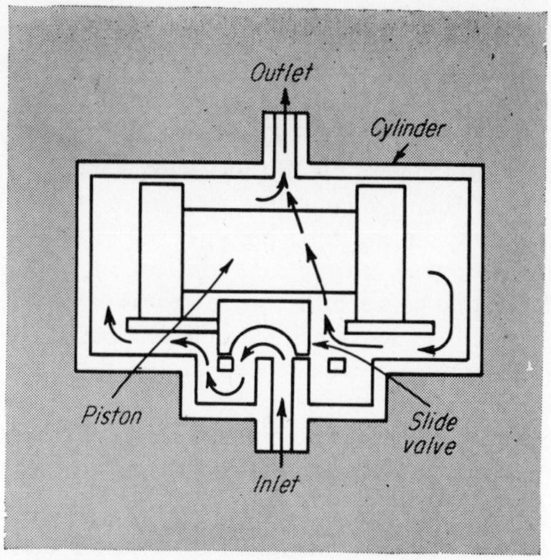

RECIPROCATING-PISTON METER

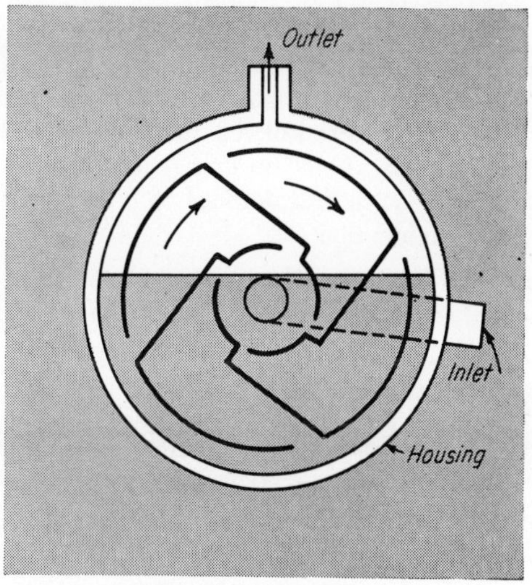

REVOLVING-DRUM METER

where turbulence is at a minimum. Flange taps, the most commonly used connection, are located 1 in. upstream and 1 in. downstream of the orifice. Because they are symmetrical, they can measure flow in either direction, providing you turn the orifice around. Also, because they are located close to the flange, they are accessible for inspection, and are located by the manufacturer of the flange, thus eliminating possible location errors. Two other arrangements for locating taps are sometimes used: the vena contracta taps, located upstream about 1 pipe diameter and downstream at the point of minimum static pressure, which are a little more accurate than flange taps because they sense maximum pressure differential; and pipe taps, located $2\frac{1}{2}$ pipe diameters upstream and 8 diameters downstream, which measure permanent pressure losses.

ADVANTAGES. Simple to install, economical, data available, accurate, wide range of materials.

DISADVANTAGES. Requires straight run of pipe, has lower pressure recovery; is inaccurate for high viscosities (below Reynolds number of 15,000), and low static-pressure applications such as blowers; handles clear fluids only (except for special segmental and eccentric types). The Reynolds number is a dimensionless index of ratio of inertia to viscosity. When number is high, inertia forces predominate and viscosity forces are insignificant; when Reynolds number is low, viscosity effects are important, causing the discharge coefficient to become a variable, seriously complicating measurement.

Flow nozzle consists of a flared inlet section followed by a cylindrical throat, whose diameter is generally between 0.25 and 0.7 of line diameter. Not as convenient or cheap as an orifice plate, the flow nozzle is often used for higher flow rates, or higher viscosities.

ADVANTAGES: Easy to install between flanges; handles dirty fluids compared with orifice type, it is more accurate for high velocity flow, gives better pressure recovery, and is less susceptible to inaccuracies due to wear.

DISADVANTAGES: Expensive to make; hard to remove from line, cannot ordinarily be modified to handle higher capacities.

Venturi is similar to flow nozzle except that the throat section is short and gradually diverges to pipeline diameter. This taper effect yields best pressure recovery of differential-pressure types. When cost permits, venturis are used where accurate flow measurement must be made with minimum loss of energy.

ADVANTAGES: Most accurate of pressure-differential types; excellent pressure recovery; good for large volumes of dirty fluids.

DISADVANTAGES. Expensive, hard to make, awkward to install, hard to modify for changes in flow range.

Pitot tube contains two pressure passages. One faces into flow, sensing total pressure; the second faces perpendicular to flow, sensing only static pressure. The difference in pressure between the two is a measure of velocity. But, because it senses velocity at only one point, the pitot tube must traverse a cross-section of the flow area measuring differential at various points. The velocities are then averaged and flow is determined by the equation: $Q = Av$, where A is cross-sectional area of line, v is average velocity.

ADVANTAGES: Very little energy consumption; simple to instal; single element can be used for most line sizes; inexpensive.

DISADVANTAGES: Relatively inaccurate; handles clean fluids only; takes time and patience to traverse.

Drag body senses the velocity of fluid flow as a function of the dynamic forces acting upon a fixed body in the flow stream. One type contains bonded strain gages in a bridge circuit, completely isolated from the fluid by a wall of stainless steel. Force on drag body deflects a rod carrying the gages. Electrical output is proportional to the square of the original flow rate.

ADVANTAGES. Can be most accurate of differential pressure meters; can operate in either direction.

DISADVANTAGES. Requires external power and meter; sensitive to shock, vibration.

Capillary type is used for very low flow ranges, and for higher viscosities—where the Reynolds number is below 15,000. Device operates on the principle that the quantity of fluid passing through a small tube in a given time is proportional to the pressure difference causing the flow times tube diameter to the fourth power over the reciprocal of tube length times viscosity. For this so-called Hagen-Poisenille law, the equation is:

$$Q = \frac{\pi \Delta p d^4}{128 \mu \ell}$$

where Δp is pressure drop, d is tube diameter, μ is viscosity, l is length.

Variable-area meters

These meters also operate by means of differential pressure, but whereas head meters contain a fixed aperture and measure differential pressure, area meters maintain a relatively constant differential pressure and measure the aperture area as it varies with flow. The most common device of this type is a rotameter, which consists of a vertical tapered glass tube containing a tapered weight. The amount of cross-sectional area is determined by height of the weight as it is suspended by the force of pressure. Differential pressure remains constant because the weight is constant—as flow is increased, the weight is forced higher in the tube, causing the area to increase. The tube is graduated so that flow rate can be read directly. Another variable-area meter is the metal tube or cylinder-and-piston type: a piston is accurately fitted inside a cylinder and is lifted by fluid pressure until enough port area in the cylinder is opened to permit passage of the flow.

ADVANTAGES. Secondary elements not needed; accurate over wide range, including low flow; handles corrosive fluids; direct reading, or can be used with magnetic following devices for remote reading; small capacity changes can be accomplished by simple changes in weight; no special upstream or downstream piping needed; pressure loss is small, constant.

DISADVANTAGES. Glass tube must be cleaned; dirty fluid requires remote readout, defeating meter's basic simplicity; though "relatively constant," differential pressure increases as flow increases because of frictional forces within meter; requires damping devices for pulsing flow.

Positive-displacement meters

Displacement meters are totalizers. They chop the flow into known volumes and count the number of chunks. Except for the diaphragm type, they share the following:

ADVANTAGES. Simply designed; accurate, even for low end of flow range; available in different materials and calibrations for measuring widely different fluids.

DISADVANTAGES. Maintenance is high because of moving parts; wear destroys accuracy; requires vapor trap upstream if entrained gas is present in liquid; relatively expensive to obtain instantaneous flow rates.

Diaphragm type is a gas flowmeter. It consists of a divided housing, two diaphragms, two sliding valves, a crank, a tangent arm, and linkage. When gas flows, the valves cause one diaphragm to fill and the other to empty, each acting as a reciprocating piston. Through linkage from the diaphragms, the crank's tangent arm picks up the reciprocating motion and turns the crank. Rotation of the crank is directly proportional to motion of the diaphragms, which in turn is proportional to the volume of gas flowing.

ADVANTAGES. Economy; linear; measures volume directly.

DISADVANTAGES. Gas only; relatively low pressure.

Reciprocating piston type is available in many variations. One common type uses two cylinders, a double-ended piston, and a sliding valve. When valve is in left position, fluid enters and is directed against piston on left, pushing entire piston assembly to right, and forcing out fluid trapped by right piston. When assembly reaches extreme right position, sliding valve moves to right. Fluid is then directed against right piston and assembly is pushed to left, forcing out fluid trapped on left side. Pistons are connected to crank arm which positions sliding valve and operates counting mechanism.

Revolving drum measures flow of gases. The measuring element is a drum partly submerged in a liquid. The gas bubbles through the liquid, pushing against partitions in drum, causing drum to rotate and release gas at the outlet.

continued, next page

HOW
TO SELECT
A
FLOWMETER

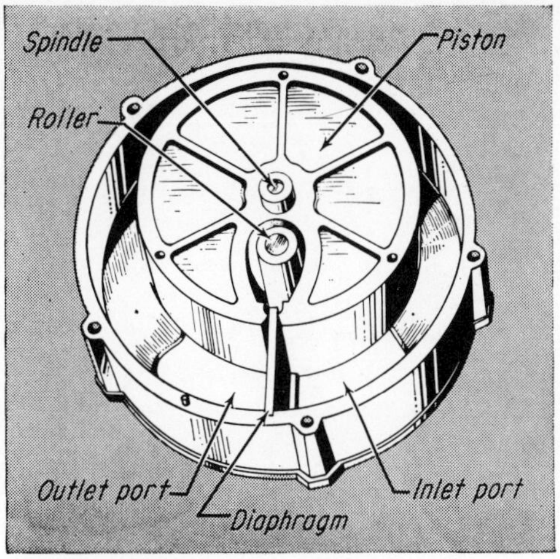

ROTATING-PISTON METER,
Neptune Meter Co

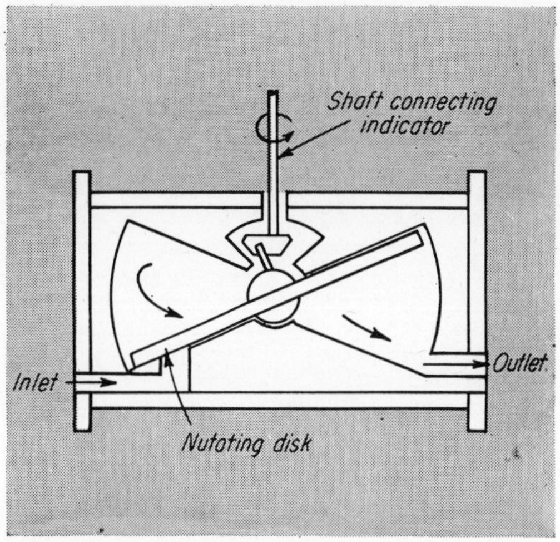

NUTATING-DISK METER

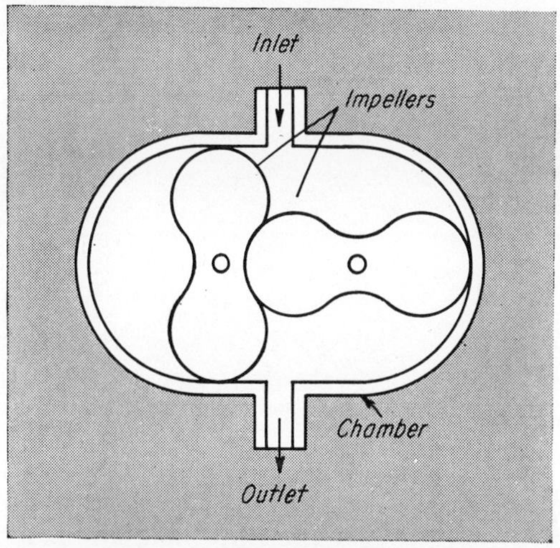

LOBED-IMPELLER METER

Rotating piston, or vane type, consists of a piston mounted eccentrically inside a cylinder. The piston contains spring-loaded sliding vanes which continually maintain contact with the cylinder wall. Fluid entering the cylinder is trapped by the vanes and causes the piston to rotate until it is carried around to the outlet port; here it is forced out by the decreasing volume of space between vanes.

Nutating disk measures liquid. The disk rocks or nutates when liquid enters the inlet. Each nutation cycle discharges a discrete volume of liquid through the outlet. Simple design, relatively low cost, and minimum maintenance make this type most commonly used displacement meter.

Lobed-impeller meter measures gas flow, and operates much like the rotating piston. Two rotors revolve in a fixed position inside the chamber. They are shaped to maintain contact between the housing and each other. As the gas enters the chamber, it is trapped by the rotors and carried around to the outlet. Flow is totalized by counting revolutions.

True-mass flowmeters

Mass flowmeters are also called transverse-momentum meters because they impart a known, constant velocity to the fluid at right angles to its direction of flow. The force required to accelerate or retard the fluid stream is equal to the rate of change of momentum in the transverse direction, which in turn is proportioned to the mass rate of flow.

ADVANTAGES. Mass flow is indicated directly, unaffected by changes in density, temperature, pressure, viscosity, compressibility; indicates rate and total flow.

DISADVANTAGES. Expensive; require rotating seals; require highly accurate, constant-speed drive.

Axial flowmeter contains an impeller and a torque-sensing wheel. The impeller is driven at a constant angular velocity. The torque wheel senses the total angular momentum, and subtracts the known momentum added by the impeller. The remaining momentum produces a torque proportional to mass rate of flow. The torque-sensing wheel is spring-restrained and magnetically coupled to an angular position pick-off.

A second version of the axial flowmeter contains two impellers, driven in opposite directions, with the torque-sensing wheel between them. The sensing wheel is restrained electromechanically instead of mechanically with a spring. Flow can be measured in either direction with this method, but pressure drop is greater because of the second impeller.

Gyroscopic type uses tubing bent in the form of a circle or square. The device is spun about the x axis while fluid is flowing through the tubing. As with a gyro, a moment acting about an axis perpendicular to both x and y is produced. The moment is a function of both angular rotation and mass flow-rate.

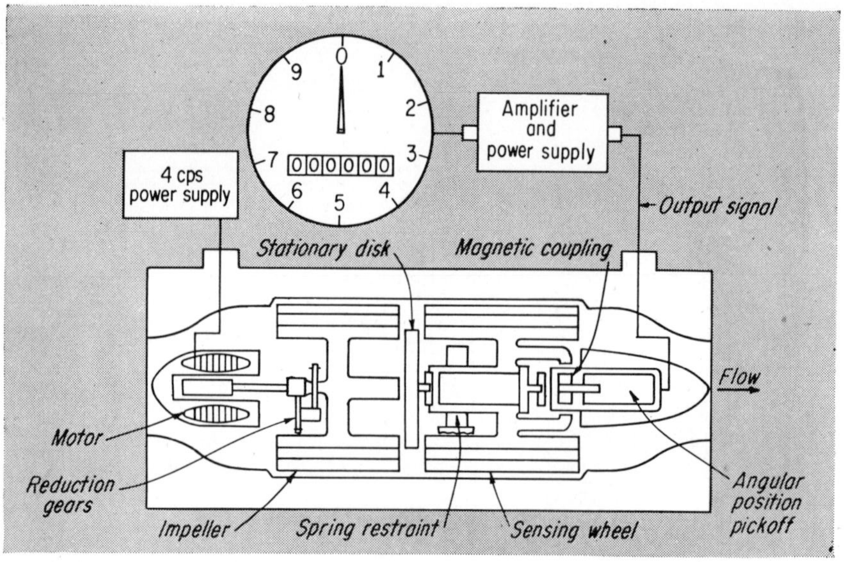

AXIAL MASS FLOWMETER, General Electric Co

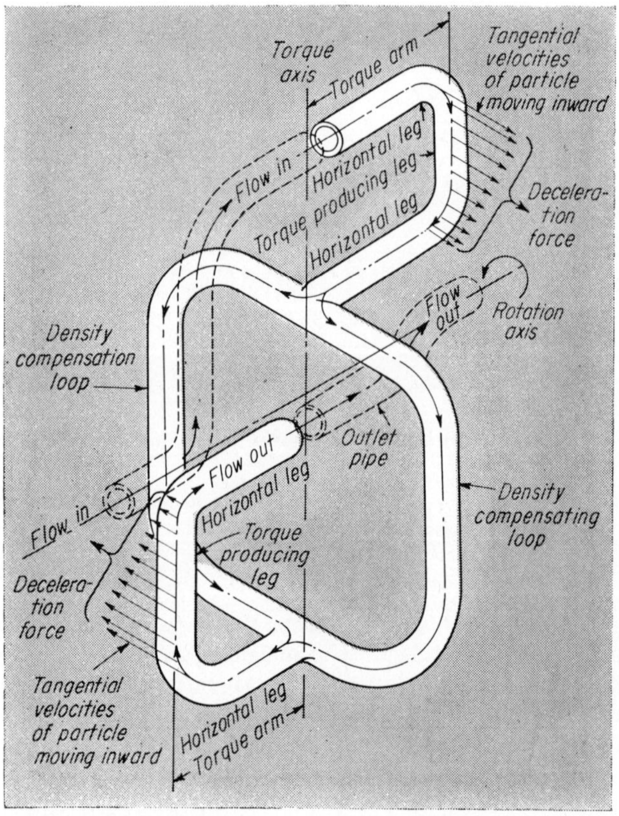

GYRO MASS FLOWMETER, Norwood Controls Div, Control Engineering Corp

A refinement of the gyroscopic meter is the vibrating gyro. Here, flow rate is indicated by the amplitude of resultant vibration. Advantages are: no flow obstruction, mass flow-rate is indicated directly, design is simple, and pressure drop is low. Disadvantages include excessive vertical-space requirements, sensitivity to turbulence, and nonlinearity at low flow rates.

Magnetic flowmeter

Based on Faraday's law of electromagnetic induction, a magnetic flowmeter consists of a piece of straight pipe, an electric coil set to produce a magnetic field perpendicular to flow, and a set of electrodes located at right angles to both magnetic field and flow. When liquid flows, the voltage induced is proportional to average velocity of liquid, strength of magnetic field, and distance between electrodes. In a given meter, the magnetic field and distance between electrodes are constant, therefore the induced voltage is directly proportional to flow velocity.

ADVANTAGES. High sensitivity and accuracy; wide flow range can be measured in either direction; output is linear; pressure loss is negligible—same as for straight pipe; handles many materials, clean or dirty; temperature and viscosity changes have little effect on accuracy.

DISADVANTAGES. Expensive; liquids must be conductive; maximum ambient temp is about 135 F, fluid to 400 F.

Sonic velocity flowmeter

Similar to magnetic flowmeter in that there is no obstruction to flow, the sonic velocity flowmeter consists of two ultrasonic transmitters and two receivers located so that one pair operates against the direction of flow (or upstream) and the other pair with flow (downstream). If both transmitters send on the same frequency, the difference in phase at the receivers between the two pairs is proportional to fluid velocity.

ADVANTAGES. Because of straight pipe, pressure loss is negligible; volumetric measurement unaffected by density change; readout can easily be analog or digital; can operate in any position.

continued, next page

HOW TO SELECT A FLOWMETER

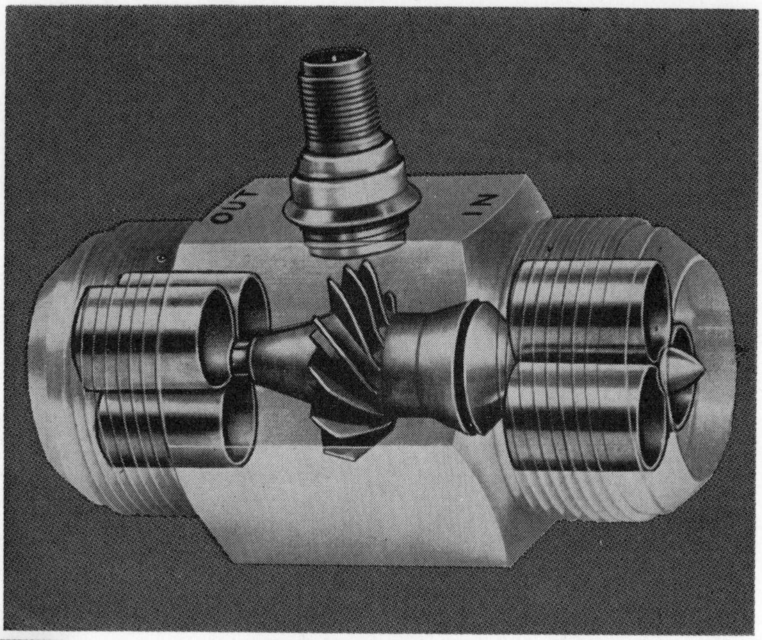

TURBINE (Pottermeter), Potter Aeronautical Corp

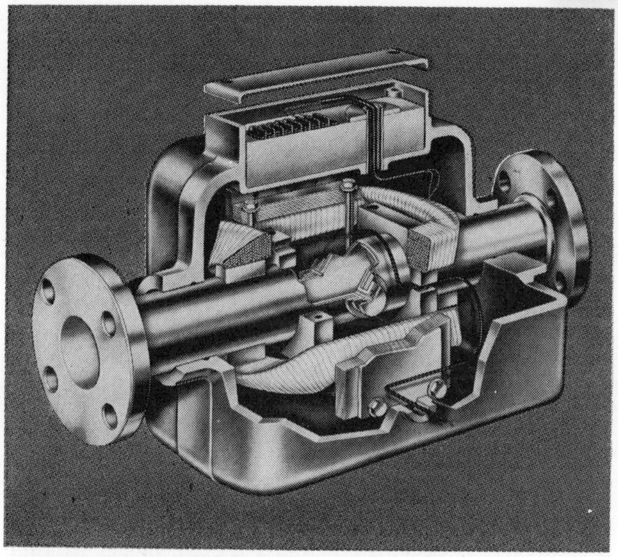
MAGNETIC METER, Fischer & Porter Co

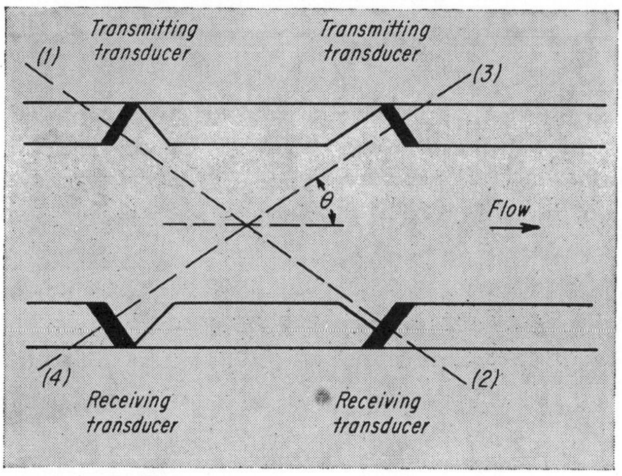

SONIC METER

DISADVANTAGES. Expensive; temperatures limited between −65 and 165 F; high viscosities cannot be measured.

Thermal meters

Used for measuring gas, thermal meters operate on the principle that when supplied from an outside source, heat is absorbed in a flowing stream proportional to the rate of flow. A thermal flow chamber contains a noble-metal thermopile. The hot junctions are heated by alternating current. The cold junctions, which sense the heat in the flow stream, are constructed to discourage heating: resistance is lower, heat conductivity is increased. Flow through the chamber tends to bring hot and cold junctions to the same temperature, which reduces voltage output of the thermopile. Thus voltage output is a measure of velocity of flow. As flow increases, voltage decreases asymptotically to zero, producing a logarithmic scale.

ADVANTAGES. Logarithmic scale permits measurement of wide range of flow; low pressure loss; fast response; flow element is relatively low cost.

DISADVANTAGES. Gas only; element must be protected from foreign particles by fine-mesh screen.

Turbine flowmeters

Turbine meter contains a turbine wheel which rotates at a speed proportional to fluid velocity. Inside the wheel is a small permanent magnet, polarized at right angles to axis of rotation. As the wheel turns, pulses are generated in a coil mounted outside the housing. (A second type uses the turbine blade tips to induce the electrical pulses.)

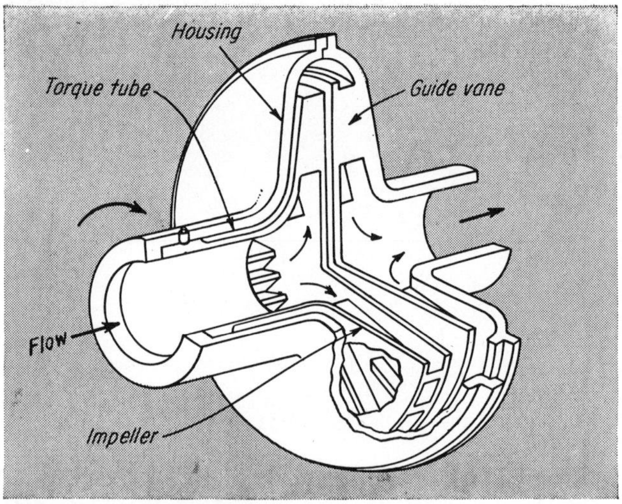

RADIAL FLOWMETER

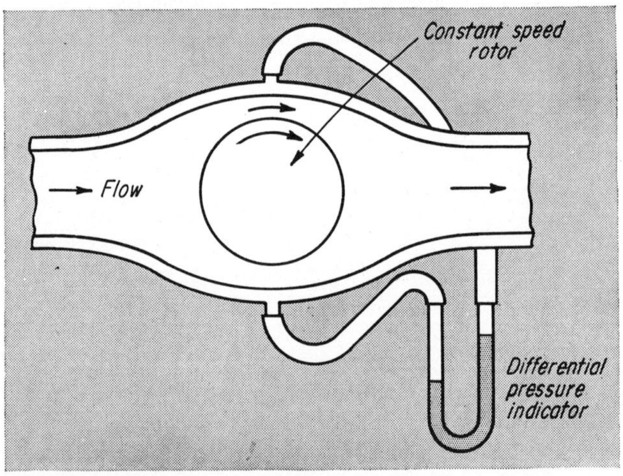

MAGNUS-EFFECT METER

Frequency of the pulsating signal gives the instantaneous flow rate, and the number of pulses totalizes the flow.

ADVANTAGES. Handles high pressures; high accuracy; quick response.

DISADVANTAGES. Abrasive media shorten bearing life; severe pulsing flow and shocks affect calibration, which must be verified.

SPECIAL-INTEREST TYPES

Radial flowmeter

The radial (Coriolis) type is a mass flowmeter, similar to a centrifugal pump. Fluid entering at the hub is deflected from its axial path by a vaned impeller. As the fluid moves away from the impeller's center of rotation, its tangential velocity increases—reflecting the effect called Coriolis acceleration. With the impeller rotating at constant rpm, the acceleration is also constant, and the force

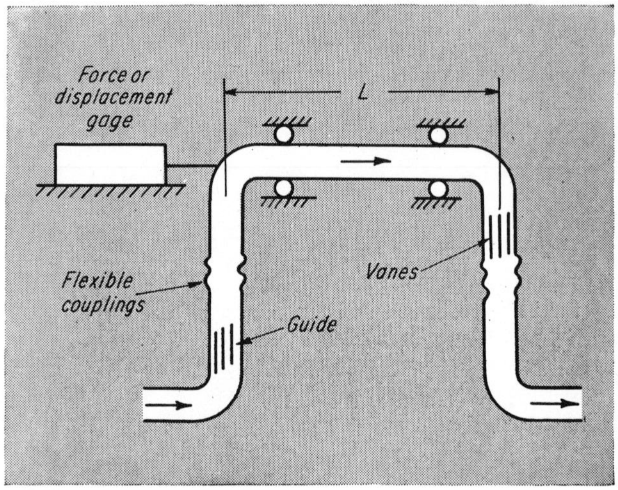

FLOW-ACCELERATION METER

required to tangentially accelerate fluid flow is proportional to mass flow-rate.

ADVANTAGES. Handles gases, liquids, foams, slurries; produces pressure rise rather than pressure drop; fast, linear response; adaptable to standard recording, indicating and controlling equipment.

DISADVANTAGES. Requires highly accurate, constant-speed drive; needs much maintenance; requires rotating seals.

Magnus-effect flowmeter

The fluid passes in two streams around a rotating cylinder as shown in the sketch. Flow in upper stream tends to be greater because the cylinder surface moves in same direction as the flow. The lower flow tends to be retarded. Net effect is to increase static pressure in the lower path, and decrease it in the upper. Pressure differential will be directly proportonal to mass flow if these assumptions are valid: flow in each passage is potential-type only (based on flowline concepts); effective "circulating" flow velocity (caused by constant cylinder rotation) is constant for any fluid; total flow is the same whether cylinder is stationary or rotating. The assumptions are not usually valid, however, and careful calibration is needed.

Flow-acceleration meter

Mass rate of flow is obtained by integrating the force required to accelerate the fluid in a given length of conduit. Flexible couplings isolate the conduit so that any horizontal force acting on it can be determined by measuring its horizontal displacement. The fluid must be assumed to be incompressible and homogeneous. Accuracy is limited only by accuracy of the integrator.

6

PUMPS, COMPRESSORS, AND BLOWERS

Performance and application of pumps, air compressors, and cooling blowers

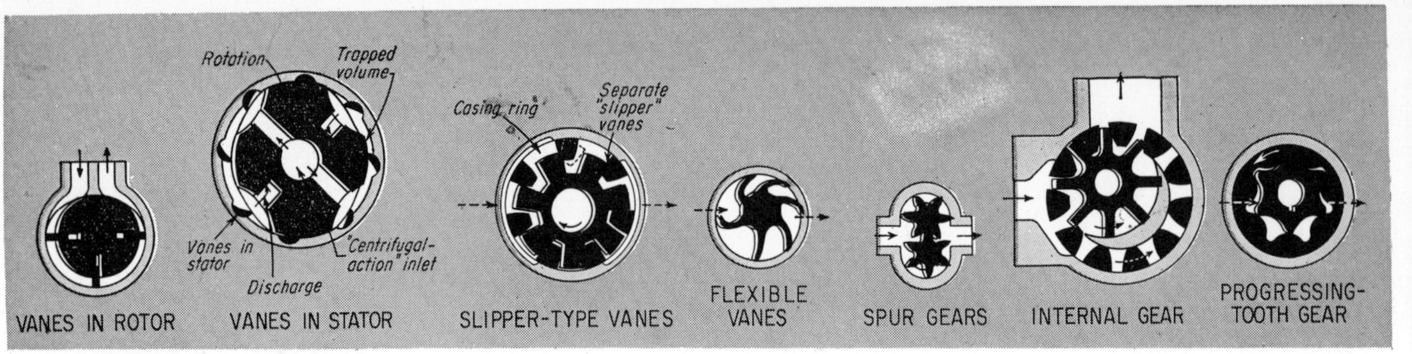

A SPECIAL REPORT

graphs simplify selection of

POSITIVE-DISPLACEMENT PUMPS . . .

these easy-to-use charts quickly give values for power, efficiency and suction. Author also compares positive-displacement pumps with other basic types such as centrifugal, listing advantages and limitations of each.

Dr Warren E Wilson, *chairman, Engineering Dept., Harvey Mudd College, Claremont, Calif*

When you need a positive-displacement pump, don't waste time searching through dozens of catalogs—just use the charts given here. They quickly give the answers needed to choose the right pump of this type, and also help predict its performance.

The charts for output power and suction head apply to all common positive-displacement pumps (gear, vane, piston, lobe and screw). Efficiency charts have been prepared for the gear, vane and piston types (detailed efficiency data not available for the other types).

Of course, before consulting the charts make sure it's a positive-displacement pump you need. For guidance, here are the advantages and limitations of all basic pump types, together with suggestions for choosing the right one.

ABOUT THE AUTHOR

Dr Warren E. Wilson, who is both engineer and educator, has been president of South Dakota School of Mines and Technology, engineering consultant to numerous pump companies, Pratt Institute's dean of engineering, and is now chairman of the engineering department at Harvey Mudd. He also organized the National Conference of Industrial Hydraulics in 1945.

Positive-displacement pumps in general have good volumetric efficiency. They can develop high pressures (to 10,000 psi), produce flow that is directly proportional to speed, and prime themselves, even at very low inlet pressure. But they must not be started or operated with the discharge valve closed; low-viscosity fluids increase the slip, and abrasive fluids cause wear. (A centrifugal pump is usually better than a positive-displacement pump at low discharge pressures.)

Of the positive-displacement pumps, the piston type has highest volumetric efficiency because leakage path between piston and cylinder can be closely controlled in manufacture. Vane, screw and lobe versions have the best suction characteristics, and are least likely to cavitate. Many unusual types of positive-displacement pumps are available with special characteristics. A few are shown among the 17 sketched above and on facing page.

Centrifugal pumps—at least, those that are most common, which are single-stage—rarely exceed 100-gpm flow or 100-psi discharge pressure. Larger sizes—single and multistage—for pipelines and other applications are available, but are often custom-made. For centrifugal pumps in this article, the product of flow and the pressure, gpm × psi, is assumed not to exceed 5000.

The output pressure of centrifugals depends on pump

text continued

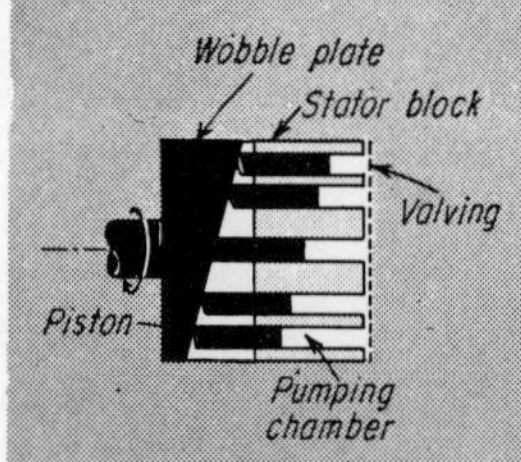

AXIAL PISTON

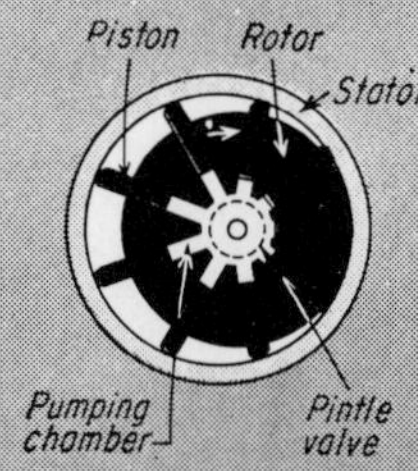

RADIAL PISTON

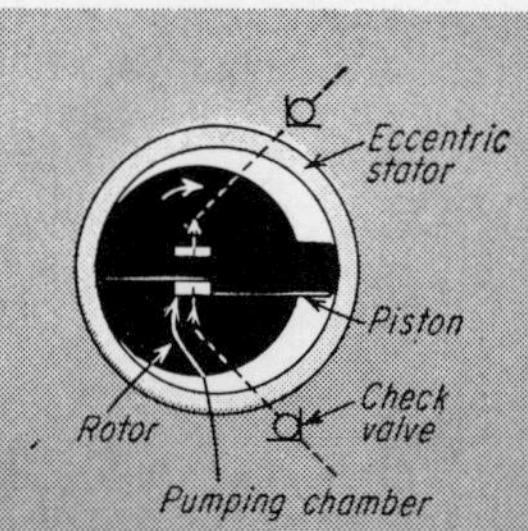

ECCENTRIC-RING PLUNGER

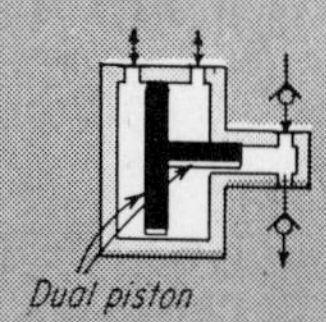

INTENSIFIER PISTON

DESIGN FEATURES OF 17 TYPES

All positive-displacement pumps—including the unusual types—have one thing in common: the liquid is forced out by reducing the volume of the pump enclosure. They differ in how the volume is reduced, and how the high-pressure liquid at the outlet is kept from leaking excessively to the inlet.

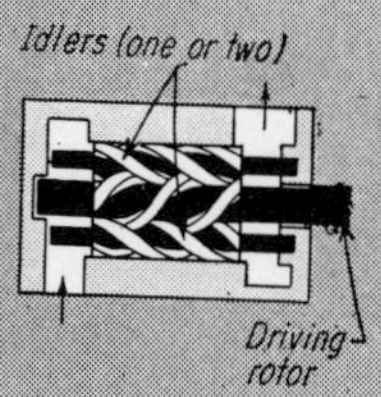

MULTIPLE SCREW

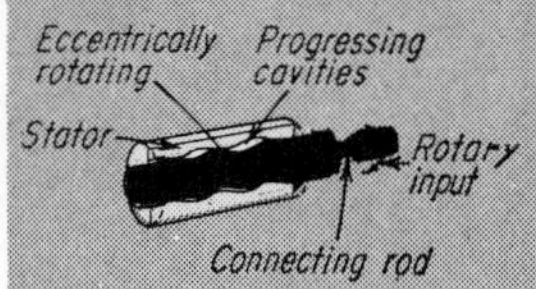

SINGLE SCREW

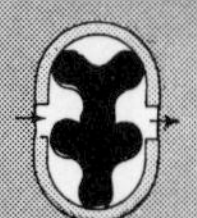

LOBE

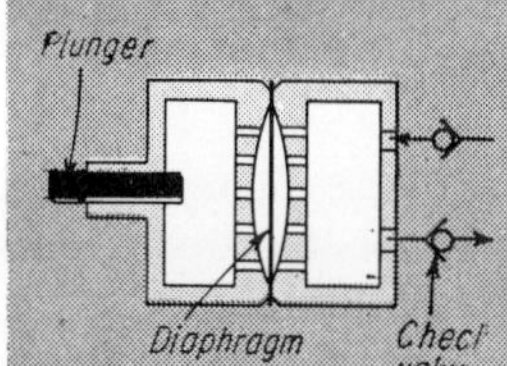

DIAPHRAGM

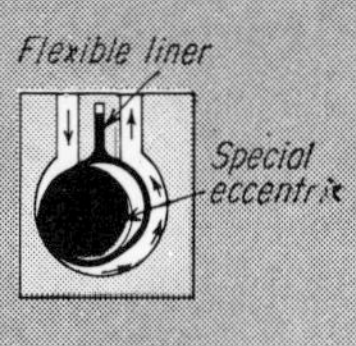

FLEXIBLE LINER

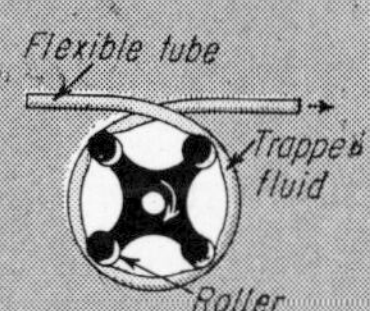

FLEXIBLE TUBE

VANE PUMPS

VANES-IN-ROTOR version is most common; vanes move radially to follow pump contour. Leakage from outlet to inlet across top is kept small because vane is completely retracted, rotor-to-stator clearance is practically zero.

VANES-IN-STATOR PUMP has hollow-shaft inlet; centrifugal action reduces cavitation. Pumping occurs as rotor turns because trapped volume of liquid is squeezed out by vanes. Outlet is through rotor—flow leaves rotor via sliprings.

SLIPPER-VANE PUMP initially depends on centrifugal force to throw slippers against casing ring. But then liquid pressure, viscous drag, and oil-film cohesion hold vanes in position. Action is similar to that in Kingsbury thrust bearing.

FLEXIBLE-VANE PUMP depends on bending of vanes to decrease trapped volume, force out liquid.

GEAR PUMPS

SPUR-GEAR PUMP carries trapped liquid between pairs of outer teeth, forces it against existing outlet pressure. Outlet-to-inlet leakage is kept small by close-meshing teeth at center.

INTERNAL-GEAR PUMP is similar to vane pump in operating principle. Inlet liquid fills expanding pumping cavities, is carried around inner and outer path as shown. Outlet-to-inlet leakage is kept small by close-meshing spider teeth and vanes.

PROGRESSING-TOOTH GEAR PUMP has one less tooth in rotor than in stator, thereby moving trapped volume one tooth width each revolution. Performance is similar to vane pump.

PISTON PUMPS

Piston pumps have reciprocating action that allows inlet liquid to fill cavity on suction stroke, forces it out on pumping stroke.

AXIAL-PISTON PUMP depends on beveled "wobble" plate or equivalent off-center linkage system to reciprocate pistons. **RADIAL-PISTON PUMP** depends on eccentric stator (or rotor) to give reciprocating action.

ECCENTRIC-RING PLUNGER PUMP is single-piston version of radial-piston pump. It is shown in suction position; 90-deg rotation of rotor will close pumping chamber, forcing out liquid.

INTENSIFIER-PISTON PUMP transfers energy from one liquid system to another. Version shown converts high-volume low-pressure fluid (large chamber) to low-volume high-pressure fluid (small chamber).

SCREW PUMPS

MULTIPLE-SCREW PUMP drives through single rotor; the other meshing rotors (there can be one or more) help trap liquid in pumping cavity that moves axially as the rotor turns.

SINGLE-SCREW OR "PROGRESSING-CAVITY" PUMP has helical rotor rotated by connecting rod. However, rotor not only rotates about its own axis, but its own axis rotates relative to axis of input. This eccentric action keeps rotor in constant contact with stator, trapping liquid in axial-moving cavity.

LOBE PUMPS

LOBE PUMPS of simplest design are similar in operation to spur-gear pumps —liquid is carried in outside pockets.

DIAPHRAGM PUMPS

DIAPHRAGM PUMP shown transfers energy from plunger-driven fluid on left to pumped fluid on right. The amount pumped is equal to displacement of the plunger. Other versions can have cam-operated diaphragms.

SQUEEGEE PUMPS

FLEXIBLE-LINER PUMP traps liquid in "squeegee" action. Pumping and sealing stops as eccentric goes through top dead center, but momentum of fluid prevents backflow. Principle advantage of squeegee is isolation of moving element from pumped liquid.

FLEXIBLE-TUBE PUMP is also squeegee type, but multiple rollers keep liquid trapped at all times.

POSITIVE-DISPLACEMENT PUMPS continued

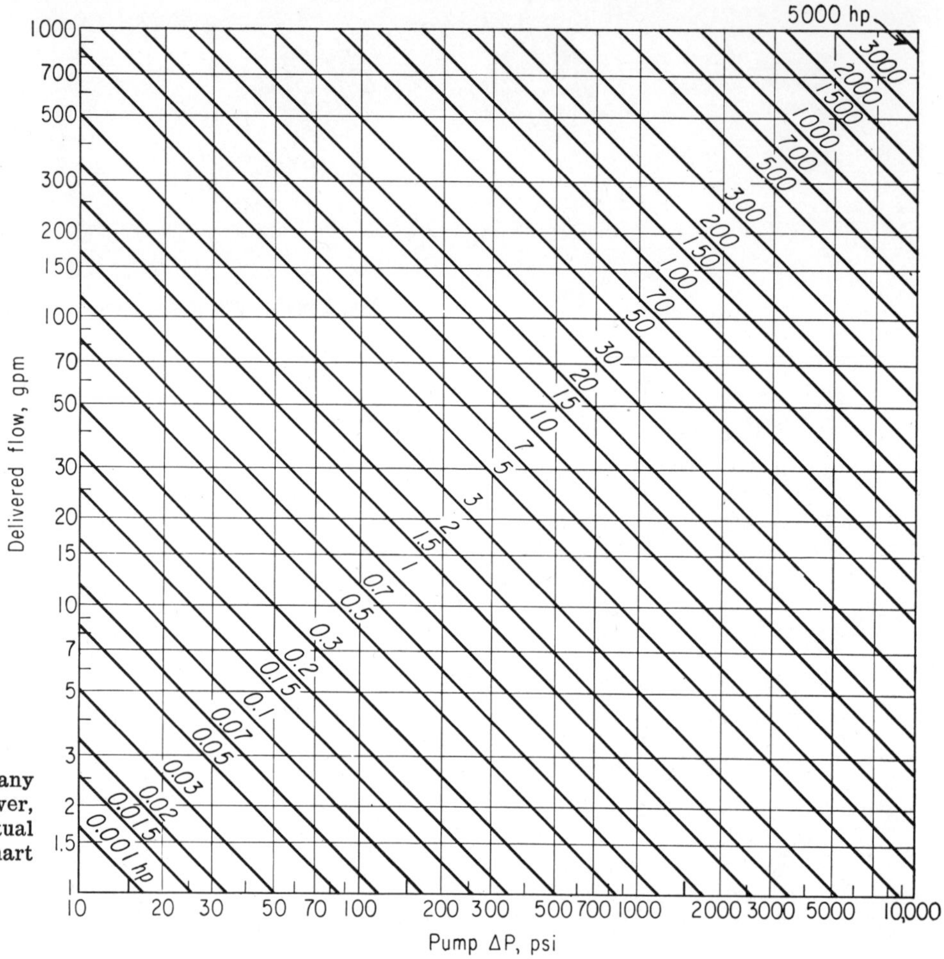

THEORETICAL POWER to drive any pump equals output fluid horsepower, $5.82 \times 10^{-4} Q_d \Delta P$, plotted here. Actual power required to drive pump is chart hp divided by over-all efficiency.

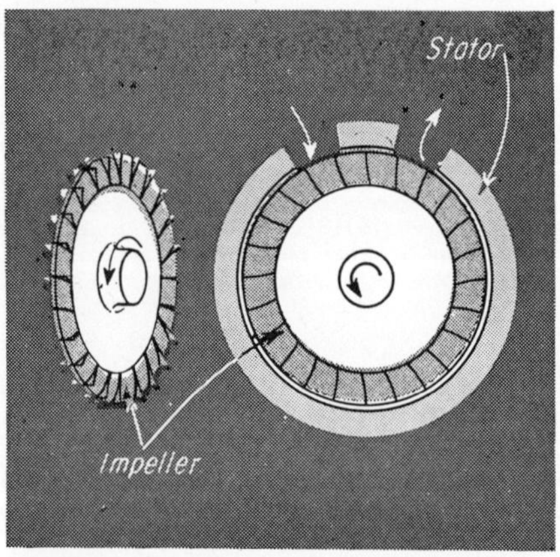

TURBINE PUMP is neither centrifugal nor positive-displacement, but has some characteristics of each. Pressure rise depends on momentum given to liquid by the spinning rotor.

geometry and speed of rotation. These pumps have a limited range of shaft speeds, and cannot tolerate as much cavitation as other types. They perform best with low fluid viscosities and are better suited for transporting fluid than building up high pressures. Abrasive or dirty fluids can be easily handled by a properly designed centrifugal pump. However, in basic hydraulic power-and-control systems, pressure is the essence of operation, and positive-displacement pumps are usually recommended.

Propeller pumps (axial flow) and mixed-flow pumps (partly centrifugal, partly axial) also have output pressure dependent on geometry and speed of rotation; have same limitations as centrifugals.

Turbine pumps combine elements of the centrifugal and positive-displacement types. The theory behind this type is not fully understood (although we do know that flow is produced by momentum transfer of turbulent liquid). But performance is well known—discharge pressure responds somewhat linearly with speed, as it does in a positive-displacement pump. Discharge pressure is usually less than 300 psi in a single stage, but this is higher than that of a single-stage centrifugal. Low-viscosity fluids, such as water, gasoline, and alcohol—even at low suction pressures—can be easily handled.

Turbine pumps are not too efficient—about 40 to 50% over-all—but their design simplicity overcomes that objection when efficiency is not a primary requirement.

IMPORTANT PUMP EQUATIONS

Pressure	$\Delta P = P_{out} - P_{in}$
Flow	$Q_d = Q_i - Q_s - Q_c$ $= E_v Q_i$ $Q_i = ND$ $Q_s = C_s D \Delta P/\mu$
Torque	$T_m = T_i + T_F$ $T_i = 36.8\ D\Delta P$
Power	$hp_{in} = 1.585 \times 10^{-5}\ TN$ $hp_{out} = 5.82 \times 10^{-4} Q_d \Delta P$
Efficiency	$E = hp_{out}/hp_{in} = E_v\ E_T$ (varies with $\mu N/\Delta P$) $E_v = Q_d/Q_i$ $E_T = T_i/T_m$

SYMBOLS FOR EQUATIONS

PRESSURE AND HEAD	
P	Pressure, psi (P'= psf, P_v= vapor pressure)
H	Head, ft of liquid
Z	Elevation of inlet above pump centerline, ft
NPSH	Net positive suction head, ft

TORQUE	
T	Torque, lb in (subscripts: m = driving; i = ideal; F = friction, dry and viscous)

FLOW AND DISPLACEMENT	
Q	Flow, gpm (subscripts: d = delivered; i = ideal; S = slip; C = cavitation)
N	Speed, rpm
V	Velocity, ft/sec (subscripts: 1 = inlet; 2 = pumping chamber)
D	Displacement, gal / rev

POWER AND EFFICIENCY	
hp_{in}	Power to drive pump, hp
hp_{out}	Power output (liquid), hp
E	Over-all efficiency, % (subscripts: v = volumetric; T = torque)

COEFFICIENTS AND CONSTANTS	
C_S	Coefficient of slip
g	Gravity constant, 32.2 ft/sec^2
μ	Absolute viscosity, slugs /(ft x sec) or lb_{force} x sec/ft^2
SSU	Kinematic viscosity, Saybolt seconds, universal
ρ	Mass density, slugs/cu ft
R	Outlet restriction (pump load), psi per gpm

Jet pumps, pitot-tube pumps, and other basic types are used for certain specialized applications. The jet type—common in well-pumping applications—entrains liquid from a low-pressure area into a jet of liquid that has high velocity head and low static head. A pitot-tube pump rotates a body of liquid in a cylinder, taps off total pressure via a stationary pitot tube inserted in the outer area where pressures are highest.

POWER AND EFFICIENCY CHARTS

Theoretical Output Power

The nomograph on previous page gives pump-output hp for any given flow at any given pressure. Actual power to drive pump will, of course, be greater than the output power; it depends on over-all efficiency of pump. The relationship is $hp_{out} = E \times hp_{in}$, where hp_{in} is power needed to drive pump, E is over-all efficiency of pump, and hp_{out} is theoretical pump-output hp based on flow and pressure.

Over-all Efficiency

Pump efficiency depends on too many variables to allow easy calculation. Typical variables are internal backflow (slip), cavitation, and friction (dry and viscous). However, pumps of similar design tend to follow the same pattern of losses and these can be estimated with "averaged" data, such as those on next page.

Volumetric efficiency—an important part of over-all efficiency—is ratio of delivered flow to ideal flow. Delivered flow $Q_d = E_v DN$, where E_v is volumetric efficiency, D is displacement, and N is pump speed. Volumetric efficiency depends on shape and clearance of rotating and stationary elements, suction characteristics of pump, viscosity of liquid, speed and pressure of pump.

If clearances are small and there is good dynamic sealing between rotor and stator, internal backflow will be small, volumetric efficiency high. But this assumes pump inlet is kept pressurized and full, so cavitation (see suction-head section, page 136) does not reduce delivery

Viscosity of the liquid has a pronounced effect on volumetric and over-all efficiency, and also on friction load and cavitation. The effects are opposing: high viscosity helps reduce slip (improving the volumetric efficiency) but increases friction load and hinders suction. Low viscosity improves suction and decreases friction load but increases slip (lowers volumetric efficiency).

Slip (internal backflow) in gpm tends to conform to this formula for viscous flow between two flat plates: $Q_s = K \times (\Delta P/\text{viscosity}) \times (\text{width}/\text{length}) \times (\text{clearance})^3$, where K is proportionality constant. Note that both width and length of clearance path can be increased without changing absolute slip flow—if width-to-length ratio and clearance are held constant.

Actual pumps, however, do not have predictable leakage-

(text continued, p. 136)

GEAR PUMPS

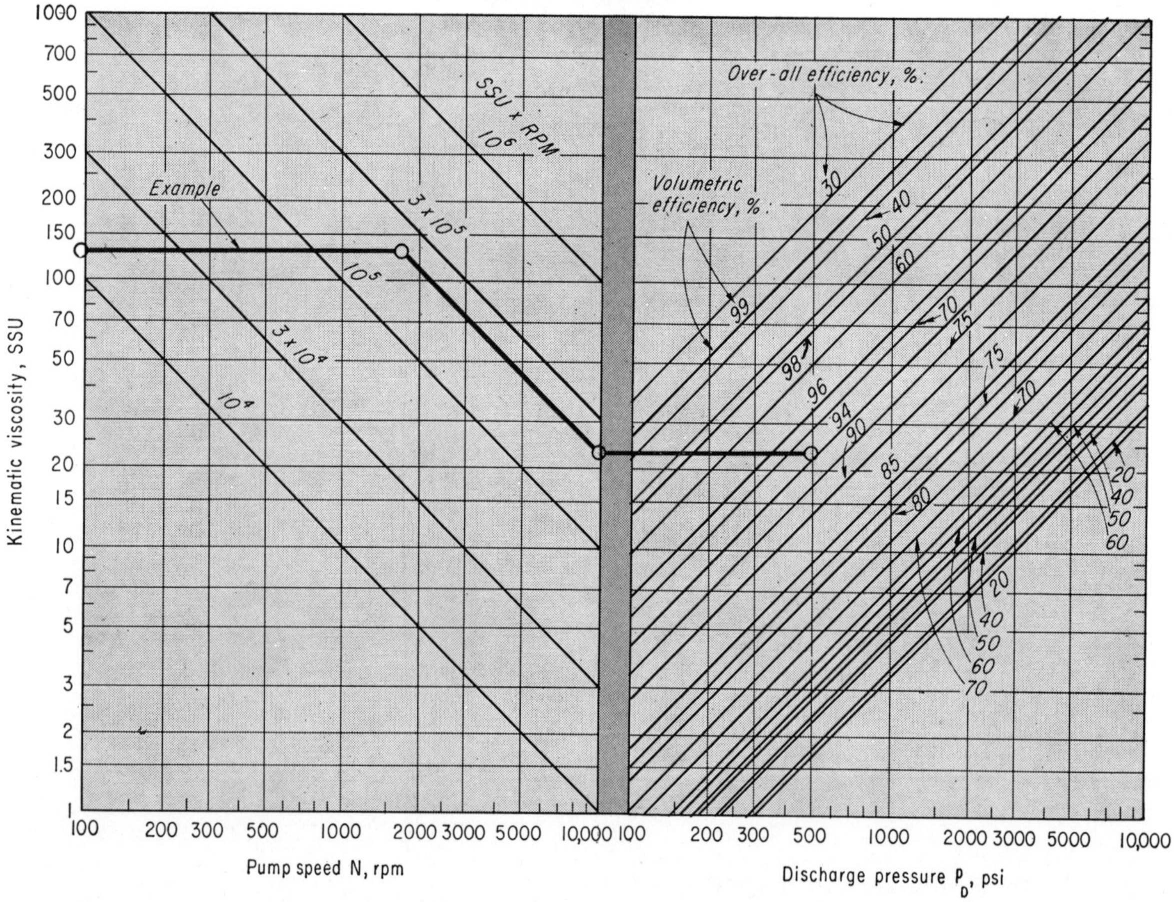

OVER-ALL EFFICIENCY and volumetric efficiency of gear, vane and piston pumps can be quickly estimated with these three nomographs—based on actual data from many pump tests. Plotting method assumes slip coefficient C_s is constant for each type pump, and $Q_s = C_s D \Delta P/\mu$. Dry friction is neglected because a well-designed pump has very little. Viscous drag is assumed proportional to speed and viscosity. Value for discharge pressure assumes inlet at atmospheric pressure. Oil used has specific gravity of 0.8, and is at normal operating temperatures.

Accuracy of these charts is probably little better than ±20% for most pumps because manufacturing and design variations are great. Better accuracy is possible if one or more performance points are known—effects of speed, viscosity and pressure can be easily predicted using charts.

Here is an example: oil at 130 SSU is pumped to 500 psi with 1800-rpm gear pump. What is probable over-all and volumetric efficiency? Answer: 94% volumetric, 72% over-all. Note: raising viscosity to 500 SSU will increase volumetric efficiency to over 98%, lower over-all to 45%. Best efficiency point can be found by trial and error.

Another example: pick gear pump to pump 1000-SSU liquid to 600 psi discharge. The best over-all efficiency is 76%—coinciding with horizontal line through SSU × rpm = 1.35×10^5. Following this line to left and upward to intersection with viscosity = 1000 SSU, best pump speed is seen to be 135 rpm. If 1000 rpm, 600 psi, and 1000 SSU are investigated, corresponding over-all efficiency is 47%, volumetric 98%. This is a decrease in over-all efficiency of 29% due to increase in speed and consequent increase in viscous drag. Cost of installing pump large enough to give required delivery at 135 rpm might offset the gain in efficiency. Compromise is intermediate speed of 600 rpm with 60% over-all efficiency.

VANE PUMPS

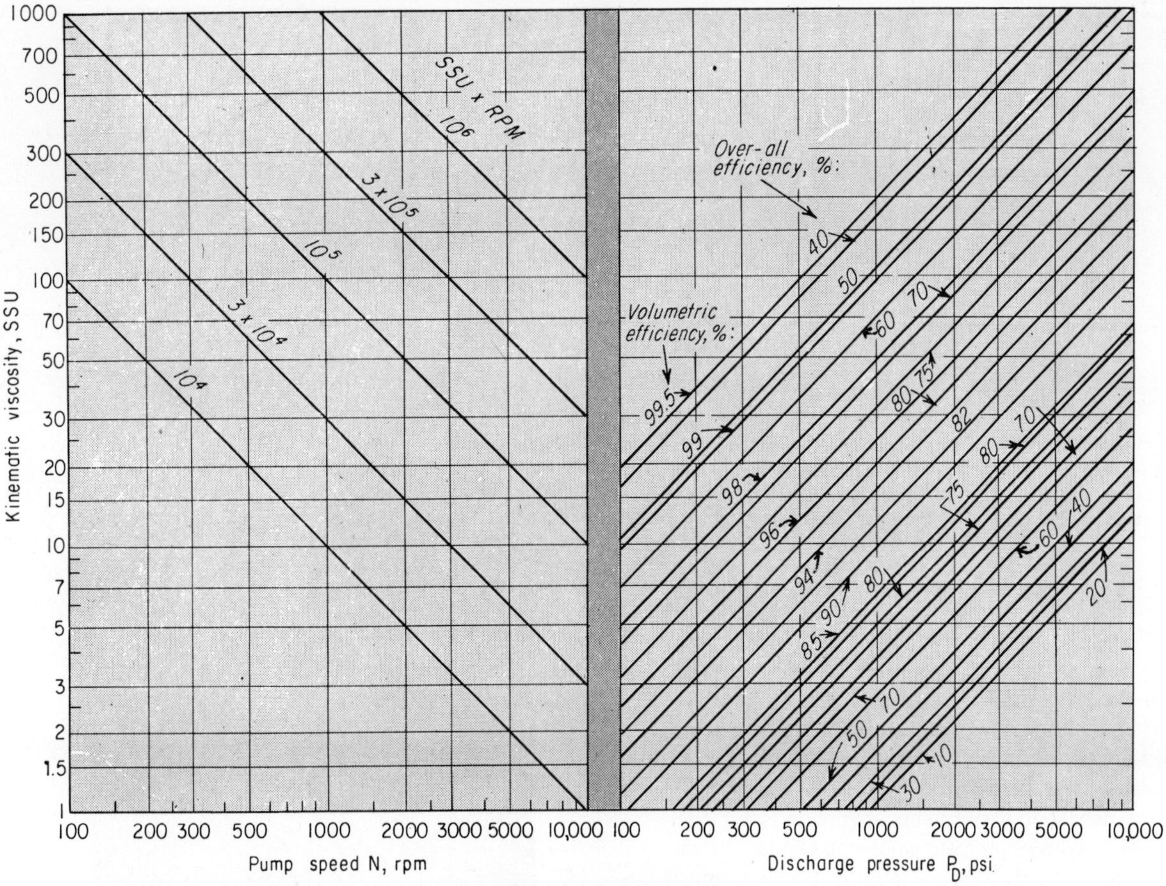

PISTON PUMPS

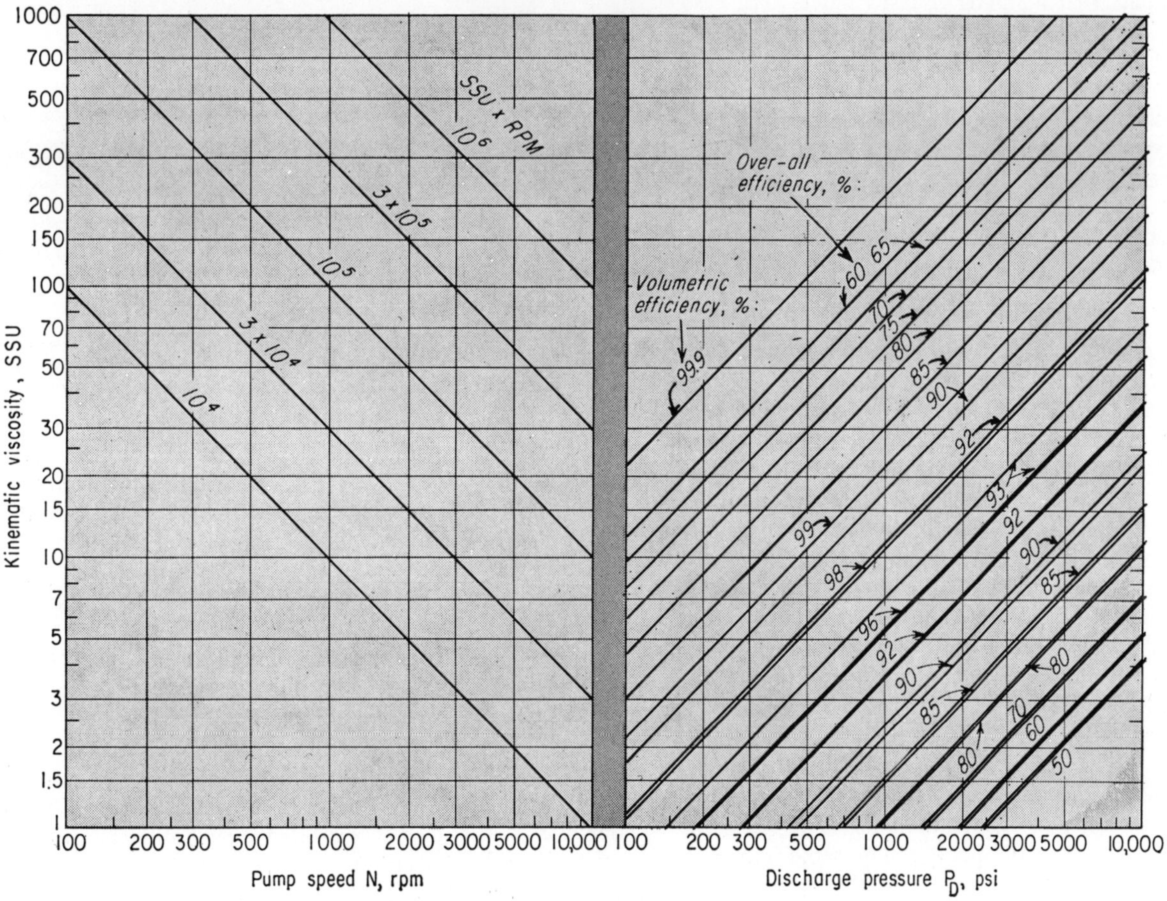

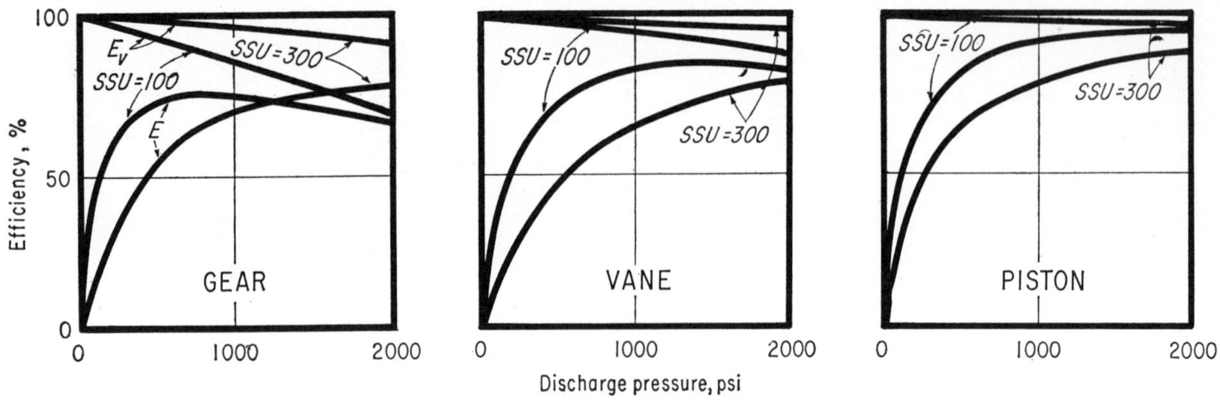

USING THE EFFICIENCY CHARTS to plot comparative performance of gear, vane, and piston pumps is easy—just follow lines on chart and read answers directly. These examples are for 1800-rpm pumps, but other speeds are evaluated just as readily. Over-all efficiency will reach peak value at a different pressure for each speed.

path dimensions, and each pump must be tested separately to determine slip and volumetric efficiency. A convenient pump-slip equation, based on flat-plate flow, is: $Q_s = (C_s D) \times (\Delta P/\mu)$, where C_s and D depend on effective width-length-clearance relationship and pump size.

Note that slip is a function of pressure, not speed. Doubling speed does not double slip—slip flow stays the same, so long as discharge pressure does not change. Percent slip—slip per gpm delivered—actually becomes one-half original value, because delivery flow doubles with speed. Also note that doubling viscosity halves slip.

Slip limits the maximum pressure of a positive-displacement pump. Assuming no cavitation, delivery flow $Q_d = DN - Q_s = DN - C_s D \Delta P/\mu$. Rearranging terms, pressure rise $\Delta P = \mu(DN - Q_d)/(C_s D)$. The maximum pressure (developed when $Q_d = 0$) is $\mu N/C_s$.

All these effects—slip, friction, viscosity, speed and pressure—have been taken into account in the "averaged" efficiency charts. But individual pumps vary in design and workmanship.

SUCTION-HEAD CHART

For each condition of flow and speed there is a minimum value of inlet total pressure (static plus velocity head) that will prevent cavitation. This minimum inlet head, based on P_v as the datum, is known as required net positive suction head (NPSH) and depends on: viscosity of liquid, square of velocity (V^2) of the pumping element, vapor pressure of liquid, design of inlet flow path, smoothness of pumping cycle, and percentage of dissolved air. Experience shows these combined effects tend to vary mostly with V^2. The simplified suction-pressure chart is based on V^2 and is fairly accurate for all positive-displacement pumps handling moderate liquid viscosities (up to 200 SSU)

To avoid cavitation, available NPSH must be greater than required NPSH. For example, if the liquid (sketch, this page) reaches the pump inlet with zero velocity and is at a pressure equal to or less than its vapor pressure, it cannot be forced into the pump. The same is true if its total inlet pressure (velocity head plus static head) is less than that of fluid adjacent to moving pump element. Here is why (see sketch):

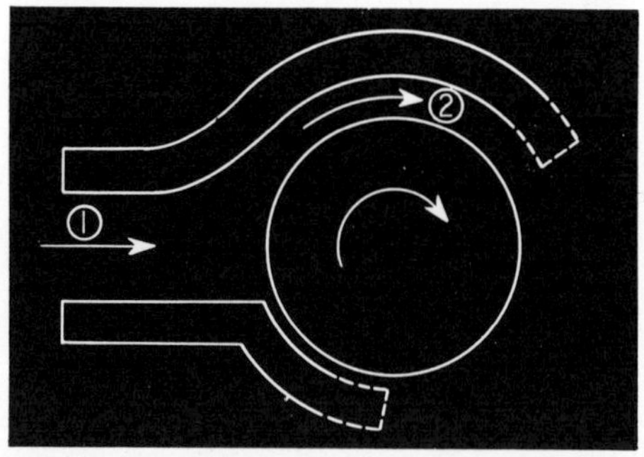

Liquid at point 1 must enter pumping chamber at point 2. Between 1 and 2 there may be gear teeth, vanes, rotor, piston, or other pump elements. At point 1 the liquid has velocity V_1 and pressure P_1. Energy (total head) per unit mass available to move liquid into pump is sum of its kinetic energy $V^2/2g$ and pressure (or potential) energy $P_1/\rho g + Z$, minus its vapor-pressure head $P_v/2\rho g$. Thus available NPSH $= V_1^2/2g + (P_1 - P_v)/\rho g + Z$.

If there were no losses of energy between points 1 and 2 and velocity V_1 was exactly equal to the required velocity V_2 at point 2, liquid would flow readily into pumping chamber even when $P_1 = P_v$. In practice, many pumps have considerable loss in energy between 1 and 2—while others actually add energy to liquid in inlet region by shearing action and other means. The required NPSH $= V_2^2/2g + H_L - H_G + P_v/\rho g$, where H_L is head loss, H_G is gain.

Available NPSH for piston pumps varies during each suction stroke, because flow pulses somewhat, depending on design. At start of stroke, fluid is at rest (or has a relatively low velocity) and available NPSH is reduced by the head needed to accelerate the fluid. At end of stroke, no acceleration is required, and available head depends on friction losses in inlet piping and passages. If inlet piping has an unusual number of fittings, the end-of-stroke (high-friction) condition is important in deter-

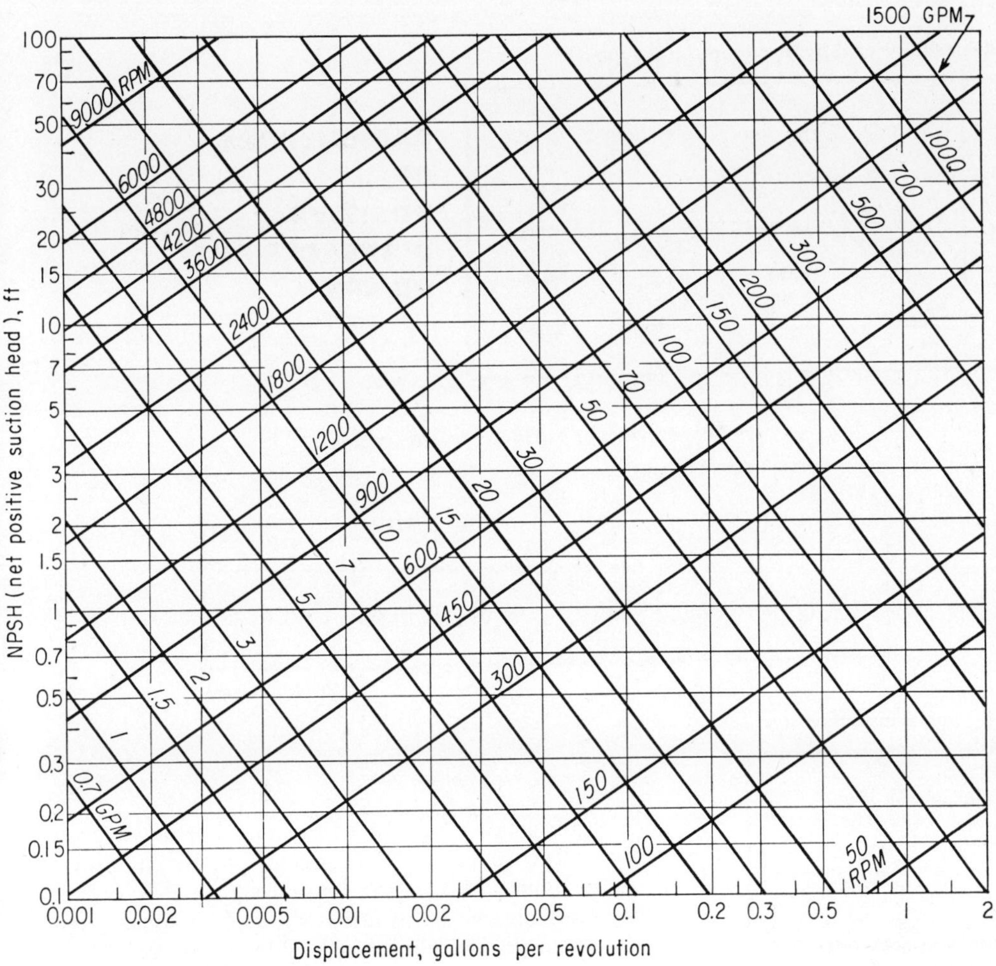

NET POSITIVE SUCTION HEAD here is based on square of velocity of pumping element, although other factors strongly influence suction requirements (see text). An example: a given pump has 10-gpm flow, runs at 1800 rpm (it can be gear, vane or piston type—the basic rules still apply.) What is NPSH? Answer: 5.2 ft of liquid pumped, absolute pressure.

mining the NPSH available. The suction-head chart, however, is based on V^2 of pumping element—the more usual criterion.

PUMP PULSATION

Freedom from pulsation—often a primary requirement—depends on geometric (shape), kinematic (velocity and flow path) and dynamic (pressure and force) considerations.

Little can be predicted about the pulsation problem in selecting a pump. Where the problem might be serious, choose a pump with very small slip (very high volumetric efficiency.) This is a good indication of fine workmanship and uniformity of clearances.

Reciprocating pumps inherently have periodic impulses. However, multipiston pumps reduce the magnitude of these variations. Gear, vane, and other positive-displacement pumps have somewhat less pulsation. Theoretically, pumps can be built which eliminate the kinematic and geometric factors contributing to pulsation, but basic design and construction problems prevent elimination of pulsation completely.

Pulsation, although appearing as a fluctuation in pressure, actually arises as a fluctuation in flow rate because pressure is a dependent quantity. Here is why:

Discharge Q_a of a positive-displacement pump $= DN - C_s \Delta PD/\mu$. The system into which liquid is pumped will have some resistance R, where $R = \Delta P/Q_a$ or $Q_a = \Delta P/R$. (This is analogous to Ohm's electrical law, $I = E/R$.)

Equating, $Q_a = \Delta P/R = DN - C_s \Delta PD/\mu$. Noting that $\Delta P = P_{out} - P_{in}$, it can be shown that this relationship between pressure, flow, and discharge resistance applies:

$$P_{out} = (DN)/(1/R + C_s D/\mu) + P_{in}$$

If N, R, C_s and P_{in} remain constant, the discharge pressure will be constant. But C_s depends on cube of pump internal clearances, and can vary during each revolution, thus introducing large flow fluctuations. Flow and pressure fluctuations are directly related: $\Delta(P_{out})/P_{out} = \Delta(Q_s)/Q_t$, where Δ indicates a small change.

EXAMPLES OF PUMP SELECTION

1—Liquid Transfer

REQUIRED: to pump 50 gpm of 100-SSU oil (70 F) to pressure of 250 psi. Motor drive will be at 1800 rpm. Inlet conditions are good, and available pressure is at least 13 psi.

SOLUTION: A large number of suitable pumps are available; however, piston pumps are somewhat expensive, and the low-pressure service does not justify them. A lower cost turbine pump can also perform in this range, but has low efficiency—less than 50%. Vane, gear or lobe pumps seem indicated.

Power-output nomograph shows that output hp is 7.2. Assuming over-all efficiency of 70%, which is reasonable

CAPABILITIES OF POSITIVE-DISPLACEMENT PUMPS

Relative desirability of typical commercial types

Pressure	Flow	Speed	hp/lb	Cavitation	Cost
piston	lobe	piston	vane	vane	gear
gear	screw	gear	gear	screw	vane
vane	gear	vane	piston	lobe	lobe
screw	vane	lobe	lobe	gear	piston
lobe	piston	screw	screw	piston	screw

Power-weight ratio* (typical values)

Type	vane	gear	piston	lobe	screw
hp/lb	0.26	0.11	0.09	0.02	0.01

*Industrial types

For the final specification—Data needed by pump manufacturer

Pressure	inlet, differential, discharge, line-and-fitting losses
Flow	gpm
Speed	rpm (range and max)
Liquid	viscosity, specific gravity, chemical inertness, operating temperature, cleanliness (solids? abrasives?)
Piping	suction conditions (flooded? lift? self-prime?)
Duty	continuous, intermittent, hours per day, days per year, driving motor, dust, moisture, fire hazard, corrosive fumes.
Misc.	seals desired, pump materials recommended, mounting.

for either a vane, gear, or lobe pump, the motor should have at least 10 hp. (A 50%-efficient turbine pump would require about 15 hp.) Gear pump would perform well and be quite economical. Its life would be long because liquid being pumped has lubricating qualities, and service is not severe. Vane pump would be somewhat more expensive, have slightly higher efficiency, and perform well. If efficiency is not the primary need, choose a gear pump.

2—High Lift

REQUIRED: transfer of viscous liquid from submerged tank 18 ft deep (lowest depth is 18 ft, but surface is essentially at ground level). Pump is to be placed at ground level. Vapor pressure of liquid at ordi-

contributing manufacturers

This article was prepared by Dr Wilson at the request of PRODUCT ENGINEERING to meet the industry's need for a quick, consistent pump-selection method.

In writing the article, Dr Wilson had the cooperation of the following manufacturers:

Adel Precision Products; American Engineering Co; Aurora Pump Div of New York Air Brake Co; Berry Hydraulics Div of Oliver Tyrone Corp; Bendix Aviation Corp, Lakeshore Div and Hamilton Div; Bijur Lubricating Corp; Blackmer Pump Co; Cessna Aircraft, Industrial Products Div; Commercial Shearing and Stamping Co; The Cornelius Co; Crossley Machine Co; Curtis-Wright Corp, Marquette Div; DeLaval Steam Turbine Co; The Deming Co; Denison Engineering Div, American Brake Shoe Co; Double A Products Co, subsidiary of Brown and Sharpe Mfg Co; Dynex Inc; Eastern Industries Inc; Eco Engineering Co; Hartmann Mfg Co; Hupp Aviation Co; Hydraulic Accessories Co; Hydraulic Products Div of Clark Equipment Co; Hydreco Div and Kinney Mfg Div of New York Air Brake Co; Jabsco Pump Co; John S Barnes Corp; Kellogg Div of American Brake Shoe Co; Lord Mfg Co; Manton-Gaulin Mfg Co Inc; Mechanisms Inc; Milton Roy Co; Moyno Pump Div, Robbins & Myers Inc; Northern Ordnance Inc, subsidiary of Northern Pump Co; Oberdorfer Foundries Inc; Oil-Dyne Inc; The Oilgear Co; Owatonna Tool Co; Parker Hydraulic Div of Parker-Hannifin Corp; Pesco Product Div of Borg-Warner Corp; Racine Hydraulics & Machinery Inc; Rockwell Mfg Co; Roper Hydraulics & Machinery Inc; Schutte & Koerting Co; SC Hydraulic Engineering Co; Sier-Bath Gear and Pump Co Inc; Simplex Engineering Co; Sundstrand Corp; Thompson Products Inc; Transfer Pump Co; Turbocraft Inc; Tuthill Pump Co; Vanton Pump and Equipment Corp; Vickers Inc; Viking Pump Co; Waldorf Instrument Co; Waterous Co; The Wayne Pump Co; Webster Electric Co, Oil Hydraulics Div; and Worthington Corp.

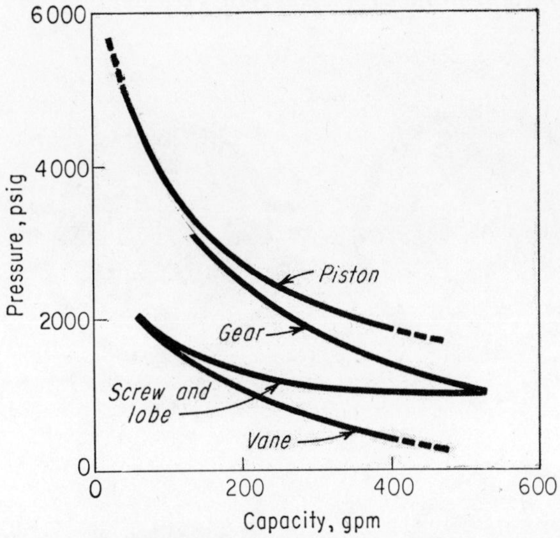

MAXIMUM PRESSURE AND FLOW values plotted here are from typical manufacturers catalogs. However, higher ratings are sometimes available for special applications.

nary temperatures is 2 psia. Flow is 150 gpm; discharge, 550 psi; viscosity, 1000 SSU at 70 F. Motors with speeds of 1800 or 2700 rpm are readily available; slower speeds will cost more.

SOLUTION: Gear, vane and piston pumps are available, but suction-head nomograph indicates that NPSH required is 32 ft of water. With an actual viscosity of 1000 SSU and an available NPSH of something in the order of 10 ft of liquid with a pump operating at 1800 or 2700 rpm, no pump is adequate. The manufacturer should be consulted, because cavitation problem is serious. There will be little competition on an economic basis—only a few pumps are available and none operate at speeds of 1800 or 2700 rpm. However, a pump with a displacement of 0.25 gal. per revolution running at 600 rpm could handle the situation.

3—Hydraulic Control Pump

REQUIRED: 1500 psi, 25 gpm, viscosity 150 SSU at 100 F. Motor to run at 1800 rpm. There are no problems regarding inlet pressure.

SOLUTION: Three types—gear, vane and piston—are available. Therefore selection must be based on some special characteristic or on costs. If pulsation is a serious problem, a piston pump with its periodic-flow characteristic might not be suitable; but a well-designed vane or gear pump will have smooth flow free from pulsation.

4—Fluid Power

REQUIRED: flow of 50 gpm at 4500 psi.

SOLUTION: Only piston and gear pumps are available—and very few manufacturers are producing pumps for this service. However, inlet conditions are good, viscosity of oil is 150 SSU at 100 F, and motor runs at 1800 rpm. Gear pump is cheaper, is less likely to cavitate. But efficiency is higher for the piston type. Either pump meets the requirements.

5—High Altitude

REQUIRED: to pump fuel at 3600 rpm, with flow of 100 gpm at 750 psi, viscosity 30 SSU. Pressure available at inlet is 3.5 inches of mercury absolute and vapor pressure is 2 inches. Many pumps will deliver 100 gpm at 750 psi, but no standard pump will deliver satisfactorily under inlet conditions prescribed.

SOLUTION: Specially designed pump, free from civitation even though energy available at inlet is extremely small. Any significant drop in pressure between the inlet and the pumping element is intolerable because vapor pockets would form and cause serious cavitation.

new method predicts

CAVITATION IN POSITIVE-DISPLACEMENT PUMPS

Inlet Configuration of 4 Different Pumps

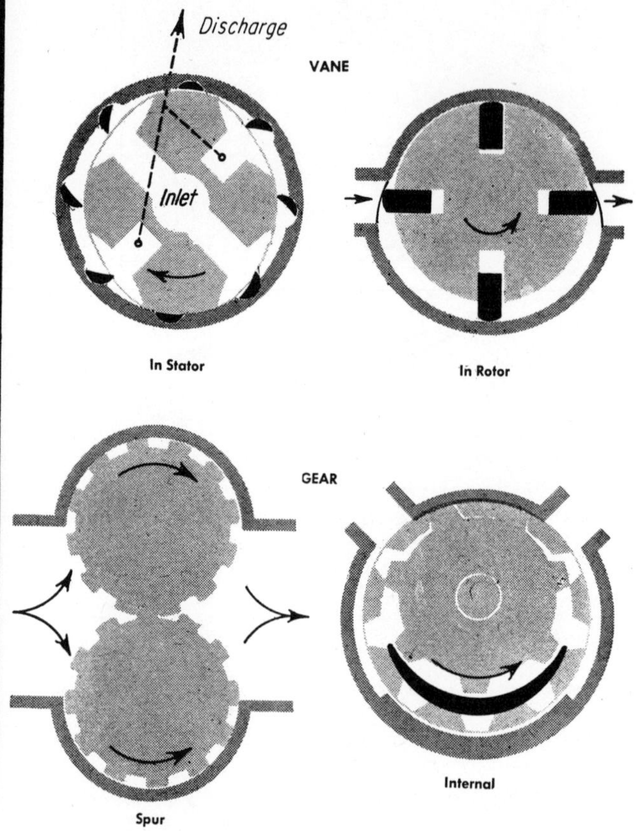

Most catalogs leave suction problems to guesswork. But a poor guess may result in a noisy pump, inadequate delivery, or damage from pitting. This article reviews the basic theory of cavitation in positive-displacement pumps; offers a consistent method of selecting pumps as well as interpreting their performance.

WARREN E WILSON

Chairman of Engineering, Harvey Mudd College, Claremont, Calif

One way to avoid cavitation is to design a pumping system that will match the suction capabilities of a particular pump. Pump manufacturers will certainly recommend that way of preventing vapor pockets behind the moving element. But the situation is usually reversed—the pump must be chosen to satisfy the system. Here is a graphical way to handle the problem.

Thorough tests on vane and gear pumps have indicated a consistent relationship between type of pump and the amount of acceleration required by the fluid inside the pumping cavity.

It is this relation that leads to the simplified equation

CAVITATION IN POSITIVE DISPLACEMENT PUMPS

Nomenclature

- **Volume flow, cu ft/sec**
  - Q_T theoretical pump discharge, $A_P V_P$
  - Q_L loss due to cavitation
  - Q_D actual pump discharge, $Q_T - Q_L$
- **Pressure, psf (1 in. hg = 70.7 psf)**
  - P_i pump inlet
  - $P_{1,2}$ pumping cavity
  - P_A atmosphere
  - P_v vapor pressure of fluid
- **Velocity, ft/sec**
  - V_P piston, vane or tooth
  - V_i pump inlet
  - $V_{1,2}$ fluid in pumping cavity
- **Other key units**
  - A area, sq ft
  - ρ mass density, slugs/cu ft
  - g 32.2 ft/sec²
  - C cavitation factor $(V_2 - V_1)/V_P$

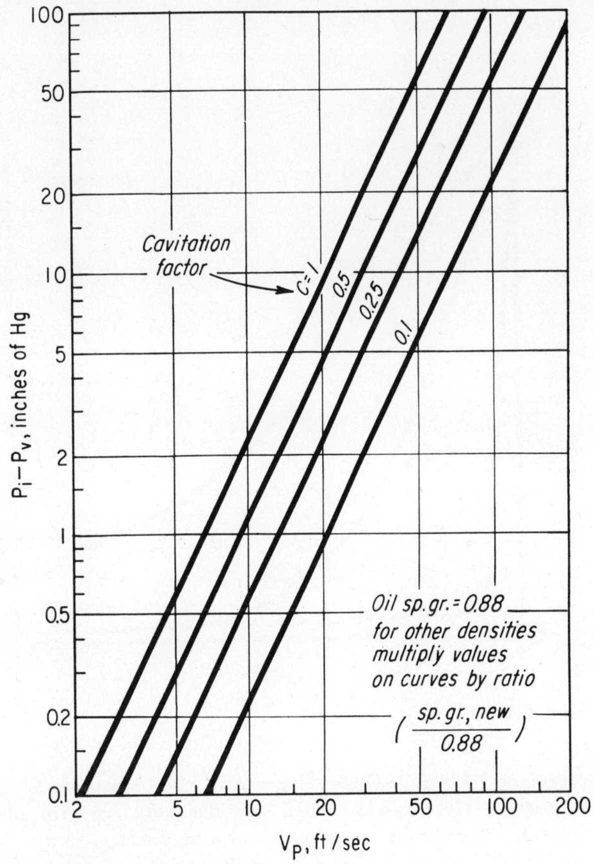

Probability of cavitation . . . when oil is the fluid can be quickly predicted with this graph that relates effective inlet pressure to pump velocity. Cavitation factor C is a measure of the acceleration required by fluid to reach velocity of the moving element. High values of C mean fluid is initially at standstill—requires max acceleration. Low values indicate fluid has initial velocity that helps get oil into pumping cavity.

Value of C depends on inlet design as well as pump type. Vanes-in-stator pump shown on page 47 gives pre-rotation to fluid because of centerline inlet—can have C as low as 0.25. Screw pumps may have value of 0.3, but spur gear and vanes-in-rotor pumps do no better than 0.4. Piston pumps will be poorer (0.7). Any pump can approach worst value (1.0) if inlet design is not correct.

EXAMPLE 1—hydraulic system has only 4 in. hg available at pump inlet; vapor pressure is 1 in. hg. Therefore, $P_i - P_v = 3$ in. hg. Plot for C = 1 shows any pump with impeller velocity less than 11 ft/sec will operate without severe cavitation. A pump that has C = 0.5 can be run with V_P as high as 16 ft/sec.

EXAMPLE 2—if pump type, speed and size are known, a velocity reading on horizontal scale can give min permissible value of $P_i - P_v$ for system on vertical scale.

for cavitation:

$$\left(\frac{Q_L}{Q_T}\right) = 1 - \left(\frac{P_A}{\rho C V_p^2}\right)\left(\frac{P_i - P_v}{P_A}\right)$$

The terms are kept dimensionless; (Q_L/Q_T) indicates loss in flow caused by cavitation, $(P_i - P_v)/P_A$ is a measure of static pressure available at inlet to accelerate fluid up to velocity of pumping element, and $(P_A/\rho C V_P^2)$ is a coefficient that includes effects of density and acceleration. This coefficient will usually remain fairly constant at fixed speed for low values of accelerating pressure. The equation is basis for all graphs with this article. The "Probability of Cavitation" curves, for example, were plotted by letting $Q_L/Q_T = 0$ (zero cavitation) and solving for $(P_i - P_v)_o$.

DERIVING THE EQUATION

A rigorous analysis of cavitation in an actual pump is not possible, but the idealized sketch given here of pumping elements shows important effects and provides groundwork for a workable equation.

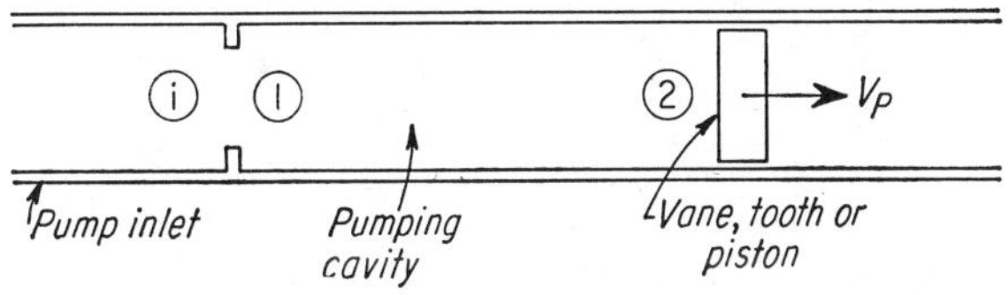

Cavitation results if the liquid cannot keep up with the moving element, and occurs between positions 1 and 2. Fluid pressure P_1 provides the only energy available to accelerate the fluid from velocity V_1 to V_2. The valuc of pressure necessary is directly proportional to the pump speed V_P. The fluid velocities inside the cavity are not readily measurable, but this general equation, based on $F = Ma$, is valid:

$$(P_1 - P_2)\,A_P = \rho\,Q_D\,(V_2 - V_1)$$

Tests have shown that each pump has a characteristic value for the ratio $C = (V_2 - V_1)/V_P$. Substituting CV_P for $(V_2 - V_1)$ gives:

$$(P_1 - P_2)\,A_P = \rho\,Q_D C V_P$$

Let $P_2 = P_v$ (nearly correct) and $P_1 = P_i$ (valid for severe cavitation because flow is low). The equation can now be expressed:

$$Q_D = \frac{(P_i - P_v)\,A_P}{\rho\,C V_P}$$

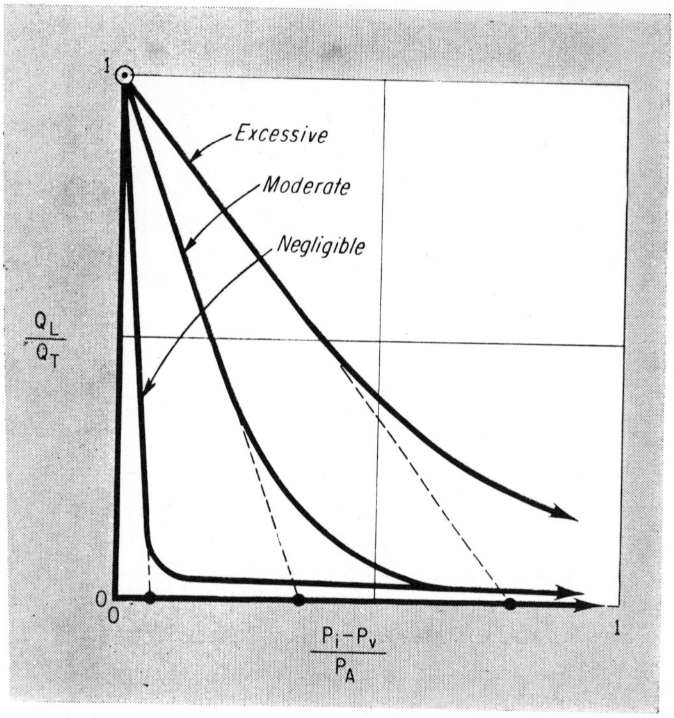

Nature of cavitation . . .
is seen in these curves plotted with dimensionless parameters. Q_L/Q_T, a measure of cavitation, varies almost linearly with available positive suction pressure in region of severest cavitation. A hypothetical pump with zero cavitation corresponds to curve on extreme left, where any pressure is sufficient to give full flow, $Q_LQ_T = 0$. Actual pumps will perform in area indicated by the three other curves. Curves are for pure cavitation (effects of entrained air and slip are not included).

From the accompanying key to nomenclature, $Q_D = Q_T - Q_L$ and $Q_T = A_PV_P$. Introducing these factors and P_A gives the cavitation equation, which is a more useful tool.

CHOOSING INLET CONFIGURATION

It is not enough to specify pump type. For any pump, suction performance will be improved by guiding and accelerating the fluid before it reaches the pumping chamber. It is much more efficient to control the fluid in a smooth inlet channel than in the pumping cavity.

Some positive-displacement rotary pumps pre-rotate the fluid by having a hollow rotor shaft serve as inlet. Radial flow through special passages guides the fluid and pressurizes the pumping cavity.

Another type of rotary pump takes leakage flow from near the discharge, controls the pressure, and reduces cavitation by feeding it back to the inlet when necessary.

LIMITATIONS TO THE METHOD

Fluid that is cavitating may not flash to vapor as completely as theorized. In such cases, the pump gives better performance than predicted because back-pressure is less. Conversely, fluids with vapor pressures higher than inlet pressure can prevent pumping entirely—and pressurization or cooling of the inlet may be required.

Air or gas entrainment will increase cavitation, particularly if inlet pressure drop is great, because trapped air expands and occupies pumping space. The simplified equation assumes air has not been entrained or dissolved, and that theoretical vapor pressure of liquid is reached.

Though temperature variations will affect vapor pressure, slope of curve Q_L/Q_T vs $(P_i - P_v)/P_A$ remains the same. Any fluid system at any temperature can be examined by this method as long as there is a positive suction head.

Viscosities up to 2000 SSU do not affect performance but above that, cavitation increases. Over 20,000 SSU, it may become necessary to pressurize or heat the fluid.

Any one or all of many separate effects can produce the family of Q_LQ_T vs $(P_i - P_v)/P_A$ curves shown. For example, if speed is increased but other factors remain the same, cavitation may become "excessive" instead of "moderate." But a severe inlet restriction or entrained air will give similar results. The method therefore is a guide only—careful analysis and testing will still be necessary.

WHAT ABOUT NPSH?

If desired, the new cavitation equation can be related to the older concept which defines "net positive suction head." This value (NPSH) is total inlet pressure less vapor pressure of the fluid:

$$NPSH = \frac{P_i}{\rho g} + \frac{V_i^2}{2g} - \frac{P_v}{\rho g}$$

The relationship can be seen in plot Q_L/Q_T vs $(P_i - P_v)/P_A$ at the intersection of straight portion of curve extended to horizontal axis. Here Q_L/Q_T is 0, and . . . .

$$\left(\frac{P_i - P_v}{P_A}\right)_0 = \left(NPSH - \frac{V_i^2}{\rho g}\right)\frac{g\rho}{P_A}.$$

Special tests help predict

Squeegee pump performance

For the first time, here is a proven technique for applying those unusual pumps with the flexible vanes, liners, or tubing.

DR WARREN E. WILSON, Chairman, Engineering Dept, Harvey Mudd College, Claremont, Calif

HOW do you predict the performance of a pump that does part of its work pumping, the rest squeezing and unsqueezing itself? And why use such a pump?

Three common pump types—flexible vane, flexible liner, and flexible tube—fit that description. For this article we'll call them *squeegee* pumps because the name is often used and seems to fit.

To answer the first question, we ran hundreds of tests. The result is the final graph (page 150) with a method (page 149) for predicting performance of any squeegee pump. All you need are two test points, usually provided by the manufacturer, and you can quickly estimate flow and efficiency under any normal conditions of speed, pressure, and viscosity.

The second question, why use a squeegee pump, is easy too. The elasticity gives each pump special advan-

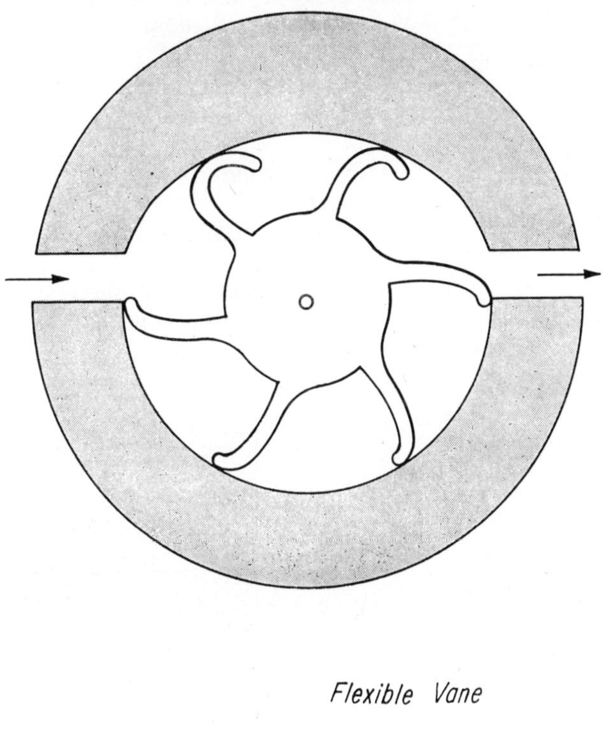

Flexible Vane

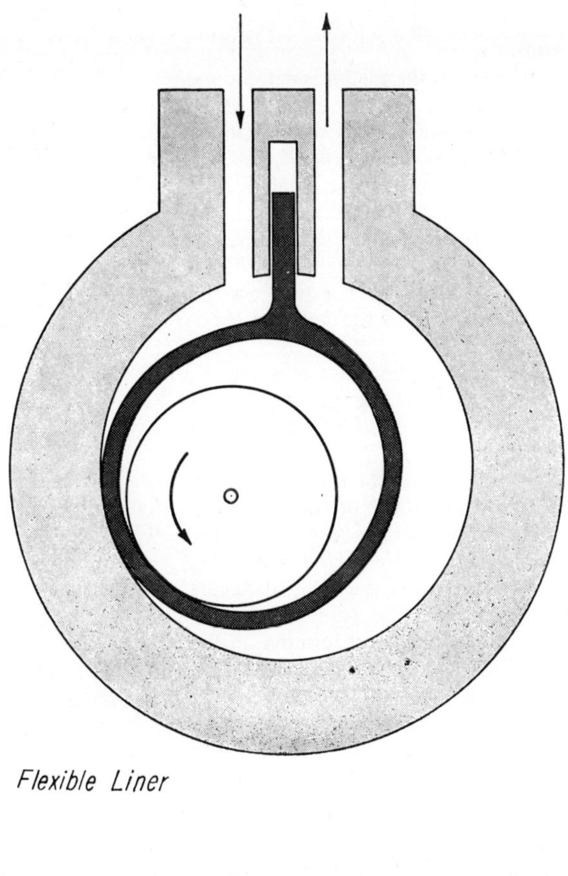

Flexible Liner

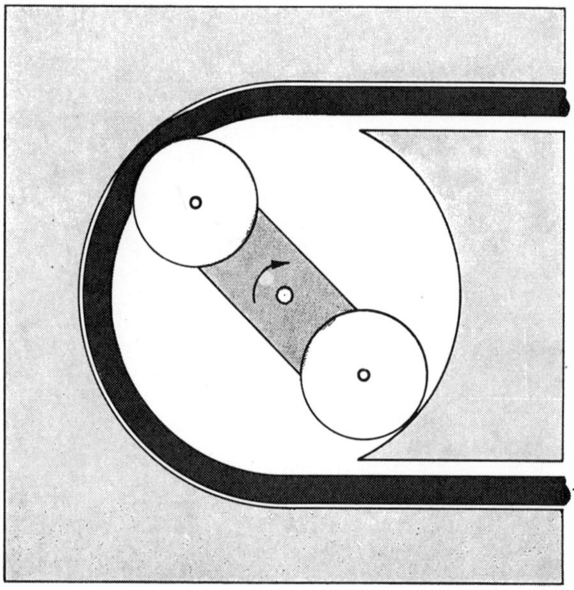

Flexible Tubing (rotary)

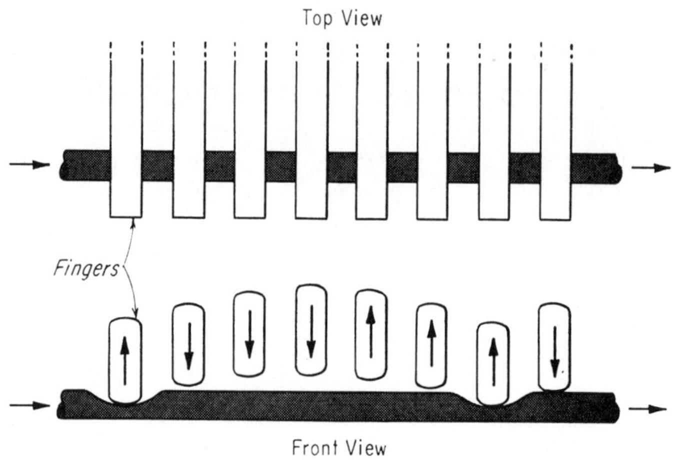

Flexible Tubing (linear)

Pump theory

It is useful to visualize a physical model from which a mathematical model can be established. The sketches *A* and *B* represent a positive-displacement pump which has the essential characteristics of most of the squeegee pumps.

In sketch *A*, the piston is acted upon by a force *F*, and is supported by springs. As the piston moves downward, as shown by velocity *V*, the motion is resisted by frictional forces, F_f, and pressure P_2 is built up in the liquid within the chamber by the resistance to flow through the outlet valve. The discharge pressure on the downstream side of the valve is P_d.

In sketch *B*, the liquid is flowing into the cavity, from inlet pressure P_i. Within the cavity the pressure is P_1. The springs will, in this case, assist the piston.

That this is a reasonable representation of at least some of the pumps to be considered can be seen in the example of a *flexible-tubing* pump. A cam or some other mechanical device applies force to the tubing and collapses it, forcing liquid to move in the tube. The level of force depends on fluid pressure, and on the elastic properties of the tubing as represented by the spring forces. The moving piston represents the cam of the real pump.

The performance equations will be similar to those for any positive-displacement pump. Here are the important ones:

Delivery

$$Q = AV \qquad (1)$$

Displacement

$$D = Q/N = AV/N \qquad (2)$$

Cam force

$$F - Kx - P_2A - F_f = Ma \qquad (3A)$$

If acceleration is small the term to the right of the equal sign can be eliminated as follows:

Cam force

$$F = Kx + P_2A + F_f \qquad (3B)$$

Cam return

$$F = Kx + P_1A - F_f \qquad (4)$$

If the piston does not follow the cam, force $F =$ zero.

Free return

$$F_f = Kx + P_1A \qquad (5)$$

where frictional resistance F_f has a viscous component and a dry-friction component.

If the springs (tubing walls) are very stiff, the hypothetical piston might move so rapidly on intake that the pump cavitates. If the springs are too weak, there might not be sufficient force to follow the cam. In this case, filling will not occur unless the inlet is pressurized. There is the necessity for careful design of the cam in relation to the operating speed, fluid, pressures, and tube design.

It might be well at this point to relate the performance of these pumps to the conventional positive-displacement pump for which there are these well-established equations:

Delivery

$$Q = DN - \frac{C_s \Delta PD}{2\pi\mu} - Q_c \qquad (6)$$

Driving torque

$$= \frac{PD}{2\pi}(1+C_f)+C_d\mu ND+T_c \qquad (7)$$

These two equations completely describe the performance of any posi-

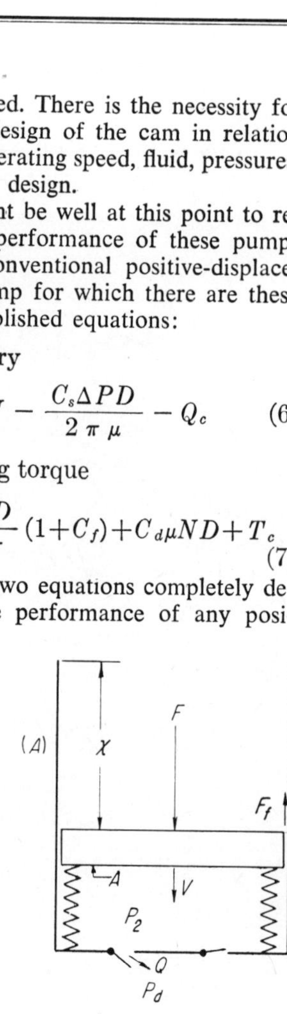

tages, even though flow calculations are less certain than those for conventional positive-displacement pumps.

For example, flexible vanes conform to the housing contour, pumping without the complexity of fitted vanes. Flexible liners eliminate vanes entirely, moving the fluid and at the same time sealing it from the mechanism. Flexible tubes are the ultimate in packless pumps: The fluid is squeezed like toothpaste and touches nothing but the inner wall. And all three types have low-cost, easily replaceable flexible elements which can pump dirty and corrosive fluids, even containing solids, without jamming the pump.

The pumps

The sketches on page 144 explain the pumping principles better than words can. The flexible-vane pump has a single-piece rotor in a shaped housing. As the vanes bend, fluid is squeezed to the outlet. A flexible-liner pump also squeezes the fluid to the outlet, but pumping and sealing stop momentarily as the rotor goes through top dead center. Momentum of the fluid prevents backflow during normal operation, but a check valve is needed if the pump stops at top dead center. A flexible-tubing pump traps fluid at inlet pressure between sets of fingers or rollers, and squeezes the trapped volume of fluid along the tubing to the outlet.

Each squeegee pump has positive-displacement action, and most of the wealth of positive-displacement theory applies (page 144). A positive-displacement pump gradually reduces the volume of the pump enclosure, thus forcing the liquid into the outlet. Differences are in how the volume is reduced and how the high-pressure liquid at the outlet is kept from leaking excessively to the inlet.

The parameters for calculating over-all performance of squeegee pumps are the same as those for positive-displacement pumps. Only the relative limits of performance differ. For example, pressure and speed will be lower because elastic parts are not as strong as rigid parts. And efficiencies are much less, but this is not usually a problem for such small pumps.

The parameters also hold for many novel types not tested in this program. The Moyno pump made by Robbins and Myers, for instance: It has a helix-angle rotor turning in an elastic stator (*PE*—Jan 8 '62, p 54). And there's the so-called peristaltic pump from Czechoslovakia, reported in *Engineering Materials and Design* magazine, April 1963, p 259. It is similar to the Randolph pump except that the squeezing is done with water pressure in a metal jacket surrounding the flexible tube.

continued, next page

By way of explanation

Step-by-step techniques for anticipating performance and applying little flexible-element pumps just haven't been available. For the centrifugals, yes—*PE*—July 23 '62, p 68. And for the conventional positive-displacement units—*PE*—June 6 '60, p 42. But the little pumps needed testing on the same stand, under the same conditions.

So PRODUCT ENGINEERING asked Dr Wilson to do the job. Of those we asked, four makers — Jabsco, Randolph, Sigmamotor, and Vanton—provided pumps and gave us counsel on each one's special characteristics. We underwrote the costs, provided encouragement, acted as liaison. The result is the graphs and selection guides in this article, based on standard tests, that fill a gap in available performance data.

---

tive-displacement pump. Note: for Eq 6 and 7 only, Q = in.³/min and D = in.³/rev.

In the case of a flexible tubing pump, one might question the existence of a slip flow, which is described in the delivery equation by the term

$$\frac{C_s \Delta PD}{2\pi\mu}$$

In the sketches the slip passageway is shown between the edge of the piston and the cylinder wall. In a flexible-tubing pump, the slip passage is the clearance between the squeezed walls. It can be zero if the cam exerts sufficient force, but it is found experimentally that there is slip flow. Cavitation is also present at higher speeds.

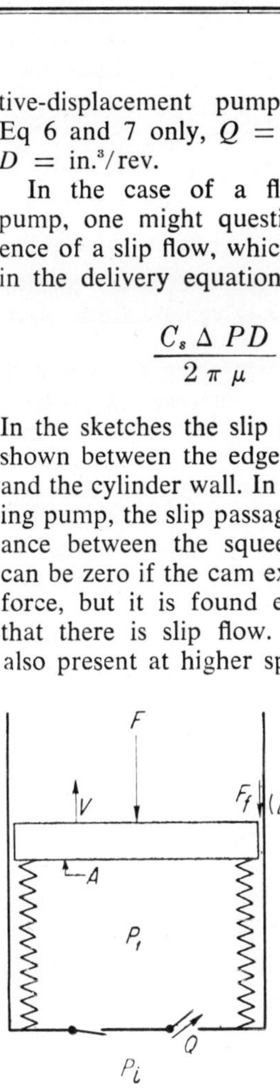

Symbols

a = acceleration, ft/sec²
A = piston area, in.²
C = coefficients: C_f = dry friction;
C_d = viscous friction; C_s = slip
D = pump displacement, gal/rev or in.³/rev
E_V = volumetric efficiency, % = actual/ideal
E_T = torque efficiency, % = ideal/actual
E = over-all efficiency, % = output hp/input hp
F_f = frictional force, lb
F = cam or piston force, lb
K = spring constant, lb/in.
L = piston stroke, in.
N = shaft speed, rpm
P_d = discharge pressure, psi
P_i = inlet pressure, psi
ΔP = pressure differential across pump, psi
P_{max} = shutoff pressure, psi, with blocked discharge
$P_{1,2,}$ = chamber pressure, psi
Q = flow, gpm: Q = delivery;
Q_c = cavitation loss; Q_s = slip loss
T = torque to drive pump, lb-in.
T_i = ideal torque, lb-in.
T_c = frictional torque, lb-in.
V = piston velocity, in./sec
x = piston motion, in.
μ = viscosity, lb min/in.²

Test results

The tests were run at room temperature and sea-level pressure with water and with glycerin as the fluids. At 70 F, water has an absolute viscosity of 1 centipoise and a kinematic viscosity of 0.000011 ft²/sec. Glycerin at 70 F has an absolute viscosity of 1400 centipoise and a kinematic viscosity of 0.0115 ft²/sec. Water is close to the "ideal" fluid, and glycerin represents an extremely viscous fluid.

Experimental data from all tests reveal clearly that with water, slip flow increases rapidly with pressure. With glycerin it increases much less rapidly. On the other hand, loss in delivery due to inlet restriction is much greater with glycerin and other high-viscosity liquids than it is with water. Torque requirements are significantly greater with glycerin than they are with water, resulting in lower over-all efficiency. However, the increase in torque is not at all proportional to the increase in viscosity of the liquid itself. If it were, it would be essentially impossible to pump viscous oils with a positive-displacement pump.

Viscosities up to 2000 cp present no apparent difficulties. Between 2000 and 4000 cp, delivery is reduced in pumps with flexible liners and hoses. Above 4000 cp, special consideration must be given to the system design.

The results: If pictures are better than words, then so are graphs. The succeeding sections include graphs based on actual tests of the typical pumps described in the performance table on the next page.

Flexible-vane pump

In the Jabsco pump tested, there is but one moving part: a Neoprene or similar impeller, which is self-priming and easily replaced. It pumps viscous or light liquids and handles solids in suspension over a wide range of speeds and in either direction. Fig 3 describes performance.

Note that although delivery decreases with pressure, it is not in a linear manner as for a piston pump. The impeller blades are flexible and the clearances through which slip occurs change in size with pressure. Consequently, slip flow is not proportional to the first power of pressure, as it would be with a viscous liquid and rigid blades, or even to the square root of pressure, as it might be with thin liquids and rigid blades.

At very high speed the delivery is somewhat less than ideal, doubtless caused by cavitation near the entrance. Note in the graphs that the experimental data were obtained without pressurization of the inlet. Pressurization usually eliminates cavitation.

Torque exceeds the ideal because of viscous and dry friction. The blades are bent with considerable force, resulting in frictional drag. There is additional drag on the ends of the rotor in contact with the endplates. As pressure increases, the difference between ideal torque and actual torque decreases, contrary to what you would expect with an ordinary positive-displacement pump of the piston, vane, or gear type. Perhaps it can be explained as a reduction in frictional drag when liquid pressures lift the flexible blades and lower the force of contact.

Volumetric, torque, and over-all efficiency are dependent on pressure and speed. The general form of the curves are as would be expected for any positive-displacement pump. Volumetric efficiency at very low pressures is high, and volumetric performance in general compares well with that of a low-pressure gear pump. However, torque efficiency is relatively low at the lower pressures, and only at the very highest pressures does it approach a value which would be considered reasonable for conventional pumps.

Over-all efficiency, even at the peak, is only slightly over 30%. This would be unacceptable in a standard positive-displacement pump, but is quite acceptable in this case because over-all efficiency has been sacrificed to obtain the desirable performance characteristics mentioned on page 145. Proof is that such pumps are being used for a wide range of applications in the in-

continued, next page

Performance of sample pumps

Pump descriptions				Test values			
	D gal/rev	Q gpm	N rpm	E_V, % at $\frac{P_{max}}{2}$	E, % (best)	Max test N rpm	Max useful P_d, psi
Jabsco *(vane)*	**0.015**	**20**	**1750**	**60**	**30**	**3600**	**26**
Randolph *(tube)*	**0.0019**	**1**	**575**	**80**	**40**	**1800**	**30**
Sigmamotor *(tube)*	**not measured**		**400**			**400**	**20-40**
Vanton 12 *(liner)*	**0.0023**	**2**	**1750**	**50**	**30**	**2900**	**30**
Vanton 60A *(liner)*	**0.0063**	**10**	**1750**	**30**	**30**	**3000**	**35**

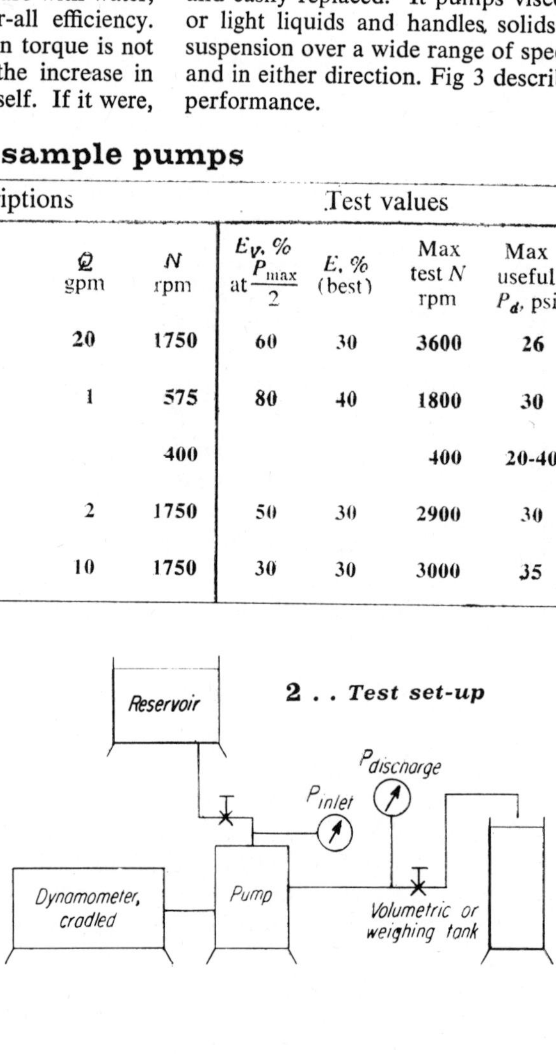

2 . . *Test set-up*

Pump inlet is several feet below reservoir. Flow is measured with a weigh tank and an electric timer.

Pump performance

(actual test data)

3 . . *Flexible vane (Jabsco): normal* P_i

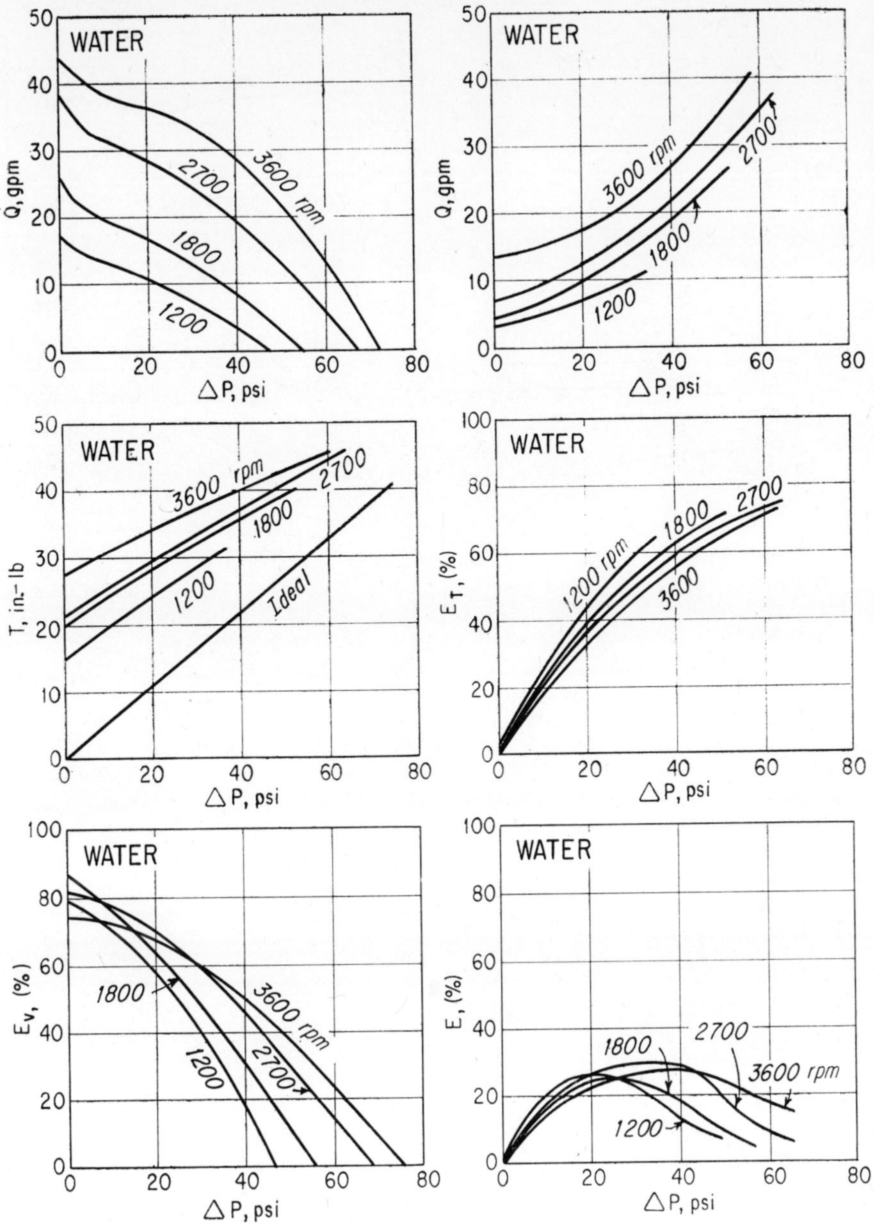

4 . . *Flexible tube (Sigmamotor): normal* P_i

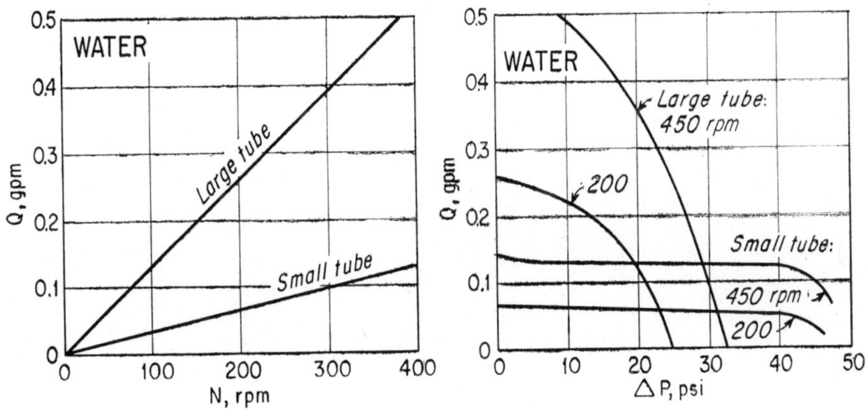

dustrial, marine, chemical, food, beverage, and dairy fields.

Flexible-tube pump

Two types were tested: a Randolph double-roller pump with a ¼-in. ID × 7⁄16-OD tube, Fig 4, and a Sigmamotor multiple-finger pump with three sizes of tubes: ¼ × ½, ⅜ × 11⁄16, and ⅜ × ⅝ in., Fig 5.

Volumetric efficiency for each is good at the recommended low operating speeds because the tube is forcefully squeezed by the rollers or fingers. At higher speeds, cavitation at the inlet reduces effective flow. Flow is roughly proportional to speed until pressures exceed 30 to 40 psi, or until high speeds cause cavitation.

Torque efficiency and over-all efficiency are somewhat low because the work done in compressing the tube is not fully recovered during the expansion. A thin-walled tube is most efficient but the operating pressures must be kept low. A thick-walled tube can hold higher pressures, sacrificing efficiency. Their relatively low efficiency is more than offset in applications such as heart pumps where strictest sanitary requirements must be met.

The roller-driven version is more efficient in the mechanical sense because the elements are simpler. The finger-driven tube and its accompanying mechanism is mechanically less efficient, but compensates by substituting simple occlusion of the cross-section for the combined occlusion and rolling action of the rotary type, thus increasing tube life. The finger arrangement also permits you to pump through several tubes of the same or different cross-sections at the same time.

Both types are available with adjustable speed. Maximum recommended speeds should not be exceeded for two reasons: Inlet cavitation reduces effective flow, and the high speeds shorten the life of the tubing considerably.

Flexible-liner pump

The pumps tested in this category were two Vanton Flex-i-liners with natural-rubber liners, Fig 6. Performance characteristics are similar to the other pump types discussed. The rolling action is more gradual than that of the tube pumps, thus stressing the liner wall less.

Tests with water for the pumped fluid show results close to calculated ideal performance, whereas tests with the more viscous glycerin are farther from the ideal. The viscosity effect is particularly noticeable at zero pressure rise where slip flow ought to be zero but cavitation reduces effective delivery.

The manufacturer recommends a maximum speed of 1750 rpm, which is borne out by tests that show a large drop in performance above 1800 rpm. Delivery at constant speed decreases as pressure increases, reaching a useful limit of 20 to 30 psi with water, and 30 to 40 psi with glycerin. However, the manufacturer can set the delivery pressure higher, to 70 psi, with some sacrifice in life.

Torque exceeds the ideal by a significant margin. Volumetric efficiency is relatively low throughout the operating range. Torque efficiency is low except at high pressures. Over-all efficiency of the smaller size pump is very low at speeds over 900 rpm, although the larger size is more efficient. The material used in the liner affects performance, the thin materials being most efficient. Thick liners are less efficient but can handle higher pressures.

Composite performance graph

Examination of all the data discloses a similarity in the shape of the performance curves. The maximum over-all efficiencies are between 20 and 40% for each pump, and seem to occur at a pressure nearly halfway between zero (open discharge) and shutoff (closed discharge) for a given rpm.

These observations and some known characteristics of positive-displacement pumps make possible a very crude but handy method of predicting performance with but the sparsest of published performance data. Fig 8, a composite performance graph, is the heart of the method. Here are the assumptions.

At any given operating speed, pump discharge pressure will vary from zero (open discharge) to shutoff pressure (blocked discharge). Over-all efficiency is highest, 33% for the composite, at rated delivery and pressure, which occurs at slightly less than half the shutoff pressure.

Volumetric efficiency is 100% when discharge pressure is zero if there is no leakage or cavitation. Volumetric efficiency decreases linearly with pressure, reaching zero at shutoff pressure.

Torque efficiency, defined as the ratio of ideal driving torque to actual driving torque, is zero when pressure is zero, and maximum when discharge pressure is maximum. The plotted curve is typical, ending at 75% shutoff efficiency.

Most important to the method is the ratio $(\Delta P/N)/(\Delta P_{max}/N)$ where ΔP is based on actual discharge pressure and ΔP_{max} is the shutoff pressure differential. See example on next page.

For example, if a pump is known to

5 . . *Flexible tube (Randolph): normal P_i*

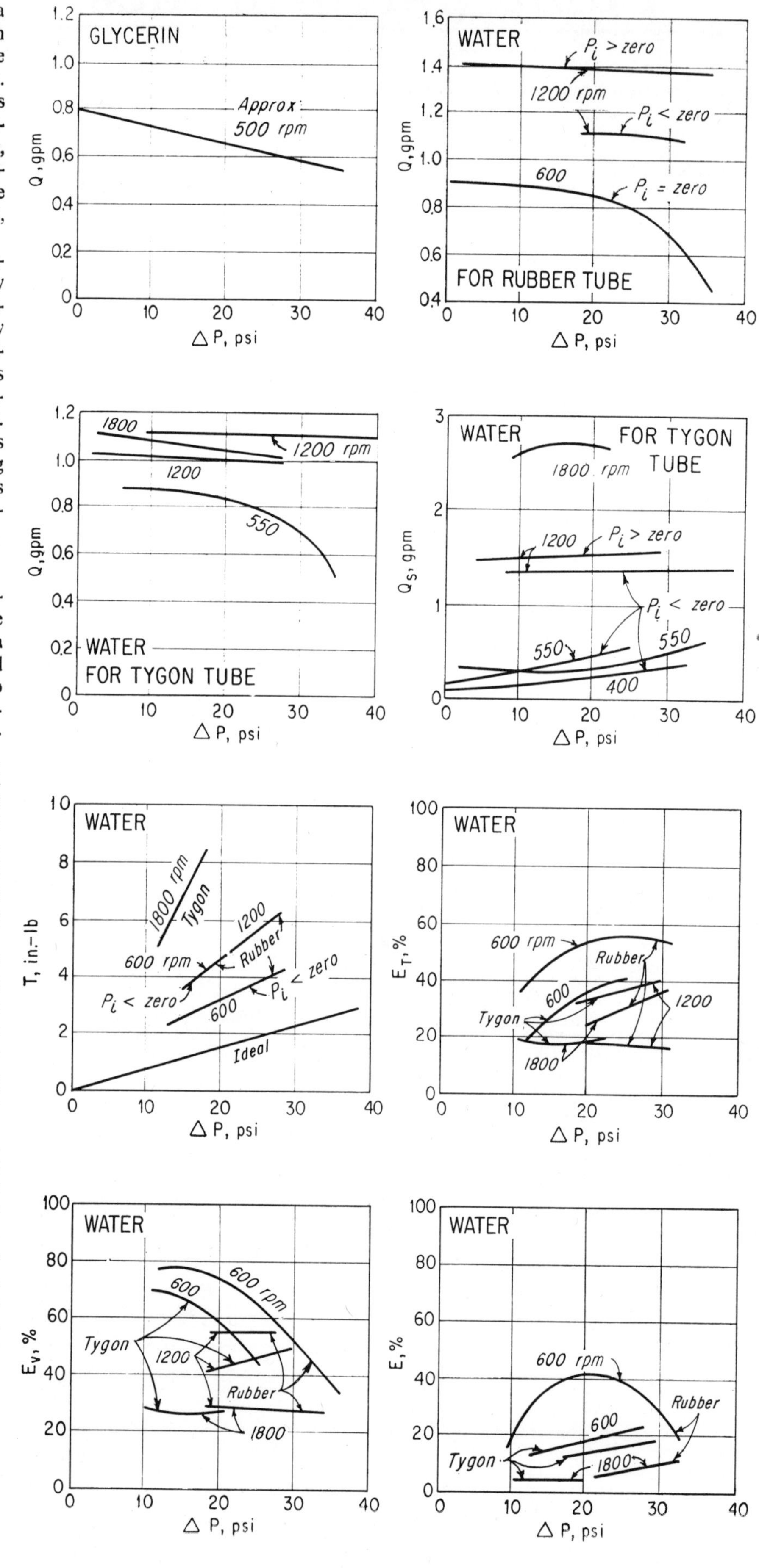

6 . . *Flexible liner (Vanton): normal P_i*

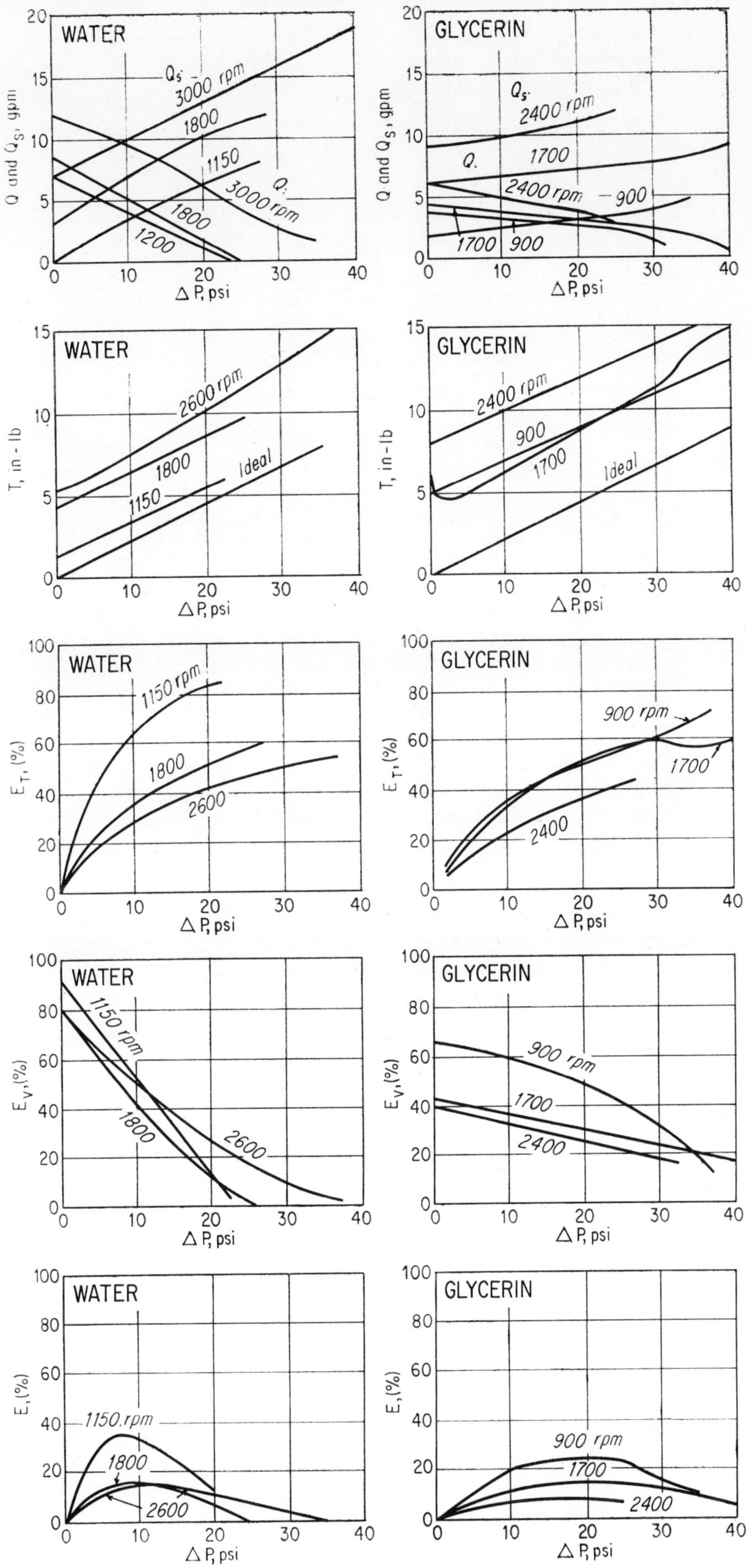

develop a maximum pressure rise of 45 psi at 450 rpm with blocked discharge, then the quotient $\Delta P_{max}/N = 45/450 = 0.10$ represents the maximum value of the abscissa in the composite graph. To determine its capabilities at 20 psi and 400 rpm, where $\Delta P/N = 20/400 = 0.05$ and $(\Delta P/N)/(\Delta P_{max}/N) = 0.5$, read the graph as follows: Over-all efficiency = 30%, volumetric efficiency = 50%, and torque efficiency = 58%. This is very nearly the peak performance. Peak performance would be at 33% efficiency, or $(\Delta P/N)/(\Delta P_{max}/N) = 0.35$. Here, torque efficiency = 50%, volumetric efficiency = 65%, and $(\Delta P/N) = 0.035$. Note that efficiency stays relatively high (over 25%) from ratios 0.15 through 0.60.

Inlet conditions are of extreme importance in predicting performance. If inlet pressure is less than 10 psia (5 psi below ambient), delivery will decrease. If inlet pressure is 5 psia or less, the pump probably will not function satisfactorily. Between 5 psia and 10 psia, you can arbitrarily assume a linear variation of the delivery with absolute pressure. In other words, at 5 psia, delivery is zero, and at 10 psia, delivery is full, all at zero discharge pressure.

Speed in excess of some reasonable value, determined experimentally, affects delivery in much the same manner as reduction in inlet pressure: Higher delivery rates cause lower inlet pressures for the same given upstream conditions. One rule is to keep inlet velocity less than 20 ft/sec.

Viscosity in excess of 500 centipoise at room temperature has an effect similar to that of a reduction in inlet pressure. A first approximation in determining the effect of viscosity is to assume a 25% reduction in volumetric efficiency for each 1000 centipoise increase in viscosity. Although this would indicate that volumetric efficiency is zero if viscosity is 4000 centipoise, it would not happen in most cases. But if viscosity is in that range, special design is in order.

Example of predicting performance

Suppose you want to pump 1.5 gpm of water against a 25 psi discharge with one of the squeegee pumps. Some information on the pump you selected indicates that shutoff pressure is 40 psi at 300 rpm, and delivery is 1 gpm at zero discharge pressure and 200 rpm. Can the pump meet your requirements of 1.5 gpm and 25 psi? And at what speed should you run it?

Use the composite performance chart (next page). The known value for $\Delta P_{max}/N = 40/300 = 0.133$. The best value for $(\Delta P/N)/(\Delta P_{max}/N) =$

0.35 based on peak over-all efficiency. Thus $\Delta P/N$ ought to be about 0.35 x 0.133 = 0.046, or $N = \Delta P/0.046$ = 25/0.046 = 540 rpm. Torque efficiency is about 50%, volumetric efficiency 65%, and over-all efficiency 33%.

Pump delivery is computed by converting known flow and speed to displacement, then applying the speed and efficiency factors. Known: Flow Q=1 gpm at 200 rpm with open discharge. $D = Q/N$ = 1/200 = 0.005 gal/rev. Delivery at 540 rpm and 25 psi discharge = ND x volumetric efficiency = 540 x 0.005 x 65% = 1.76 gpm, which is more than the requirements. Speed can be reduced slightly if you want, at some sacrifice in efficiency.

Torque to drive the pump is computed as follows:

$$T = \frac{\Delta PD}{2\pi} \times \frac{1}{E_T} = \frac{25}{2\pi} \times 0.005 \times \frac{231}{0.5}$$

$$= 9.2 \text{ lb-in.}$$

Correspondingly, input power is:

$$Hp_{in} = 2\pi \times 540 \times \frac{9.2}{(12)\,(33{,}000)}$$

$$= 0.079 \text{ hp}$$

Power output is:

$$Hp_{out} = 1.76 \times 231 \times \frac{25}{(12)\,(33{,}000)}$$

$$= 0.026$$

yielding an over-all efficiency of 33%, which is as assumed.

7 . . *Effects of viscosity and inlet restriction*

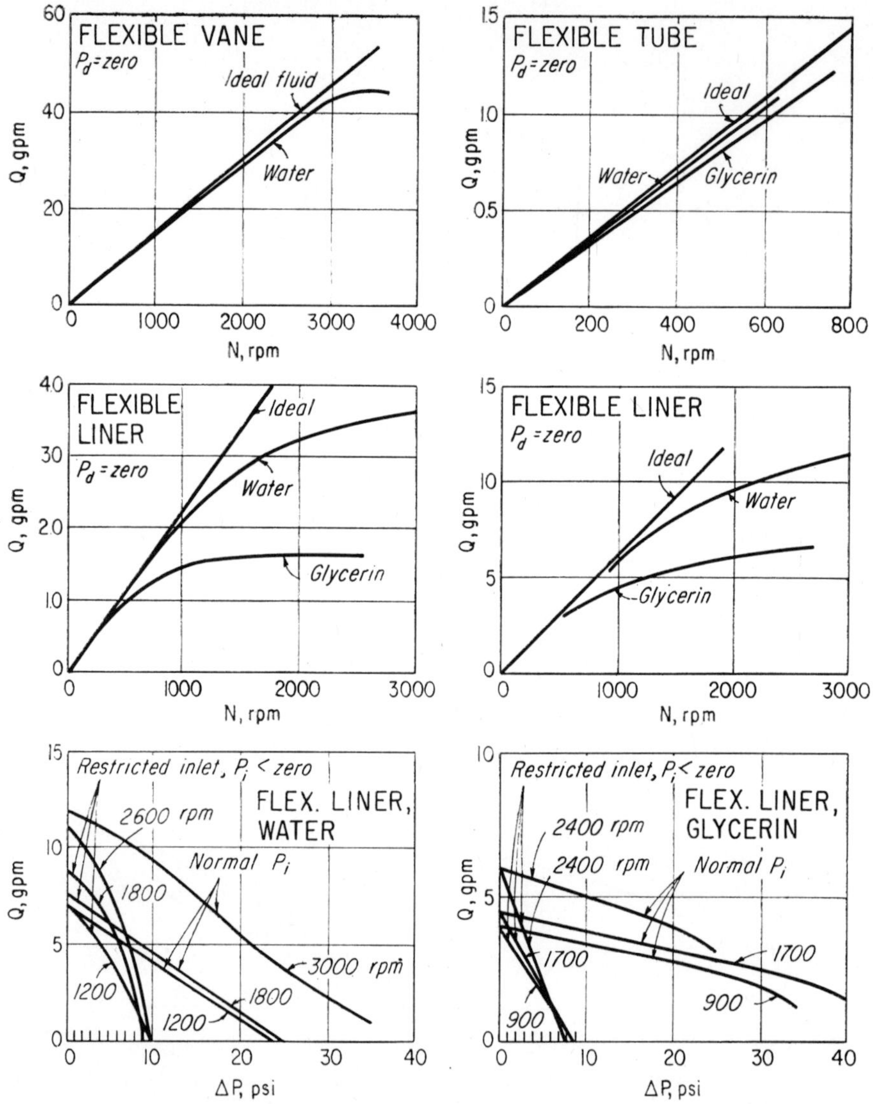

8 . . *Composite performance (See text for method)*

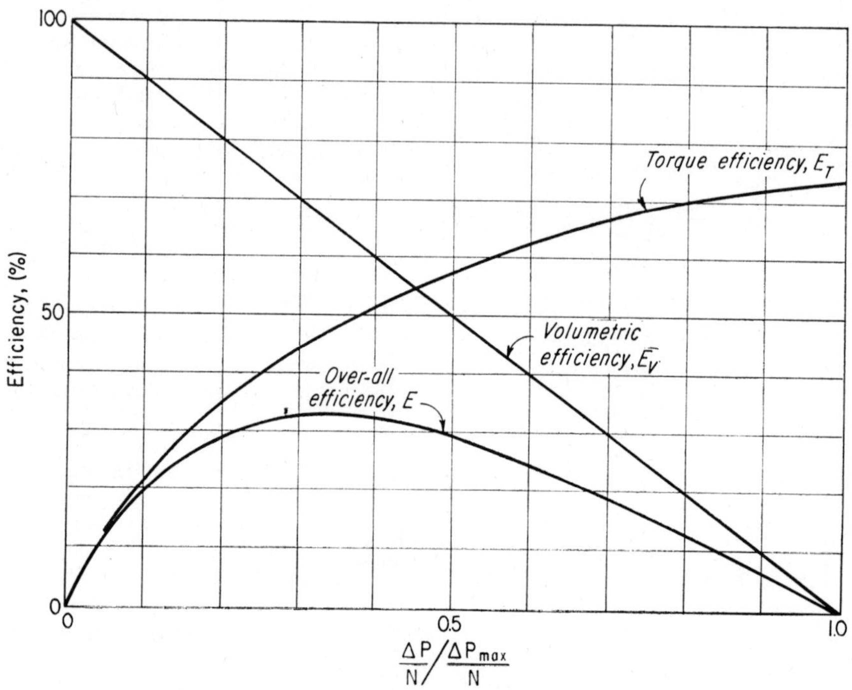

400-cycle motor drives . . .

New centrifugal pump

High-speed single-impeller pump develops high pressure at low flow, competing with more-complex pumps operating at conventional speeds.

FRANKLIN D. YEAPLE, Associate Editor

WHEN flows are less than 50 gpm at discharge pressures over 200 psi, you need more than an ordinary centrifugal pump. Such jobs in the past have been handled by positive-displacement pumps, turbine-type regenerative pumps or, in some instances, multistage centrifugals.

Now you can buy a single-stage high-speed centrifugal pump (photo below) good for high pressure and low flow. It can be driven directly by a 400-cycle motor at speeds up to 24,000 rpm.

This new pump is much smaller than ordinary centrifugals. The impeller is full-open; the casing is concentric instead of volute; the casing and impeller rings are eliminated; and the diffuser is small, tangential to the casing bore.

Conventional centrifugal pumps often have double-shrouded impellers to guide the flow and support the vanes. And the vanes and impeller casing are shaped to partially convert the velocity energy into static pressure by diffusion within the casing. This improves efficiency somewhat, but the static pressure inside the casing forces leakage flow back to suction; therefore "wearing" rings are necessary for sealing between the impeller and casing near the hub.

The new design has several basic advantages over conventional centrifugals for operation at high speeds and pressures:

- Reduced axial thrust. Conventional shrouded impellers have considerable pressure unbalance at high pressures.
- Better radial balance. Pressure is

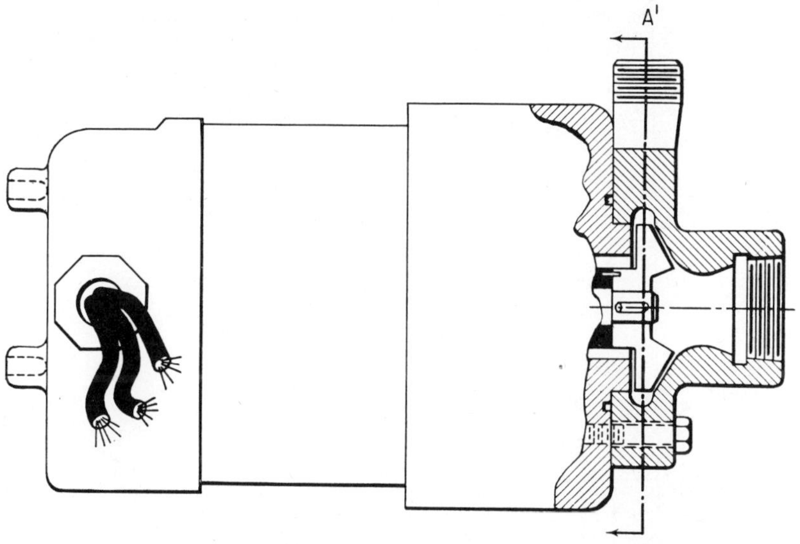

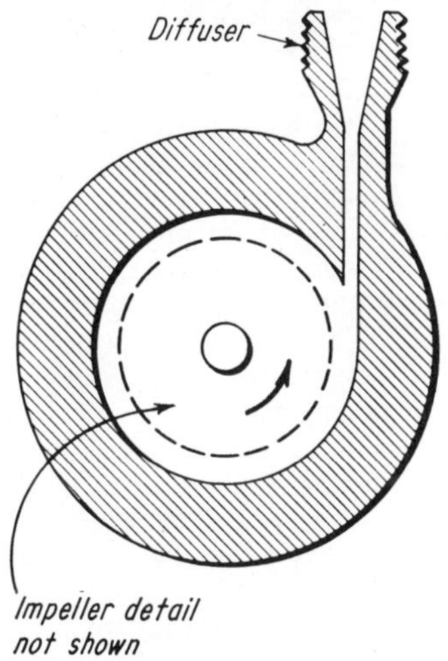

Section A-A'

HIGH-SPEED IMPELLER . . . achieves high pressure with relatively low flow because of its small size and special design. Impeller is full-open. Pump casing is circular and the radial clearance is about equal to the width of the impeller. Rating of version shown is: 12,000 rpm, 5 hp, 32 gpm, 320-ft head, 52% efficiency.

uniform because casing is concentric instead of volute.

• No rubbing wear. There are no wearing rings, therefore no rubbing at any speed.

• Simplified machining and casting. Elimination of shrouds and close-clearance wearing rings reduces manufacturing costs.

New Pump's Performance

Conversion of kinetic energy of the fluid into static pressure energy occurs entirely in the diffuser (sketch below). Part of the pressure comes from the ram effect of the rotating mass of fluid in the casing as it enters the diffuser. But most of the pressure is generated in the expanding section of the diffuser, where velocity head is converted to pressure head.

For a given impeller at a given speed, the discharge pressure stays constant for all flows up to the maximum. This is because the pressure level is set by the velocity energy in the rotating mass of fluid in the casing, and the diffuser section merely taps this energy. Pressure begins to drop rapidly when flow through the impeller cannot keep up with the discharge flow. Interestingly, the pump may be operated stably even in this extended flow region.

Flow is adjusted in three ways: by changing the diameter of diffuser section (best method); modifying vane design; or altering clearances between impeller and casing. Recirculation or throttling are not used at all.

Efficiency of this new pump is higher than that of low-speed pumps in the same range of specific speed N_s (see Theory on last page of article). Versions already built have N_s values from 400 to 800, based on flows less than 50 gpm. Comparative efficiencies are plotted in the performance graphs. One reason for good efficiency is that the entire liquid mass is brought into rotation—the only loss is by skin-friction, which is low. Another reason is that the clearance from the face of the impeller to the casing wall (suction side) is only a few thousandths of an inch.

Pumps of this new high-speed design have been operating for over a year, and it appears certain that minimum life will be at least 12,000 hr for bearings, seals, and other mechanical parts. Impeller and casing will last longer than 12,000 hr if ordinary, clean hydraulic fluids are used. The electrical parts—motor and controls—will outlast the pump.

High-speed motor

The 400-cycle motor shown in the sketch is available; the pump is close-coupled to the motor. Unfortunately,

60-cycle vs 400-cycle pumps

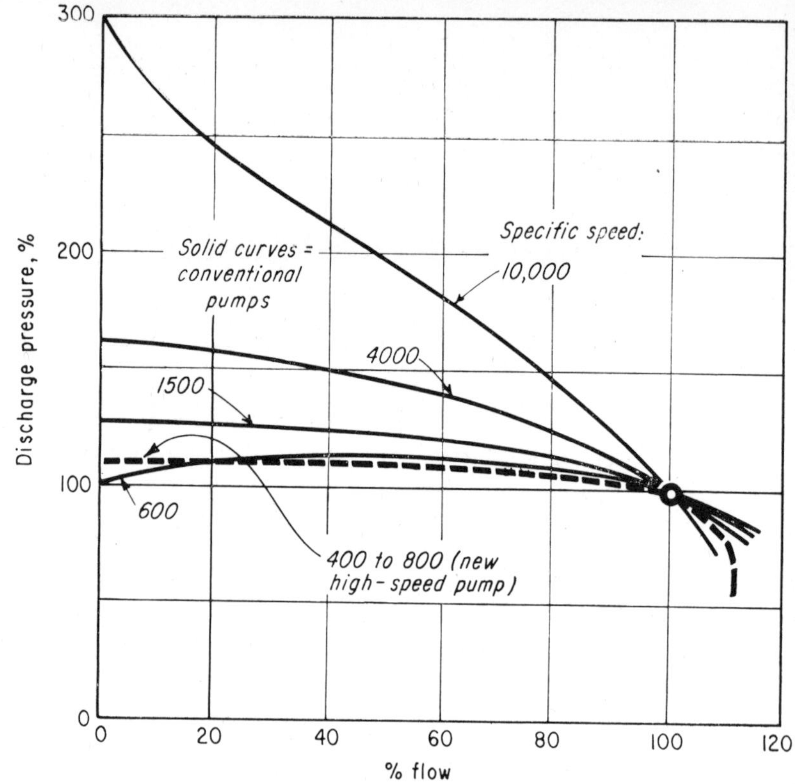

EFFECT OF SPECIFIC SPEED ON HEAD-VS-CAPACITY

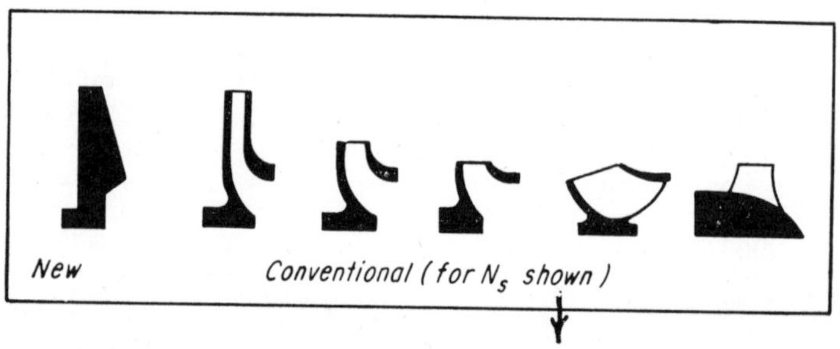

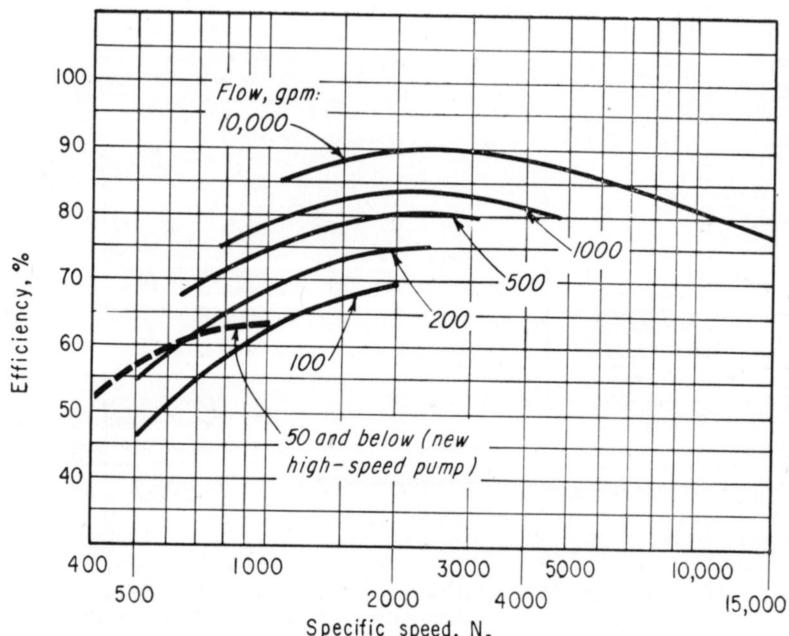

EFFECT OF SPECIFIC SPEED ON MAXIMUM EFFICIENCY

Performance of new high-speed pump

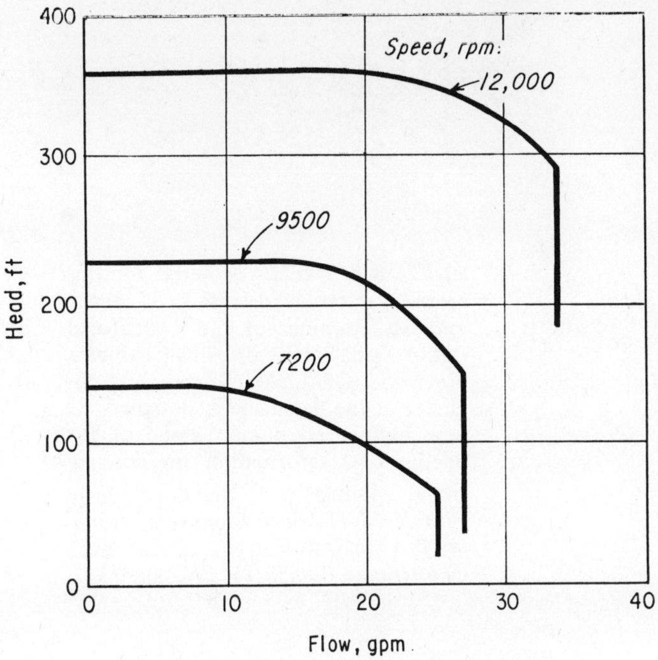

EFFECT OF IMPELLER SPEED ON HEAD-VS-CAPACITY

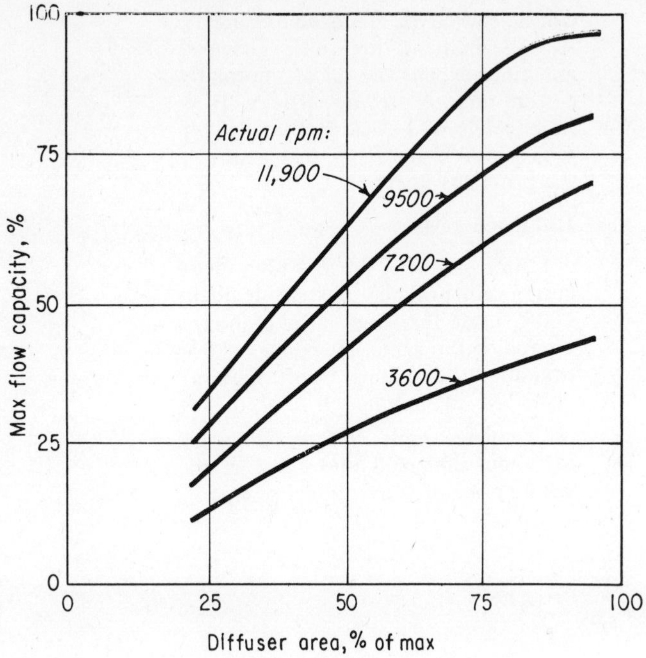

EFFECT OF DIFFUSER INLET AREA ON MAXIMUM CAPACITY

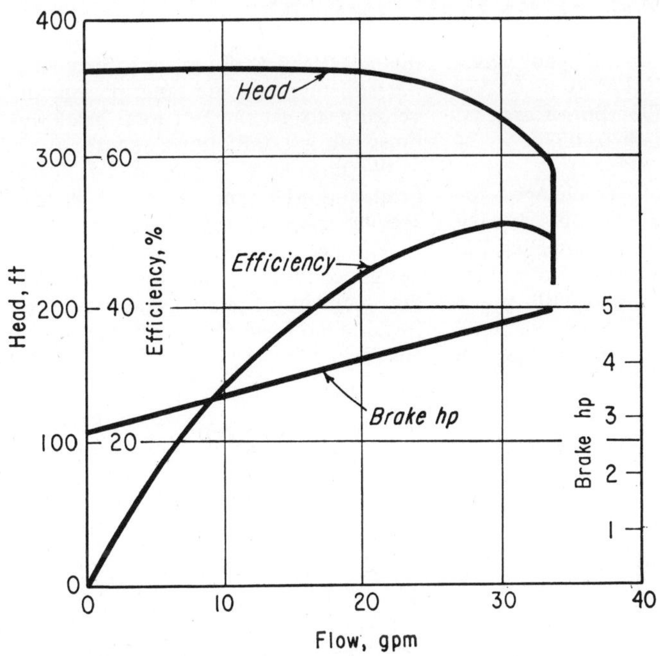

TEST PERFORMANCE AT 12,000 RPM
(5-hp pump, 2.85-in. impeller)

IMPELLER DIA., IN.	2.625		3.04	
Speed, rpm	12,000	24,000	12,000	24,000
Flow, gpm	23.5	47	36	72
Head, ft	300	1200	400	1600
Brake hp	3.5	27.4	7.1	54
Efficiency, %	49	52	51	54
Overall pump: length, in. dia., in. weight, lb	 11 6 35	 14 8.5 95	 12 6 45	 16 8.5 125

CALCULATED PERFORMANCE FOR TWO IMPELLER SIZES
(specific speed=800 for each example)

400-cycle power is not supplied in most plants, and conversion equipment will have to be purchased. Rotating-type converters are sold by many manufacturers, and soon there will be static-type frequency changers. At the moment, the static changers are quite expensive and are normally custom-built. A previous article, "How Practical Is 400-Cycle Power Today" (*PE*—March 6 '61, p 54), outlines the problems involved.

Estimated costs

Motor-pumps of the new high-speed design will probably cost from 50 to 200% more than conventional motor-pumps of the same horsepower. For example, the pumps (with motors) described in columns 1 to 4 of the performance table will probably sell for about 300, 200, and 150% of the selling price of conventional pumps.

EDITOR'S NOTE: Material for this article was obtained from L. H. Sence and M. Kammer of the Centrifugal Pump Department of Allis-Chalmers, Norwood, Ohio. Mr Sence, chief engineer of the department, designed the new high-speed pump. Some of the background information on conventional pumps was obtained from Mark's *Mechanical Engineers' Handbook* (McGraw-Hill) and Karassik's *Centrifugal Pumps* (F. W. Dodge).

A way to visualize specific speed

Specific speed N_S of a centrifugal pump is the speed at which a geometrically similar pump would have to run to deliver 1 gpm against a head of 1 ft. Its formula is:

$$N_S = \frac{N\sqrt{Q}}{H^{3/4}}$$

where: Q = gpm pumped
N = pump speed, rpm
H = total head, ft

To visualize this, picture a pump reduced in size until it delivers a flow of 1 gpm against a total head of 1 ft.

Remember the fundamental rules of centrifugal-pump hydraulics: Capacity change is directly proportional to speed change, and head change is directly proportional to the square of speed change.

Now let's take a pump and shrink it down to get its specific speed. Assume a pump rated at 24 gpm, 300-ft total head, and 12,000 rpm. First we slow it down until it produces only 1-ft total head. We must reduce speed by the square root of the head, which means a cut to 17.31/300 = 0.0578 times 12,000, or 693 rpm. Capacity, too, will go down, to 0.0578 × 24 or 1.386 gpm.

But we have not yet changed the size of the impeller, merely slowed it down. Now shrink the impeller but maintain its proportions. To reduce capacity to 1 gpm and still pump against the 1-ft total head, keep all velocities the same. So reduce all areas down to 1/1.386 and all dimensions down to $1/\sqrt{1.386} = 0.85$. However, if we reduce the impeller diameter to 0.85 times its former size but still maintain the same peripheral velocity, to get the 1-ft total head we must run it 1/0.85 times as fast.

We now have 1 gpm at 1-ft total head and 816 rpm. This 816 is the specific speed of the original pump.

As a check, take the original pump-design conditions, substitute them in the centrifugal-pump specific-speed formula and see it we get the same answer:

$$N_S = \frac{12{,}000\sqrt{24}}{300^{3/4}} = 816$$

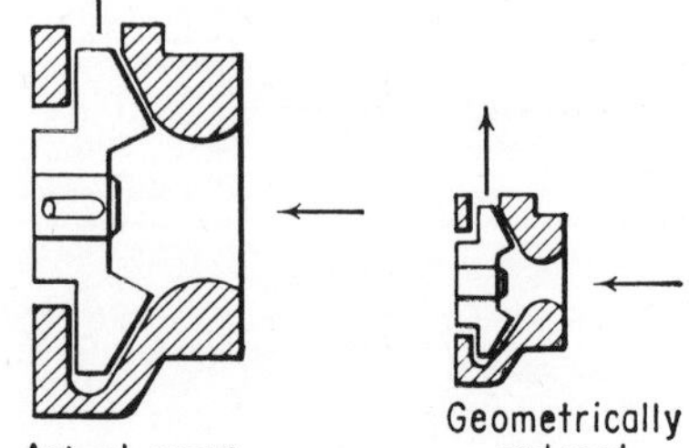

from thermodynamics, equation for . . .

ADIABATIC HEAD selects compressor type

ROTARY POSITIVE-DISPLACEMENT COMPRESSORS

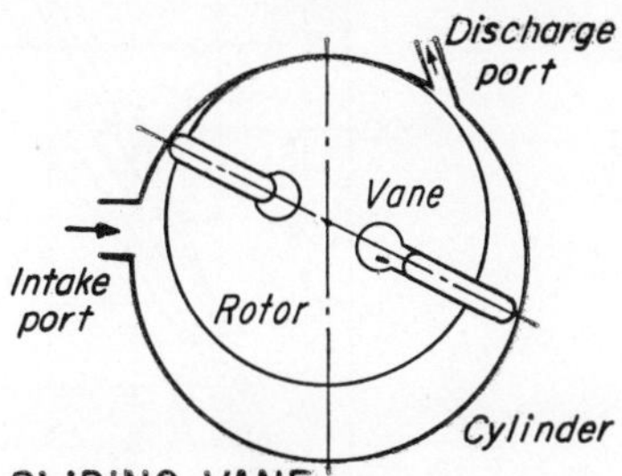

SLIDING VANE

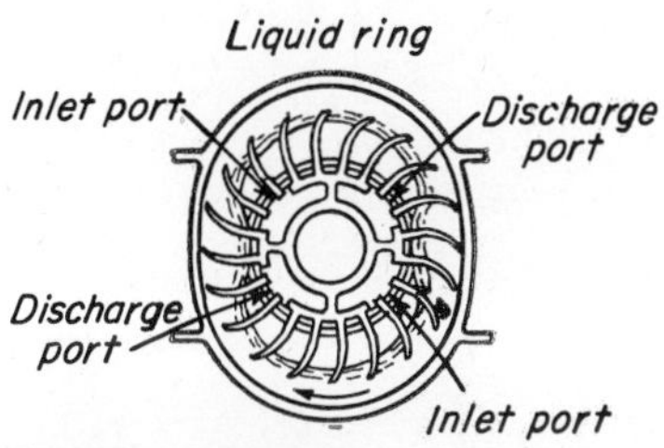

LIQUID PISTON

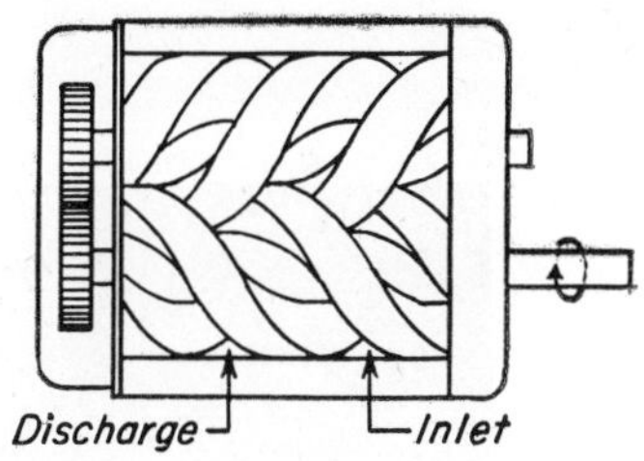

ROTARY SCREW

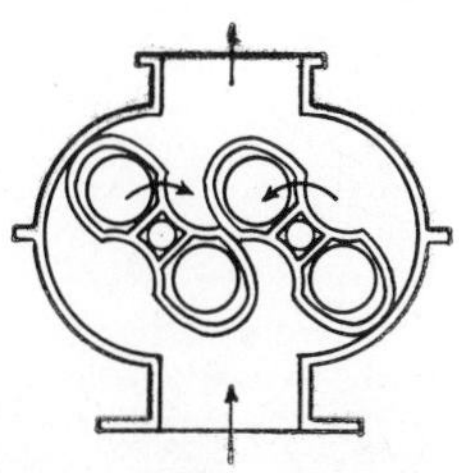

ROTARY LOBE

If you know the pressure ratio, required flow and shaft speed, the "adiabatic head" concept can lead directly to the choice between a positive-displacement and centrifugal compressor—for all gases. Here's a step-by-step method, including the needed reference data.

WILBUR A SPRAKER, and ROBERT A STEIN
Aeronautics and Thermodynamics Div
Battelle Memorial Institute, Columbus, Ohio

The word "compressor" is loosely defined in practice. For many years, "blower" meant any machine that compressed gas with a pressure rise from 1 to 35 psi; a "compressor" produced a pressure rise greater than 35 psi. This definition ignored a number of single-stage centrifugal and axial machines, and today the distinction between blower and compressor has vanished; compressor now means any machine producing a pressure rise greater than 1 psi. Under 1 psi it is a fan.

A "small" compressor for our discussion here has a flow of less than 1000 cfm. While this is an arbitrary dividing point, the selection problems change character at around this capacity.

In this area there are two basic types of compressors: positive displacement, and turbocompressor. Main distinction is in the energy transfer method: in positive displacement the fluid is compressed by the varying volume of the container; in turbo machines the pressure comes from dynamic effects such as centrifugal force.

Depending on the principal direction of the gas flow through the machine, turbocompressors are axial-flow and radial or centrifugal-flow machines, with intermediate mixed-flow types employing varying combinations of axial and centrifugal flow.

continued, next page

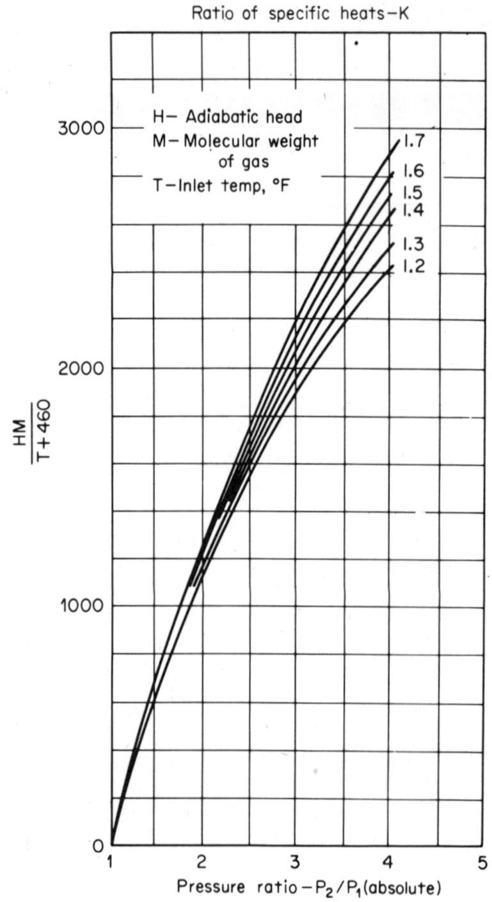

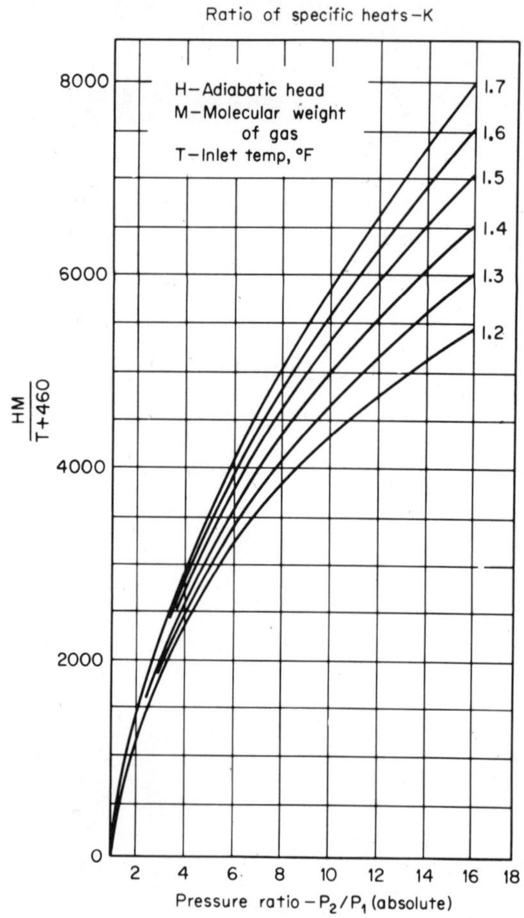

ADIABATIC HEAD for the compressor, when pressure ratio is known, is computed from the values read on the vertical scale.

Positive Displacement

Positive-displacement compressors include reciprocating machines such as piston and diaphragm types, and rotating machines of the vane, gear, screw and lobe types. Pressure rise in such machines requires small clearances plus some method of lubrication, and the compressed gas therefore may be subject to contamination by the lubricant. However, some special reciprocating or vane-type compressors with carbon piston rings or carbon vanes are designed to run without lubricating oil.

Piston-type reciprocating compressors are available in great variety, including special machines for unusual requirements. Maximum compression ratio is 10 per stage; two-stage compressors are available for compression ratios greater than 8; multistaging can produce discharge pressures up to 5000 psi.

Diaphragm-type compressors are limited to a flow of only a few cfm; low initial cost is offset by limited life and frequent maintenance under severe service.

Rotary sliding-vane compressors can produce a com-

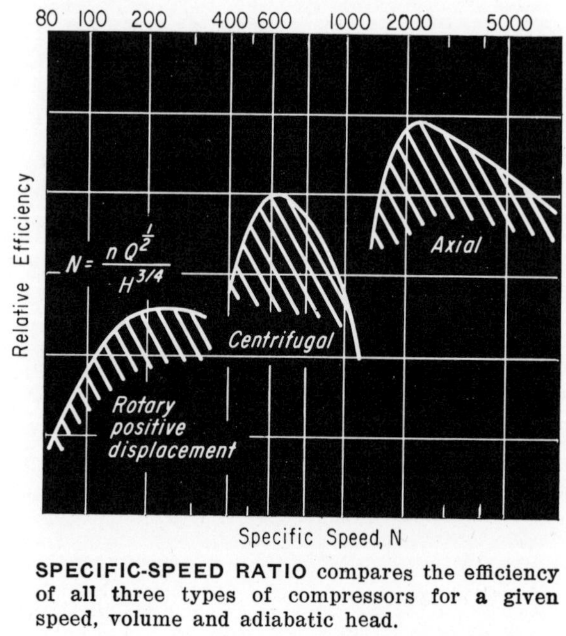

SPECIFIC-SPEED RATIO compares the efficiency of all three types of compressors for a given speed, volume and adiabatic head.

pression ratio of about 8 per stage. They perform efficiently over a wide range of pressure and flow rates, particularly for pressure ratios of 2 or greater.

Liquid-piston compressors employ water or other low-viscosity liquid to compress gas trapped between the blades and casing wall. There is no metal contact between blades and wall, therefore no lubrication or wear at these points. Single-stage compressors are available for a wide range of flow capacities, with compression ratios up to 5.

Gear-type compressors have two meshing-screw rotors to produce internal compression. Timing-gears maintain a fixed clearance between the rotors and no internal lubrication is needed. Pressure ratios of 3 are possible with inlet at atmosphere.

Rotary two-lobe compressors have mating lobed impellers that push the gas from inlet to discharge line. No internal compression takes place, but the impeller forces the gas through the discharge opening against back pressure of the system. As in the screw-type rotor, clearance between impellers is fixed by timing gears, thus eliminating need for internal lubrication. Compression ratios are limited to about 1.7 in a single stage.

Usual limiting pressure ratios per stage for various types of positive-displacement compressors in capacities up to 1000 cfm are:

Type	Max Pressure Ratio
Piston	10
Vane	8
Liquid piston	5
Diaphragm (10 cfm max)	5
Screw	3
Lobe	2

Adiabatic-head Equation

Operating characteristics of all types of compressors can be compared by the combined use of two parameters—specific speed and adiabatic head—each a function of the geometry of the machine. Specific speed is a dimensionless ratio first developed for liquid pumps and then applied to compressible fluids:

$$N = nQ^{1/2}/(gh)^{3/4}$$

For liquids, h is simply pressure × density. But for compressible fluids, h must be integrated for density change in the compression process, using the thermodynamic description of a reversible adiabatic process, and becomes the "adiabatic head" H:

$$H = \frac{R\,(T + 460)}{\frac{K-1}{K}M}\left[\left(\frac{P_2}{P_1}\right)^{\frac{K-1}{K}} - 1\right]$$

or

$$\frac{HM}{T + 460} = f\left(K, \frac{P_2}{P_1}\right)$$

This relationship, presented graphically in the "adiabatic head" curves (facing page) expresses compressor performance in terms of: pressure ratio of the machine, fluid inlet temperature, molecular weight and ratio of specific heats of the fluid. The relationship can be used to determine the adiabatic head needed for a given application, or to evaluate a specific compressor from catalog performance data.

In a centrifugal or axial-flow compressor, H is determined by the speed, impeller diameter and vane geometry. In a positive-displacement machine, H is determined by the volume change in the fluid space during the compression stroke. Thus, this factor is a useful constant in

GAS PROPERTIES FOR COMPUTATION OF ADIABATIC HEAD

	Molecular weight	Ratio of specific heats*
Hydrogen	2	1.4
Helium	4	1.7
Methane	16	1.3
Ammonia	17	1.3
Steam	18	1.3
Nitrogen	28	1.4
Carbon monoxide	28	1.3
Air	29	1.4
Oxygen	32	1.4
Carbon dioxide	44	1.3
Freon 12	121	1.1

*** Values are for input temperatures in the normal ambient range.**

SYMBOLS

g = gravitational acceleration
h = pressure head, ft
H = adiabatic head, ft
K = ratio of specific heats of gas for ambient temperatures
M = molecular weight of gas
n = rotational speed, rpm
N = specific speed
P_1 = compressor inlet pressure, psia
P_2 = compressor outlet pressure, psia
R = gas constant, 1544 ft lb/lb/deg Rankine
T = inlet temperature, °F
Q = volume, cfm

evaluating the capability of any compressor under given operating conditions.

Further, the adiabatic-head concept helps evaluate the effect of varying the fluid properties, inlet pressure and inlet temperature on compressor performance. For a given machine,

- An increase in inlet temperature decreases the pressure ratio
- Change in inlet pressure produces a proportional change in discharge pressure
- Change to a gas with higher molecular weight increases the pressure ratio

Since this relationship applies to both positive-displacement and turbocompressors, it offers a direct method of comparing performance in a specific application. When specific speed is plotted as:

$$N = nQ^{1/2}/H^{3/4}$$

as shown in the diagrams, the relative efficiency of positive-displacement, centrifugal and axial-flow machines is readily evaluated for a specific speed, flow and head.

In using this method to select a machine for a specific application, the first step is to define the system requirements:

- Molecular weight and ratio of specific heats for the gas
- Pressure ratio required, including system losses
- Inlet gas temperature
- Flow rate

Next determine the optimum type of compressor by these steps:

- Calculate adiabatic head H from curve
- Decide on speed
- Calculate specific speed
- Select compressor type from chart

If chart indicates positive displacement, use pressure ratio to narrow the choice.

While pressure ratio, flow and speed are the primary factors in selecting a compressor, the decision must also consider pressure losses in the system, and uniformity

SOME EXAMPLES

Now, let's carry through a simple problem. Suppose we need 50 cfm of room air at 2 psi, and plan to use an ac universal motor at 20,000 rpm with a life of 1000 to 2000 hr. Molecular weight of air is 29, ratio of specific heat is 1.4. Pressure ratio (absolute pressure) is 16.7/14.7 or 1.136. From the adiabatic head curve, HM/T is 200, and H is 200 x 530/29 or 3680.

Substituting this value for H in the equation for specific speed gives $N = 20{,}000\ (50)^{1/2}/3680^{3/4}$ or 300. Turning to the chart for specific speed, this value falls between the displacement and centrifugal types. A centrifugal type can be made at relatively low cost from aluminum stampings, while a positive-displacement compressor needs more parts, closer tolerances and therefore costs more.

If the application were industrial, requiring a service life of 5,000 or 10,000 hr, problems of brush wear and bearing life in a universal motor could be eliminated with a 3600-rpm induction motor. At this lower motor speed the specific speed would be 55, and the choice would now swing to a vane-type compressor, which would also be much quieter.

To show how this design approach works with another gas, consider a machine to compress 400 cfm of carbon dioxide from 5 psi and 100 F to 85 psi, with a compressor rotating at 3600 rpm. Molecular weight of carbon dioxide is 44 and ratio of specific heats is 1.3. Pressure ratio (absolute) is 5. Entering the adiabatic head curve from the pressure ratio, the value of 3000 is read on the ordinate. Adiabatic head H is 3000 x 560/44 = 38,200. Specific speed is $3600(400)^{1/2}/38{,}200^{3/4}$ or 26, at the low end of the positive compressor range. Since the pressure ratio is 5, either a piston or vane type can be considered.

The flow would have to be increased by a factor of 500 or the rpm multiplied by 10 before a centrifugal compressor would be effective.

Performance data in catalogs is generally given in terms of air at room temperature and atmospheric pressure. The adiabatic head chart can be used to find the equivalent performance for air instead of carbon dioxide. Adiabatic head H will be the same, 38,200. Molecular weight of air is 29, inlet temperature 520R, inlet pressure 14.7 psi, and ratio of specific heats 1.4. Therefore

$$HM/T = 38{,}200 \times 29/520 = 2130$$

Entering the chart for adiabatic head at 2130, and K at 1.4, the pressure ratio is 3.2, and outlet pressure is 47.0 psia. Therefore, a compressor designed to deliver 400 cfm of air at 47 psia with 60-F inlet air temperature would compress 400 cfm of carbon dioxide from 20 psia to 100 psia at an inlet temperature of 100 F.

Or take an extreme example, helium with a molecular weight of 4 and a ratio of specific heats of 1.66. Assume the compressor is to take helium at 80 F and 50 psig and deliver 75 psig. Pressure ratio (absolute) is 1.39. From the adiabatic head chart, the vertical scale value is 520. With $T = 540$ R, H is calculated 70,200.

Translating these values to air as a base ($T = 520$ R, $K = 1.4$, and $M = 29$) using the value of H calculated for helium; the new ordinate is

$$70{,}200 \times 28.95/520 = 3910$$

This value, read down from the curve for $K = 1.4$, gives a pressure ratio of 6.8. Therefore, a compressor designed to deliver a pressure ratio of 6.2 in air with an inlet temperature of 60 F would compress the same volume of helium to a pressure ratio of 1.39 with an inlet temperature of 80 F and inlet pressure of 50 psig.

of load; corrosiveness, dirt and moisture content of fluid; service life and maintenance problems.

A compressor system must provide the required flow and pressure at the discharge of a pneumatic system, which includes pipe, valves, and fittings. Methods for estimating system pressure losses are described in reference texts, and articles such as "Pressure Losses in Pneumatic Systems" (*PE—Oct 26 '59, p 50*).

Provision should also be made for additional flow capacity for potential growth. Additional components are almost invariably added—an allowance of at least 25% over immediate needs, depending on the situation, will call for little increase in investment. Also, compressor performance decreases with age and use, primarily from wear of vanes, valves or piston-rings. This shows up as a decrease in discharge pressure, rather than a decrease in flow.

While there is no hard-and-fast rule, in general, the system should be designed for one additional pressure ratio. For example, a system discharging at 100 psi (gage) with atmospheric inlet (a pressure ratio of approximately 8) would need a pressure ratio of 9 and a discharge pressure of approximately 120 psi. Again, the additional investment is minor.

For a system with an intermittent load—the more common case—the compressor capacity should be less than the system capacity, with the difference made up by an accumulator in the system. If some variation in supply pressure is acceptable, a relatively small compressor can be used with a large accumulator. If the supply is used for long periods of time with only short idle periods, compressor capacity should nearly equal the supply. If a relatively constant discharge pressure is required, use a small accumulator and a large compressor. For regulated pressure, use an oversize compressor with a pressure regulator.

For Specialized Requirements

Other requirements include corrosive gases, wet gases, high temperatures, high pressures or extremely low pressures, as well as gases of differing physical properties. For high pressures and very low vacuums, a range of heavy-duty high-pressure compressors and close-clearance high-suction compressors is available. For corrosive applications, the compressor surfaces in contact with fluid may be glass, stainless steel, or graphite.

Wet gases, and gases other than air, influence selection. Compression of wet gases leads to a more nearly isothermal compression with a consequent change in compressor performance; lubrication may present problems not present with dry gases. Corrosion-resistant construction may be needed for high-temperature gases; or the choice may be between an expensive specialized compressor or use of cooler to decrease the inlet temperature. But these spcialized compressor conditions often present problems that only the compressor manufacturer can solve, and the requirements should be discussed with him.

Compressors—in Series or Parallel

When additional flow or pressure is needed beyond the capacity of an available compressor, within certain limitations compressors can be operated in parallel or in series. In series, matching problems arise in positive-displacement machines—piston, rotary vane, rotary lobe.

If two of these compressors are operated in series the discharge flow rate of the first must match inlet flow rate of the second. If not carefully matched, a series system may show an appreciable vacuum or excessive pressure between the two compressors rather than the intermediate pressure desired. Mismatch may also introduce seal problems in the second compressor, since inlet and discharge pressure, and the compressor-case pressure, will be higher than design values.

With turbocompressors in series, a flow mismatch will cause surging of one or the other, and flow instability.

Positive-displacement compressors can be operated in parallel if the discharge pressures of the two machines are about the same; flow through one compressor has little or no effect upon the other. However, turbocompressors operated in parallel do present a matching problem; one compressor will frequently pick up the entire flow while the second idles or becomes unstable. Consequently, an automatic flow-control system is required.

Turbo and positive-displacement compressors can be combined; a turbocompressor can be used as a supercharger ahead of a positive-displacement machine, or vice versa. Or a turbocompressor could be used in parallel with a positive-displacement compressor as a flow booster.

to move air—EQUATIONS FIND THE RIGHT IMPELLER

Once the airflow, static pressure and rpm are determined, just a few simple calculations will show the type and size of impeller needed. This sharply reduces time spent searching through manufacturers' catalogs.

J M WATSON
The Torrington Mfg Co
Torrington, Conn

After you calculate the specific speed needed, it's easy to settle on the type of impeller for a particular application. But making the final specific choice with only this information involves a long search of manufacturers' catalogs. However, most of this waste time can be eliminated if you also know the required impeller size; and the method described in this article shows how to find it.

All you need are a few simple calculations involving coefficients of pressure and flow. Typical values for these factors, for all major impeller types, are listed in Table II.

IMPELLER TYPES

Table I shows the three major impeller types, together with subgroups and typical applications. The most frequently used commercial types are: forward-curved-centrifugal, mixed-flow, and propeller-type-axial impellers. Forward-curved centrifugal impellers are divided into two subgroups—normal blade area, where total blade surface is about equal to peripheral surface of the impeller; and high blade area, where total blade surface is 30 to 40% greater than peripheral area. This division is important because of the effect of blade surface on pressure and flow coefficients. Propeller fans are subdivided into groups with high, medium, and low pressure coefficients. Vane-axial and tube-axial impellers are usually designed to customers' specifications, and so should be selected by working directly with the manufacturer.

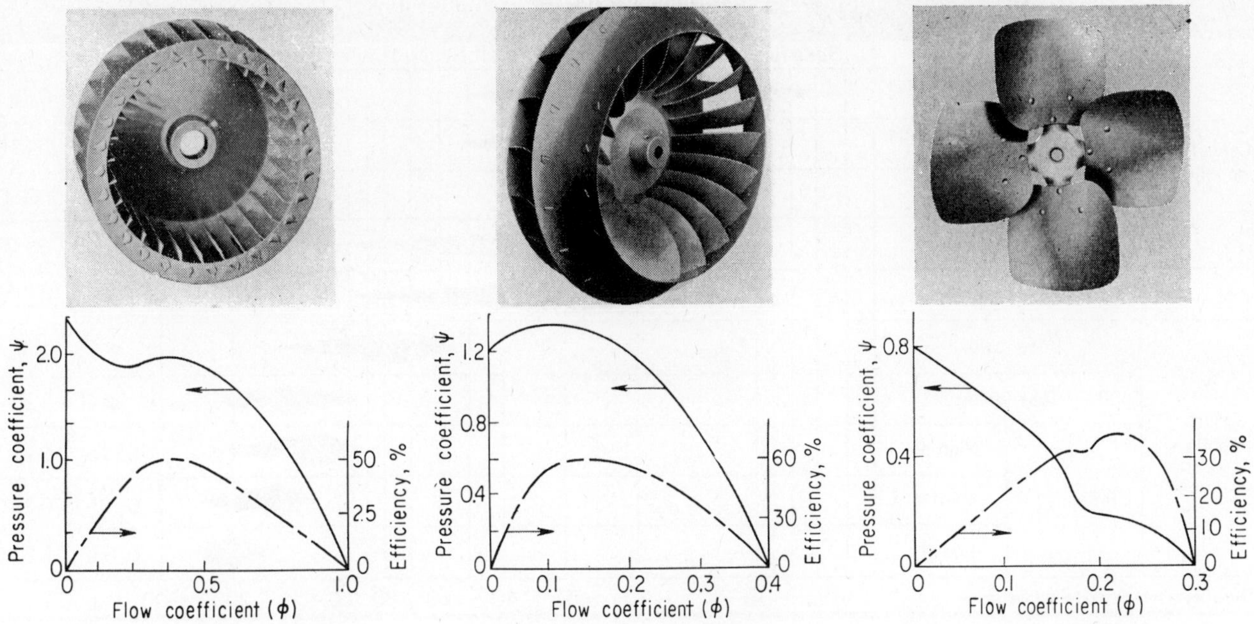

Typical centrifugal impeller | Typical mixed-flow impeller | Typical axial impeller

THREE BASIC IMPELLERS—Solid curves relate pressure coefficients to flow coefficients for typical impellers of each type. Dotted curves indicate relative efficiency.

Table I—Characteristics of Various Impeller Types

	Type			Pressure coef ψ	Flow coef φ	Typical applications
Centrifugal impellers	Forward normal curved	High blade area		1.5–2.0	0.20–0.75	Room air conditioners Central air conditioning and heating Electronic-equipment cooling
		Normal blade area		1.0–2.0	0.15–0.65	
	Radial blade			1.0–1.4	0.002–0.09	Washing machines Clothes dryers Assorted household appliances
	Backward curved			0.60–1.10	0.09–0.30	Industrial exhaust systems Electronic-equipment cooling
Mixed-flow impellers				0.60–1.10	0.10–0.25	Room air conditioners Roof ventilators
Axial impellers	Vane axial			0.20–0.60	0.10–0.40	Aircraft air conditioning Pressurizing aircraft compartments Shipboard air conditioning
	Tube axial			0.10–0.30	0.10–0.40	Electronic-equipment cooling Aircraft air conditioning
	Propeller	High ψ		0.15–0.20	0.25–0.40	Air conditioning condensing units Light industrial exhaust fans
		Med ψ		0.10–0.15	0.15–0.35	Unit and space heaters, Air circulators
		Low ψ		0.05–0.10	0.10–0.25	Assorted portable appliances

HOW TO CHOOSE

Here is the simple step-by-step procedure for finding the specific impeller needed for any application.

Find the right type

List your application requirements: airflow Q, in cfm; static pressure p_s, in inches of water; and impeller speed, N, in rpm. Airflow rate is determined by established heat-transfer rates and unit performance required for the application. Static pressure is determined by air-path configuration, coil resistance, filters, louvers, and other elements in the flow system. Impeller speed is determined by the choice of motor (standard four- and six-pole shaded-pole and split-capacitor types are available) and also by the drive—direct or belt.

Calculate specific speed. Use equation:

$$N_s = \frac{NQ^{1/2}}{p_s^{3/4}}$$

or find specific speed, using nomograph,

Table II—Performance Data for Impellers

Type			Specific speed $N_s \times 1000$ (Peak efficiency at widest part of bar)	Typical ψ	Typical ϕ
Centrifugal impellers	Forward curved	High blade area	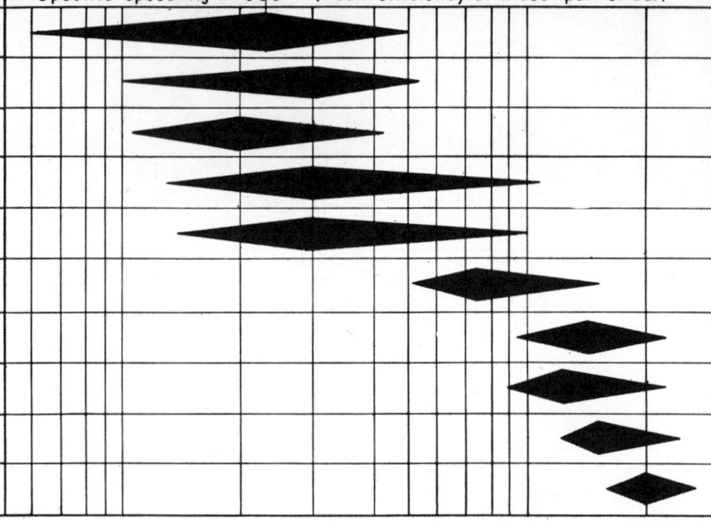	1.70	0.254
		Normal blade area		1.40	0.176
	Radial blade			1.0 – 1.4	0.002-0.09
	Backward curved			0.60–1.10	0.09–0.30
Mixed-flow impellers				1.00	0.112
Axial impellers	Vane axial			*	*
	Tube axial			*	*
	Propeller	High ψ		0.175	0.263
		Med ψ		0.125	0.250
		Low ψ		0.075	0.238
*Wide deviation possible within range			6 8 10 20 30 40 60 80 100 200 300		

Check the tables. Table II shows the type (or types) of impellers that operate at high efficiency at the calculated specific speed. The widest portion of the bar indicates the point of peak efficiency. If more than one impeller type is appropriate, base your final selection on such other factors as cost, size and shape of space available, and nature of flowpath. Generally, one type of impeller will be a clear choice for a particular application.

Find the right standard diameter

Using appropriate pressure coefficient ψ, from Table II, calculate theoretical impeller diameter from this equation:

$$D = \frac{1.53 \times 10^4}{N} \sqrt{\frac{p_s}{\psi}}$$

(For derivation of this equation and the others that follow, see p. 164.)

Consult a catalog to find a standard diameter that most closely matches the theoretical diameter you have calculated. Use this value for the remaining steps.

Find other dimensions, if needed

Propeller type—find the pitch angle by substituting in this equation (see Table II for flow- coefficient ϕ):

$$\alpha = \sin^{-1} \frac{350\,Q}{\phi_1\,ND^3}$$

Forward-curved centrifugal or mixed-flow impeller—find impeller width:

$$W = \frac{175\,Q}{\phi_1\,ND^2}$$

Radial-bladed or backward-curved centrifugal impeller—consult a manufacturer's catalog for additional information, because they may be carried as special products.

Make the final choice

Consult a manufacturer's catalog for final selection. It should be easy because type and approximate size are known.

SYMBOLS

N_s —specific speed
N —impeller speed, rpm
Q —airflow rate, cfm
p_s —static pressure, in. water
ψ —pressure coefficient
ϕ —flow coefficient
g —acceleration of gravity, ft/sec/sec
ρ —weight density of air, lb/cu ft
U_t —blade tip velocity, ft/sec
D —impeller diameter, in.
V_o —airflow velocity, ft/sec
ΔA—cross-section of airflow, sq ft
α —fan pitch angle, deg
K —constant for given class of fans
W —width of impeller, in.

OTHER APPLICATION REQUIREMENTS

Although this selection method has been presented with airflow rate, static pressure and rotational speed as the application requirements, it is not limited to impeller selection where these are the only known quantities. For example, if you must use either a particular type of impeller or a particular impeller diameter and you must find the best impeller, the selection procedure described above can still be used.

If diameter is known, for instance, the equation for D can be rearranged to solve for one of the other factors.

Even where there are too many unknowns you can accurately estimate a solution by varying a particular parameter within reasonable limits and calculating the other parameters until a satisfactory impeller is found. After a little practice, solutions can be found in this

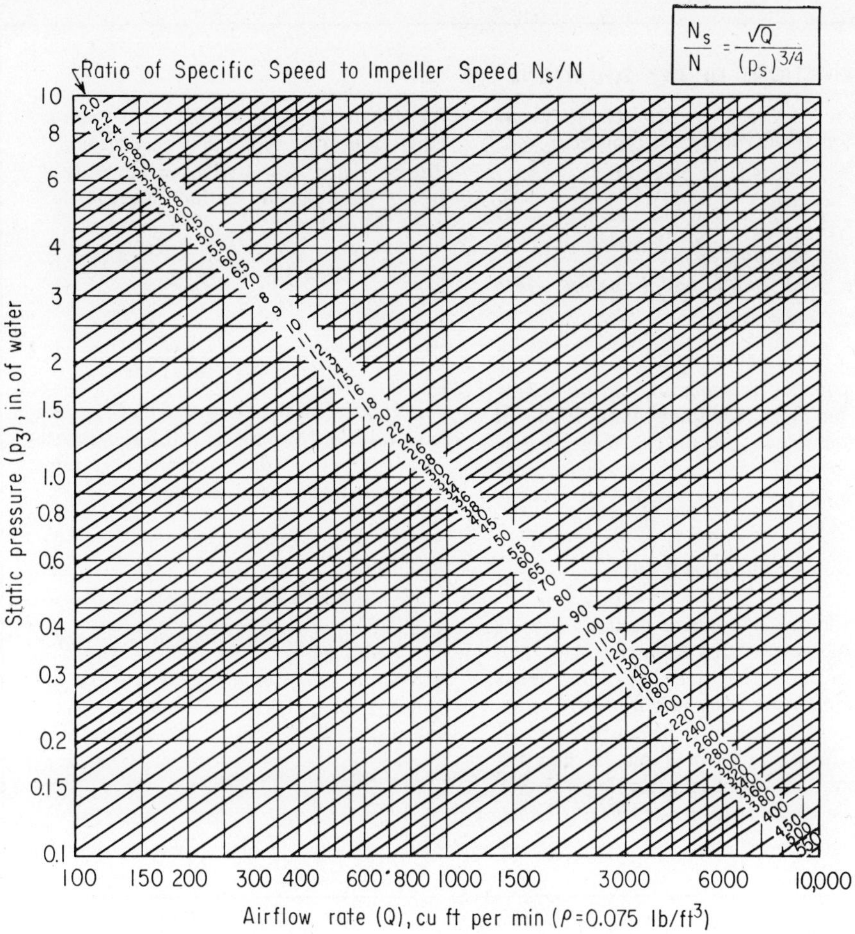

Alternate method for finding specific speed. Because the result (diagonal column) is a ratio of $\frac{N_s}{N}$, mutiply by impeller speed N to get specific speed, N_s.

manner almost as quickly as if all the requirements were given.

HOW THIS METHOD WORKS

Two typical problems will show how simply and quickly you can find the right impeller for a given set of conditions.

Example I

Assume the following requirements:

$$Q = 500 \text{ cfm}$$
$$p_s = 0.10 \text{ in. of water}$$
$$N = 1550 \text{ rpm}$$

To find the impeller type required, calculate specific speed:

$$N_s = \frac{NQ^{1/2}}{p_s^{3/4}} = \frac{(1550)(500)^{1/2}}{(0.10)^{3/4}} = 194{,}500$$

In Table II, the most efficient impeller for a specific speed of 194,500 is a propeller with a low-pressure coefficient.

Table II gives 0.075 as the typical pressure coefficient for the low-ψ propeller. Diameter then is calculated:

$$D = \frac{1.53 \times 10^4}{N}\sqrt{\frac{p_s}{\psi}}$$

$$D = \frac{1.53 \times 10^4}{1550}\sqrt{\frac{0.10}{0.075}}$$

$$D = 11.4 \text{ in.}$$

To find fan pitch angle, find a standard diameter closest to 11.4 in. One catalog lists a standard fan with an 11-in. dia.

Table II gives 0.238 as typical flow coefficient, ϕ, for the low-ψ propeller. Pitch angle α then is:

$$\alpha = \sin^{-1}\frac{350Q}{\phi_1 ND^3}$$

$$= \sin^{-1}\left[\frac{(350)(500)}{(0.238)(1550)(11)^3}\right]$$

$$= \sin^{-1}(0.357)$$

$$\alpha = 21°$$

The catalog lists performance data for 16° and 24° pitch angles for the 11-in.-dia fan. The 24° fan delivers 660 cfm at 0.10 in. of water and 1550 rpm, which is higher air delivery than necessary. With 16° pitch angle, the fan delivers about 515 cfm at 0.10 in. of water and 1550 rpm. This is very close to the requirements, so final selection is an 11-in.-dia., low-ψ propeller fan with a 16° pitch angle.

Example II

Assume the following requirements:

$$Q = 100 \text{ cfm}$$
$$p_s = 0.25 \text{ in. of water}$$
$$N = 1000 \text{ rpm}$$

The specific speed is

$$N_s = \frac{1000(100)^{1/2}}{(.25)^{3/4}} = 28{,}200$$

Table II indicates that any of four impeller types may be efficiently applied:

DERIVATIONS OF KEY EQUATIONS

Pressure and Flow Coefficient

Here's how the coefficients listed in Table II were calculated.

Pressure coefficient ψ, a dimensionless quantity, is the ratio of potential energy developed by an impeller to pressure energy corresponding to blade-tip velocity:

$$\psi = \frac{2\,gp_s}{\rho U_t^2} \tag{1}$$

It can also be expressed as:

$$\psi = \frac{2.35 \times 10^8\, p_s}{N^2 D^2} \tag{2}$$

By rearranging, an equation for impeller diameter becomes:

$$D = \frac{1.53 \times 10^4}{N}\sqrt{\frac{p_s}{\psi}} \tag{3}$$

Flow coefficient ϕ, also a dimensionless quantity, is the ratio of airflow velocity to blade-tip velocity:

$$\phi = \frac{V_o}{U_t} \tag{4}$$

In order to maintain a uniform approach for all axial- and centrifugal-impeller types, ΔA is defined as $\pi D^2/4$. Therefore flow coefficient can be expressed:

$$\phi = \frac{700Q}{ND^3} \tag{5}$$

where $Q = \Delta A \times V_o$

Blade pitch angle and impeller width

Pitch angle can be determined by:

$$\sin\alpha = K\phi$$

This equation is empirical, and holds true only for flow rates from 70 to 100% of maximum flow. But for this range of flow rates, propeller fans are most efficient and are most effectively used. If, in the equation

$$\frac{\sin\alpha_1}{\phi_1} = \frac{\sin\alpha_2}{\phi_2} \tag{6}$$

the ratio $\sin\alpha_1/\psi_1$ for a particular fan is known, the sine of the pitch angle α_2 of another fan of the same class can be found by finding its flow coefficient, ϕ_2. If, for example, a pitch angle of 30° is assigned to α_1, then $\sin\alpha_1 = 0.5$. In Table II, flow coefficients are given out for fans of 30° pitch in each class of propeller fan. Therefore, pitch angle for any fan can be calculated from the application requirements, diameter, and flow coefficient, by substituting the value of ϕ from Eq (5) for ϕ_2 in Eq (6).

$$\sin\alpha_2 = \frac{350Q}{\phi_1 ND^3} \text{ or} \tag{7}$$

$$\alpha = \sin^{-1}\frac{350Q}{\phi\, ND^3} \tag{8}$$

For centrifugal and mixed-flow impellers, width of the impeller must be known. The ratio of width to diameter is proportional to the flow coefficient:

$$\frac{W}{D} = K\,\phi \tag{9}$$

where K is constant for a given impeller design $\frac{W}{D} = 0.25$ is assumed, and corresponding values of ϕ_1 are determined in Table II for each class of impellers.

Substituting Eq (5) in Eq (9) gives:

$$W = \frac{175Q}{\phi\, ND^2} \tag{10}$$

- Forward-curved centrifugal impeller of high blade area
- Forward-curved centrifugal impeller of normal blade area
- Backward-curved centrifugal impeller
- Mixed-flow impeller

Typical pressure coefficients from Table II for these types of impellers start from 1.7 for high-blade-area forward-curved centrifugals and drop down to 1.0 for mixed-flow impellers.

Because diameter is an inverse function of pressure coefficient for a fixed speed and a fixed static pressure, it is evident that for a higher pressure coefficient the impeller diameter will be smaller. Thus, a forward-curved centrifugal impeller of high blade area will require the smallest impeller diameter in this particular application, and is the most desirable.

The required diameter will then be:

$$D = \frac{1.53 \times 10^4}{1000}\sqrt{\frac{.25}{1.7}} = 5.85 \text{ in.}$$

The closest diameter listed in the manufacturer's catalog for a standard forward-curved centrifugal blower wheel of high blade area is $D = 5\frac{19}{32}$ in.

Table II indicates 0.25 for a typical value of ϕ for a forward-curved centrifugal impeller of high blade area, resulting in a blade width:

$$W = \frac{(175)(100)}{(.25)(1000)(5.59)^2}$$

$$W = 2.25 \text{ in.}$$

The closest standard width listed in the catalog is $W = 2\ 16/32$ in.

The catalog performance-curve for this particular impeller shows it will deliver 98 cfm against a static pressure of 0.25 in. of water when operated at 1000 rpm. A slight increase in speed, such as from 1000 to 1020 rpm, will raise the airflow rate to at least the 100 cfm originally required. This increase in speed is well within normal motor capabilities, so the final selection is a forward-curved centrifugal impeller of high blade area, $5\frac{19}{32}$ in. dia. and $2\frac{1}{2}$ in. wide.

Ex. I

Ex. II

7

VALVES AND CONTROLS

Construction and performance of pressure-control, flow-control, directional, servo, manifold, and fluid-amplifier valves and devices

After five years in the incubator

Fluid amplifiers go commercial

Major corporations today are about to take the wraps off products controlled with fluid amplifiers. Here's a first-hand report of what's coming in '66

FRANK YEAPLE, Associate Editor

Over five years in development, and some no bigger than a button, these amplifiers can respond to a whisper of air, yet dispatch instant orders for crushing a complete auto frame into a 1-ft cube!

There are no moving parts other than the fluid (gas or liquid), therefore nothing to wear out. The devices themselves can be made of any material, including plastic, metal, and ceramic, so temperature need not be a limiting factor.

They seem to have no limitations other than the problems men have in understanding them. They are so difficult, in fact, that the only fluid devices appearing in the new commercial products are simple on-off elements. These, at least, are predictable and can be grouped to perform logic functions.

The rest of the fluid-amplifier art is still experimental, though on a grand scale. Industry and Government are spending millions to realize the promise of the new technology.

Fluidics is born

The newly coined word has caught on and is the rallying cry for this scampering technology. It means *fluid logic,* principally of the no-moving-parts sort exemplified by the fluid amplifier.

Fluidics refers to control of fluid circuits with a whole new family of logical devices having no (or a minimum of) moving parts. The pioneer device, and the impetus for this whole business, is the so-called wall-effect amplifier. Its principle was discovered years ago (circa 1930) by Henri Coanda, and was developed as a practical control device about 1960 by Ray Bowles, B. M. Horton, and Ray Warren, who at that time worked at the Army's Diamond Ordnance Fuse Lab in Washington.

In his vigorous eighties now, Henri Coanda discovered the wall-effect phenomenon (the tendency of a stream to cling to a wall under certain conditions) while piloting the world's "first" jet plane in 1910. He had attached an engine-driven airjet to each wing, and succeeding in igniting additional fuel in a sort of afterburner. He got the thrust he wanted, but the streams clung stubbornly to the fuselage, destroying the plane and nearly Coanda. He couldn't understand why until about 1930, while taking a bath. The water from the faucet clung to his arm—defying logic—and he reasoned that it was the entrainment of trapped air that caused the phenomenon. As the air next to his arm was carried along, a partial vacuum was created. Atmospheric pressure, then, forced the water stream to cling to his arm (as it had forced the hot gases to cling to the fuselage).

A variety of other fluidic devices followed quickly—the turbulence amplifier invented by Ray Auger; the impact modulator invented by Bjorn Bjornsen and Tom Lechner of Johnson Service Co; and the vortex amplifier, developed over the years by many experimenters.

Do-it-yourself?

It would seem possible to stack fluid amplifiers like pancakes to get all the power and control needed. But it isn't so simple. There are a few natural phenomena to consider—shock waves, compressibility, viscosity, turbulence, resonance, and entrainment. As soon as the output of one device is made into the input of another, the complexity begins.

Users—engineers trying to breadboard the amplifiers—admit that great improvements have been made in wall-effect and turbulence amplifiers, but no one is willing to say system design is easy. The problems arise when you try to set up fluid amplifier sequencing circuits to replace electrical relay "trees." The loading effects are hard to predict. It is even worse if you choose proportional-type amplifiers.

The keenest minds in industry and in top universities are trying to analyze the streams and swirls with math models, but it will be years before proportional amplifiers can be cataloged like transistors.

Interface blues

For all the fluidic amplifiers, and in particular the turbulence amplifier, some sort of output amplifier is needed to boost pressure levels. It would be very useful also to raise the working pressure of the turbulence amplifier to at least a few psi. All the other fluidic logic elements can operate at line pressures where needed—only the turbulence amplifier is restricted somewhat by its need to maintain laminar-type flow.

Why fool around with low pressures at all? Efficiency. Whether on or off, even the tiniest fluid amplifiers require a constant flow of supply air, and when there are numerous devices, the flows mount. It is agreed that pressures must be kept as low as possible, meaning less than 3-psi supply, unless air supply is no problem. Only for the final output should the pressure level be raised.

EDITOR'S NOTE: PRODUCT ENGINEERING was first to publish an article on fluid amplifiers (Mar 14 '60, p 17). Since then the proliferation of knowledge and paper has been both amazing and helpful, including about a dozen heavy volumes from Harry Diamond Lab symposiums.

DIGITAL (bi-stable)

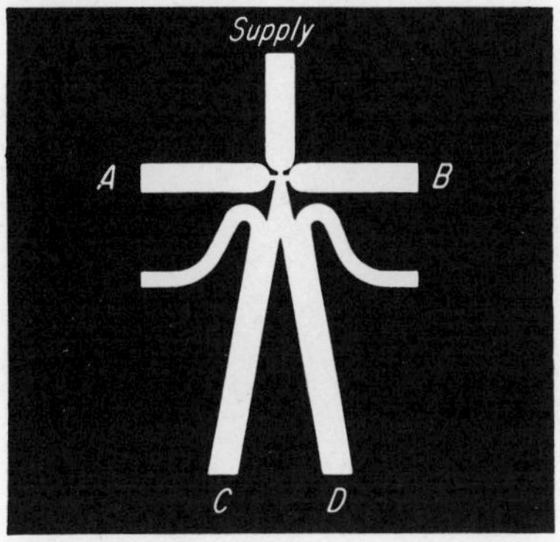

Wall-effect flip-flop is one of the pioneer devices. The supply jet will tend to hug one wall (or the other), thus emerging from outlet **C** (or **D**). Once established, the jet continues through the one outlet until a control signal at A or B deflects the jet to the opposite outlet. The remaining two outlets are vents.

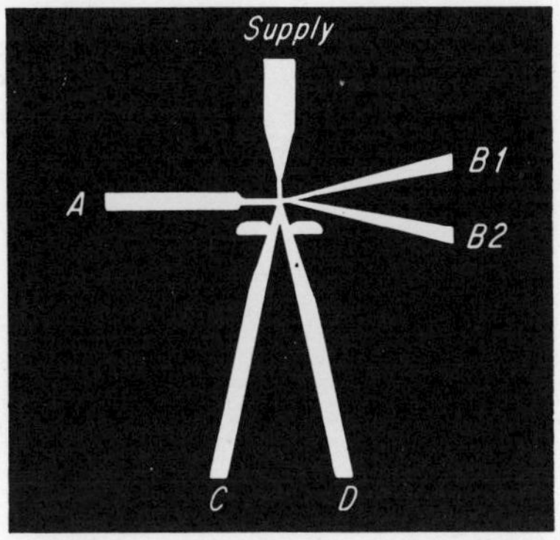

Wall-effect OR-NOR is a variation on the flip-flop. Control ports **B**[1] and **B**[2] constitute the **OR** effect, in that either one **OR** the other can switch over the device. If neither **B**[1] **NOR B**[2] are energized, this produces an output at **D.** This example is only one of the many logic functions that can be performed by fluidics.

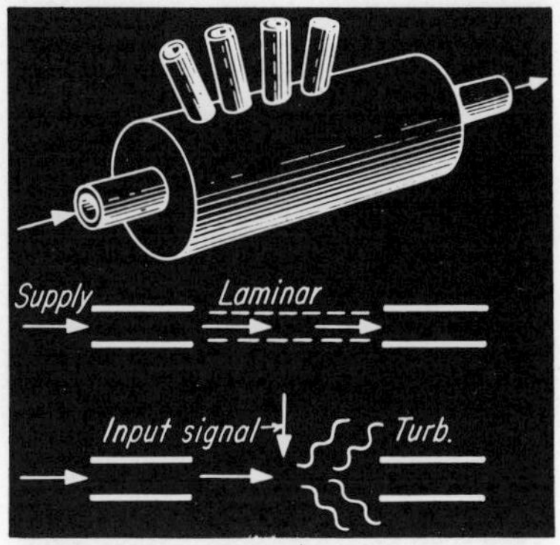

Turbulence amplifiers are the simplest in concept. The supply jet is aimed at the output and will continue to reach it until the laminar characteristic of the jet is disturbed. It takes only a whisper of air to disturb the jet, thus destroying the output signal. The logic mode is **NOT** or **NOR** (no output when any input).

ANALOG (proportional

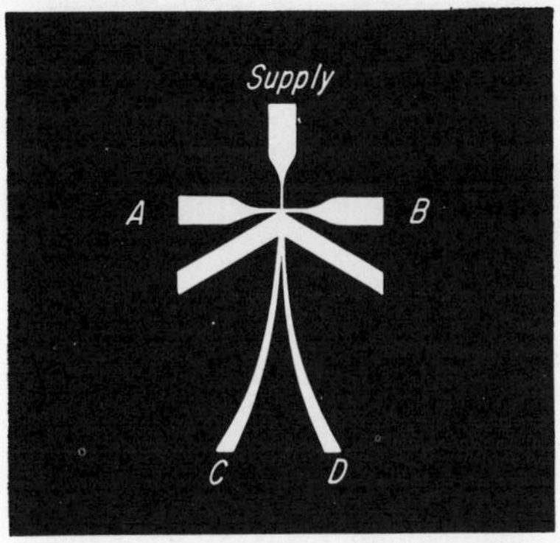

Momentum exchange fluid amplifiers depend on the interaction of control jets A and B with the main supply jet, which aims a portion of the jet to one output or the other **(C** or **D)**. The other legs are the vents, as in the wall-effect digital amplifiers. The appearance of the proportional and digital devices are similar.

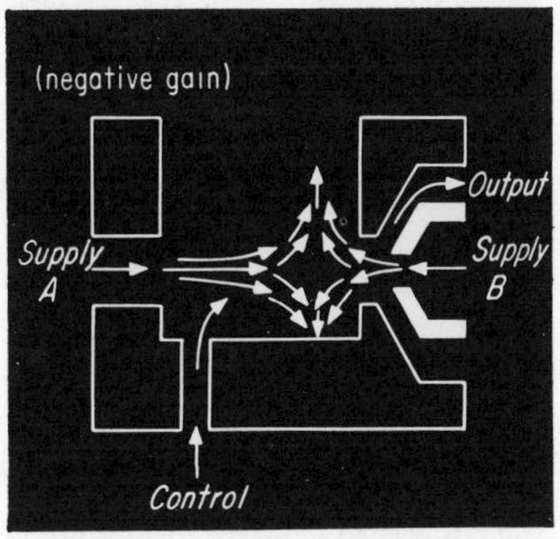

Transverse impact modulators can vary output by moving the area of impact axially with a force balance between the opposing jets **A** and **B.** The control jet deflects the supply-**A** stream and tends to negate its effect, thus moving the impact area to the left, reducing output. The device therefore has negative gain.

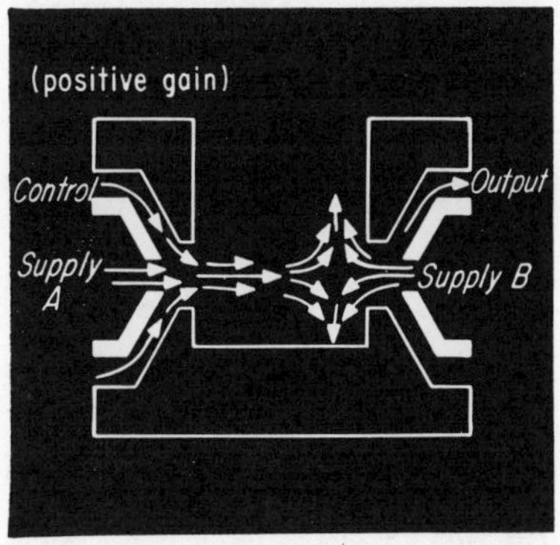

Direct impact modulators have the same basic function as the transverse variety, but use the control jet to strengthen, rather than to weaken, the supply-**A** jet. There are many performance charts plotted to show where the spread-out region occurs, but it is practically impossible so far to compute it mathematically.

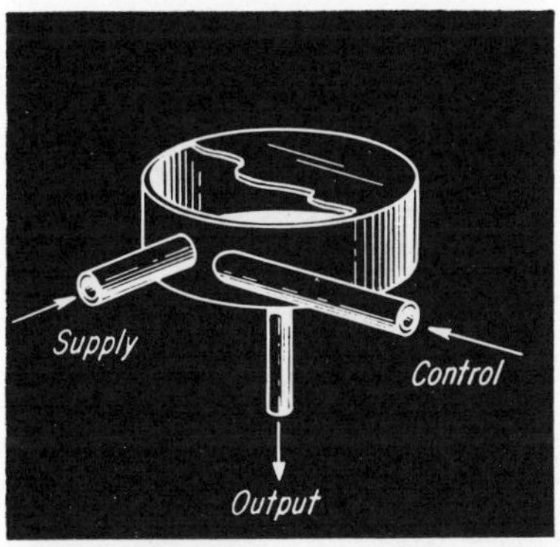

Jet-controlled vortex amplifier has least resistance to flow when the control jet is not operating, because the flow goes directly from supply to output without vortexing. The control jet causes the supply jet to vortex, thus lengthening the effective flow path in proportion to the amount of control flow.

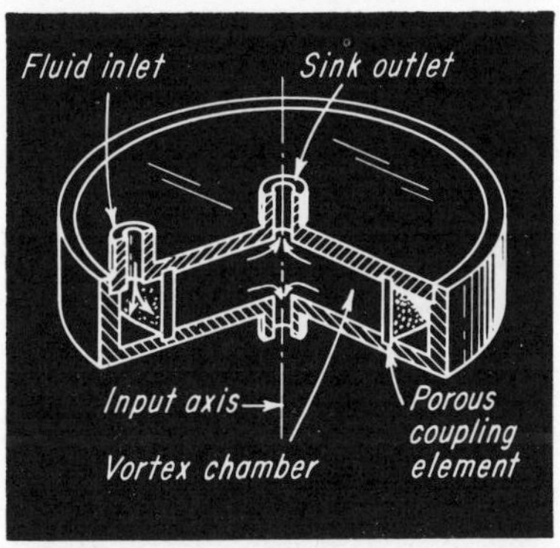

Rate-sensitive vortex device actually rotates in operation, imparting vortex-producing forces which serve to lengthen the effective flow path from supply to output. The principle is similar to that in the jet-controlled version. The rate-sensitive vortex is used in some experimental missiles and aircraft.

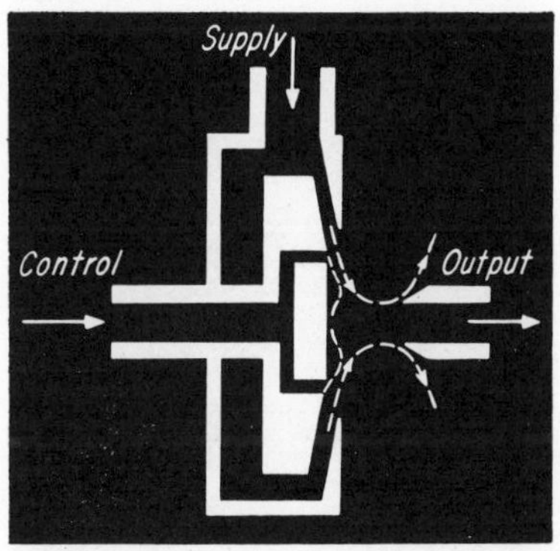

Focused-jet amplifier is designed to aim the supply jet directly into the output pipe except when a control signal tends to deflect the jet away from the target. In some versions, there can be several output pipes clustered in the target area, with several control inputs to steer the main jet selectively.

DIODES

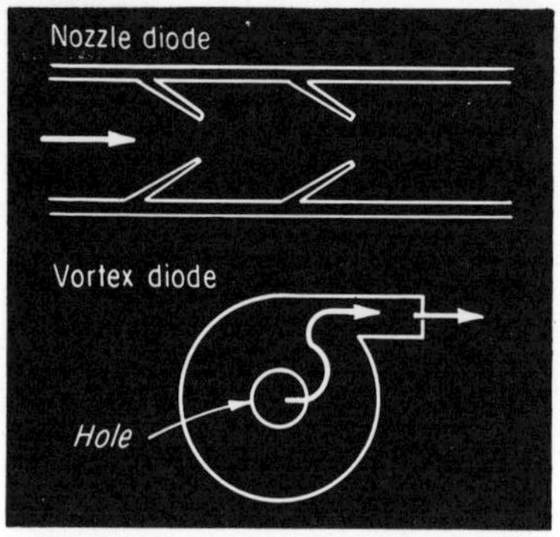

Easy flow forward and restricted flow in the reverse direction characterize all the pure fluid diodes (so-called). The vortex device delivers flow easily in the direction of the arrows, but much less efficiently in the reverse direction. If you want zero reverse flow, then pick a conventional check valve . . . not these.

RESISTORS

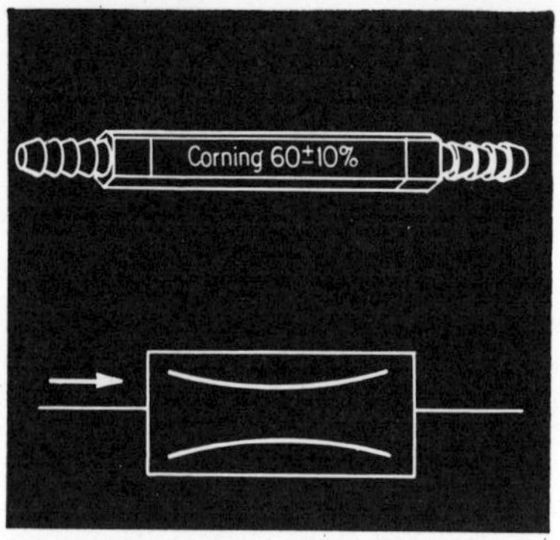

Simple capillaries seem to make the best resistors because laminar flow exists and the pressure loss then is directly proportional to flow. Corning makes the device shown. Resistance **R** is expressed in lb-sec/ft^5. A resistor with **R** $= 40$ has a pressure drop of 3.5 psi for a flow of 0.09 in.^{3}/sec with atmos exhaust.

INPUTS

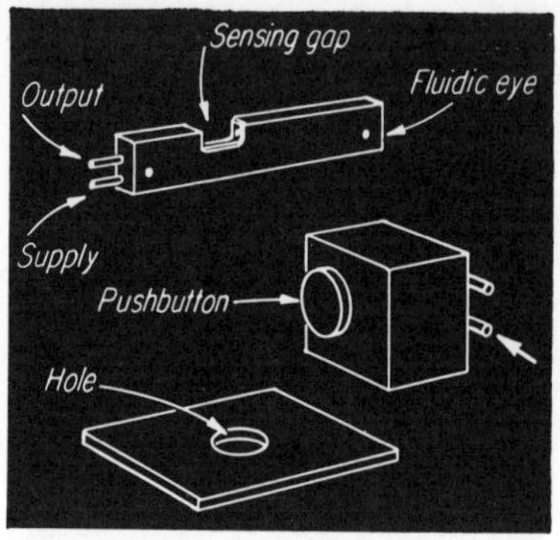

Every input mode from the Fluidic Eye (**PE**—Nov.8,**1965**), to a hole in a plate that the operator covers with the tip of his finger, is being tried. Howie Corp just announced a pushbutton that looks like an electrical pushbutton but acts instead like a bleed-off valve: When pushed, it blocks exhaust, raises pressure.

OSCILLATORS

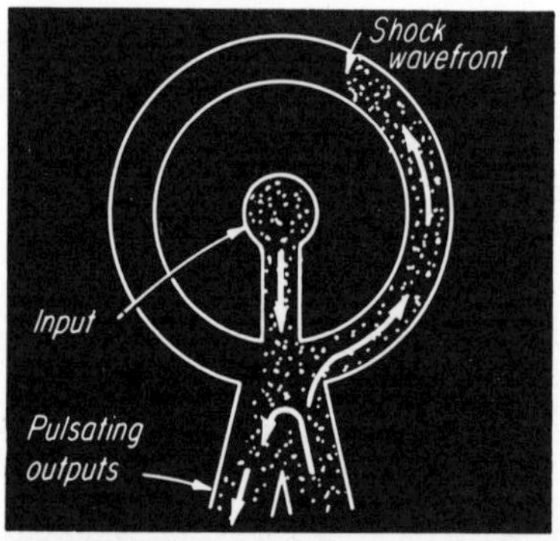

The wall-effect oscillator is the most common type. The speed of sound is the key to the frequency. The flow leaving the input goes to one output leg or the other, but soon gets switched because air entrained from the opposite leg speeds around the by-pass and signals the stream to change to the other leg . . . etc.

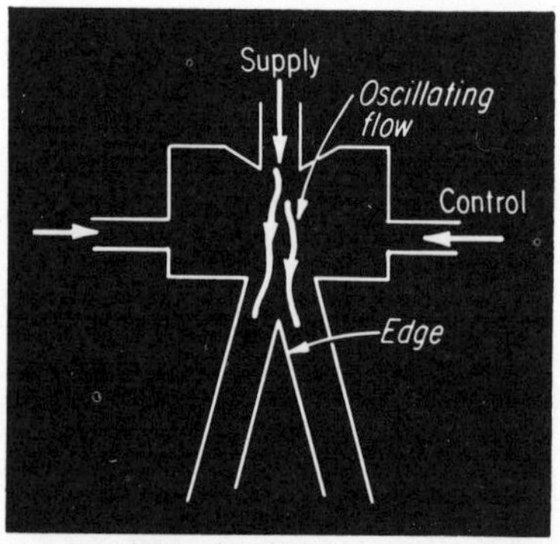

The edgetone oscillator has its origin in a child's whistle, according to H. H. Unfried of Mattel Co (and they should know). Flow oscillates at the knifedge until a side signal shifts it to one leg or the other. Response is fast: Takes only one oscillation, and the period of oscillation can be less than a millisec.

OUTPUT AMPLIFIERS

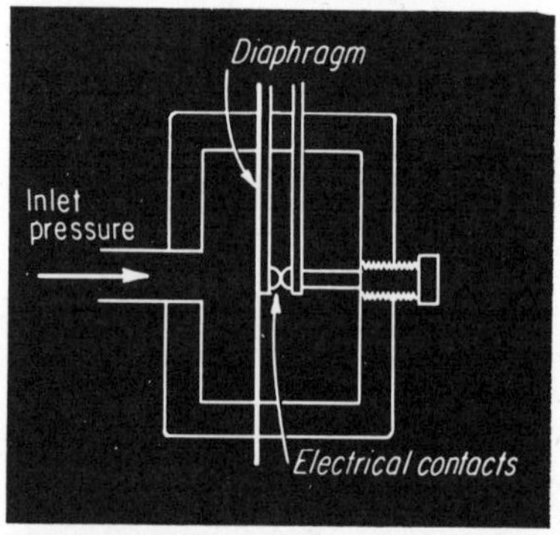

A scramble for ideas is inevitable. Here is the interface between the usually-weak output of the fluidic devices and the demanding input of the power devices. Diaphragms and pistons are one approach. Force-balance pneumatic amplifiers are another. Shown is a transducer. In any concept, the outputs must be cheap.

MECHANICAL DIGITAL LOGIC (for comparison)

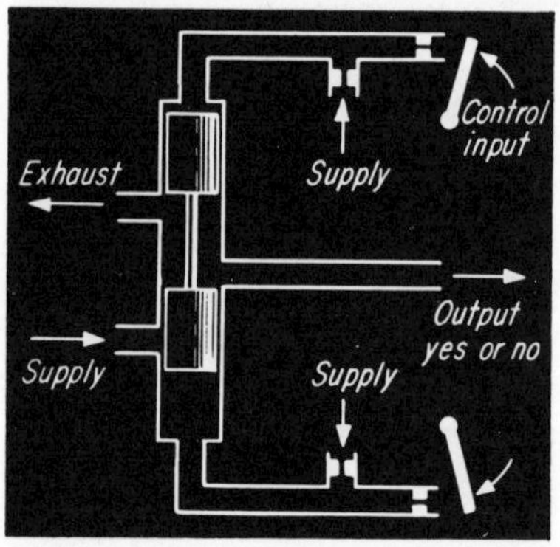

Spool valves for logical control are not new, but there is renewed interest in low-cost arrays to perform all the logic functions such as **OR, NOT, AND, MEMORY,** and so forth. Except for the additional space and the need for moving parts, digital spool logic is as flexible as digital fluid amplifier logic. Also see **Ball logic.**

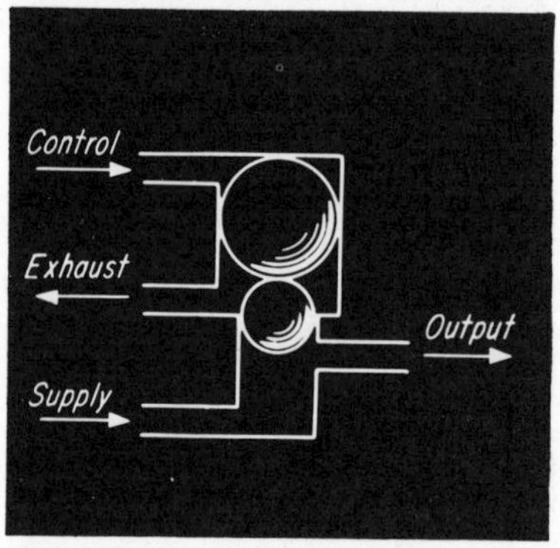

Ball valves for logical control are newer in concept than spool-valve logic, but can do the same job. It might be that purely fluid logic will bypass the mechanical variety, but then again the balls and spools are easier to comprehend and might be a stepping stone to understanding the wall-effect and turbulence types.

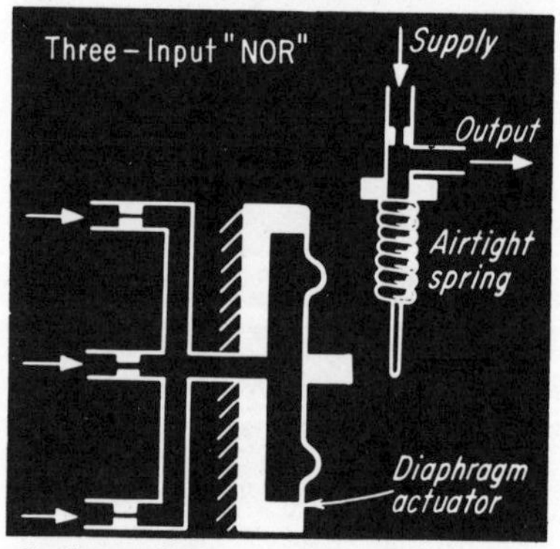

Spring-valve amplifiers are new in this country (Techne-England). The principle is bleed-off. A pressure signal is fed through a restriction into internal space surrounded by the tightly-wrapped turns. No air can escape until the spring is tilted slightly. Then the pressure is instantly lost because of restriction.

HYDRAULIC CONTROL VALVES

I—Relief Valves

Basic academic theories of fluid flow and Pascal's laws relative to the behavior of contained fluids are familiar. The mathematical relationships do not change. Pumps or pressure generating devices are generally well described and rated. as are cylinders and fluid motors.

Valve operating characteristics are less well known, however. Capacity and pressure ratings may not mean the same things to manufacturer and user. The misapplication of hydraulic valves has cost industry untold thousands of dollars in inefficient heat generation and lost power.

This series of articles is intended to cover the identification, function, construction, installation and maintenance of valves in relation to the circuit as a whole and to help in analyzing and predicting malfunction and making satisfactory correction. In some instances it may be of assistance in tailoring a valve to a specific application.

The three major valve types—pressure control, flow control and directional control—will be con sidered in that order. A wind-up article is planned to include material about certain specialty items like the hydraulic fuse, multiple control valves that are really valves in a common housing and special systems like high-low and regenerative circuits.

JOHN J. PIPPENGER
Double A Products Co., Manchester, Mich.

THE RELIEF VALVE is a device that automatically creates an orifice sufficiently large to divert fluid back to the reservior at a predetermined value of pressure. The ideal relief valve would permit pressure to rise to the set value and then create an orifice to maintain that pressure. This ideal is not economically feasible so units used as relief valves are designed to operate at some degree less than the ideal condition according to the economic requirements of the circuit involved.

Simplest type of true relief valve is a sliding spool normally closed two-way valve which has one end of the spool connected to the input line and the other end loaded with a simple coil spring as shown in Fig. 1. The pocket in which the spring is enclosed is drained to the downstream or tank connection and back to the reservoir. The spring positions the spool so that the input line is blocked from the exhaust line. As input line pressure increases, oil pressure against the end of the spool tends to move the spool against the opposing spring. The primary, or pressure port of the valve is gradually opened to the secondary or tank port so that an orifice is created directing the oil back to the reservoir.

Size of orifice created is directly dependent on effective area on the end of the valve spool, spring rate, and amount of fluid flowing through the orifice.

Some refinements to this basic relief valve include angles

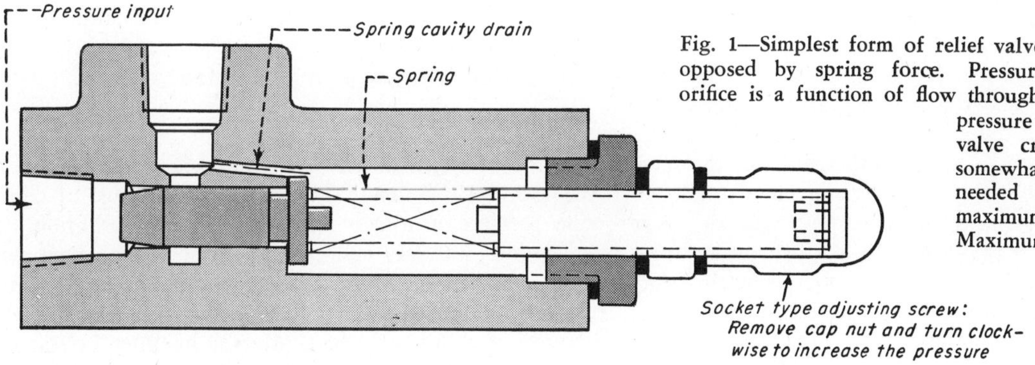

Fig. 1—Simplest form of relief valve. Fluid pressure is opposed by spring force. Pressure drop across the orifice is a function of flow through the orifice so that pressure at which the relief valve cracks open will be somewhat lower than that needed to dump a rated maximum flow of fluid. Maximum flow is 1 gpm.

A Short Course in Hydraulic System Design

Many papers, articles and books have been published in an effort to familiarize the reader with efficient hydraulic circuits. There are scores of more or less standard circuits that have performed excellently in specific machines. Certain basic segments of these circuits are common to many and often may be purchased as a multiple component incorporating pump, control valving and even fluid motor in a single package.

Standard circuits are convenient but they, too, may be misapplied. Each new problem should be approached with the following questions.

Does the job economically justify use of fluid power? A thorough knowledge of cost versus performance of oil hydraulic, pneumatic and mechanical transmissions is required to make this decision. It is necessary to be extremely cautious to overcome the natural temptation to apply new found knowledge uncritically.

What forces and speeds are required in the operation. Force is pressure times area. Small spaces (and areas) require higher pressures. Speed is equivalent to flow (gpm or in.^{3}/min). Pumps and fluid motors or cylinders may now be specified.

Control valves are chosen in a similar way. Pressure ratings must match maximum pressures generated by the pump unless rapid movements will involve shock waves. Flow capacity through the valve will be based on capacity of pump with the exception of regenerative circuits or those using accumulators.

Valves may be divided into three major categories. Pressure control valves are chosen to meet the needs of the pressure generating equipment in relief and safety functions, sequencing, unloading, counterbalancing, etc. Flow control valves control speed of fluid flow and are selected according to the degree of sensitivity required and quantity of fluid handled. Directional control valves are chosen to suit the operating signal (manual, cam, solenoid, pilot) and the type of control desired such as two-way, three-way, four-way or any other multiplicity of control needed. Many variations such as spool notches, taper grinds, reliefs, etc., are available as refinements of the direction control function.

Having chosen the circuit and its components, design proceeds to plumbing, reservoir and physical placement of components for maximum efficiency and ease of maintenance. Improvements in high pressure plumbing and connectors in the form of flanges and subplates have eased maintenance and increased customer acceptance to a very satisfactory level.

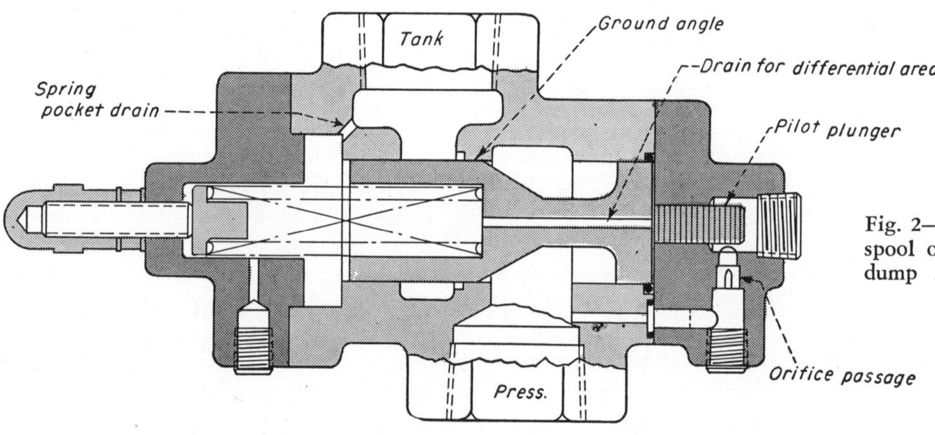

Fig. 2—A pilot piston actuates the spool of this relief valve used to dump larger quantities of fluid.

ground on the spool, orificed passages to the pocket at the end of the spool, orifices in the drain line from the spring pocket, and pilot pistons instead of pressurized fluid to work against the end of the spool as shown in Fig. 2. Use of pilot pistons is often necessary to keep the effective pilot pressure area within the physical range of practical spring rates. When a pilot piston is used, it is necessary to drain the pocket between pilot piston and end of the spool because of the difference in areas. Sometimes this pocket is drained through a hole in the center of the spool which conducts fluid to the drained spring chamber.

Pressure range is usually quite narrow with this type relief valve because of the inherently narrow spring ranges that can be accommodated in the relief valve housing. In some applications it is necessary to nest several springs to obtain the desired spring rate.

Relief valves may be constructed with a stepped spool of two different diameters in a housing with both end pockets drained as shown in Fig. 3. The larger diameter is toward the spring pocket. The larger diameter is also between the input or pressure port and the outlet or tank port. The net difference in area between the two concentric diameters acts like a pilot piston.

Operating Characteristics

Direct spring loaded relief valve design, while simple in construction, has many drawbacks. The range of adjustable pressure is usually quite narrow. In addition, the valve does not fail safe. If an orifice is plugged with foreign matter the valve fails to open permitting an overpressure condition. There is a tendency, unless highly damped, for the valve to chatter or hammer in response

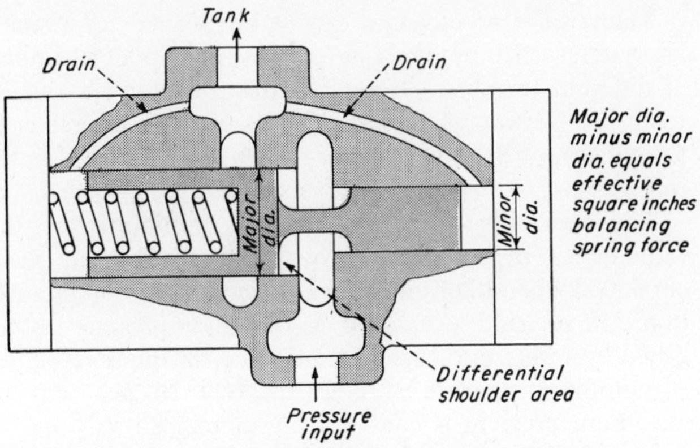

Fig. 3—Net area between minor and major diameters is the effective area against which pressure acts in this relief valve. Both ends of the spool must be drained to tank.

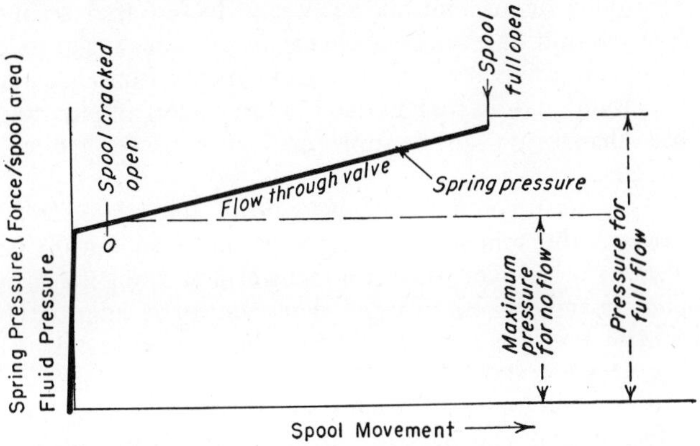

Fig. 4—If full pump capacity is dumped through the relief valve with the spool full open it may take a considerable drop in applied pressure to reduce the tank flow to zero.

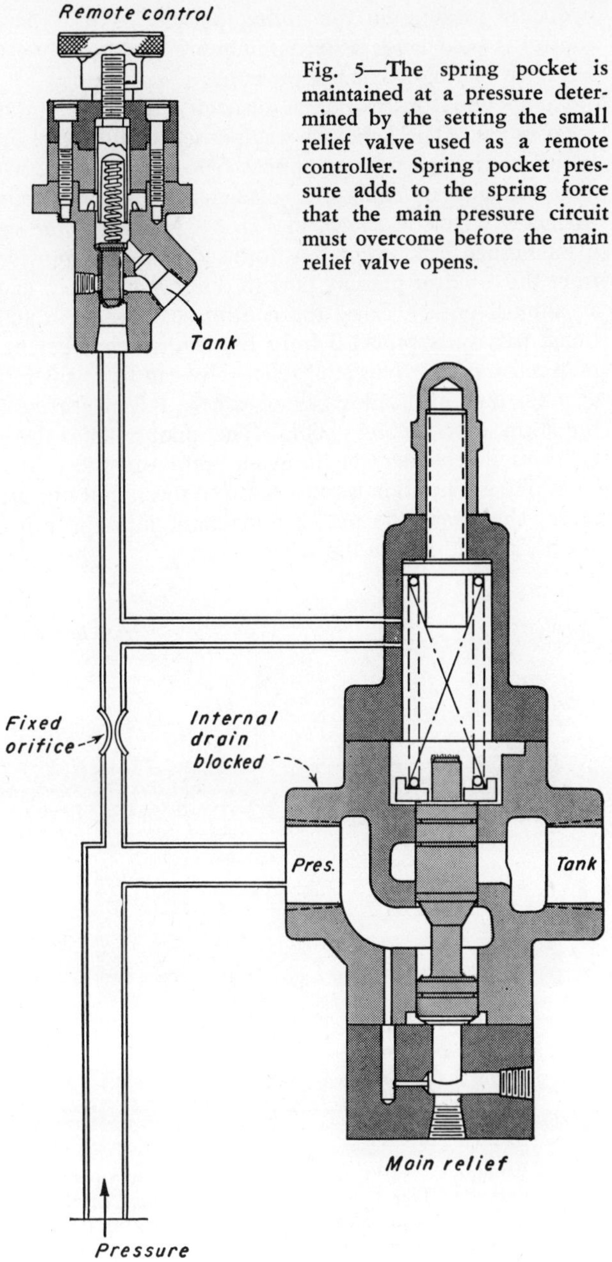

Fig. 5—The spring pocket is maintained at a pressure determined by the setting the small relief valve used as a remote controller. Spring pocket pressure adds to the spring force that the main pressure circuit must overcome before the main relief valve opens.

to rapid changes in circuit requirements. Sometimes this is the result of harmonics set up in the spring.

Any restriction in the line to the reservoir will increase the effective spring values so that the operating pressure will be higher. In certain applications it is common to connect returns from several different devices to a common return header. Pressure in this header will increase with large tank flows which, when reflected back to the relief valve, cause an increase in the entire circuit pressure. Therefore, individual and independent lines to the reservoir are desirable where economically feasible. Independent tank lines are also convenient in tracing down circuit losses.

Spring loaded relief valves are inherently variable orifice devices as shown in Fig. 4. The spring requires a definite increment of pressure between barely cracked and full open. An 18 gpm pump might direct its full volume through the valve in certain parts of a machine cycle. At 1000 psi the relief valve has established a maximum orifice. However, if work resistance during some other part of the cycle created a pressure of 900 psi a portion of the flow may still be by-passed to tank through a partly open orifice. Some valves do not entirely close until pressure drops to 50 per cent of the relief valve setting.

Often this loss is unimportant. Some machines use the effect to intentionally slow operation as work resistance increases. When pressure reaches the set value the valve is full open and the machine stops entirely. However, a penalty in efficiency is paid for this type of bleed-off speed control.

The piloted poppet relief valve is perhaps the most accurate and efficient relief valve for all around use on hydraulic applications. In some cases it is higher priced on initial applications but usually saves in operating costs.

Pilot Operated Relief Valves

Operating pressure of the spring loaded valves may also be controlled remotely as shown in Fig. 5. If a small amount of oil is diverted from the pressure port through an orifice of proper size into the spring chamber, and the drain from the spring chamber is diverted through a small pilot type relief valve, an addition to the spring force is created. The pilot valve can be placed remotely from the main valve. An established minimum pressure is assured by the main spool spring, any higher value within the scope of the equipment may be reached by the added hydraulic pressure in the spring pocket. This type of control is used when a fixed minimum pressure is necessary with varying maximum pressures.

Compound construction is characteristic of pilot type relief valves. The basic two-way portion of the valve may be either sliding spool or poppet type construction, with a multiplicity of concentric diameters to create desired balancing factors as shown in Fig. 6. Minimum pressure is established by a spring. An orificed passage is provided from the input or pressure port to the spring pocket, or to an annular pocket, called the control chamber. An additional passage is provided from the control chamber to a pilot relief valve. This pilot relief valve can be in the form of a small spring loaded ball or cone. It can also be in the form of a sliding spool. The pilot relief valve is relatively independent of the main portion of the valve as far as physical position is concerned but most pilot operated relief valves have the pilot section built into the cap or directly manifolded to the valve.

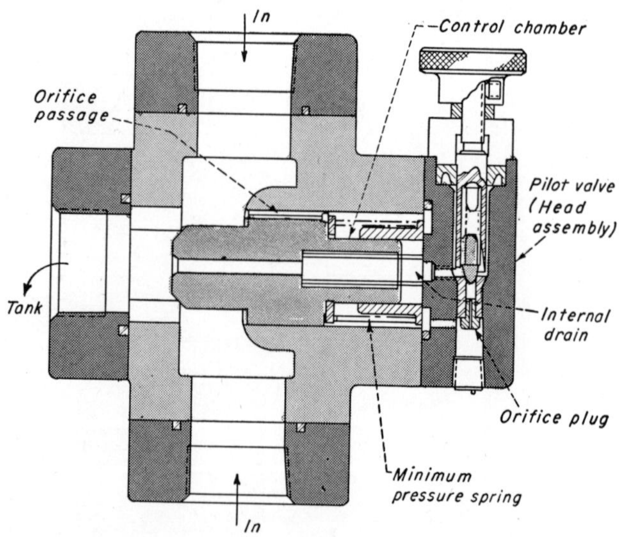

Fig. 6—Compound valve in one housing whose operation is similar to that of Fig. 5. Only an annular portion of the main piston area is used for added hydraulic pressure in the spring pocket. Note two orifices; between main circuit pressure and spring pocket, and between spring pocket and pilot valve cone.

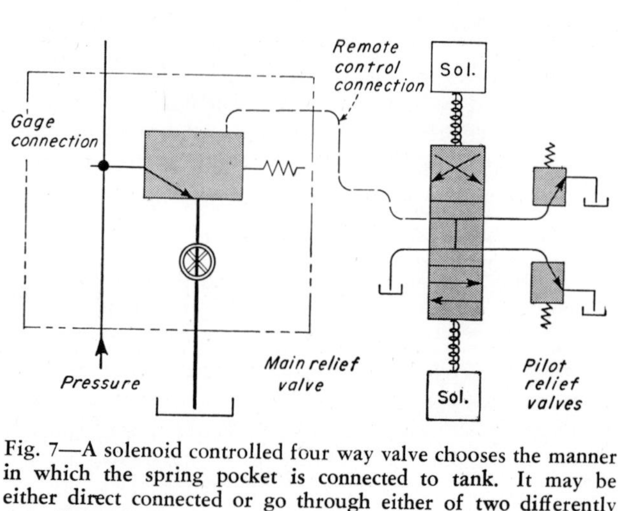

Fig. 7—A solenoid controlled four way valve chooses the manner in which the spring pocket is connected to tank. It may be either direct connected or go through either of two differently set pilot relief valves. Any one of three maximum relief settings may be chosen.

Many different pressures can be established for a single basic relief valve by directing oil to various pilot chambers at different pressure settings. A definite program can be set up in circuits where varying pressures are desirable by the process of diverting the oil to different pilots as shown in Fig. 7.

The basic valve body usually has a minimum pressure value of less than 100 psi. This pressure will dump main circuit oil when pilot oil is released to tank without restriction. At pressures over 2000 psi the initial pressure (established by spring rates) may be increased for quick response. A minimum pressure of approximately 10 per cent of maximum pressure is common in commercial valving.

In addition to the relatively quiet and dependable service that can be expected of the piloted type relief valve, versatile controls can be tailored to a multitude of applications as shown in Fig. 8. Small pilot relief valves can be controlled by an adjusting screw and locked after setting, levers permit quick manual change of pressure, a cam roller type unit can sense pressure requirements from machine operations and diaphragm units can sense and amplify minute changes in air controls or low pressure hydraulic circuits.

The most common remote control for piloted relief valves is the adjusting screw type as shown in Fig. 8 (E). Two or more pilot valves can control flow from the main pilot chamber. An integral pilot control is adjusted to control pressure between control chamber and tank and is set for any desired maximum pressure. A remote valve will control for any pressure less than maximum. The integral pilot control valve is in parallel with auxiliary remote control valves. The unit with least resistance to fluid flow is the effective control.

Physical location of remote valves should be within 20 ft of basic relief valve because of frictional losses in passage of pilot oil through the lines. Special units and adjustments can increase this maximum distance.

In some circuits needle type two-way valves are used as remote control for a relief valve. Pressure can be controlled accurately in this manner. In all cases discharge from the remote control valves must be connected back to the reservoir without resistance if the pressure control is to be consistent.

Rapid drop in operating pressure to the established minimum is possible by allowing the control chamber to drain without restriction to the tank. A two-day valve in parallel with the maximum pressure adjustment pilot control valve will perform this function. When the two-way valve is open to the reservoir, pilot oil is released and the pressure drops to the established minimum. When the two-way valve is closed the oil is diverted from the control chamber through the pilot pressure control unit to tank. Maximum pressure will then be a function of the established setting of the remote or integral control pilot valve or the resistance of the work being performed.

Fig. 9 illustrates a standard lever operated four way valve with detents to hold the spool in one of three positions at the will of the operator. The end cap has been replaced with a two way valve housing. The spool for the two way portion is fastened to the end of the four way directional spool so that the two will operate simultaneously.

When the valve is in the neutral position the in and

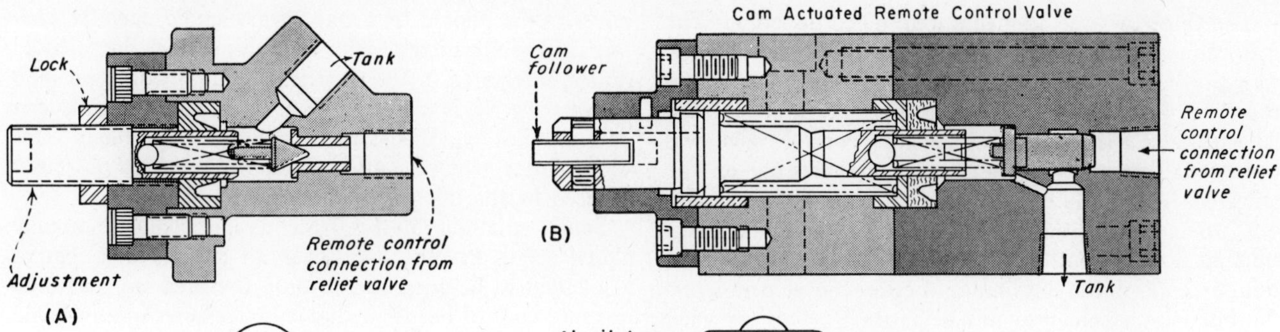

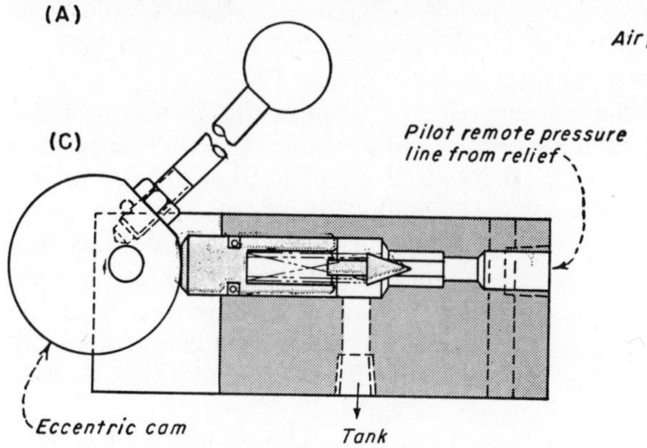

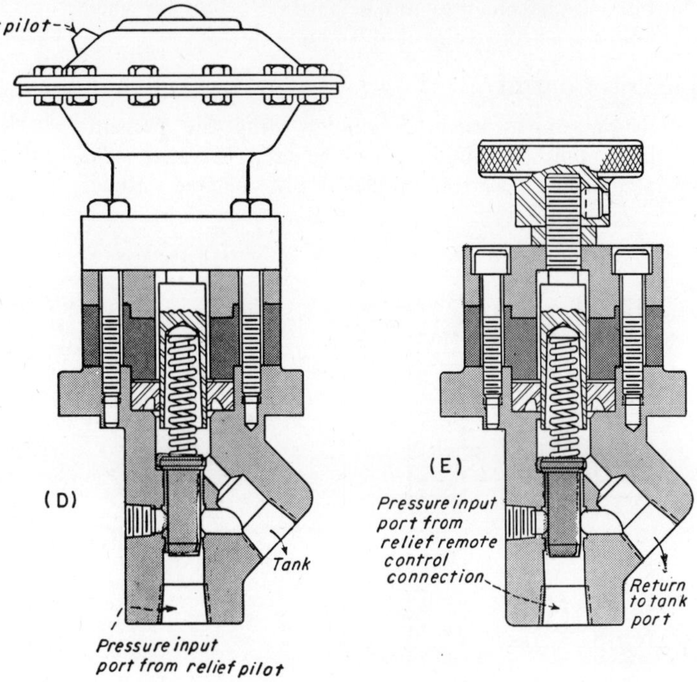

Fig. 8—Three methods of changing the pressure settings of remote pilot relief valves. (A) Spring pressure is adjusted by a screw and locked. (B) A cam operated valve establishes spring pressure on the relief spool. (C) A lever with eccentric cam establishes spring pressure on relief cone. (D) Air pilot pressure increases spring force on the relief spool. (E) Adjusting screw is most common form of pressure adjustment on the pilot relief valve.

out ports of the auxiliary two way valve are connected. The out port is connected to the reservoir and the in port is connected to the remote control port of the relief valve. Thus the control chamber in the pilot operated relief valve would be directed back to tank or relaxed. This diversion of the control oil back to the reservoir will drop main relief valve pressure down to the minimum as established by the main spool spring.

When the four way valve is shifted in either direction, flow through the two way valve is stopped. The relief now functions in the normal manner, raising pressure to the resistance value of the circuit or to the maximum pressure setting of the integral pilot in the event that the entire volume is not being used to move a machine member. Auxiliary two way valve sections are available in many different types and sizes of directional control valves.

It is possible to connect several valves in series as illustrated in Fig. 10. The directional control valves can be in a parallel circuit and the relief valve pilot lines connected in series through the two way sections. This permits using a valve of a limited size for directional control without minimizing the flow during the unloaded portion. All unloading is done through the piloted relief valve.

If a pilot valve is installed in the remote control line between two valves as illustrated in Fig. 10 it will determine maximum pressure while valve *B* alone is being used. When valve *A* is used the system will have the full pressure as set on the main relief valve. In this manner it is possible to cascade pressure values down through several settings.

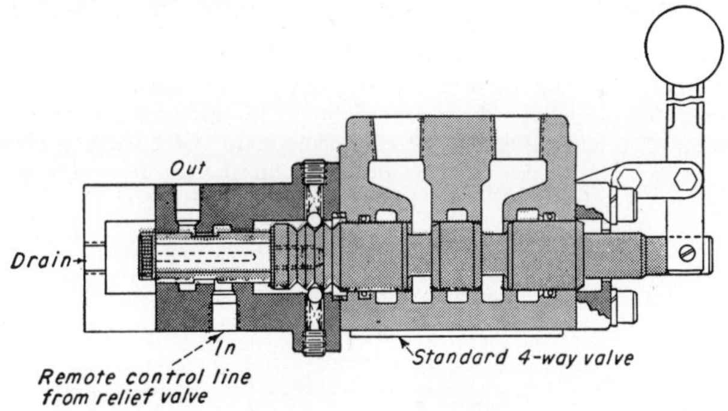

Fig. 9—A two-way valve with open center is here integral with a standard four-way valve. The two-way valve may be used to pilot a relief valve spring pocket to tank, effectively lowering pressure setting during idling of a hydraulic circuit.

One large installation used a 90 gpm pump with a 2 in. relief valve to supply oil to the circuit. The valve controlling the main winch was of a two inch pipe size. Maximum pressure was needed for the fluid motor driving the winch. A tailgate mechanism needed a lesser pressure and only a $\frac{3}{4}$ in. directional control valve. An auxiliary bulldozer needed only a minimum pressure but $1\frac{1}{4}$ in. pipe.

The three two-way segments were piped in series. The first connection from the relief valve remote control line was to the winch control valve. The discharge from this two way section was directed to the input of the tailgate valve. In this line a remote control valve was connected that controlled the maximum pressure on the tailgate. The out line from the tailgate two-way valve was connected to the bulldozer valve assembly with a suitable remote control valve ahead of the two-way section to control the maximum pressure developed while the bulldozer mechanism was being used. This provided a simple and effective circuit at a minimum cost.

Electric Control

Two-way valves used to remotely drop the pressure to the established minimum can be lever operated, cam operated, stem operated, or electrically operated. Relief valves of the piloted type may have solenoid operated two way valves built integral with the valve so that the solenoid can be integrated in the electrical circuit to drop pressure between cycles as shown in Fig. 11. Many circuits can be simplified and made much more dependable in this way. Space requirements and piping costs are obviously reduced by the integral solenoid pilot valve.

Some variations of this basic design include normally closed spools in the pilot valves for circuits that require a solenoid to be energized to drop the pressure, and dual pressure control heads so that two pressures are available with the solenoid acting as the shifting medium between the two pressures as shown in Fig. 12.

If the pressure control chamber is allowed to bleed back to the reservoir without resistance and faster than oil can enter the control chamber through the orifice, the pressure is equal to the force created by the spring on the main spool. If this bleed is blocked and oil is directed to the normal integral pilot, operation is conventional. If the two-way valve is made a part of the relief valve assembly piping and pressure drop are reduced.

These valves are usually provided normally open so that pressure controlled by the relief valve is at its minimum value until the solenoid is energized.

Sample Applications

Accumulators that operate with a weight or spring offer a physical movement to sense the charged and discharged position. It is simple and economical to have limit switches at the desired extreme positions to actuate a solenoid operated relief valve and control the recharge cycle Usually it is much less expensive to run electrical conduit and limit switches rather than pilot piping. An important factor is the normally greater protection from mechanical injury to the connecting

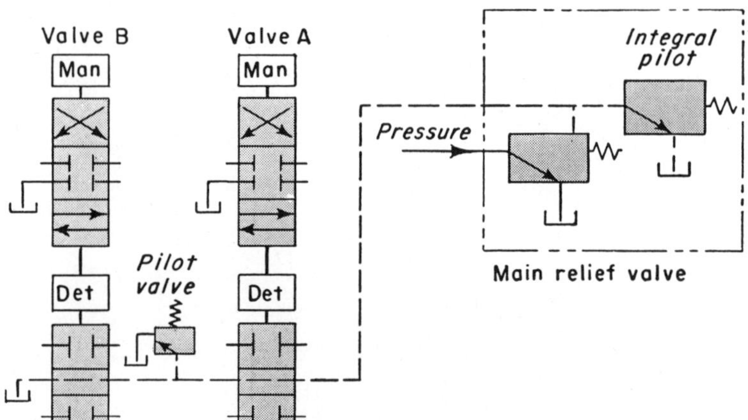

Fig. 10—A second small pilot relief valve may be added in parallel to the one built into the main relief valve. When four-way valves *A* and *B* are centered, lowest relief valve setting is made by dumping the relief spring pocket to tank. When valve *B* is off center the main relief valve is controlled by the auxiliary pilot valve. When both *A* and *B* are off center the relief valve operates normally with its integral pilot relief valve.

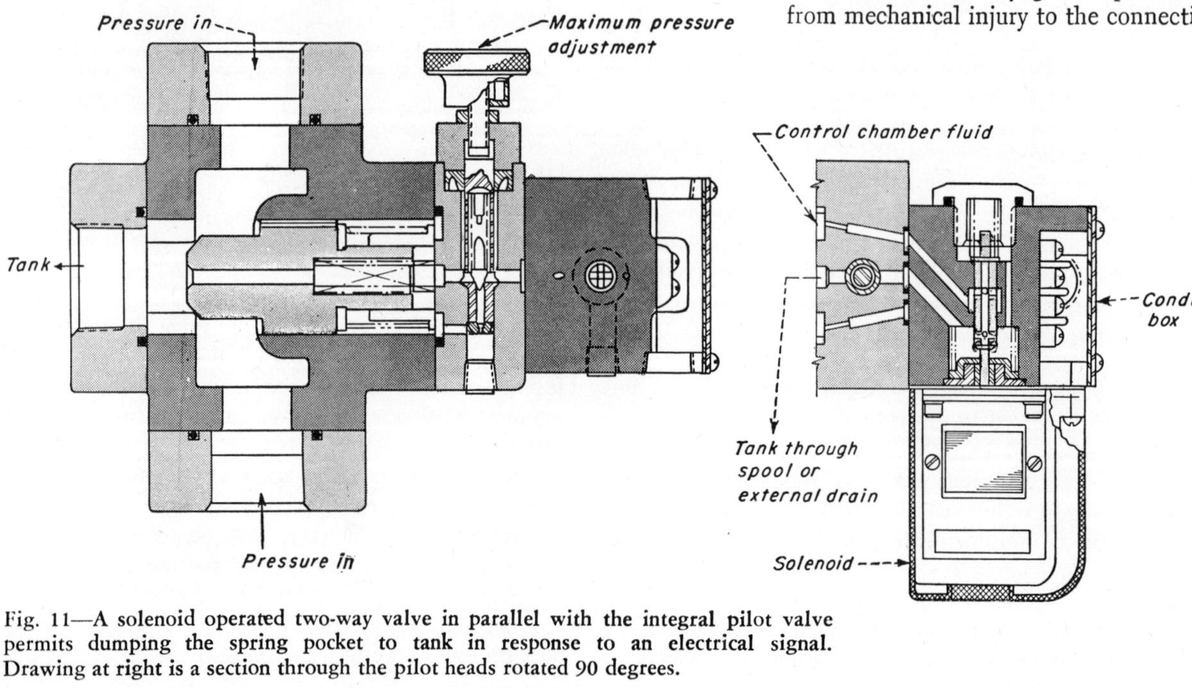

Fig. 11—A solenoid operated two-way valve in parallel with the integral pilot valve permits dumping the spring pocket to tank in response to an electrical signal. Drawing at right is a section through the pilot heads rotated 90 degrees.

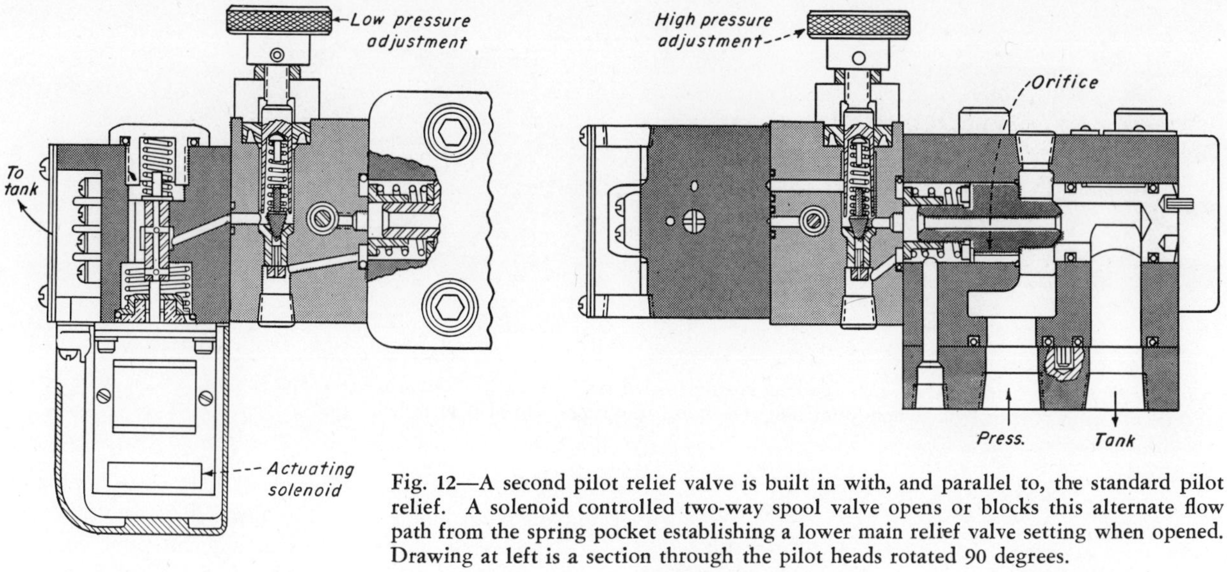

Fig. 12—A second pilot relief valve is built in with, and parallel to, the standard pilot relief. A solenoid controlled two-way spool valve opens or blocks this alternate flow path from the spring pocket establishing a lower main relief valve setting when opened. Drawing at left is a section through the pilot heads rotated 90 degrees.

media when electrical devices are used.

Accumulators using gasses as a balancing medium or so constructed that physical movement cannot be used to sense the amount of charge must, of necessity, depend on pressure for a signal. In this application a pressure switch with an adjustable maximum pressure as well as an adjustable minimum pressure can be used to actuate the solenoid operated relief valve to control the pumping cycle for recharging the accumulator.

In some circuits the solenoid control is used for establishing two different adjustable pressures. These models have two independent pilot relief valves manifolded to the main relief valve body incorporating a solenoid operated diversion valve for control as shown in Fig. 12. With the solenoid de-energized, the primary pressure control head establishes a certain pressure at the desire of the operator. When the solenoid is energized the primary head is removed from control and the secondary head takes over as the control medium. The secondary head must always be set at a higher pressure than the primary head or the control will be nullified. On certain types of spinning and metal forming machines this valve provides a medium or low pressure for delicate operations and a higher pressure for coining or other operations during the cycle when additional pressure is necessary. If there is some question as to the amount of back pressure that may be encountered in the secondary or tank port of the relief valve it is usually advisable to specify externally drained models. If the tank line is relatively short and the volume is well within the rating of the valve, external drains are not essential.

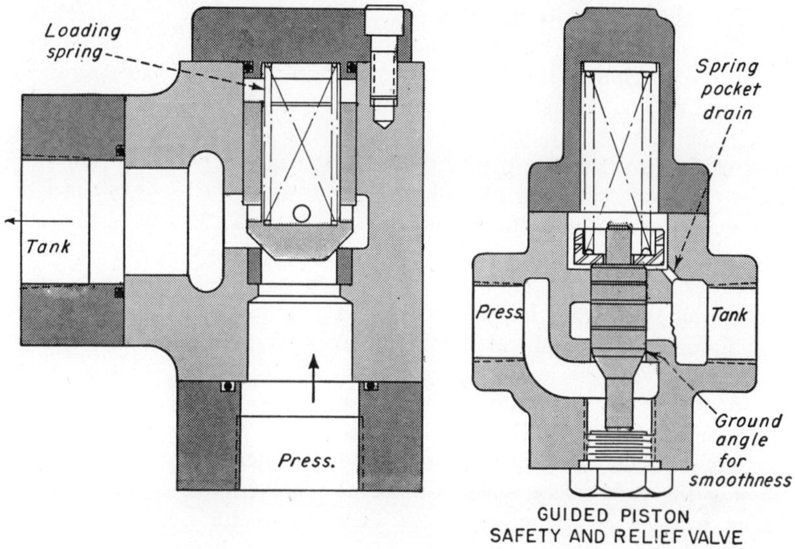

Fig. 13—Simplest safety valve (left) in which cone lifts off seat when pressure equals spring force. Holes in cone are to drain the spring pocket. (right) A guided piston safety valve. The spring pocket drains to the tank port through an orifice used to damp out spool oscillations without adversely affecting speed of pressure actuation.

Many automation devices operate at relatively low pressures during a large portion of the time cycle. As an example, the index, positioning and locking portion of a machine cycle may only require 200 psi. The automatic insertion of a sleeve bushing may require as high as 1000 psi. In this cycle a dual pressure solenoid operated relief valve might be used. During all portions of the cycle with the exception of the pressing operation the pressure is held at a maximum of 200 psi by the dual pressure relief valve. As soon as the circuit is actuated to bring in the pressing cylinder the solenoid operating the relief valve pilot diversion spool is energized. Actuating or energizing the solenoid on the relief valve provides a pressure setting of 1000 psi instead of the previous 200 psi.

Other reasons for limiting the pressure may be as safety precautions, power saving, easier control, preventing intensification during certain dynamic operations or to limit forces on certain tooling. At the desired point in the circuit the pressure can be changed by an electrical signal to any higher value desired, within the scope of the equipment involved.

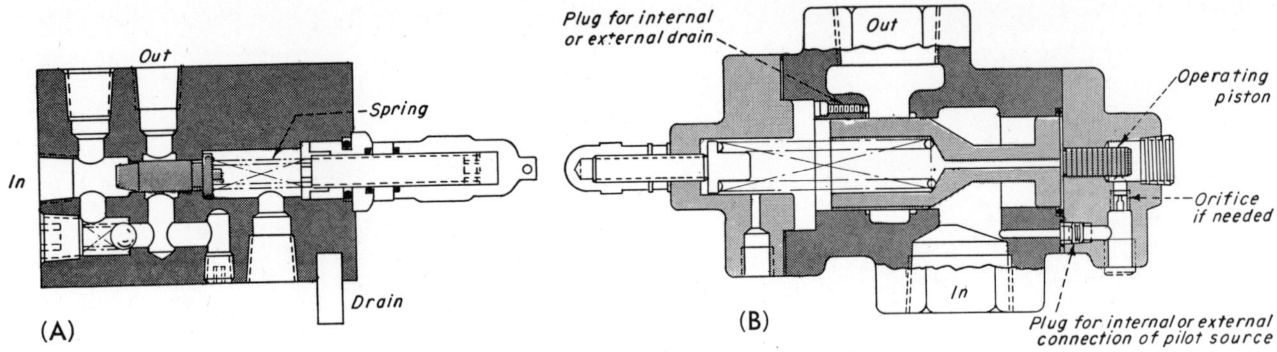

Fig. 14—Two forms of bypass valve. (A) is direct spring loaded and operated by the pressure at the in port. (B) may be actuated by pressure at the in port or by a remote pilot line. Damping provisions are also included.

BYPASS VALVES

If the relief valve were only in the circuit for protection against undesirably high pressures and operating pressure were considerably lower than the setting of the valve it might be considered a safety valve. As a safety valve, operation only occurs because of some improper functioning of the machine. However, construction of a safety valve might be considerably simpler than a true maximum pressure relief valve. It might also be desirable to have the safety valve create noise in the form of a chatter or whistle to alert the operator when the valve is in operation.

In many installations the safety valve is a simple ball check with a heavy spring to hold ball against seat. At pressures sufficiently high to unseat the ball a path is created back to the tank for this excess fluid. No orificing or damping action is provided so that the operation is erratic and often very noisy. Since the device only operates in an emergency the noise is relatively unimportant. Two variations on this design are shown in Fig. 13. Low first cost and dependability are the primary criteria in choosing the true safety valve. Choice of a relief valve usually involves many more factors such as quietness, dependability, repetitive accuracy, versatile remote control, smooth passage through valve and metals which will last a long time in severe service.

There are two basic types of bypass valves. Probably the most widely used is the direct spring loaded type shown in Fig. 14 (A). This is a normally closed two way valve with a spring on one end of the spool balanced by a piston on the other. This piston may be the end of the spool or some separate actuating medium such as an integral piston of a different diameter or a diaphragm for amplification of pressure signals as in (B). This actuating piston may be connected to the input line internally with or without a restricting or damping device. If operating signal is taken from the input or upstream side, position of the spool is directly related to input pressure; relief, safety or internally piloted sequence valves operate in this manner.

If this pilot piston device is connected to some external fluid pressure signal point it is an externally piloted unit. Pressure at the in port of the valve has no effect on operation of main spool. This type of pilot connection is used in most unloading circuits, some counterbalancing circuits and some sequencing operations. Pilot oil losses through normal clearance flows are low. In some critical installations pilot pistons are provided with seals and a diaphragm type pilot is virtually "dead end" service.

Second type of bypass valves are the pilot controlled variety shown in Fig. 15. The main spool is normally closed with a relatively soft spring. Operation of the valve results from fluid pressure plus spring force. A chamber of trapped oil is created in the same general area as the main spool spring. Usually, oil under pressure, is allowed to enter this control chamber from the input of the valve through a fixed orifice. At the set value of the integral pilot relief mechanism the pilot oil is allowed to escape back to the reservoir at a larger flow rate than that at which it enters the control chamber. The amount of pressure created by this integral pilot relief valve is added to the initial spring tension against the main spool of the valve. The total of the two is the system pressure ahead of the bypass valve. This means that as an internally controlled unit it operates in the same manner as a pilot type relief valve. As an externally piloted unit a small piston is included to upset the integral pilot control relief valve so that the main spool can rise at the pressure created by the main spool spring.

Fig. 15—When pilot operated the small control piston lifts the integral pilot control relief valve and back pressure in the spring pocket drops to zero. Any pressure above the spring valve will then operate this valve.

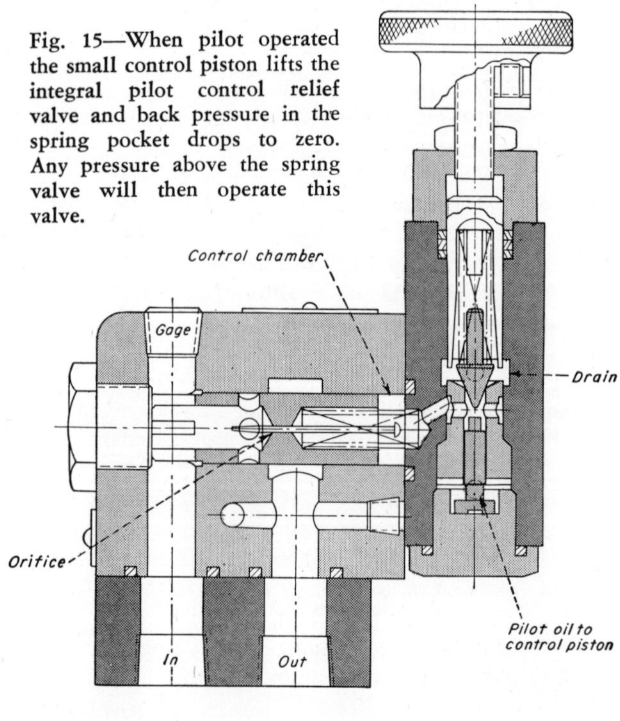

HYDRAULIC CONTROL VALVES

II—Pressure Control Valves

JOHN J. PIPPENGER
Double A Products Co., Manchester, Mich.

ALL PRESSURE CONTROL VALVES, including relief valves, are basically two-way valves held normally open or closed by a spring or equivalent device (counterweights or pneumatic balance). Either input, output or external pilot pressure may be used to control the valve by aiding or inhibiting spring force depending on the desired function.

Variations on the basic pressure control valve include compounding with check valves, providing alternate flow paths, or inverting the pilot connections to achieve specialized functions. Each application requires special consideration of the consequences of failure of the valve or of other components in the system. Every valve must be properly drained of clearance flow or it will not function as planned.

COUNTERBALANCE TYPE

Quoting from the glossary of terms for fluid power as published by the National Fluid Power Association, "VALVE, COUNTERBALANCE—A valve which maintains resistance against flow in one direction but permits free flow in the other. Usually connected to the outlet of a double acting cylinder to support its weight or prevent uncontrolled movement." Fig. 1 is an example of this type of circuit.

When the piston is lifted the counterbalance valve offers no obstruction to flow since the integral check valve permits free flow in this direction. When lowering, fluid coming out of the cylinder is controlled at a set minimum pressure by the pressure actuated segment of the valve. This means that the piston must be "pumped" down.

Flow from the counterbalance valve is usually connected through the directional control to tank. This permits the spring pocket to be internally drained saving piping and eliminating potential external leakage as shown in Fig. 2. Note that when the cylinder is lifting pressure is applied in the spring pocket of the pressure actuated segment of this valve closing it off tight. Reverse flow must go through the integral check.

When lowering, the check valve closes, so that flow is controlled by the pressure actuated segment which will not open until a set pressure is built up under the piston. The downstream or outlet port of the counterbalance valve is directed to tank so that the internal drain is also to tank and the pressure actuated segment of the

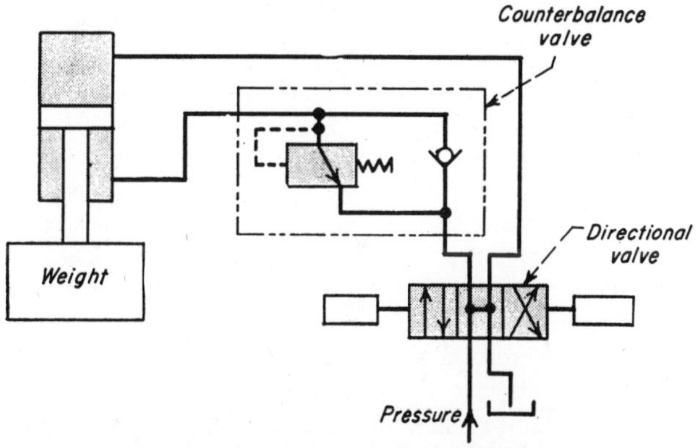

Fig. 1—Typical application of a counterbalance valve. Pressure actuated segment is set to open at or above the pressure generated in the trapped fluid under the piston by the load weight. To lower the weight the piston will then have to be "pumped" down.

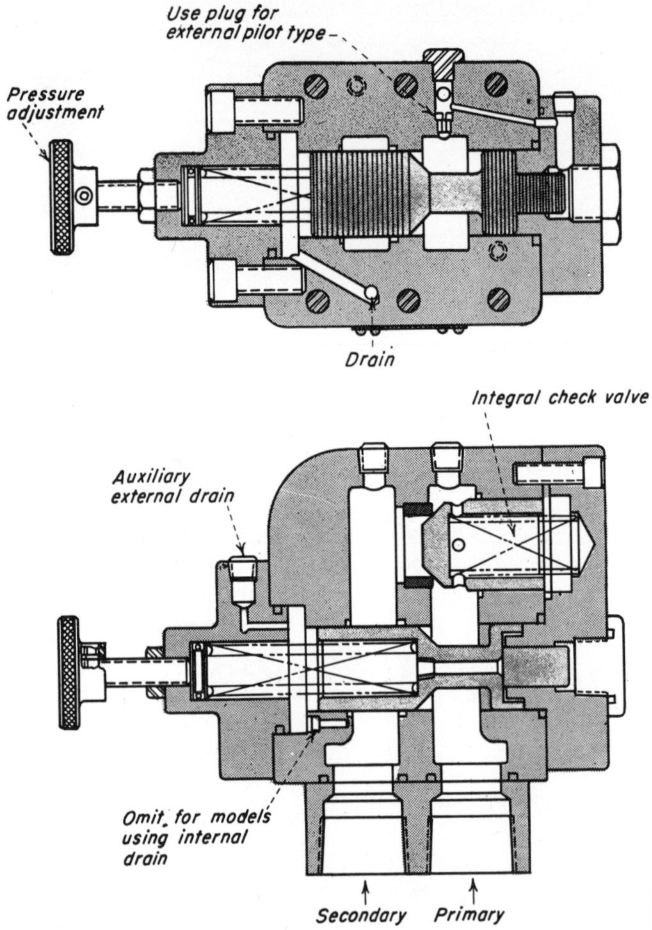

Fig. 2—Two sections of a counterbalance valve using internal drain. External connection to spring pocket is plugged and internal connection goes to secondary.

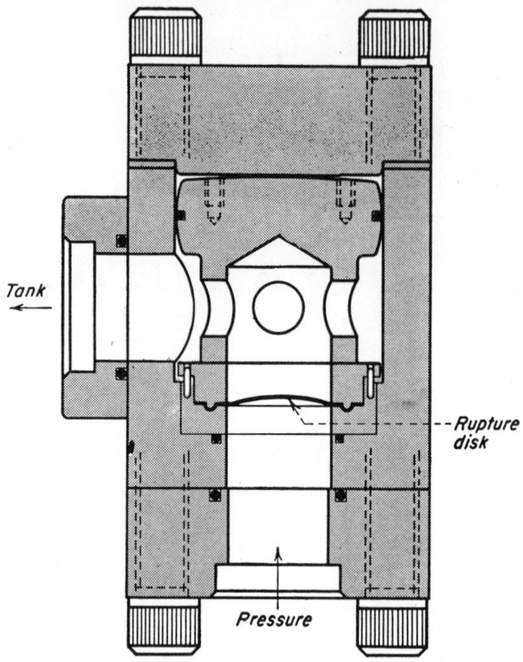

Fig. 3—A hydraulic fuse in which flexible disc will rupture above specified design pressures permitting free flow to tank.

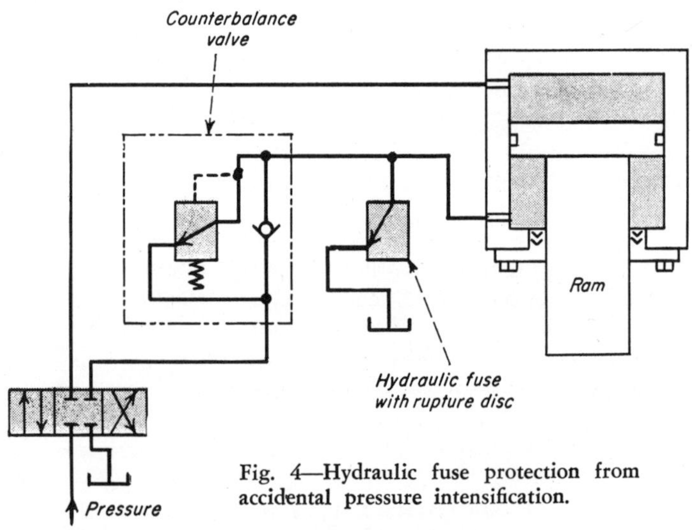

Fig. 4—Hydraulic fuse protection from accidental pressure intensification.

valve operates in a normal manner.

The actuating mechanism of the counterbalance valve may be operated from a control line internally connected to the primary port or from an external source. The direct operating type wherein the pressure is connected internally is generally used where it is desired to prevent uncontrolled movement of the cylinder.

A relatively light circuit may depend on the counterbalance valve as the only mechanism preventing cylinder drop during idle cycles. If loading on the cylinders is likely to change in value, as in the case of a control gate on a power dam covered with ice in the winter time, the pilot line from the counterbalance valve may be taken from the opposite end of the cylinder so that there would have to be pressure on the top of the piston before the counterbalance valve will open.

This connection has the further advantage of preventing a runaway condition that may develop in closely counterbalanced cylinders. When lowered, a high inertia load can maintain pressure under the piston even if pump pressure is cut off. If the counterbalance valve is piloted from this side of the cylinder it will remain open. If it is piloted from above the piston any drop in supply pressure will drop the pilot pressure so that the counterbalance valve will restrict movement to the speed of pump flow. A fast uncontrolled gravity drop is avoided.

In many circuits, especially press circuits, the counterbalance valve may be controlling oil coming out of jack rams or the lip of the main ram. If the pilot line is damaged or the counterbalance valve accidentally jammed with foreign matter there is a possibility of high pressure amplification factors. Recognizing this problem, provisions should be made to install a hydraulic fuse like the one shown in Fig. 3. Burst pressure of the hydraulic fuse should be based on maximum pressure that the packing on the lip of the ram or the jack ram or that mechanical linkages in the press can withstand.

The hydraulic fuse is always connected between cylinder and counterbalance valve with a suitable "tee" connection as shown in Fig. 4. In most installations the hydraulic fuse need not be large in physical size because high velocities are expected through the orifice after the disc has

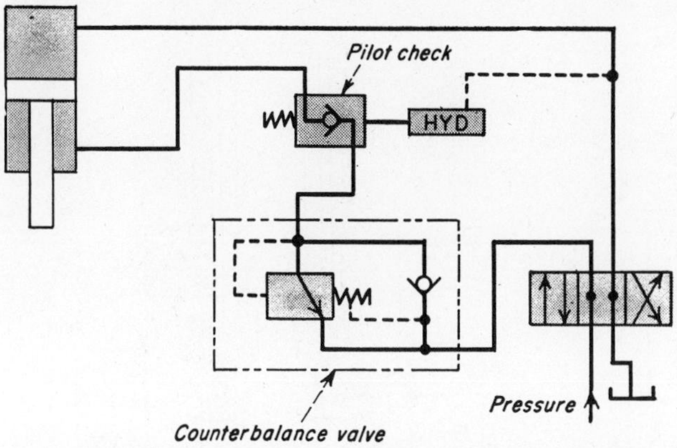

Fig. 5—The pilot check placed in series with a counterbalance valve prevents downward drift of piston until the upper face is pressurized.

burst. Contained pressures need not be dropped to zero but only to a safe limit established by the machine components. The extremely accurate burst values of the hydraulic fuses make it possible to choose one close to the maximum pressure allowable on the machine components. This factor makes it practical and safe to have a wide range between normal operating pressure and burst pressure of the disc.

When a mechanical counterbalance is used with a vertical cylinder the counterbalance valve would only be installed to prevent uncontrolled movements. If the counterbalance valve is the only mechanism to control the static position of a vertical cylinder perfect functioning becomes extremely important. Clearance flow through the valve in the static condition of the machine must be minimized since gravity drift of the cylinder will be a function of total clearance flow through valve plus clearance flow past cylinder rings. If no drift is permissible it may be necessary to provide suitable lock pin devices or ratchet mechanisms to make certain the cylinder does not drop during long static periods.

In vertical cylinder installations a pilot operated check valve and a counterbalance valve are sometimes installed in series as shown in Fig. 5. The check valve is normally piloted from the opposing cylinder line. The counterbalance valve may be either internally or externally piloted. The pilot operated check provides a tight seal without clearance flow during the static portion of the machine cycle. The counterbalance valve creates sufficient resistance during downward movement to prevent the pilot check from closing and reopening which would create chatter and shock waves in the machine cycle.

In some installations it is possible to use a poppet type pilot controlled relief valve as a counterbalance valve to minimize clearance flow during the static cycle. The balancing sleeve in the poppet type relief valve is a sliding fit so that even in this type of valve there will be a certain clearance flow.

SEQUENCE VALVES

From the glossary of terms for fluid power as published by the National Fluid Power Association: "VALVE, SEQUENCE—A valve which directs flow to more than one portion of a fluid circuit in sequence. Flow is directed only to that part of the circuit which is connected to the primary or inlet port of the valve until pressure setting of the valve is reached. At this time the valve opens and pressure in the secondary port may vary from zero to near the setting of the primary side with no variation in the primary pressure. If pressure in secondary equals primary setting, valve opens fully." Fig. 6 is a sectional drawing of a typical sequence valve.

Practically all sequence valves must be externally drained since the secondary portion of the circuit is pressurized during some functions. Excessive back pressure on the external drains would act as an additive to the spring adjustments creating an abnormal operating range.

Pilot signal for a sequence valve is generally from an internal source. However, if the sequencing operation must be interlocked with other portions of the circuit it may be necessary to use an external pilot.

If the sequence valve is "teed" between directional control valve and a second cylinder to provide a sequence of operations, the valve will normally be equipped with an integral check valve for a free flow in the opposite direction as in Fig. 2. If the sequence valve is used ahead of the directional control valve to insure a source of pressure to certain portions of the circuit the check would not be necessary. As shown in Fig. 7, any severe drop in main circuit pressure will close the sequence valve and guarantee a minimum pilot pressure to the pilot circuit.

There are installations in which it is necessary to prevent an uncontrolled low pressure movement of the cylinder in either direction. Fig. 8 illustrates a dual pilot

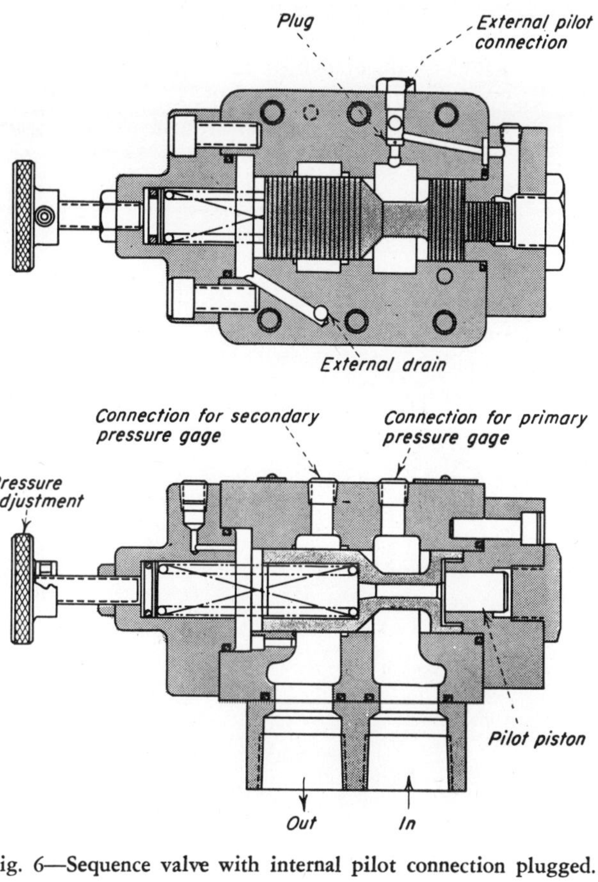

Fig. 6—Sequence valve with internal pilot connection plugged.

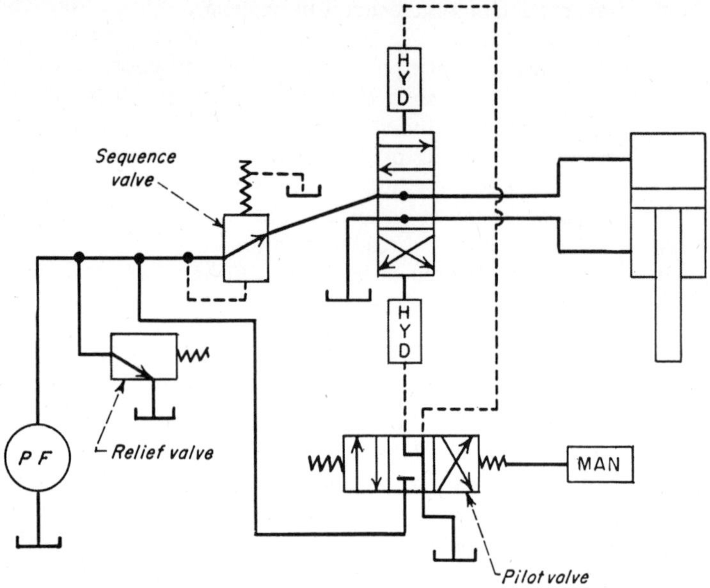

Fig. 7—Sequence valve is here used to guarantee a minimum pressure to the control circuit. Set to a low value the sequence valve closes only under unusual conditions in the main power circuit.

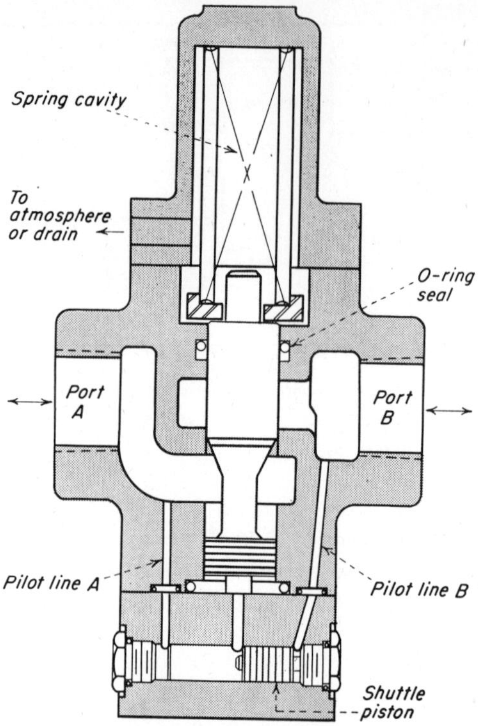

Fig. 8—This sequence valve operates identically for both directions of flow. Either way a minimum pressure is required to open the valve.

sequence valve which can be installed in one or both cylinder lines to keep the cylinder under control at all times. When port A is pressurized the shuttle piston is shifted as illustrated and a pilot line directs fluid to the bottom of the spool. This fluid under a predetermined pressure will cause the spool to rise against the spring and connecting ports A and B. If pressure diminishes in port A the spool will be closed by the spring. When port B is pressurized the shuttle piston will be shifted and the operation is reversed. The spring cavity may be drained to the tank or sealed tight with an O-ring on the spool as illustrated so that the spring chamber can be open to atmosphere.

This same dual pilot operated sequence valve can be used as a direct spring loaded dual relief valve between the cylinder lines of a double rod cylinder or fluid motor as a protection against over pressures. However, this unit makes no provision for replenishing fluid lost in certain functions of a fluid motor as some of the more complex dual relief and replenishing valve units do.

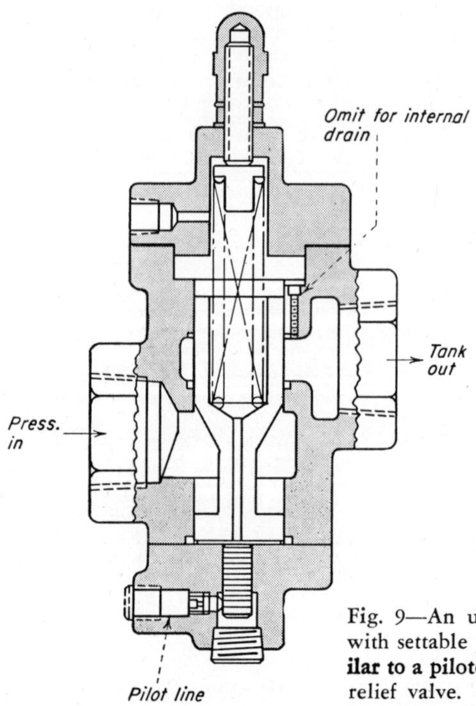

Fig. 9—An unloading valve with settable pressure is similar to a piloted low-pressure relief valve.

UNLOADING VALVES

From "A Glossary of Terms for Fluid Power" prepared by the National Fluid Power Association: "VALVE, UNLOADING—A valve which allows pressure to build up to an adjustable setting, then by-passes the flow as long as the present pressure is maintained on the pilot port by a remote source."

From this explanation the unloading valve can be a spring offset two-way valve which usually has an adjustment to regulate the tension on the spring that controls the pressure setting at which the valve opens. Fig. 9 is a typical unloading valve.

Since the unloading valve is generally used to by-pass a portion of the circuit back to the reservoir at a negligible pressure internal drainage of the spring pocket is possible. An exception to this would be the circuit wherein the by-pass type unloading valve is used to unload a high pressure circuit into a lower pressure circuit. In this case the pressure from the secondary, or lower pressure circuit might adversely affect the spring adjustment because of the additive effect of spring pocket pressure.

An unloading valve directly actuated by a pilot piston and balanced by spring compression has some basic disadvantages. Spring rate determines the range from cracking value to full open which is usually quite wide. In

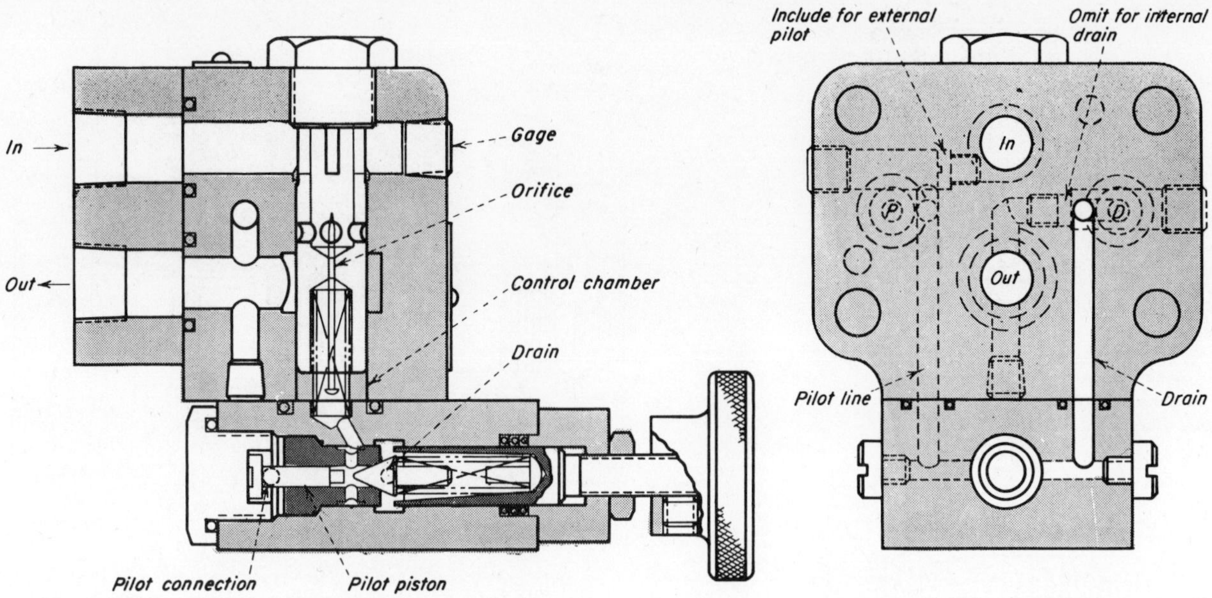

Fig. 10—Auxiliary pilot control unloading valve. Pressure is equalized across the main spool orifice. When pilot pressure is applied the pilot cone is lifted off its seat dropping spring pocket pressure to a low value. Main circuit pressure then overcomes the spring and flows to the outlet port.

addition, repetitive reset accuracy is limited. Where these factors become troublesome, a pilot operated sequence type unloading valve may be used in which the unloading valve is similar to a piloted relief valve and the pilot cone is forced off the seat by a small piston as shown in Fig. 10. Small physical motion is required from full open to full closed and the unloading cycle and the repetitive reset value are very accurate.

A disadvantage of the pilot chamber control type is that it will operate if inlet port pressure is high enough to lift the cone off its seat. In some applications this may be an advantage but it must be taken into consideration when choosing circuit components.

The pilot chamber control type has another possible disadvantage in that the minimum operating pressure required at the inlet port is somewhat higher than that of the direct spring balanced type after the valve is fully unloaded by the pilot source. The pilot chamber control type is built similar to a pilot operated relief valve and has its main spool relief spring effective during the unloading cycle. This low value, (usually less than 50 psi) may be high enough to cause overheating in certain circuits. Smaller valves are not usually affected by the high minimum pressure. A larger valve must drop to as low a value as possible so most are manufactured with direct spring loading. Fig. 11 is an example of this construction.

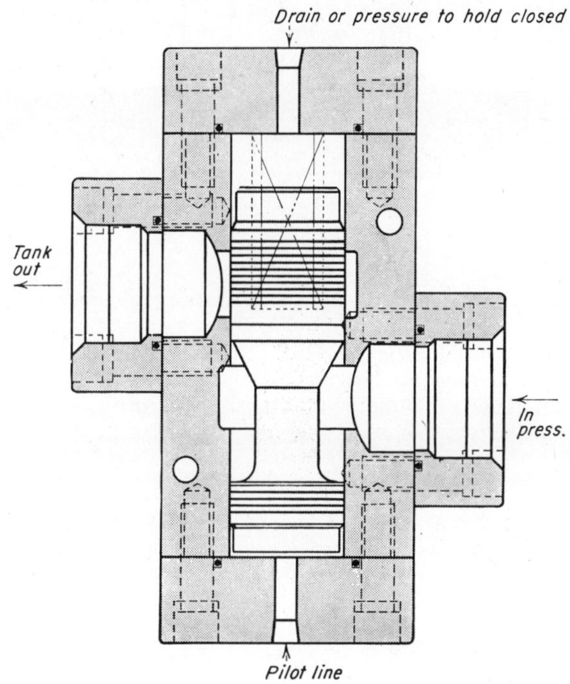

Fig. 11—Large capacity unloading valve with direct acting spring and pilot passages.

PRESSURE REDUCING VALVES

A pressure reducing valve is almost the exact opposite of a relief valve. A relief valve is a spring offset, normally closed, two-way valve with the sensing line to the inlet port. A reducing valve is a spring offset, normally open, two way valve with the sensing line taken from the outlet, or downstream line as shown in Fig. 12. Both valves may include certain damping devices as well as bleed orifices and other controls for most satisfactory service over a wide variety of applications.

The spool is held normally open with a balancing spring until the secondary, or reduced pressure, rises to the value of the spring and gradually lifts the spool. This lifting operation closes the passage between high pressure inlet and low pressure outlet and high pressure oil is prevented from reaching the secondary, or low pressure area.

Since the spool in this series is of the sliding type a certain amount of clearance flow can be expected between high pressure source and low pressure area. To prevent this clearance oil from increasing pressure in the secondary area, an adjustable bleed spigot may be provided. This bleed spigot is in effect a variable orifice. When the desired amount of bleed is established for satisfactory

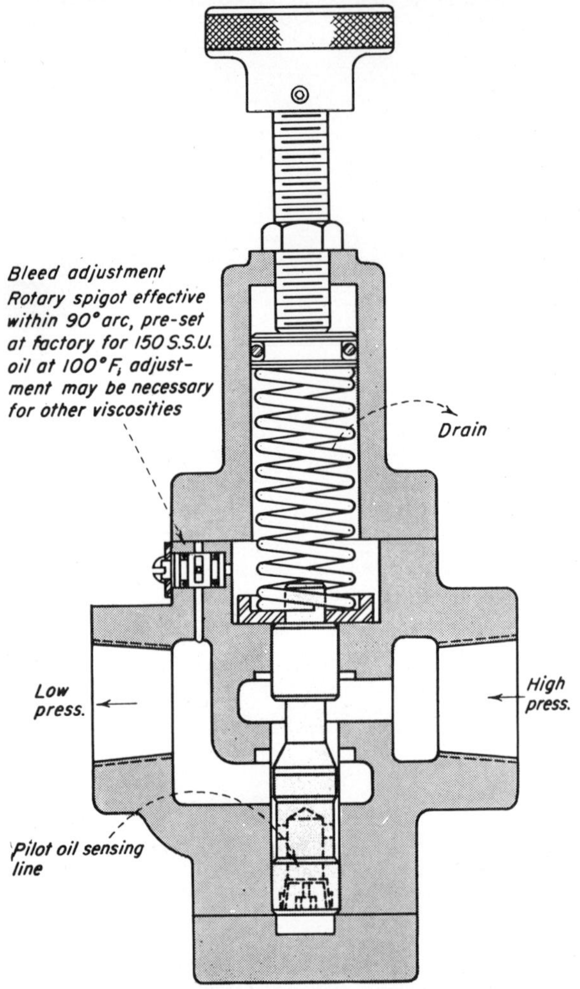

Fig. 12—A direct spring operated reducing valve in which oil from the low pressure connection is directed through an orifice to the back of the spool. As secondary pressure increases the valve is closed.

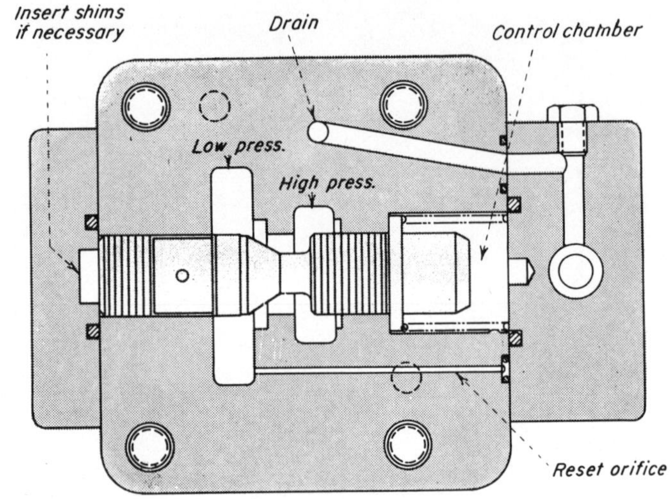

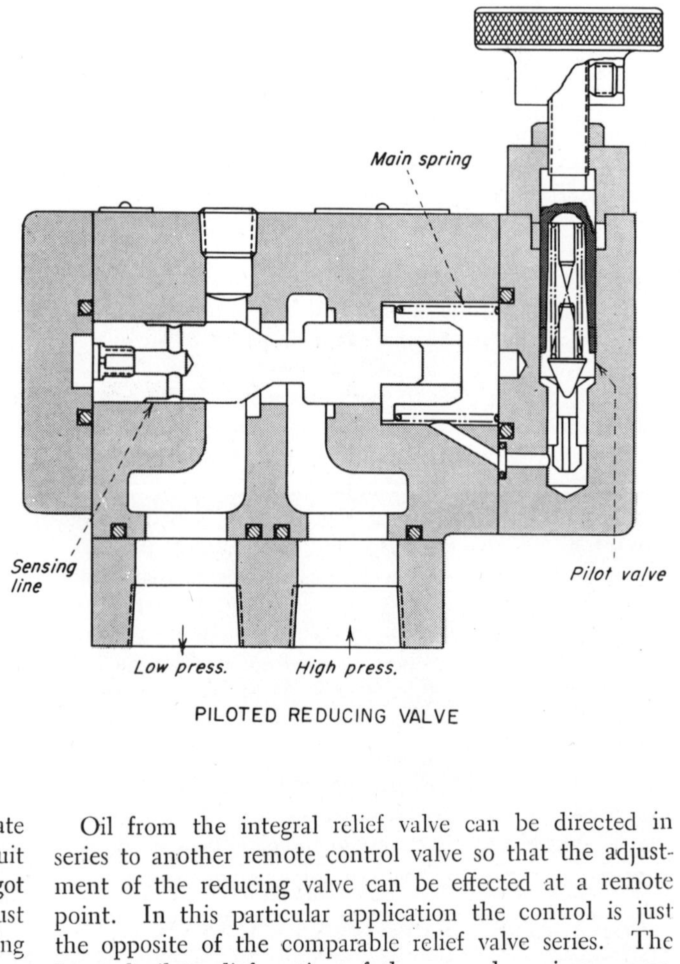

Fig. 13—A piloted reducing valve with integral pilot to control oil pressure in the spring pocket. Tightening the pilot valve spring permits higher pressure to be developed in the spring pocket requiring higher secondary pressures to close off the main spool passage.

operation, it is locked in position. If there is adequate clearance flow in the low pressure portion of the circuit it may be possible to close off the spigot. The spigot drains into the spring pocket of the valve which must itself be drained to prevent clearance flow oil from adding to the spring value.

Larger valves of this type are similar in operation with the exception of a hydraulic additive to the main spring as shown in Fig. 13. The main spring is relatively weak, usually equivalent to around 50 psi. Bleed oil from the secondary, or low pressure side, is introduced into the spring chamber at the top of the spool. A small pilot relief valve holds pressure in this spring chamber which is additive to the main spool spring value. Higher pressure oil passing over this small relief valve is then directed back to the reservoir. Rate of pilot flow is usually low.

Oil from the integral relief valve can be directed in series to another remote control valve so that the adjustment of the reducing valve can be effected at a remote point. In this particular application the control is just the opposite of the comparable relief valve series. The integral pilot relief section of the normal maximum pressure relief valve controls the maximum pressure, and the remote stations are in parallel and control the pressure to lesser values. In the case of the reducing valve pilots, the integral pilot relief segment controls the minimum setting and the pilots in series control to higher pressures. A program of varying pressure is possible with the pilot type reducing valve by directing the oil to various remote control pilots or by putting the desired number of control pilots in series and teeing off lines to tank through suitable directional control valving to establish the desired degree of control.

A word of caution is in order when remote control stations are involved with a piloted reducing valve. If the integral pilot control is set at the desired minimum and secured with a lock wire and seal, it will only control MINIMUM pressures. The remote control device must be set to a desired maximum when connected in series. Special parallel remote control circuit models are available when necessary.

Installations that are critical to possible over-pressure conditions can be accommodated by tailoring the spool movement closely to the volume requirements. Insertion of shims under the main spool will reduce the opening between the high pressure inlet and the low pressure outlet to the minimum necessary for maximum flow conditions. Less pilot oil is required to move the spool to the closed position and reaction time is proportionally reduced. Where inherent shocks are expected, shimming valves is a common practice.

It is sometimes possible to setup the circuitry so that a return check is not necessary. In the average application of a reducing valve in a cylinder line where it is only desirable to reduce pressure in one direction, the return check provides a dependable method of getting the oil back past the valve with a minimum of resistance. If the pilot segment of the reducing valve can be drained into the opposite cylinder line it will act as a pressurizing line when the cylinder is reversed. Application of pressure to the drain on a piloted reducing valve prevents the spool from closing so that a return flow through the normal valve is possible without the use of an integral check valve.

However, this is not a dependable method when there are large area differentials between head and rod end of the cylinder. Excessive flows because of the differential operation or intensification factors may preclude the use of the drain into an opposite cylinder line and require the model with integral check valve.

Most piloted models are provided with an orifice between secondary line and control chamber that contains a small pin. This pin controls the rate of flow into the control chamber at the reduced pressure value. In certain circuits it is possible to increase efficiency and degree of control by varying this orifice. To change the orifice the diameter or length of pin is varied. Variation is rarely required but in the event the valve is to be used with fluids with either extremely high or low viscosity, a method of making the valve more compatible with the fluid media is readily available.

Special connections are sometimes made into the control chamber of pilot operated reducing valves to connect pressure switches so that they will register or operate on the pressure additive created by the integral pilot relief valve. The switch is protected to a degree by the fact that the oil must come through the orifice spigot, or pin, prior to acting upon the switch. The switch must be set at a value lower than the adjustment of the reducing valve or it will not function.

A pressure gage connected at this point will not provide an absolutely accurate reading, but it is protected both by orifice and integral relief. The gage will provide a relative reading which is often satisfactory. The life of the gage is lengthened by the absorption of surges in the orifice mechanism. The immediate response of the integral pilot relief protects the gage from over pressures due to shock. This is, of course, not true of excessive or prolonged shock conditions which most often indicate faulty circuit components.

In developing circuits containing one or more reducing valves, it is usually the practice to allocate approximately 80 to 100 cubic in./min of operating oil or fluid. Each valve in the circuit would be figured at this value. Normally the requirement is about 25 per cent of the figure stated above, but the larger quantity takes into consideration inherent circuit changes such as wear on the pumping medium and nuisance losses in other parts of the circuit.

When reducing valves are used from a central source of hydraulic pressure to actuate molding presses or similar devices, it is usually considered good practice to install a pilot operated check valve, or similar poppet type shut off valve, ahead of the reducing valve to block the line while the press is open during the loading cycle. Often the loading portion of the cycle is relatively long, and the poppet type pilot check valve avoids losses from the central system while this portion of the circuit is involved. Special pilot operated check valves can be provided with large piston areas operated from a low pressure pilot source to open a relatively high pressure line. Practically any type of pilot valve operating from the low pressure source can actuate the pilot check valve. Sometimes the pilot check is operated through the pilot valve upon receipt of an electrical signal from limit switches on the machine, or in some more complex circuits from an electric eye or air gap sensing device.

Automatic devices of the type mentioned above appear at first glance to be "gilding the lily," but a close analysis of production costs over a relatively short period of time will indicate savings that make the cost of the equipment involved seem negligible.

Actual wearing parts on a reducing valve are not normally critical and long service life can be expected. As long as the circuitry is compatible with the inherent operating requirements of the valve, few difficulties need be expected.

HYDRAULIC

III. Flow Control Valves

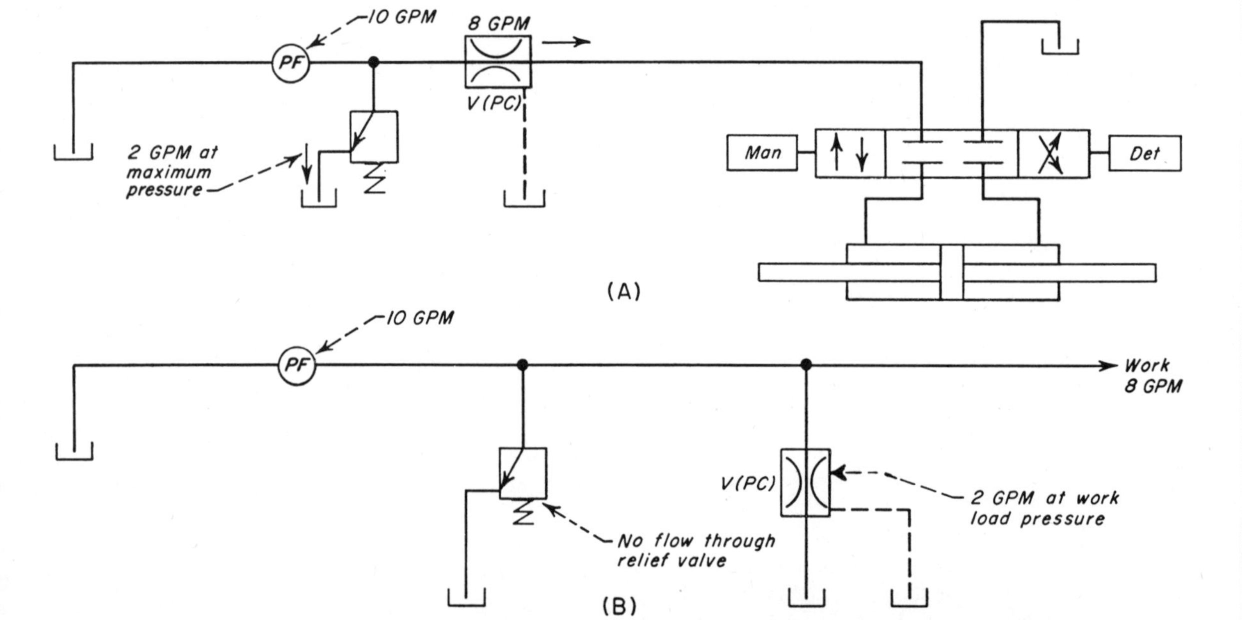

Fig. 1—(A) Flow to tank is by-passed by a pressure sensitive valve while flow to the hydraulic actuator passes through a controlling orifice in this meter-in circuit. (B) Bleed off flow control conducts bypass oil through the orifice to achieve flow control at the hydraulic actuator.

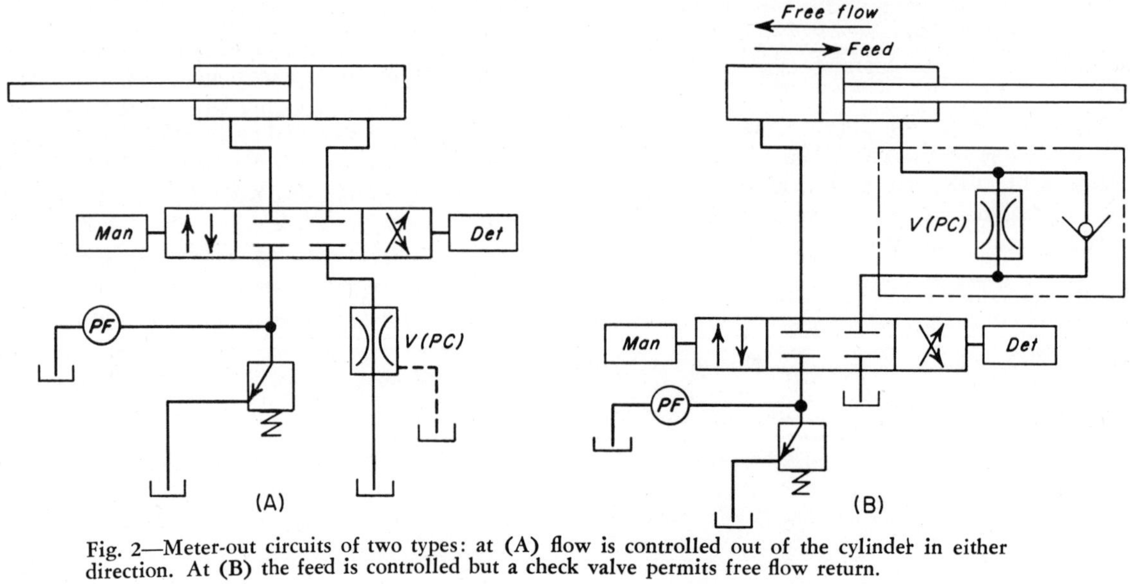

Fig. 2—Meter-out circuits of two types: at (A) flow is controlled out of the cylinder in either direction. At (B) the feed is controlled but a check valve permits free flow return.

JOHN J. PIPPENGER
Double A Products Co., Manchester, Michigan

CONTROL VALVES

These valves control and limit flow to, therefore speed of, a hydraulic actuator. Included are details of construction, operation and maintenance of direct and pilot operated flow control and check valves.

MANY advances in the art of metal working can be traced to the fine feeds made available by hydraulic transmission of power. Variable displacement pumps are excellent for this type of service but they may be too expensive to warrant installation in some circuits. Flow from a constant delivery pump may be controlled by an orifice effectively and at reasonable cost. This orifice may bypass part of the pump output to tank or it may be used directly in the line with an alternate flow path to tank provided by an additional valve. Fig. 1 shows two examples of these circuits.

The basic problem with orifices is that flow is a function of pressure drop as well as area. When pressures at the cylinder or fluid motor fluctuate in response to load changes, flow rate is affected. Some additional pressure-sensitive device must be used to compensate.

Simplest form of compensator is a reducing valve installed between high pressure line and the orifice diverting fluid to tank. Control of orifice area and the secondary pressure of the reducing valve insures a fixed flow of bypass oil. Total pump delivery minus bypass flow determines flow to the cylinder or fluid motor. Work load determines maximum pressure.

Combined orifices and reducing valves are available and widely used when large volumes of fluid are involved as in broaching machines, large planers and shapers.

A variation on this circuit makes use of a needle valve as a combination flow control and relief valve as shown in Fig 1(B). Various types of die tryout presses, leather molding presses, and certain types of straightening presses, as well as laboratory presses incorporate this circuit which is simple and satisfactory. The pressure generating device is protected with a relief valve set at maximum operating pressure. The whole circuit consists of a pump, relief valve used as a safety device, operating ram, large needle valve bypassing oil to tank, and piping.

The pump delivers oil whose minimum pressure is a function of the resistance of the needle valve when wide open. As the needle valve is closed resistance to flow increases, and pressure rises to a point at which the ram will start to move. At this point the needle valve is a combination relief and flow control valve.. As the orifice through the needle valve is further decreased, ram movement speeds up. When the ram meets heavier resistances pressure develops according to the size of orifice at the needle valve. Ram speed drops since more fluid is diverted through the orifice. When the needle valve is completely closed pressure will rise to that value set on the maximum pressure relief valve. A check valve may be installed in the line between pump and ram to hold the oil in the ram with pump not operating.

Meter-out Circuits

When more precise control is required two basic methods are used. Oil may be metered directly into or out of the cylinder. The meter-in circuit is generally used as shown at (A) in Fig. 1, when the valve controls rate of flow to a directional valve so that speed will be uniform in both directions (assuming a double end rod cylinder or motor) or when back pressure on the directional control valve would impair circuit function.

The meter-out circuits shown in Fig. 2 are most widely used in milling, drilling, reaming, and spot facing circuits. Their basic advantage is that they act as a check on the tool. When climb milling and drilling there is a tendency for the cutting tool to pull through the work at some points. The meter-out circuit prevents this action. The meter-out circuit also takes up all hydraulic play by keeping a back pressure on the piston at all times.

An adequate supply of control oil is vital for proper operation of a flow control circuit. A minimum of 4 or 5 cu in./min is specified on many flow control devices. This is an absolute minimum based on clearance flow requirements of the valve components. As such, it should be like a red flag to the circuit and machine designer. Recognizing that wear is difficult to estimate on hydraulic equipment, hard and fast rules cannot be made as to how much consideration should be given for normal wear. For new components 5 cu in./min may be sufficient; a minimum of 25 cu in./min is reasonably conservative when wear is taken into account.

Many circuits do not operate anywhere near this low a flow, and it is not a problem. In other installations the force required is low and small cylinders seem adequate and desirable. Valves are then asked to function on flows that could not be controlled by the finest needle valve.

Other components in the circuit must be taken into consideration; such as, the clearance flow past rings on a cylinder, any spool type valves in the circuit, or other packings which may drain oil out of the control circuit. Most control panels are based on a meter-out circuit. If the oil is clear and components kept in good order, wear factors are insignificant. However, foreign matter in the fluid medium can cause rapid wear. Over a period of time the clearance flow may increase from 5 to 10 cu in./min. If the feed is based on a rate between 5 and 10 cu in./min, this wear is critical. If the feed were based on flows of 20 or 25 cu in./min, the wear would do no harm.

Of course economics fit into this picture and cylinders of adequate size often offer space problems. However, good engineering practice dictates the requirement for adequate control oil. Larger cylinders will operate satisfactorily at lower pressures and circuits can be designed so that lower pressures decrease clearance flows so the advantages pyramid.

FLOW CONTROL VALVES

Flow control valves fall into several basic categories. For machine feeds with varying pressures and loads, a compensated flow control mechanism is usually required for satisfactory operation. Machine feeds with relatively constant loads and pressures may operate satisfactorily with a precise, adjustable-orifice flow control valve without any compensating mechanism. Certain other mechanical movements controlled and actuated by fluid power may also operate satisfactorily with a rather coarse adjustable orifice without integral pressure compensation ahead of the orifice to control the flow of fluid.

Non-Compensated Flow Control

Simplest form of flow control valve is basically a check valve poppet held off the seat by a mechanical screw jack adjustment as shown in Fig. 3. The adjustment is sealed from pressure so that the valve does not leak when adjusted. Reverse flow to the under side of the poppet lifts it and passes through the valve in free flow. Flow in the opposite direction closes the poppet against the screw jack so that the resultant orifice determines the flow rate through the valve. A relatively low number of turns of the screw jack will alter the flow from a stopped position to wide open. The valve poppet can be lapped to the seat if it is desired to use the valve as a high pressure globe valve with a free flow check in one direction.

Another group of flow control valves performs a similar operation but in a more precise manner as shown in Fig. 4. The adjustment requires many more turns of the adjusting screw from full open to full closed. This relatively fine adjustment is further improved by using a sliding fit spool rather than a seated poppet. Minimum flow is controlled by the physical clearances between spool and valve bore necessary for thermal expansion and normal manufacturing operations. From this minimum value the valve passes through three degrees or ranges of control. First, the flow is controlled by a notch in the spool for small flow. Second, a ground angle is encountered which increases the flow to a larger amount than that passing through the notch. Third, after coming to the end of the ground angle a full opening is created. This opening is increased until the spool stop is encountered. Reverse flow is the same regardless of the position of the adjustment.

Fig. 5 illustrates how the sliding spool non-compensated flow control valve can be built into a manifold unit to eliminate high pressure piping. In this unit two sliding spool, non-compensated flow control units have been manufactured in a housing that fits between sub plate and body of a solenoid controlled pilot operated four-way valve. They are so ported that they meter the oil out of the cylinder from both ends. Oil enters the cylinder in either direction of flow through the free flow check mechanism.

With the meter-out configuration it is possible to control cylinder movement with a relatively high degree of accuracy. The unit illustrated is widely used in automated conveyor mechanisms. Two independent valves manufactured in a common body permit independent control of the speed of cylinder travel in each direction. The savings in piping and installation are apparent.

Non-compensated flow control and check valves may be used in a meter-out circuit to control the movement of press rams as long as the valve is in no danger of being

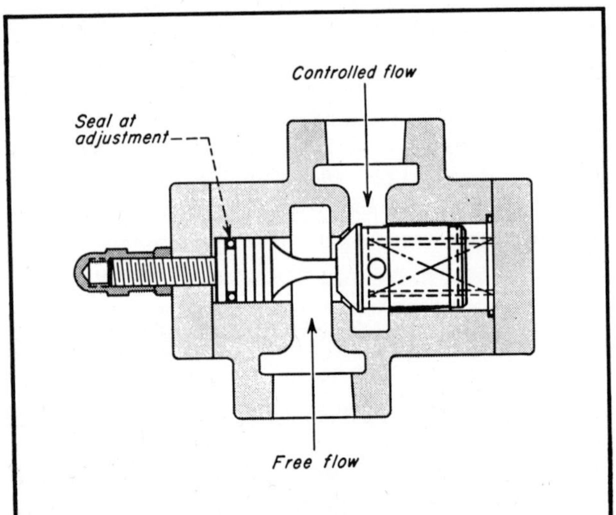

Fig. 3—Poppet type non-compensated flow control and check valve controls flow with a settable orifice in one direction and permits free flow in the reverse direction.

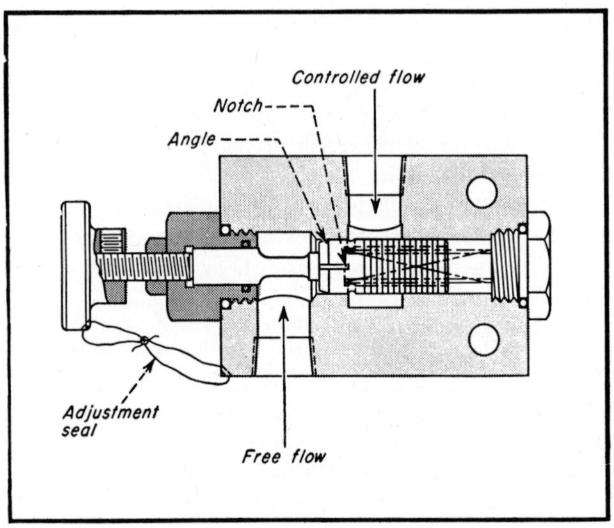

Fig. 4—Finer lead screw and a sliding spool in this flow control valve allows closer flow control. Reverse flow is still free.

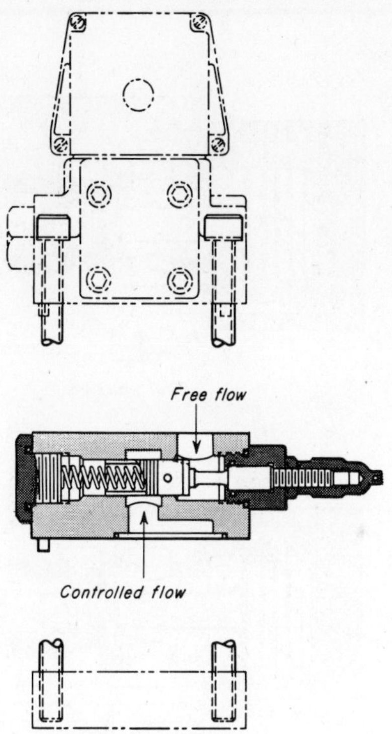

Fig. 5—Sandwich type non-compensated sliding spool flow control and check unit is a dual unit (two valves are side by side) whose housing has been designed to fit between a directional valve and mounting pad.

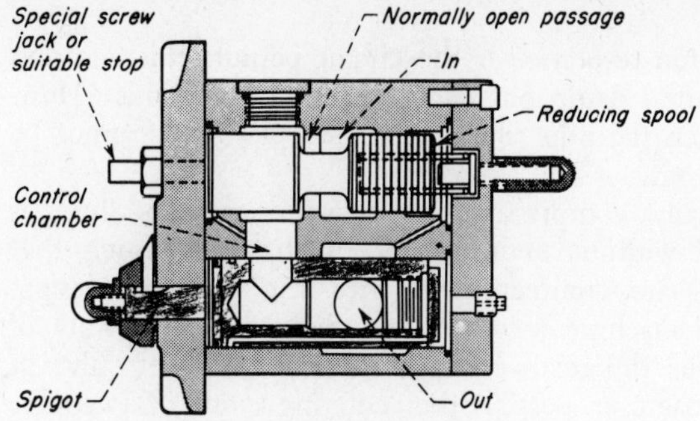

Fig. 6—A compensated flow control valve in which pressure in the control chamber is conducted to one side of the spool and bucks spring pressure plus outlet pressure on the opposite side. Desired flow is set by turning the shut-off spigot to close off the flow path.

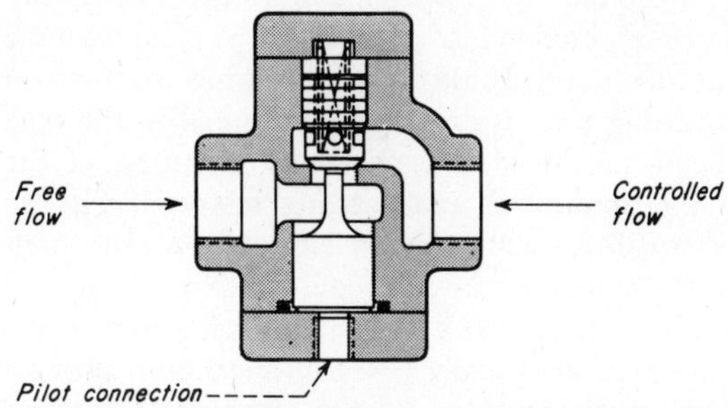

Fig. 7—Basic pilot check valve permits free flow in one direction. Flow is completely stopped in opposite direction until pilot pressure is applied at pilot connection. Pilot piston touches but is not attached to poppet.

accidently shut off. In many installations the ratio between the main ram of a press and the jack rams can be 100:1. The intensification factor is such that piping, packing glands, and cylinders themselves may be destroyed. If flow control valves are used in an application of this type, they should be protected by the insertion of a hydraulic fuse of sufficient value to blow prior to any possible damage to the equipment or operators.

Pressure Compensated Flow Control

Inherent versatility of a commercial, compensated flow control valve will at times create problems. For an applicable range of capacity it is necessary to make the reducing valve of sufficient size to pass maximum flow. This may mean that a relatively small amount of fluid is being controlled by a large valve like the one shown in Fig. 6.

Since the reducing spool is normally open, at small flows the spool may have to travel a considerable distance before it is fully effective. If the machine is stopped during a feed cycle and then started again it may result in a jump of varying magnitude while the reducing spool is moving into a position to effectively control the pressure drop across the orifice. In precision machining operations this can be very serious.

Special models have been manufactured with an adjustment in the compensator spool to limit the amount of opening to a value just slightly larger than the maximum required to satisfy the volume being controlled in the feed operation. In other installations a suitable stop, sometimes a slug, is inserted in the cavity at the end of the spool, opposite from the spring, to prevent the valve from opening to the full extent when the flow through the valve is relaxed. Fig. 6 shows a screw jack for this function.

PILOT CHECK VALVES

Pilot operated check valves are made in three different basic designs. The simplest type are poppets similar to an angle check valve. The poppet is pushed off the seat by a conventional piston when pilot pressure is applied to the rear of the piston. Pilot pressure times pilot piston area must be somewhat greater than contained pressure times poppet seating area to lift the poppet. As shown in Fig. 7, force of the light spring tending to hold the poppet closed must also be overcome by pilot pressure. Ratio of piston area and check seat area varies in different models.

Most single acting pilot check valves are abrupt in opening and throttling is limited. If throttling is required some other valve should be used.

Adequate pilot pressure must be maintained on the pilot of the check to guarantee that it is held open for the requisite length of time. A check valve piloted from a line in which the pressure varies with load will flutter, setting up detrimental shock waves.

A needle valve in the pilot line can be used to control

pilot piston response. If the circuit permits this addition it will often damp out harsh machine operations. However, when the pilot check is a safety device it cannot be damped out.

Normally compression of cylinder or ram fluid is achieved without undue shock or stress, assuming that pressures are commensurate with the machine design. Yield of machine members and slight compressibility of fluid, plus the relatively soft action of a relief valve at peak pressure, or possibly the centering action of a variable delivery pump, will help to avoid shock loading on the compression cycle of a machine. Further, a gradual pickup of work load will often soften the compression cycle. Operations such as coining, shearing, or piercing are exceptions to this characteristic.

In many machines the pressure never rises any higher than the force required to move the work. A good example of this type of circuit would be a broaching machine in which the cutting force is directly proportional to the type of metal being machined, the degree of sharpness of the tools, and the amount of metal being removed. If the tools are sharp, force will be relatively low. As the tools become worn, higher forces are required to push or pull the cutters through the work. Maximum pressures of machines of this type are usually based on maximum possible loading that can be tolerated by the mechanical design.

Decompression Valves

Sudden decompression of these systems may damage or wear circuit components. When abrupt action is undesirable, controlled pilot check action can assist in decompression. A ball or a small piston is first unseated by the pilot piston. A portion of the oil under pressure is released through a small orifice at high velocity to lower the contained pressure to a secondary value. The main poppet then releases at a reduced shock rating. A reduced shock rating is specified because this is usually a relative matter. The higher the pressure with respect to the speed of release, the higher the shock condition. By first releasing oil through a relatively small orifice at high velocities movement of the load mass is slowed until all stresses are diminished.

A two-step decompression cycle is usually adequate for the average machine tool application as achieved by the valve in Fig. 8. In larger press applications it is often desirable to incorporate a three-step decompression cycle.

Three-step decompression checks are made in two basic series. The standard type are manufactured with a pilot piston whose area is smaller than the effective area of the poppet as shown in Fig. 9. A molding press circuit might incorporate a simple ram type cylinder for application of pressure against the closed platens. The press is opened and closed with one or more jack rams, or cylinders. As the jack rams close the press, the main ram is filled by creating a vacuum satisfied through a check valve. After the press is closed, fluid under pressure is diverted into the main ram so that the main ram will build up the desired tonnage on the platens. Area ratios between jack rams and the main ram may be in the order of 100:1. In this case 1000 psi on the jack ram would only create 10 psi pressure on the main ram if oil were locked in the main ram by a pilot operated check.

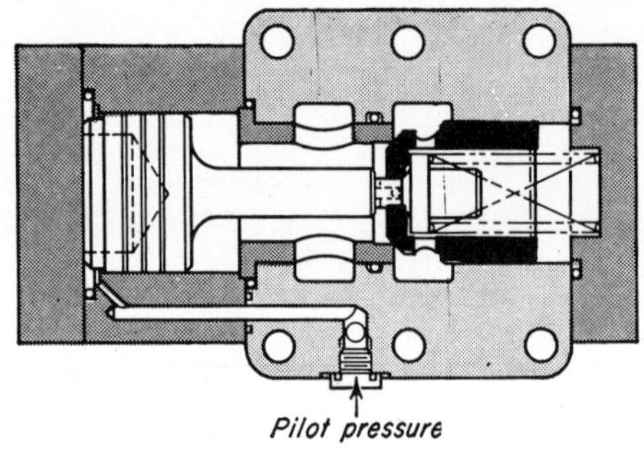

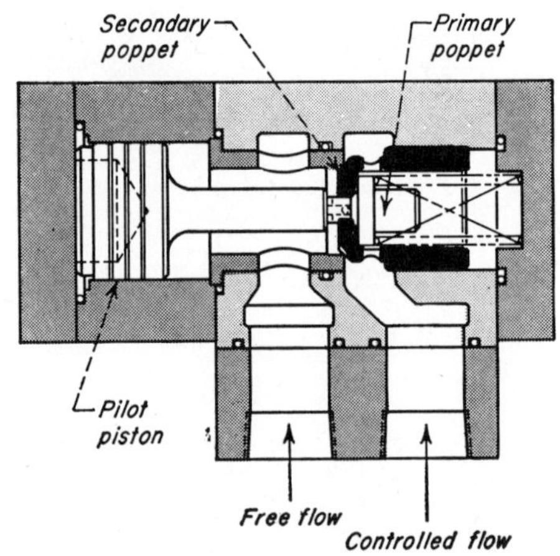

Fig. 8—Pilot check with two step decompression. Movement of the pilot piston lifts the primary poppet opening a small orifice between ports. Further motion of the pilot piston lifts the secondary poppet and opens a large area for flow between ports.

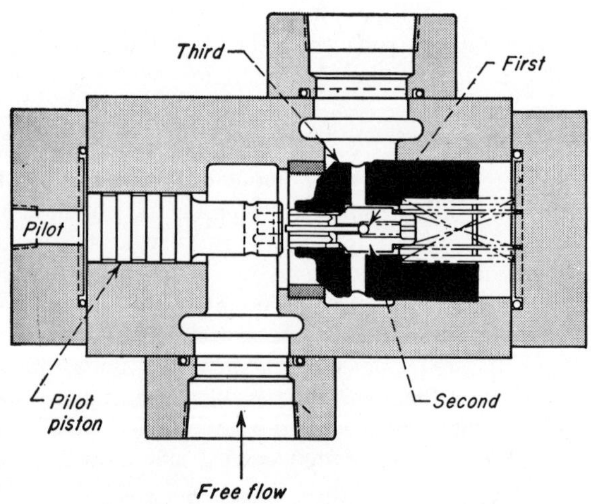

Fig. 9—Three step decompression valve. Initial movement of piston lifts a small ball check off its seat. Additional motion lifts the second poppet widening the reverse flow orifice. Still further pilot piston motion lifts the entire poppet assembly off the main seat. Reverse flow is free.

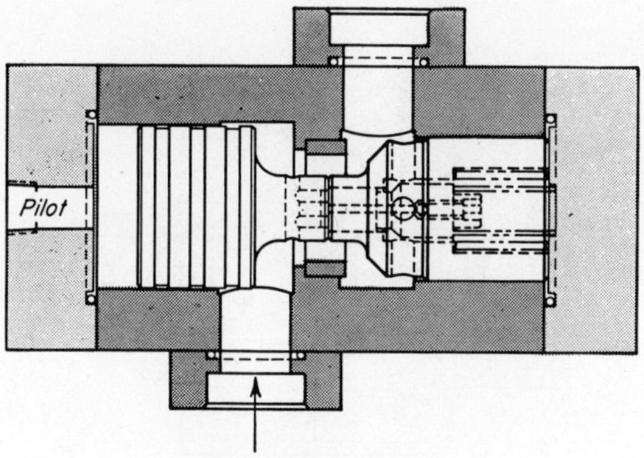

Fig. 10—Large piston pilot check with three step action used for pressurizing a low pressure circuit rather than decompression functions. This valve is shown open under full pilot pressure.

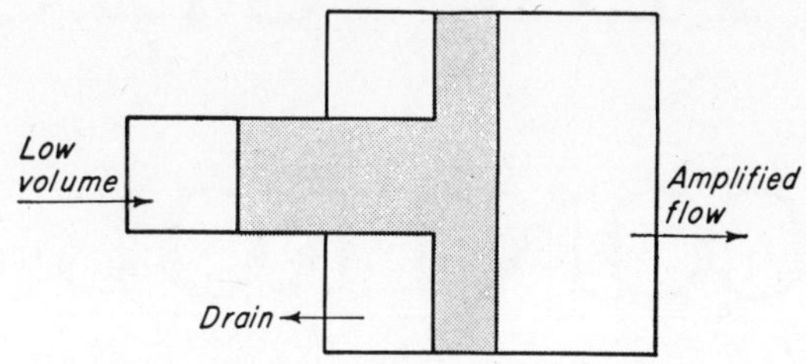

Fig. 11—An intensifier used in reverse will increase flow volume making metering a simpler problem. Low-volume, high-pressure fluid is piped to the small piston. Low-pressure fluid leavs the large side at high flow rate.

Application of 3000 or 5000 psi to the main ram sets up stresses in all of the machine members. When pilot pressure is applied to the piston of a typical pilot check valve for the release of this pressure it first encounters an area ratio of 64:1. Very little oil volume is released, but the contained pressure drops from 5000 psi down to 500 to 1000 psi. The second step unseats the secondary poppet spool at an area ratio of 4:1. This operation normally brings the pressure down to less than 100 psi. The third step completes the decompression.

Some checks have a ratio on the third step of 0.40:1. This means that 1000 psi pilot pressure is necessary to release 400 psi contained pressure. The reason for the negative ratio is to insure smooth operation of the first two steps. If the first two steps do not correctly decompress, the pilot check cannot be opened. By the time the first two steps have been completed opening of the main poppet is not much of a problem because in circuits where the check is used, jack rams do not create large forces or pressures in the main ram. Available pilot pressure is well in excess of the pressures exerted on the check. The first step might open a flow area of 0.0153 sq in.; the second step a flow area of 0.0552 sq in.; the third step a flow area of 1.77 sq in.

A pilot check valve may be equipped with a pilot piston having greater area than the main poppet. In this unit the decompression factor may be less important than the function of opening up a flow in a relatively smooth manner. Units with a large pilot piston are intended to open against a continuing flow of fluid wherein the opening of the pilot check will only momentarily lower the pressure value.

For example, in opening a line from an accumulator into another part of the circuit. The gradual opening through three steps softens the sudden shock of fluid hitting the downstream head at high velocities. Fig. 10 illustrates the relationship of the pilot piston to the poppet seating area. In this particular configuration the values may be 235:1 on the primary opening, 15:1 on the secondary opening and as high as 2.5:1 on the ratio of piston to poppet seat areas. The additional pilot fluid permits use of throttle valve arrangement if it is desirable to control the speed at which the valve opens.

Fine Feed Control

Circuits have been developed in which pilot operated check valves are the key to precise control of machine movements. For small flows in some feed circuits, directional control function is taken over by several pilot operated check valves. The check valves are controlled by conventional pilot type four-way or three-way valves. The check valve either blocks direct flow of fluid from cylinder for stopped position or opens up to a desired orifice to permit uncompensated feed. A second pilot operated check is "teed" in between the blocking pilot operated check and the cylinder that can be opened back to reservoir to permit a rapid traverse when needed. This permits fine feed rates when load and pressure are uniform and a compensating factor is not essential.

Another method of providing an extremely fine feed is to use an intensifier in reverse as shown in Fig. 11. This is then an amplifier of volume rather than an intensifier of pressure. With the assistance of pilot operated checks to channel the fluids in the proper manner, meter-out oil from the cylinder would be directed to the small diameter piston in the intensifier. Oil from the large diameter piston would be directed through an orifice (either compensated or not compensated) at a rate much greater than the fluid entering the same unit on the small end. The quantity of oil to be controlled would be a function of the length of feed required. The degree of control would determine the amount of amplification.

Note that a limited stroke is involved and an intensifier must be chosen of sufficient size to handle maximum length of feed. In addition, circuit operation must permit resetting the intensifier to its start position before further feeds are attempted.

HYDRAULIC CONTROL VALVES

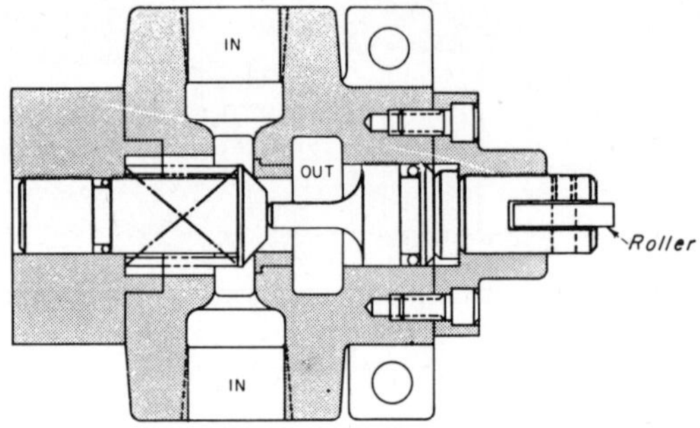

Fig. 1—A cam operated two-way valve with two manifolded input ports and one output port.

IV. Directional Valves

Four-way valve spool design and typical circuit applications when ports are blocked, open, throttled or variously interconnected in the neutral position.

JOHN J. PIPPENGER
Double A Products Co., Manchester, Michigan

THE TERMS TWO-WAY, three-way, four-way or diversion valve indicate the function that is expected from a component. A two-way valve is expected to either block or permit free flow from one port to another or from one group of ports to another group of ports. A two-way valve may have a plurality of ports as shown in Fig. 1. A two-way valve with twc input ports may be inserted in a pipe line to permit through flow. When the spool is shifted it will connect the two common ports to a third port.

Three-way valves add another function to these units. Porting in a three-way valve includes a pressure port, one cylinder port and a tank port. Again, the three-way valve often has additional porting that is merely a convenience in the plumbing or for gage connections. Pressure port to cylinder port connection is similar to the two-way valve function, either open or closed. Cylinder and tank ports are also similar to two-way valve function in that they are either open or closed as the valve is shifted from one extreme position to the other.

When pressure is directed to cylinder port, tank port is blocked. The opposite shift of the valve connects cylinder port with tank port and pressure port is blocked. This three-way operation is most commonly used as directional control valve with gravity return rams, spring loaded cylinders or gage checking stations.

Diversion valves are sometimes constructed with three ports and designated three-way diversion valves. In this unit the pressure port is directed to one port while the second port is blocked and vice versa. Three-way diversion valves are often of the open center type.

Four-way valves usually have four ports though there are a large number manufactured with five ports. A five port two-pressure valve is shown in Fig. 2. Basic operation is usually the diversion of pressure to one cylinder port in an extreme shifted position with the fluid returning from the other end of the cylinder and directed through the valve to a tank port. This function is reversed in the opposite extreme position of the valve. Neutral position flow will vary as explained below. The four-way valve may also be used as a double two-way valve. In this kind of operation, the unit would be shifted from neutral position to one extreme position. In neutral position all flows are stopped. In the shifted position pressure is connected to one port, the other port is connected to tank. The tank port would here also be suitable for pressure service.

Drain ports are an addition to ports mentioned in the various configurations. Multiple ports are also common. A four-way valve may have two tank ports to minimize resistance to flow through the valve particularly when it is designed as a part of a machine. Catalogued four-way valves that are manufactured to cover a multitude of applications usually have the tank lines manifolded within the valve and then connected to one external port.

In switching between alternate flow paths a three or four-way valve must pass through a center position. If the spool is not held in center for any length of time,

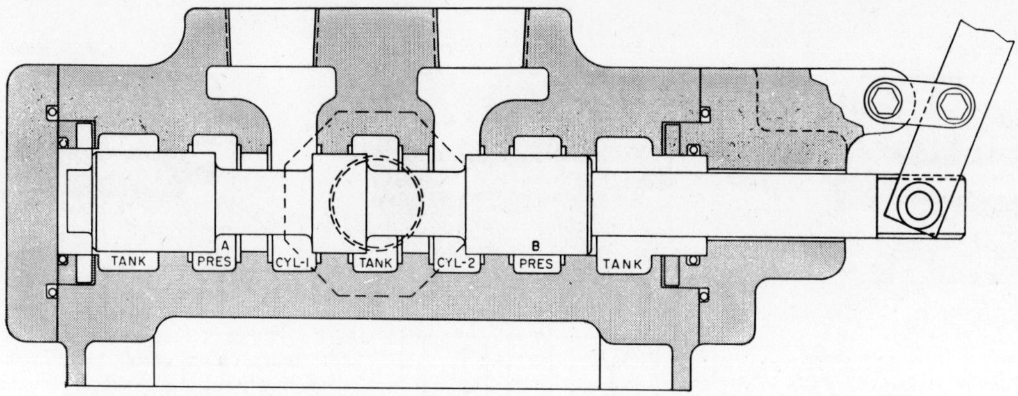

Fig. 2—A five-way solenoid actuated water valve. Ports open to two different pressures, A and B, two cylinder connections and tank. In extreme position shown pressure A is connected to cylinder 1, pressure B is blocked, cylinder 2 goes to tank. When reversed, pressure B is connected to cylinder 2 and cylinder 1 is dumped

it is designated a two position valve. A two position valve is only expected to reverse a fluid motor. When the spool may be held on center by a ball retainer or balance of spring forces it is called a three position valve.

Differences among four-way directional valves are confined to the various interconnections set up when the spool is centered and the speed and smoothness of switching between alternate flow paths. Since by far the majority of directional valves are of the sliding spool type, and since the three-way valve is, in most applications, a simplified (one port blocked) version of the four-way valve, this discussion will be confined to four-way valve spool designs.

Basic Construction

Many four-way directional control valve main bodies are made symmetrical so that the operating spool can be assembled from either end. This construction normally uses a loop core in the casting as the method of connecting the end valve ports together and to the proper exterior connection. To decrease physical size of directional control valves it is necessary to eliminate the loop coring in the valve body casting. The center of the valve spool is hollowed out with suitable connecting passageways to cover the requirements formerly served by loop coring. Fig. 3 illustrates the common closed center spool design when hollowed out.

Closed Center Valves

The term closed center indicates that flow through the valve is momentarily stopped as the spool travels from one extreme position to the other. At the midpoint of travel the cessation of fluid flow locks pressure in the several lines connected to the valve. In a two position

Fig. 4 — Closed center spool with throttling grooves to smooth acceleration and deceleration during reversal. Flow is cut off gradually but longer lands limit maximum flow for the same size of valve envelope.

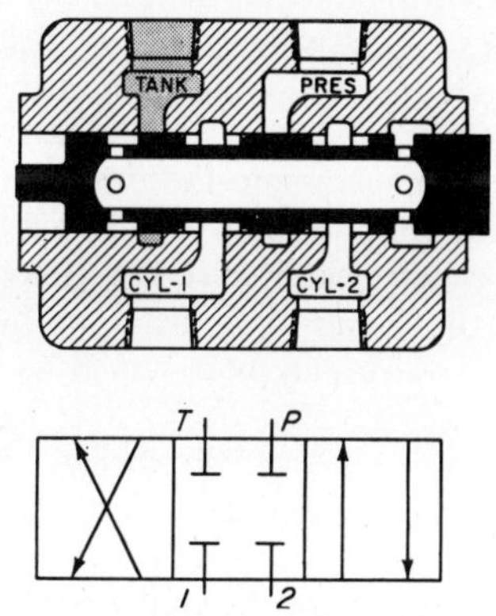

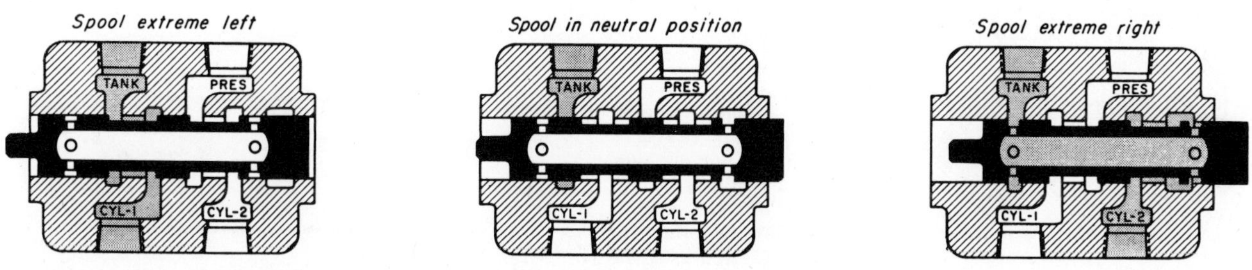

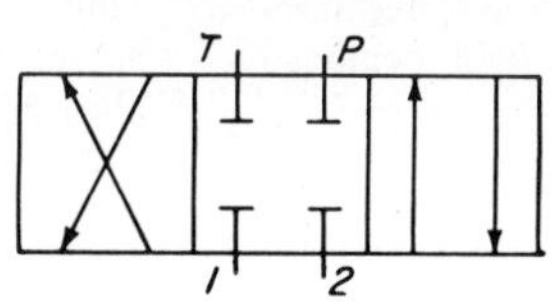

Fig. 3—Typical closed center valve switches flow paths to the two cylinders ports in extreme left and right positions. All ports are blocked when the spool is in the center position.

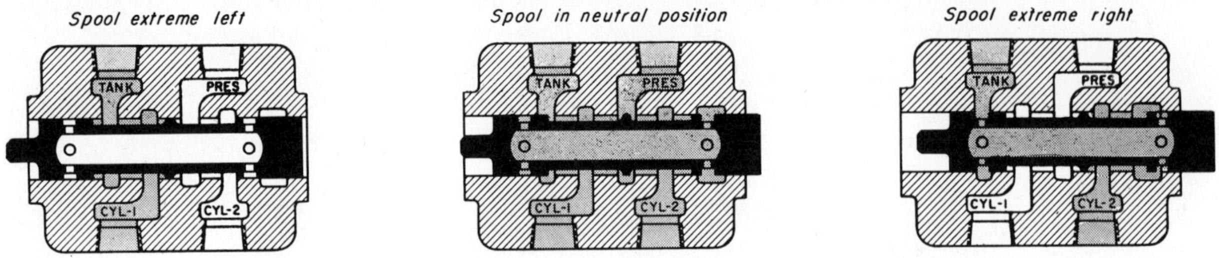

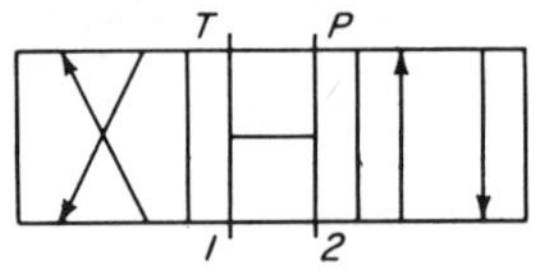

Fig. 5—Open center valve operates the same as the closed center valve in the extreme positions. All ports are connected when the spool is in the center position.

valve this midpoint of spool travel is only of importance if the open or closed center condition will affect the machine operation. If an accumulator is used in a circuit a closed center spool design will prevent undue loss of fluid from the accumulator during the brief period of time necessary to shift the spool from one extreme end position to the other.

A closed center spool may be modified to assist in controlling acceleration, deceleration, compression and decompression by including notches or ground angles on the edges of the spool to open and close the flow passages in a more gradual manner. A closed center spool with throttling devices incorporated is illustrated in Fig. 4. Usually this modified spool will not have quite as much flow capacity as the square ended spool. Longer spool lands are necessary for sufficient linear travel to incorporate the throttling construction.

Only machine movement that may occur on center of a closed center valve is a function of clearance flow past the lands of a metal to metal sealed valve. A packed valve may be assumed to be dead tight.

Open Center Valves

An open center spool design is illustrated in Fig. 5. In two position valves this construction offers built in shock protection. Assuming the circuit does not include accumulators and a momentary drop in system pressure may be tolerated, the open center spool precludes any sudden pressure build up and release as the spool shifts across center. At the exact center position all parts are interconnected and dumped to tank.

In addition, pressure release as the valve crosses center permits all machine members to relax momentarily before taking up the reverse load. As the valve centers all flow paths are connected together and to the reservoir. Pressures drop to zero until the spool has nearly completed its shifting operation. Normal friction forces in machine slides will help in decelerating the load and accelerating it in the reverse direction during the valve shifting operation.

Since in the use of two position valves there is no particular need to dump full flow to tank as the spool traverses center, considerable latitude is possible in spool design. Angles or notches on an open center spool permit transient flow control without losing the open center condition entirely. Fig. 6 is an example of a throttled open center valve often referred to as a partially open center spool. Flow capacity may be restricted during cross over to allow a relatively small pressure drop and firm acceleration and deceleration. In some installations throttling also may be desirable for compression or decompression purposes. (*Turn page*)

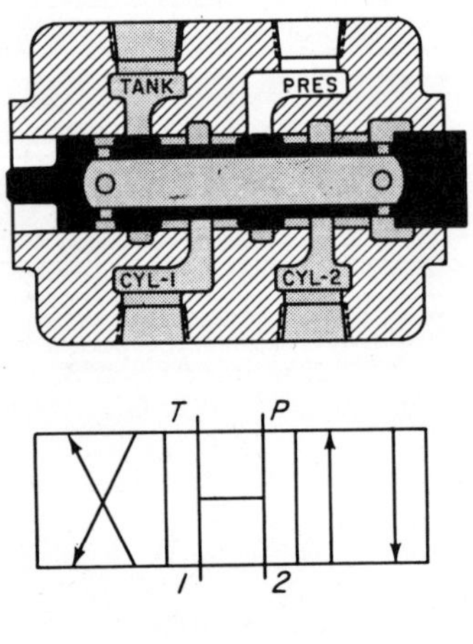

Fig. 6—A throttled open center valve is incapable of dropping line pressure to zero, but it reduces line pressure to a value sufficiently low to eliminate shock waves and assist in acceleration and deceleration of the load members.

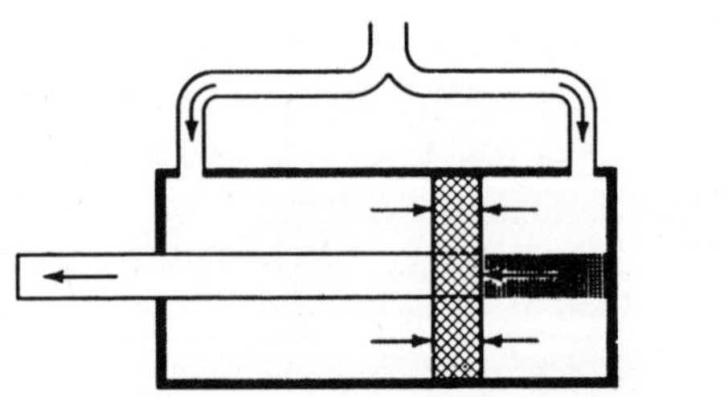

Fig. 7—A low pressure acting equally on both sides of a single rod piston works against two different areas. Result is a simple ram which will move if the pressure times rod area is greater than frictional resistance of piston and machine member and associated loads.

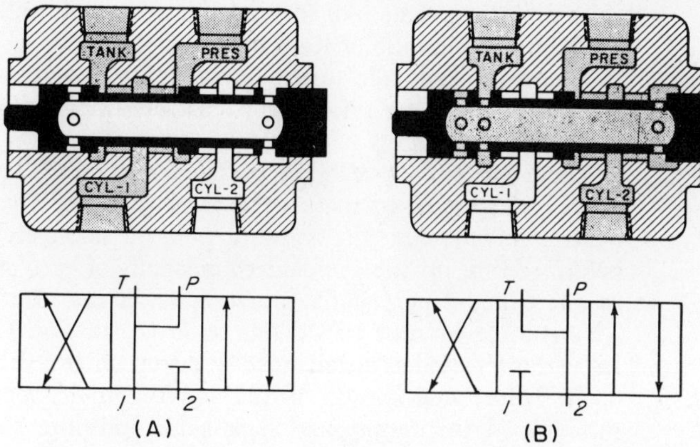

Fig. 8—Unsymmetrical design sets up a longer flow path to the tank connection in B than in A. Pressure drops will not be identical for the two valves though their JIC symbols and function are the same.

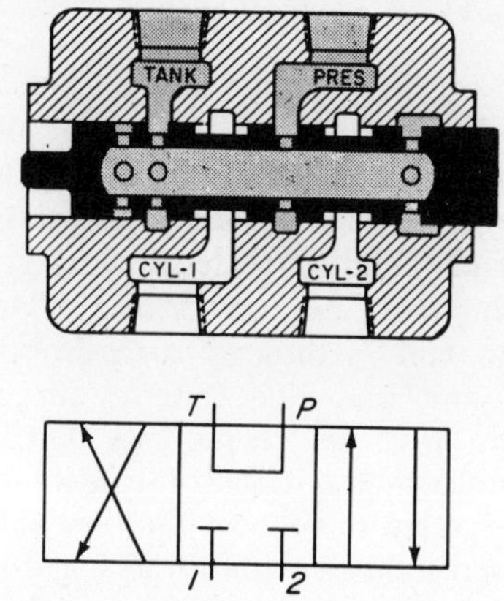

Fig. 9—Flow from pressure to tank ports is fairly direct and pressure drop in neutral position will be low. Both cylinder ports are blocked and piston movement is limited to clearance flow.

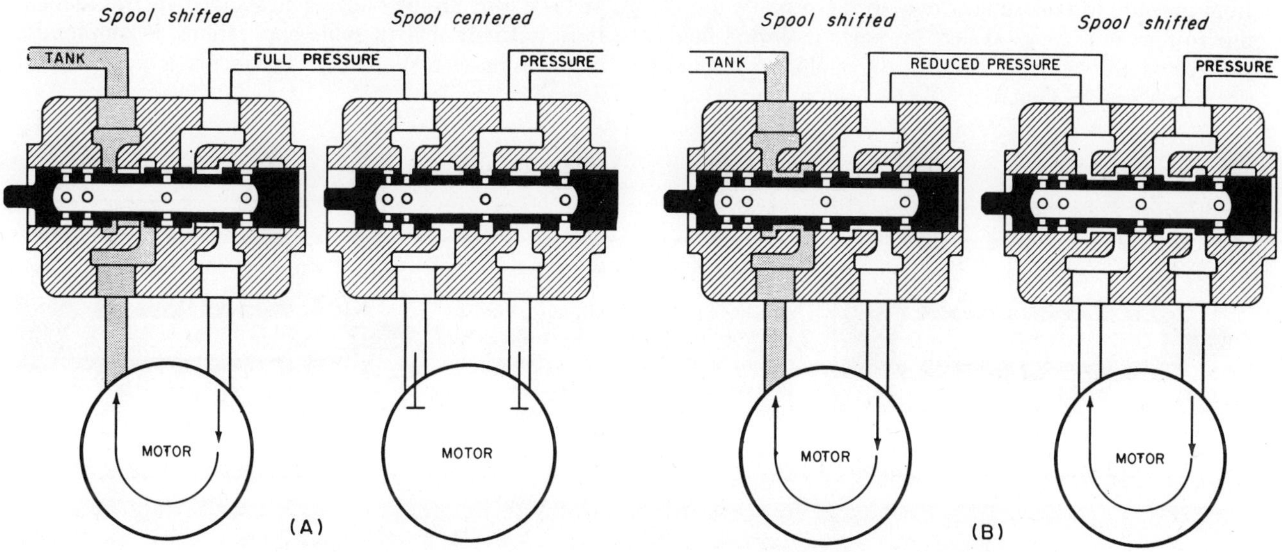

Fig. 10—Series circuit using the valve of Fig. 7 permits actuation of two or more devices from a single pressure scource. If only one valve is shifted (A) its motor receives full pressure. If both are shifted (B) a reduced pressure is generated.

Three Position Valves

More complex spool configurations may be required by three position valves whose center position is used in an operational rather than transient mode. Extreme end positions still provide straight and reversed flow channels for the two connection pairs but the center position may be the basic open or closed center with or without throttling, or it may be any combination of interconnected or blocked ports.

An open center valve connecting all four channels is undesirable in some circuits. If the line to the reservoir is long or small the high flow resistance may start a cylinder moving as shown in Fig. 7. All four channels at the valve will be at some low pressure and the differential area on the piston permits creation of an unbalanced force. Rotary fluid motors or double end rod cylinders are immune to this problem unless the valve has not centered exactly and pressure drops are unequal across the cylinder channels.

When an open center valve has been connected to a cylinder or fluid motor the piston is in a free floating condition when the valve is centered. No hydraulic obstruction other than fluid friction prevents piston movement if external forces are applied. To eliminate this, one of the two cylinder ports may be blocked while the second is permitted to flow freely to tank as shown in Fig. 8. This construction reduces centered flow capacity through the valve roughly to half of the flow through symmetrical loop core designs. In the bored out spool of Fig. 8, the two methods of design do not have equal pressure drops to tank because of the differences in flow paths.

Tandem Spools

Another method of preventing cylinder drift in neutral position yet still diverting pressurized fluid to tank is shown in Fig. 9. In neutral position both cylinder ports are blocked and input pressure is connected to tank. This spool design usually directs fluid to tank with a minimum of resistance. Flow from pressure port to tank is through large drilled holes into the spool center and fluid traverses roughly half the spool length on its way to the tank port.

This construction is called a tandem spool design and is widely used to connect the flow path for valves piped in a series circuit as shown in Fig. 10. In a series circuit pressure enters the first valve, and is discharged into the input of the second and so on as required by the circuit. Maximum allowable pressure drop determines the number of valves that may be connected in series.

When all valves are in neutral flow goes directly from pressure to tank subject only to the pressure drops due to flow resistance. When any valve is shifted fluid is directed through the valve to the motor in normal directional valve function. When both valves are shifted, total pressure divides itself across each valve according to the ratio of loads on the motors.

Tandem spools are often provided with ground notches or angles for smooth flow characteristics during the shifting operation.

There are two basic types of tandem spools. The first is constructed so that as the valve is shifted from neutral it will close the pressure port off completely before setting up the standard connections of pressure to one cylinder port and return cylinder port to tank. This is a closed cross over design as in (A) of Fig. 11. In vertical cylinder installations a piston might drift down before starting up if the lower cylinder port was opened to pressure and tank before pressure and tank connections were blocked.

If a cylinder is operated horizontally or is counterbalanced an open cross over tandem spool may be used as shown in (B) of Fig. 11. All ports are connected to the reservoir until the spool is either in an extreme position or definitely in neutral. Shock of pressure rise during shifting is avoided in this design. It is apparent that in some applications a compromise between those two spool arrangements is possible and desirable.

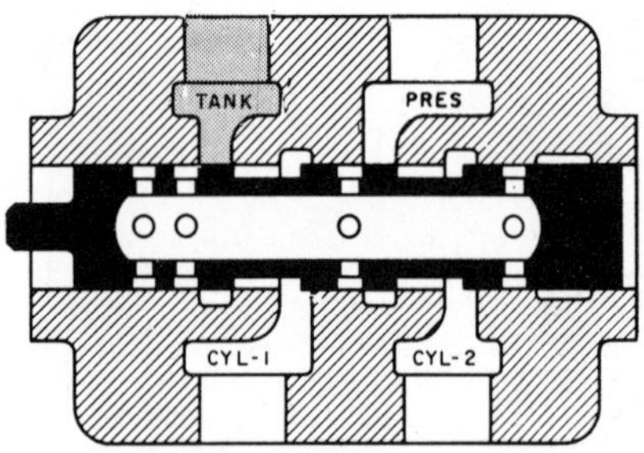

(A)

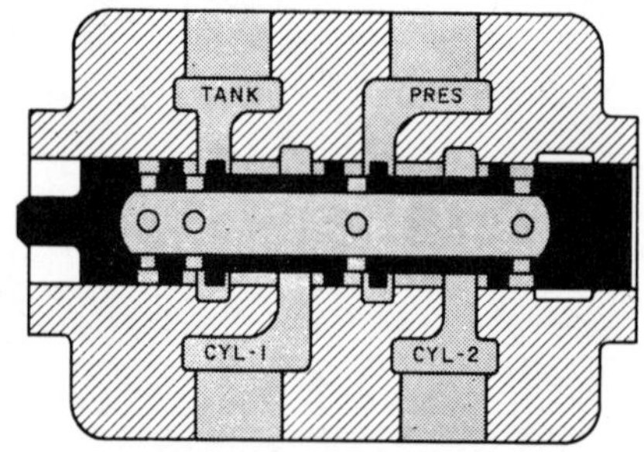

(B)

Fig. 11—Two tandem spool designs drawn in crossover position between neutral and extreme shift to the left. In (A) pressure is blocked before connection to cylinder 2. In (B) pressure is relaxed to tank before connection to cylinder 2.

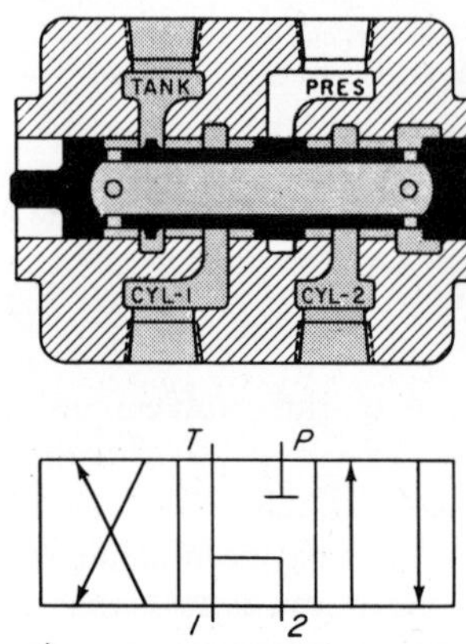

Fig. 12—Pressure is maintained in an accumulator when the spool is in neutral but the piston may be mechanically driven since free flow is permitted between cylinder ports and tank.

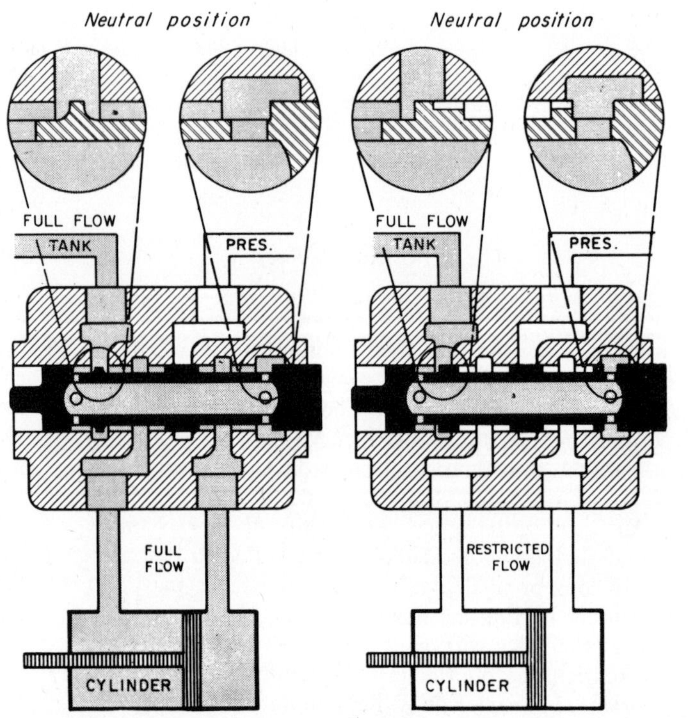

Fig. 13—Comparison of a standard pressure port blocked spool with a notched spool to severely restrict flow between cylinder ports and tank.

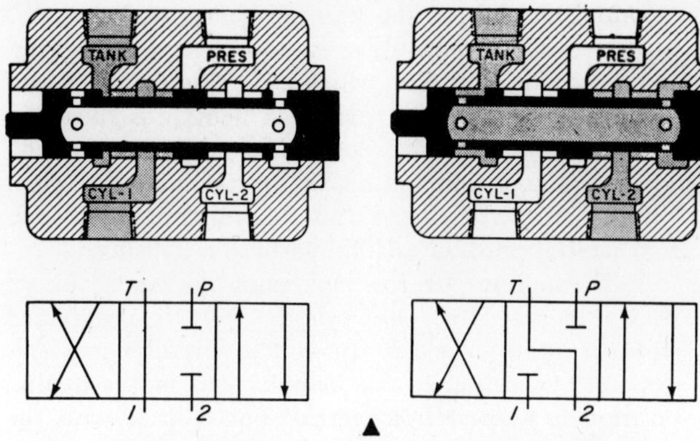

Fig. 14—Alternate forms of a valve spool with pressure port and one cylinder port blocked in neutral position.

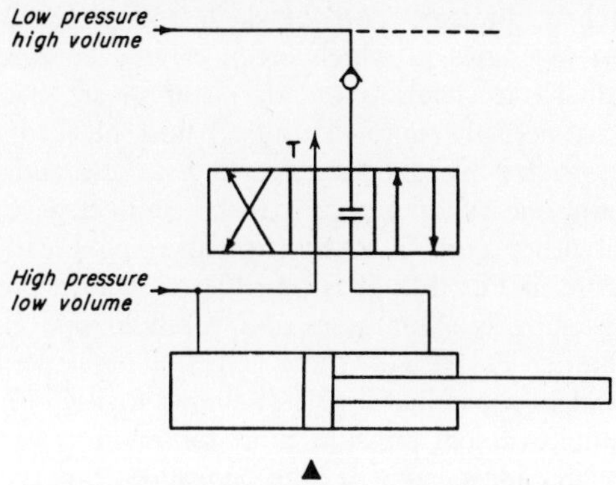

Fig. 15—Low pressure oil establishes rapid traverse until work resistance is met. Pressure rise then closes off check valve and high pressure along finishes work cycle. Low pressure oil establishes high speed reversal. High pressure is dumped. When spool is in neutral, low pressure port is blocked so that other valves and cylinders may be operated from this supply. High pressure dumps to tank at low resistance.

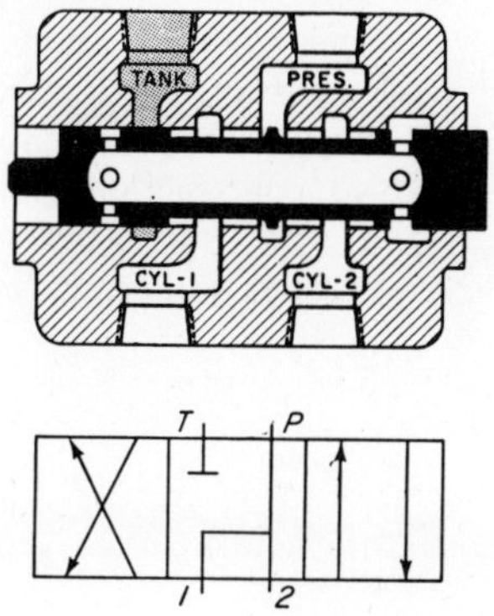

Fig. 16—In neutral position, as shown, a single rod and cylinder will work regeneratively setting up high speed forward. This spool configuration is required in pressure centering devices for definite neutral positioning.

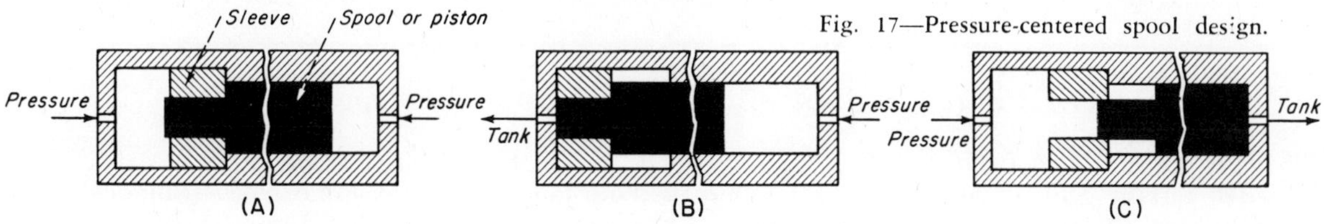

Fig. 17—Pressure-centered spool design.

(A) Pressure at both ends forces sleeve against housing; in reverse, forces piston against sleeve.

(B) Pressure at right end permits piston to pick up sleeve and carry to left end plate.

(C) Pressure at left end pushes sleeve to shoulder and forces piston to the limit of its travel.

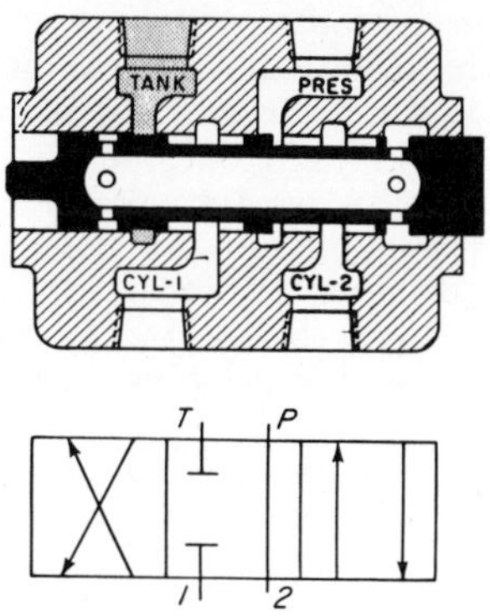

Fig. 18—Two variations on tank port blocked spool configuration close off one of the cylinder ports as well as the tank port. Pressure is applied to the second cylinder port in neutral position. Extreme end positions' are standard.

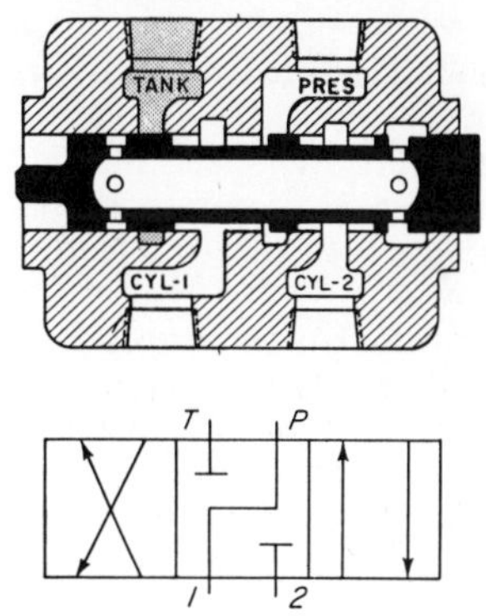

Pressure Port Blocked

Cylinder lines are relaxed but the pressure port is blocked by the valve spool shown in Fig. 12. The piston may be moved mechanically when the valve spool is in neutral as may be required when a cylinder parallels a lead screw for fast traverse and precise lead screw positioning.

Practically all pilot sections of pilot operated four-way valves are designed in this fashion. The pilot spool is spring centered when the solenoids are de-energized. Springs on the ends of the main spool urge fluid out of the pockets through the pilot body.

If only sufficient pressure relaxation is required to make certain a piston will not move because of area differentials, small angles or notches may be ground on the spool lands to relax cylinder lines in neutral and throttle flow as the spool is shifted. Fig. 13 compares complete relaxation of cylinder ports in the neutral position to the restricted flow possible with ground notches. Pressure port is blocked in both valves.

Notched, pressure port blocked, spools are widely used in machines in which creep cannot be tolerated but smooth acceleration and deceleration are essential. They are actually more like an all port blocked spool with restricted bleed between cylinder ports and tank.

If only one cylinder port is to be connected to tank and all other ports blocked the valve spool will look like those in Fig. 14. It is possible to remove pressure from packing lines or hoses and yet hold one end of the cylinder locked with these valves. These spools are also used in a parallel circuit as shown in Fig. 15 with one source of low pressure fluid for rapid traverse of several cylinders and a high pressure, low volume pump for each cylinder to apply large forces.

Notched spools to relax pressure slightly set up a controlled bleed to tank when the spool is in neutral. Pressure build up from clearance leakage is prevented where it would be detrimental to machine operation.

Tank Port Blocked

Pressure may be applied to both cylinder ports and the tank port blocked in neutral position as shown in Fig. 16. This spool intentionally causes a single end rod cylinder to travel until the valve is shifted in the conventional manner to an extreme position. This is one method of causing a regenerative action. Oil from the rod end of the cylinder is directed both through the valve porting to the head end combining with the incoming fluid from the pressure generator.

This spool configuration is widely used with pressure centering devices in which a shouldered piston and sleeve establish an area balance in neutral as shown in Fig. 17. The piston or spool is moved by maintaining pressure on one end and relaxing pressure on the other. When pressure is applied to both ends the spool comes to a neutral position established by the housing.

Two variations on the tank port blocked spool are shown in Fig. 18. One cylinder port is pressurized and the other blocked. In certain installations it is desirable to pressurize one cylinder end in the neutral position to keep it from getting out of pace with the rest of the machine.

Summary

Spool configurations described above do not cover all of the spools used in hydraulic industry today. However, diversion spools, dual two-way, three-way, servo construction, auxiliary pilot control spools and the many modifications of the spools illustrated are simpler to understand once basic types are known.

Oil Piloted Valves—A note on spool movement

Whatever the function of two-way, four-way or relief valve certain general considerations apply when pilot operated by pressurized oil. Whether the pilot valve is remote fro mthe main valve or integral with it an auxiliary needle valve or orifice is usually installed to control the rate of spool movement. A check valve is sometimes installed in parallel with the needle valve or orifice for effecting different speeds for the two directions of spool movement.

By controlling the speed of spool movement as illustrated a degree of deceleration and acceleration control is added to the simple directional control of fluid flow.

Degree of control is highly dependent on configuration of spool. Square ended, open or closed center spools are least effective in controlling transient flow, but they offer largest flow passage through the valve. Ground angles and notches machined on the edges of the spool subtract somewhat from maximum flow. If maximum flow is to be kept the same, length of spool travel must be increased with attendant slower valve action.

Additional factors affecting degree of control include configuration of needle valves, pilot pressure, viscosity of fluid medium, and the manner in which pilot pressure is applied. The last item refers to differences between spring offset, spring centered and no spring methods of spool actuation.

Regardless of valve function, a uniform shifting force can be expected from pilot operated valves in a conventional circuit. With constant pilot pressure, reasonable repetitive accuracy can be expected without introducing pressure compensating devices like those used in most flow control valves. Note, however, that changes in fluid temperature and pressure must affect speed of spool movement.

With spring centered or spring offset valves certain differences in function must be anticipated. For example, a spring centered valve will have pilot pressure plus spring force urging the spool to move away from its extreme position. Beyond the neutral position pilot pressure continues but the spring force reverses. With no pilot pressure the spring will shift the spool to neutral but only the potential energy stored in the spring is available for this function. Response is bound to be slower under spring force alone than under combined spring force and pilot pressure.

This difference is even more noticeable in a spring offset valve. Spool movement is a function of pilot pressure versus spring in one direction and of spring alone in the return. Note, too, that pilot pressure is relatively constant but spring force increases with deflection.

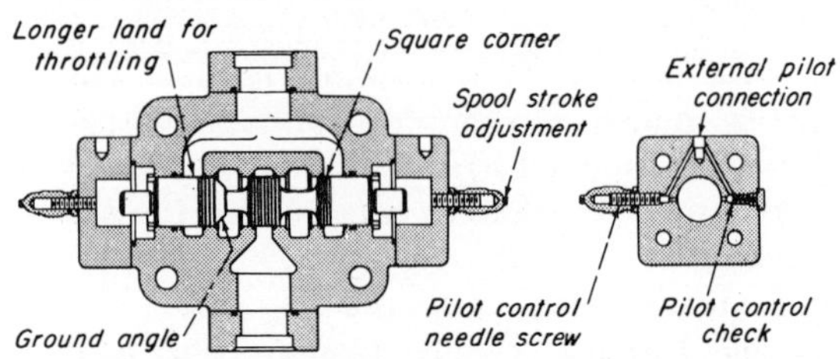

HYDRAULIC CONTROL VALVES

V—Valve Circuits

Beyond serving in their obvious capacities as pressure, flow or directional control valves, standard combinations of these components result in configurations designed to perform specialized functions. Regenerative, tandem and high-low circuits are among the most familiar forms. Factors governing the choice of valve, its function in the circuit and the problems that may be anticipated and avoided are discussed in this last article in the valve series.

JOHN J. PIPPENGER
Double A Products Co. Manchester, Michigan

Regenerative Circuits

A typical regenerative circuit would operate thus: (a) Rapid approach of cylinder with rod and blind ends of cylinder tied together for regeneration. (b) Rapid return (pressure at rod end and blind end open to tank).

This requires a four-way valve wherein pressure is connected to both cylinder ports in neutral position. As shown in Fig. 1, the spool will operate conventionally in either extreme position.

Assume a cylinder so constructed that the rod area is one-half of the blind-end area. With a 15-gpm pump and the valve in neutral 15 gpm enters at the pressure port. Tank port is blocked. Considering cylinder port two as the blind end, 15 gpm enters this port. Even though pressure is connected to cylinder one port, or the rod end, 15 gpm is delivered out of the rod end to join with the incoming 15 gpm from the pump because of the differences in area, creating a flow of 30 gpm to cylinder two port during the regenerative cycle. A 30-gpm valve sounds large when used with a 15-gpm, but there are several advantages to this system.

Saving in cost of piping. The largest flow through the cylinder lines would be 30 gpm from cylinder two port to the blind end of the cylinder. This might be piped with 1-in. pipe or tubing. The opposite cylinder line would carry 15 gpm, requiring either ¾- or 1-in. pipe or tubing. The pressure line would only need be ½- or ¾-in. pipe or tubing, and the tank line which would only carry a possible 30 gpm could be either 1- or 1¼-in. pipe or tubing. Thus, the valve body could be purchased with 1¼-in. pipe taps even though the larger capacity was being used in the internal parts. The 1¼-in. pipe taps would be bushed to the desired size. If a flange-connected valve were used, it would be possible to obtain reducing flanges so that bushings would not be required. Only one valve is used in this installation. In other regenerative circuits, more valves are required and, consequently, more piping and associated fittings.

Simplicity of maintenance. The fact that all of the valving is contained in one package certainly lends itself to easy maintenance.

Over-all cost. In some models the cost of one large valve is equal or less than the cost of several small valves

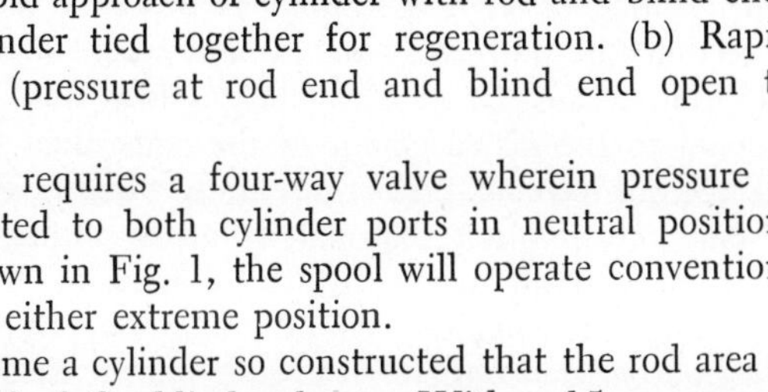

Fig. 1—Simplest form of regenerative circuit has only three functions: rapid forward, full-pressure forward, full-pressure return.

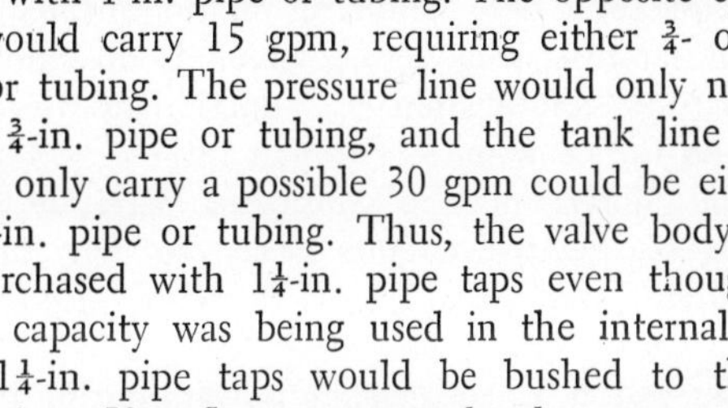

Fig. 2—With four-way valve in neutral, return flow is shut off and by intermittent operation of the two-way valve the piston may be inched forward. With two-way valve open, the circuit operates exactly like that of Fig. 1.

to obtain the same results. Consequent space economies can laso be expected inasmuch as one big valve will often go in a much smaller space than several smaller valves.

However, there are some disadvantages. With the circuit previously described, it is difficult to inch the circuit for setup purposes. There are only three functions: regenerative rapid forward, full-pressure forward, full-pressure return.

With the addition of another two-way valve between the cylinder ports for the regenerative function, as shown in Fig. 2, four functions are obtained. It is now possible to use a partially closed center four-way valve so that "stop" position is available and the cylinder can be stopped at any convenient point in the circuit for setup, safety or stroke-limiting operations. Stop is achieved by centering the four-way valve and blocking with the two-way valve.

Acceleration or deceleration requirements are also more easily controlled with the use of a combination of valves. The single-valve circuit offers considerable resistance to flow and throttling is difficult.

The manner in which the cylinder is to be controlled is arbitrary. If it is primarily a manual operation wherein the operator is the controlling influence, then four-way valve may be lever operated, pilot operated with a remote manual pilot valve or solenoid controlled with the operator actuating pushbuttons. Valves are also manufactured with various air piloting mechanisms. Whether control is by lever, solenoid, oil pilot, air pilot, or some combination is a function of the application and the economics of the circuit.

Several points must be considered carefully when designing regenerative circuits. Even though the following items may seem basic, they are often overlooked:

(1) The piston, when traveling forward during the regenerative cycle, has only a driving force created by the effective area of the rod passing through the packing.

(2) When the full effective force of the head end is desired, rod end must be connected to tank.

(3) When the cylinder is returning, twice as much oil is ejected from the head end as that being introduced in the rod end (assuming a 2:1 rod).

(4) If flow of oil is restricted from the rod end by a metering device, pressure intensification may be severe because of the area ratio on the two piston faces. If the restrictive device on the rod end were closed accidentally, 3000 psi could be intensified to 6000 psi. This is not a remote possibility; it happens with unfortunate regularity. If any restriction in the form of a metering or deceleration device is used in a regenerative circuit, consideration should be given to the introductio of a relief valve on the rod end ahead of the restrictive device which will limit the pressure to a safe value.

(5) Careful consideration should be given to the use of relatively high velocity through the pipe lines during the regenerative cycle. During this part of the cycle, the work being performed is usually the rapid movement of the machine part from one position to another. A slight increase in pressure because of high-velocity flows through the piping at this time may not cause difficulty and save expense and space.

(6) There must be an adequate supply of fluid in the reservoir so the displacement of the larger-than-normal rod will not lower the oil level unduly when it is in its extreme out position. Several cylinders pulling from the same reservoir may create this condition.

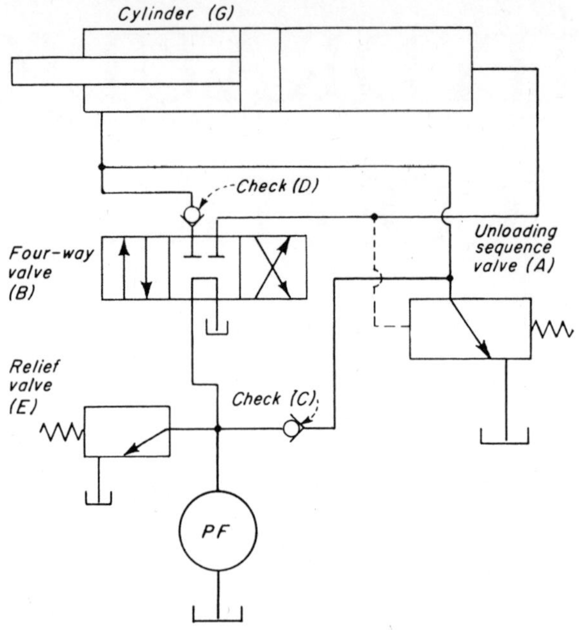

Fig. 3—Regeneration is set up on the forward piston stroke of this circuit and is maintained until work resistance is met. Valve *A* then unloads the rod end to tank and full pressure is applied at the head end. Piston return is standard.

A simple regenerative circuit made up entirely of standard "garden variety" valves is shown in Fig. 3. Assume first that valve *B* is a two-position valve. With the valve shifted so that pressure is on the rod end of G, a static condition will prevail wherein the flow path is through the check value D, to the rod end. There is no pilot pressure to actuate valve A so this line is essentially a dead end. Check valve C will be closed. Head end flow is to tank through valve *B*.

When the four-way directional control valve is reversed so that pressure is directed to the head end of G, the piston will advance. Oil from the rod end cannot go through check valve D nor can it go through valve A until pressure rises to a point equivalent to the setting of valve A in the head end line. There is a path through check valve C so that the oil from the rod end will join that coming from the pump and provide the regenerative action.

When the piston rod on cylinder G reaches sufficient resistance to build up the pressure in the pilot line to valve A, valve A starts to open and oil will be diverted from the rod end of the cylinder through valve A back to the reservoir rather than through check valve C. This will provide maximum force on the cylinder by providing a flow path from the rod end of G so that there is no back pressure.

When valve *B* is again shifted to return the rod of cylinder G, pilot pressure will be released from valve

A and it will close. Oil flows through check valve *D* to the rod end, and the head end is connected to tank. The piston will return to the end of its stroke and the pump will dump over the relief valve *E*.

If valve *B* were a closed-center three-position, four-way valve, the cylinder could be inched in either direction but pump flow would have to dump over the relief value. If valve *B* were an open-center valve, the pump would dump in neutral. If there is any tendency for cylinder G to drift out on the differential pressure, check valve C may be loaded somewhat to prevent the rod end oil from leaving the cylinder. In extreme cases, valve C could be supplemented with a low-pressure sequence valve to lock the oil in the rod end of cylinder G and prevent drift.

If valve *B* were a tandem-type valve as shown in Fig. 3, there would still be a flow passage through valve C from the rod end cylinder line to tank in neutral position so that the rod end cannot be considered locked.

Valve *B* must be of sufficient size to handle the pump flow plus the flow from the rod end of the cylinder without creating a pressure drop incompatible with the circuit requirements. Valve A only needs to be sufficiently large to handle the pump delivery minus the displacement of the rod. For example, if *P* delivered 10 gpm and cylinder G were a 2:1 cylinder, then valve A need handle only 5 gpm. The same flow characteristics are true of valve C. Valves *D* and *E* need only handle pump capacity.

The manner in which valve *B* is actuated is unimportant. It will work equally well manually, electrically, or through some remote piloting method. Likewise, the manner in which the signal reverses the action of valve *B* could be through limit switches, cam valves, pressure switches, or any other device that provides a signal compatible with the type of controls used. The change from regenerative action to conventional action is automatic by the pressure signal to valve A. The piston will have a return speed equal to the regenerative fast forward. Standard forward speed under full load is half the fast forward speed.

Tandem Circuits

Tandem circuits are series circuits, and normally only one valve is used at a time. Since flow through the valve is in series, if the first valve in the series is operated the exhaust oil under pressure passes through the remaining valves in the circuit as shown in Fig. 4. Oil passing through subsequent pipes and valves will cause back pressure. A second valve should not be operated simultaneously with the first or a loss of force will result due to additional back pressure.

Tandem circuits are considered foolproof circuits because of their inherent sequencial operation. Only one valve can be used effectively at one time, and there is only one common path for the fluid through the circuit. Another bonus is the fact that when all valves are in neutral, the pump is idling at a minimum pressure value, depending only on frictional losses of fluid passing through the circuit.

If a vertical cylinder is used in a circuit with a tandem valve with a partially open crossover, the cylinder may drop as the valve is shifting prior to starting in the up direction. If the cylinder is counterbalanced, this is not a problem. In some installations it is not economically feasible to control the cylinder with a counterbalance valve and the closed-center crossover is necessary.

In other installations a partially open center condition is desirable to prevent a rise in pressure as the valve shifts. If the partially open center spool is used and is compatible with the other circuit requirements, it is often possible to minimize inherent shock.

When one tandem-type directional valve is used it is usually satisfactory to use the valve with an integral drain for the shaft packing. (Shaft packing drained into the tank port of the valve.) If several tandem valves are to be used in series with internal drains, packing glands must be included on the end of the spool or a balanced spool design used with high-pressure packing mounted directly on the spool, Fig. 5. In most cases an externally drained valve is preferable, Fig. 6,

Fig. 4—Tandem circuit permits operation of several cylinders from one pressure source. Only one cylinder may be operated at a time without major pressure loss.

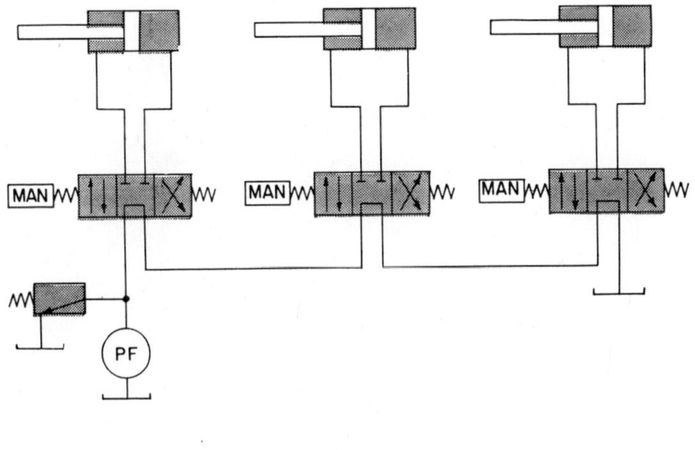

Fig. 5—Tandem valves with internal drain require high-pressure packings to prevent oil leakage.

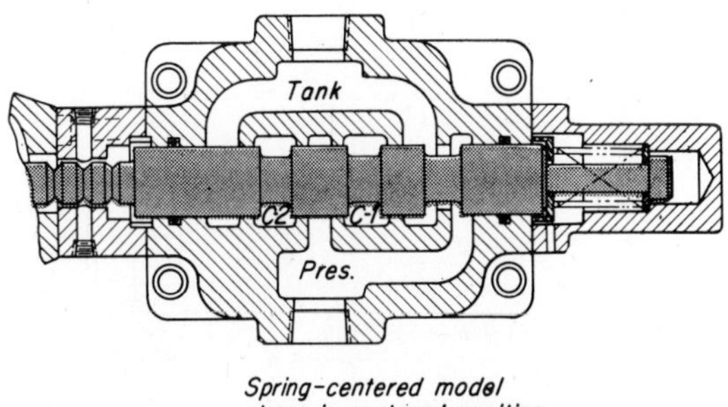

Spring-centered model shown in centered position

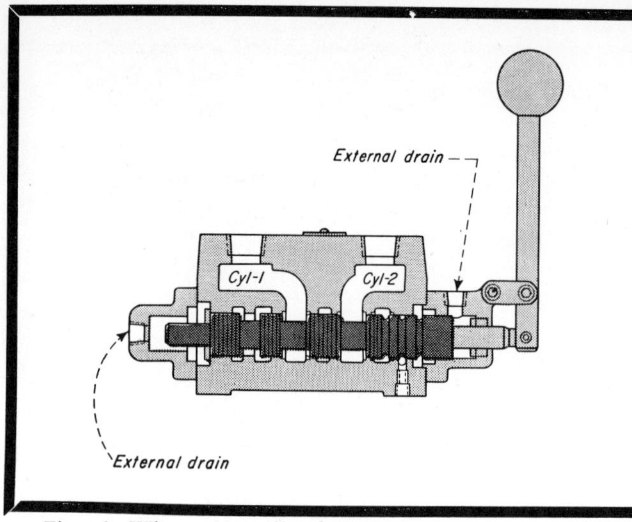

Fig. 6—When externally drained, a four-way tandem valve does not require packings. Drains eliminate sensitivity to back pressures which may unbalance the spool.

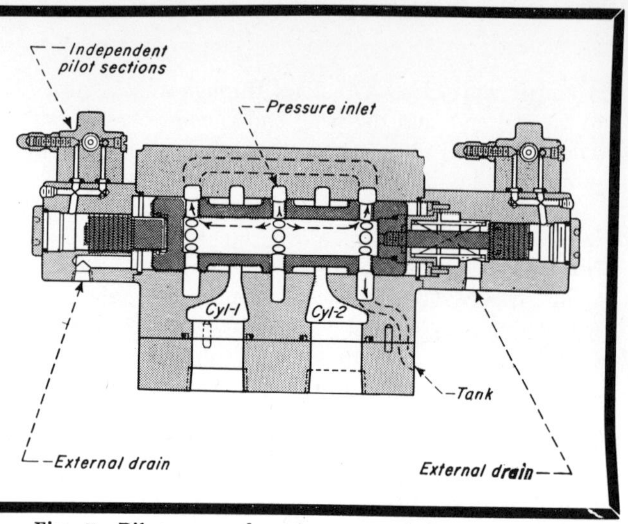

Fig. 7—Pilot-operated spring-centered four-way tandem valves normally are provided with external drains for the spring and pilot sections.

so that there is no tendency for the spool to shift as a result of back pressures generated if the spool is not fully balanced. High-pressure packing usually creates considerable friction on the operating spool.

When solenoid-operated valves are used, they are usually of the solenoid controlled, pilot-operated type that normally has provisions for an external drain of the pilot section, Fig. 7, and spring pockets.

High-Low Circuits

A high-low circuit usually consists of one relatively high-pressure pump of small volume with flow characteristics based on circuit requirements when the machine is either feeding (through some flow control mechanism) or holding work (such as in a clamping operation). In a feeding operation, volume required must provide adequat flow for the fastest feed rate desired with sufficient additional oil to maintain clearance flows through the components under pressure. In a clamping cycle, pump volume would be based on clearance flows only in circuit components.

The high-pressure pump usually feeds directly into the machine circuit, Fig. 8. The low-pressure pump feeds into the machine circuit through an angle check valve. This angle check valve provides a free flow path into the circuit with a minimum of resistance and at the same time prevents a return flow of high-pressure oil back to the low-pressure pump when the pump is unloaded. An unloading valve is connected in the circuit between the outlet of the low-pressure pump and the angle check valve. The unloading valve is closed as long as the pump is in its operating portion of the cycle and the pilot line has not been activated. Outlet of unloading valve is generally connected to reservoir. The pilot connection comes from the high-pressure portion of the circuit beyond the check valve.

In general there is no relief valve on the low-pressure pump. An unloading valve of two possible types serves this function. Simplest unloading valve is controlled by pilot oil whose action against a pilot piston is shown in Fig. 9. Because of the relatively large pilot-pressure differential between full closed and full open, when using unloading valves of this type there should be at

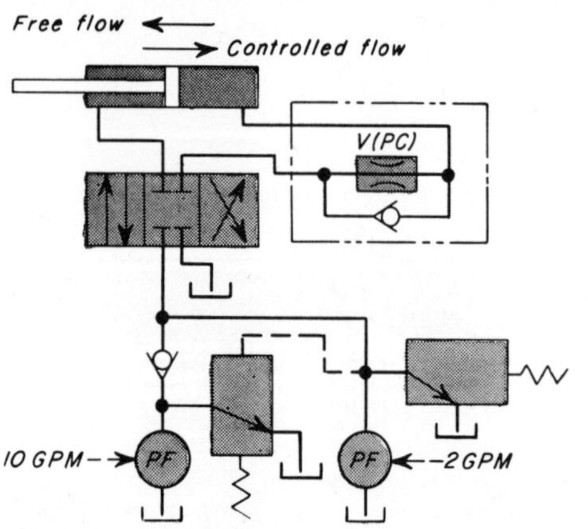

Fig. 8—Typical high-low circuit with two pumps of different pressure-volume characteristics. The high-pressure pump is protected by a relief valve. The low-pressure pump is controlled by a pilot-operated unloading valve.

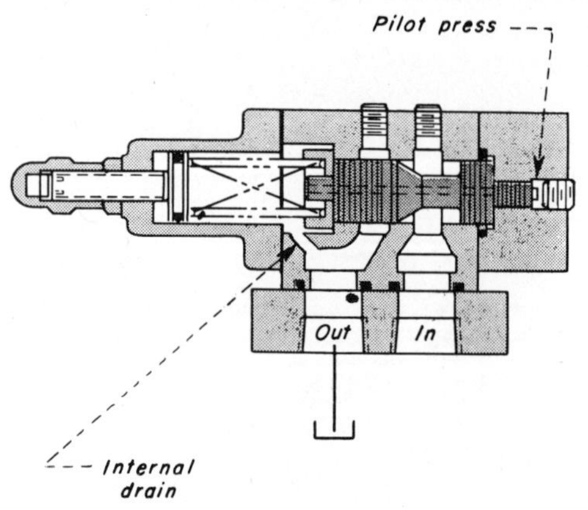

Fig. 9—Pilot-operated unloading valve has pilot pressure directly applied to the pilot piston. A fixed pilot pressure is required to open the valve.

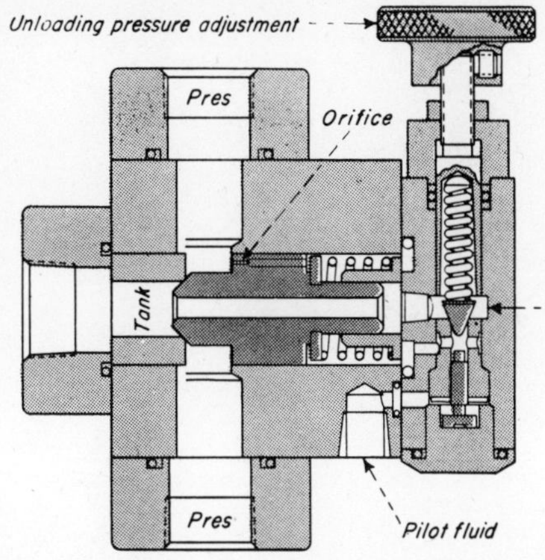

Fig. 10—A more complex pilot-operated unloading valve in which pilot pressure releases oil that has reached the spring pocket through an orifice.

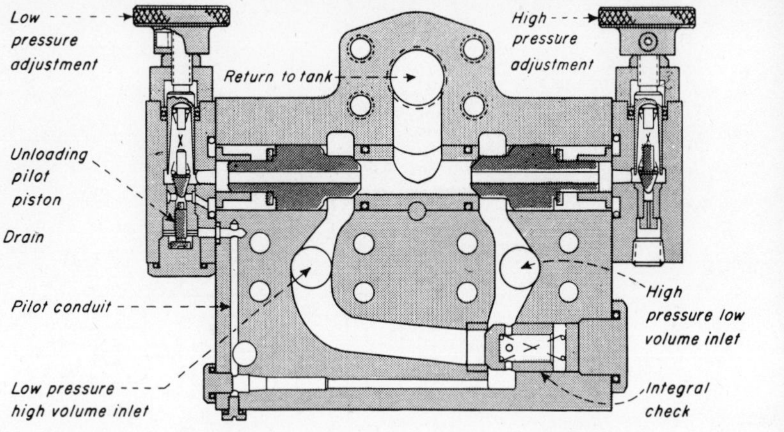

Fig. 11—Valve combination in one housing includes relief valve for high-pressure pump, unloading valve for low-pressure pump, check valve and internal pilot porting.

least 200-psi difference between maximum pressures of the high- and low-pressure pumps.

In later models, the unloading valve takes the form shown in Fig. 10. Here the spring is very light and pressure holding the spool closed comes from the pressure port and leaks through an orifice in the spool to the spring pocket. A tiny relief valve closes off flow between the spring pocket and drain. The relief valve is unseated by pilot pressure on a plunger and more abrupt control of the entire unloading function is made possible. With this type of unloading valve high and low pressures may be set as close as 100 psi.

The direct-acting or pilot-relief types are both actuated by a pressure signal from the high-pressure portion of the circuit. The spring on a direct-acting unloading valve can be adjusted for various physical forces which, in turn, control the value of the pressure required to divert delivery of the high-volume pump back to reservoir. This pilot mechanism is independent of the low-pressure or diversion portion of the two-way spool. With the direct-acting two-way valve the high-pressure section reaches a predetermined pressure value sufficient to lift the two-way spool and the high-volume pump delivery is diverted by this action back to the reservoir at frictional resistance values only. With a piloted-relief-type unloading valve, oil is diverted back to reservoir at a slightly higher pressure setting but insignificantly higher in most installations. The value is established by friction factors plus spring force on the main spool.

Unloading valve, relief valve and check valve may all be separate and individually piped in standard fashion. However, combination valves are available, Fig. 11, including the three valves in one housing and internally ported with consequent better leakage insurance and installation convenience.

Power requirement of this circuit is the sum of low-pressure pump power computed at the point at which the unloading valve becomes fully effective plus the power needed to drive the high-pressure pump to the relief value of the unloading valve. During the high-pressure part of the cycle, power consumption is usually so low that it can be disregarded. This may not be entirely true in gargantuan jobs, but the run of the mill system will follow these general rules.

Power dissipated during a holding action will appear in the form of heat. These losses are unavoidable and can be considered a normal cost of doing business. However, there is no point in making this high-pressure flow any greater than absolutely necessary.

Therefore, the temptation to use high-pressure oil as an addition to the low-pressure pump flow to increase rapid-traverse rate should be resisted. From a first-cost standpoint it may seem desirable to stay in the smallest possible pump-housing series. Thus if 18-gpm rapid-traverse flow is required, and only 2-gpm feed and holding fluid, it may seem logical to use an 11-gpm low-pressure pump and an 8-gpm holding pump. However, power loss to heat in the 8-gpm pump would cause difficulty throughout the life of the machine. It would be far better to pay the higher original cost to buy an 18- or 20-gpm volume pump and a correct 2-gpm holding pump even if there is more oil volume available than necessary and a sacrifice of small size is involved.

For the same reason the unloading valve must be adequate for the flow it must handle. Valve sizing should always be on the plus side. Excessive pressure drop through this valve is a total loss and will create additional power losses and heating problems.

Most circuits in use in industrial applications incorporate sliding spool-type valves which depend on a metal-to-metal fit for sealing between directional control ports. Recognizing that there must be a certain drain loss of the fluid media with this equipment to provide adequate lubrication and provision for thermal expansion, the high-pressure pump is chosen with extra capacity. Excessive wearing of circuit components can, after a period of time, make it necessary to either add more volume capacity to the high pressure pump or repair the circuit component in which excessive loss is occurring. If this is not done, the high-pressure pump may be unable to create sufficient pressure to actuate the unloading valve or valves. This condition makes itself apparent by loss of high operating pressures, ex-

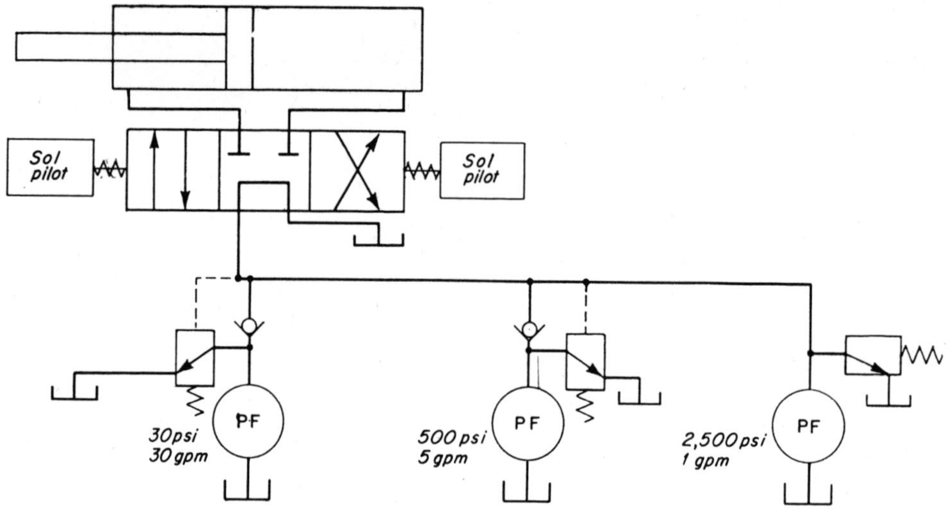

Fig. 12—Three-pump system for three pressure ranges required in a bearing assembly operation. Forward piston speed depends on load resistance, return speed has all pumps acting simultaneously.

cessive heat, and usually higher over-all operating noise level of the power unit.

When circuits require fairly long periods of high-pressure operation but flow is limited to clearance losses in the directional valving a relatively small capacity accumulator may be added. Full charge of the accumulator signals the high-pressure pump to unload so that excess flow is not forced to spill over a relief valve.

In some applications there may be several steps of pressure and volume requirements, as in the circuit of Fig. 12. A machine for inserting bearings into a large housing may require a large volume of fluid to provide sufficient speed to bring the bearing applicator to the inserting point. This may be a 30-gpm pump operating at a maximum pressure of 100 psi. When the bearing is against the bore, a pressure of 300 to 500 psi may be necessary to force it into place, supplied by a 5-gpm pump. During this time the 30-gpm is unloaded because the pilot line opens the valve controlling the unloading cycle on this pump at the 300-psi pressure setting.

After the bearing has been pushed into a portion of the bore at a relatively low speed, a final bore, perhaps of a very close or interference fit, may require a very slow push at 2500 psi. The 5-gpm pump is unloaded when the pressure exceeded 500 psi, and a 1-gpm high-pressure pump completes the operation. Return stroke, as well as initial rapid traverse, would be with all three pump flows, or a total of 36 gpm. Variations of this circuit are widely used in all types of industrial applications.

In many installations a high-volume, low-pressure pump is used in conjunction with a relatively low-volume variable-delivery pump. An unloading valve controls maximum pressure of the low-pressure pump. Any further rise in pressure will automatically reduce delivery of high-pressure pump until a certain minimum value is reached. This minimum is often just sufficient to make up for clearance flows at maximum pressure value in the circuit.

A small-capacity, variable-delivery pump will often introduce further economies in the circuit design of the high-low type by avoiding the necessity of passing oil over a relief valve at the maximum value. The economic factors involved in each particular circuit will dictate the correct combination. High pressure oil delivered by a variable delivery pump is tailored to instantaneous circuit needs and will vary to suit the condition of components.

how to evaluate . . .

HYDRAULIC SERVOVALVES

Today there are dozens of types and combinations of industrial servovalves, made by over two dozen manufacturers. However, choosing among them is not just a matter of comparing catalogs . . . there are these important design considerations, based on the valve characteristics.

WALTER ERNST, *vp and director of engineering*
The Commonwealth Engineering Co of Ohio
Dayton, Ohio

New demands for controlled hydraulic power have made the electrohydraulic servo—developed originally for aircraft—an important addition to the commercial market. Simple in concept, a **servovalve is a fluid valve that varies output flow or pressure in response to an input control signal.** The signal can be electrical, mechanical, fluid-pressure, force-reaction of a gyroscope, or even thermal-expansion of a trapped liquid. The most commonly chosen servovalve is electrohydraulic–the type discussed in this article. Its electrical input signal is easy to transmit and modify; and any other signal or response, including speed, acceleration, and light intensity, can be converted to a suitable electrical input.

In evaluating an industrial servovalve, **flow, pressure, accuracy** and **sensitivity** are usually the essential considerations.

Flow and pressure values are given in catalogs. But accuracy and sensitivity under actual operating conditions, even with complete catalog information, are difficult to predict–some knowledge of servovalve design is needed. Here are design tools that can make the evaluation simpler.

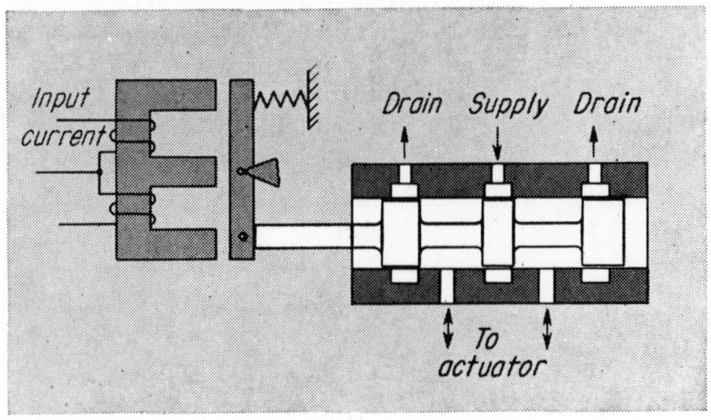

Single-stage servovalve controls oil flow directly in proportion to input current received by electric actuator, called torque motor.

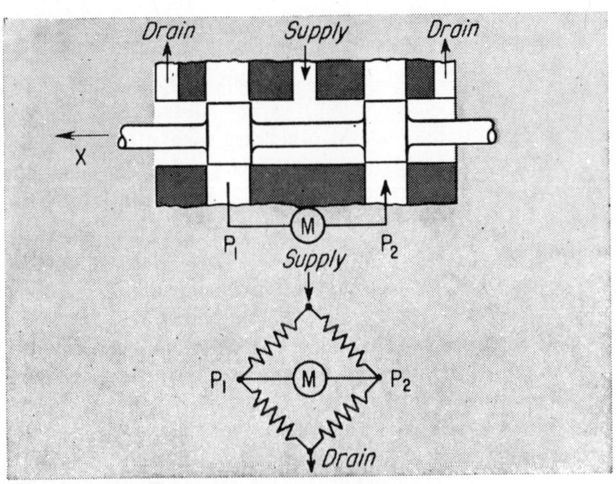

Fig 1 Hydraulic-electric analogy . . .
between 4-way valve and Wheatstone bridge is key to accompanying method for calculating valve response. *M* is the hydraulic-motor load. Branches resemble electrical resistances, and fluid flow tends to obey Ohm's law.

VALVE DYNAMICS

What happens inside as well as outside a valve should be kept in mind when choosing servovalves for any system. These basics–circuitry, balance of forces on moving parts, output vs input, leakage–are important.

Wheatstone Bridge Analogy

Flow-pressure relationships inside a valve can be approximated by assuming that fluid resistance is akin to electrical resistance, then solving the resulting linear-resistance circuit by analogy.

The circuit in Fig 1—a 4-way valve—is from Transactions, ASME 75 (no. 6), 1953. One or more of the branches can have variable resistance to flow.

Oil flow, Q, cu in./sec, = about $100A\sqrt{\Delta P}$ for this or any valve. P is pressure in psi, and A is area of opening, sq in., for valve-spool position selected. The constant, 100A, may be defined as "conductance" (an electrical analogy) because it is a measure of ability of valve to "conduct" fluid.

An ideal valve of type shown has no flow if spool is centered—conductance is zero. For any position away from center, conductance is zero on the closed side of bridge. Flow from open side goes through motor, then returns through diametrically opposite arm to drain. Conductances of opposite arms are equal, and equation for pressure drop through motor is:

$$P_M = P_S \times \frac{2Q_M^2}{(100A)^2}$$

Flow Forces on Spool

Most servo valves have spools for the main valve element, therefore the discussion will be based on spools. However, poppets, plates and other valve elements will have similar problems.

Actual flow through a spool valve is proportional to minimum cross-sectional flow area × $\sqrt{\text{pressure drop}}$. But flow causes side loadings and other disturbances that decrease accuracy and repeatability.

A perfect valve would be in perfect force balance for all flows, and a low-force input signal would be sufficient. Some spool designs come close to achieving this perfection, but special, costly contouring of the valve elements is needed. The following explains why.

Loadings on the valve during operation are both **hydrodynamic** and **hydrostatic.** When the loading is **hydrodynamic,** so-called "Bernoulli" forces are caused by the reaction of the jet of fluid escaping through the valve orifices. This jet–owing to the skew-symmetrical configuration of the orifice–is inclined to the valve axis at an angle of about 70°, Fig 2.

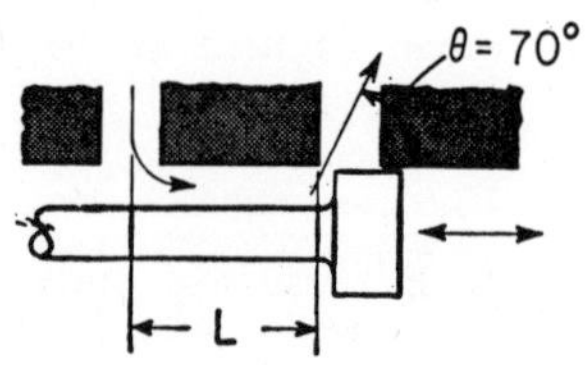

Fig 2

Two forces are generated. One, a "spring" force, F_T, lb, depends on total distance that spool moves. For an ideal valve (or an underlapped valve beyond the region of underlap):

$$F_T = -2Q \cos\theta \sqrt{\rho P_V}$$

where Q is total flow, cu in./sec, ρ is density, lb × sec²/in.⁴, and P_V is total pressure drop across valve (two orifices in series), psi. The negative sign indicates that this force, proportional to spool displacement, tends to center the valve. Combined with the equation for flow, Q = constant × $\sqrt{P_V}$ the final equation for F_T becomes

$$F_T = 0.47A\ P_V \quad \text{(for oil)}$$

In addition there is a velocity-responsive damping force having the magnitude

$$F_D = CwL \frac{dx}{dt} \sqrt{2 \Delta P \rho}$$

where F_D is the force on one orifice only, and ΔP is the pressure drop across one orifice. L is the "damping length" or distance between centers of incoming and outgoing flows of the orifice. C is discharge coefficient, w is peripheral width of the valve opening, x is distance the valve moves . . . all in consistent units.

If L is positive, F_D will be positive and valve will be unstable, because of an increasing force in direction of valve travel. L will be positive when flow through the orifice is inward (toward the piston axis and vice versa). Since a conventional four-way valve has two orifices with damping lengths opposed in sign, an effective damping length is the algebraic sum of the two L's for the two chambers. If a valve has negative effective damping length, it will be stable.

The "Bernoulli forces" (preceding paragraphs) can be used for gain compensation of the second stage of a two-stage servo valve. The forces are proportional to the pressure drop across the valve and thus cancel the increase in gain that would otherwise take place.

Hydrostatic forces on a stationary valve can lock the spool—a well-known problem. Circumferential piston grooves can reduce such forces.

In theory, there is no lateral force (at right angles to axis) for any piston configuration having axial symmetry or for a decentered parallel piston. However, for a decentered conical piston, a cocked parallel piston, or one with burrs, lateral forces occur. The lateral force developed by a tapered piston will be:

$$F = \frac{\pi L r b \,\Delta P}{2e}\left[1 - \frac{2c + b}{\sqrt{(2c+b)^2 - 4e^2}}\right]$$

where L = axial length of piston, in.; r = radius of piston, in.; b = radial taper, in./in.; ΔP = pressure drop across the piston, psi; e = eccentricity of piston, in., and c = radial clearance of large end of centered piston.

TYPICAL UNITS

Q	oil flow, cu in./sec
P	pressure, psi
P_S	supply pressure, psi
P_M	pressure drop in hydraulic motor, psi
P_V	pressure drop in valve, psi
A	flow area, sq. in.
F	force, lb
ρ	density, lb × sec²/in.⁴
L	damping length between orifices, in.
C	discharge coefficient
w	peripheral width of valve opening, in.
x	distance valve spool moves, in.
r	radius of piston, in.
b	taper, in./in.
e	eccentricity, in.
c	clearance, in.
u	underlap, in.
K_Q	flow gain, cu in./sec/in.
K_P	static stiffness, psi/in.
K_{QvsP}	slope of pressure-flow curve cu in./sec/psi

If the higher pressure is applied to the large end of the taper, the piston will be unstable and will be forced against the cylinder wall. Conversely, a tapered piston with the small end exposed to the higher pressure will tend to center itself . . . a stabilizing effect. However the small taper is difficult to make, and the piston is suitable for one-directional pressure application only. Non-tapered pistons with good dimensional accuracy, fine finish, and circumferential grooves are still the best answer for combatting undesirable hydrostatic forces.

Output vs Input

An important qualification for a servo valve is that large outputs be obtainable with small inputs. The ratio of output to input, called "gain," is usually measured with output in volume-flow units, and input in electrical current to the torque motor or other actuator. Input is also given as movement of pilot spool, in. There is a separate value of gain for each element of a servo—torque motor, first-stage amplifier, main spool—as well as for the overall valve.

Referring to the Wheatstone-bridge analogy, Fig 1, the over-all output-input relationship for an ideal 4-way valve can be readily analysed by partial differentiation:

- Flow gain

$$K_Q = \frac{\partial Q_M}{\partial x}, \quad \frac{\text{cu in.}}{\text{sec in.}}, \quad \text{for } P_M = \text{constant}$$

- Static stiffness (centered spool)

$$K_P = \frac{\partial P_M}{\partial x}, \quad \text{psi/in., for } x = 0$$

- Slope of pressure-flow curve

$$K_{Q\ vs\ P} = \frac{\partial Q_M}{\partial P_M}, \quad \frac{\text{cu in.}}{\text{sec psi}}, \quad \text{for } x = \text{constant}$$

For the general case, these final equations apply:

$$K_Q = 100w\sqrt{\tfrac{1}{2}(P_S - P_M)}$$
$$K_P = \text{infinity}$$
$$K_{Q\ vs\ P} = \frac{\text{conductance, cu in./sec/psi}}{4 \times \sqrt{\tfrac{1}{2}(P_S - P_M)}}$$

The oil supplied to an underlapped valve (the typical design) can be at either **constant pressure** or **constant flow.** Valve and load resistance determine flow in a constant-pressure valve.

Constant-pressure valve. All arms of the bridge are conducting; diametrically opposite conductances are equal. Denoting the underlap as u, then motor flow is:

$$Q_M = 100w(u + x)\sqrt{P_S - P_1} - 100w(u - x)\sqrt{P_1}$$

It follows that:

$$P_M = P_1 - P_2; \quad P_S = P_1 + P_2;$$
$$P_1 = \tfrac{1}{2}(P_S + P_M);$$
$$P_2 = \tfrac{1}{2}(P_S - P_M)$$

And:

$$Q_M = 100w(u + x)\sqrt{\tfrac{1}{2}(P_S - P_M)} - 100w(u - x)\sqrt{\tfrac{1}{2}(P_S + P_M)}$$

- Flow gain:

$$K_Q = 100w(u + 1)\sqrt{\tfrac{1}{2}(P_S - P_M)} - 100w(u - 1)\sqrt{\tfrac{1}{2}(P_S + P_M)}$$

$$\text{For } P_M = 0, \quad K_Q = 100w\sqrt{2P_S}$$

- Static Stiffness:

$$K_P = \frac{2P_S}{u}$$

- Slope of pressure-flow curve:

$$K_{Q\ vsP} = 100w\left[\frac{u + x}{2\sqrt{2(P_S - P_M)}} + \frac{u - x}{2\sqrt{2(P_S + P_M)}}\right]$$

$$\text{For } P_M = 0, \quad K_{Q\ vs\ P} = \frac{100wu}{\sqrt{2P_S}}$$

Constant flow valve. Motor pressure drop is:

$$P_M = P_1 - P_2$$
$$= \frac{Q_2^2}{[100w(u - x)]^2} - \frac{Q_1^2}{[100w(u + x)]^2}$$

It follows that:

$$Q_1 = \tfrac{1}{2}(Q_S + Q_M); \quad Q_2 = \tfrac{1}{2}(Q_S - Q_M)$$

And:

$$P_M = \frac{(Q_S - Q_M)^2}{4[100w(u - x)]^2} - \frac{(Q_S + Q_M)^2}{4[100w(u + x)^2]}$$

Flow gain can be obtained by differentiation of the implicit function, $F\ (x, Q_M)$:

$$\frac{\partial Q_M}{\partial x} = -\frac{\partial F/\partial x}{\partial F/\partial Q}$$

$$\frac{\partial Q_M}{\partial x} = \frac{(Q_S - Q_M)^2}{2Q_S(u - x)} + \frac{(Q_S + Q_M)^2}{2Q_S(u + x)}$$

- Flow gain:

$$K_Q = \frac{Q_S}{u}, \quad \text{for } P_M = 0, \quad Q_M = 0, \quad x = 0$$

Flow gain is now independent of P_M and a nonlinear function of valve displacement x; thus its characteristics are totally different from those of the constant-pressure valve.

- Static stiffness:

$$K_P = \frac{Q_S^2}{(100w)^2u^3}$$

- Slope of the pressure-flow curve can again be determined by differentiation of the implicit function, giving:

$$K_{Q\ vs\ P} = \frac{2(100w)^2(u - x)^2(u + x)^2}{(u + x)^2(Q_S - Q_M) + (u - x)^2(Q_S + Q_M)}$$

$$K_{Q\ vs\ P} = (100wu)^2 \div Q_S \quad \text{for} \quad x = 0, \quad Q_M = 0$$

A shunt is sometimes put across the load to improve stability. Such a shunt can be treated as a reduction of Q_M:

$$Q_{shunt} = Q_M - C_S\sqrt{P_M},$$

where C_S is conductance of shunt.

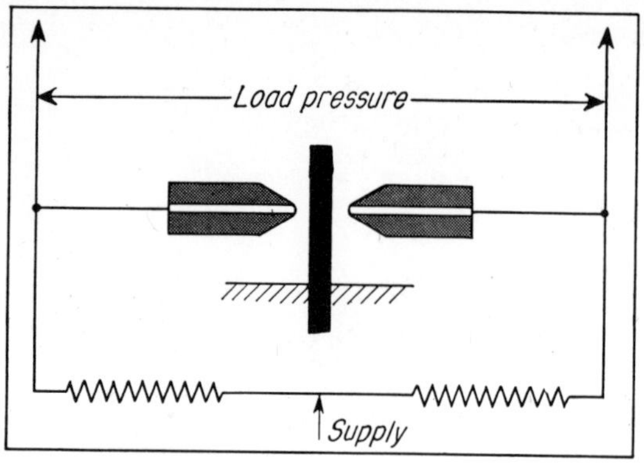

Fig 3 Flapper valve

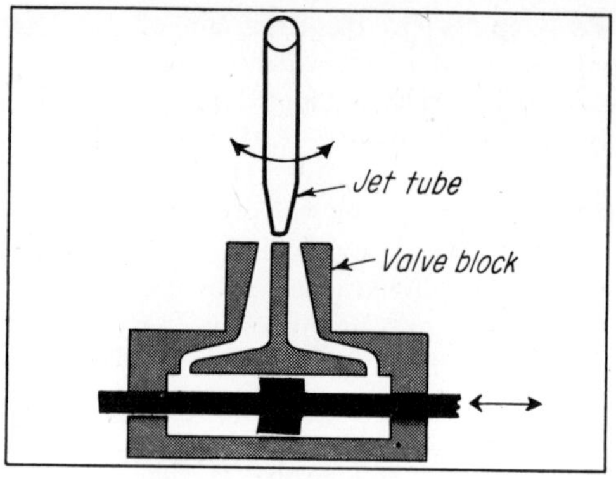

Fig 4 Askania nozzle

Leakage

In neutral position, leakage depends on spool underlap. Even with "zero" lap (± 0.000, 2 in.) some leakage exists and up to 5 percent of the rated horsepower may be dissipated. Hysteresis is the loop formed by the flow-vs-current curves for increasing and decreasing electrical current. Hysteresis of a good commercial valve should be below 2% of rated flow. "Zero shift" is a condition where zero flow does not occur at zero input current . . . a common condition for most servo valves. Therefore, some means for readjusting null should be provided.

VALVE STAGES

The simplest electrohydraulic servo valves have a single stage of amplification—the input signal drives the main spool directly. More common, however, are two-stage valves, which have these parts and functions:

Torque Motor Input

The electric actuator controlling a servo valve is often a "torque motor" whose thrust output is proportional to differential current supplied to it. The thrust is converted to spool or flapper position, controlled with restraining or centering springs, and the fluid output is proportional to differential current. A single-stage valve is directly actuated by the torque motor, while a two-stage valve uses the first stage–generally a nozzle or flapper–as a preamplifier. There has to be feedback between first and second stage, which can be done mechanically, with force or position feedback, or electrically with potentiometers. More often feedback is first-stage flow converted to differential pressure through a hydraulic Wheatstone-bridge arrangement, and applied to the spring-restrained second-stage spool valve.

The electrical input-power requirements for single-stage servos is high–1 to 6 watt–because the torque motor must drive the power spool. Two-stage servos can usually be controlled with less power. For example, simple flapper-nozzle 2-stage servos with no feedback operate on less than 0.1 watt. Three-stage designs also operate on less than 0.1 watt, but handle higher flows.

Differential currents of about 40 milliamps for single-stage and 15 milliamp for 2-stage valves are typical.

Preamplifiers

There is no preamplification in a single stage servo valve–the input force signal moves the main spool directly. But for better response, particularly for oil flows greater than about 4 gpm, more than a single stage is desirable. In a 2-stage servo, the input controls a smaller pilot valve, which in turn positions the main spool. For highest flows (100's of gpm), 3-stage servo valves–preamplifier, amplifier, and main valves–are also available.

The pilot (preamplifier) valve has many possible designs. It can be a spool, slide valve, flapper, nozzle-jet, or variations of each. Two common forms are the flapper valve and the Askania nozzle.

A **flapper valve**, Fig 3, has a continuous flow to tank from two opposing orifices. It is modulated by movement of the flapper between the orifices, thus forming a Wheatstone bridge with two fixed and two variable arms. The fixed orifices of the first stage are about 0.005 in. in diameter; the null clearance between nozzle and flapper is about 0.001 in.

The **Askania nozzle**, Fig 4, has a pivoted tube which discharges a jet of fluid. The kinetic energy of the jet is reconverted into pressure energy in the tapered passages, the amount depending upon the relative angular displacement of the jet from the respective passages.

Internal Feedback

The response–movement or pressure–of the main spool is sometimes fed directly back to the preamplifier inside the valve housing. In this way, the main spool motion can be stopped at the desired point by effectively neutralizing the output signal of the preamplifier. The feedback can be a mechanical linkage, a pressure signal, or a proportional spring bias of some sort.

TYPES OF VALVES

There are two main types of servo valves–**flow-control** and **pressure-control.** The **flow-control** type is used in systems the most. It provides an output flow proportional to input displacement at constant load pressure. It has a high impedance and is most suitable for noninductive (non-time-dependent) load. The **pressure-control** type, useful too, provides a pressure proportional to input displacement, has low impedance, is highly load-sensitive.

An **"ideal"** valve has high static stiffness in terms of psi/in. It will accelerate inertia loads quickly, resist disturbances and has low static error. Undesirable

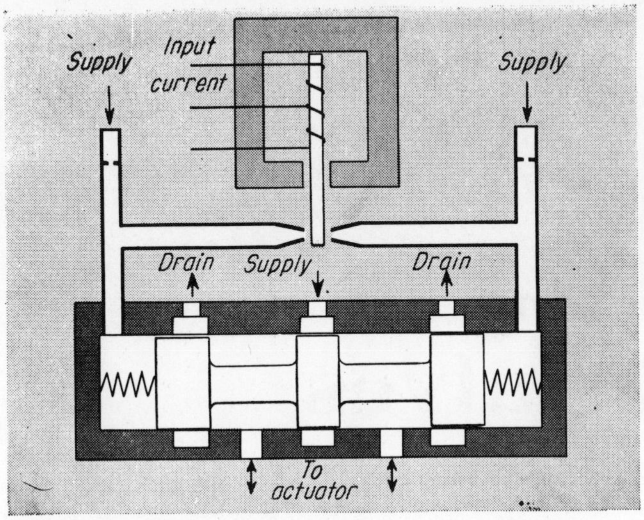

Fig 5 Flow-control type . . .
servovalve varies output flow in proportion to input electrical current. This example is 2-stage, with flapper preamplifier.

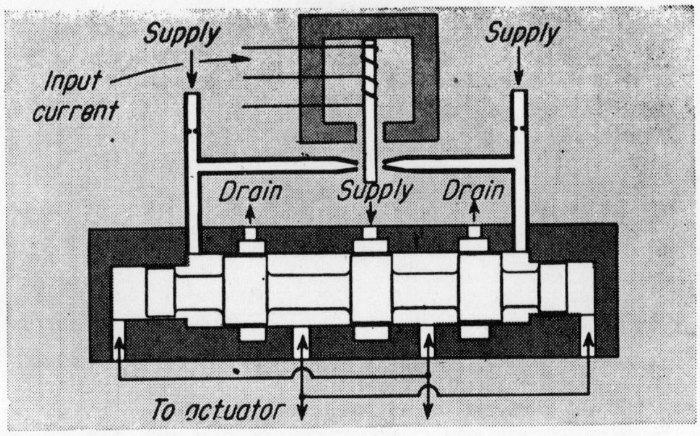

Fig 6 Pressure-control type . . .
servovalve varies output pressure in proportion to input electrical signal. Output pressure is fed to balancing piston at appropriate end of spool, giving feedback needed to control pressure.

valve characteristics are poor linearity and zero pressure-flow slope at neutral, which has a destabilizing effect.

A constant-pressure underlapped valve has a large flow gain, stable pressure-flow slope, and good linearity over most of the flow range. It has zero static error for inductive loads only. The error for resistance or static loads increases with static stiffness—some movement of the valve off neutral is required to produce the pressure that sustains a static load. This movement, Δx, = approx $(u/2)(P_M/P_S)$ for small ratios of P_M/P_S.

A **constant-flow underlapped valve** has poor linearity and is not suitable for precision servomechanisms; however, it works for less exacting jobs. An example is power steering for vehicles, where pump is practically unloaded when the valve is in neutral position. If pressure is from an accumulator or variable-displacement pump with pressure-compensating control, an ideal valve will minimize power consumption. In most cases a constant-displacement pump with relief valve is used, and it does not matter whether power is wasted through the underlaps of the valve or through the relief valve. An important point: choose pump to meet requirements of the servomechanism at maximum displacement rate or else the constant-pressure system will become a constant-displacement system. Maximum efficiency is obtained when $P_M = (2/3) \times P_S$.

Flapper-type valves, although often used as preamplifiers, are rarely used alone for high-power applications. As preamplifiers, they need a constant-pressure supply. For power applications, a slightly underlapped valve is a good compromise.

For any servo valve, dynamic response is very high. Some small valves can operate at frequencies as high as 900 cps—100 cps is common. A few have max capacities to 45 gpm, frequency response to 80 cps; one is able to pass 160 gpm, can respond to 30 cps. Time constants are normally from 2 to 5 millisec. A typical output-spool deadband is about 0.5% of rated output.

The well-known book by Walter Ernst, "Oil Hydraulic Power and Its Industrial Applications," is scheduled for a new printing—completely revised—in January.

Some of the background information on servos for his book and for this article has been obtained from papers written for the ASME Transactions in 1952 and 1953 by J F Blackburn and S Y Lee; and from studies made by R E Bojar, B A Johnson, and Lee Schmid of the Cook Research Laboratories under Air Force Contract AF 33 (616)-2447. For this help, the author is grateful.

How to calculate response time of

Pilot-operated valves

Based on solenoid-energized, pilot-operated spool valves, this quick method examines the five forces that slow down valve operation.

LOUIS DODGE, Hydraulics Consultant, New Richmond, Ohio

A familiar sight in almost any hydraulic automatic control system is a four-way flow-control valve with a solenoid pilot valve perched on top. There are more of these four-way valves than all other remotely operated flow-control valves combined.

How fast will they respond? It's not just a matter of computing the flow rate through a known restriction, although that is part of the answer. Speed of response also depends on the characteristic force-vs-airgap curve of the solenoid; the shape, size, clearance, and displacement of each spool; and the fluid viscosity.

The method outlined in this article relates those parameters for the sketched valve, Fig 1, and can be applied to any other pilot-operated spool valve. A special technique is worked out for a large spool valve actuated by a small auxiliary piston.

The valve in action

Before applying the response equations, let's establish the sequence of

1. . TYPICAL SOLENOID-ENERGIZED PILOT-OPERATED VALVE

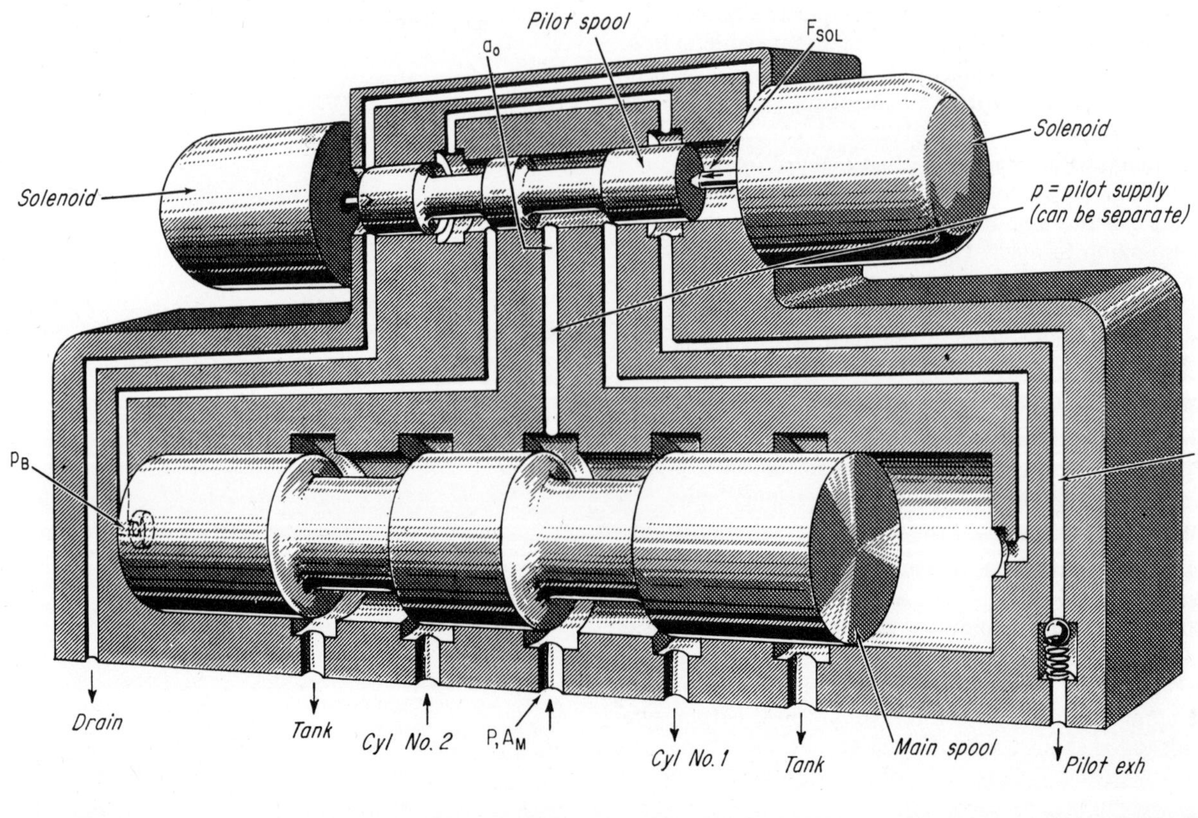

2. TYPICAL RESPONSE OF PILOTED VALVE

CALCULATED, VERIFIED BY TEST

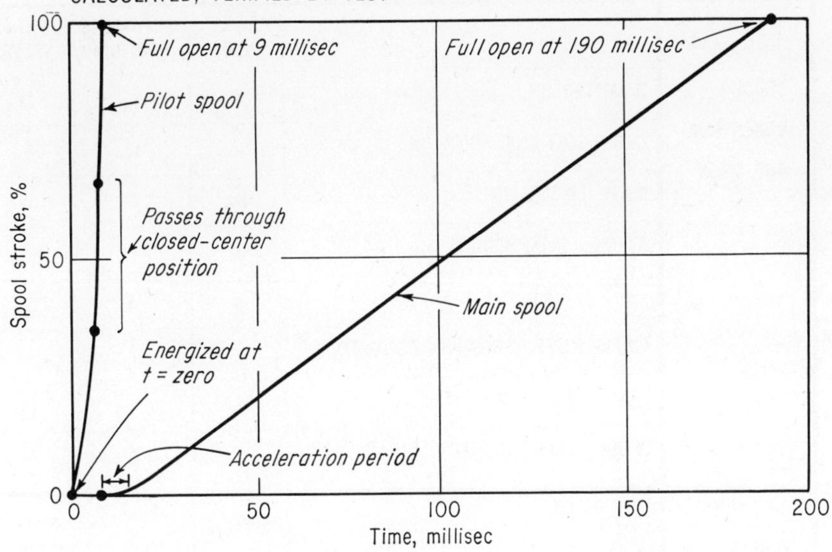

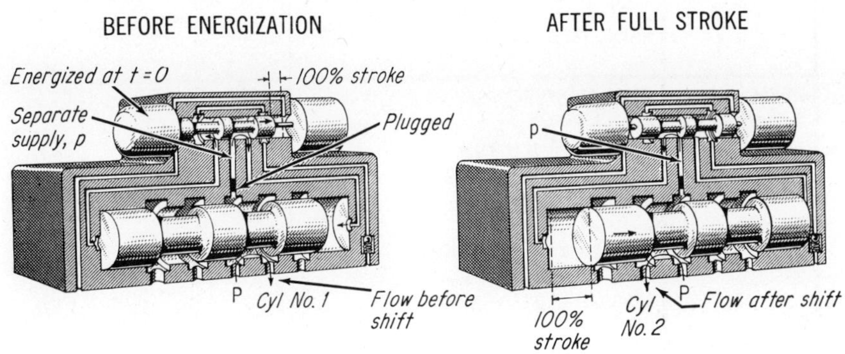

DIMENSIONS AND OPERATING CONDITIONS

	PILOT SPOOL	MAIN SPOOL
DIAMETER, IN.	$d=0.25$	$D=2.5$
MASS, LB-SEC^2/IN.	$m=0.0002$	$M=0.05$
STROKE, IN.	$s=0.375$	$S=1.5$
LAND LENGTH, IN.	.....	$L=6.0$
RADIAL CLEARANCE, IN.	.....	$C=0.0003$
COEFFICIENT OF FRICTION	.....	$F_R=0.04$
SOLENOID FORCE, LB (initial; final)	$F_{SOL}=1$; 8.5	.....
BACK PRESSURE, PSI	.....	$p_B=20$
SUPPLY PRESSURE, PSI	$p=100$	$P=500$
DIFFERENTIAL PRESSURE, PSI	$\Delta p=70$ (approx)	$\Delta P=450$ (approx)
PORT AREA, IN.^{2}	$a_o=0.05$	$A_M=1.2$
FLOW COEFFICIENT	$f=0.6$	$f=0.6$
VISCOSITY, CP	$\mu=80$	$\mu=80$
DENSITY, LB-SEC^2/IN.^{4}	0.000,085	0.000,085

operation. Briefly, here is what happens: The solenoid is energized, the pilot spool moves quickly to the full-open position, and the main spool is shifted at a rate determined by the amount of flow that can move through the pilot ports against these five resisting forces:

- Pilot system back pressure, psi
- Viscous damping force, lb
- Radial jet force, lb
- Axial jet force, lb
- Acceleration force, lb

Starting with the energization of the solenoid, here are the steps to take in calculating individual and total response times.

Solenoid and pilot

First obtain the force-vs-airgap curve of the solenoid from the manufacturer. From this you can determine the accelerating force available at each point of the pilot-spool stroke. The approximate linear relationship is

$$F_{SOL} = A + Bx \tag{1}$$

where F_{SOL} = solenoid force, lb; A = solenoid force at start of pilot-spool stroke; B = force gradient, lb/in., based on the force-vs-airgap curve; and x = pilot-spool displacement, in.

Next, calculate the time-displacement curve. Damping forces can be neglected because the stroke is short, velocities are relatively low, oil is of low viscosity, and the pilot spool is lapped to a loose fit. These assumptions are usually valid, and therefore time vs pilot-spool displacement can be approximated in this relationship (see *PE*—June '56, p 158):

$$t = \sqrt{\frac{m}{B}} \ln\left[\frac{x + r + \sqrt{2rx + x^2}}{r}\right] \tag{2}$$

where t = time, sec; m = mass of pilot spool, lb-sec^2/in.; and $r = A/B$, the ratio of solenoid starting force to force gradient.

Finally, read the time-displacement graph to find out how long it takes for the spool shift to the new position. The curves in Fig 2 are typical. They are for a ¼-in.-dia closed-center pilot spool controlling flow to a 2½-in. main spool.

Pilot pressure chosen for this example is 100 psi, which is less than main supply pressure, and a separate supply for the pilot is required. There are no hard and fast rules for establishing pilot pressures, but if possible keep intensity down to a few hundred pounds per square inch if that will do the job, to avoid possible distortion or leakage in the pilot system.

The graph shown reveals this: The pilot spool starts at the far left position, moves ⅛ in. to the closed center position in about 0.006 sec, and opens to the final position in another 0.003 sec. The average velocity during the last 0.003 sec is 0.125/0.003 = 42 in./sec.

In the open position, the pilot-valve flow rate depends on effective orifice diameter a_o and the pressure drop through the pilot valve:

$$q = fa_o \sqrt{\frac{2\Delta p}{\rho}} \qquad (3)$$

where q = flow, in.^{3}/sec; f = flow factor, dimensionless, ranging from 0.55 to 0.70 depending on valve type; a_o = cross-sectional area, in.^{2}, of the minimum port opening—usually the drilled port hole; $\Delta p = p - p_1$ = differential pressure, psi, measured across the pilot inlet and outlet ports; and ρ = fluid mass density, lb-sec^2/in.^{4}, normally 0.000085 for oil.

The value of Δp, measured at the pilot spool, depends on the performance of the main spool, including the effects of the back pressure reflected from the pilot exhaust.

Main spool

The five hydrodynamic resisting forces of back pressure, viscous damping, radial jet, axial jet, and acceleration determine shifting speed of the main spool.

Some simplifying assumptions must be made; otherwise there is no practical mathematical solution. For one, assume that back pressure of the pilot system, set by the pilot exhaust valve, is constant. Ignore line resistances, because connecting lines are short. Neglect viscous damping except at the full-velocity portion of the stroke. Then the five dynamic resistances, explained in detail below, can be handled with simple equations.

Pilot-system back pressure: Back pressure is usually 5 to 7% of pilot pressure p. Select the higher value if operating pressures are over 200 psi, because it adds a margin of safety that compensates for spool rubbing friction. The friction is from metal-to-metal contact at points where the oil film is partially destroyed.

Above 400-psi operating pressure, a separate pilot supply usually is provided. Pilot pressure in these instances ought to be at least 7% of the main operating pressure to ensure adequate force to move the main spool. For example, when the main operating pressure is 3500 psi, pilot pressure is 0.07 x 3500 = about 250 psi, and back pressure will be 0.07 x 250 = about 20 psi. Back pressure of the main system will range from 50 to 250 psi, depending on the load.

Viscous damping: For average

VALVE SYMBOLOGY

		PILOT	ACTUATING PISTON	MAIN VALVE
SPOOL DIMENSIONS AND MASS	DIAMETER, IN.	d	d_x	D
	CROSS-SECTIONAL AREA, IN.^{2}	—	a_p	A_S
	MASS, LB-SEC^2/IN.	m	—	M
	STROKE: INTERMEDIATE	x	—	—
	FULL	s	S	S
	ENGAGEMENT (LENGTH IN CONTACT), IN.	—	l_p	—
	LAND LENGTH, IN. (TOTAL)	—	—	L
	SPOOL-TO-BORE RADIAL CLEARANCE, IN.	C	C	C
SOLENOID FORCES	INITIAL, LB	A	—	—
	GRADIENT, LB/IN.	B	—	—
	FINAL, LB	F_{SOL}	—	—
	RATIO, A/B	r	—	—
DRAG FORCES	BACK PRESSURE, PSI	p_B	p_B	p_B
	VISCOUS DRAG, LB (or PSI)	—	d_V	D_V
	RADIAL JET, LB	—	P_{rad}	P_{rad}
	COEFFICIENT OF FRICTION	—	F_R	F_R
	AXIAL JET, LB	—	P_{ax}	P_{ax}
	ACCELERATION FORCE, LB	$F=ma$	—	$F=Ma$
OIL PRESSURE, FLOW, AND PORT SIZE	PRESSURE: SUPPLY, PSI	p	p	P
	PILOT DOWNSTREAM, PSI	p_1	p_1	p_1
	DIFFERENTIAL, PSI	Δp	Δp	ΔP
	PORT AREA, IN.^{2} (EFFECTIVE ORIFICE)	a_o	—	A_M
	FLOW COEFFICIENT (0.55 TO 0.70)	f	f	f
	VISCOSITY, CENTIPOISE	μ	μ	μ
	OIL DENSITY, LB-SEC^2/IN.^{4}	ρ	ρ	ρ
	FLOW RATE, IN.^{3}/SEC	q	q	—
	OIL VELOCITY, IN./SEC (THROUGH PORT)	—	—	V_o
	OIL MASS FLOW, LB-SEC/IN.	—	—	M_f
	OIL-JET DEFLECTION ANGLE, DEG	—	—	a
VALVE RESPONSE	ACCELERATION TIME, SEC	t_a	T_a	T_a
	SHIFTING VELOCITY, IN./SEC	v_p	v	V
	SHIFTING TIME, SEC (AFTER ENERGIZATION)	t	T	T

3. JET-FORCE DRAG

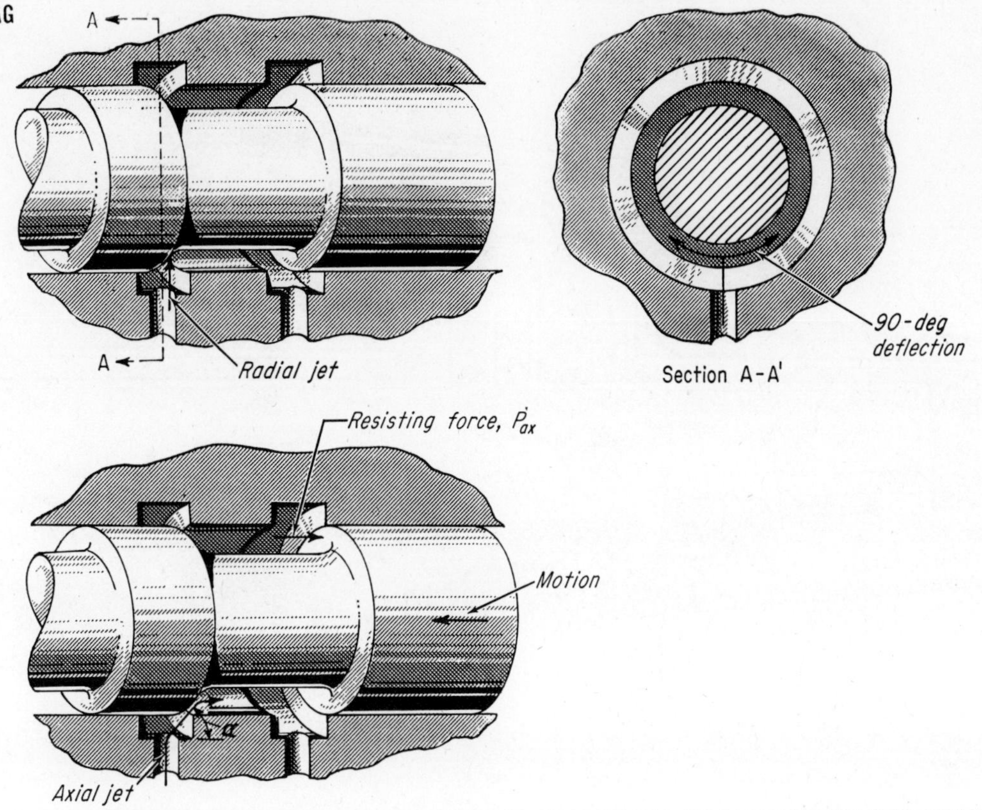

conditions of constant velocity and constant temperature, the damping force D_V, lb, can be expressed with this approximate relationship (Petroff's law):

$$D_V = \frac{D\pi L V \mu}{C \times 6.9 \times 10^6} \qquad (4)$$

where D = spool diameter, in.; L = length of the spool lands, in.; V = main spool velocity, in./sec; μ = absolute viscosity, centipoise; and C = spool to bore radial clearance, in.

If temperature varies more than 30 or 50 deg, it is nearly impossible to compute viscous resistance.

Radial jet pressure: The spool is forced radially against the far side when oil flow impinges against the near side of the spool (Fig 3). The radial force itself would be of no consequence except that it increases friction in the longitudinal direction in accordance with this expression:

$$P_{ax} = P_{rad} F_R \qquad (5)$$

where F_R = coefficient of friction and P_{ax} and P_{rad} = force, lb. The value of F_R will range from 0.02 to 0.06.

Formulas for the value of P_{rad} given in Marks's *Mechanical Engineers' Handbook* (McGraw-Hill) and other references depend on the number of degrees the jet is deflected. The deflection is assumed to be 90 deg for this article, giving the equation

$$P_{rad} = M_f V_o = Q\rho V \qquad (6)$$

where M_f = mass flow, lb-sec/in.; V_o = oil velocity through port, in./sec; and V = spool velocity, in./sec.

Then, because $V_o = \sqrt{2\Delta P/\rho}$ and $Q = fA_M \sqrt{2\Delta P/\rho}$, this useful relationship is derived:

$$P_{ax} = 2F_R f A_M \Delta P \qquad (7)$$

where A_M is the port opening of the main valve.

For low system pressures, P_{ax} is small and can usually be neglected. Above 2000 psi, don't ignore P_{ax}.

Axial jet pressure: The axial component of the incoming jet of oil will resist or augment spool motion, depending on spool design. Most spool valves have conventional lands: The jet resists motion during opening of a pressure port and assists motion during closing (Fig 3). To be on the safe side, neglect forces assisting motion.

The jet in effect acts against the spool shoulder away from the port and is effective only in proportion to its axial component, cos α:

$$P_{ax} = 2fA_M \Delta P \cos \alpha \qquad (8)$$

where α normally varies from 70 deg at initial opening to 90 deg at full opening. In calculations, use axial jet pressure during initial opening, and the axial component of radial pressure during remainder of travel.

Mass acceleration: The force necessary to accelerate the spool is

$$F = Ma \qquad (9)$$

where M is spool mass, lb-sec²/in., and a is dV/dt, in./sec². This force can be neglected in most instances unless very high speeds are expected.

Special case: piston-operated spool

Some pilot-operated valves have main spools too large to shift quickly with small pilot-oil flows unless special actuators are added. With special actuators such as the small-diameter pistons shown in Fig 4, an ordinary solenoid pilot is plenty large enough to shift very large main spools.

The actuating piston size must be carefully selected to give optimum response. If the diameter d_x is too large, spool velocity will be low. If the diameter is too small, spool friction and back pressure will be too hard to overcome.

Compute the best size by writing a differential equation containing velocity and area of the actuating piston and setting the first derivative of velocity vs area equal to zero—the peak value of velocity. Start with the pilot flow (Eq 3):

4. SPECIAL CASE: PISTON-OPERATED SPOOL

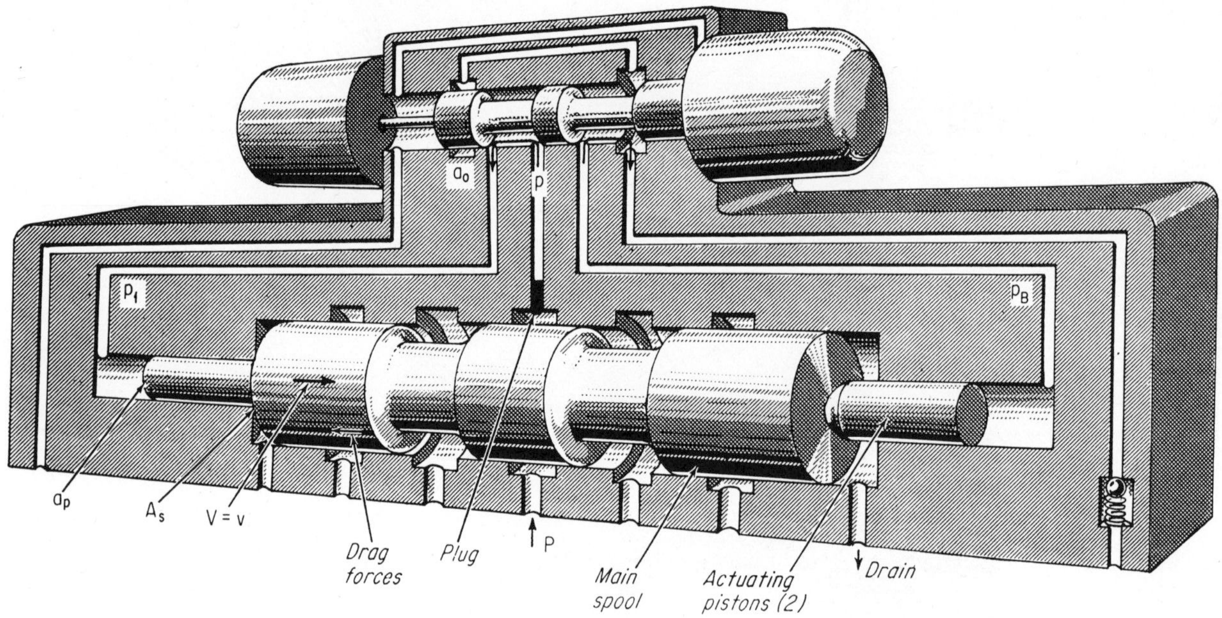

$$q = fa_o\sqrt{\frac{2\Delta p}{\rho}} \qquad (10)$$

Substitute constant k_1 for $fa_o\sqrt{2/\rho}$ and va_p for q. Term v is actuating piston velocity and a_p is its cross-sectional area. Eq 3 becomes

$$va_p = k_1\sqrt{\Delta p} \qquad (11)$$

Pressure drop $\Delta p = p\text{-}p_1 = p\text{-}p_B\text{-}(P_{ax} + \Sigma D_V)/a_p$, which is pilot pressure p less the aggregate of the hydrodynamic resistances discussed before. To simplify, set $k_2 = (p\text{-}p_B)$ and $k_3 = (P_{ax} + \Sigma D_V)$. Then:

$$va_p = k_1\sqrt{k_2 - k_3/a_p}$$

or

$$v = k_1\sqrt{\frac{k_2}{a_p^2} - \frac{k_3}{a_p^3}}$$

$$= \sqrt{\frac{k_1^2\,k_2}{a_p^2} - \frac{k_1^2\,k_3}{a_p^3}} \qquad (12)$$

The first derivative is

$$\frac{dv}{da_p} = \frac{1}{2}\,\frac{\left(\dfrac{3k_1^2\,k_3}{a_p^4} - \dfrac{2k_1^2\,k_2}{a_p^3}\right)}{\sqrt{\dfrac{k_1^2\,k_2}{a_p^2} - \dfrac{k_1^2\,k_3}{a_p^3}}}$$

Let the first derivative = zero, and multiply both sides of the equation by the denominator:

$$0 = \frac{3k_1^2\,k_3}{a_p^4} - \frac{2k_1^2\,k_2}{a_p^3}$$

Solve for optimum area a_p:

$$a_p = \frac{3k_3}{2k_2} = \frac{3(P_{ax} + \Sigma D_V)}{2(p - p_B)} \qquad (13)$$

Some approximations must be made before a_p can be computed. For instance, a_p appears as a piston diameter in the expression for damping force, Eq 4, thus changing Eq 13 into quartic form with four roots. Accuracy is not improved enough to warrant such a rigorous approach.

The simplification is to use trial and error. First assume a value for the actuating piston diameter, then substitute it in the viscous-drag equation (following) and Eq 13 to see how close you've guessed.

Viscous drag includes the main spool and the actuating piston, and Eq 4 still applies. The combined expression can be simplified to

$$\text{total } D_V = \frac{4VA_s\,\mu}{C\times 6.9\times 10^6\times d_x} \times\left(\frac{DL}{d_x} + 2S\right) \qquad (14)$$

where D_V = damping force, lb; V = piston and main-spool velocity, in./sec; A_S and $a_p = \frac{\pi D^2}{4}$ and $\frac{\pi d_x^2}{4}$, the cross-sectional areas of the main spool and the actuating piston, in.^{2}; C = radial clearance of piston and spool (assumed same), in.; L = combined lengths of lands of main spool, in.; and S = main spool stroke, in. Term $2S$ is for two actuating pistons and S is assumed equal to the average piston engagement l_p, in. Example 3 at the end of the article uses this equation.

Special case: spring-centered spool

A spring-centered main spool has the additional resistance of the springs, but otherwise response is calculated in the same way without the spring. When actuated from the center position, spool travel is one half that of the free spool, and acceleration is somewhat less. Return speed is faster because of the spring return force.

Total time of spool travel for a spring-centered valve won't differ much from that for a non-spring valve. The photograph and performance curves on page 213 are from actual tests made by Vickers Inc on spring-centered valves, showing the effect of pilot pressure on valve response. The 3-in. and 2-in. notations in this case refer to port size rather than spool diameter. Note that the total shift

SPECIAL TESTS OF TYPICAL MAIN SPOOL
(courtesy Vickers Inc. Division of Sperry Rand Corp.)

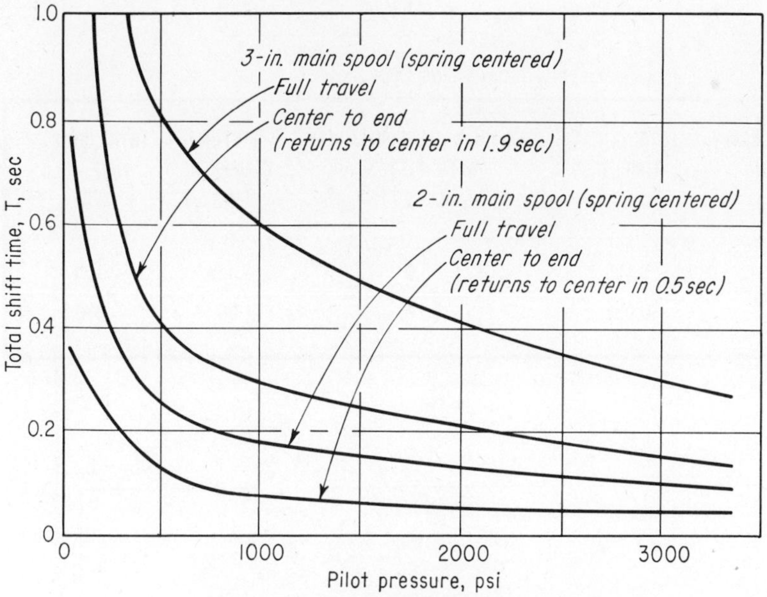

times are of the same order of magnitude as the times for the non-spring-centered valve on page 209.

CALCULATION EXAMPLES

1—Velocity of main spool

Dimensions, operating pressures and performance are given on page 214. Pilot supply pressure is 100 psi; main supply pressure is 500 psi.

The procedure is to compute maximum velocity first, then the acceleration time. Final computation is for total time.

The forces acting upon the main spool at maximum velocity are: pilot back pressure p_B, viscous damping force D_V, and radial jet force P_{rad}. From Eq 7, $P_{ax} = 2 \times 0.04 \times 0.6 \times 1.2 \times 450 = 26$ lb; $26/4.9 = 5.3$ psi. Provisionally, estimate that D_V is equivalent to 3.2 psi and $p_B = 20$ psi. The combined hydrodynamic resistance is $5.3 + 3.2 + 20 = 28.5$ psi.

Hence, pilot-valve pressure differential $\Delta p = 100 - 28.5 = 71.5$ psi, and the flow rate is

$$q = 0.6 \times 0.05\sqrt{2 \times 71.5/0.000085} = 40 \text{ in.}^3/\text{sec}$$

The maximum velocity of the main spool, $V = 40/4.9 = 8.2$ in./sec. With V known, $D_V = (2.5\pi 6 \times 8.2 \times 80)/(0.0003 \times 6.9 \times 10^6) = 3.05$ psi. The provisional estimate of $D_V = 3.2$ was close enough and recalculation is not necessary.

The forces acting upon the spool during acceleration are: p_B, P_{ax}, D_V and F, where $F = Ma$. Assuming a mean value for initial port opening $A_M = 0.4$ in.^{2}, then from Eq 8, $P_{ax} = 2 \times 0.6 \times 0.4 \times 450 \times 0.26 = 56$ lb; $56/4.9 = 11.4$ psi, where $\alpha = 75$ deg, and $\cos\alpha = 0.26$. Viscous drag will be the average: $D_V = 3.2/2 = 1.6$ psi. Back pressure is still $p_B = 20$ psi, so the total is $11.4 + 1.6 + 20 = 33$ psi.

Therefore accelerating pressure $= 100 - 33 = 67$ psi; $67 \times 4.9 = 328$ lb. From Eq 9, acceleration time $t_a = MV/F = 0.05 \times 8.2/328 = 0.0013$. The displacement of the main spool during the acceleration period is negligible, being less than 1% of the total stroke. The total stroke of 1.5 in. takes $1.5/8.2 = 0.182$ sec. The time interval from energization of the solenoid to completion of the main valve stroke, $T \simeq 0.190$ sec.

2—Higher pilot pressure

Dimensions and main operating pressure are unchanged from Example 1. However, pilot pressure is now made equal to main operating pressure: 500 psi. Much higher flow will pass through the pilot valve port. Here are the resulting response figures:

Maximum velocity period: $P_{ax} = 5.3$ psi (same as before); $D_V = 7.7$ psi (higher estimate, proportional to anticipated velocity); $p_B = 20.0$ psi (same). Total is 33.0 psi.

New $\Delta p = 467$ psi, and $q = 4.7\sqrt{467} = 102$ in.^{3}/sec; $V = 102/4.9 = 20.8$ in./sec; $D_V = 1.82 \times 20.8 = 37.8$ lb $= 7.75$ psi (proves out assumption).

Acceleration period: $P_{ax} = 11.4$ psi (same); $p_B = 20.0$ psi (same); $D_V = 3.6$ psi (increased). Total is 35.0 psi. Accelerating force $F = 465 \times 4.9 = 2280$ lb.

Accelerating time $t_a = (0.05 \times 20.8)/(2280) = 0.0005$ sec. The 1.5-in. stroke takes $1.5/20.8 = 0.072$ sec. Total time $T = 0.081$ sec.

The flow rate of the pilot oil is more important than pressure intensity in obtaining a fast-acting valve. A slightly larger pilot valve and enlarged pilot porting have a marked effect on the operational speed of the main valve.

Note that increasing the pilot pressure fivefold, from 100 to 500 psi, only doubles the speed of response, from 0.19 to 0.08 sec. Increasing the port area can result in a nearly proportional gain in speed, and no additional pressure is necessary. Costs of producing a ⅜-in. pilot spool are not much greater than those for a ¼-in. spool, and the increase in capacity is 50% without the additional heat losses entailed by an increase in pressure.

3—Special piston actuators

Dimensions and operating pressures are unchanged from Example 1. However, a small actuating piston is placed at each end of the main spool to increase the longitudinal velocity for a given pilot-fluid flow rate. Trial and error would normally be used to calculate the most effective diameter for the actuating piston. To simplify this

EFFECT OF ADDING ACTUATING PISTONS

PILOT PRESSURE p, PSI	MAIN VALVE DIAMETER, IN.	MAXIMUM VALVE VELOCITY, WITHOUT PISTON IN./SEC.	TOTAL SHIFT TIME T, SEC	MAXIMUM VALVE VELOCITY, WITH PISTON IN. SEC	PISTON DIAMETER, IN.	TOTAL SHIFT TIME T, SEC
100	2.50	8.2	0.190	15.0	1.425	0.109
500	2.50	20.8	0.081	55.0	1.06	0.036

EFFECT OF VARYING THE PISTON DIAMETER

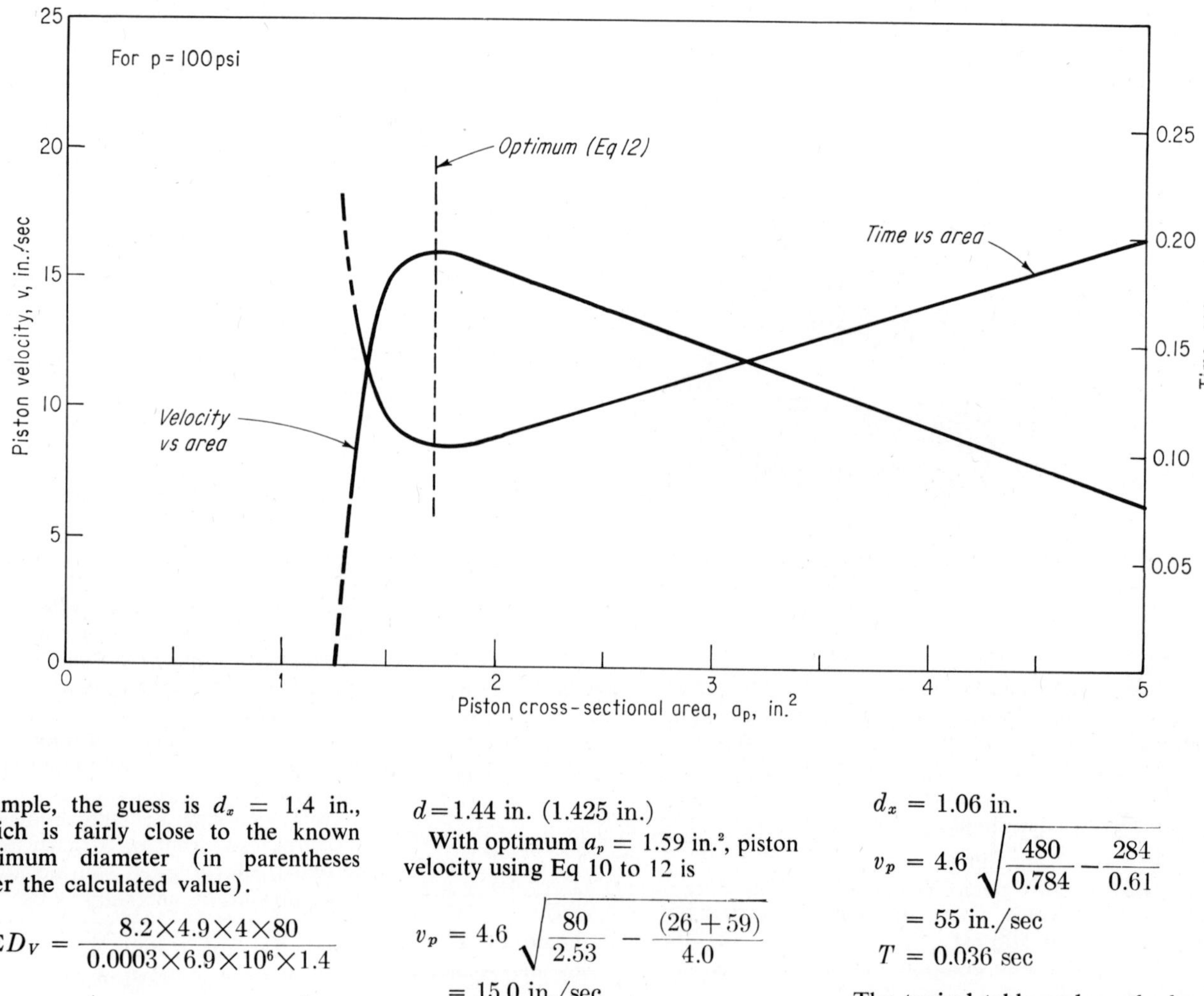

example, the guess is $d_x = 1.4$ in., which is fairly close to the known optimum diameter (in parentheses after the calculated value).

$$\Sigma D_V = \frac{8.2\times4.9\times4\times80}{0.0003\times6.9\times10^6\times1.4}$$

$$\times\left(\frac{2.5\times6}{1.4}+2\times1.5\right)$$

$$=\frac{12800}{2075\times1.4}\left(\frac{15}{1.4}+3\right)$$

$$= 60.5 \text{ lb } (59.0 \text{ lb optimum})$$

Introducing the value for ΣD_V in Eq 13:

$$a_p=\frac{3(26+60.5)}{2(100-20)}=1.62 \text{ in.}^2\ (1.59 \text{ in.}^2)$$

$d = 1.44$ in. (1.425 in.)

With optimum $a_p = 1.59$ in.^{2}, piston velocity using Eq 10 to 12 is

$$v_p = 4.6\sqrt{\frac{80}{2.53}-\frac{(26+59)}{4.0}}$$

$$= 15.0 \text{ in./sec}$$

Total time $T = 0.100 + 0.009 = 0.109$ sec. Using the pilot pressure of example 2 ($p = 500$ psi), $V = 20.8$ in./sec, and $d = 1.06$ in. Then:

$$\Sigma D_V = \frac{20.8\times4.9\times4.80}{0.0003\times6.9\times10^6\times1.06}$$

$$\times\left(\frac{15}{1.06}+3\right) = 258 \text{ lb}$$

$$a_p = \frac{3(26+258)}{2(500-20)} = 0.886 \text{ in.}^2$$

$d_x = 1.06$ in.

$$v_p = 4.6\sqrt{\frac{480}{0.784}-\frac{284}{0.61}}$$

$$= 55 \text{ in./sec}$$

$$T = 0.036 \text{ sec}$$

The typical table and graph shown above are of Eq 12, depicting the effect of actuating-piston area upon spool shifting velocity and shifting time.

Ball valves of plastic

"Cost too much" is the charge often leveled at metal ball valves. Here is a plastic ball.

ROBERT E. SANCTUARY, Product Design Manager, Worcester Valve Co, Inc, Worcester, Mass

It is expensive to make industrial quality ball valves out of metal. In sizes up to six inches, the cost of tooling, precision machining to close tolerances, and the special metals and synthetics (for seals), make conventional ball valves more expensive than gate, globe, and plug valves.

Our problem was to develop a ball valve to compete with 125-lb bronze gate valves, semi-steel lubricated plug valves, bronze and iron plug cocks, and iron body gate valves—but without sacrificing quality or performance.

. . .

WHY A BALL VALVE?

A ball valve (a plug valve too) goes from full open to full closed in a quarter-turn of the stem. That is the obvious principal advantage. Let's look at some others:

- **Visual clarity.** When the handle is parallel to the pipe, the valve is open. When perpendicular, the valve is closed.
- **Size.** The overall vertical mounting dimension required (from the bottom of the valve body to the top of the handle and stem) for a 1-in. ball valve varies from 2¼ in. to 4 in. depending upon the manufacturer. A comparable gate valve in the open position requires from 8 to 14 in.; a globe valve takes 5 to 12 in. And here is one of the ways that a ball has advantage over a plug: The vertical dimension of a 1-in. plug valve is 4 to 7 in. See sketches.
- **Weight.** A 1-in. ball valve weighs between 1.6 and 4.5 lb; whereas comparable conventional valves weigh between 4 and 43 lb.
- **Lubrication.** A ball valve does not have to be lubricated, so is easier to operate automatically.
- **Flow loss.** Straight-through flow in a round passage substantially reduces turbulence, which minimizes head loss for a given port diameter.

However, ball valves have their limitations: When operating temperatures exceed 450 to 500 F, and when operating pressures become extremely high, ball valves cannot generally compete with gate or globe valves. And they are not recommended for throttling service on high pressure steam lines: They are more ideally suited to full-open, full-closed service.

Envelopes for typical 1-inch valves (open position)

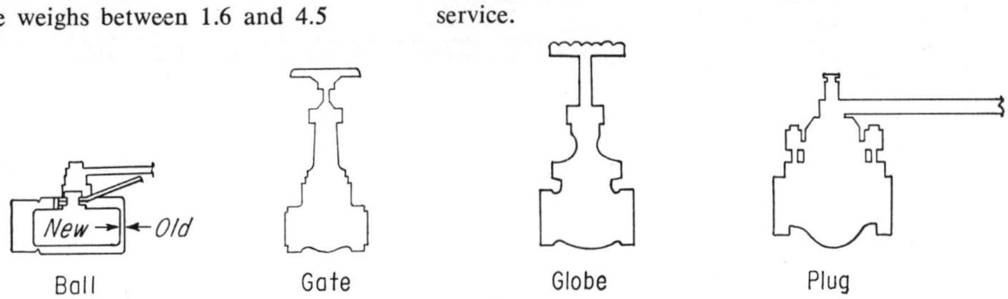

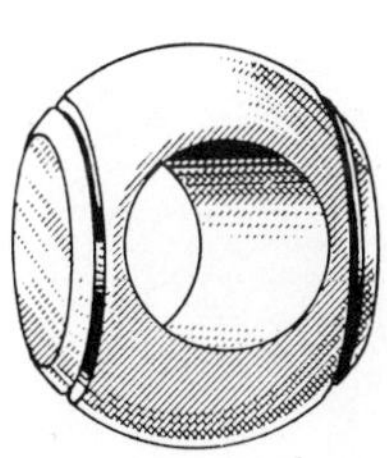

Plastic one-piece ball

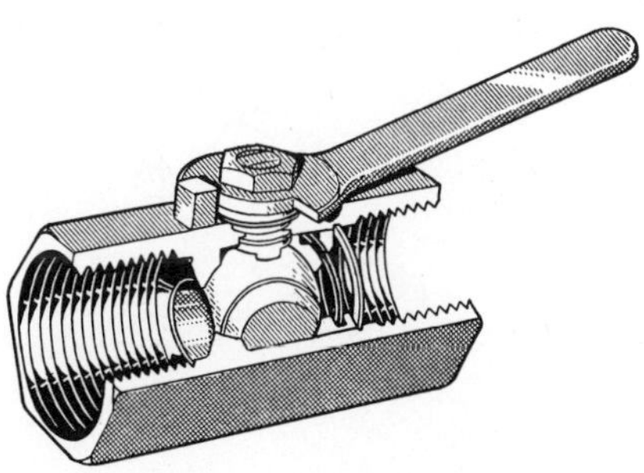

Integral body

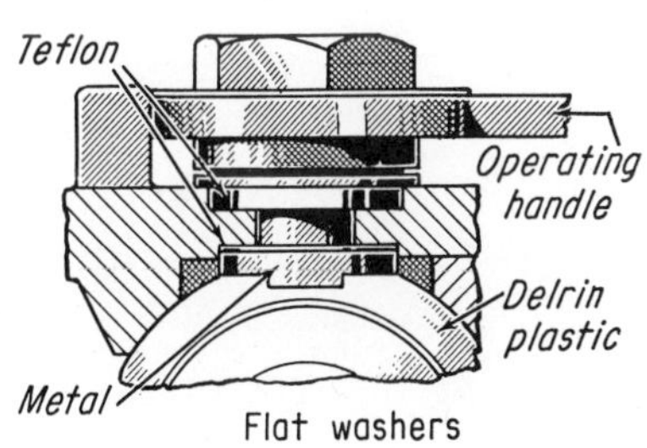

Flat washers

For uncluttered controls . . .

Manifold control valves

Manifolds, mounting plates, and stacked valves are the building blocks of compact hydraulic and pneumatic control circuits. Here are design tips.

LOUIS DODGE, Consultant, New Richmond, Ohio

THE maze of piping and fittings usually associated with hydraulic and pneumatic valves can be eliminated by manifolding. Also referred to as stacking or plate-mounting, manifolding reduces leakage in the system, saves assembly time, lowers maintenance costs, and improves appearance. But the initial cost might be higher because the manifolds are somewhat special. And the design will be less flexible than if you piped up non-manifolded valves because the manifold passages are internal and difficult to alter. However, the advantages usually outweigh the disadvantages.

The five basic manifolding techniques, sketched below, are: stacked valves, valve block, stacked subplates, drilled plate, and sandwich plate. The article sections immediately following give the specifications; the last section gives some general tips on manifold materials and valve arrangement.

Stacked valves

Mated ports with O-ring or gasket seals eliminate interconnecting piping in stacked valves. Usually there is a common supply passing through all the valves, and a common drain. Any additional porting can be specified, and each valve can be actuated manually, mechanically, or electrically.

The space saving is considerable because each separate valve needs a minimum of five line connections: one pressure line, one or two exhaust lines, two service lines, and one drain line. For example, four individual valves have about 20 line connections; four stacked valves have only 11.

Two-way, three-way, and four-way valves are available, in any standard spool size. Special mounting brackets on the machine frame are usually needed; line connections should not be used as supports.

Five manifolding techniques

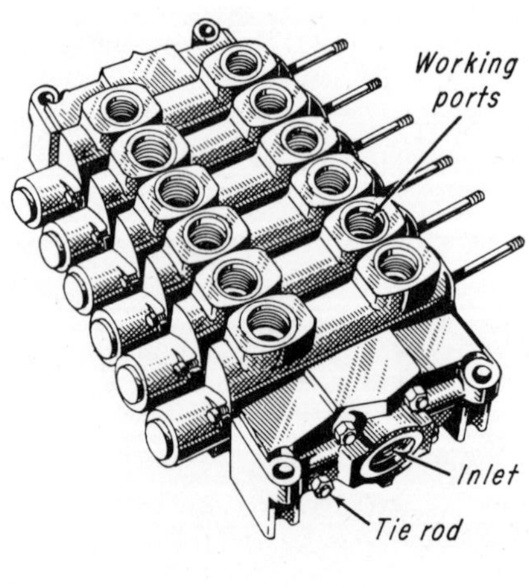

Stacked valves

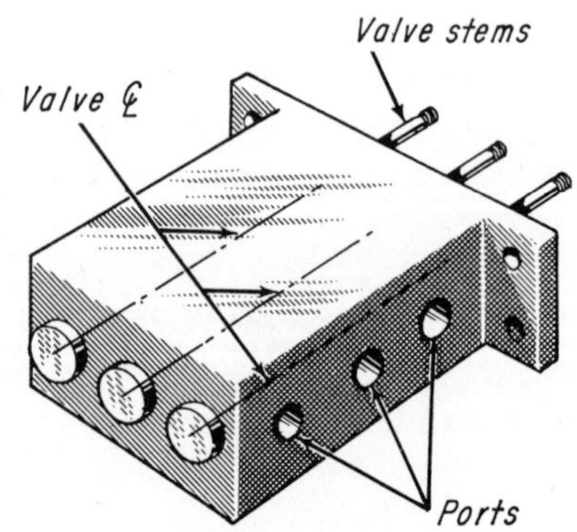

Valve block

Valve blocks

A variant of stacked valves is a single valve block in which several valve-spool assemblies are contained. Almost all types of valves can be accommodated and the passage spacing can be closer than for any other combination of valves. Pressure drops are lower because passages are shorter; speed of response is improved. Fine-grained cast iron is normally selected as the block material to reduce porosity and provide mechanical strength.

One disadvantage is that flexibility is severely curtailed when many valve functions are combined in one compact block. But the method is invaluable in crowded locations and where multiple valves must be very close to the controlled devices.

Valve blocks are usually custom built to suit the application, and it is worth a great deal of engineering time to obtain an optimum arrangement of passages and valves. Costs aren't necessarily high, however, if production quantities for each design are large. The parts and machining steps required in valve-block assembly are fewer than those for separate valves stacked together, because there are no interconnection seals or joints.

Design practices: The tables and sketches on pages 218 and 219 summarize the key design standards. Many of them apply equally to all manifold types.

Follow this machining sequence: Make valve bores parallel where possible, square the block, finish valve bores, then interdrill. To prolong wear life and reduce maintenance, it is desirable to use press-fitted replaceable valve sleeves in the bores. Provide them with grooves and drilled ports. Harden the sleeves to 32-35 Rockwell C; harden the spools to 58-62 Rockwell C.

If you do not use sleeves, machine the valve bores and lap them right in the block. You must provide internal grooving and drilling, but don't do this machining until after you've lapped the bores. Stone the edges of the grooves and clean them thoroughly of compound residual. Afterwards, relap the bore. The advantages of omitting sleeves are that the over-all assembly can be smaller, valve stroke reduced, and response time decreased. The stroke is less because the annular opening in the bore has more port area for the same axial displacement than the drilled holes in the sleeve.

Stresses produced by cover screws can distort the block and the valve bores, interfering with valve operation. The problem is lessened if the ends of the valve bores are relieved $\frac{1}{64}$ to $\frac{1}{32}$ in.

Pipe connections can develop leaks if the threaded joints are distorted when the cover screws or fittings are tightened. For that reason NPT-threaded ports should not be located close to a valve bore or to the block edge where mechanical strength is lower.

Straight-thread JIC ports are preferable to tapered threads, but they require O-ring seals or equivalent. Where tapered threads cannot be avoided, a Tru-Seal nut (Flick-Reedy Corp) with a Teflon sealing ring will seal even with light torque. A squared, finished surface at the top of the hole is required.

Stacked subplates

By interconnecting valves via subplates, you keep the connections independent of the valves. This offers several advantages: You can replace valves without disturbing the piping; you can modify the interconnections in the subplates without altering the valves; you can accommodate different valve sizes and porting in a single stack of matched subplates; and you can allow for a limited amount of distortion in the subplate without distorting the valve. The design practices given previously for valve blocks apply here too.

Subplates aren't limited to multiple valves; there are probably more individual subplates, used for single valves, than there are stacked subplates. And they offer many of the same advantages.

Drilled plates

Plate mounting of valves puts a wall between the valves and the piping and is desirable for these reasons: Appearance is far better because the panel conceals the piping; the valves are mounted on a flat surface and are not subjected to strains from piping connections; maintenance is simple because the front of the panel is mechanically independent of the back; and valves can be simply arranged and

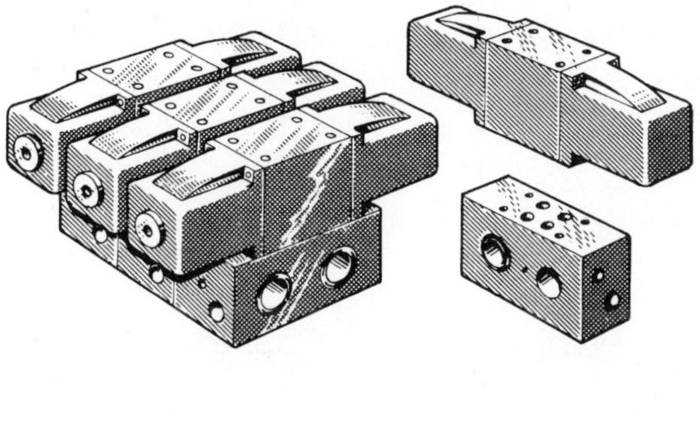

Stacked or single subplates

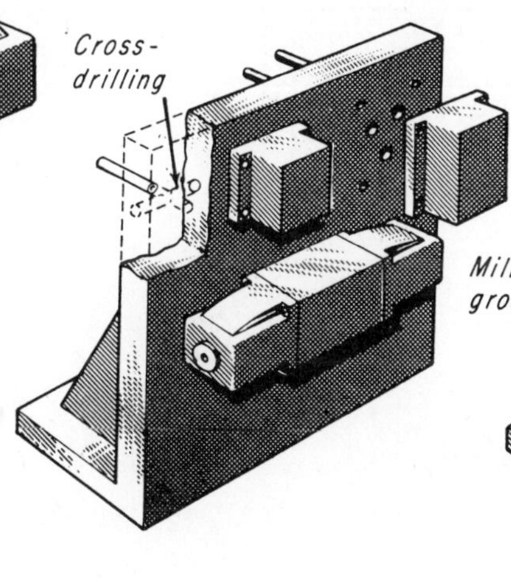

Drilled plate

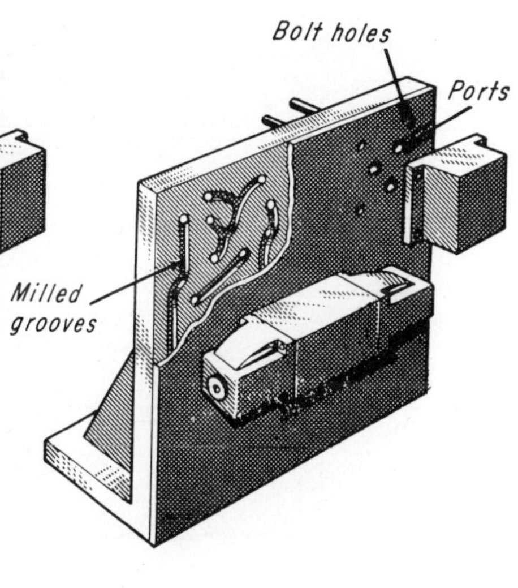

Sandwich plate

Stacked valve interconnections

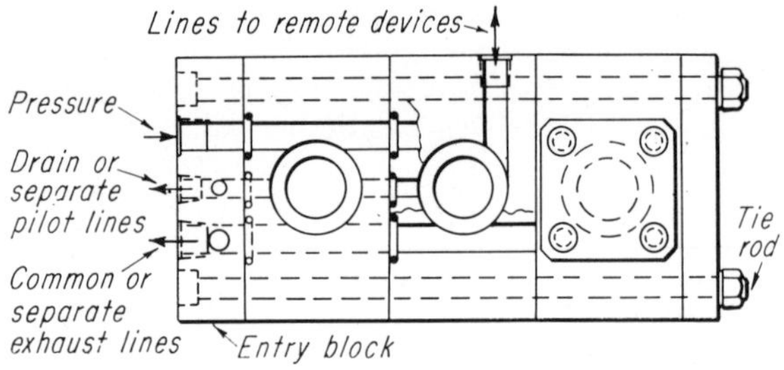

Typical stack

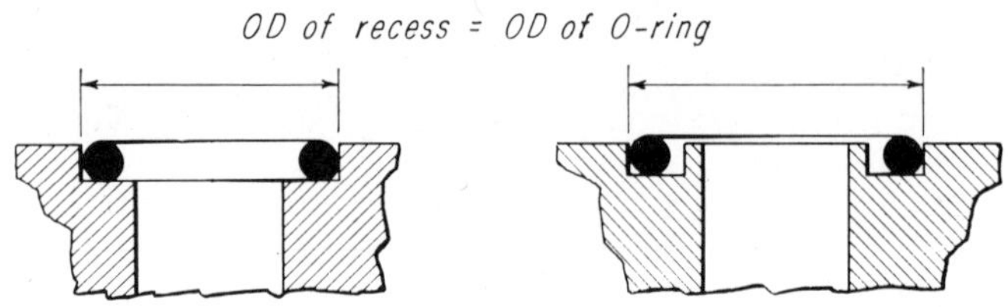

Counterbore for O-ring

Grooved for high pressure

labeled so that their functions will be easily understood.

The simplest plates are drilled straight through and the piping connection made directly to the back. In some instances the hole is used for clearance only, and the piping connection is made right to the valve. However, for the purpose of this discussion, it is assumed that 1) the plates are drilled and cross-drilled to make interconnecting passages in one or more planes within the plate, 2) piping connections are made on the edges or back, and 3) the valves are mounted on the plate and ported directly to it. A special advantage is that needle valves and simple check valves often can be inserted inside the plate, thereby eliminating extra lines and valve bodies.

Plates up to 2-in. thickness are usually made of cold-rolled steel, strain-relieved before machining. Plates up to 5-in. thickness should be made of hot-rolled steel, flame-cut from mill block. All surfaces must be cleaned, squared, and finished. After the plate is drilled, all mounting surfaces for the valves must be ground.

Plates can also be made from cast iron, but it must be fine-grained to avoid leakage paths along large flakes of graphite, particularly in the thin sections. Meehanite is one suitable material. Hydraulic-grade cast iron is another.

Design practices: Plate thickness and hole spacing depend on tap size and the number of planes in which the plate is cross-drilled. The table on next page gives minimum dimensions for plates drilled with a single layer and double layer of holes.

Leave enough metal between holes in adjacent layers: 1/8 in. for pipe tap size up to 1/2-in. dia, 1/4 in. for taps up to 1-in. dia, and 5/16 in. for taps above 1-in. dia. Be sure that runout in deep holes does not diminish the margin. You can reduce runout by

Valve block details

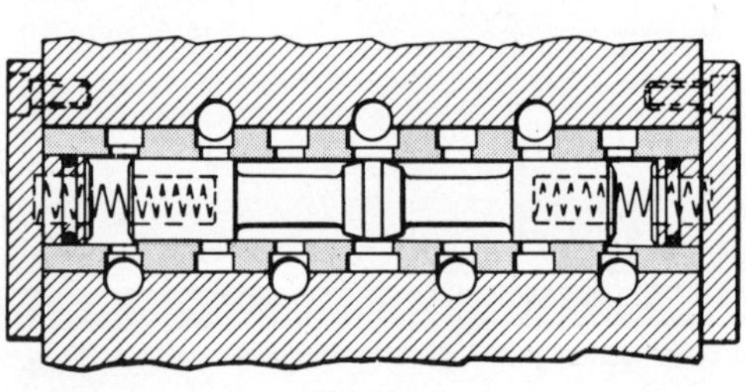

With sleeve

Valve symbol

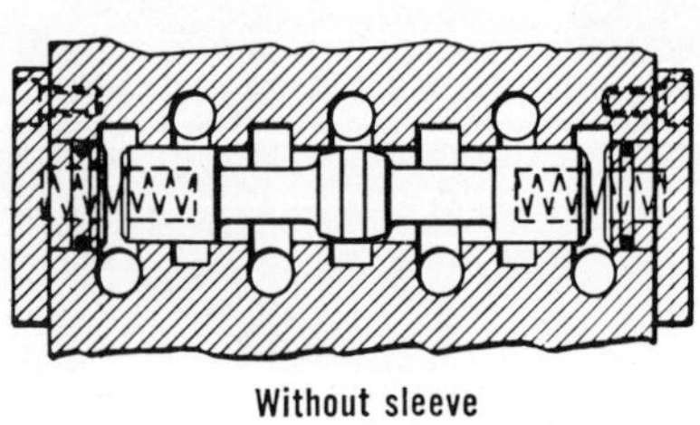

Without sleeve

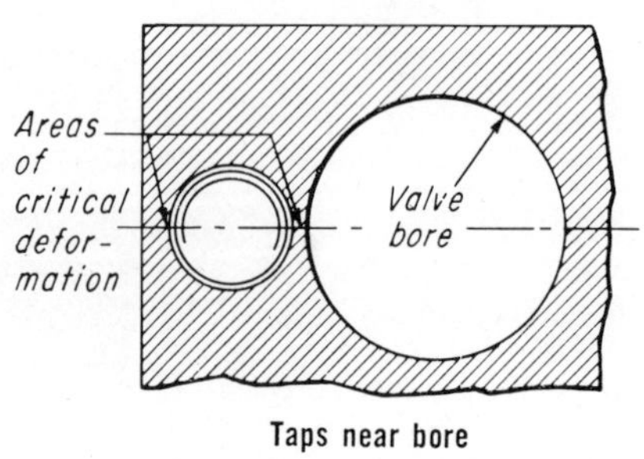

Taps near bore

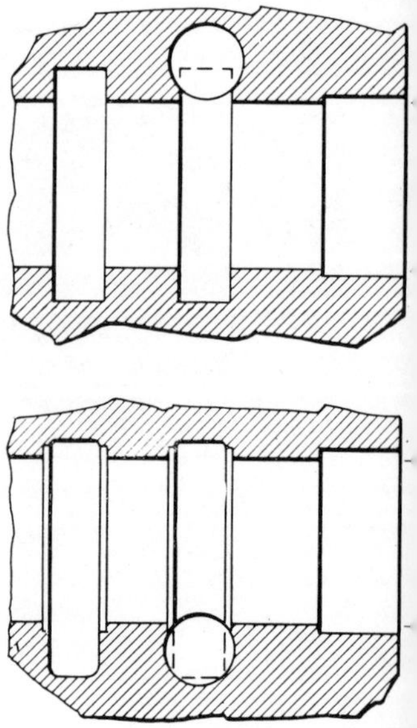

Stacked subplate drilling

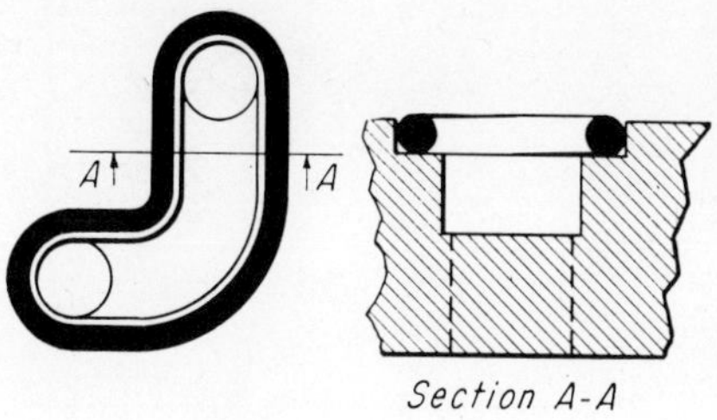

Irregular groove

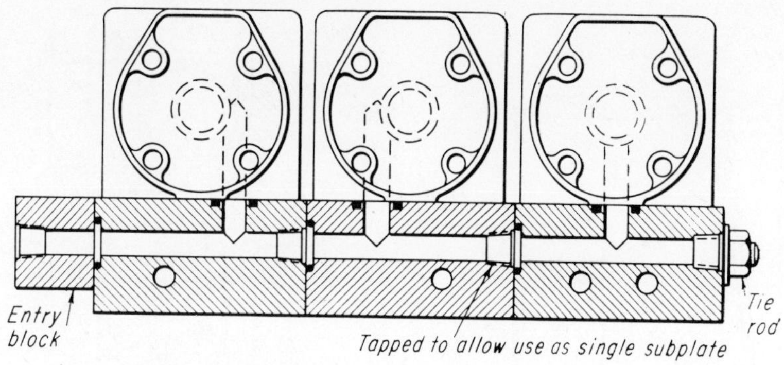

drilling the holes from both edges of the plate.

Maximum practical drill depth is about 1 ft. On such deep holes you can facilitate chip removal by specifying the tap drill for the first 4 in., then reducing the drill size by 1/64 in. for each successive 4-in. depth. However, use the tap drill for the entire hole where possible.

When a hole is drilled to intersect another, the drill point should penetrate 1/16 in. beyond the far wall into solid metal. The centerlines of the two holes need not intersect exactly, but they should not be offset more than the radius of the smaller hole. An exception is where only a small opening or tap is needed; just make sure the breakthrough is complete.

Holes to be tapped should be taper-reamed before tapping. Relieve upper threads of the tapped hole about 1/16 in. below finished surface.

Avoid unnecessary interdrilling in the main plate by adding a junction block to the back (see sketch on this page) and combining similar functions. All the supply ports can be combined, and so can the drain lines and pilot lines. It will be cheaper than using standard fittings. Exhaust lines can usually be combined too, but if a valve is sensitive to back pressure, provide a separate exhaust for that valve.

Make the junction block—also called terminal block—of steel and dimension it to match standard stock sizes. Except for drilling and tapping, it need not be machined.

Enclose the tubing in a sheet-metal box in the back of the main plate. Over-all size depends on tubing diameter, as shown by the sketches.

Front of manifold plate and the mating surfaces of valves must be extremely flat. Avoid setting up stresses in valve body and mounting surface during clamping for grinding. Remember that the metal is flexible and the

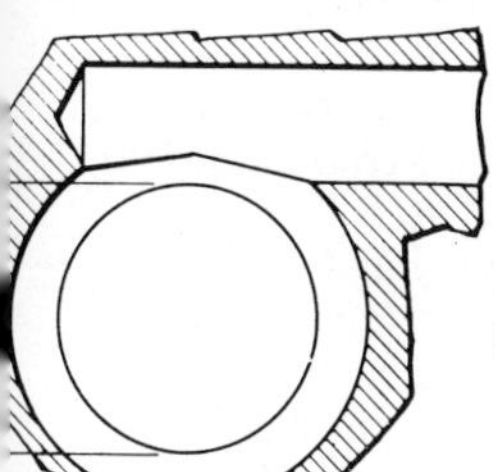
Tangential port to avoid jet pressure

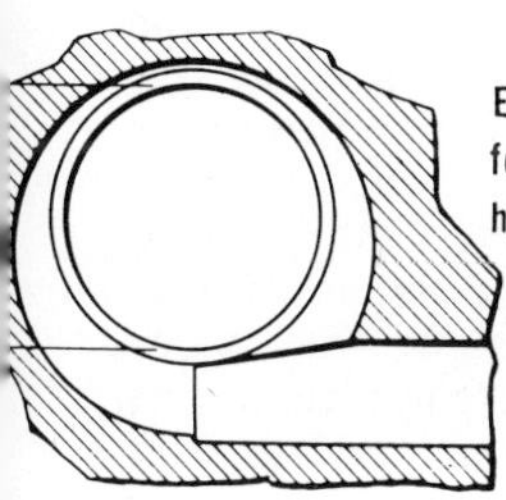
Eccentric groove for better hydraulic balance

PREFERRED DIMENSIONS FOR DRILLING (BLOCK, SUBPLATE, AND PLATE).

		SINGLE ROW	DOUBLE ROW	
PIPE TAP, NPTF	TAP DRILL d	PLATE THICKNESS A	HOLE SPACING B	PLATE THICKNESS C
		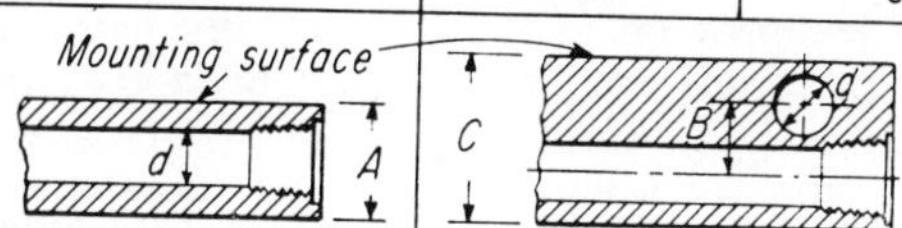		
1/4	7/16	7/8	9/16	1 7/16
3/8	19/32	1	3/4	1 3/4
1/2	23/32	1 3/8	7/8	2 1/4
3/4	15/16	1 5/8	1 1/8	2 3/4
1	1 5/32	2	1 3/8	3 3/8
1 1/4	1 1/2	2 3/8	1 3/4	4 1/8
1 1/2	1 23/32	3	2	5
2	2 3/16	3 1/2	2 1/2	6

Drilled plate details

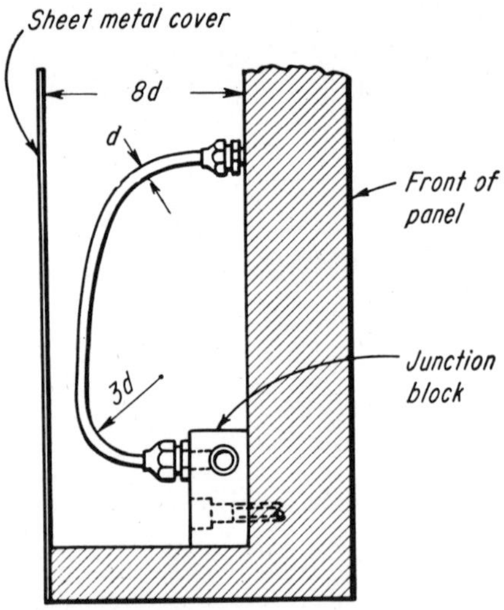

End view

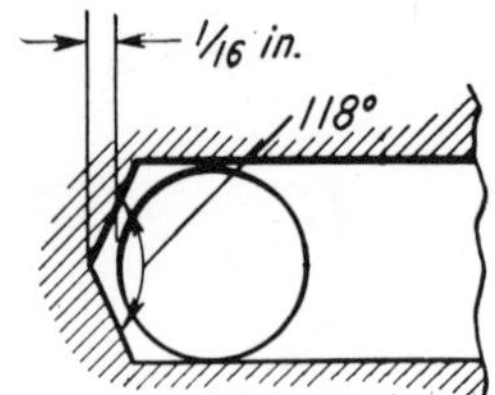

Full intersection

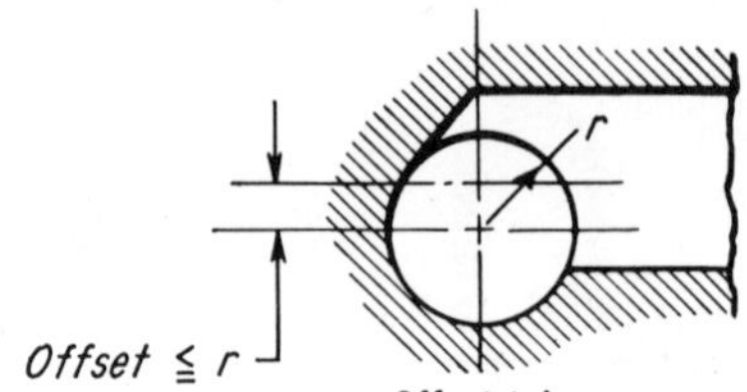

Offset tolerance

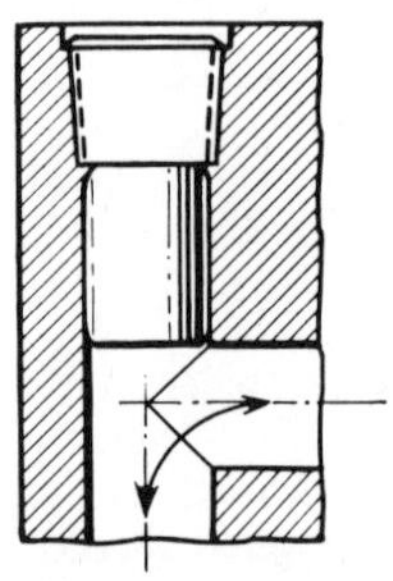

Plug to eliminate dead space

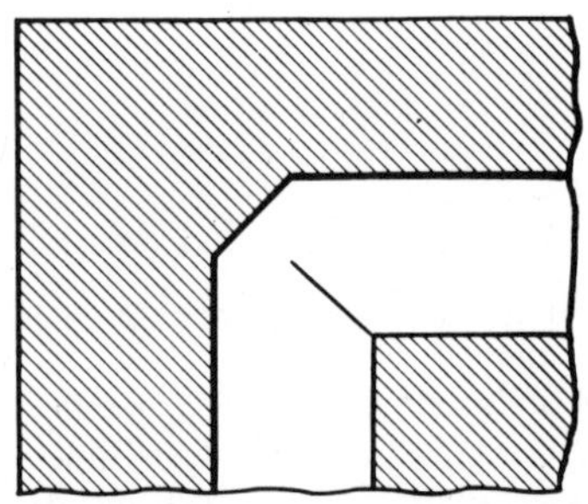

Right way to avoid air pocket

surface will not stay as flat when the holding clamps are released.

To seal across-the-joint ports, use O-rings or gaskets. O-rings can be placed in a simple counterbore or in a groove as shown in the sketches on page 218. The counterbore is less expensive and usually adequate. A groove is suggested for pressures higher than 1000 psi.

Outside diameter of counterbore or groove should be equal to the nominal diameter (OD) of the O-ring. This helps to keep the O-ring from expanding under pressure which would cause a pumping action during pressure rise and fall. A good O-ring technical bulletin will give detailed information on clearance and compression.

Port spacing depends on O-ring dimensions or on minimum gasket width as specified in the following section on sandwich plates. In any event, staggered arrangement is best for high port density.

Sandwich plates

The sandwich manifold, like the drilled-plate manifold, eliminates interconnecting piping because the connections are internal. The difference is that the sandwich type has milled or cast passages that can be made in any convenient shape, and more valves can be mounted in a given space.

Materials for the plates can be steel, fine-grained cast iron, or hard aluminum. The advantages of each are discussed under *Over-all Design,* the last section of this article. If steel is the material, the passages are milled. If cast iron is chosen, the passages can be cast and need no further processing other than sandblasting to clean them. Casting costs less too, if the manifolds are made in quantity.

Plates can be brazed, bolted, or bonded together to form one or more layers of grooved passages. The example on next page is an Almo Manifold and Tool Co patented design. Previous articles (see Editor's Note on next page) give other examples.

The following discussion is based on ordinary design practice and does not necessarily apply to the products of any particular companies.

Brazed: Plates can be copper-brazed under pressure and heated in an electric furnace. Capillary action causes the copper to flow to all contact surfaces, and the same action forces out the liquid flux. The excess copper will flow out with the flux through the open ports. A good joint must be flux-free, because trapped flux will block copper flow during brazing and create leakage paths.

Gasketed: For low pressure (up to 500 psi) vellumoid gaskets will work, For higher pressures a zinc-coated copper gasket 0.005 to 0.010 in. thick is better. Lands between adjacent ports should never be less than ¼ in. Plates must be bolted securely and the bolt spacing must be adequate to provide clamping force at all points despite the separating force of the pressurized fluid.

Metal-to-metal: If slight leakage is permissible the plates can be lapped together and used without a gasket, provided that bolts are spaced closely enough. To eliminate leakage, apply a thin coat of sealing compound before bolting.

O-ring: Bolted plates with O-ring-sealed ports are used occasionally. Design practices are outlined in the previous section on drilled plates.

Bonded: Epoxy resins, adhesives, and other bonding methods are used in some applications. A recent development of General Electric Co is metallurgical bonding. The sandwich plates are made of type 410 stainless steel and are diffusion-bonded by the vacuum hot-pressing technique. Such metallurgical bonding needs no flux or adhesive and therefore eliminates the problem of clogged passageways and flux inclusions.

Over-all design

Materials: Manifolds of all types are usually made of steel or fine grained cast iron, because these materials are readily machined and brazed, have good hardness and strength, have thermal coefficients of expansion similar to coefficients of the valves mounted on them, are resilient enough to withstand high bolting loads without distortion, and can be readily mounted on a machine.

A hard aluminum alloy is some-

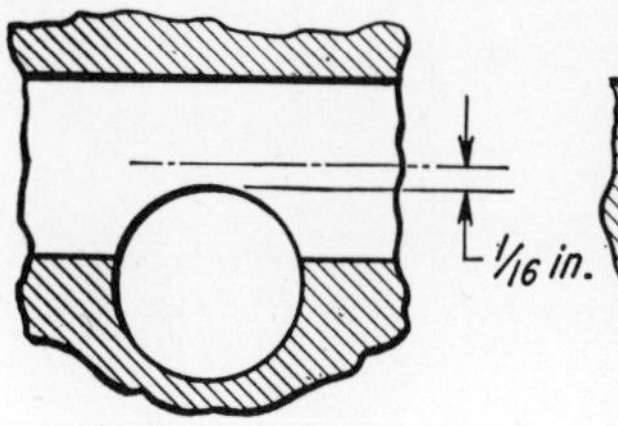

For breakthrough only

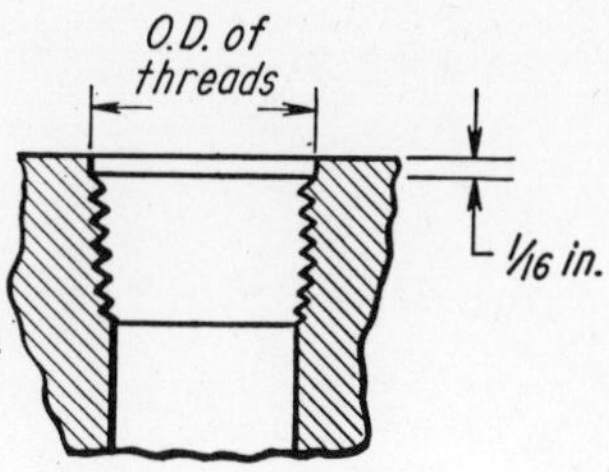

Thread relief

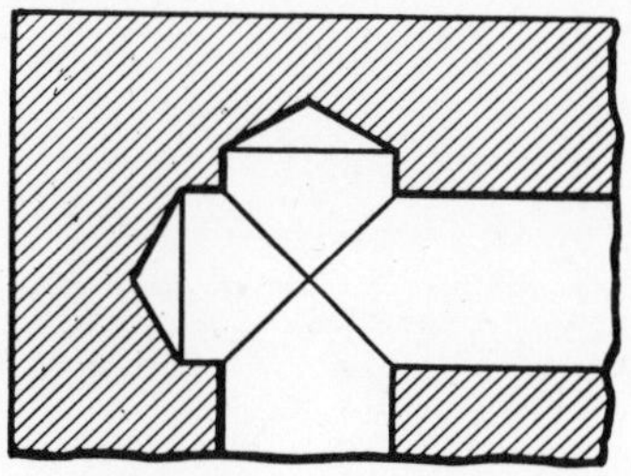

Wrong way (has air pocket)

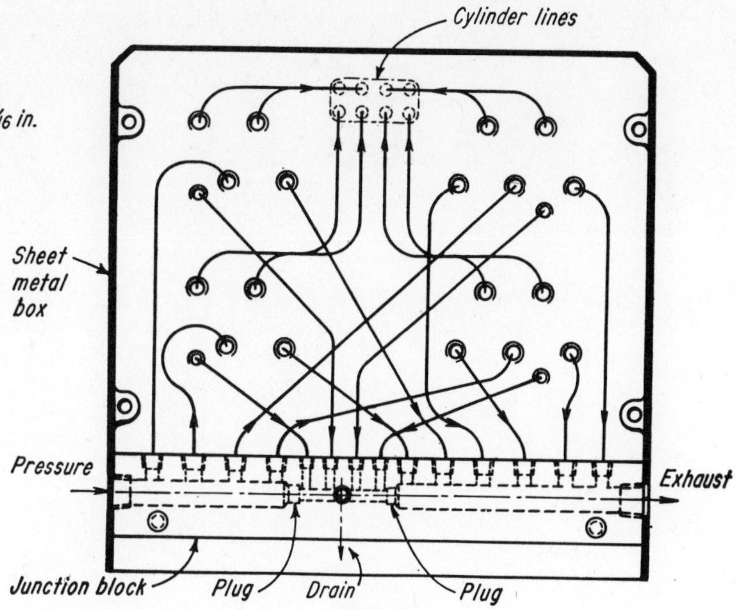

Rear view

times used if weight is a problem. However, aluminum has lower strength; it is more difficult to braze and more easily scratched; and it has a much higher coefficient of expansion than steel and iron. The Editor's Note describes one technique that has worked.

Valve arrangement: In spacing valves, provide ample room for electric lines and conduits and avoid interference with hydraulic lines. Locate the terminal box on top of the manifold for easy access. Remember, the electric hook-up is the last step in assembly before the unit becomes operative.

Double-solenoid valves are mounted in horizontal position. Spring-return valves are mounted in vertical position. Manifolds are frequently mounted on reservoirs as hydraulic-power-unit accessories. Use straight fittings instead of elbows for hydraulic lines leading to other units; it is more convenient for the fitter and saves space.

EDITOR'S NOTE: Modular hydraulics has always been of interest. These previous articles have followed its progress:

Sandwich manifolds, Feb 17 '58, p 80—Brazed aluminum plates with cast grooves save space and weight.

Today's film adhesives, Dec 11 '61, p 86—Bonded sandwich panels for hydraulic circuits use nitrile phenolic film to seal manifold passages and mount parts. —*Franklin D. Yeaple*

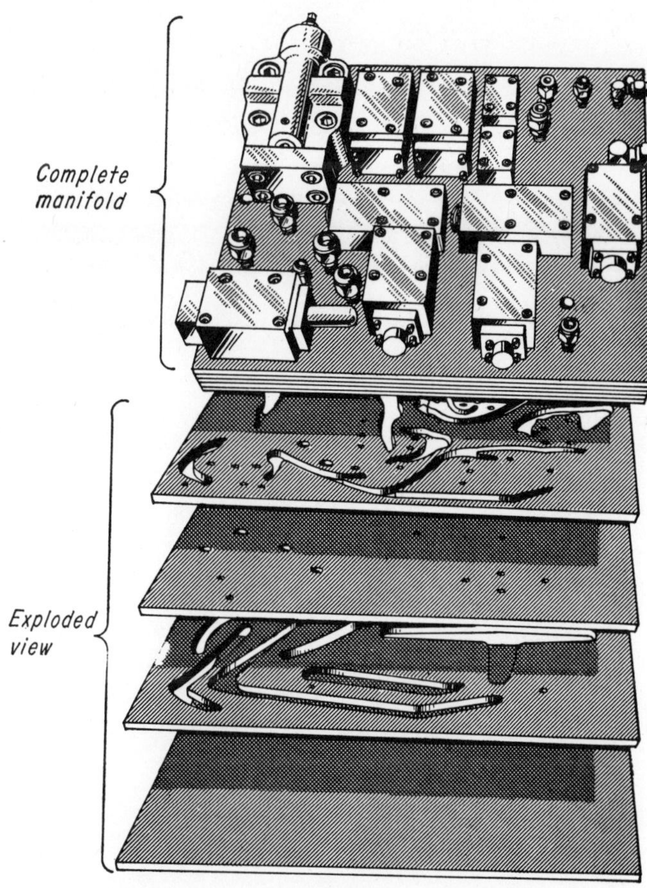

Sandwich plate grooves

8

CYLINDERS AND ACTUATORS

Stroke, strength, and performance of oil-operated and air-operated cylinders

POWER CYLINDER MECHANISMS—

full stroke

KURT HAIN *Institute For Fundamental Research, Brunswick, Germany*
Translated by Prof. F. R. E. Crossley, Yale University

Choosing the right proportions for mechanisms requires balancing opposing factors. This discussion and the charts tell how.

For efficient transmission of forces from power-driven cylinders, a mechanism must be designed so the moment of the output load on the piston is as constant as possible. But a number of forces may act on the mechanism's links, independent of the normal output load. These forces must be analyzed and evaluated in order to determine the proper proportions of each link. When more than one output load is connected to the mechanism, each one must be considered independently. To minimize peak loads on the piston, the mechanism must be designed so that its variable velocity ratio will give minimum side thrusts. Where a number of linkages can fulfill the requirements, two parameters are particularly useful in evaluating the choice: the transmission angle μ, and the side-thrust angle σ (see Fig 1).

Comparing the two cylinders in Fig 1 shows that the best transmission angle μ is 90 deg; the worst, either zero or 180 deg. Fig 1(B) shows that the best possible side-thrust angle σ is zero, and that any increase in this angle creates a transverse-force component which results in friction losses in the power-driven cylinder. The most advantageous combination of transmission and side-thrust, angles will be between these extremes.

Two sets of curves, Fig 2 and Fig 3, show how to pick the right combination of angles. The curves are more easily understood by considering typical examples.

Angle of swing without intermediate positions

In Fig 1, cylinder d with a given stroke is required to give an output swing of 120 deg. Two possible linkages are shown. Linkage 1(A) gives advantageous transmission angles of μ_{min} and μ_{max}, but the side-thrust angle is too large. Linkage 1(B) represents an improvement; transmission angles were selected from Fig 2, from which the side-thrust angle and linkage proportions were also determined. From curves in Fig 2, angle of swing $\phi = 120$ deg gives $\mu_{min} = 20$ deg and $\sigma_{max} = 10$ deg. Intersection of these two curves, gives the ordinate ratio $b/s = 0.83$, which determines the length of the connecting rod b for a given cylinder stroke s. The ordinate for angle $\phi = 120$ deg intersects the curve a/s to give 0.57, and the curve e/s to give 0.42. Thus, the length of connecting rod a and the offset e are readily determined for a given cylinder stroke.

Even if the force produced at the piston in the linkage were constant, it would still be impossible to make an exact prediction concerning the variable moment produced to turn the driven connecting rod a. We can, however, pose this problem: For a 120-deg output angle and a constant piston force, let the turning moment of the output lever be as nearly constant as possible. Because the mechanism has an inherently variable mechanical advantage (velocity ratio), it will never be posssible to produce a moment at the output lever which is mathematically constant. But the best linkage will be the one that produces a turning moment with an acceptably small variation from a desired constant.

For this selection the curves in Fig 3 are used. Each curve δ in the series indicates the percentage above or below the

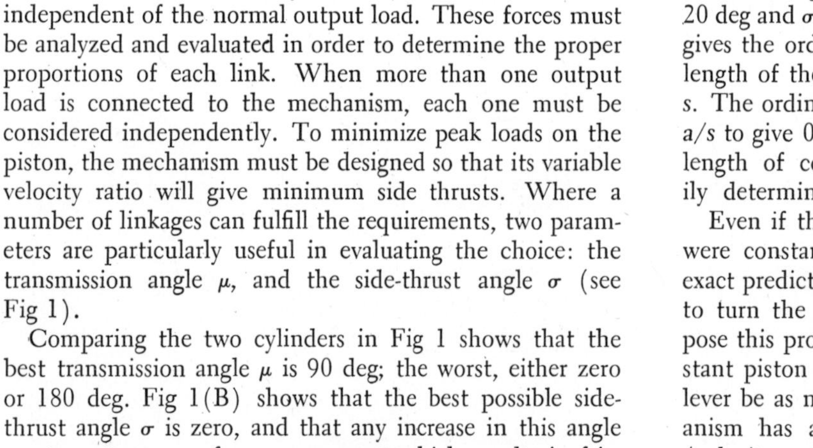

1. **THE OPPOSING PARAMETERS** must be balanced to make the most efficient use of levers with power-driven cylinders. Best combination would be high transmission angle μ, and low side-thrust angle σ; but this is not possible. In (A) both of these angles are large; in (B) both are small. Best result lies somewhere in between; charts on facing page show how to compromise.

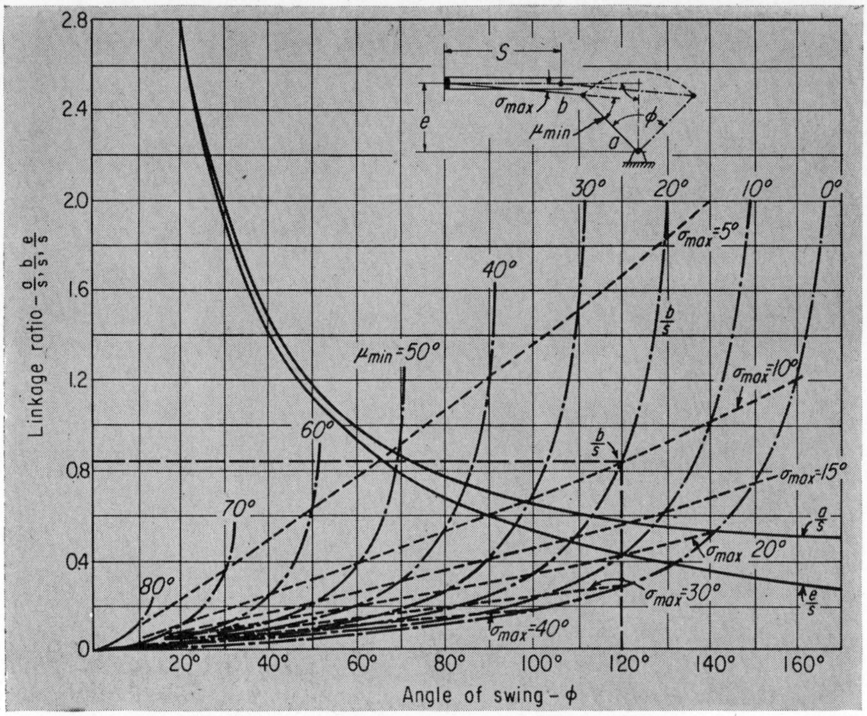

2. LINKAGE RATIOS for the various output angles of swing can be determined from this graph for various transmission and side-thrust angles.

constant moment desired for various values of ϕ and μ_{min}. For any given angle of swing ϕ any number of transmission angles μ_{min} are possible, but at the same time we also want to get the minimum deviation from a constant output turning moment. For example, if we allow a maximum deviation $\delta = \pm 5\%$ from the desired constant turning moment, then for an output swing of 120 deg, the worst transmission angle $\mu_{min} = 31.5$ deg. On the other hand, if the transmission angle is greater, the deviation δ must be allowed to increase. This accents the opposing characteristic of these two parameters and requires a suitable compromise, which will give a maximum transmission angle for a minimum deviation δ. The curves in Fig 3 show what choices are possible in this respect.

To overcome this limitation on angle ϕ, the arrangement in Fig 4(A) shows the cylinder pivoted at C_o and the piston rod attached to connecting rod a_1. The maximum practical angle of swing ϕ is 100 deg, which gives a corresponding transmission angle μ_{1min} of 40 deg. However, this angle of swing can be used to produce a much greater output angle of swing ψ by compounding a second four-bar linkage with links a, b, c and d. The output lever c swings about the fixed journal B_o, and the secondary transmission angle μ_{2min}, can be made equal to μ_{1min}, the angle between the coupler b and the lever c. This results in an output angle of 220 deg for an input angle of 100 deg.

Fig 4(B) shows how this four-bar linkage can be applied to a mobile crane. The load arm is integral with the output lever c of the mechanism, and its large output angle makes it possible to handle loads in front as well as behind the tractor.

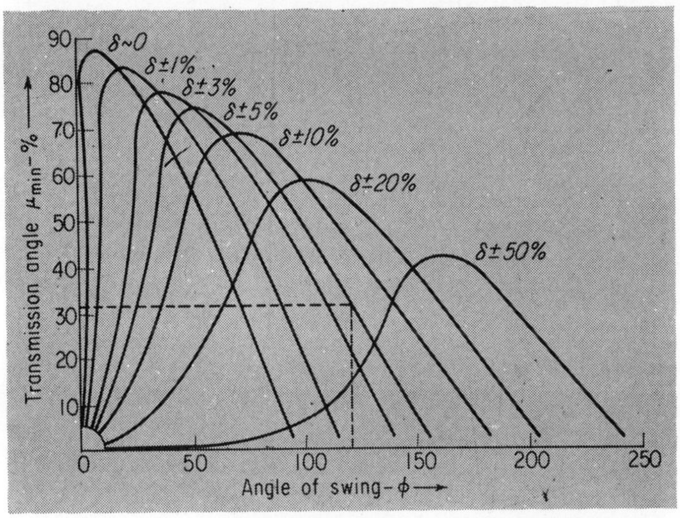

3. THE TRANSMISSION ANGLE determined from previous graph can be checked with this graph for minimum deviation when a constant turning moment is desired.

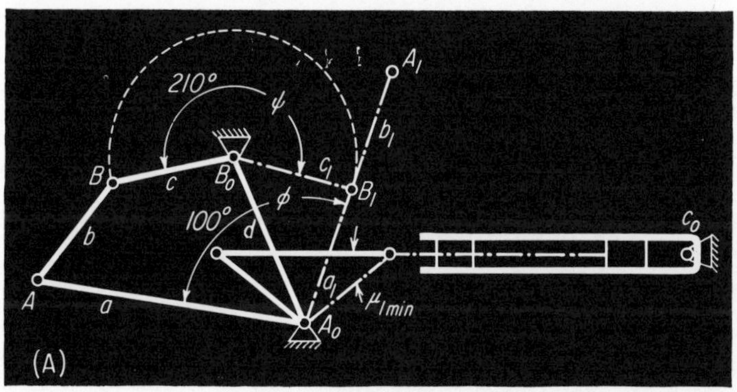

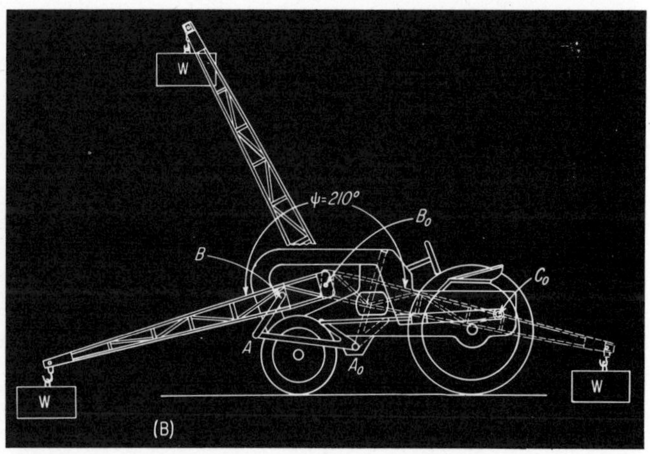

4. FOUR-BAR LINKAGES as developed in (*A*) can increase the range of output angle from 100 to 210 deg with a favorable mechanical advantage of the power transmission. This linkage is incorporated in the mobile crane (*B*).

POWER CYLINDERS—

intermediate positions

KURT HAIN, *Institute for Fundamental Research, Brunswick, Germany*
Translated by F R Erskine Crossley, Yale University

Graphical methods solve the coordination of linear motion of piston with angular motion of mechanism.

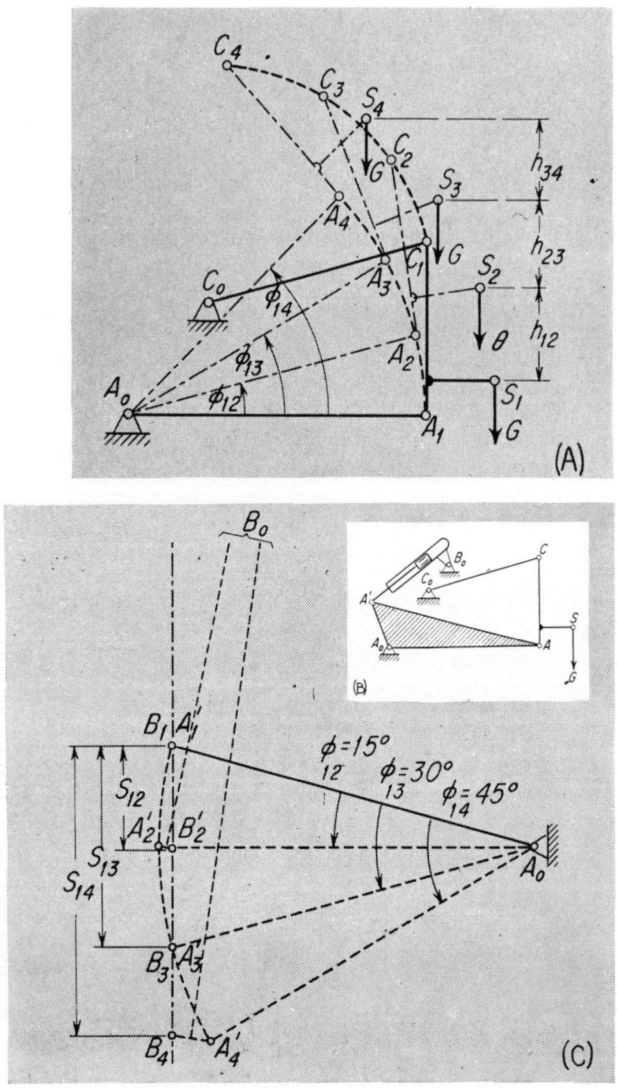

In power-cylinder mechanisms it is sometimes desirable to divide the output angle into several equal portions which are predetermined and controlled by partial stroke lengths of the cylinder piston. The problem here is to coordinate the linear motion of the piston with the angular motion of the mechanism. Consider, for example, the mechanism in Fig 1(A) which is designed to lift a weight G of 200 lb through an angle $\phi_{14} = 45$. Dividing this angle into 3 equal parts gives $\phi_{12} = 15$, $\phi_{13} = 30$, $\phi_{14} = 45$. Assume the corresponding lifts to be $h_{12} = 11$ in., $h_{23} = 10\frac{1}{4}$ in., and $h_{34} = 9\frac{1}{4}$ in. Amounts of work required are:

$$A_{12} = Gh_{12} = 2200 \text{ lb-in.}$$
$$A_{23} = Gh_{23} = 2050 \text{ lb-in.}$$
$$A_{34} = Gh_{34} = 1850 \text{ lb-in.}$$

If the piston stroke is assumed to be $s = 10$ in., partial piston displacements for intermediate positions are:

$$s_{12} = s \cdot \frac{A_{12}}{A_{12} + A_{23} + A_{34}} = 3.44 \text{ in.}$$

$$s_{23} = s \cdot \frac{A_{23}}{A_{12} + A_{23} + A_{34}} = 3.34 \text{ in.}$$

$$s_{34} = s \cdot \frac{A_{34}}{A_{12} + A_{23} + A_{34}} = 3.03 \text{ in.}$$

Pivoted cylinders

One choice of mechanism for lifting this weight G with center of gravity S is shown in 1(B) with a side-pivoted cylinder. While this type of arrangement permits a compact design, the offset pivot creates a side thrust which requires cylinder guides capable of withstanding the turning moment. A compound linkage, however, offers more efficient transmission of the piston force.

Fig 1(C) shows how a compound linkage can be laid

1. LAYOUT OF A LIFTING LINKAGE (A) for predetermined positions. (B) Pivoted power-driven cylinder can be applied to lifting linkage to give a compact design as shown here, but at expense of high side thrust. (C) This layout coordinates intermediate positions of piston with corresponding positions of output swing determined in (A).

From the text for the Refresher Course in Mechanism 1959/1960, arranged and sponsored by the Chamber of Commerce & Industry of the City of Brunswick, Germany, in cooperation with the Division of The Science of Mechanism of the Society of German Engineers (VDI). Course director: Kurt Hain.

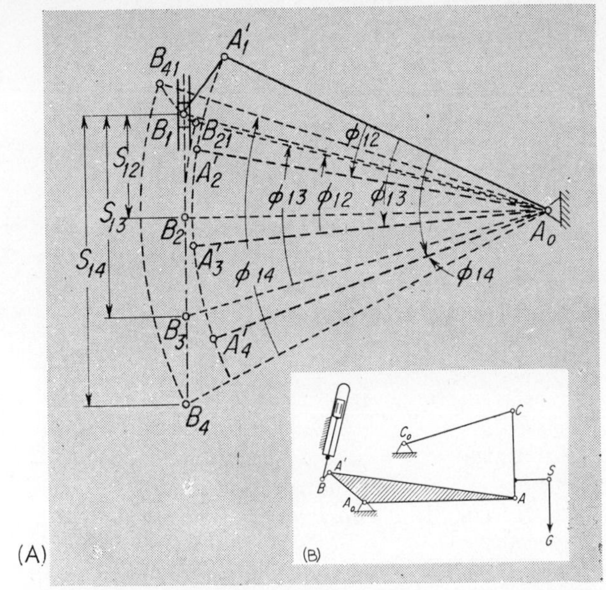

2. FOR FIXED CYLINDERS (A) coordination of intermediate piston and output-arc positions is determined by the point-position-reduction method. This layout gives one solution, and its application is shown in (B).

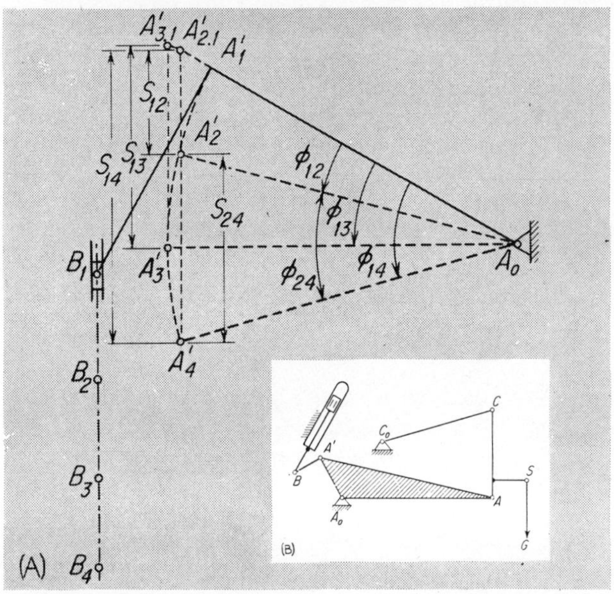

3. ANOTHER SOLUTION (A) by the point-position-reduction method may be desirable for some applications. Resulting linkage is shown in (B).

out for a pivoted-cylinder arrangement with 3 partial strokes, s_{12}, s_{13}, and s_{14}, and the corresponding angular displacements $\phi_{12} = 15$, $\phi_{13} = 30$ and $\phi_{14} = 45$ already determined. The distance s_{13} is chosen as the base of an isosceles triangle, in which the angle of swing ϕ_{13} is the apex angle. This establishes point A_o and also the length of arm $A_oA'_1 = A_oA'_3$, then from angles ϕ_{12} and ϕ_{14} the points A'_2 and A'_4 can be found. Now if the perpendicular bisectors of A'_2B_2 and A'_4B_4 are drawn, their point of intersection will locate the point B_o, which is the pivot point for mounting the power cylinder. After this, the given dimensions of the four-bar linkage such as determined in Fig 1(A) can be laid out. Dimensions of the particular power cylinder chosen will determine the distance AA', which can be usually adjusted to suit at assembly.

Fixed cylinders

For fixed cylinders there are two fundamentally different solutions as shown in Fig 2 and 3. In Fig 2(A) the three given partial strokes s_{12}, s_{13}, and s_{14} are shown together with the corresponding angular displacements $\phi_{12} = 15$, $\phi_{13} = 30$, and $\phi_{14} = 45$. The distance s_{13} is used as the base line of the isosceles triangle $B_1A_oB_3$, with ϕ_{13} as the apex angle. With A_o as a turning point, line A_oB_2 is swung through an angle ϕ_{12}, and the line A_oB_4 through an angle ϕ_{14}, in the opposite sense of direction to that in which the link A_oA' turns from its station 1. These rotations move point B_2 around to $B_{2.1}$ and B_4 to $B_{4.1}$. If point B_3 were similarly turned back through the angle ϕ_{13}, B_3 would be found to coincide with point B_1. This coincidence is a result of the construction with the isosceles triangle, and is an example of the so-called point-position-reduction method. The center of the circle, which can be drawn through the three points B_1,$B_{2.1}$, and $B_{4.1}$ is the point A', which determines the crank length $A_oA'_1$. A circular arc is next described through A'_1 with center at A_o, and it is cut with arcs with centers at B_2, B_3, and B_4 and radii equal to the length $B_1A'_1$. This establishes points A'_2, A'_3, and A'_4, which form the equal angles, $A'_1A_oA'_2 = A'_2A_oA'_3 - A'_3A_oA'_4$, and completes the desired coordination of the partial piston strokes with the crank angles ϕ.

Fig 2(B) shows how the four-bar linkage thus developed can be coupled to the lifting linkage A_oACC_o and the fixed cylinder. Once again, the whole four-bar linkage can be swiveled around point A_o to provide any adjustments that may be required by the particular power cylinder that is chosen. When the relative positions are settled, the distance AA' will be established.

Fig 3(A) shows the second method for developing the design of a four-bar linkage actuated by a piston in a fixed cylinder. Again, the isosceles triangle is constructed with partial stroke $s_{24} = s_{14} - s_{12} = 8 - 3 = 5$ in., used as the base in this example. This corresponds to the crank angle $\phi_{24} = 30°$. With center at this apex point A_o, an arc is drawn to pass through A'_2 and A'_4, and with the angles ϕ_{12} and ϕ_{13}, points A'_1 and A'_3 are located. In the direction of the baseline $A'_2A'_4$ and parallel to it, we now move the point A'_2 through a distance s_{12} to $A'_{2.1}$, and the point A'_3 through a distance s_{13} to $A'_{3.1}$; these displacements must be in opposite sense of direction to which piston wrist pin moves when passing through corresponding stations.

Similarly, point A'_4 can be shifted through the distance s_{14}, and it would coincide with point $A'_{2.1}$; this is another example of the so-called point-position-reduction method. The circle which passes through the three points A'_1, $A'_{2.1}$ and $A'_{3.1}$ is drawn, and its center is the point B_1 of the linkage. The line of action of the piston is the line through B_1 parallel to the base line $A'_2A'_4$ of the isosceles triangle. The succeeding three points B_2, B_3 and B_4 are located by striking off arc from the centers A'_2, A'_3 and A'_4, all having a radius equal to $B_1A'_1$. This means that:

$$B_1B_2 = s_{12},\ B_2B_3 = s_{23},\ \text{and}\ B_3B_4 = s_{34}.$$

In Fig 3(B) the four-bar linkage thus developed is shown coupled to linkage A_oACC_o, and again linkage can be adjusted relative to point A_o to establish point A^1.

CHARTS give SAFE PISTON-ROD SIZE

for strokes of 10 to 300 in. in 5 different mounting styles

Z. J. LANSKY, *Chief Engineer, Hannifin Company, Div of Parker-Hannifin Corporation, Des Plaines, Ill.*

HOW THE CHARTS WORK

Suppose you want to exert 25,000-lb thrust over an 80-in. stroke, and the cylinder has a fixed-style mounting with a pinned rod-end.

On chart 2, appropriate for those mounting conditions, 25,000-lb thrust and 80-in. stroke intersect at a point between the 2- and 2½-in. rod-diameter lines. These lines actually represent maximum safe conditions at which rods will not buckle. Therefore, **always choose rod size to right of thrust-stroke intersecton**—in this case 2½-in. dia.

Stop-tube must be used if intersection lies within stop-tube length-bands. In this example a 1-in. stop-tube will be necessary—therefore a cylinder with an 81-in. stroke will be limited to 80-in. stroke.

• For a cylinder pressure of 2000 psi, piston dia works out in this example to be 4 in. **Always reduce pressure if a larger-size cylinder is used without increasing piston-rod dia.**

• Charts are based on 100,000-psi yield-point for rod material, and on standard rod-extensions that are general throughout most of the square-type, power-cylinder industry. For longer extensions, increase the stroke accordingly.

Fixed-cylinder charts

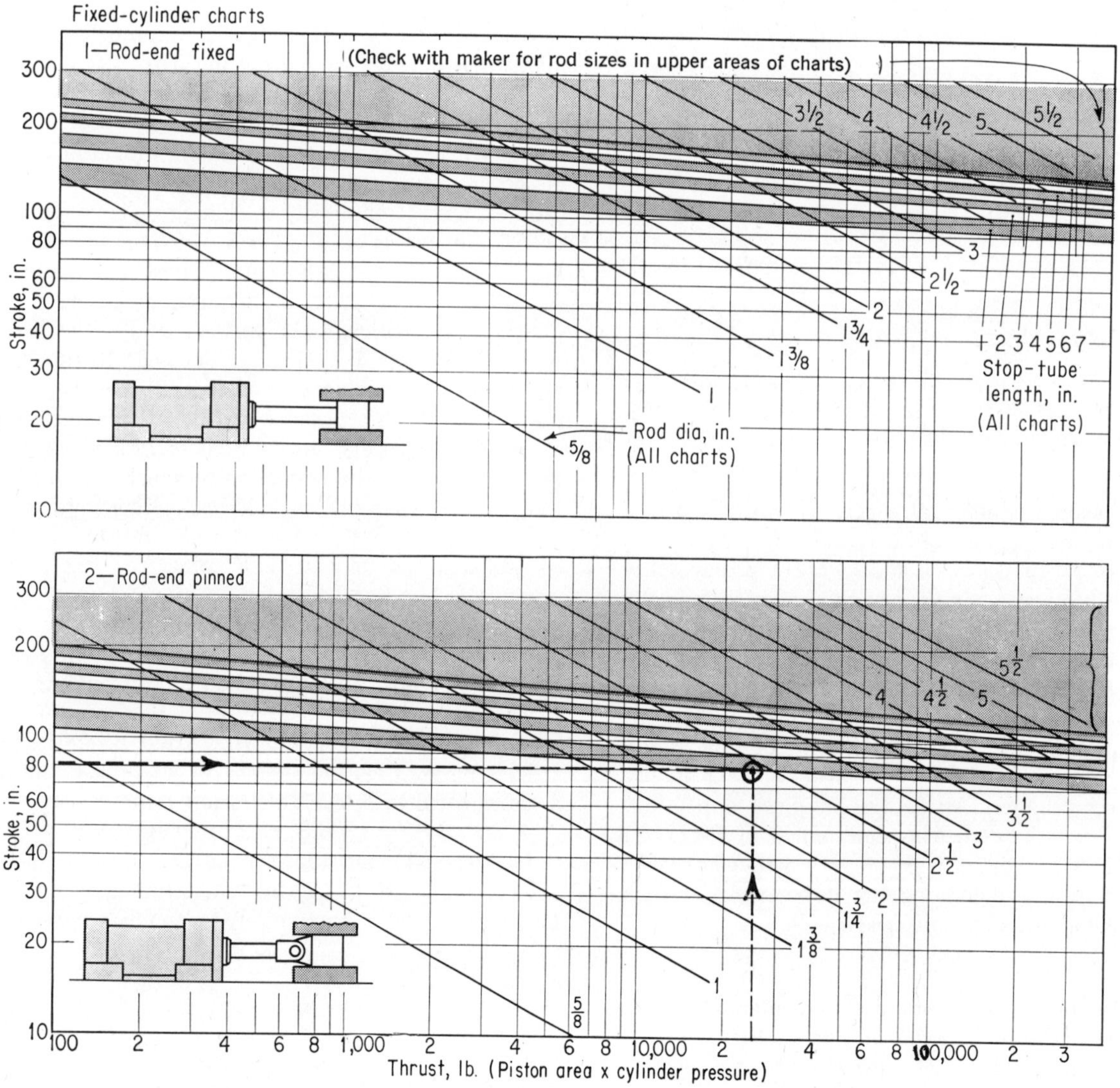

CHARTS give SAFE PISTON-ROD SIZE (Cont.)

Pivoted-cylinder charts

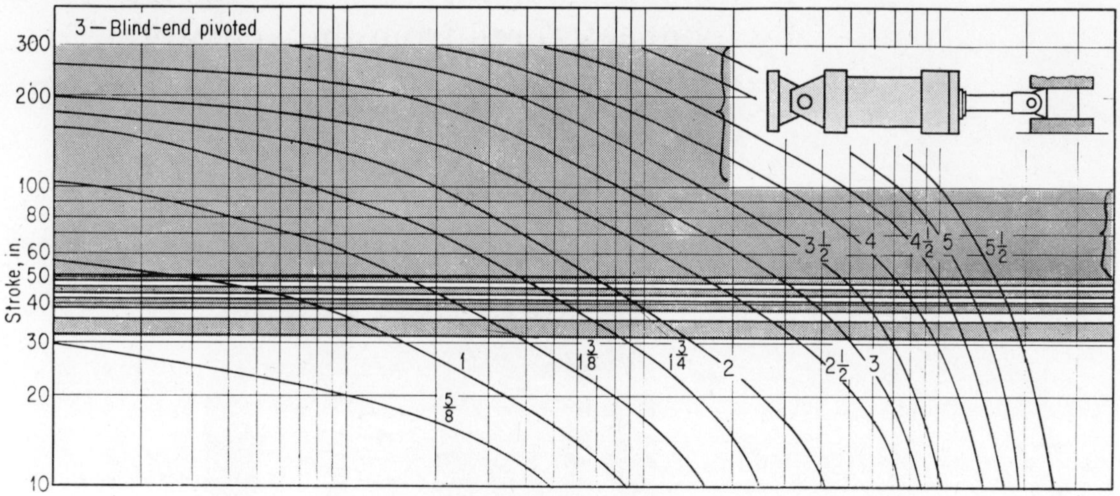

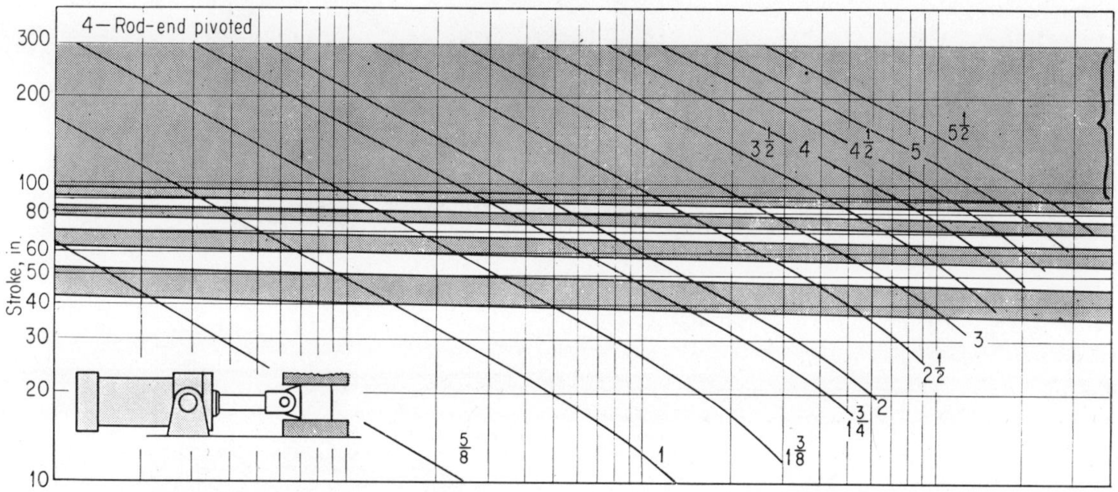

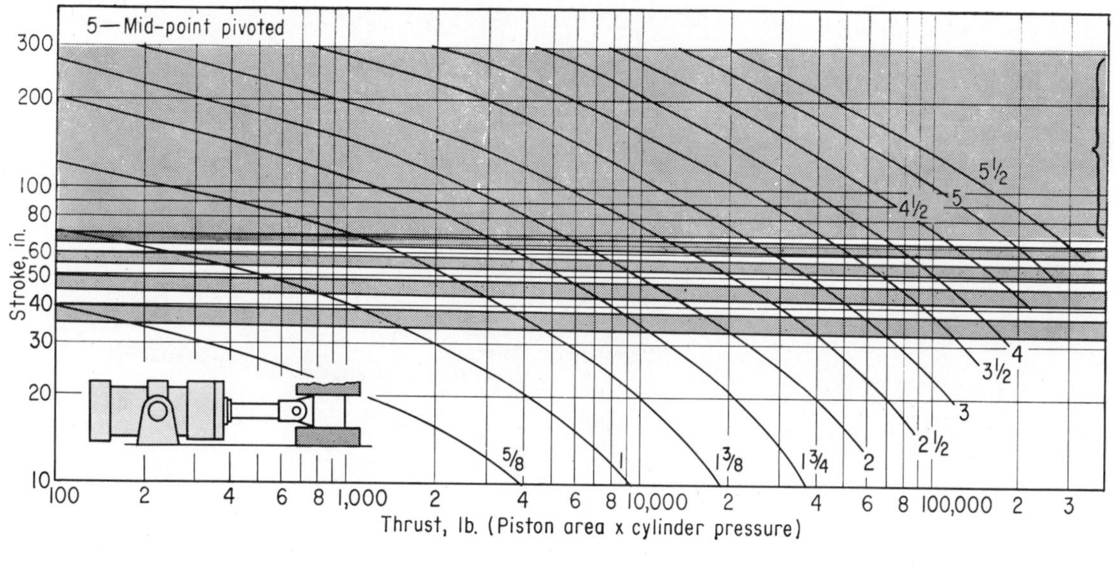

6 fluid-power force

When the actuator in these amplifiers is in position, input force is balanced by feedback force from the actuator.

GERALD BLOOM, Project Engineer, Simmonds Precision Products Inc, Tarrytown, NY

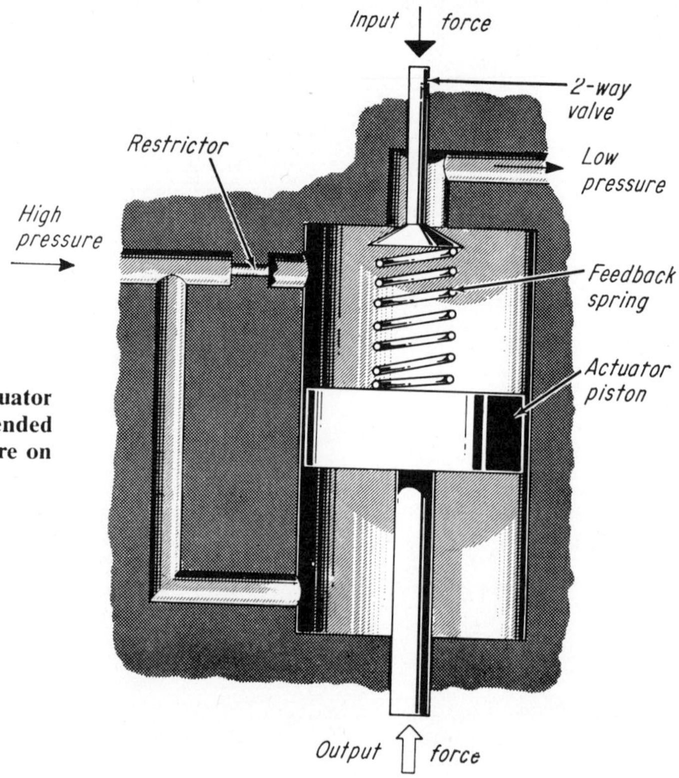

1 **IN SIMPLEST AMPLIFIER** the actuator is retracted by pressure force, extended by feedback spring-force plus pressure on the piston rod area.

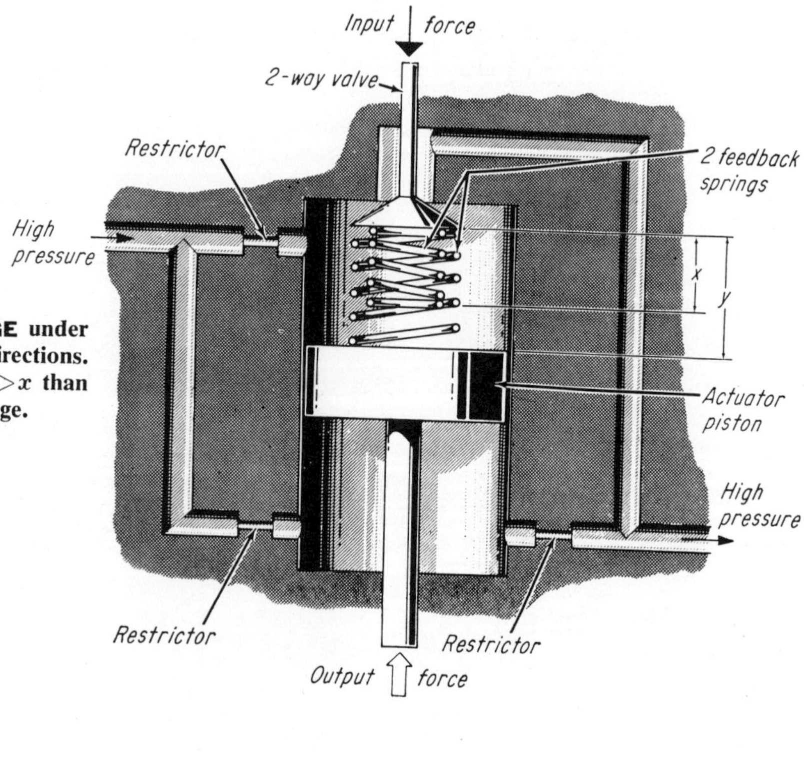

2 **RESTRICTED FLOW PASSAGE** under the piston provides force in two directions. Actuator moves farther when $y>x$ than when $y=x$ for a given input change.

amplifiers

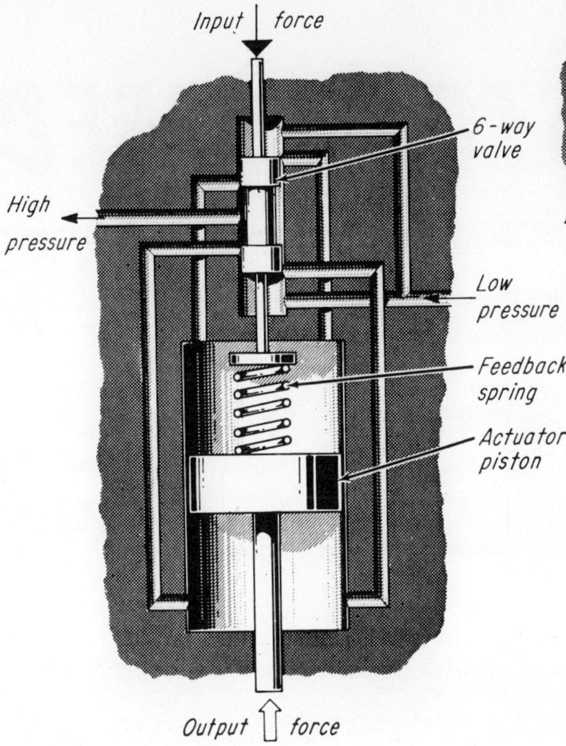

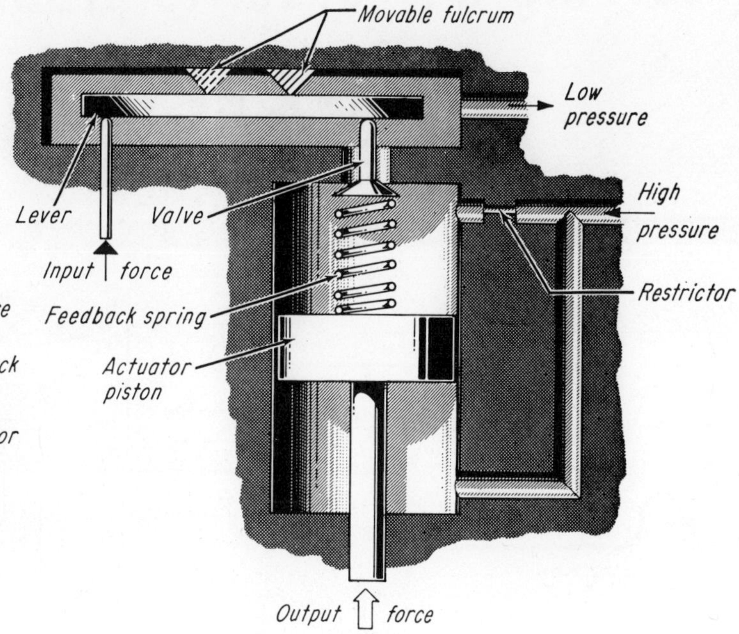

3 **MULTI-PORT VALVE gives full pressure actuation in both directions; therefore it provides high force and fast response for a given pressure level.**

4 **LEVER-ACTUATED VALVE may be advantageous in some applications. Here input force can be adjusted by changing the position of the fulcrum.**

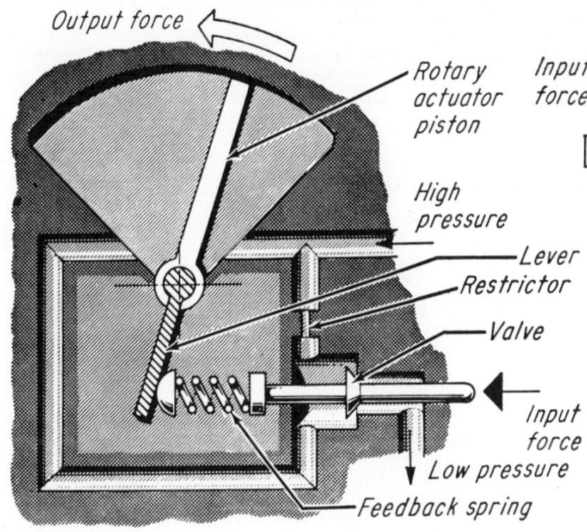

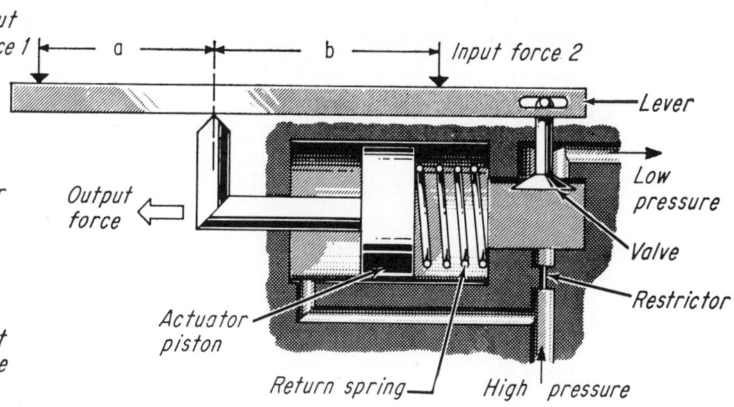

5 **ROTARY OUTPUT for large or multiple rotations, including those greater than 360 deg, uses fluid motor as actuator, with feedback through a torsion spring.**

6 **FULCRUM ON ACTUATOR moves until input forces balance. This amplifier, therefore, is useful for dividing forces so that Input 1 $\times$ *a* = Input 2 $\times$ *b*.**

solving a hydraulics problem . . .

WHEN PISTONS NEED

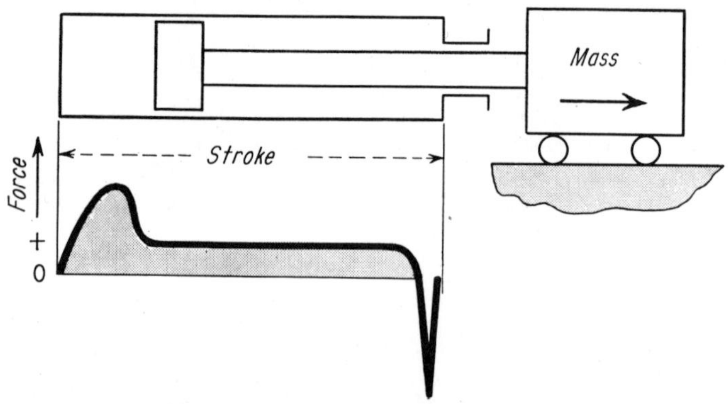

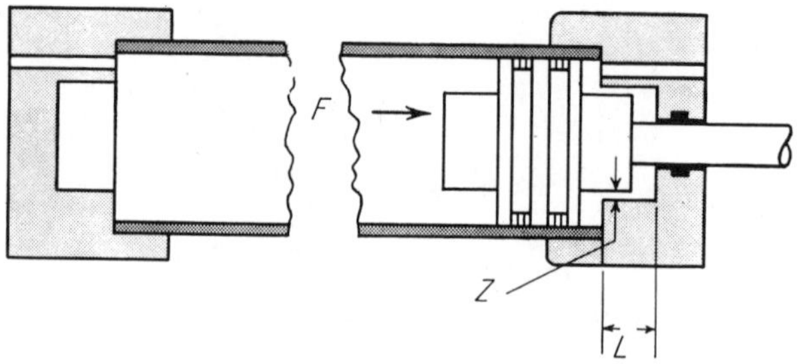

1 Undamped cylinder . . . has high forces to accelerate mass, at start of stroke, then reduced force to hold constant velocity, and finally experiences shock forces at end during reversal of mass.

2 Annular clearance . . . in cavity of cylinder cap serves as flow restriction for this dashpot. Oil then escapes at outlet port.

It's necessary to take the bang out of end-of-stroke forces that damage hydraulic cylinders and other components. The author describes braking and reversing methods that allow a variety of displacement-time responses.

LOUIS DODGE

Cincinnati, Ohio

SYMBOLS

a	acceleration and deceleration, ft/sec²
M	mass, lb-sec²/ft
F	mean dashpot resistance, lb
H	hydraulic head, ft
P	mean dashpot pressure, psi
P_B	back pressure, psi
ΔP	differential pressure, psi
L	dashpot depth, ft
R	radius of dashpot, in.
Z	annular clearance, in.
V	velocity, ft/sec
Q_G	discharge, gpm
Q'	discharge, ft³/sec
Q''	discharge, in³/sec
B	capacity of dashpot, in.³
S	specific gravity
C_D	coefficient of discharge
T	time, sec
A'	area, ft²
A''	area, in.²
G	32 ft/sec²

BRAKES

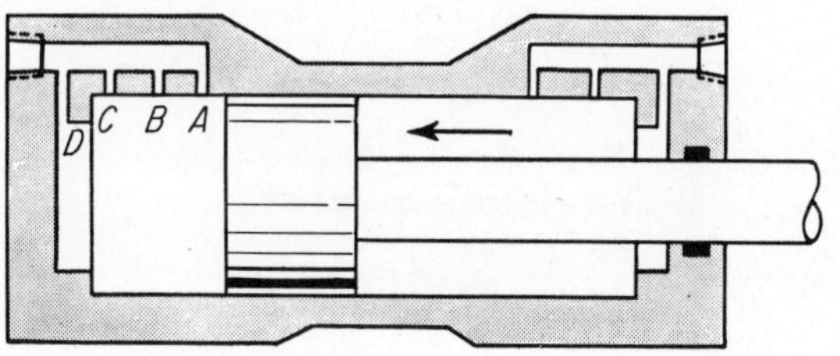

3 Communicating orifices . . .
provide increasing resistance to oil flow as dashpot piston moves to left and covers holes. On return stroke, induced flow is restricted less and less.

Pistons, like people, are hard to stop once they get going. With hydraulic pistons, however, the control need not be difficult. Some of the same oil-flow principles that push the piston can also be put to work as a brake.

Result is less shock caused by end-of-stroke dynamic forces—the shock that creates problems of noise, wear and high stresses. When created by a suddenly applied load, these forces give a deflection which may be twice as much as when the same load is applied slowly. But that slowness can itself be a drawback. The answer is skillful compromise—decelerate with a controlled force strong enough to reverse the stroke swiftly, but not so fast that it shocks the operating components.

Hydraulic damping and shockless reversal can be had with flow-restriction or pressure-reducing devices. These include servo-controlled variable pumps, flow-control valves, cylinder modifiers (orifices, special pistons), and other power cutouts or reducers.

ANNULAR-CLEARANCE METHOD

A familiar component for this work is the hydraulic shock absorber (dashpot). It limits axial piston velocity where and when desired by trapping oil ahead of the piston, then releasing it through a restriction, slowly and with predetermined control.

Here is an example of how to calculate an annular-clearance restriction, Fig. 2. Assume that at the start of dash-pot action, inertia forces alone are dissipated through the ejection of dashpot oil. Hence, kinetic energy of moving parts equals work done during penetration of the dashpot

$$\frac{1}{2}MV^2 = FL$$

The acceleration (and deceleration) is

$$a = \frac{F}{M}$$

Assuming a parabolic velocity-time curve, then

$$V = aT/2$$

Mean dashpot pressure is

$$P = \frac{F}{\pi R^2}$$

Differential pressure is

$$\Delta P = P - P_B$$

Standard equation of flow issuing from an orifice is

$$Q' = C_D A' \sqrt{2GH}$$

Assume that annular clearance of dashpot is equivalent to an orifice, and then convert to conventional units

$$Q_G = 238\ RZC_D \sqrt{\frac{\Delta P}{S}}$$

An assumed value of the coefficient C_D may be checked against the Reynolds number of the calculated flow and adjusted if it deviates too much from what experience has shown to be reasonable. For Reynolds numbers below 100, C_D may vary from 0.1 to 0.7.

The time required to eject the dashpot oil is

$$T = 0.26\frac{B}{Q_G}$$

Suppose that a piston velocity V of 1.1 ft/sec is selected, and that the following data are known: $M = 64$ lb-sec²/ft; $L = 4$ in.; $R = 1$ in.; $B = 12.5$ in.³; $S = 0.85$; $C_D = 0.62$, and $\Delta P = 30$ psi. Then,

$$F = \frac{MV^2}{2L} = \frac{(64)(1.1)^2(12)}{(2)(4)} = 116 \text{ lb}$$

$$a = \frac{F}{M} = \frac{116}{64}\ 1.81 \text{ ft/sec}^2$$

$$T = \frac{2V}{a} = \frac{(2)(1.1)}{(1.81)} = 1.22 \text{ sec}$$

$$Q_G = 0.26\frac{B}{T} = 0.26\frac{12.5}{1.22} = 2.66 \text{ gpm}$$

$$Z = \frac{Q_G\sqrt{S}}{238RC_D\sqrt{\Delta P}} = \frac{(2.66)(0.92)}{(238)(1)(0.62)(5.5)} = 0.003 \text{ in.}$$

ORIFICE METHOD

Orifices in the cylinder wall, rather than annular clearance, can serve as the restriction, Fig. 3. Flow resistance increases in steps as the piston covers each hole, and

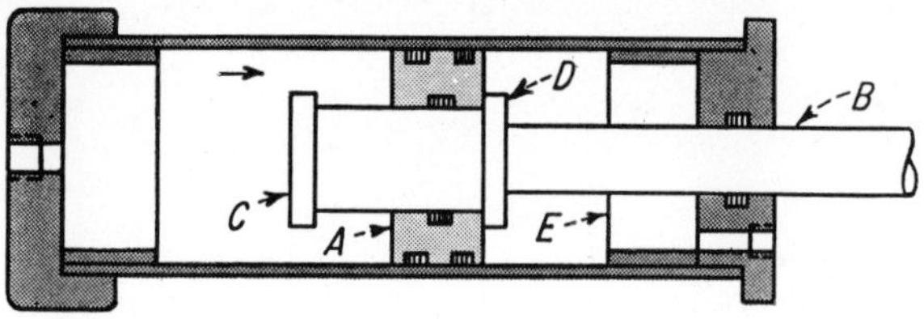

4 Open binary . . .
dashpot reduces force of oil pressure against piston at end of stroke because outer piston *A* hits stop *E* and only area of C is effective.

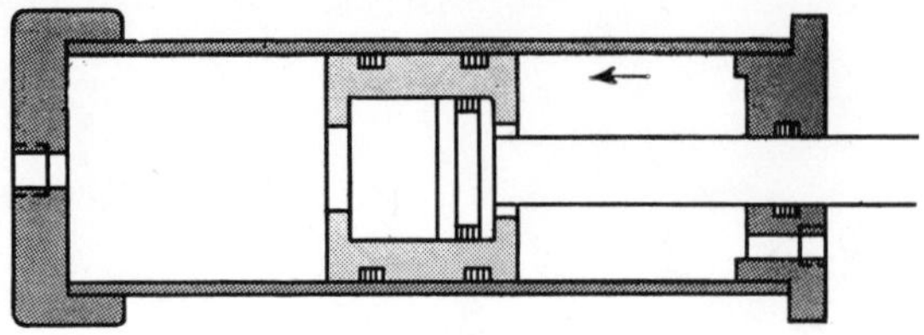

5 Caged binary . . .
dashpot works same as open binary: Outer piston reaches end of travel first, and inner piston completes travel at reduced force.

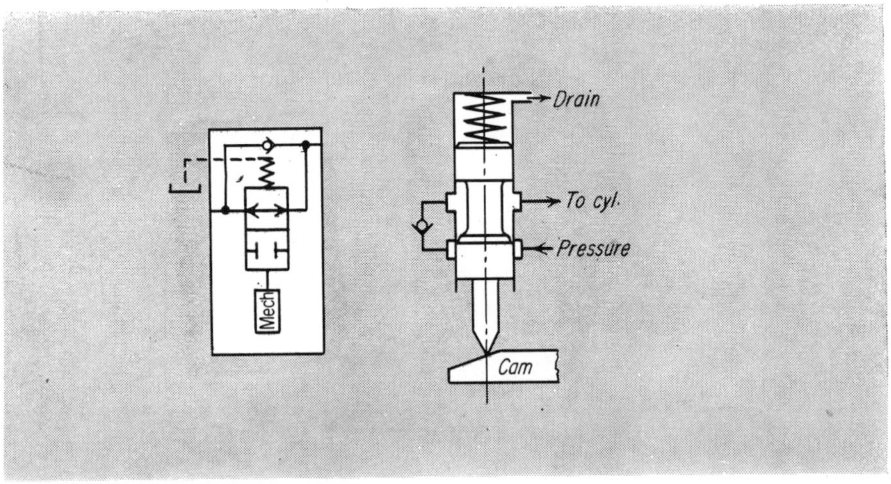

6 Deceleration valve . . .
throttles oil flow to cylinder at will of cam, making possible a variety of velocity-time responses. Cam may be operated by actual cylinder travel.

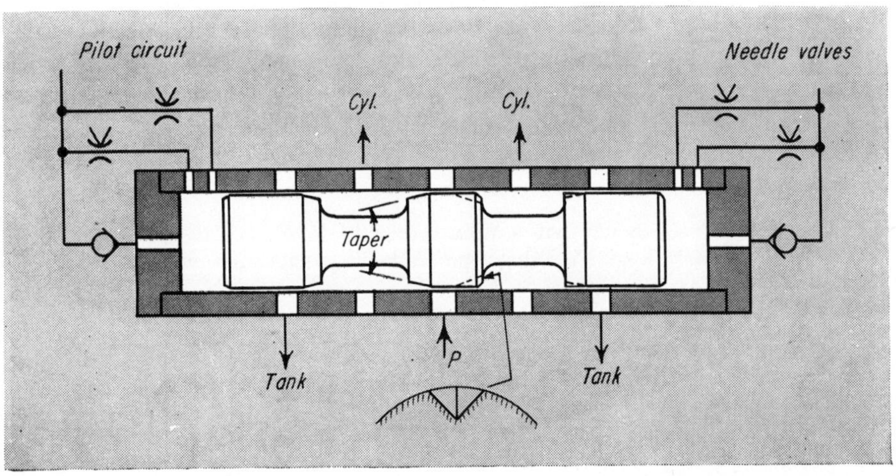

7 Pilot-operated . . .
flow-control valve uses carefully tapered anc notched piston to provide accurate restriction for both loading and unloading working cylinder. Also, pilot flow is separately throttled for finer control.

8 Reversing valve . . .
is spring-centered to stop flow in both directions if pilot pressures are zero or equal. Slowness of movement to ends depends on difference in pilot pressure and design of springs.

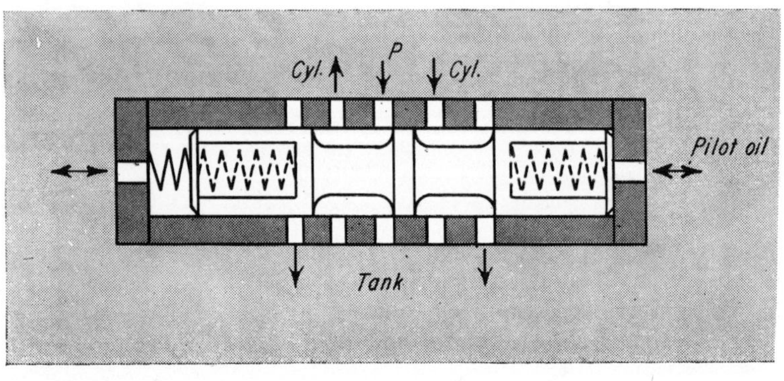

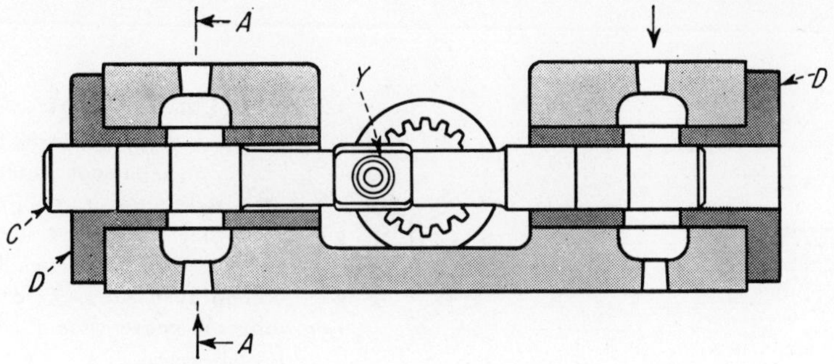

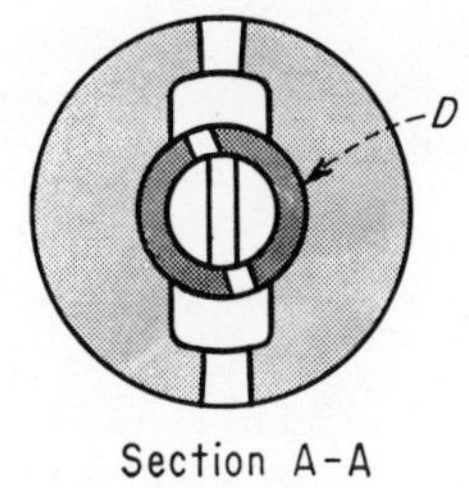

9 Sinusoidal flow-control . . . system uses gear train and hydraulic valve to tie in cylinder piston position with oil pressure, matching sine and cosine curves.

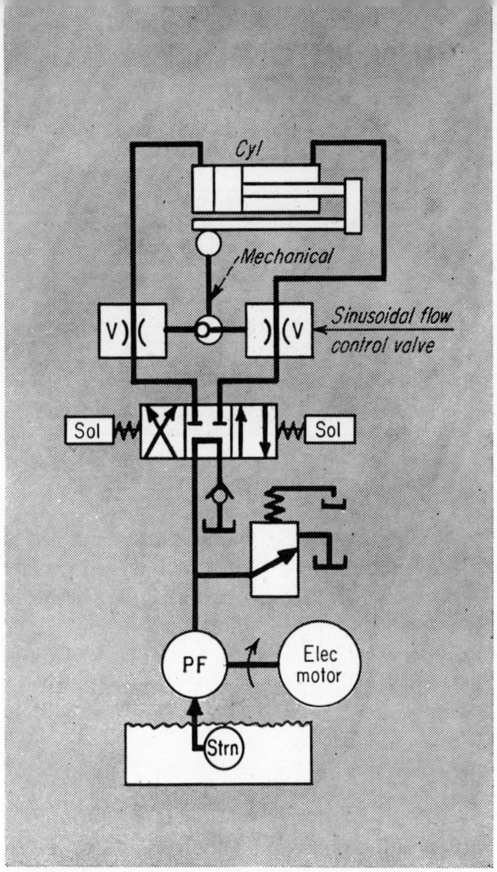

10 Symbolic picture . . . of sinusoidal flow-control (Fig. 9) shows how gears are driven by piston extension to vary valve position and maintain sine-cosine response.

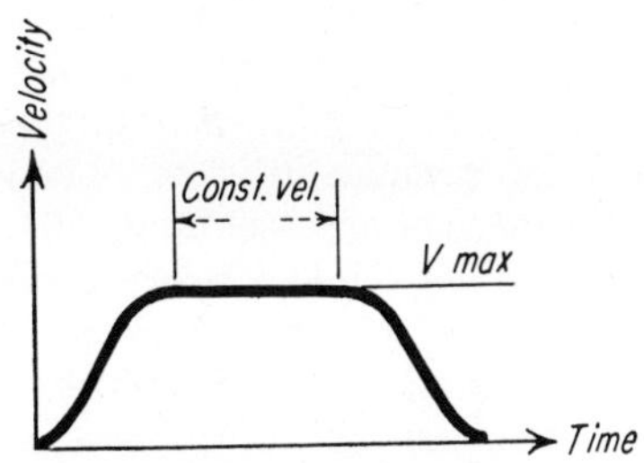

11 Trapezoidal . . . response maintains max velocity at center stroke where needed, yet damping prevents end-of-stroke shock.

decreases in steps as the piston moves to uncover the holes. In the uncovering direction, oil flow is induced rather than pressurized.

Location and spacing of orifices can be used to give a suitable basic displacement curve. Orifice area can be made variable for the purpose of greater flexibility. Hydraulic balance of piston can be attained by arranging orifices in diametrically opposed pairs. When A_P represents piston area and Q_t the total oil input,

$$V_{max} = \frac{Q_t}{A_P}$$

Recalling the equation for orifice flow, $Q = C_D A \sqrt{2GH}$, and assuming that mean pressure differential ΔP is constant then Q is a function of effective portion A_E of total orifice area A_T, hence,

$$V_X = \frac{V_{max} A_E}{A_T} = \frac{V_{max}(A_T - A_A - A_B \ldots)}{A_T}$$

where A_A, A_B, etc., are individual orifice areas.

SLOWDOWN WITH BINARY PISTON

The binary piston, Figs. 4 and 5, is a third way to slow down. It is primarily a free-floating piston *A*, Fig. 4, confined in its movement between collar *C* and *D* of piston rod *B*. Its working cycle differs from that of the conventional cylinder in one important respect. Initially when cylinder is pressurized, piston *A* will slide until it contacts either collar *C* or *D*. Hence, piston rod *B* will commence to move under full pressure until piston *A* is stopped by sleeve *E*. Remainder of the stroke is under reduced power proportional to area of the piston bore; on reversal, the same action occurs, except that the piston-rod area reduces power output further. Binary-piston design meets frequent demands for max power output at start of piston stroke and diminished output at end of stroke. It needs no flow-control device.

FLOW-CONTROL VALVES

Simple deceleration valves, cam-controlled, can throttle the pressure of oil to the cylinder and provide the smooth reversals needed (Fig. 6). The pressure port is normally open to the cylinder and can be closed through cam action, resulting in a desired rate of deceleration at end of stroke. This form of damping works for dynamic loads, and can also interrupt long strokes with periods of dwell at desired intervals by a series of adjustable dogs or cams spaced as needed. Conventional cylinders can be employed.

Piloted flow-control valves, Fig. 7, allow better control of cylinder operation. To avoid abrupt starting, the spool section governing the port openings may be provided with tapered lands or V notches, as shown in the illustration. The smaller the included angle of taper or V notch, the more gradual the flow rate when the spool is shifted. Also, the pilot-operated valve can be made to reverse slowly by metering the oil from pilot circuit through adjustable needle valves.

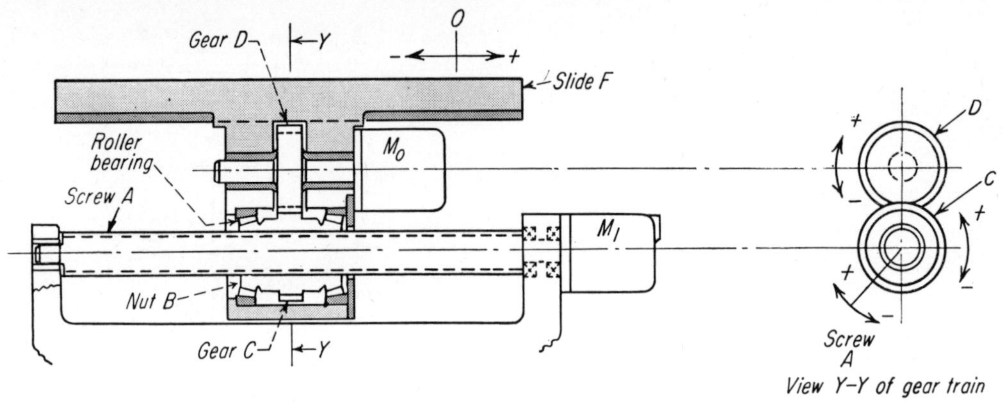

12 Differential . . . drive is a hydro-mechanical arrangement using the difference in rpm of hydraulic motors M_0 and M_1 to move nut *B* axially along feed-screw *A*, and thereby move slide *F*.

Other ways to dampen movement are available. A spring-centered pilot-operated reversing valve (Fig. 8) attains a measure of damping depending on the ratio of spring power to pilot oil pressure. A coil of small copper tubing is common practice when an inexpensive yet dependable method is required, and when oil volume is not large.

SINUSOIDAL RECIPROCATION

Simple harmonic motion is an example of reciprocation without abrupt variation. Gear mechanisms are available to coordinate the movement of a flow-control valve with a hydraulic cylinder, producing true sinusoidal travel. These mechanisms are designed to match the typical sine-cosine curves where displacement, velocity, and acceleration have sine or cosine response when plotted against time.

The valve spool C in Fig. 9 is actuated by a sine-cosine gear-type mechanism (not shown) which engages the spool at fulcrum Y. Fig. 10 shows how the sine-cosine mechanism can be coupled to a suitable sliding member of the cylinder mechanism, providing the required speed ratio; piston stroke and speed will correspond to linear motion of fulcrum Y. (An alternative is with a feedback system using a timing device that works with a reversing valve in the hydraulic system).

When the mechanism is set in motion, fulcrum Y will describe an oscillatory linear motion matching the cylinder piston motion. The ports of the sleeve D, serving alternately as pressure and exhaust ports, are slots of uniform width, as are the corresponding openings in spool C. The sleeves are individually adjustable for volume control, and timing of the cylinder stroke can be conveniently arranged.

The exposed area of each port is a function of spool displacement and the spool motion is characteristic of the cosine curve; therefore flow Q into and from the cylinder is represented by the area under the cosine curve (in units of velocity and time) multiplied by piston area A_p.

When mean velocity equals unity,

$$Q = A_P \int_0^{2\pi} (1 - \cos X)\, dX = A_p 2\pi$$

For practical reasons, oil pump delivery should exceed by 20% the peak demand of the system. Therefore, stroke length $L = 2\pi$ in., T_s = time of stroke in sec, and required pump capacity is

$$Q_P = \frac{2.4\, LA_p}{T_S} = \text{in.}^3/\text{sec}$$

Oil compressibility, ambient temperature and the variation of the Reynolds number during oscillation have some effect but may be neglected in most problems.

Intermittent gearing can be employed in the drive mechanism of the sinusoidal flow-control valve if shockless reversal in an optimum velocity program, plus a constant feed-rate level, is required. A trapezoidal velocity-time response, for example, may be obtained (Fig. 11). In transfer mechanisms requiring sequential cyclic operation (i.e., total cylinder stroke divided in periodic intervals of dwell and fast operation) the cosine mechanism can be geared to suit. Standard four-way solenoid actuated valves can aid control too.

DIFFERENTIAL DRIVE

The differential hydro-mechanical drive actuator, Fig. 12, offers some advantages not found in conventional feed-screw designs. Machine tool slide *F* moves in proportion to the sum (or difference) of hydraulic motor movements. Here is how it works.

Zero feed results when both nut *B* (driven by gear *D*) and screw *A* rotate in the same direction and at equal angular velocity. When they rotate at different angular velocities, then the nut will travel axially along the screw, carrying the whole slide assembly (slide, motor M_0, gear *D*, nut *B*). It is possible to reverse linear slide motion by a negative or positive change in differential rotation. Furthermore, since the differential causes slide movement, the absolute speeds of the motors can be held near the most efficient range. This avoids low-speed (below 10 rpm) servomotor operation which tends to be erratic.

Velocity-time response is determined by inputs to servomotors M_0 and M_1, and can be varied to suit actuator requirements. Max feed rate occurs when nut and screw rotate full speed in opposite directions, and is twice that in conventional feed-screw design.

for cushioning and control of air cylinders,

REGULATE THE EXHAUST

Regulating the exhaust of an air cylinder creates a pressure which resists the forward movement of the piston. If you control this back pressure, you can greatly increase the usefulness of the actuator, especially for many high-speed applications. Here are three simple ways to use back pressure.

B A BOGGS
Lucasville, Ohio

CUSHIONED STOPS

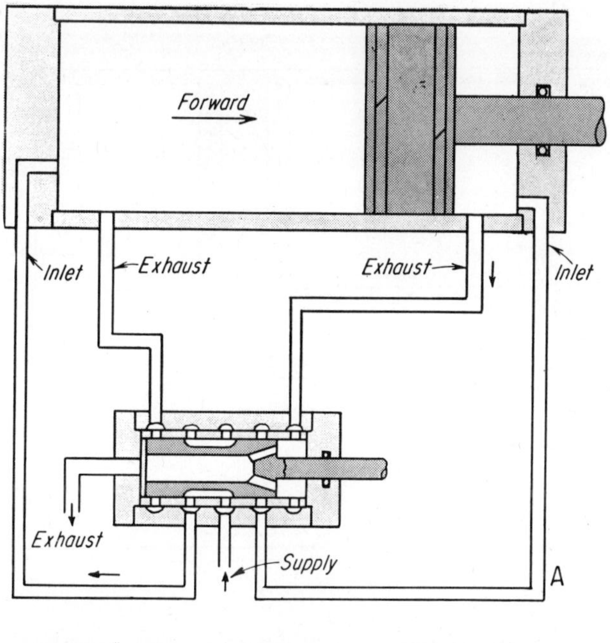

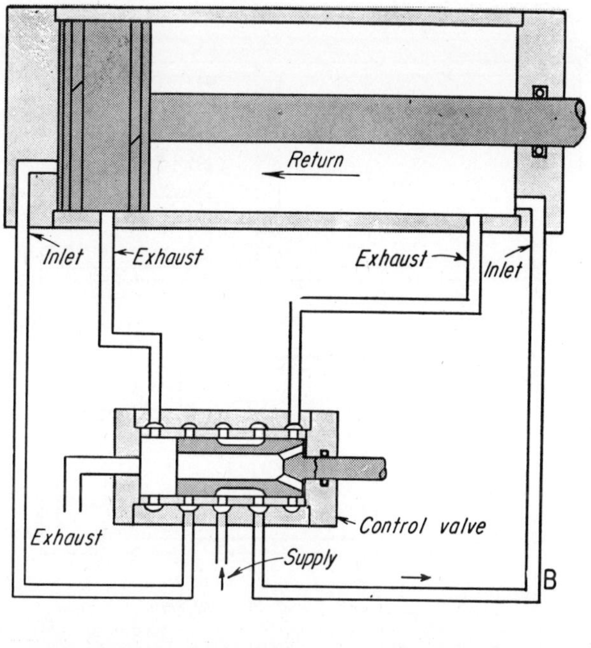

AIR COMPRESSED AT END OF STROKE cushions high-speed pistons. On forward stroke (A) piston moves at full force. At the end of the stroke, piston covers exhaust port to compress air remaining in the cylinder. Return stroke (B) shows piston blocking exhaust port for similar cushioning action. Exhaust ports can be located any desired distance from end of cylinder as long as distance does not exceed three-quarters of piston length. Because of pressure build-up in inlet line during cushioning, control valve should be as close as possible to the cylinder. Volume of this inlet line must be considered part of cylinder volume when calculating cushioning pressure.

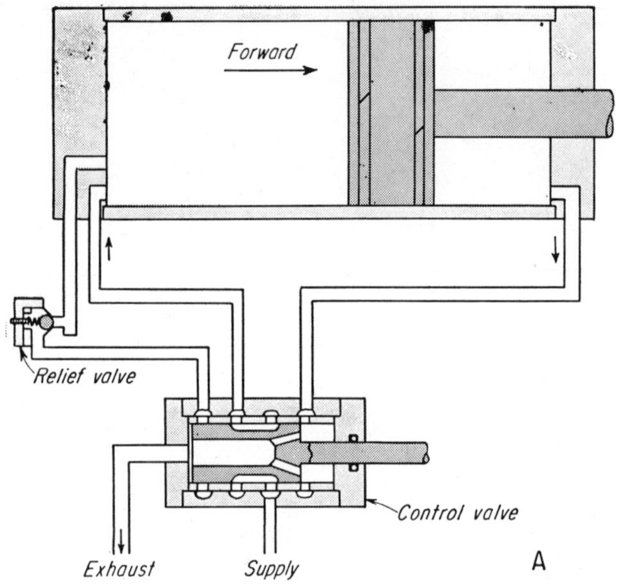

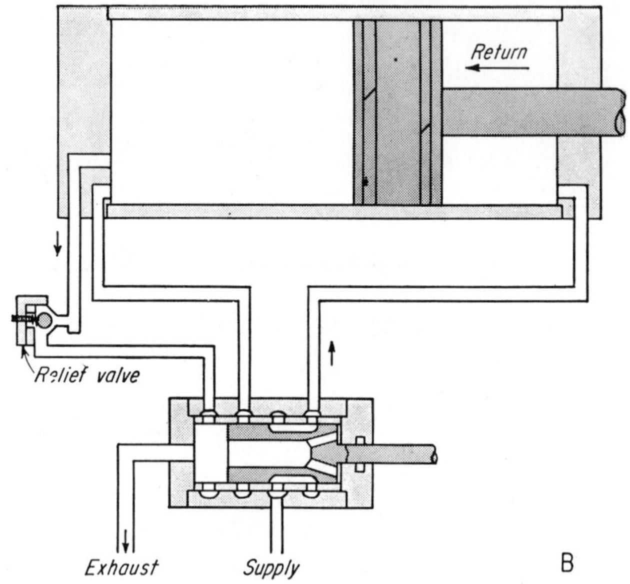

ADJUSTABLE DIFFERENTIAL FORCE for linear movement with a specified force in one direction and a different force in the return direction. Difference in the two forces can be controlled by relief-valve setting.

Relief valve is set so that return stroke will be resisted by known back pressure. On forward stroke (A), no resistance is met because exhaust through control valve is unrestricted, and piston moves forward with force equal to supply pressure times piston area. When piston reaches end of forward stroke, valve is moved to second position (B) which opens special exhaust port through relief valve. Supply is shifted to opposite side, and piston moves on return stroke.

Force in return direction is

$$F_2 = (A-a)\,P - AP_b$$

where A is area of piston, sq. in.; a is area of piston rod, sq. in.; (A—a) is differential area; P is inlet pressure, psi; and P_b is back pressure, psi. Both supply and back pressure can be varied, so either stroke is adjustable.

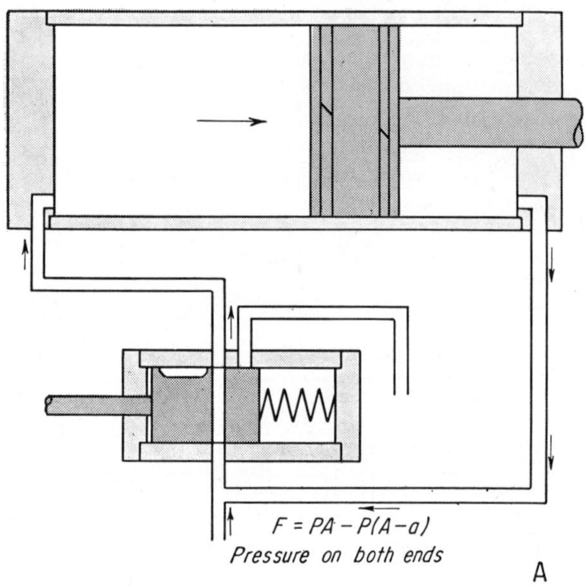

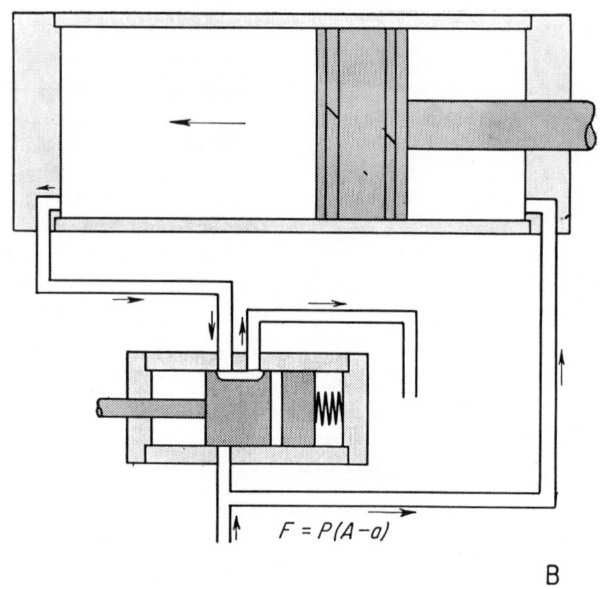

FIXED DIFFERENTIAL FORCE. As above, different forces are needed in each direction. In this circuit, however, no relief valve is required. Same pressure is applied to both sides of the piston at once causing forward stroke (A) to be a function of piston rod diameter. For return stroke (B) left side of cylinder is exhausted to atmosphere through control valve. Piston rod must be proportioned to produce desired force differential, and no adjustment is possible.

9

FLUID MOTORS AND DRIVES

Design and performance of adjustable-speed hydrostatic drives, air motors

When the speed has to be adjustable

Which hydraulic drive?

Each type of hydraulic drive, including torque converters and split-torque transmissions, is evaluated here on the basis of its unique advantages and limitations.

WERNER HOLZBOCK, Staff Engineering Specialist, Vickers Inc, Div Sperry Rand Corp

THERE is a story about a designer who wanted to indicate on his drawing that a certain part had to be knurled. So next to the part he wrote a note: *Nurl, nerl, gnerl, knerl*—oh, forget it, leave it plain.

Chances are that something similar happens at times when you try to select the right adjustable-speed hydraulic drive. The variety can be bewildering—unless you categorize them first by basic principle, then by function.

This article does just that. You'll find the first step below: categorization of the thirteen basic types.

Almost any specific need can be met with one of the thirteen drives. For a low-cost drive where efficiency is not important, such as that for a small winch, a simple four-way valve controlling a hydraulic gear motor will do the job. Where efficiency is essential and wide adjustability is secondary, a planetary gearbox combined with an adjustable-speed hydrostatic drive is available—it's called a split-torque drive. If a controllable-slip clutch or an adjustable-torque coupling is adequate, then a viscous clutch or a hydrodynamic drive will be the answer.

But for applications where efficiency, wide adjustability, and the ability to hold constant speed at all loads are equally important, the hydrostatic drive—a positive-displacement pump driving a positive-displacement motor — is generally selected. Almost all adjustable-speed hydraulic drives, other than those generally used in automobiles, are hydrostatic.

Any hydraulic drive has certain advantages over electrical or mechanical drives, simply because it is hydraulic. For example, there are no electrical parts to arc or fail, and expensive explosion-proofing and the associated need for maintenance are eliminated. The fluid motor being driven can be located a distance from the prime mover, with simple piping in between. The drives can absorb almost any shock or cycling disturbance because hydraulic fluids in conjunction with ample pipe sizes and accumulators have great resilience. And torques can be extremely high.

continued, next page

Categories of adjustable-speed hydraulic drives

HYDRODYNAMIC:	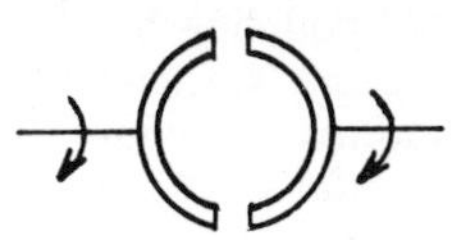	**Scoop-type fluid coupling Special torque converters**
HYDROVISCOUS:		**Magnetic-particle clutch Electroviscous slip clutch Silicone slip clutch**
HYDROSTATIC: P = pump M = fluid motor F = fixed displacement V = variable-displacement	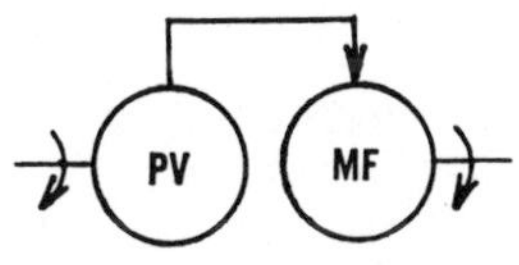	**1) PF-MF (variable-speed prime mover) 2) PF-MF (with flow-control valves) 3) PF-MV 4) PV-MF 5) PV-MV**
HYDROMECHANICAL:	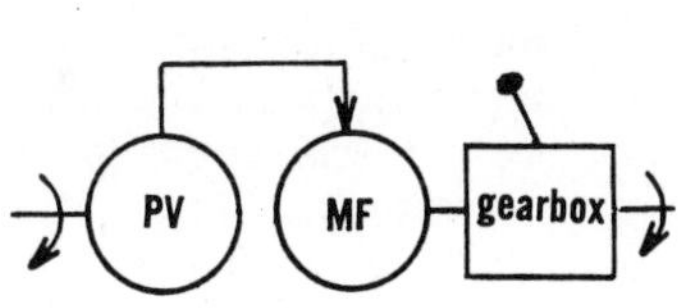	**Split torque Shiftable gearbox Hydraulic-planetary gearbox**

Hydrodynamic drives

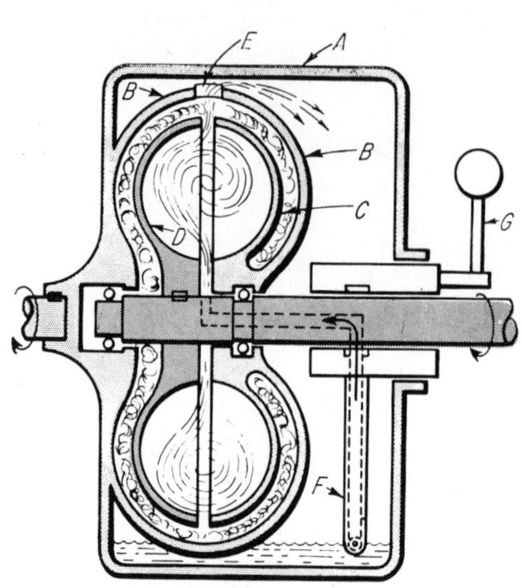

SCOOP-TYPE COUPLING

Principle: The kinetic energy of a hydraulic fluid, put in motion by an input impeller, drives an output impeller. *Hydrodynamic* is synonymous with *hydrokinetic*. The simplest adjustable-speed version is the single-stage scoop-type hydraulic coupling, shown at right, used primarily for driving large cooling fans.

When impellers are at rest, the drive housing is partly filled with oil. As the input shaft rotates, the oil is put into motion by the input-shaft impeller, and kinetic energy drives the output impeller, called the runner. There must be a speed difference between impeller and runner: at least 3% slip is required. Torque depends on oil volume and slip.

As shown in the figure, outer casing *A* together with drive casing *B* and impeller *C* are driven by the input shaft. The output shaft carries runner *D*. As the impeller rotates, it loses oil through nozzle *E*, tending to empty the drive housing and reduce the coupling between impeller and runner. However, scoop tube *F* prevents this by scooping up the oil and returning it to the drive casing.

The volume of oil in the drive casing and the speed of the runner are controlled, via the scoop tube, with lever *G*. Torque at 100% slip is very high, excellent for starting up a still shaft. But speed droop is very great, not suitable for constant-speed operation with varying load.

The main advantage of a simple hydrodynamic drive is that it combines the characteristics of a fluid clutch with those of an adjustable-speed drive. A natural application is for frequent starting of heavy loads. Heavy shocks are smoothly absorbed with relative ease, and reliability is

Hydroviscous drives

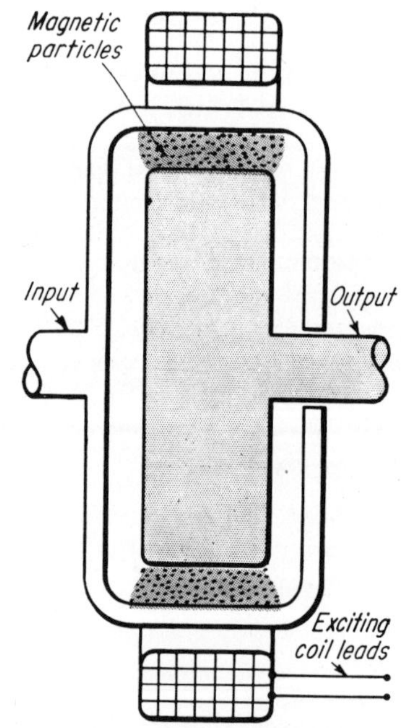

MAGNETIC-PARTICLE CLUTCH
("dry" fluid)

Hydroviscous is a coined word to describe drives that depend on viscosity change for torque and speed control. The most common is the magnetic-particle clutch. The particles act as a fluid when unenergized. Lesser known, but also a true viscous drive, is the electroviscous hydraulic clutch. A truer viscous drive is the liquid-slip clutch. Although they are clutches, each in a limited way can control speed.

Magnetic-particle clutch: The space between the two rotating members in the sketch at right is loosely filled with metallic particles, spherical in shape and screened to a uniform size. The electrical coil is stationary and produces a magnetic field which acts through an air gap; slip rings are not needed.

Unenergized, the magnetic particles lie at random and do not couple mechanically. When the coil is energized with dc, the resulting field aligns the magnetic particles in "chains," linking one rotating member with the other. The linking force can be increased or decreased by varying the current flow through the energized electrical coil.

One of the outstanding characteristics is that this force does not decrease with increasing speed as it does with friction clutches. Output torque is constant throughout the speed range.

Electroviscous hydraulic clutch: It is possible to change the viscosity of some liquids by applying an electrical field across them. In the second sketch at right the space between the impeller and runner is filled with such a liquid. The amount of coupling between the input and output surfaces is a function of the electric field intensity. At a field potential of 75 v per mil of fluid thickness, measured between the plate surfaces, the viscosity is as much as 200 times that of the nonenergized oil.

This concept of electroviscosity, called the *Winslow effect,* is not new, but it recently aroused new interest when Armour Research Foundation and Colorado Dynamics Corp announced progress in developing a suitable liquid. They are not releasing much information on their liquids, but one type that has worked well in the past is a suspension of about 50% by volume of something like silica gel in a carrier liquid. It resembles oil and has no metal particles.

Liquid-slip clutch: Silicone oil, available in almost any desired viscosity and extremely stable at all normal temperatures, is the coupling

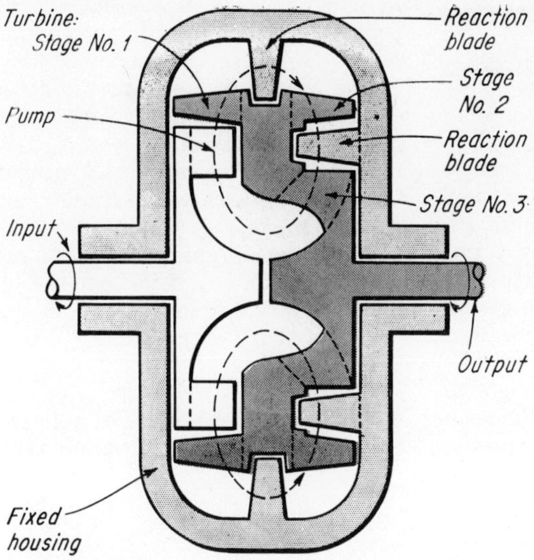

THREE-STAGE TORQUE CONVERTER

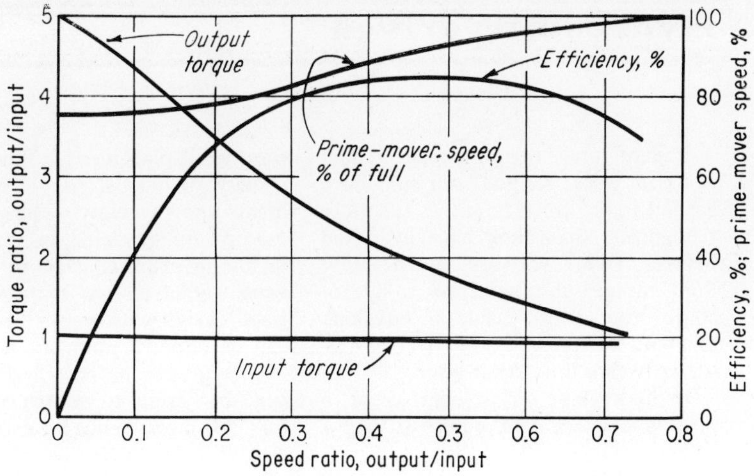

TYPICAL TORQUE-CONVERTER PERFORMANCE

excellent. It is bulkier than positive-displacement drives such as the hydrostatic drives described later.

Torque converter: A hydrodynamic drive is also a torque converter if it has a third element, the stator, between impeller and runner. Some torque converters have three stages. The output torque at stall can be made many times the input. As output speed approaches input speed, output torque approaches zero.

In theory, speed can be changed by making the stator blades adjustable, but this principle has so far been used only experimentally and torque converters are not sold as adjustable-speed drives.

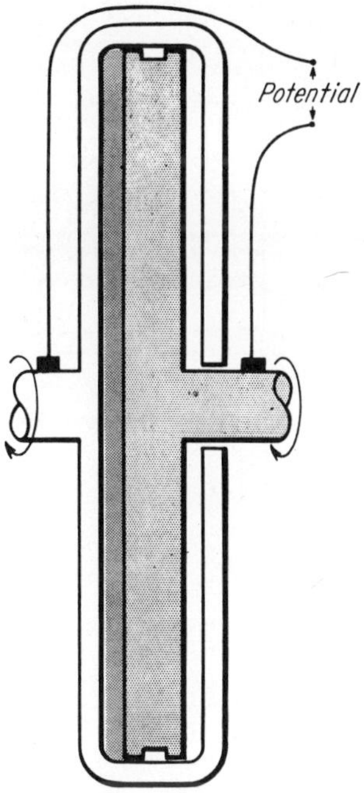

ELECTROVISCOUS CLUTCH

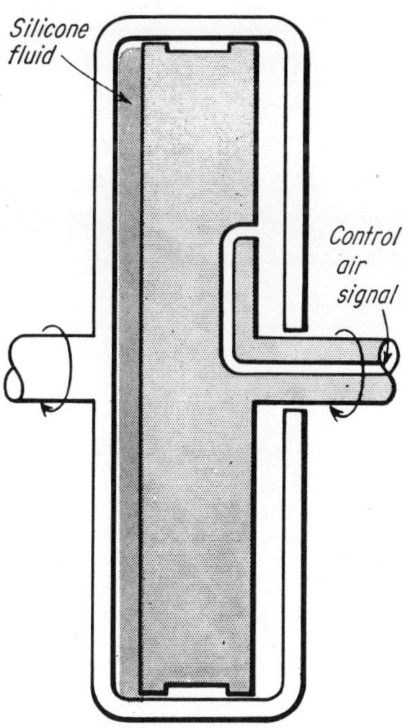

LIQUID SLIP CLUTCH

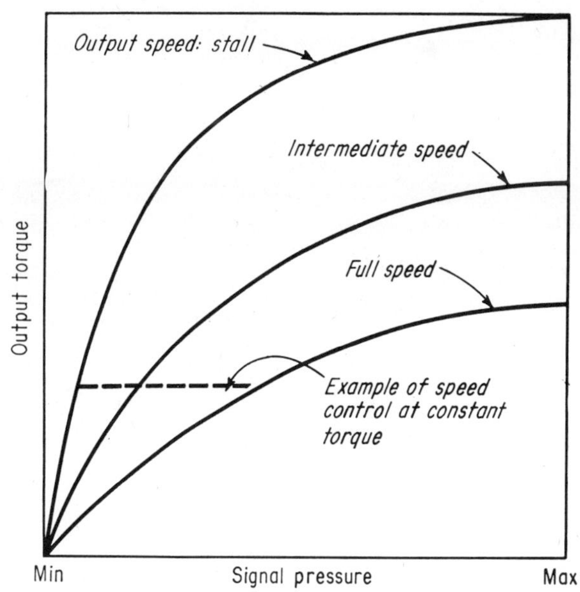

TYPICAL SLIP-CLUTCH PERFORMANCE

medium in several clutch types. Transmitted torque and therefore speed can be varied by changing the fluid thickness or its shearing effect.

In one technique, sketch and graph at far right, the distance between the rotating surfaces is adjusted pneumatically, lending itself to feedback control. In another, centrifugally actuated valves permit some leakage of fluid, increasing slip and thus maintaining preset speed.

As in eddy-current clutches and simple hydrodynamic couplings, efficiency is highest when slip is lowest. Heat must be removed, usually by fins on the rotating parts, or with auxiliary cooling fans.

Hydrostatic drives

Recent and revolutionary reductions in price, weight, and size have earned for hydrostatic drives far more recognition than they have ever had before. Today you can buy a 20-hp fluid motor—the heart of a hydrostatic drive—to fit inside an envelope of about 0.2 ft^3. A 20-hp electric motor is twelve times that large.

A hydrostatic drive consists of a positive-displacement pump driving a positive-displacement fluid motor, usually in one of the basic arrangements shown below. The pumps and motors are available in many forms, including external gear, internal gear, vane, radial piston, and axial piston. Each design with minor modifications can be either a pump or a motor.

The graphs of cost and weight on page 246 apply to pumps or fluid motors. Oil flow, hence size, is expressed in cubic inches per revolution. To convert to gallons per minute, multiply by $4.34 \times 10^{-3} \times$ rpm.

The wide variation in cost and size emphasizes the rapid development in hydraulic pumps and motors. Although the old types are still around, lower-cost, lower-weight units are entering the market rapidly. This also explains the abrupt price break for units slightly above 5 in.^{3} per revolution; the new developments have not yet covered the larger sizes.

Pump-motor operation

The pump or the motor or both can be adjustable, that is, of variable dis-

HYDROSTATIC DRIVE

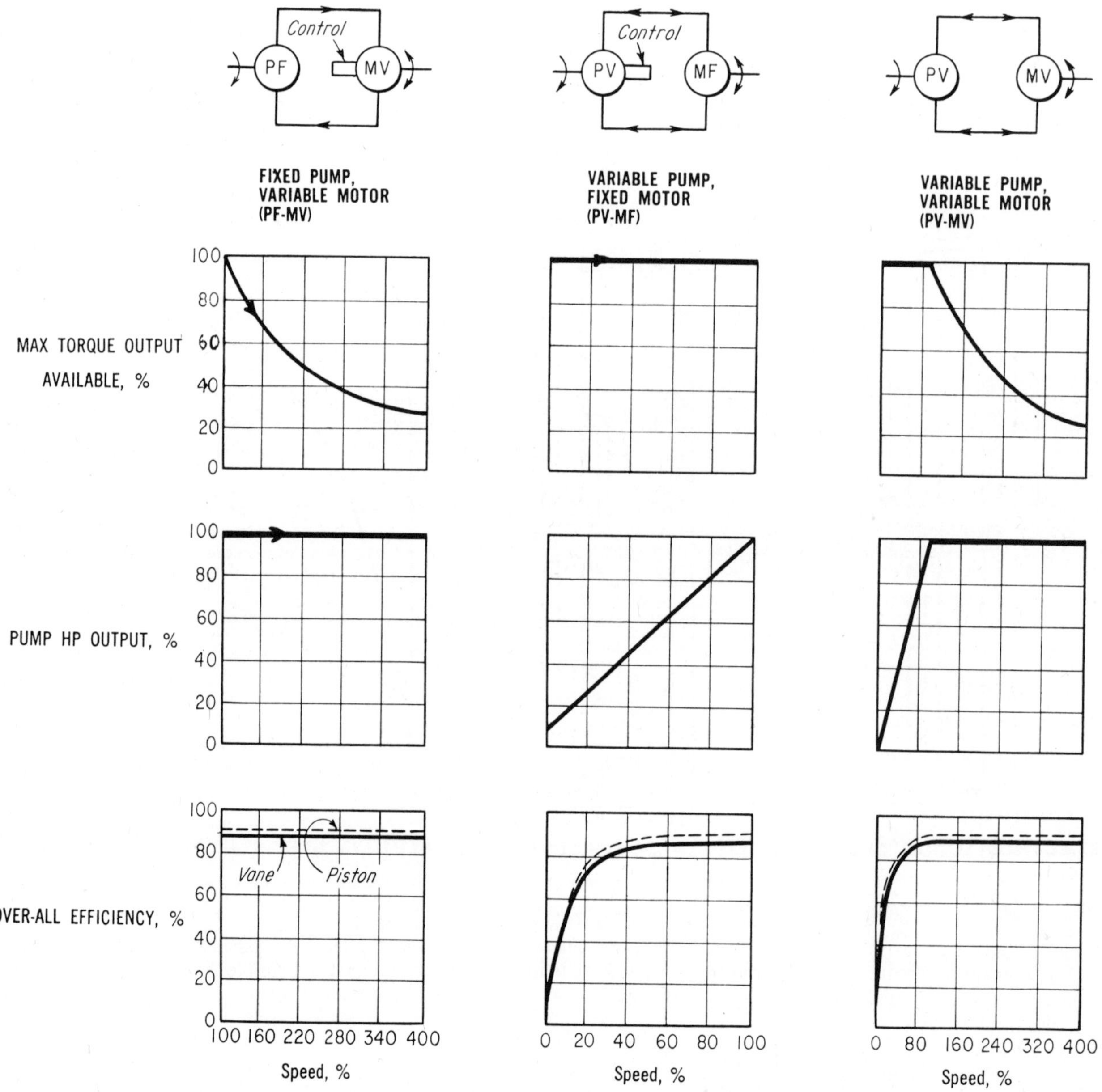

placement. Both can be constant displacement, with a valving system to vary the flow to the fluid motor or a variable-speed prime mover to drive the pump.

Only vane and piston-type pumps and motors can have variable displacement. Gear pumps have constant displacement and must be rotated at variable speed to adjust flow.

Adjustable radial-piston and vane pumps and motors have movable pressure rings that change the eccentricity of the vanes or pistons relative to the shaft, thus changing the displacement. Adjustable axial-piston pumps and motors usually have swash plates or wobble plates that determine the effective stroke of the pistons—or the valve block itself can be tilted. But the principles are the same. Typical examples are shown in the sketches on the next page.

Fluid motor speed is dependent on output flow of the pump divided by the displacement per revolution of the motor. Input torque to the pump (for a constant-speed electric motor) and output torque of the hydraulic motor for a given efficiency depend on pressure and displacement. In a positive-displacement pump-motor drive, pressure rises to whatever level is required to move the load at the set displacement (speed), unless load torque exceeds the capacity of the fluid motor, the pump, or the prime mover. If the pressure exceeds the setting of any of the relief and safety devices, the drive slows down or stalls without damage because fluid is bypassed. As soon as the load returns to normal, the motor operates as before.

Pump and motor performance are defined in the graphs shown below and the table on the next page. The most efficient units are the piston type; vanes are slightly less efficient.

Safe, smooth braking: Hydrostatic drives can be slowed down and stopped more smoothly and rapidly

PERFORMANCE

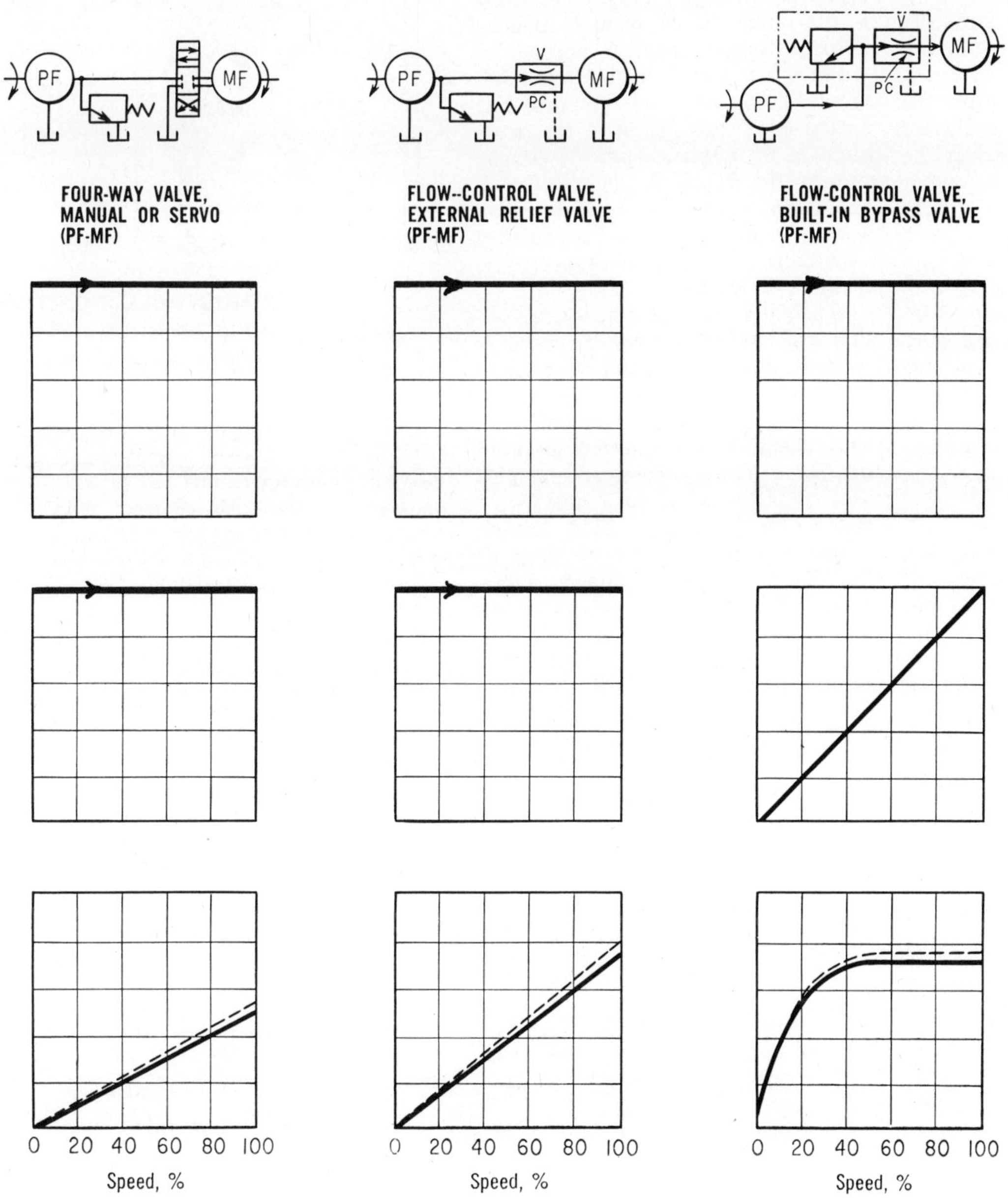

than most other drives. This is due in part to the overload capability of the fluid motor. There are two other important reasons: 1) The fluid motors are smaller and have considerably less inertia than equivalent electric motors, eddy-current clutches, or mechanical drives. 2) The motive fluid is easy to control or restrict at any of several flow points.

Two inherent braking systems are ideal for fluid drives: regenerative and dynamic.

With *regenerative* braking pump delivery is purposely reduced to slow down the fluid motor. The regenerative action occurs when the inertia of the driven machinery begins to drive the fluid motor as a pump, which in turn drives the pump as a motor. The kinetic energy is quickly absorbed, stopping the fluid motor and its connected equipment.

In *dynamic* braking, flow from the pump to the fluid motor is cut off by a valve. Load inertia drives the fluid motor as a pump, and the output is forced through a relief valve, cushioning the braking effect.

Conventional external braking with a friction clutch is often used, usually on the driven machinery. Inherent and external braking are combined in some applications. One example is a hydraulic-powered winch, where regenerative braking action is supplemented by a friction brake to hold the load.

Adjusting the displacement

There are many conventional ways to change the position of the yoke or stroking ring of adjustable-speed pumps and motors, including hydraulic, pneumatic, electric, and mechanical. The pressure compensator at the right is of particular interest because it is a feedback system in itself.

Pressure compensator: This refers to a control feedback device that senses pressure in the line to the fluid motor and uses it to adjust pump or motor displacement to correct or improve the existing operating conditions (compensate).

This pressure signal is useful for a variety of reasons. For a fixed-displacement fluid motor, you can determine torque by reading a pressure gage. And you can use the pressure as a feedback signal to reduce pump delivery, hence speed, if this has raised torque above a pre-set maximum.

In the illustration, the compensator is attached to the pump. Hydraulic pressure is applied at upper port. If pressure — representing torque — exceeds the setting of pilot valve screw *F*, the pump displacement is reduced until the torque returns to normal.

Initial pressure is below the minimum setting of the compensator. Spring *B* extends piston *A* for maximum pump displacement. Hydraulic pressure at *R* acts through orifice *X* and passage *V* to impose a thrust on spool *H*. Piston *A* remains extended when this pressure cannot overcome spring *G* to shift spool *H*.

When the pressure reaches the setting of pilot valve screw *F*, spool *H* shifts to direct fluid from chamber *Q* and passage *V* through interconnected holes *N* of spool *H*, to chamber *Y*. This chamber is connected to drain through orifice *Z* and to the annulus area of piston *A* through hole *W* of piston *J* and passage *U*. Flow through orifice *X* causes a pressure drop so that the pressure in chamber *R* becomes greater than in chamber *Q*. The pressure difference unseats piston *J* and fluid is throttled from chamber *R* to passage *U*. Fluid in *U* moves piston *A* to decrease pump delivery until it equals system requirements.

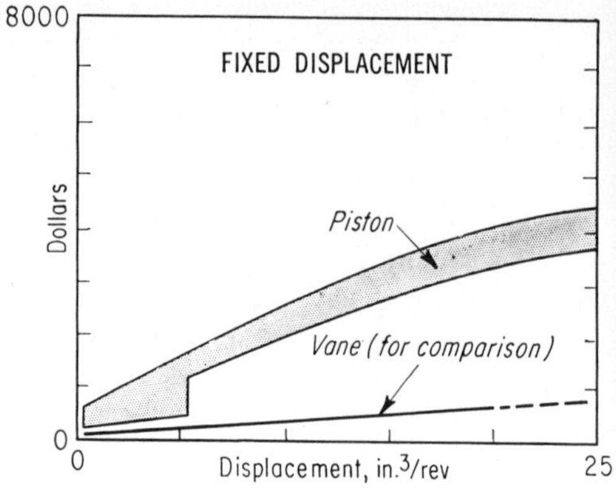

TYPICAL COST RANGE OF STANDARD

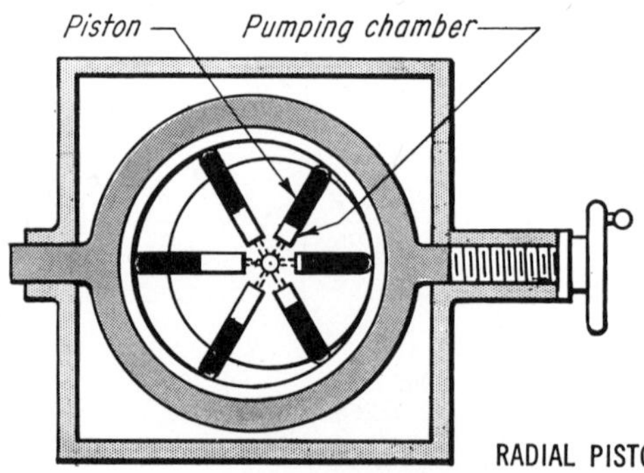

RADIAL PISTON

ADJUSTABLE POSITIVE-

One application of the pressure compensator is to assist a flow-control valve in a meter-in circuit, p 250, driving a fixed-displacement fluid motor. If the motor runs only at intervals and the flow-control valve has to throttle the flow frequently or cut it off completely, a pressure compensator can efficiently readjust pump delivery. Otherwise the excess flow from the pump must be dumped through a relief valve.

CHARACTERISTICS OF TYPICAL PUMPS AND FLUID MOTORS

TYPE	MAX PRESSURE, PSI	MAX DELIVERY RATE, GPM	MAX SPEED RPM	OVER-ALL EFFICIENCY	MAINTENANCE COSTS	LIFE EXPECTANCY	INITIAL COST
GEAR	**2500**	**150**	**7000**	**Poor to good**	**Low to medium**	**Good**	**Low to medium**
VANE	**2500**	**250**	**4000**	**Good**	**Low**	**Very good**	**Medium**
PISTON	**5000**	**450**	**12000**	**Very good**	**Very low**	**Excellent**	**Medium to high**

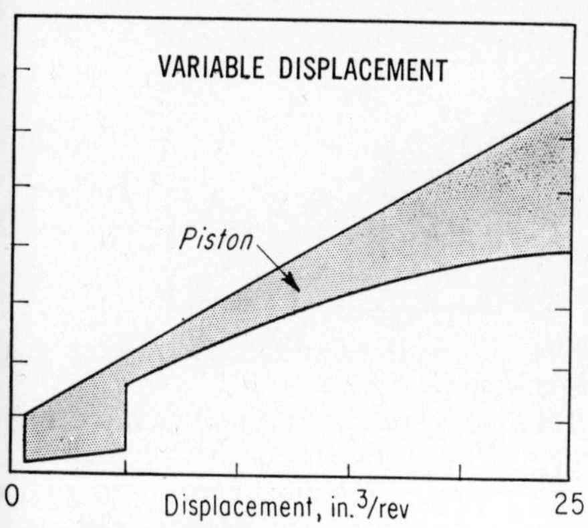

PISTON PUMPS

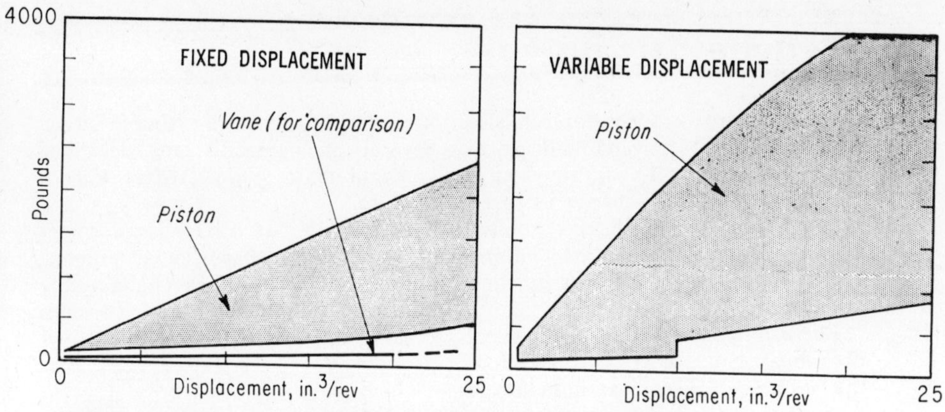

WIDE RANGE OF WEIGHTS FOR PISTON PUMPS

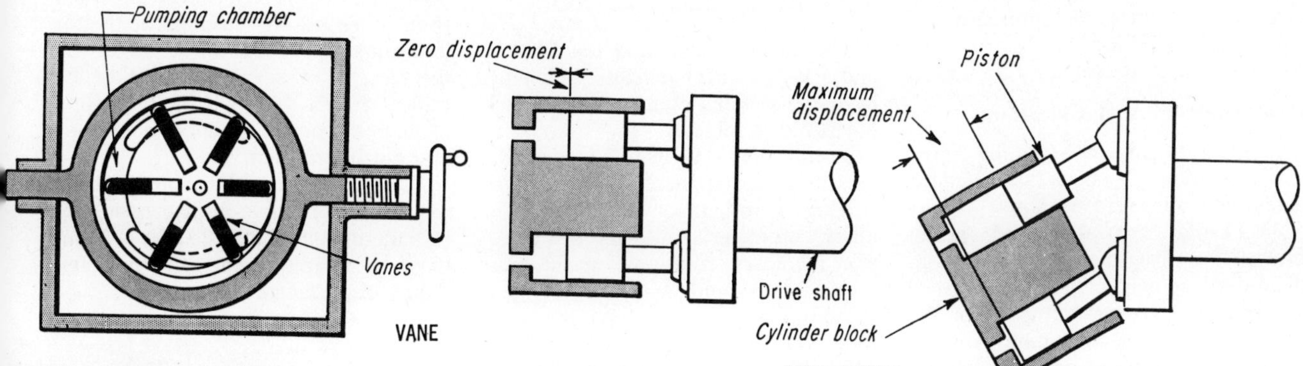

ISPLACEMENT PUMPS

PRESSURE-COMPENSATOR CONTROL

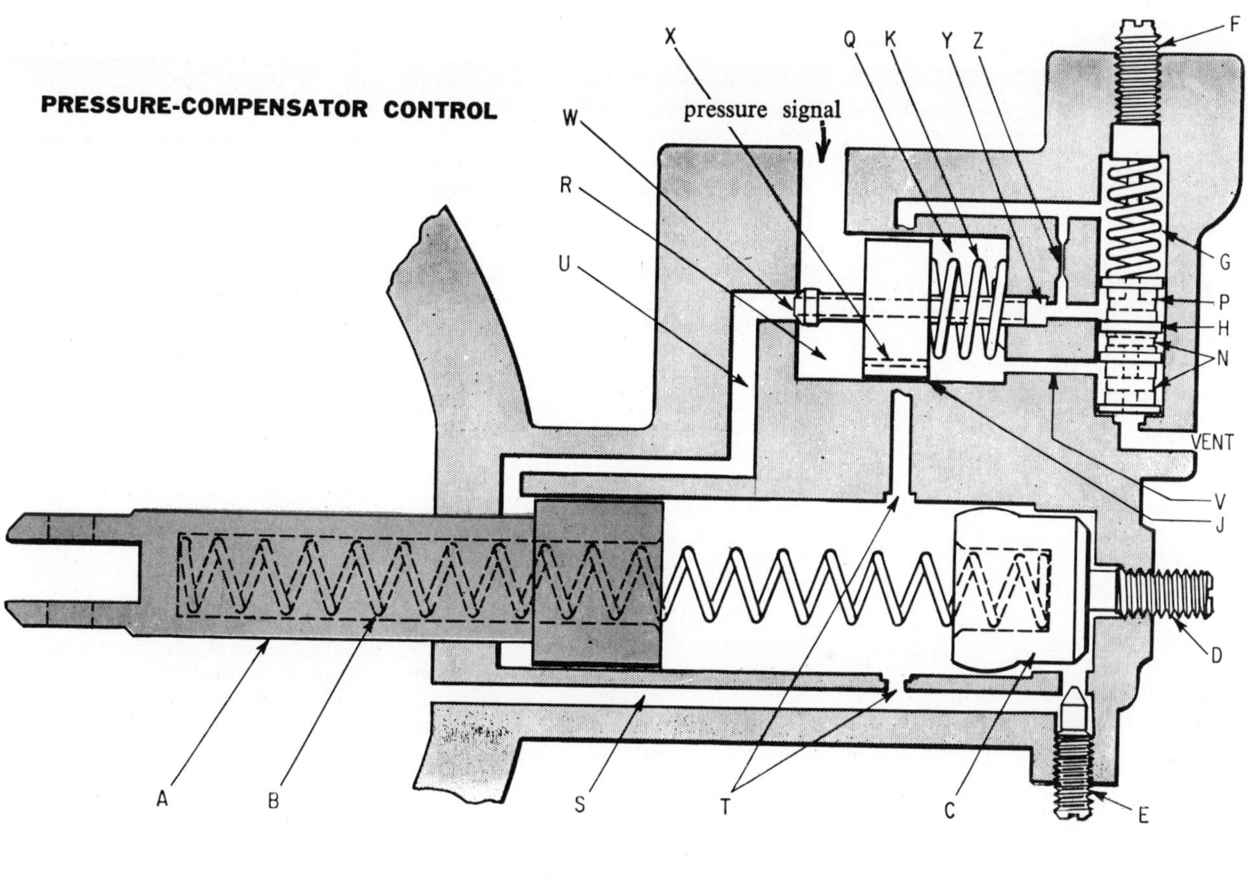

The pressure compensator can also be mounted on a variable-displacement fluid motor. By adjusting motor displacement you can change its speed to cope with a load change, holding the same hydraulic pressure as before. Thus speed is decreased as a function of load.

For example, in the slasher control described on the next page, speed of the variable-displacement fluid motor that drives the loom beam can be easily controlled by a compensator. As the diameter of the roll grows, the torque on the fluid motor increases and its pressure goes higher. Because speed is to be reduced as the diameter increases, the increased pressure signal can be applied to the pressure compensator to decrease motor speed.

Hydrostatic-drive classifications

Hydrostatic drives can be most easily classified in these five standard categories, all based on flow:

1) PF-MF (variable-speed prime mover) = Fixed-displacement pump, fixed-displacement motor; motor speed adjusted by varying the speed of the prime mover (electric motor or internal-combustion engine) to control pump output.

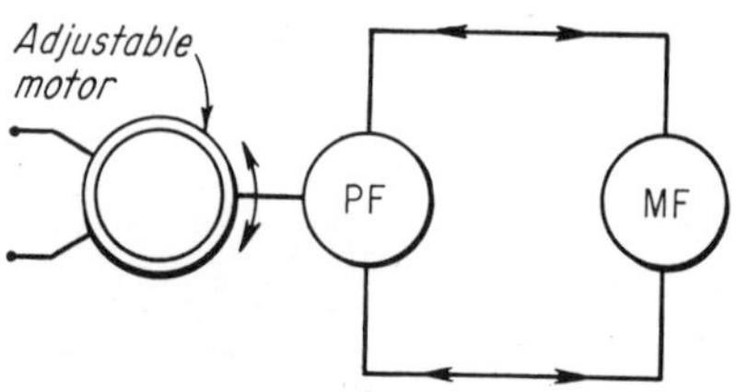

Prime mover characteristics and not pump-motor performance are involved here. See Editor's Note on page 254 for previous article on variable-speed prime movers.

2) PF-MF (with flow-control valves) = Fixed-displacement pump, fixed-displacement motor; motor speed adjusted with valving that controls the input or output flow of the motor.

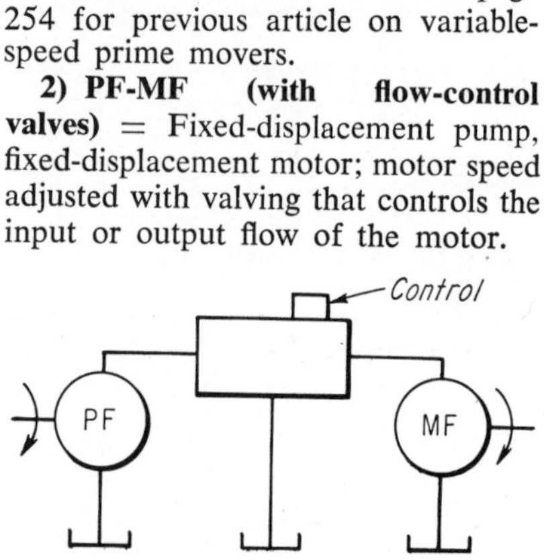

In its simplest form the PF-MF (valved) is the lowest in cost of all hydrostatic drives. Efficiency is relatively low because control is by throttling, but there are many important uses for this basically simple drive. Several arrangements are described in detail under Flow-Control Valves, page 252.

3) PF-MV = Fixed-displacement pump, variable-displacement motor; motor speed adjusted by changing the effective stroke of the pistons or vanes (gears not applicable).

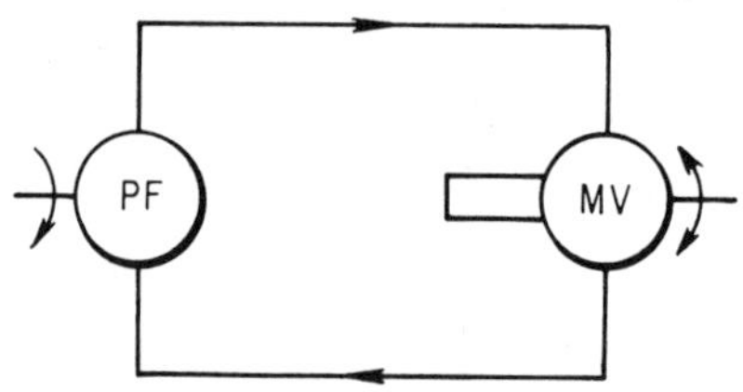

This arrangement, and the PV-MF and PV-MV arrangements following, have well-defined characteristics, each portrayed graphically.

In the PF-MV drive, torque decreases as speed increases, useful for driving a roll that winds paper or other web material. As the roll diameter increases, rotational speed must decrease to hold a constant linear velocity of the paper. At the same time, mass and resisting torque of the roll increases, matching the PF-MV performance characteristics.

Motor displacement is gradually increased by the control to reduce motor speed. The speed drops because the motor takes more fluid per revolution. Motor torque rises because the effective moment arm of the vanes or pistons is increased. The horsepower remains relatively constant over the entire speed range.

Using the constant torque drive described in category 4, below, would be uneconomical here because in such a drive the motor would have to be sized to withstand maximum torque at maximum speed—yet it would not normally run at that point.

For example, suppose the load has a torque of 40 lb-ft at 300 rpm and 10 lb-ft at 1200 rpm. A constant-torque drive would have to be sized to produce 40 lb-ft and 1200 rpm, or 9.6 hp, neglecting losses. A variable-torque drive would be sized for 10 lb-ft and 1200 rpm, or 2.4 hp, only one-fourth the size of the constant-torque drive.

4) PV-MF = Variable-displacement pump, fixed-displacement motor; motor speed adjusted by varying pump displacement to vary pump flow.

Torque remains constant over the entire speed range because motor displacement is fixed and the pump is able to maintain full pressure at all

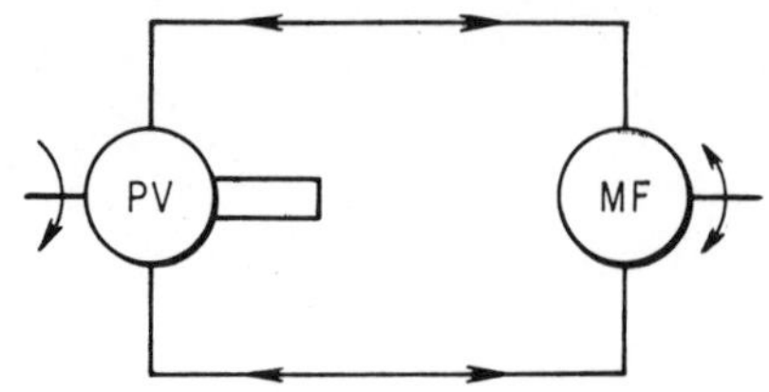

flows, that is, if you disregard changes in efficiency.

Motor speed is proportional to pump delivery rate, and pump power is proportional to motor speed. Pump speed is constant, even though delivery rate is adjustable.

Special drives have been built with fixed-displacement motors and variable-volume pumps with accurately controlled speed ranges of 1:120,000,000. The slowest speed is 1/60,000 rpm, achieved by pulsing the motor intermittently to follow a target—in this case a missile being tracked by radar. Speeds from 0.1 to 2000 rpm, a 20,000:1 ratio, are servo-controlled and not intermittent.

5) PV-MV = Variable-displacement pump, variable-displacement motor; motor speed adjusted by a combination of adjustments of pump and motor displacements.

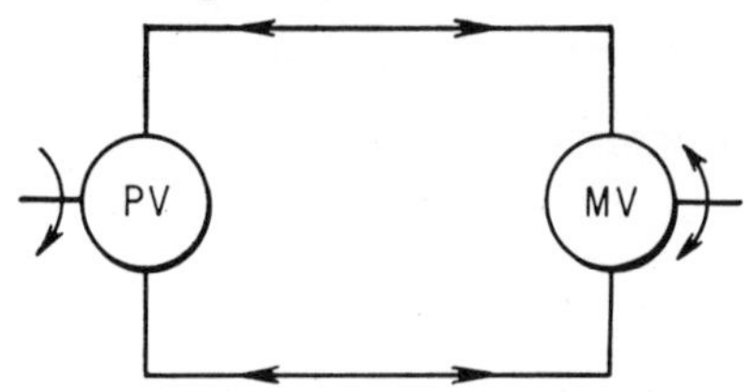

If neither the constant-torque nor the constant-horsepower characteristics of the PF-MV or the PV-MF drives is exactly suitable, then combining them will add versatility—and cost. Either the pump or the motor can be adjusted, or both can be adjusted together.

Speed range is greatly extended. For example, assume that the minimum speed of your driven equipment must be 50 rpm (a typical value) and maximum speed 3600 rpm.

Assume that the pump selected has a maximum displacement which is 75% of the full displacement of the fluid motor, and that the pump speed is held constant at 1200 rpm. Thus the fluid motor will operate at 1200 x 0.75 = 900 rpm. If pump displacement is reduced to about 5% of its full stroke, the fluid motor speed will drop to about 50 rpm. With pump stroke at full displacement, and fluid-motor displacement at about 25% of full stroke (a recommended minimum), output speed will increase to 900 x 4 = 3600 rpm. Total range = 3600/50 = 72:1, and much larger ranges are possible.

Where two machines or two separate parts of the same machine must be driven at different speeds, and one of the speeds varies in a different manner from the other, the answer again is a PV-MV drive, but with two fluid motors.

Take as an example a textile-mill slasher, sketch at right, which carries multiple strands of yarn through a chemical sizing solution and winds them onto a loom beam prior to weaving. The control conditions are:

1) Speeds of main drive roll and loom beam must change with a control signal from a moisture detector.

2) Speed of loom beam relative to main drive roll must change with the wound diameter of the loom beam.

Condition 1 is constant torque, satisfied by a variable-displacement pump and a fixed-displacement motor. Condition 2 is variable torque, calling for a variable-displacement motor.

The answer is to turn the main drive roll with a fixed-displacement motor and the loom beam with a variable-displacement motor. A single variable-displacement pump can drive both motors. Moisture-control signals are applied to the pump, changing the speed of both fluid motors in unison. An additional signal is applied to the variable-displacement fluid motor, changing only the speed of the loom beam as its diameter builds up.

Speed control with valves

The second category of hydrostatic drives—PF-MF drive with control valves—depends on the valving for speed control and is discussed separately here for that reason.

One versatile pump-motor-valve arrangement has a so-called double or split-flow pump, each half having fixed displacement but of different magnitude. Valving can provide three basic flow rates to the fluid motor by utilizing one, the other, or both pump outputs. Operation is depicted in the circuits at the right. A variant is a multiple-output pump, where each of numerous outputs can have a different displacement and can drive its own fluid motor.

But most conventional valved hydrostatic drives have a single output pump driving a single fluid motor. The main differences are in the way the valves are arranged in the circuit.

Four metering configurations

Motor speed is controlled by the valving in one of four ways: bleed-off, meter-in, meter-out, or a combination of these. *Bleed-off* control diverts a portion of pump delivery to the oil tank, the remainder drives the fluid motor. *Meter-in* control adjusts input flow to the motor; *meter-out* control

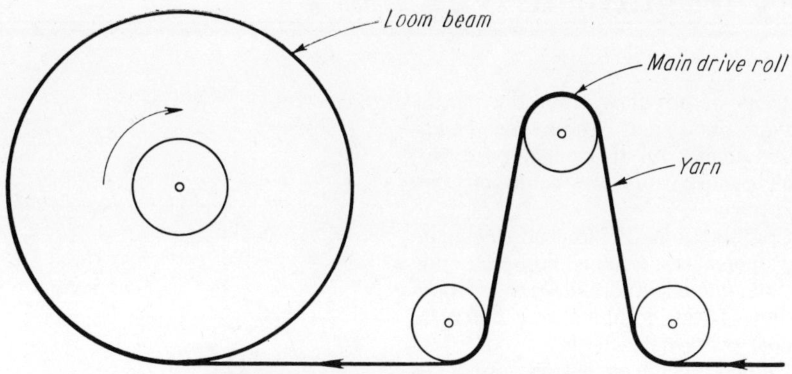

APPLICATION OF PV-MV DRIVE
(Textile-mill slasher)

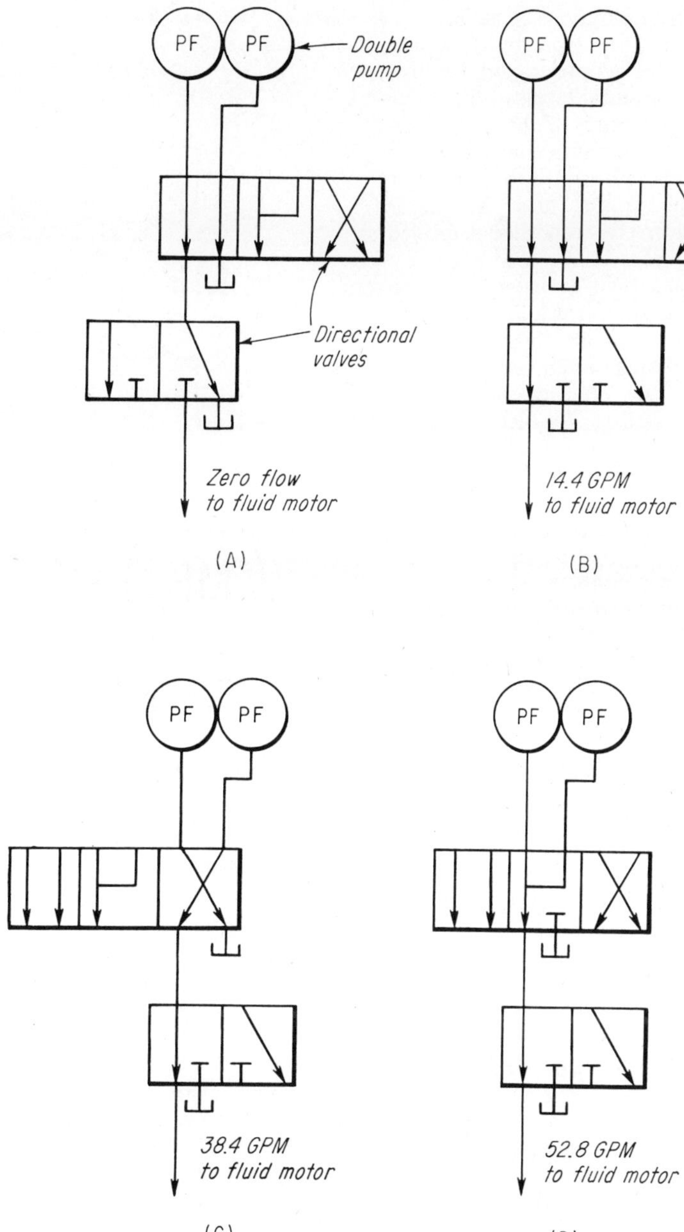

DOUBLE-PUMP DRIVE (Directional valves)

restricts output flow from the motor to vary its speed. These are shown schematically on this page; performance comparisons are made on the next page.

Combinations of bleed-off, meter-in and meter-out circuits comprise the art of valved hydraulic-transmission design. These combinations differ in almost every new circuit.

Choice of the metering circuit is especially important. Drive performance is often based more on the circuit than on the pumps, motors, and valves you might choose to place in the circuit. Let's examine inherent advantages and limitations of each.

Bleed-off: Flow from the pump to the fluid motor is unrestricted. Control is with an adjustable bleed-off valve located in a branch of the pump discharge line. Assuming constant pump flow, whatever oil is bled to tank never reaches the motor; thus the motor speed is inversely proportional to bled flow.

There are two advantages: pump pressure adjusts to load, thus reducing pump effort, and the throttling losses are in the branch and not in the motor circuit. If an efficient flow-control valve is selected for the bleed-off, part-load efficiency is reasonably good.

Furthermore, if the fluid motor speed never goes to zero, the bleed-off valve never has to pass full pump flow and can be proportionately smaller. The other two configurations, meter-in or meter-out, pass full pump flow.

Meter-in: Flow from the pump to the fluid motor must pass through the control valve, which adjusts flow by automatic control of its valve port area. Pump excess flow is returned to tank through a relief valve. Thus the pump must work against full pressure or higher at all times. This extra work is avoided in the integral bypass throttling valve.

Meter-out: All flow through the fluid motor must pass through the meter-out valve located downstream of the motor. The circuit is similar in performance to the meter-in circuit, in which the pump excess flow is discharged through a separate relief valve.

One advantage of meter-out control is its excellent shock-absorbing capability. The back pressure against the motor outlet damps sudden load changes.

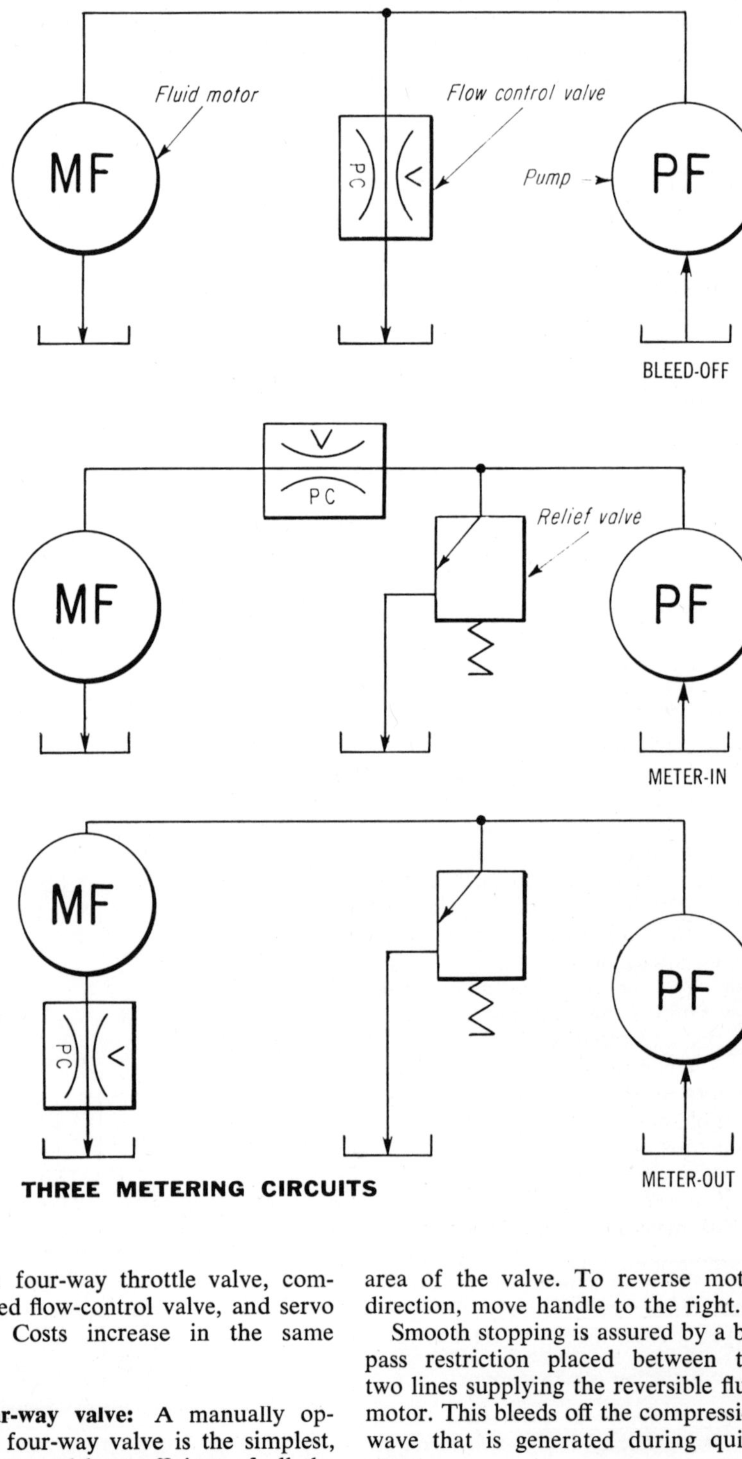

THREE METERING CIRCUITS

Valve types

There are three valve types used in most hydrostatic PF-MF (valved) drives: four-way throttle valve, compensated flow-control valve, and servo valve. Costs increase in the same order.

Four-way valve: A manually operated four-way valve is the simplest, cheapest, and least efficient of all the controls for hydrostatic drives. It operates basically as shown in the control circuit on the following page.

Move handle *C* to the left to drive the fluid motor in one direction at a speed determined by the throttled flow area of the valve. To reverse motor direction, move handle to the right.

Smooth stopping is assured by a bypass restriction placed between the two lines supplying the reversible fluid motor. This bleeds off the compression wave that is generated during quick stops.

The pump output is fully utilized only at full opening of the four-way valve; at partial flows the relief valve bleeds the excess flow to the oil tank. At the center position, fluid flow is blocked, stopping the motor.

METERING CONFIGURATIONS COMPARED

METERING CONFIGURATION	SPEED REGULATION	EFFICIENCY	SHOCK ABSORPTION
BLEED-OFF	**Good**	**Good**	**Good**
METER-IN, SEPARATE RELIEF	**Very good**	**Poor**	**Good**
METER-IN, INTEGRAL BYPASS	**Very good**	**Good**	**Good**
METER-OUT	**Very good**	**Poor**	**Very good**

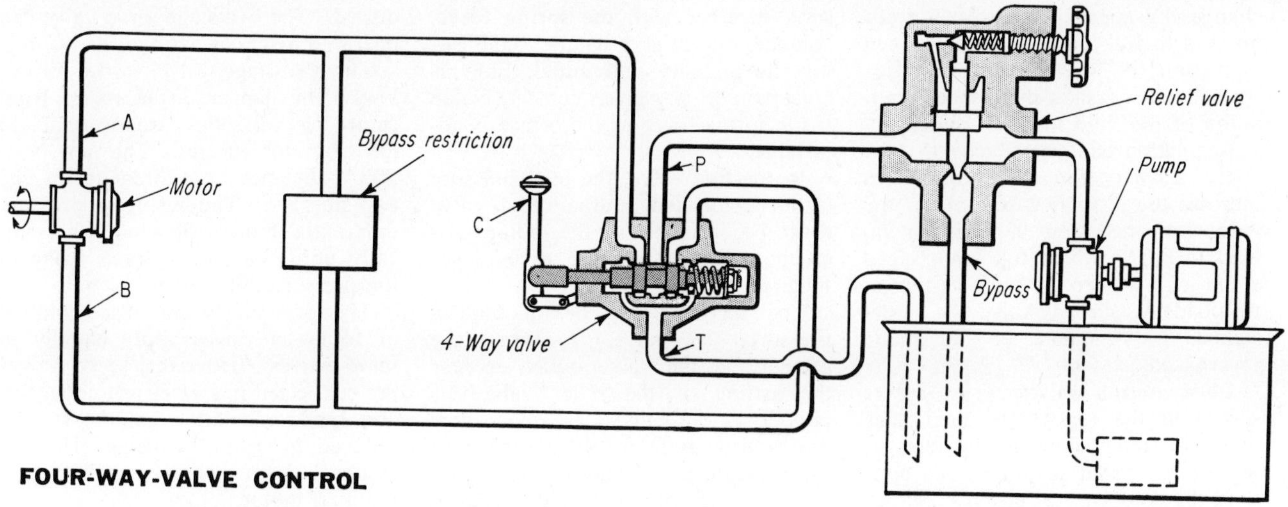

FOUR-WAY-VALVE CONTROL

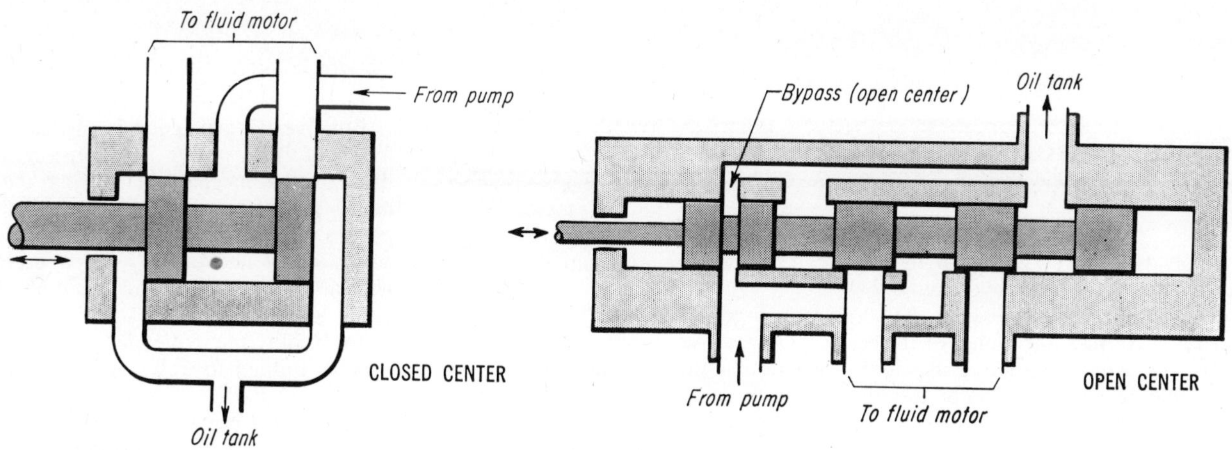

FOUR-WAY VALVES

A variant of the closed-center valve is shown below the control schematic. An open-center four-way valve, it allows flow of the pump delivery back to the tank at greatly reduced pump-discharge pressure.

One disadvantage of the open-center valve is that the spool must be moved a considerable distance before the bypass closes off to allow flow to the motor. This ineffective motion is called *deadband.*

The biggest disadvantage of any four-way-valve control is that maximum pump power is needed whenever the valve is throttling the flow. This lost power is dissipated in heat which must be removed. For small drives, say ½ hp or so, the power consumption and temperature rise probably will be insignificant.

A further limitation of the four-way control illustrated is speed droop, called *regulation.* It is the change in fluid-motor speed with load, and can exceed 30% of normal speed. Here's an example:

Let the pump flow rate in the four-way valve control circuit, above, be 30 gpm and the relief-valve setting be 1000 psi. With a valve opening that provides a flow area of 0.056 in.^{2} from P to A and from B to T, respectively, the flow that passes at 1000 psi will just about equal the pump flow rate.

The basic flow equation is

$$Q = KA\sqrt{\Delta P}$$

Where A is area, ΔP is pressure drop, and K is a constant. ΔP is 500 psi, not 1000 psi, because there are two pressure drops, P to A and B to T.

Hydrostatic drives *continued*

This is true as long as there is no load on the fluid motor.

Suppose the motor is loaded and requires a pressure drop of 500 psi. For the same valve setting, the flow will now drop to 21.2 gpm because the available pressure drop at the valve has been halved. The excess is bled off by the relief valve through the bypass.

The speed of the fluid motor thus changes by approximately 30% from no load to full load. Hence the speed regulation of this circuit is 30%. Actually, it is greater if the speed regulation of the fluid motor is added.

Regulation can be reduced to zero if the valve opening is readjusted to increase the flow. For example, if the valve is opened from 0.056 in.^{2} at no load to 0.079 in.^{2} at full load, speed at no load will be the same as that at full load. A four-way valve can be designed for manual or automatic (servo) adjustment.

Flow-control valves: The valves shown at the top of the next page maintain steady flow despite changes in inlet or outlet pressures. It's done with the compensator piston *C*, which automatically adjusts its position to hold the pressure differential constant across the valve orifice. The equation is

$$Q = KA\sqrt{\Delta P}$$

and flow Q will stay constant if orifice area A and differential ΔP do not change.

With the spring, the compensator piston will move until the outlet pressure, together with the spring force, balances the inlet pressure, establishing the pressure differential. Flow is adjustable to any given constant value if the spring force or the orifice is adjustable.

In the first valve, the inlet pressure is not controlled, and a relief valve must be installed at the pump discharge to bleed off the excess (see meter-in circuit).

The second valve has a built-in pressure-relief valve, *H*. When the pressure at the valve outlet exceeds the setting of the relief valve, its poppet is unseated. A small flow passes through the restriction, lowering the pressure at *E*. This further opens the compensator valve and increases the flow to the tank. Part-load efficiency is improved because pump pressure is reduced as bypass flow increases.

Servo valves: A servo valve can be a simple four-way valve, shown at right, or any other kind. But it is controlled electrically by a magnetic coil that positions the spool in response to a remote electrical signal, the feedback. It is called a *servo valve* because it is used in a servo system.

The spool of a single-stage servo valve is moved by the torque motor directly, and up to 2-w input is required. The two-stage servo valve can operate on signals as small as 40 mw.

The two-stage valve works as follows: The flapper displacement from center is directly proportional to torque-motor current. The new position unbalances the pressure in the two nozzles. The resulting pressure differential moves the spool automatically until the spring force balances the pressure differential.

The advantages and disadvantages of four-way valves apply equally to servo valves. However, servo valves are connected into electrical networks, and speed regulation can be readily reduced to 0.1% by using electrical feedback from tachometers driven by the fluid motor.

Hydromechanical drives

Really two drives in one, a hydromechanical drive is a hydrostatic or hydrodynamic drive combined with a gearbox. It can take one of three forms: 1) *split-torque* drive in which a variable-displacement motor adjusts the output speed of a mechanically driven differential planetary by varying the speed of one of the gears; 2) *mechanically shiftable* but separate gearbox driven by a hydrostatic or hydrodynamic drive; and 3) *hydraulic-planetary* gearbox in which the gears double as gear pumps and can be speed-controlled by varying the restriction at the pump outlets.

Split-torque drive: The first figure at right shows the most common arrangement, but many others will work. In any arrangement, the prime mover is connected to the load through two paths: directly to the differential planetary gearing and indirectly off the input shaft and through the hydrostatic drive. The planetary sums the two inputs to produce output N_o.

The speed relationship is

$$N_o = \frac{N_1 \pm 2N_C\,(1 + a)}{1 + 2a}$$

where a is the ratio of radii of planet gears to sun gear, and N_1, N_C and N_o are input, planet-carrier, and output speeds respectively.

Carrier speed N_C depends on the displacement of pump-motor *A*. If the displacement is zero, neither *A* nor *B* passes flow, and N_C is zero. Output speed N_o in this case equals $N_1/(1 + 2a)$, called the base speed.

To run above base speed, adjust *A* to rotate the planet carrier in the same direction as N_o, shown in the figure. To run below base speed, reverse N_C.

Pump motor *A* operates as a pump while N_o is above base speed but runs as a motor when N_o is below base speed. In either case, the pump is driven from the mechanical system, and the motor returns the energy to the same system. Theoretically, there is no energy loss.

But over-all efficiency is highest if most of the power is transmitted through the mechanical gearing, because gears are more efficient than pumps or motors. Typical efficiency is considerably above 90% for a drive with an output speed ratio (max N_o)/(min N_o) of 2:1. A split-torque drive is not usually recommended unless the ratio is limited to 3:1. For top efficiency—over 95% is possible—select the narrowest possible speed range.

Size of the hydraulic portion of a split-torque drive is small compared with a straight hydrostatic drive because the gears do most of the work. The pump-motors only speed up or slow down the planet carrier as needed.

Initial cost of a split-torque drive is higher than that for a straight hydrostatic drive. So when simplicity or low first cost is more important than operating efficiency, as in low-horsepower applications, choose the straight drive. Where operating economy and small size are paramount, choose split-torque.

Where both operating economy and wide speed range are important, neither the split-torque nor a straight

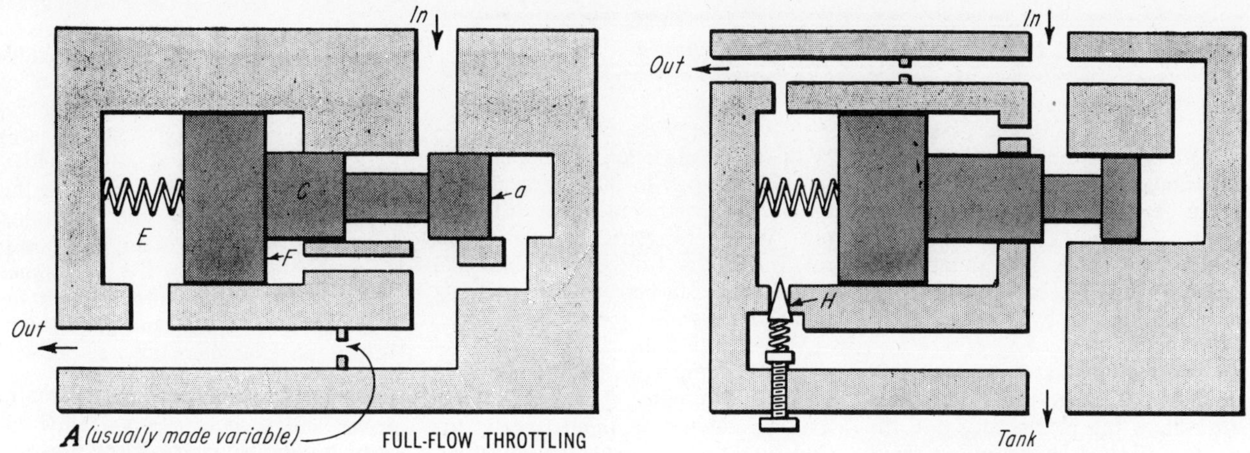

FULL-FLOW THROTTLING

INTEGRAL-BYPASS THROTTLING

FLOW-CONTROL VALVES

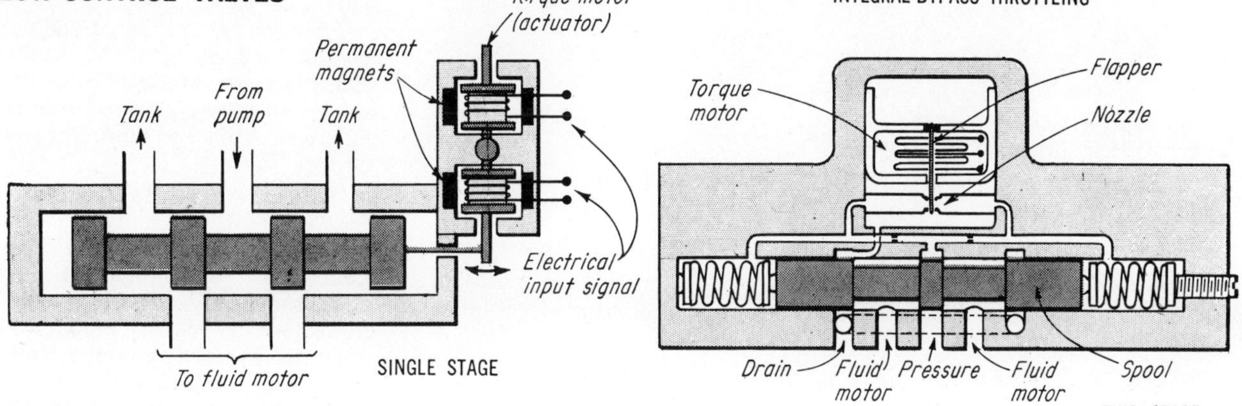

SINGLE STAGE

TWO STAGE

SERVO VALVES (four-way)

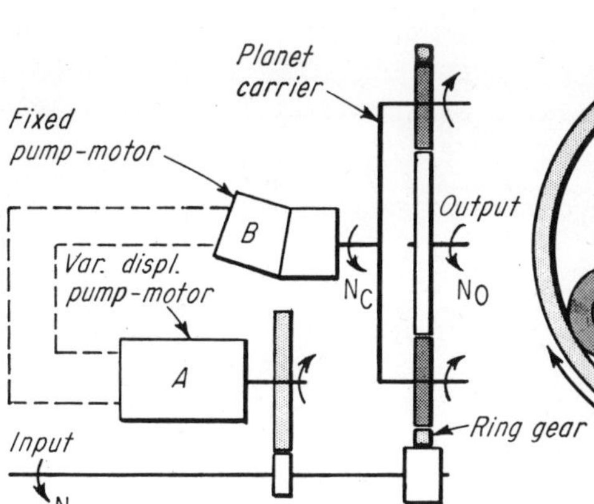

SPLIT-TORQUE DRIVE

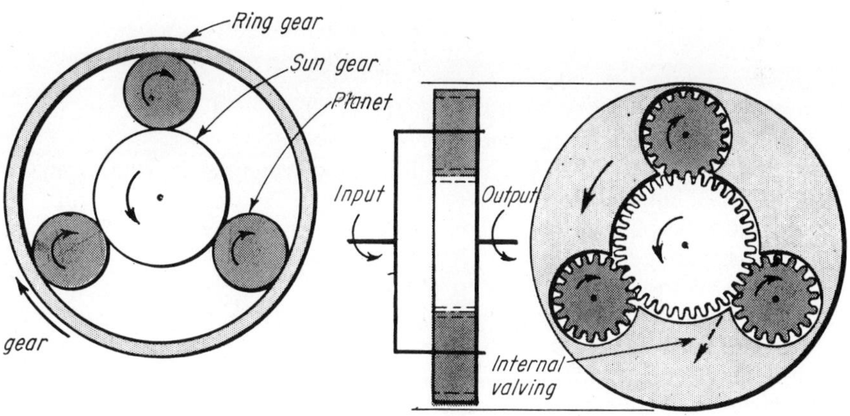

HYDRAULIC-PLANETARY DRIVE

drive is readily applicable. However, a special split-torque drive with several planetaries and gear-shifting devices will work. First cost will be high.

One application of a split-torque drive is for an aircraft alternator: alternator speed is held constant even though the input-shaft speed varies. This is done by regulating the speed of the hydrostatic portion of the drive with a feedback signal from the alternator to correct for changes of output speed. Efficiency is high.

Shiftable gearbox: The speed range of any hydrostatic or hydrodynamic drive can be extended by adding a gearbox with adjustable ratios. For example, a truck with a torque-converter hydrodynamic drive coupled to a five-speed gear transmission can utilize nearly full engine horsepower and full drive efficiency in an output-speed range of over 6:1.

Without the gearbox, output torque, power, and speed depend on the characteristic performance curve of the torque converter, and although the drive can be readily matched to almost any load, speed is not truly adjustable.

Hydraulic-planetary gearbox: If the sun gear and planets of a simple planetary—without ring gear—are designed to double as gear pumps, the rotation of the planets around the sun can be controlled by restricting the pumped flow.

In one commercial version of this, shown on previous page, the input shaft rotates the planet cage and the sun gear is keyed to the output shaft. With 100% restriction of pumped flow, the planets lock to the sun gear, and the output speed is identical to the input. With zero restriction, the planets rotate freely and the output shaft can be held still. In between, any output/input speed ratio is possible.

Valves inside the drive housing restrict the pump flow, and their adjustment is external. The entire drive housing rotates at input speed; the centrifugal action maintains an annular bath of oil to supply the valves.

Performance is not unlike an eddy-current clutch in that a decrease of output speed requires slip and therefore lowers efficiency. Slip in this case is flow through a valve. However, by mixing air with the oil at the high-slip conditions, the pumping losses are reduced. The gear-pump drive also has some of the advantages of an eddy-current clutch, which include compact design and good shock absorption characteristics.

QUESTIONS TO ASK

Answers to these four important questions will help pinpoint the drive type you need.

- How much output power must the drive produce, and how much input power do you have available?
- What torque does the load require at each speed, and what input torque can you provide at each of these speeds?
- How much variation is there in the load speed and torque, and can you accept large speed droop (poor speed regulation) during load changes?
- How efficient must the drive be, and can you dissipate the heat?

The performance graphs on page 244 are sufficient for matching the proper drive to your load, if you can answer the four questions. Keep in mind that the hydraulic drive is simply an adjustable coupling between two pieces of your own equipment.

Efficiencies are lowest at low output speeds for most hydraulic drives, although the decrease is usually relatively small. If the load torque decreases too, the total heat to be dissipated at low speeds rarely exceeds that at full speed.

For example, a 100-hp hydrostatic drive running at 1000 rpm with 90% efficiency has to dissipate 10 hp. The same drive running at 100 rpm will have about 55% efficiency, developing 10 hp and dissipating 5.5 hp.

Suppose instead that a different type of drive were selected, and the efficiency dropped directly with speed. Then 55% efficiency would coincide with 600 rpm instead of 100 rpm, developing 60 hp and dissipating 33 hp—more than three times the losses at full speed!

Heat-dissipating capacity of a hydraulic drive can be defined in terms of its oil-tank surface area. A rule of thumb is

$$A = \frac{20{,}000\,P}{\Delta T}$$

where A is the surface area of the oil tank, in.^{2}, P is horsepower not converted into useful work, and ΔT is the temperature difference, degrees Fahrenheit, between oil and atmosphere.

For example, a 20-hp drive with 80% efficiency must dissipate 4 hp. If the maximum temperature differential between oil and atmosphere is 40 F, then

$$A = \frac{20{,}000 \times 4}{40} = 2000 \text{ in.}^2$$

EDITOR'S NOTE: Specific hydraulic drives and specific components used in hydraulic drives are covered in detail in:

Flow-control valves: **Fluid-power controls,** a book by R. M. Koff and J. J. Pippenger, McGraw-Hill, 1959—Covers all valve types used in hydraulic-drive circuits and illustrates typical circuits.

Hydrodynamic theory and circuit design: **Handbook of fluid dynamics** book edited by V. L. Streeter, McGraw-Hill, 1961; **Design of hydraulic control systems,** a book by E. E. Lewis and H. Stern, McGraw-Hill, 1962; **Oil hydraulic power,** a book by W. Ernst, McGraw-Hill, 1960; and **Fluid power control,** a book edited by J. F. Blackburn, G. Reethof, and J. L. Shearer, Technology Press and John Wiley & Son, 1960.

—*Franklin D. Yeaple*

A comprehensive study of flow through vane air motors in which pressure drops, torque losses and compression and expansion efficiencies are derived in explicit form. These relationships are then made non-dimensional and plotted in two graphs to permit easy design analysis of new and existing motors and a quick determination of the effect of changing the various motor parameters.

analysis and design of air motors

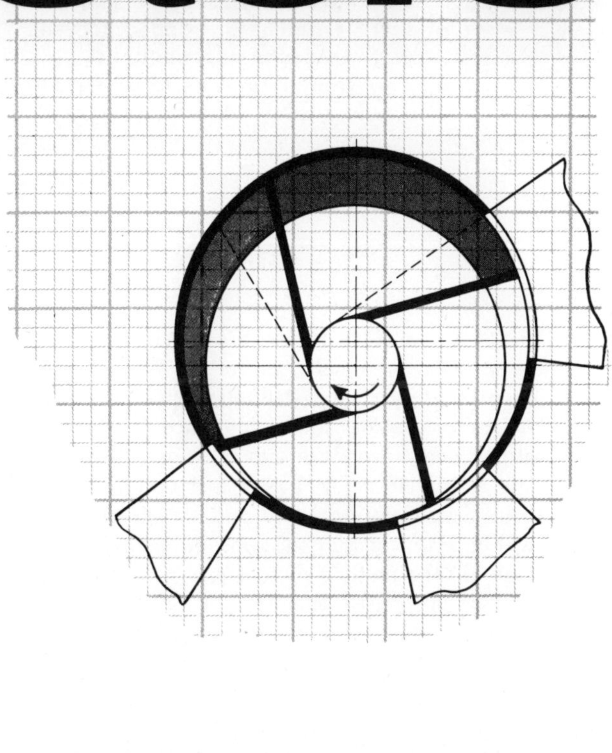

OTMAR E. TEICHMANN
Manager, Product Research
Borg-Warner Corp.

Formerly Manager of the Heat-Power Research Department, Armour Research Foundation.

air motors

symbols

A — Area, in^2.
A_H — Equivalent flow area of lead-in passages, in^2.
A_L — Leakage flow area, in^2.
b — Vane breadth, in.
b' — Projected vane breadth, in.
C — Design constant determined by location of inlet and outlet orifices and number of vanes.
C_1 — A_m/eL
C_2 — Ratio of mean pressure drop to inlet pressure drop.
C_3 — Ratio of mean torque radius to housing radius.
d — $(k-1)/k$
e — Eccentricity of rotor in housing, in.
E_{mech} — Mechanical efficiency.
G — Total weight rate of air flow to motor, lb/min.
G_L — Leakage weight rate of air flow, lb/min.
G_{th} — Theoretical air consumption, lb/min.
g — Acceleration of gravity, ft/sec^2.
h — Oil film thickness, in.
hp — Horsepower.
k — Adiabatic expansion coefficient.
K — Supply air factor, min/in.
n — Polytropic expansion exponent.
N — Rotor speed, rpm.
N_s — Rotor speed, rps.
p_o — Pressure of ambient air (atmospheric), psi.
p_1 — Pressure of air in reservoir, psi.
p_a — Pressure at inlet to space between vanes, psi.
p_b — Pressure of air trapped between vanes after expansion, psi.
p_m — Average pressure to do equivalent work of expansion cycle, psi.
Q — Total air consumption, cfm.
Q_L — Leakage flow, cfm.
Q_{th} — Theoretical air consumption, cfm.
r — Rotor radius, in.
R — Housing radius, in.
R_g — Gas constant.
R_t — Torque radius, in.
S — Motor size factor, in.^{3}.
t — Vane thickness, in.
T — T_{in}/Sp_o
T_F — Friction torque, ft-lb.
T_{in} — Input torque, ft-lb.
T_{out} — Output torque, ft-lb.
T_v — Viscous friction torque, ft-lb.
U — Vane velocity, in/sec.
v_a — Specific volume of air at intake port, ft^3/lb.
V_o — Volume of air at atmospheric pressure, ft^3.
V_1 — Volume of air after compression, ft^3.
V_a — Volume of pressurized air trapped between vanes at instant inlet port is closed, in^3.
V_b — Volume of air between vanes just prior to opening of exhaust port, in^3.
W — Compression or expansion work.
z — Number of vanes.
α — Non-dimensional area, A/R^2.
ε — Non-dimensional eccentricity, e/R.
μ — Dynamic viscosity, slug/ft-sec.
ξ — Non-dimensional pressure ratio, m/r.
ξ_{cr} — Critical pressure ratio.
ρ — Air density, lb/ft^3.
τ — Viscous sheer stress, psi.
Φ — Leakage — inlet ratio, in.
Ψ_L — Leakage flow variable.
Ψ_{max} — Non-dimensional flow variable.
Ω — Motor-inlet ratio, in.

Basic advantages in the use of air motors for portable power are: low weight per horsepower, small size, indifference to overload and stall, cool operation, the availability of speed control, and indifference to dirty or explosive atmospheres. Disadvantages as compared to electric power are: lower efficiency, higher first cost (nearly double the equivalent electric motor) and a maintenance cost which will be approximately the same and perhaps more expensive than electric motor maintenance.

Poor efficiency is the result of the need to charge the air motor for the cost of compressing air to operating pressure as well as the inefficiencies in expanding it during the work cycle. In this article efficiency of air compression and expansion in vane motors will be analyzed and design principles derived that may be used as a basis of synthesizing new designs and predicting their performance.

Power for Air Compression

Actual air compression follows a polytropic curve as shown in Fig. 1. Work required for this compression is given by the equation

$$W = \frac{n}{n-1} p_o V_o \left[1 - \left(\frac{p_1}{p_o}\right)^{\frac{n-1}{n}}\right] \qquad (1)$$

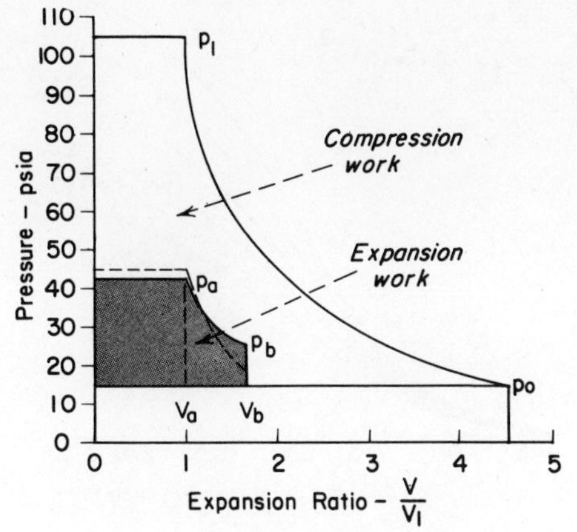

Fig. 1 — Absolute pressure versus expansion ratio for compressor (in color) and air motor (shaded). Note that for this instance the motor is only approximately 25 per cent efficient.

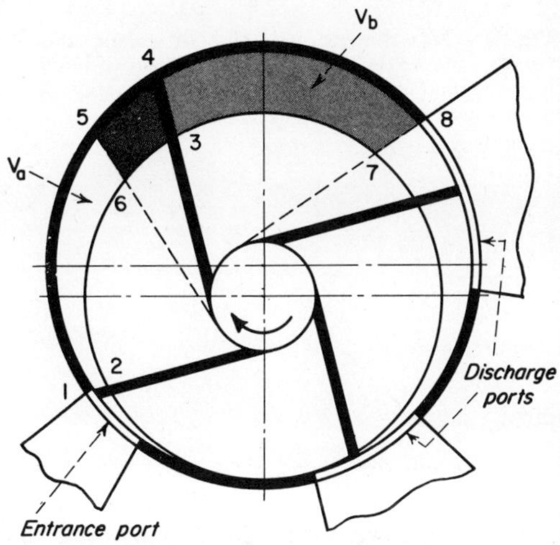

Fig. 2 — Schematic cross-section of air motor. Intake air fills space 1234 at which time entrance port is closed off. This air expands to volume 5678 when discharge port is opened.

If the compression were isothermal less work would be needed and this difference increases with increasing pressure ratio p_1/p_o. Therefore when pressure ratios are higher than 7, compression is carried out in two or more stages with intercoolers between stages to more nearly approximate isothermal conditions. Table I lists the theoretical horsepower requirement for compression of 100 cfm of free air (p_o = 14.7 psia) to 90 psig for isothermal, adiabatic and polytropic conditions. Actual brake horsepower as estimated by the Compressed Air Handbook is approximately 18.0 hp/100 cfm for these conditions.

Power from Compressed Air

If the sequence of operations in a compressor is reversed and the valves properly governed a compressor can be operated as an air motor. The formulas given in Table I are still valid for idealized conditions. Actual air motors are far from ideal as shown in Fig. 1. In this graph the abscissa is plotted as an expansion ratio with the compressed volume assume as unity. To achieve a compression of 7.12 (104.7/14.7) the compressor volumetric change must have been 4.52 for a polytropic compression in which $n = 1.3$.

Compressed air is then piped to the motor. If the motor could receive the air without pressure loss and expansion of 4.52 times were possible within the motor then the motor diagram would be nearly the same size as the compressor diagram, indicating high efficiency. Unfortunately practical limitations in design prohibit this and a typical vane motor is designed with an expanson ratio of only 1.67. Compressed air reaches the entrance port at only 42 psia and the darkened area represents an effective use of the compressed air with an efficiency of only 25 per cent.

This low efficiency may be improved by reducing pressure drop between supply reservoir and entrance port of motor. Further improvement is possible by increasing the expansion ratio in the motor.

Vane Motor Torque

A diagrammatic sketch of a vane motor is shown in Fig. 2. A cylindrical rotor with sliding vanes is set eccentrically in a cylindrical housing. Compressed air is admitted through the inlet ports and fills the cavity between vanes, housing and rotor. Pressure unbalance acting on the vanes causes the rotor to turn. Torque, T, developed at the shaft is proportional to the product of effective vane area, A, pressure unbalance, Δp, and torque radius, R_t, measured from the center of pressure to center of rotor shaft.

$$T = \Delta p A R_t \tag{2}$$

Effective vane area is the projection of the actual vane area on the radial plane. If axial vane length is L, and actual and projected vane breadth b^1, and b respectively, instantaneous torque is the product

$$T = \Delta p b L R_t \tag{3}$$

Volumetric Expansion in Vane Motors

Neglecting leakage, the space between two consecutive vanes is filled with high pressure air when the space is open to the entrance port. Filling stops when the trailing vane

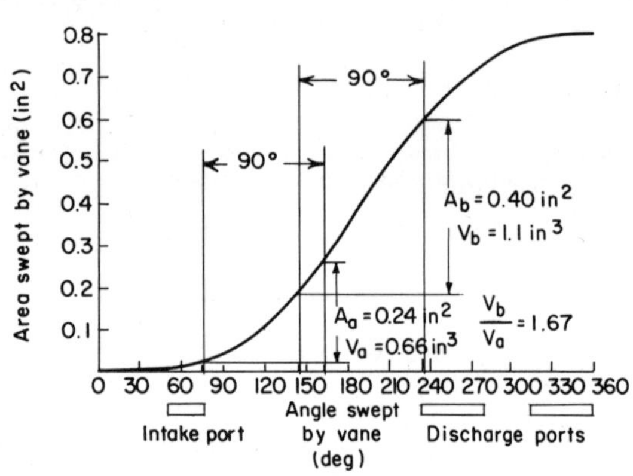

Fig. 3 — Area swept by vane plotted versus angular position of vane. Locations of intake and discharge ports permit reading initial and final volumes and the computation of expansion ratio.

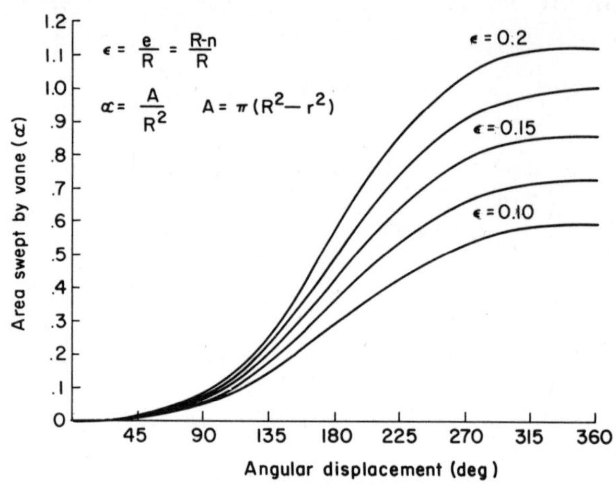

Fig. 4 — Non-dimensional plot of swept area versus vane angle. Having chosen radius of housing and eccentricity, area versus angle may be determined for any motor.

covers the port and the volume of air, V_a, trapped between the two blades expands as the rotor turns eccentrically. Just before the leading vane opens to the exhaust port maximum expansion has taken place to volume V_b. Expansion ratio V_b/V_a is determined by the motor dimensions, the location of entrance and discharge ports and the number of vanes.

A graph of accumulated area swept by the vanes of a typical grinder motor is shown in Fig. 3. Area is plotted as a function of angular position of vane and the locations of intake and discharge ports are indicated. Cross sectional area enclosed between two vanes 90 deg displaced is determined by reading the difference in areas on the ordinate between any two points 90 deg apart on the abscissa. Since swept area is proportional to volume, the volumetrical expansion ratio may be quickly determined from this plot.

An ideal air motor would expand the compressed air to atmospheric pressure within the motor. Physical size limitations in eccentricity and number of vanes restricts the compression ratio severely.

Derivation of curves like that of Fig. 3 requires laborious computation or careful graphical layout. A convenient non-dimensional set of curves has been developed as shown in Fig. 4 which will handle all sizes of vane motors of varying eccentricity. In this plot non-dimensional eccentricity, ϵ, is

$$\epsilon = \frac{e}{R} = \frac{R-r}{R} \tag{4}$$

and non-dimensional area, a, is

$$a = \frac{A}{R^2} \tag{5}$$

Power Output

Instantaneous torque developed by a vane was derived in Eq. (3) above. Projected vane breadth, b, and torque radius R_t, change much less during the working cycle than does the pressure and so average values may be used as scaled off a cross-sectional drawing of a motor.

Pressure follows a polytropic expansion

$$p_b = p_a \left(\frac{V_a}{V_b}\right)^n \tag{6}$$

This pressure may be plotted as shown in Fig. 5 in which the space between adjacent vanes is filled at a constant pressure p_a to volume V_a. It then expands to p_b, exhausts at atmospheric pressure, p_o, and then the volume of air remaining is evacuated in preparation for refill at p_a. Superimposed on this graph is a rectangle (6745) equal in area to the expansion curve (12345). Average pressure, p_m, that will produce the same work over the cycle may be calculated by the following equation.

$$\frac{p_m}{p_a} = \frac{V_a}{V_b}\left\{1 + \frac{1}{n-1}\left[1 - \left(\frac{V_b}{V_a}\right)^{1-n}\right]\right\} \tag{7}$$

$$\Delta p_m = p_m - p_o \tag{8}$$

This equation has been solved for several expansion ratios and for values of $n = 1.4$ and 1.3 in Table II.

When substituted in Eq. (3) the values of Δp_m, b, L, and R_t will solve for an average torque per vane over a full

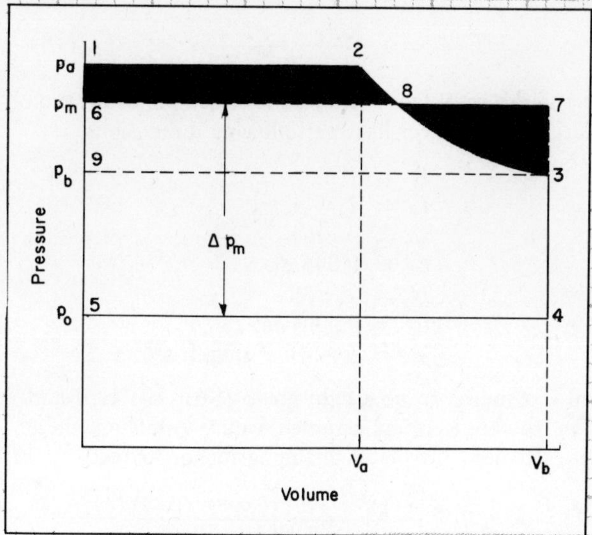

Fig. 5 — Pressure-volume plot of air expansion in motor. Curve 12345 may be approximated by an equivalent area 4567 making the computation much simpler.

TABLE I — Compression Work

	Formula	Theoretical hp/100cfm
Isothermal	$W = p_0 V_0 \ln \left(\frac{p_0}{p_1}\right)$	12.6
Adiabatic	$W = \frac{k}{k-1} p_0 V_0 \left[1 - \left(\frac{p_1}{p_0}\right)^{\frac{k-1}{k}}\right]$	17.0
Polytropic	$W = \frac{n}{n-1} p_0 V_0 \left[1 - \left(\frac{p_1}{p_0}\right)^{\frac{n-1}{n}}\right]$	15.9

Table II — p_m/p_a for Typical Expansion Ratios

V_b/V_a	n	p_m/p_a
1.50	1.4	0.917
1.67	1.4	0.877
2.00	1.4	0.803
1.50	1.3	0.919
1.67	1.3	0.882
2.00	1.3	0.812

revolution. When multiplied by the number of vanes z, the speed, N, and a conversion constant, horsepower output appears in the familiar form

$$hp = \frac{TzN}{63{,}000} \tag{9}$$

Flow

Weight rate of air consumed by a motor without leakage will be called theoretical weight rate of air consumption G_{th}.

$$G_{th} = \rho_a \frac{V_a}{1728} zN \tag{10}$$

where ρ_a is the density of air (lb/ft³) trapped in volume V_a (in³). N is speed (rpm) and z the number of vanes. When expressed in terms of cfm of free air

$$Q_{th} = \frac{G_{th}}{\rho_o} \text{ (cfm free air)} \tag{11}$$

where ρ_o is the weight density of atmospheric air (ρ_o = 0.076 lb/cu ft).

To determine ρ_a, pressure p_a must be known. From the $p-V$ diagram of the motor Fig. 5, if the power developed by the motor is known, the area (12345) may be calculated by

$$\Delta p_m V_b = \frac{396{,}000 \times hp}{E_{mech}\, zN} \text{ (in-lb)} \tag{12}$$

where E_{mech} = mechanical efficiency.

Since V_a and V_b are known Δp_m and p_m ($= \Delta p_m + p_o$) can be determined and p_a may be established by referring to Table II.

$$\rho_a = \rho_1 \left(\frac{p_1}{p_a}\right)^{\frac{1}{n}} \tag{13}$$

where ρ_1 and p_1 are reservoir density and pressure. Thus Q_{th} is evaluated and compared with total air consumption, Q, of the motor.

Leakage

Difference between the actual and the theoretical air consumption is leakage flow.

$$Q_L = Q - Q_{th} \tag{14}$$

Leakage flow occurs mainly at the edge of the vanes and at the end plates. Vane edge leakage occurs only during the time when the air inlet port and the discharge port are straddled by two consecutive vanes. End plate leakage from entrance to discharge ports of the motor is constant and of a larger magnitude than vane edge leakage. For the purpose of this analysis, vane edge leakage will be lumped with end plate leakage.

In addition to Q_L there is a second kind of leakage which takes place after the air has been trapped between the motor vanes. This leakage does not affect Q_L but causes a change in pressure on the vanes. Flow from a trapped volume around a vane to the exhaust port decreases the mean torque acting on the vanes and in order to correct for the corresponding decrease in area of the p — V diagram, a higher p_a is needed as shown in dotted outline of Fig. 1. Area lost due to lowered p_b has to be compensated for by a higher p_a. In this analysis no special correction will be made for this leakage because its effect is similar in nature to the change from adiabatic to polytropic conditions and can be taken care of by a proper choice of the exponent **n**.

Torque

Torque input exerted by compressed air on the vanes of the motor is used to overcome external torque load and internal friction torque losses.

$$T_{in} = T_{out} + T_F \tag{15}$$

There are two main sources of torque losses: bearing friction and vane friction. Bearing friction is low for a motor without speed reduction gears and somewhat larger with them. Experimental data indicate that over-all torque losses are linear with speed for low speeds and increase more rapidly at higher speeds. Assuming a thin film of oil between end face of vane and motor housing, viscous shear may be estimated and compared with experimentally determined data.

Viscous shear stress, τ in the oil film between end face of vane and housing is

$$\tau = \mu \frac{U}{h} = \mu \frac{2R\pi N_s}{h} \tag{16}$$

where h is the oil film thickness, U the vane velocity, N_s the speed of rotor in rps and R the housing radius used as a mean torque arm for all vanes.

$$T_v = \tau t L Rz \tag{17}$$

$$T_v = \mu \frac{2R^2\pi N_s t L z}{h} \tag{18}$$

where t is the vane face thickness. As a sample illustration a typical grinder motor has the following dimensions

$R = 1$ in.
$L = 2.75$ in.
$z = 4$ vanes
$t = 0.096$ in.
$N_s = 90$ rps
$h = 0.0003$ in.
$\mu = 5 \times 10^{-4}$ slug/ft-sec

The oil is assumed to be a light grade (SAE 10) type and the film thickness is an assumed value. Applying these quantities to Eq. (18) and changing inches to feet

$$T_v = 5 \times 10^{-4} \frac{2(1)^2 \pi (90)(0.096)(2.75)(4)}{0.0003 \times 12^3}$$

$$T_v = 0.575 \ ft\text{-}lb$$

Written in more convenient terms viscous friction torque appears as

$$T_v = \frac{\mu \pi z}{20{,}736} \times \frac{D^2 t L}{h} \times N \ (ft\text{-}lb) \tag{19}$$

where all dimensions are in inches, speed in rpm and dynamic viscosity in slug/ft-sec.

Viscous friction torque is here assumed to be proportional to speed. Experimental data on some motors support this assumption. On others in which a more rapid increase of friction torque is noted at higher speeds it may be assumed that film thickness, h, decreases as centrifugal force increases.

10

DASHPOTS, DAMPING, AND FLUID HAMMER

Calculation and selection of dashpots, snubbers, accumulators, and shock absorbers for both oil and air systems

Pressure losses absorb energy

Dashpots and damping

Volumes have been written on dashpot theory. But where can you quickly look up an equation to solve a practical problem? The answer: here

LOUIS DODGE, Hydraulic Consultant, New Richmond, Ohio

WHATEVER your dashpot function is, one of the equations developed here will be appropriate. Mass, inertia, friction, pressure, restriction flow area, piston area, and mechanical force are the parameters. The damping mechanism is assumed to be trapped oil forced by a plunger through an orifice or annular clearance, throttling the flow for a specified stopping distance and time.

An orifice restriction in a plunger develops a resisting pressure proportional to v^2 of the plunger, thus responding nonlinearly with speed. An annular restriction (the clearance between plunger and bore), develops a resisting pressure proportional to v if the flow is laminar, and to v^2 if the flow is turbulent. Orifice restrictions rapidly lose their damping effect as the plunger slows down, in contrast to laminar restrictions. The equations developed here cover both types.

Don't expect to find the exact equation for your exact design, however. You've got to make simplifying assumptions before you can write solvable equations. The calculated example of a cylinder snubber (following) illustrates some of the practical assumptions that can be made. You'll still have to build a prototype, test it,

1. TYPICAL DASHPOTS

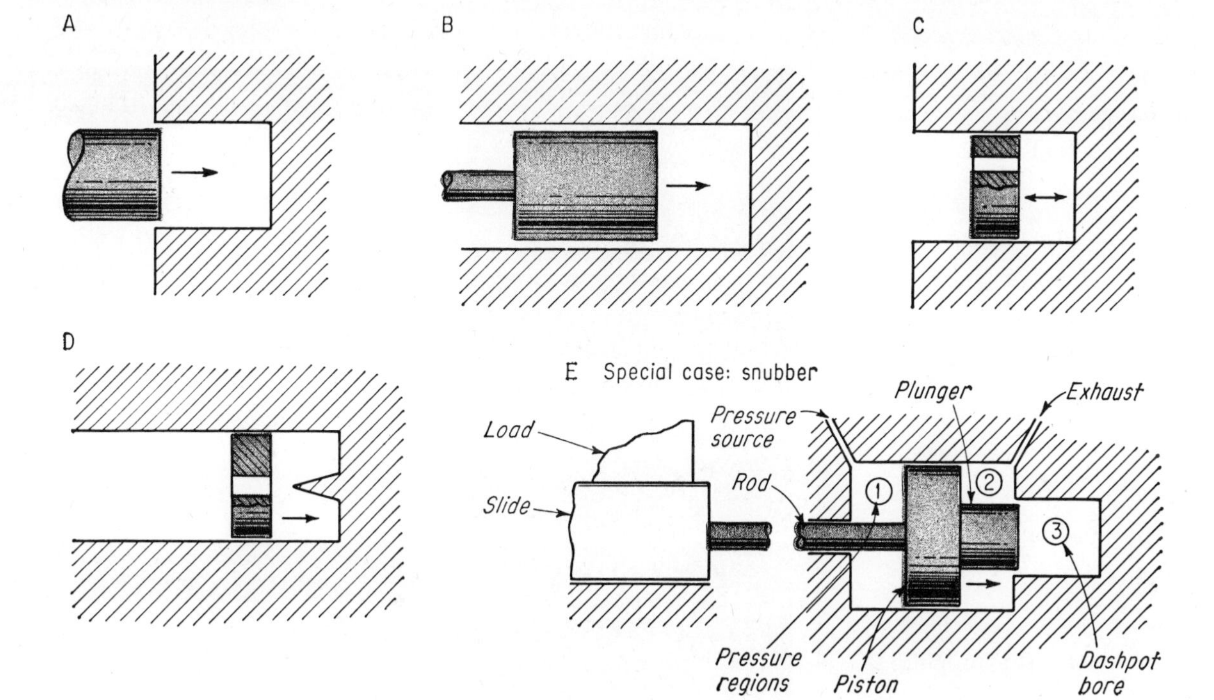

SYMBOLS

a = area, in.^{2}

c = clearance, in.

C_f = coefficient of friction, lb/lb

d = diameter, in.

E_k = kinetic energy, in.-lb; $E_k = \frac{1}{2}mv^2$

F = force, lb

f = friction factor for viscous flow; $f = 64/R_e$

g = acceleration caused by gravity, 386 in./sec^2

H_L = head loss, ft

K = K-factor; $H_L = KV^2 2g$

l = length, in.

m = mass of load, lb sec^2/in.

p = pressure, psi

Δp = differential pressure, psi

q = flow, in.^{3}/sec

R_e = Reynolds Number

t = time, sec

v, V = velocity, in./sec, ft/sec

w = work, in.-lb

x = travel, in.

μ = absolute viscosity, lb sec/in.^{2}

ρ = weight density, lb/in.^{3}

then adjust clearances, backpressures, and spring constants. But at least you can come closer on the first try.

DERIVE PRACTICAL EQUATIONS

With the right equation, and enough assumptions, you can design or analyze a dashpot or damper that will perform any of the three conventional functions:

- Stop or retard an inertial load
- Resist an applied force
- Absorb shock or vibration

The table of equations below summarizes some useful combinations of parameters. Actually, with the seven parameters plus two types of restrictions, you can write dozens more equations, and design numerous styles of dashpots. For detailed information on specific dashpots and damping devices, see the Editor's Note on the last page of the article.

Annulus restriction

Let's take this practical problem as an example: a close-clearance dashpot snubber on the end of a conventional cylinder for a heavy slide (Fig 1E). The equations in the third row of the table apply. Some of the conclusions also will apply to the plain dashpot in Fig 1B (see comments after Eq 10).

Set up an energy balance between input and resisting forces:

work of input = work of resisting forces

Input includes kinetic energy, mechanical force, and pressure (region 1). Resistance includes friction, back pressure (region 2), and dashpot pressure (region 3). Here is the math, based on Fig 1 symbols and Fig 2 assumptions:

$$E_k + w_{mech} + w_1 = w_f + w_2 + w_3 \quad (1)$$

Where

$$E_k = \tfrac{1}{2}mv_i^2 \quad (2)$$

$$w_{mech} = (F_{mech})l_b \quad (3)$$

$$w_1 = \int_o^{l_b} p_1 a_e dx \quad (4)$$

$$w_f = F_f l_b \quad (5)$$

$$w_2 = p_2 a_e l_b \quad (6)$$

$$w_3 = w_{pl} = \int_o^{l_b} (\Delta p) a_{pl} dx \quad (7)$$

Now examine each of the six terms separately. Kinetic energy, E_k, is clear. Mechanical work, w_{mech}, can also remain as is. Input work of source pressure can be handled easiest by carrying it as $F_1 l_b/2$, based on the assumption (Fig 2) that pressure drops linearly to zero during stroke.

Friction work is impossible to express in simple terms because so much depends on the seal and slide design. A properly lubricated slide has a coefficient of friction of $C_f = 0.1$. A piston has negligible resistance in most instances. Rod-seal friction depends on contact pressure between rod and

SUBSCRIPTS ONLY

a = actual

b = bore of piston or dashpot

e = effective

f = friction

i = initial

l = load

m = mean or average

o = orifice

p = piston

pl = plunger

rad = radial

t = total (to stopping)

1,2,3 = pressure regions

TYPICAL DASHPOT EQUATIONS

	Stopping time, t_t, sec
A – Annular restriction (laminar case) inertia *resisted by* p_3	$t_t = \dfrac{1.18\mu(d_m l_b/c_{rad})^3}{mv_i^2/2}$
inertia *resisted by* p_3, friction	$t_t = \dfrac{1.18\mu(d_m l_b/c_{rad})^3}{mv_i^2/2 - w_f}$
inertia, p_1 mechanical force *resisted by* p_2, p_3, friction	$t_t = \dfrac{1.18\mu(d_m l_b/c_{rad})^3}{mv_i^2/2 + (F_1/2 + F_{mech} - F_2)\, l_b - w_f}$
	Stopping time to $v_p = 0.1\, v_i$, sec
B – Orifice restriction (based on oil) inertia *resisted by* p_p	(Eq 15) $t \text{ @ } 0.1v_i = \dfrac{0.9ma_o^2}{v_i a_p^3 K 0.42 \times 10^{-5}}$

seal, and on fluid pressure in the cylinder. Seal C_f = 0.06 to 0.11 for lubricated leather and 0.08 to 0.1 for metal. Initial contact pressure will remain constant up to about 750-psi cylinder pressure in most designs. Rather than go into detail here, carry the friction work term as w_f, to be computed for each design application.

Region 2 work, w_2, is simple to compute because back pressure is assumed constant. Carry this term as $F_2 l_b$ to be consistent with w_1. Be careful with your computation of F_1 and F_2, because the effective areas involve several diameters. Remember that backpressure p_2 must be assumed

text continued, next page

ASSUMPTIONS (Fig 1E and 2)

- Dashpot restriction is an annulus with close clearance and laminar flow (also see orifice restriction sketch)
- Dashpot is centered (if not, damping force will be much less)
- Dashpot action doesn't begin until plunger starts into bore
- Dashpot is used to full depth (otherwise use l_a instead of l_b, and all equations still apply)
- Plunger velocity reduces linearly with x from initial to zero (actual curve is S-shaped)
- Pressure in region I reduces linearly to zero (valid if shutoff valve closes at about same rate as dashpot moves)
- Back pressure is constant
- Friction force is constant
- Mechanical force is constant

2 . . ANNULAR RESTRICTION

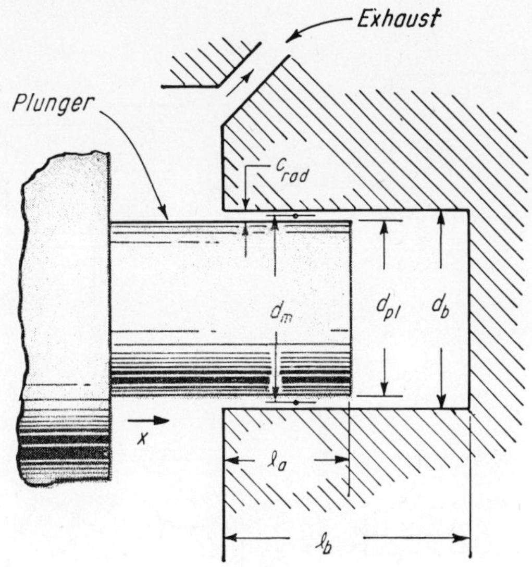

Typical response

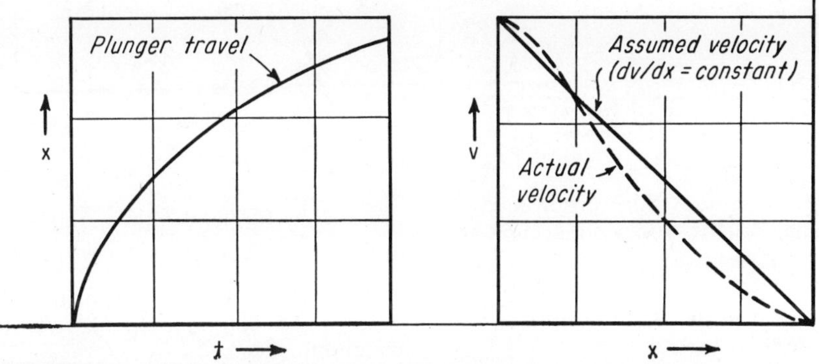

Annular clearance, c_{rad}, in.

$$c_{rad} = d_m l_b \sqrt[3]{\frac{1.18\mu/t_t}{mv_i^2/2}}$$

$$c_{rad} = d_m l_b \sqrt[3]{\frac{1.18\mu/t_t}{mv_i^2/2 - w_f}}$$

(General Eq 10)

$$c_{rad} = d_m l_b \sqrt[3]{\frac{1.18\mu/t_t}{mv_i^2/2 + (F_1/2 + F_{mech} - F_2)\, l_b - w_f}}$$

Bore length needed, l_b, in. (solve by trial and error where necessary)

$$l_b = \frac{c_{rad}}{d_m}\sqrt[3]{\frac{mv_i^2 t_t}{2.36\mu}}$$

$$l_b = \frac{c_{rad}}{d_m}\sqrt[3]{\frac{\left(\frac{mv_i^2}{2} - w_f\right)t_t}{1.18\mu}}$$

$$\frac{1}{l_b^3}\left[\frac{mv_i^2}{2} + (F_1/2 + F_{mech} - F_2)\, l_b - w_f\right] = \frac{1.18\mu d_m^3}{t_t c^3_{rad}}$$

Orifice area, a_o, for special case

(Eq 16)

$$a_o = \sqrt{\frac{(t \; @ \; 0.1v_i)(v_i a_p^3 K0.42 \times 10^{-5})}{0.9m}}$$

Piston travel, x (General equation)

(Eq 14)

$$x = \frac{ma_o^2}{a_p^3 K4.2 \times 10^{-5}}\left[\ln\left(\frac{m}{v_i} + \frac{a_p^3}{a_o^2} K4.2 \times 10^{-5} t\right) - \ln\frac{m}{v_i}\right]$$

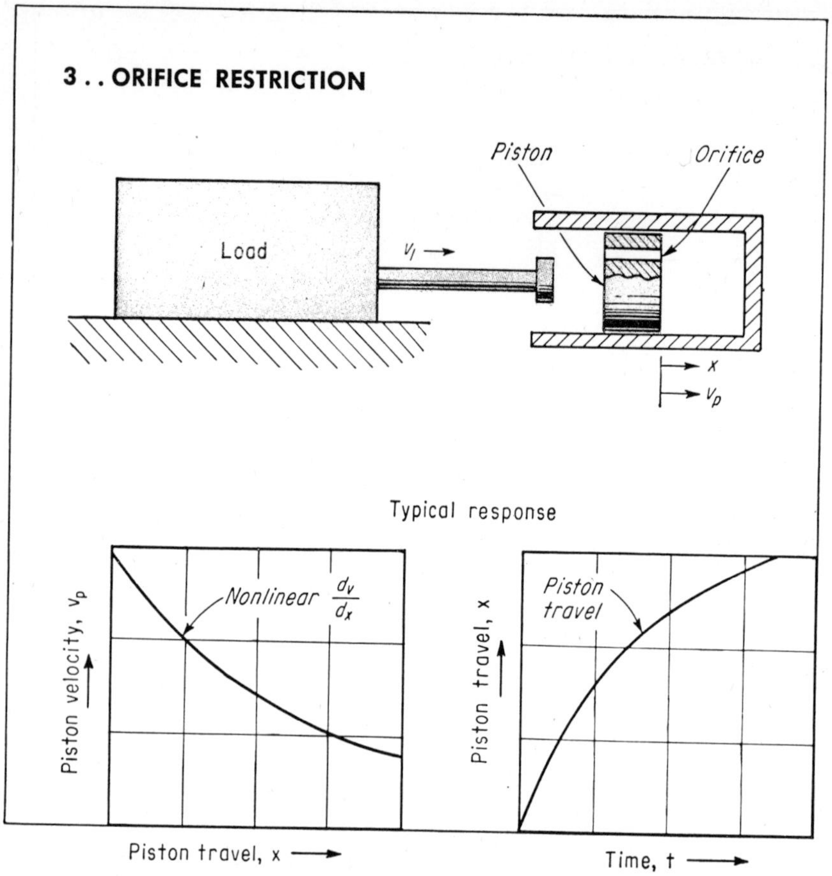

3 .. ORIFICE RESTRICTION

to act over the entire piston area, including a_{pl}, if you also assume that $\Delta p = p_3 - p_2$.

Plunger work, w_{pl}, involves integral calculus, and is derived as follows:

$$w_{pl} = \int_o^{l_b} (\Delta p) a_{pl} dx \qquad (7)$$

where

$$\Delta p = p_3 - p_2 = \frac{12 q_m \mu l_m}{c^3_{rad} d_m \pi} \qquad (8a)$$

Eq 8 is the laminar-flow equation from the combined reprint *Flow resistance*, offered in the Editor's note. Subscript m is mean, so that $l_m = l_b/2$ and $q_m = v_i a_{pl}/2$. Thus Δp is a mean value based on mean flow along the mean plunger length, as if the plunger were moving at a constant rate with plunger length fixed. Eq 8A then becomes

$$\Delta p = \frac{12 v_i a_{pl} \mu l_b/2}{2c^3_{rad} d_m \pi} \qquad (8b)$$

Substitute $\pi d^2_m/4$ for a_{pl} in Eq 7 and 8, and here is the result:

$$w_{pl} = \int_o^{l_b} \frac{0.59 v_i \mu l_b d_m^3}{c^3_{rad}} dx$$

$$= \frac{0.59 v_i \mu l_b^2 d_m^3}{c^3_{rad}}$$

Get rid of v_i by recognizing that $v_i/2 = v_m$ and that $v_m = l_b/t_t$ by definition. Thus:

$$w_{pl} = \frac{1.18 \mu l_b^3 d_m^3}{c^3_{rad} t_t} \qquad (9)$$

Eq 9 is the basic relationship used in the annular-restriction equations in the table. Going back to Eq 1, and substituting the known terms:

$$\tfrac{1}{2} m v_i^2 + F_{mech} l_b + \frac{F_1}{2} l_b =$$

$$w_f + F_2 l_b + \frac{1.18 \mu l_b^3 d_m^3}{c^3_{rad} t_t}$$

Solve for c_{rad}:

$$c_{rad} = l_b d_m \times \sqrt[3]{\frac{1.18\mu/t_t}{\tfrac{1}{2} m v_i^2 + (F_{mech} + \frac{F_1}{2} - F_2) l_b - w_f}} \qquad (10)$$

You can manipulate the individual forces to match your own situation. The mechanical force, for instance, can be negative (resistive) thus combining with the friction force. Also, back pressure in region 2 can be variable, instead of fixed. There even can be a spring involved. And the dashpot restriction can be an orifice (text below) instead of a laminar flow restriction.

An interesting variation is to assume that the plunger is not a snubber, but is a dashpot with a fixed plunger length (Fig 1B). Thus l_m in Eq 8 is not $l_b/2$, but is l_{pl}. The final Eq 10 will be the same, except that the (l_b^3 1.18) term becomes ($2.36 l_b^2 l_{pl}$).

Not included in the dashpot equations are terms to account for acceleration of the fluid as it is forced from the dashpot cup into the annulus. These forces can be considered negligible in all designs except those where the load and piston inertia are very small, as perhaps in an instrument dashpot with a small piston and wide clearances. The fluid accelerating force comes from the dashpot fluid pressure, thus increasing the damping effect and lengthening the total stroke time.

Orifice restriction

Any restriction can be equated with an orifice if flow depends on $\sqrt{\Delta p}$. Thus a wide annular clearance qualifies, and a drilled hole. Just be careful to note the corrections and deviations.

A good example for analysis is a shock absorber with a drilled hole in the piston (Fig. 3). Assume here that there is zero friction, and no spring. Newton's law of motion is sufficient to equate force and acceleration:

$$F = m \frac{dv_p}{dt} \qquad (11)$$

where $F = a_p \Delta p$ is the resisting force of the dashpot, m is load mass (piston is assumed weightless), and v_p is the velocity of the load and piston after they collide and move together.

Orifice relationships and values of pressure-loss coefficients were developed in a series of articles, recently made available as the combined reprint "Flow resistance" (Editor's Note). Here is a key equation:

$$H_L = K V^2/2g, \text{ or in our units}$$
$$\Delta p = K v_o^2 7.5\text{x}10^{-7} \rho \text{ psi (for any fluid)}$$
$$\Delta p = K v_o^2 4.2\text{x}10^{-5} \text{ psi (for oil)} \qquad (11a)$$

where K is the effective K-factor based on inlet, local, and exit restrictions of the passage, and v_o is the flow velocity through the orifice, in./sec. Velocity $v_o = v_p a_p / a_o$, therefore:

$$\Delta p = K(v_p a_p/a_o)^2 4.2\times10^{-5} \quad (11b)$$

Now complete the differential equation:

$$-m\frac{dv_p}{dt} = F$$

$$= a_p K v_o^2 4.2\times10^{-5}$$

$$= \frac{a_p^3}{a_o^2} K 4.2\times10^{-5} v_p^2 \quad (12)$$

For easier manipulation, gather together the constants. Let:

$$C_1 = \frac{a_p^3}{a_o^2} K 4.2\times10^{-5} \quad \text{Thus:}$$

$$C_1 v_p^2 = -m\frac{dv_p}{dt}$$

Rearrange and integrate:

$$\int C_1 dt = \int -m\frac{dv_p}{v_p^2}$$

$$C_1 t + C_2 = \frac{m}{v_p}$$

And finally:

$$v_p = \frac{m}{C_1 t + C_2} \quad (13)$$

To find C_2, let t = zero, so $v_p = v_i$:

$$C_2 = \frac{m}{v_i}$$

The relationship for piston travel for a given orifice size can be developed with another integration:

$$v_p = \frac{dx}{dt} = \frac{m}{C_1 t + C_2}$$

$$\int dx = \int \frac{m}{C_2 + C_1 t} dt$$

$$x = \frac{m}{C_1} ln(C_2 + C_1 t) + C_3 \quad (14)$$

If $x = 0$ at $t = 0$

$$x = \frac{m}{C_1} ln(C_2 + C_1 t) - \frac{m}{C_1} ln C_2$$

The graphs in Fig. 3 show the general shape of v_p vs x and x vs t. In theory, the plunger slows but never stops. For all practical purposes, the dashpot is stopped when $v_p = 0.1\ v_i$, so to determine stopping time use this relationship from Eq 13:

$$t@0.1v_i = \frac{m - 0.1v_i C_2}{0.1 v_i C_1}$$

$$= \frac{0.9 m a_o^2}{v_i a_p^3 K 0.42\times10^{-5}} \quad (15)$$

Here is the expression for orifice area, found by solving Eq 15 for a_o:

$$a_o = \sqrt{\frac{(t@0.1v_i)(v_i a_p^3 K 0.42\times10^{-5})}{0.9m}} \quad (16)$$

which means that for a given desired time to reach a velocity of 0.1 v_i, it requires an orifice area = a_o.

A more useful relationship, based on travel x rather than time t, can be developed too. For this, start with an energy balance instead of Newton's law:

$$(E_k)_{initial} - (E_k)_{@x} = \int_o^x F dx \quad (17)$$

into which Eq 2 and 11 are substituted, letting v_p represent v_x:

$$\tfrac{1}{2}m(v_i^2 - v_p^2) = \int_o^x \frac{a_p^3 K v_p^2 4.2\times10^{-5}}{a_o^2} dx \quad (18)$$

Let y = first term. Then:

$$\frac{dy}{dv_p} = -m v_p, \text{ and } dy = m v_p dv_p$$

Now differentiate Eq 18 and separate the variables:

$$-\frac{dv_p}{v_p} = \frac{a_p^3 K 4.2\times10^{-5}}{m a_o^2} dx$$

Integrate from v_i to v_p, and from 0 to x:

$$ln\frac{v_i}{v_p} = \frac{a_p^3 K 4.2\times10^{-5} x}{m a_o^2}$$

$$a_o = \sqrt{\frac{a_p^3 K 4.2 \times 10^{-5} x}{m ln\frac{v_i}{v_p}}} \quad (19)$$

which means that for a given dashpot to reach a velocity of v_p at point x in its travel, the orifice area must = a_o.

Maximum pressure is at impact, where $v_p = v_i$:

$$\Delta p_{max} = \frac{a_p^2 K v_i^2 4.2 \times 10^{-5}}{a_o^2} \quad (20)$$

Variable orifice

A special case (Fig 4) is an orifice that varies with x to give constant Δp, thus constant deceleration. In order for Δp to be constant, v_o must be constant. But $v_o = a_p v_p / a_o$, therefore: v_p/a_o = constant, meaning that the orifice area must vary directly with piston velocity.

4 . . VARIABLE—ORIFICE RESTRICTION

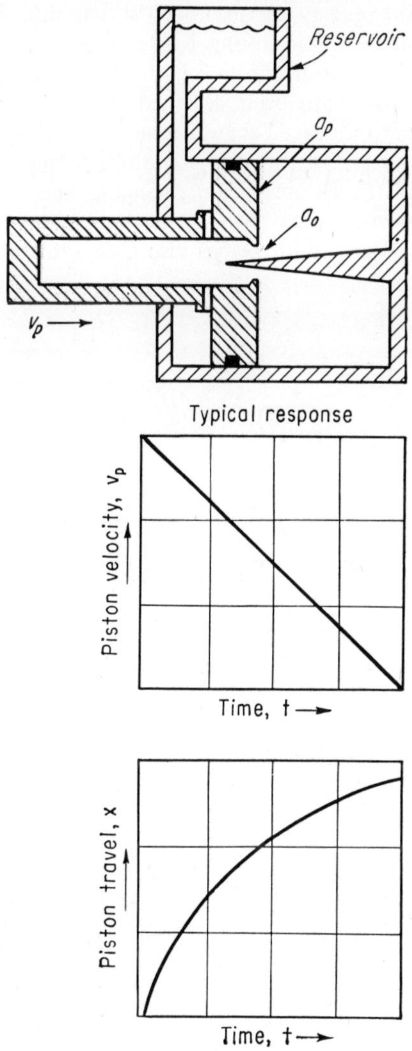

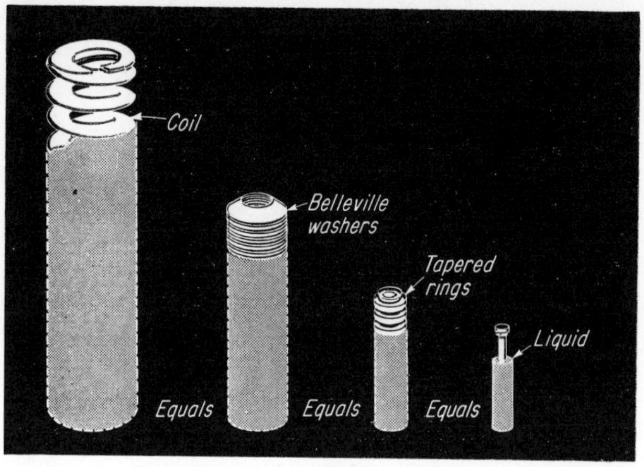

Each of these springs, drawn to same scale, would theoretically handle the same 25-ton load with the same 5-in. deflection. But free length for coil is an impractical 68 in.; for liquid spring, only 17 in.

SAVING SPACE

The spring action given by a liquid has proved itself as a dependable workhorse of the airlines. Now it's ready to move into trucks, cars, buses and many other industrial products. Here is a summary of what it can do, and how to use it.

LLOYD M POLENTZ
Engineering consultant, Whittier, Calif

How much spring action can you expect from liquids? Some are available that will compress to less than 90% of their original volume, with application of 20,000-psi pressure. And the action is elastic: when force is applied, the energy is absorbed, stored and fully released. A single liquid spring can provide as much "springiness" as 30 coil springs of the same length and diameter. Some have been designed to reciprocate at 1000 cycles per minute. Yet by restricting fluid flow with orifices in the piston, there is a controlled slow-down effect useful for shock absorption, damping and timing. The liquid springs below show various ways this is done.

But liquid springs also have some drawbacks too; the table (p. 270) points out limitations as well as advantages. High costs eliminate their consideration for any but those jobs that lower priced springs cannot handle. Loads must be high before a liquid spring design is practical . . . and the resulting high pressures of the compressed fluid must be contained.

Fewer Problems Now

Springiness, size, sealing and **temperature stability** have become satisfactory—a result of much development. **Springiness in small sizes** depends on good fluid compressibility, high pressures, and zero leakage. If a fluid is too "stiff," larger cylinder volumes are needed for the

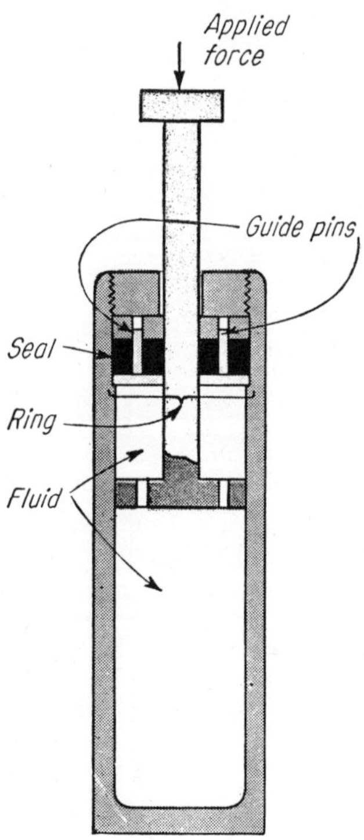

◀ Any liquid spring . . . operates from compressibility of the contained fluid. It is the piston rod entering the fluid chamber and not the piston itself that builds up fluid pressure. Restoring force for this example (a direct-compression type) equals fluid pressure times cross-section area of piston rod. Orifices or check valves in piston can speed or restrict motion to any desirable degree in either direction—useful when shock-absorbing action is needed.

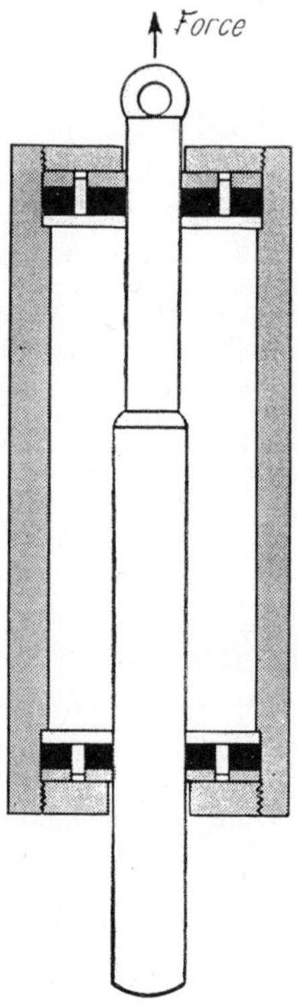

◀ Tension-type . . . builds fluid pressure when rod moves upward because larger-dia, lower half of rod displaces more fluid than upper half. Restoring force, in downward direction, equals fluid pressure times difference in cross-section areas of rod.

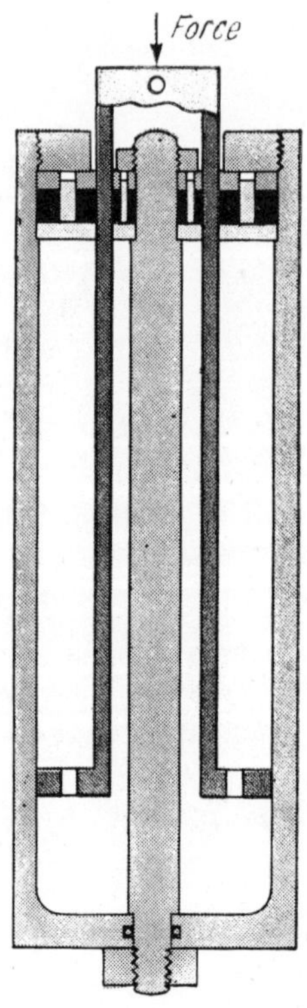

with the LIQUID SPRING

same springiness. But cylinder-wall thickness will increase in proportion to diameter for a given fluid pressure, further increasing total size and weight. Better fluids, better steels and better ways to stress-relieve the cylinders have already resulted in good springiness in very small sizes.

Sealing must be absolutely tight, or fluid will be lost and spring characteristics will change. This problem was solved with the "unsupported-area seal" developed by P. W. Bridgeman of Harvard in his research on high pressures. With these seals, pictured in each of the illustrations, liquid springs have been run through thousands of cycles without adding fluid. The sealing is so complete that the rod is wiped clean of fluid on each stroke, and reservoirs of grease must sometimes be added for lubrication.

A temperature variation—whether ambient or self-generated—will tend to expand or contract the fluid and change the preload. It does not affect the slope of the load-deflection curve, however, because compressibility stays constant over a reasonable temperature range. Natural cooling of the cylinder surfaces will usually prevent the temperatures from rising too high if heat generation is small. (See "When Air Is the Coolant," Feb. 16 '59, p 54). If heat is generated rapidly during a work stroke (such as in a shock-absorber), but cooling intervals between strokes are long, then the mass of the spring and foundation can act as a temporary heat sink. Helpful is the fact that both the compression and expansion of a fluid are elastic and adiabatic, so that the only energy not restored is frictional energy generated in the rod packing, and throttling energy that is purposely generated to absorb shock or slow down movement.

More About Heating

An example—the 50,000-lb, 5-in.-stroke liquid spring in the table—shows that frictional heat energy is not likely to be a problem in normal applications:

Assume 5000-lb average friction load for the entire stroke. Energy absorbed per stroke is 5000 x 5/12 or 2080 ft-lb (2.65 btu). The assembly weighs about 35 lb, has an average over-all specific heat of about 0.15 btu/lb/°F. Therefore temperature rise per stroke is 2.65/(0.15 x 35) or about 0.5°—an amount easily dissi-

◀ **Long-stroke type . . .** has hollow shaft to increase structural rigidity. Restoring force equals fluid pressure times cross-section area of hollow-shaft wall. Strokes up to 5 ft are possible.

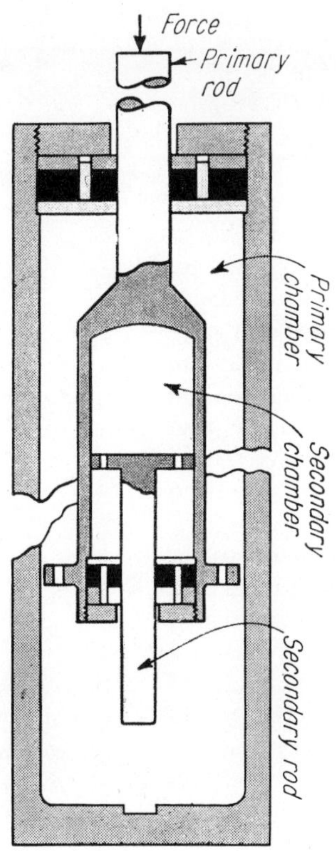

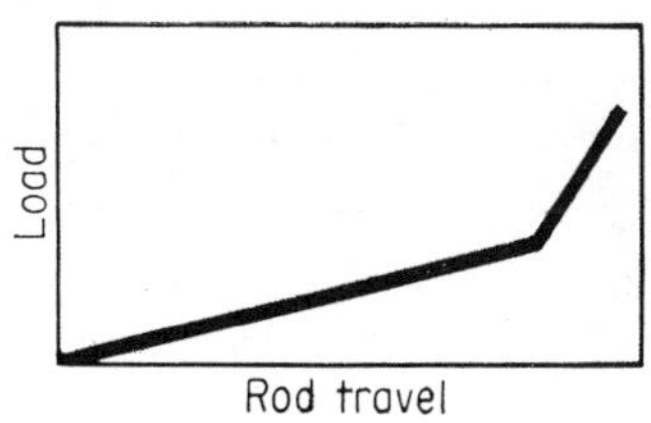

▶ **Compound spring . . .** has dual spring rate, shown by graph. Zero load finds primary rod in upward position; secondary rod resting on bottom of secondary chamber. As load is applied, primary rod moves downward, compressing fluid in primary chamber; while secondary rod moves small distance into its chamber because of differential pressure against its lower end. It will not float in balance until pressures are about equal in both chambers. When load is increased to change-over point (knee of curve), secondary rod will finally be touching bottom of outer cylinder; still more load mechanically forces this rod into its chamber. Slope of load-vs-travel curves is greater under these conditions because contained volume of secondary chamber is relatively small and pressure rise is rapid for small movements of rod.

DESIGNING A LIQUID SPRING

The 50,000-lb, 5-in.-stroke liquid spring design shown in the table was based on a fluid pressure of 50,000 psi and a liquid compressibility of 18% (82% volume remaining at 50,000 psi). Rod diameter is 1 in.; therefore total volume of fluid required (5 x 0.785)/0.18, or 21.9 cu in. Ignoring volume change of steel parts under stress, satisfactory cylinder dimensions, (in inches) are 1.8 ID x 3 OD x 8.6 long. Allowing 3.4-in. total for cylinder ends and seals, and 5 in. for the stroke, the total length is 17 in.

For comparison, the equivalent coil spring was designed for a 65,000-psi maximum stress, including the Wahl correction factor. The dimensions, in., are: wire diameter, 3; coil OD, 11.5; and length, 68. There are 19 active turns and 2 inactive ones. The Belleville washers and tapered ring springs were estimated from published design data.

Performance of Four Typical Spring Types

	Coil	Nested: Belleville washers	Nested: Tapered rings	Liquid
Useful range:				
low load	1 oz	20 lb	2 ton	100 lb
high load	10 ton	100 ton	150 ton	200 ton
force vs deflection	low to high	high	high	med. to high
stroke	short to long	short	short to med.	short to long
Damping ability	low	low	low	low to high
Relative cost	low	low	medium	high
An example: for 50,000-lb load, 5-in. stroke:				
Size, in: length	68	37	24	17
dia	11.5	8	5	3

Compressibility of liquids (approximate)

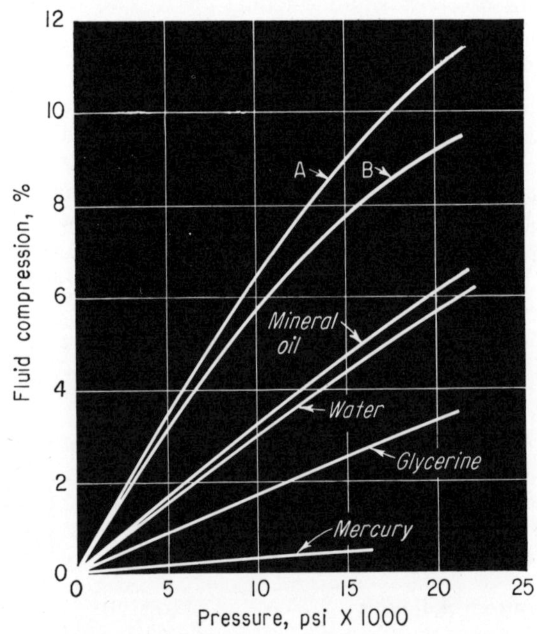

Common fluids . . . for liquid springs are Dow-Corning type F-4029 (Curve A) and type 200 (Curve B). Each is silicone-based, has viscosity of 100 centistokes at 77F. Other four curves are for reference only—these fluids would not ordinarily be selected.

pated. But a smaller liquid spring under rapidly fluctuating loads may experience higher temperature rise; each design must be evaluated on its own merits.

If heat generation and ambient temperature combine to significantly overheat a liquid spring, then cooling means must be provided, or the spring must be compensated for temperature. One compensating design allows the fluid to expand (or contract) into a constant-pressure reservoir during the no-load condition; controlled valving can prevent backflow during the operating stroke. Another answer is to make the rod stroke adjustable—or design the product to be insensitive to small variations in stroke.

This example of a non-compensated design shows what happens if temperature rise is too high: The coefficient of thermal expansion of a silicone fluid is about 6.5 x 10^{-4}/°F. A 100° temperature rise in any contained volume corresponds to a volume compression of 6.5%, and raises the pressure about 10,000 psi. Such a spring would have poor repeatability.

Future for Liquid Springs

The first and still most popular application for liquid springs is in aircraft landing gear. Response times are almost instantaneous, repeatability excellent, maintenance problems nearly nonexistent. And even with high pressures, total stored energy is small, and liquid springs are safe—unless brittle materials are chosen. Other applications are also attractive, particularly where space is at a premium and cost is secondary. Such springs have been successfully used for railway buffers, presses, dies, and draw benches.

Automobiles have been operated experimentally for periods of over five years with the conventional leaf springs replaced by small liquid springs. Costs are higher, but inherently a single dual-rate liquid spring can—in a single package—replace a conventional spring, overload spring and shock absorber. And mass production would certainly make the price competitive. Trucks and station wagons as well as automobiles can take advantage of a dual-rate liquid spring that gives a ride like a passenger car, yet carries loads like a truck.

For greater versatility

Design with air springs

Developed for cushioning shock and maintaining constant height in vehicles, air springs are finding applications as machine mounts and linkage actuators.

J. E. GIECK, Product Manager, Airide-Airmount Div, Firestone Industrial Products Co, Noblesville, Ind

Vehicle suspensions: Air springs make possible a leveling system which keeps the vehicle at constant height regardless of load. A typical suspension system is shown.

Vibration mounts: Both in vehicle suspension and in industrial application where the air springs are used as machine vibration mounts, the fundamental requirement is to keep the natural frequency of the mass-spring system substantially below the disturbing frequency (or forced frequency).

ROLLING-DIAPHRAGM

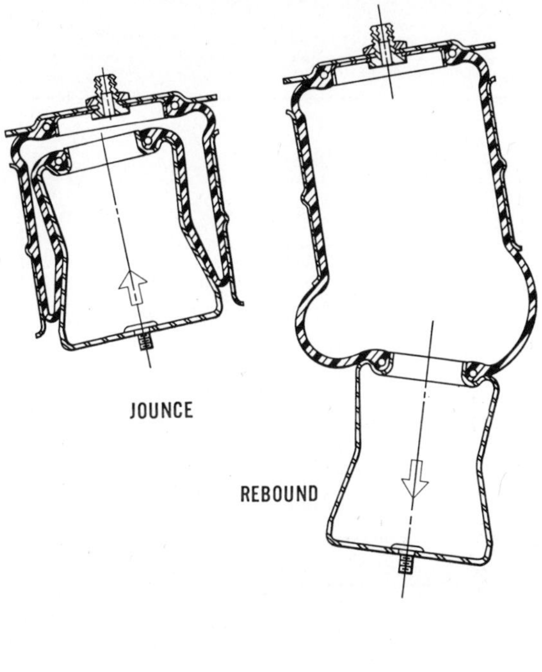

TWO-CONVOLUTION BELLOW

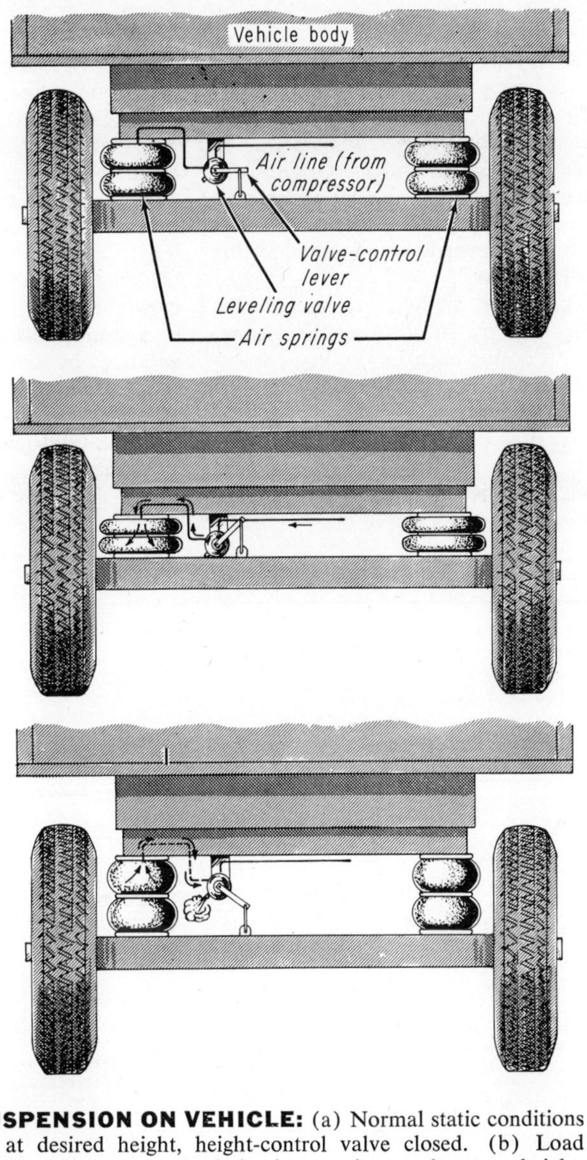

AIR SUSPENSION ON VEHICLE: (a) Normal static conditions—air springs at desired height, height-control valve closed. (b) Load added to vehicle—valve opens to admit air to springs and restore height, but at higher pressure. (c) Load removed from vehicle—valve permits bleeding off excess air pressure to atmosphere and restores design height.

Calculate dash pots without calculus

Air snubber graphs

Nondimensional parameters open the door to a shortcut method for computing orifice area, maximum pressure, and stroke time, of any snubber. An analog computer produced the design data

TOM CAREY, Senior Design Engineer, and T. T. HADELER, Senior Design Engineer, Aerospace Division, Walter Kidde and Company, Belleville, NJ

FORMERLY, the chief difficulty in designing air snubbers for pistons (Fig 1) has been that the equations are nonlinear, thus cumbersome to solve without computers. Original efforts to develop graphical techniques failed, largely because the parameters were too numerous, and crossplotting too laborious.

This new method overcomes the difficulties by substituting dimensionless ratios for each of the key parameters. For instance, pressure is replaced by the dimensionless ratio P/P_i, which is actual pressure vs initial pressure. And piston velocity becomes v/v_i, which is actual vs initial velocity.

Eight of these dimensionless ratios, plus four simple relationships (Equations Table on next page), and the charts in Fig 2, enable you to quickly calculate orifice size for any given combinations of initial pressure, load mass, external constant force, snubber length and diameter, initial and final velocities, initial temperature, and gas constant.

Assumptions. The method has been validated by actual tests, and will work for any snubber from pea-size to several feet in diameter. However, it must be used within these limitations:

- Only the piston-orifice type of dashpot is considered
- The piston is slowed throughout the stroke and is firmly stopped when it reaches the end of the stroke. It does not rebound before the end, nor oscillate or bounce, nor increase in velocity during the stroke because of excessive external force
- Friction is zero
- The external force is constant. If the force will vary, then you must work out solutions for several force levels (say at zero, midpoint, and maximum), and select the calculated orifice size that is the best under the circumstances.

1 . . The snubber model

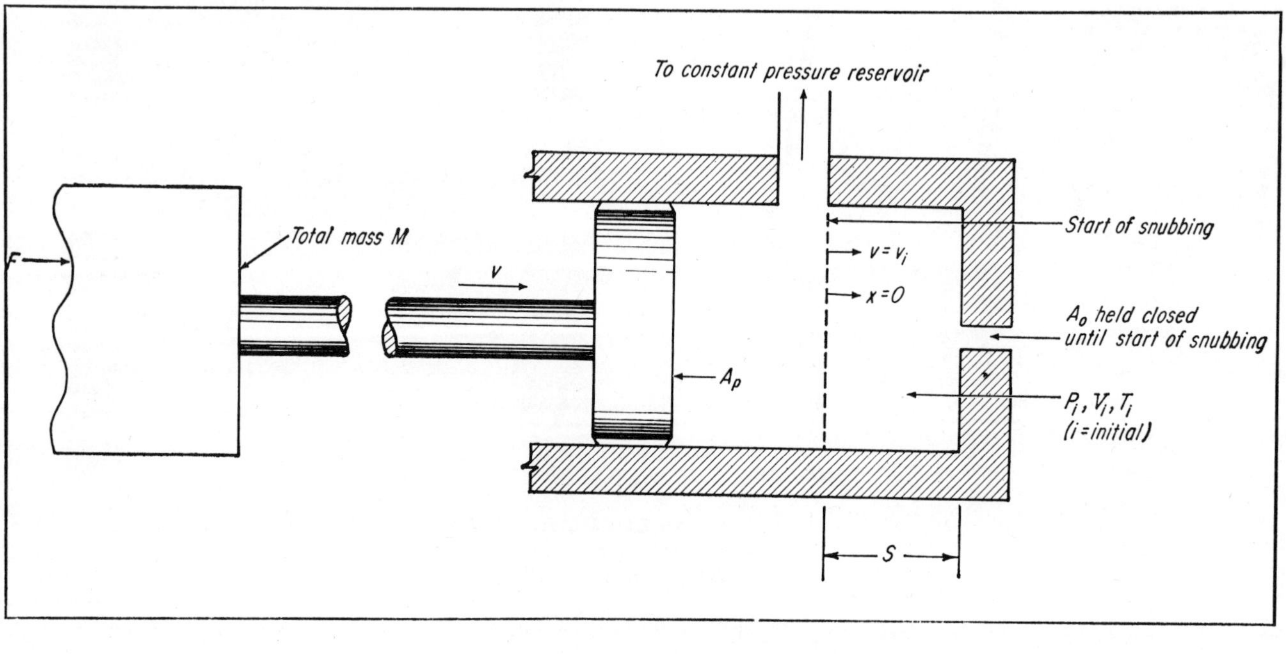

SYMBOLS

a = acceleration, in/sec²

A_o = orifice area, in.²

A_p = piston area, in.²

C_D = orifice discharge coefficient (dimensionless)

F = constant force, lb (includes force of atmospheric pressure)

g = gravitational constant, in/sec²

k = ratio of specific heats (dimensionless)

m = weight of gas in actuator, lb

M = total load mass, lb-sec²/in. (includes piston and rod)

P = snubber pressure, psia

R = gas constant, in. lb/lb −°R

S = snubber active length, in.

t = time, sec

T = gas temperature, °R

V = snubber volume, in.³

v = piston velocity, in/sec

w = orifice flow rate, lb/sec

x = piston travel, in.

ρ = gas density, lb/in.³

SUBSCRIPTS

i = initial (at start of snubbing action) such as in v_i

p = piston

o = orifice

I—Basic equations

Force and motion

1 $$(F - PA_p) = Ma$$

2 $$v = v_i + \int_o^t v\,dt$$

3 $$x = \int_0^t v\,dt$$

Perfect gas law

4 $$P = P_i\left(\frac{m/m_i}{V/V_i}\right)^k = P_i\left(\frac{T}{T_i}\right)^{\frac{k-1}{k}}$$

Gas flow through bleed orifice

5 $$w = C_D P A_o \sqrt{\frac{kg}{RT}\left(\frac{2}{k+1}\right)^{\frac{k+1}{k-1}}}$$

II—Dimensionless parameters

6) $K_E = \dfrac{P_iV_i}{Mv_i^2}$	$\dfrac{\text{stored energy}}{\text{kinetic energy}}$
7) $K_F = \dfrac{F}{P_iA_p}$	$\dfrac{\text{constant external force}}{\text{initial pressure force}}$
8) $K_P = \dfrac{P}{P_i}$	$\dfrac{\text{pressure}}{\text{initial pressure}}$
9) $K_w = \dfrac{w_i}{\rho A_p v_i}$	$\dfrac{\text{initial orifice flow}}{\text{initial displacement flow}}$
10) $K_x = \dfrac{x}{S}$	$\dfrac{\text{piston travel}}{\text{snubber length}}$
11) $K_v = \dfrac{v}{v_i}$	$\dfrac{\text{piston velocity}}{\text{initial piston velocity}}$
12) $K_a = \dfrac{Ma}{P_iA_p}$	$\dfrac{\text{acceleration force}}{\text{initial pressure force}}$
13) $K_t = \dfrac{v_it}{S}$	$\dfrac{\text{travel at initial velocity}}{\text{snubber length}}$

III—Parametric relationships

14) $$K_P = \left[\frac{1 - K_w\int_o^{K_t}(K_P)^{\frac{k+1}{2k}}\,dK_t}{1 - K_x}\right]^k$$

15) $$K_x = \int_o^{K_t} K_v dK_t$$

16) $$K_v = 1 + K_E\int_o^{K_t} K_a dK_t$$

17) $$K_a = K_F - K_P$$

- Air, hydrogen, nitrogen, oxygen and any other gas with a specific-heat ratio of about $k = 1.4$ can be computed with this method. Gases with other than $k = 1.4$ cannot
- Contained air or gas is ideal
- Compression is adiabatic
- Flow through bleed orifice is critical. This is a good assumption except when dashpot initial pressure = atmospheric, such as in snubbers for screen doors

In summary, this is the problem that is solved: A load mass with a constant external force is slowed by a pneumatic snubber until it bottoms at full stroke. Velocity is positive throughout the stroke, but reaches a minimum value at, or near, impact.

The accompanying charts give nondimensional velocity at the point of impact, as a function of nondimensional external force, pressure-volume stored energy of the snubber cylinder, and weight flow of the air or gas.

No data for nondimensional time are given because the choice of parameter ($K_t = v_i t/S$) proved fortunate in that it held a constant value of 0.95 ± 0.15 for every combination of variables. Thus, within reasonable limits of accuracy, there was no need to graph the minor differences.

The maximum value of pressure occurs just before impact at the end of the stroke, and these points are recorded in Fig 3. It is rarely necessary to know the intermediate points from P_i = initial to P_{max} = final, because cylinder wall dimensions are based on P_{max}.

How the data were obtained. Dimensionless equations 11 through 17 comprised the mathematical model for an analog simulation. The dimensionless parameters were varied, and the analog output was recorded and plotted (Figs 2 and 3).

Because the plotted parameters are dimensionless, each point represents an unlimited number of solutions which can be converted to your actual conditions by simple algebra.

Calculated example: 3-in.^{3} snubber

What is the required orifice area A_o when: $M = 0.1$ lb-sec^2/in., $P_i = 100$ psia, $A_p = 3$ in.^{2}, $V_i = 3$ in.^{3}, $S = 1$ in., $v_i = 100$ in./sec, v at end of travel (impact) = 20 in./sec, F = 150 lb, $T_i = 530°$ R, $R = 639.6$ in.-lb/lb-°R (air), and $C_D = 0.9$.

Also determine peak actuator pressure, peak acceleration (negative), and approximate time of stroke.

Solution. From Eq 6, 7, and 11: $K_E = (100 \times 3)/(0.1 \times 100^2) = 0.3$; $K_F = 150/100 \times 3 = 0.5$; K_v at end of stroke = $20/100 = 0.2$.

Now find the value for K_w at $K_E = 0.3$ and $K_v = 0.2$ (Fig 2). Note that there is no chart for $K_F = 0.5$. Therefore you must interpolate between the charts for $F = 0$ and $F = 1$. Do it as follows: Prepare a chart of K_w vs K_F as in Fig 4, by reading values of K_w from each plot in Fig 2. Read off $K_w = 0.375$ from Fig 4.

Solve for true flow w_i in Eq 9 : $w_i = K_w \rho A_p v_i = 0.375\ (P_i/RT_i) \times 3 \times 100 = 0.0332$ lb/sec. Solve for A_o in Eq 5, initial flow:

$$A_o = \frac{w_i}{P_i C_D \sqrt{\dfrac{kg}{RT_i}\left(\dfrac{2}{k+1}\right)^{\frac{k+1}{k-1}}}}$$

$$= 0.016 \text{ in.}^2$$

Read off K_P (max) from Fig 3, knowing that $K_F = 0.5\ K_w = 0.375$: K_P (max) = 10.2. Then true P_{max} (at end of stroke) = $10.2 \times 100 = 1020$ psi, from Eq 8. K_a (max) = $K_F - K_P$ (max) = $0.5 - 10.2 = -9.7$, from Eq 17. Thus $a_{max} = (P_i A_p/M)\ (-9.7) = -29{,}100$ in./sec^2, from Eq 12.

Travel time, based on K_t = about 0.95, is $t = (S/v_i)\ 0.95 = 0.0095$ sec. Slight increase in K_t can be expected in a system with friction.

2 . . Impact velocity vs orifice flow (dimensionless)

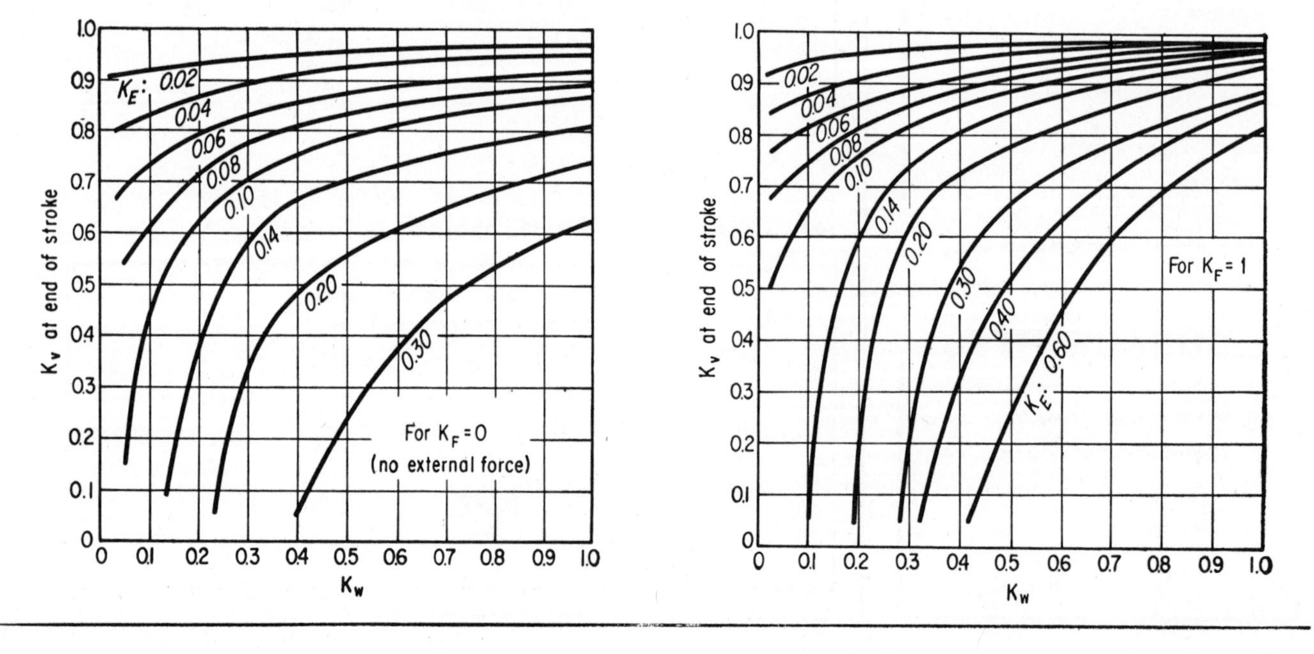

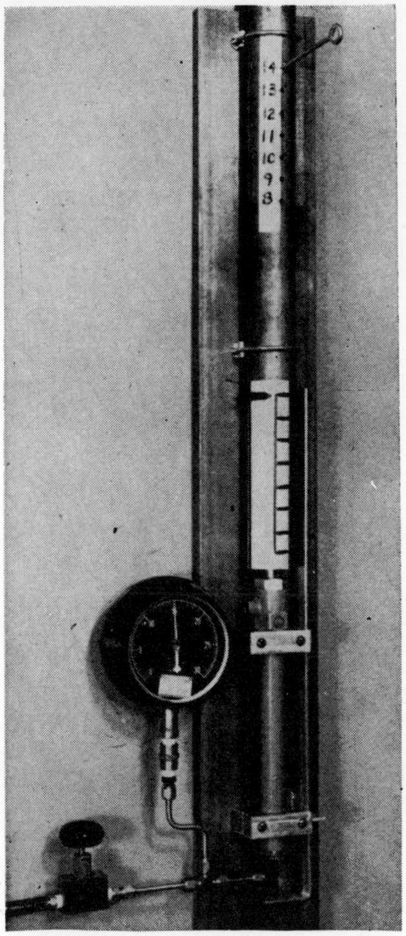

Test rig used to verify method has free-falling weight to simulate mass plus constant force. Actuator is charged to 15 psi initially. When weight strikes actuator rod, the flow orifice is opened instantly (see Fig. 1).

3 . . Max. pressure at impact (dimensionless)

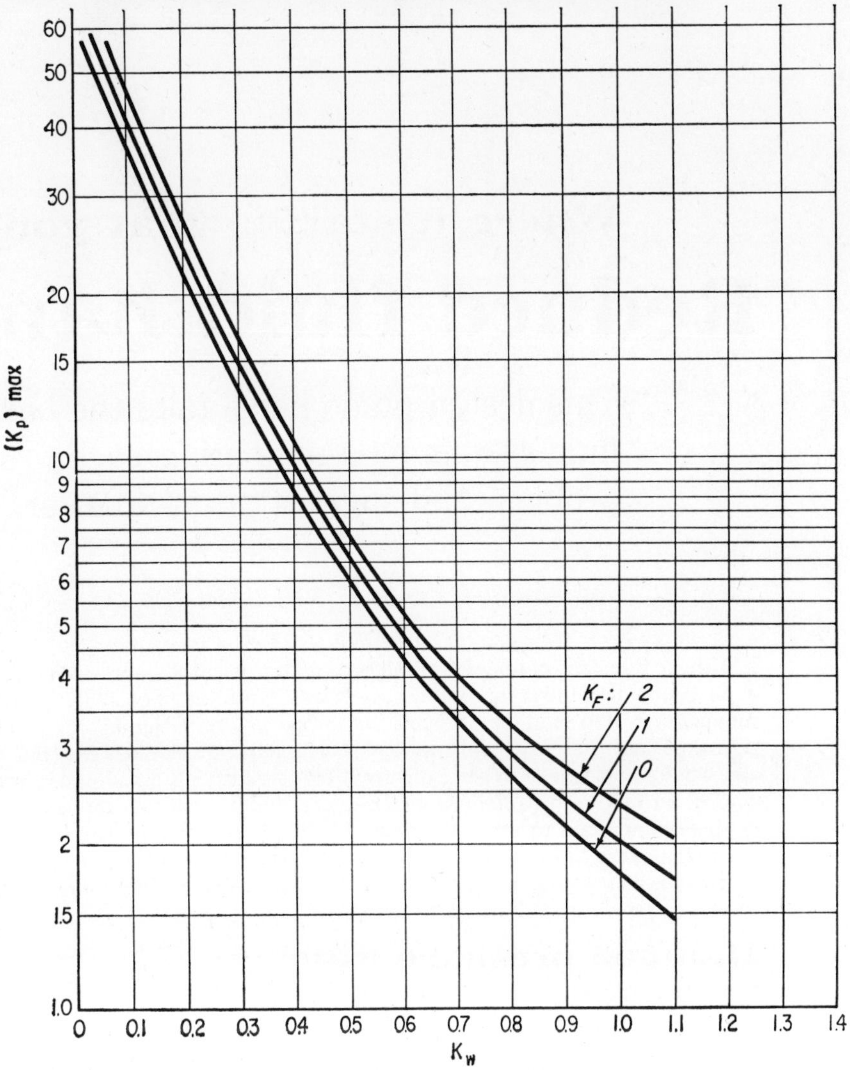

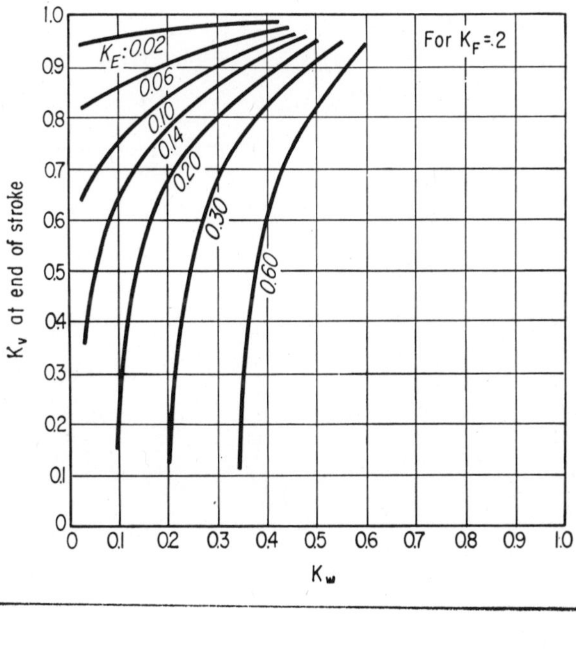

4 . . K_w vs K_F crossplot (the example)

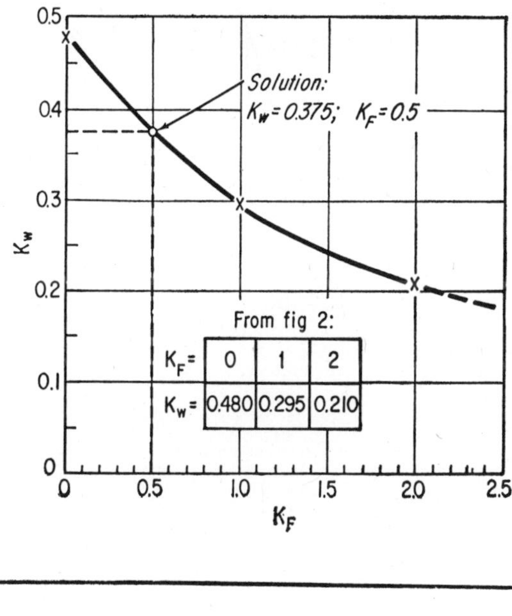

Where it starts, what you can do

Reduce fluid hammer

This design guide helps take the bang out of a fluid system by anticipating each pressure-transient problem and suggesting an answer.

LOUIS DODGE, *Consultant, New Richmond, Ohio*

IN theory, instant closing of a valve or piston will burst the piping, assuming that flow can be stopped in zero time and that fluid and pipewall are wholly inelastic. What actually does occur is known as *fluid hammer;* its theory and ways to avoid the damage it can cause have been the subject of treatises as far back as 1775.

The scientific approach (see Editor's Note on page 290) is awkward when you're designing for production and need quick answers, but with a few simplifying assumptions researchers' findings can be applied easily:

- Single run of pipe (no branches).
- Constant equivalent pipe diameter.
- Elastic system (damping neglected).

The simple reservoir-piping-valve circuit on the next page typifies the circuit, and each pressure-transient problem should first be reduced to that form. The calculation procedure on page 278 shows how to do this. Results will be conservative, because damping in an actual system helps lessen the shock.

Damped pressure waves

(opening and closing shown)

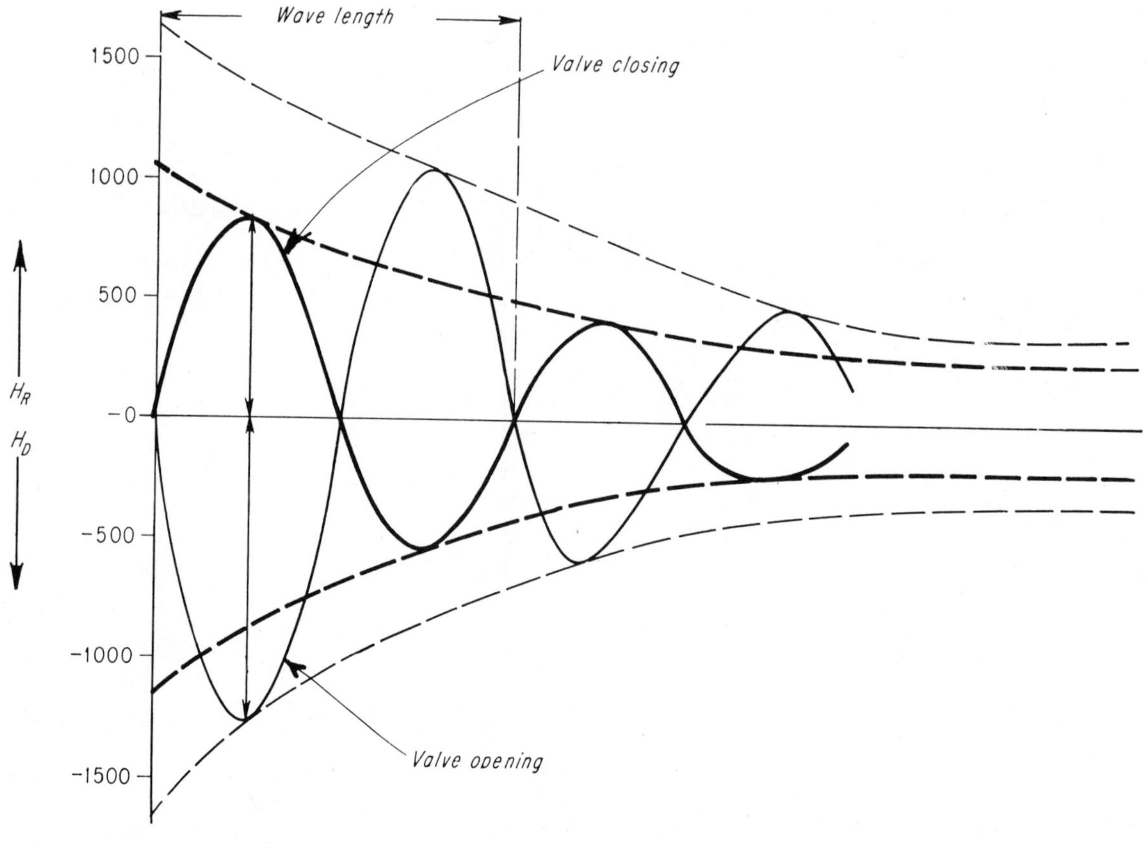

What fluid hammer is

Fluid hammer is the term for the over-all effect of pressure transients in a piping system. More specifically, it is fluid shock, accompanied by noise, resulting from the opening or closing of a valve or other flow-control component. It includes effects of initial system pressure, fluid inertia, piping dimensions, pipewall elasticity, fluid bulk modulus, and valve operating time (to open or close).

During the closing cycle of a valve the kinetic energy of the moving fluid is converted into potential energy. (During the opening cycle the opposite occurs.) Elasticity of fluid and pipewall produces waves of positive and negative pressures which are superimposed upon the original pressure and travel from the valve, through the fluid in the pipe, to the reservoir, and back to the valve. The process would keep repeating indefinitely except that internal friction and other energy losses gradually damp out the waves.

The initial shock of suddenly stopped flow can induce transient pressure changes that exceed the static pressure in both the positive and negative directions. So-called negative pressures in reality cannot be less than absolute zero, but the pipewall contracts during the negative wave and the effect is the same as a negative pressure.

An analogy to a positive-negative pressure wave is a coil spring that is extended and then released: the stresses reverse from tension to compression. Pipe failure could occur from negative pressure but rarely does, because the pipe is stronger in the inward direction.

Burst piping is not the only damage possible from excessive pressure surges. Any component in the circuit or mounted on the same structure is subject to impact stresses induced by the pressure wave. And even if there is no physical damage, the noise alone can be intolerable.

Critical period: The time T_c of one round trip of the pressure wave from valve to reservoir and back is the critical time (one period) that establishes the severity of the fluid hammer.

If the valve closes within the critical time (called fast closing) the pressure rise is the maximum. It doesn't matter how much faster than T_c the closing time is as long as the fluid is stopped completely before the first pressure wave completes the round trip.

If the valve does not close within critical time T_c, the maximum pressure rise will be less. Special equations are needed.

Hypothetical pressure wave

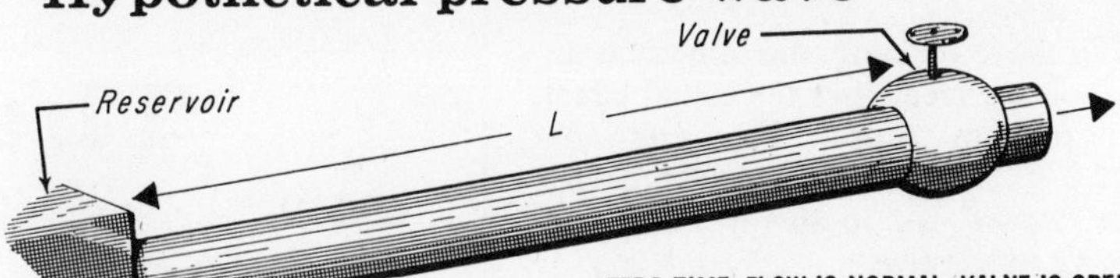

ZERO TIME: FLOW IS NORMAL, VALVE IS OPEN.

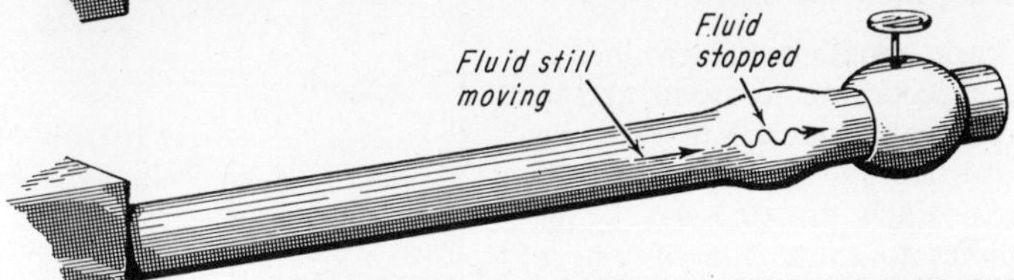

NEXT INSTANT: VALVE SLAMS ALL THE WAY CLOSED, FLUID "PILES UP."

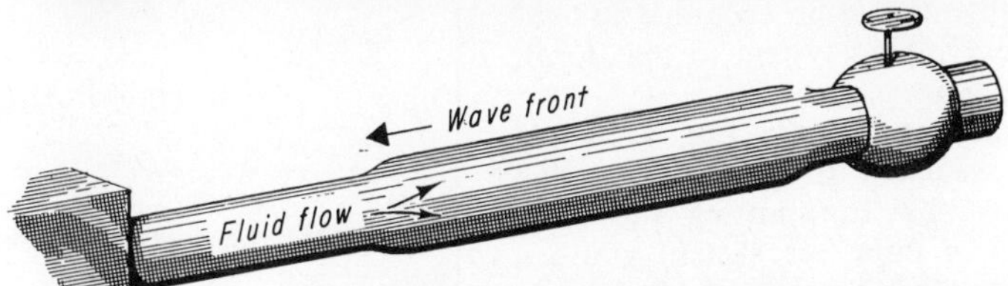

PILE-UP TRAVELS ALMOST AT SPEED OF SOUND, FED BY FLUID INERTIA.

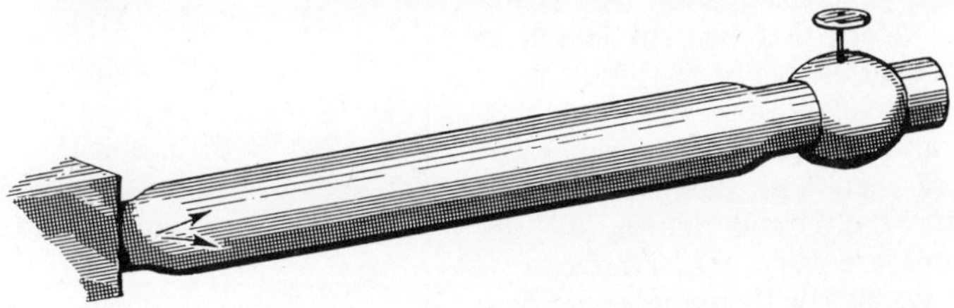

INERTIA OF FLUID INSIDE PIPE IS USED UP (RESERVOIR FLUID HAS NEGLIGIBLE VELOCITY).

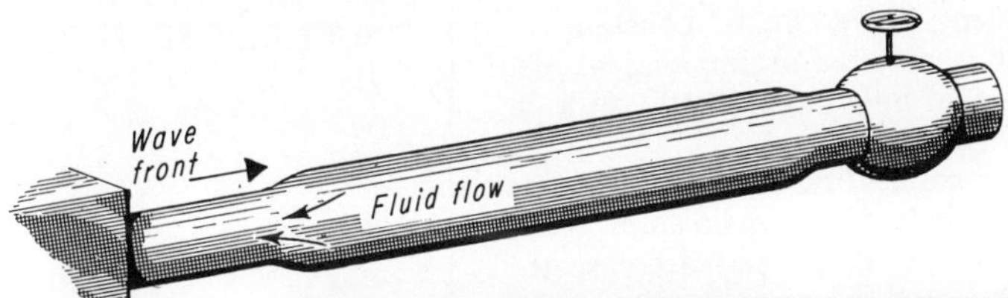

EXPANDED PIPE CONTRACTS AND SQUEEZES FLUID BACK TO RESERVOIR.

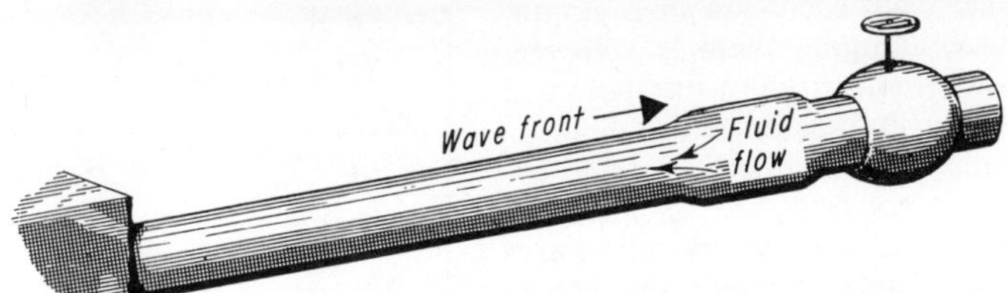

ALMOST ALL FLUID IS SQUEEZED OUT AND PIPEWALL RAPIDLY CONTRACTS.

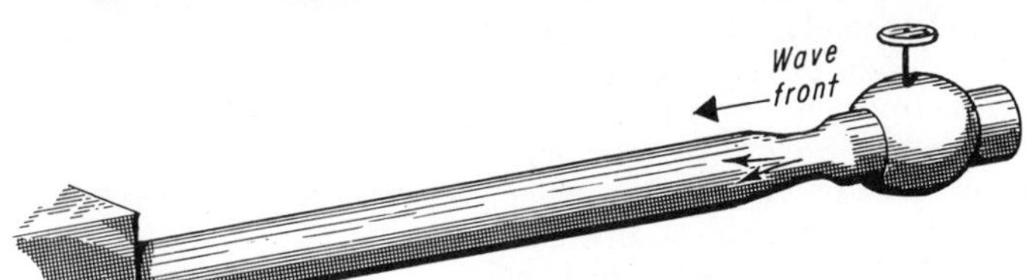

PIPEWALL OVERSHOOTS AND GOES INTO COMPRESSION.

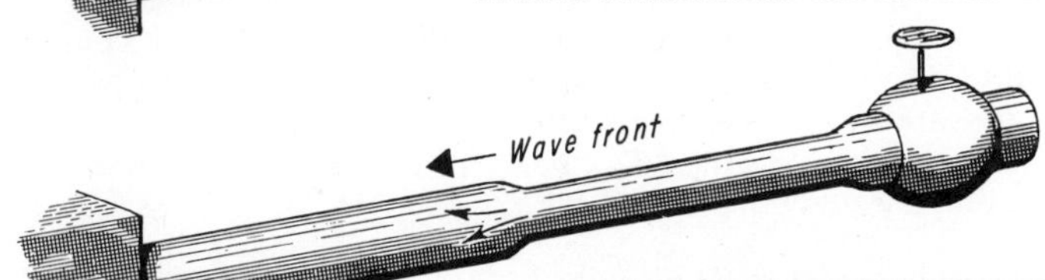

COMPRESSION CONTINUES TOWARDS RESERVOIR.

Note: Pipewall alternately contracts and expands until the energy is dissipated. The entire process takes place in less than a second in most cases.

For valve opening, the same reasoning holds, except that the initial effect is a pressure drop. The graph on page 276 compares the damped pressure sure waves for both opening and closing of a valve.

Calculating fluid hammer

The basic equations are grouped on this page according to speed and direction of valve operation: fast closing, slow closing, fast opening, slow opening. Each pressure-wave equation applies to single run of pipe of constant diameter, so the first step in the mathematical procedure is to convert your piping system to that form. Here is how:

Equivalent circuit: For a simple series system with pipes of varying diameter but transporting the same amount of fluid (see sketch) you can calculate the average fluid velocity using the general equation (Eq 1) for $V_{F(avg)}$. Note that over-all length L in the equation is not modified.

Sharp bends (more abrupt than $5 \times d_i$) and elbows lessen the shock because of their high damping effect—thus for the circuit shown, all the equations are very conservative.

For systems with branches or parallel circuits the conversion to an equivalent single run of constant diameter is difficult or impractical, depending on how much math you want to tackle.

In general, branches add volume and elasticity to the main line. They diminish the shock and decrease the magnitude of the pressure wave. The more branches there are in a circuit, the more damping there is. Hence, complex circuits in most instances will not have a fluid-hammer problem.

Parallel non-identical circuits are impossible to calculate in a simple procedure because the pressure waves will travel in a different manner in each line and will not return to the valve simultaneously. Parallel identical circuits can be calculated individually and pose no special problem.

Safe pipe size: Eq 2 to 10 determine maximum surge pressures. For initial sizing of your system piping, use simplified Eq 7, $P_R = 50V_F$, which is accurate enough to estimate pressure rise. This equation is based on the following: oil weighing 55 lb/ft³; steel pipe; pressure-wave velocity of 4300 ft/sec; closing time less than T_C. For better accuracy use Eq 5 and the V_P graph.

It is simple to calculate pipe stress from Eq 4:

$$S = (P + P_R)d/2e$$

where S is hoop stress, psi; $P + P_R$ is calculated surge pressure, psi. Typical steel tubing has an allowable stress of 85,000 psi. For other materials, use these figures: soft copper, 36,000 psi; half-hard copper, 40,000 psi; annealed stainless steel, 85,000 psi; 316 or 314 stainless steel (¼ hard), 12,000 psi. All are based on temperatures up to 100 F.

But be careful how far you trust simple arithmetic. Remember these facts:

- Stresses and deformations caused by impact pressures are double those caused by slow application of the same pressure.
- Safety factors from 4 to 8 are usually required in rigid piping. JIC (Joint Industry Conference) recommendations for rigid piping have set safety factors of 8 for a working pressure of 0 to 1000 psi, 6 for 1000 to 2500 psi, and 4 for pressures above 2500 psi.

Ordinary surges won't harm the small-diameter tubing and piping common in most hydraulic and pneumatic

The equations

GENERAL

$$V_{F(avg)} = \frac{l_1v_1 + l_2v_2 + l_3v_3 + \ldots}{L} \quad (1)$$

where

l_1 = **length of first section of pipe in a series of such sections**

v_1 = **velocity of fluid in first section**

$$V_P = \frac{12}{\sqrt{\frac{\rho}{g}\left(\frac{1}{\beta} + \frac{d_i}{eE}\right)}} \quad (2)$$

T_C = $2L/V_P$ **sec, which is time of pressure-wave round trip** (3)

S = $(P + P_R)\, d_i/2e$ **(hoop stress in pipewall)** (4)

FAST CLOSING ($T \leq T_C$)

$H_{R(max)}$ = V_PV_F/g, **ft** (5)

H_R = $V_P\Delta V_F/g$ **ft, for fast partial closing** (6)

ΔV_F = **initial** V_F **– final** V_F

P_R = **50** V_F, **psi (Simplification of equation 5 for "average" conditions)** (7)

SLOW CLOSING ($T > T_C$)

$H_R = M - H - \sqrt{M^2 - m_1^2}$, **ft** (8)

$M = m_1 + m_2$

$m_1 = H + H_{R(max)}$

$m_2 = V_F^2\,(V_P - 2L/T)^2/2g^2H$

FAST OPENING ($T \leq T_C$)

$H_D = \sqrt{n_F\,(2H + n_F)} - n_F$, **ft** (9)

$n_F = H_{R(max)}^2/2H$

SLOW OPENING ($T > T_C$)

$H_D = \sqrt{n_S(2H + n_S)} - n_S$, **ft** (10)

$n_S = 2V_F^2L^2/g^2T^2H$, **ft**

THE SYMBOLS

P = system static pressure, psi

P_R = pressure rise, psi

H = system static head, ft (H for oil $= 2.52\,P$; H for water $= 2.31\,P$).

H_R = pressure rise, ft

H_D = pressure drop, ft

V_F = velocity of fluid, ft/sec

V_P = velocity of pressure wave in fluid, ft/sec (depends on pipe dimensions and fluid)

E = modulus of elasticity of pipewall, lb/in.^{2} (E for steel $= 30 \times 10^6$; copper $= 15 \times 10^6$; cast iron $= 12 \times 10^6$; aluminum $= 10 \times 10^6$; asbestos cement $= 3.4 \times 10^6$)

β = bulk modulus of elasticity of fluid, lb/in.^{2} (β for water $= 294{,}000$; β for oil $= 250{,}000$)

L = pipe length, ft

d_i = pipe inside diameter, in.

e = pipewall thickness, in.

S = hoop (circumferential) stress in pipewall, psi

T = valve closing or opening time, sec

T_C = critical period, sec. $T_C = 2L/V_P$

g = acceleration of gravity, 32.2 ft/sec^2

ρ = fluid density, lb/ft^3

TERMINOLOGY

Fluid hammer — total effect of pressure transients. Also called water hammer.

Pressure rise — increment above system pressure.

Pressure surge—peak pressure in absolute units during a pressure rise.

Negative pressure—term for the hypothetical fluid pressure sometimes used to explain pipe contraction resulting from elastic springback of a pipewall after a positive-pressure surge.

Equivalent single run of piping

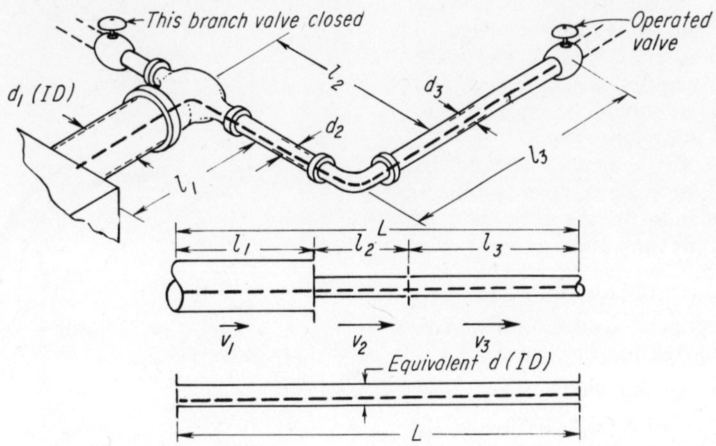

Pressure-wave velocity V_P

(average values for oil or water)

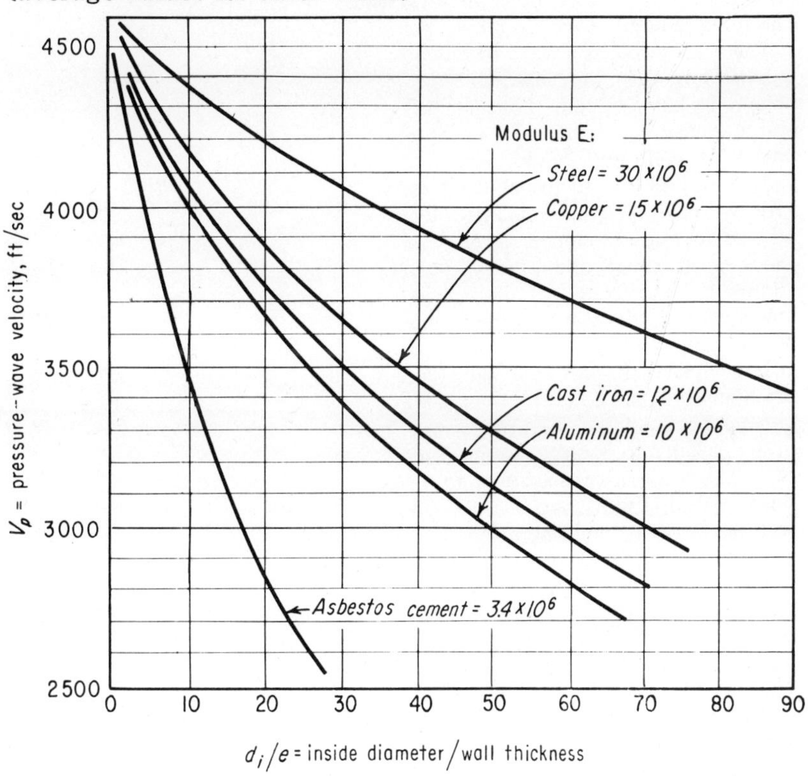

systems. But the ability to withstand surge goes down as tubing size goes up.

If calculated surge pressure is too high for the maximum wall thickness available from stock (very likely if tubing diameter is more than 2 in. and pressures are high) there is a simple answer: use two or more lines of smaller diameter.

For example, assume that flow velocity is 20 ft/sec through 2-in.-dia. steel tubing with a 0.120-in. wall. Allowable stress is 85,000 psi. Let system static pressure equal 2000 psi.

Using Eq 7, $P_R = 50 \times 20 = 1000$ psi. Surge pressure $P + P_R = 2000 + 1000 = 3000$ psi. Hoop stress $S = 3000 \times 2.0/(2 \times 0.12) = 25{,}000$ psi. Safety factor at 3000 psi must be 4 or more, so the 85,000 allowable stress is inadequate.

One answer is to select heavier-walled tubing, but a search will reveal that there's not much available in large tubes with heavy walls.

A better answer is to parallel three 1¼-in. lines—they have greater total flow area and also can handle surges better. Hoop stress at 3000 psi for a 1¼-in. tube with a 0.109-in. wall is $3000 \times 1.25/0.218 = 17{,}200$ psi, which gives a safety factor of more than 4. And because the larger flow area reduces velocity V_F, the effective safety factor is even greater. Cost is less too: three small tubes are cheaper than one large tube, and they are easier to bend and flare because of the smaller outside diameter and thinner wall thickness.

Pressure lines should be rigidly supported at short intervals to prevent vibration and subsequent damage to flared tubing ends, fittings, and seals. Accompanying cylinders, pumps, and fluid motors should be mounted to withstand vibration.

Even if surge damage is unlikely, there still may be a noise problem. Noise depends on the time rate of change of pressure and can occur even at low absolute pressures. Noise is not eliminated merely by making the piping stronger. You must damp or smooth out the pressure wave.

Reducing fluid hammer

The discussion will be based on noise, because this is the best indicator of a serious condition. Diminishing noise solves the shock-damage problem. Fast-acting valves are the primary source of fluid-hammer noise, but there are other sources such as cylinders, pumps, and fluid motors. The same remedies apply to all.

There are two basic ways to reduce noise: 1) modify the valve or component to have smooth motion, preferably sinusoidal, within the time allowed for operation; 2) add capacitive volume or elasticity to absorb the pressure waves. Sketched on this page are devices that reduce noise and shock. Sometimes, of course, you can eliminate the problem by reducing system flow, or by increasing pipe diameter to reduce the velocity.

Inexpensive changes to a valve can often smooth out flow. Examples are valve spools tapered to throttle flow, nested springs to control closing force of a solenoid valve, dashpots to reduce end-of-stroke shock, and cams to time valve or pilot operation.

Cylinders should have cushions or dashpots to ensure shockless reversal, particularly if the load has high inertia and is controlled by fast-reverse valves. A previous chapter on pistons (see pg 232) describes a dozen helpful techniques, including a sine-cosine gear train to produce sinusoidal cylinder and valve motion.

For example, suppose a cylinder driving a load of 5000 lb at 1.4 ft/sec is controlled by a valve closing in 0.08 sec. Normal system pressure is 200 psi and effective piston area is 12.5 in.^{2}. Further assume that the frictional load during stopping just balances what's left of the 200 psi after being throttled by the valve. Thus the shock pressure principally will be caused by inertia of the load and can be calculated easily:

Acceleration $a = dv/dt = 1.4/0.08 = 17.5$ ft/sec^2

Force $=$ mass $\times\ a = 5000 \times 17.5/32.2 = 2720$ lb

Pressure $P = 2720/12.5 = 218$ psi

The actual pressure surge will be smaller than 218 psi because the equations neglected the compressibility of the oil trapped in the cylinder and the elasticity of the metal parts surrounding the oil. However, this is more than offset by the fact that sudden pressures are twice as damaging as gradual pressures. Suggested: Assume an effective peak pressure of 400 psi and allow a safety factor of 8. Better: Add a dashpot or a decelerating valve to reduce the shock.

Typical pulsation

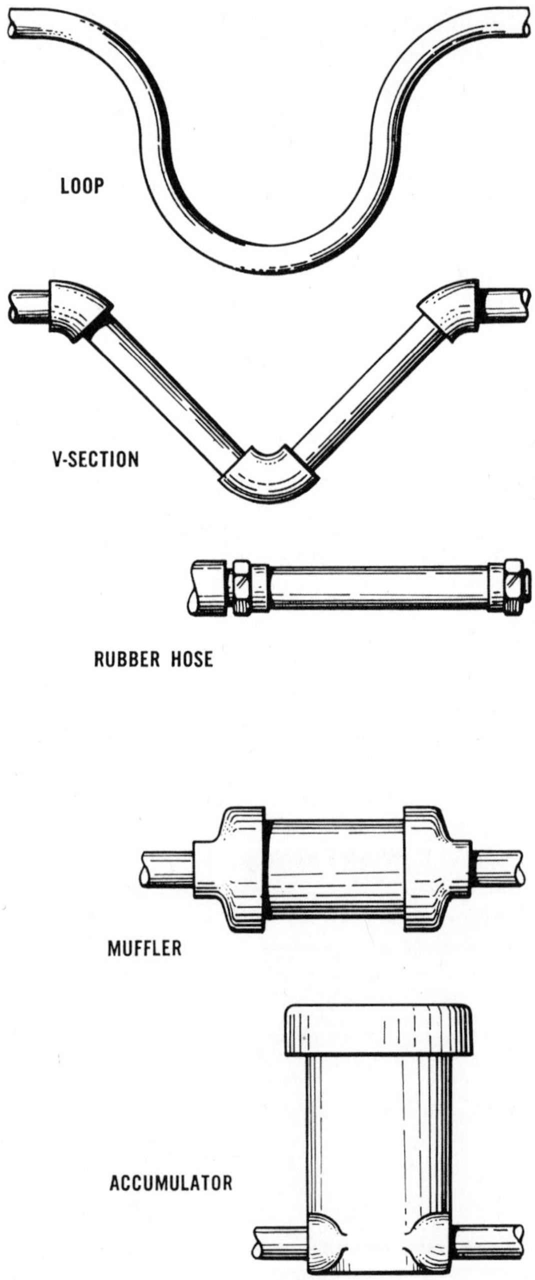

dampers

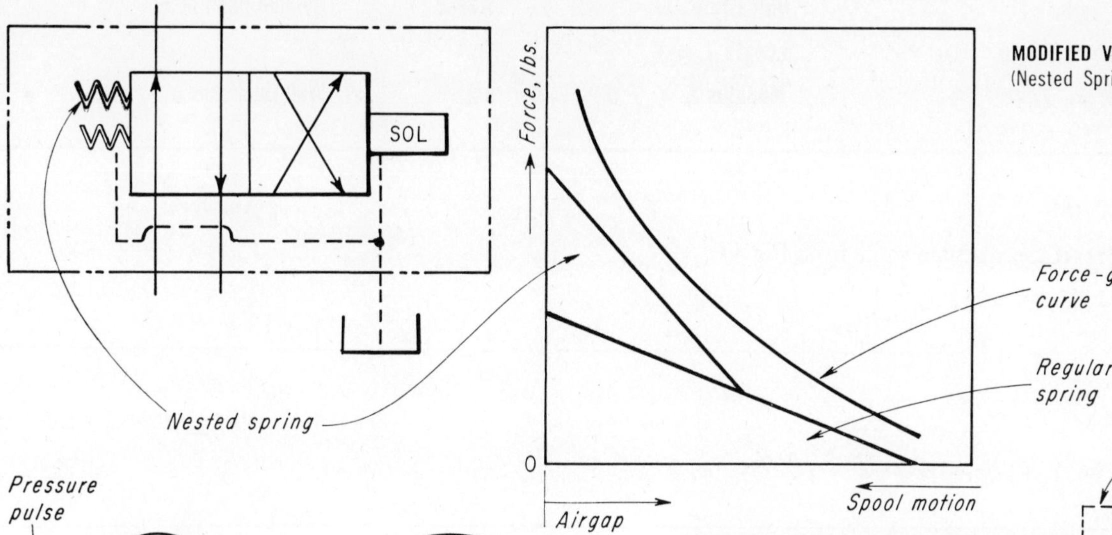

MODIFIED VALVE
(Nested Spring)

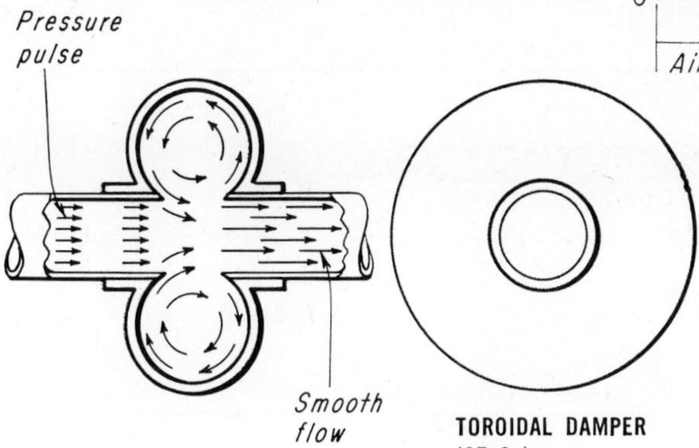

TOROIDAL DAMPER
(GE Co)

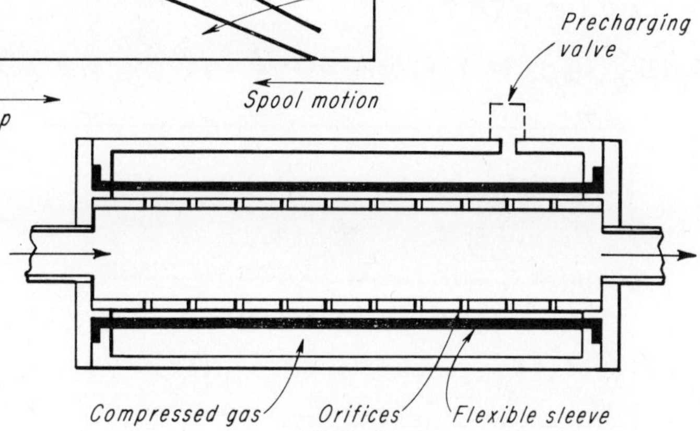

CUSHIONED SLEEVE
(Wabco)

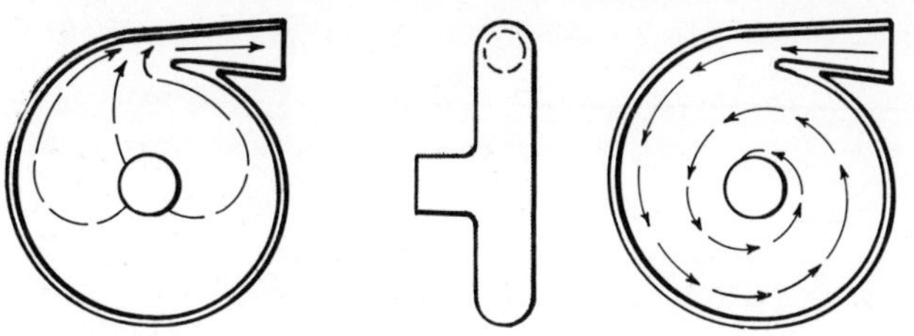

Low damping effect

High damping effect

INVOLUTE DAMPER

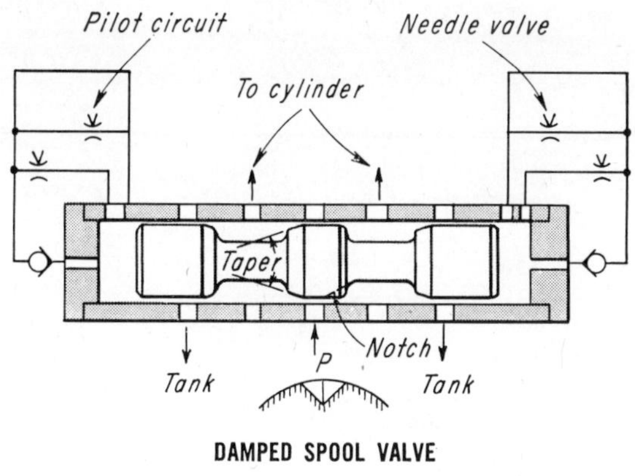

DAMPED SPOOL VALVE

Where the pulsation is continuous and in resonance with the piping (possible when the source is a pump) then the remedy may be to redesign the pump. Resonance can amplify noise up to annoying or dangerous levels. A previous article (*PE*—Feb '52, p 178) shows how to make the modifications.

Without altering the valve, cylinder, or pump, you can reduce fluid hammer by adding volume, inertia, or elasticity to the pipeline adjacent to the source of the hammer. Most of the sketches on this page show such in-the-line pressure-wave absorbers. They are called by many names, including *isolators, capacitive-load chambers, mufflers, surge dampers, suppressors, snubbers,* and *accumulators.* A common name for all of them is *pulsation damper.*

The sketches are self-explanatory, but remember the following:

• A half-loop of copper or steel tub-

EXAMPLE 1 . . Oil line to hydraulic press

Fluid = oil		Bulk modulus β = 25×10^4 psi		Density ρ = 55	lb/ft³
Flow = 10 gpm		Velocity V_F = 16.4	ft/sec	System head H = 17,000	ft
Pipe dia, d_i = 0.5 in.		Length L = 40	ft	= 6000	psi
Pipe material = steel		Modulus E = 30×10^6	psi	Wall thickness e = 0.11	in.

GENERAL EQUATIONS:

(2) $V_P =$ (read approximate value from Fig 4)

(3) $T_C = 2L/V_P = 2 \times 40/4540$

Answers:

$V_P =$ 4540 ft/sec

$T_C =$ 0.0174 sec

FAST CLOSING ($\leqslant T_C$):

(5) $H_{R\,(max)} = V_P V_F / g = 4540 \times 16.4/32.2$

$H_{R\,(max)} =$ 2300 ft

= 875 psi

SLOW CLOSING: (For example, 0.1 sec)

(8) $H_R - M - H - \sqrt{M^2 - m_1^2} = 19{,}407 - 17{,}000 - \sqrt{19{,}407^2 - 19{,}300^2}$

$m_1 = H + H_{R\,(max)} = 17{,}000 + 2300 = 19{,}300$

$m_2 = V_F^2 (V_P - 2L/T)^2 / 2g^2 H = \frac{16.4^2}{34{,}000 \times 32.2^2}\left(4540 - \frac{80}{0.1}\right)^2 = 107$

$M = m_1 + m_2 = 19{,}300 + 107 = 19{,}407$

$H_R =$ 447 ft

= 164 psi

FAST OPENING ($\leqslant T_C$):

(9) $H_D = \sqrt{n_F(2H + n_F)} - n_F = \sqrt{158(34{,}000 + 158)} - 158$

$n_F = H_R{}^2{}_{(max)}/2H = 2300^2/34{,}000 = 158$

$H_D =$ 2163 ft

= 800 psi

SLOW OPENING: (For example, 0.1 sec)

(10) $H_D = \sqrt{n_S(2H + n_S)} - n_S = \sqrt{4.9(34{,}000 + 4.9)} - 4.9$

$n_S = 2V_F^2 L^2 / g^2 T^2 H = 2 \times 16.4^2 \times 40^2 / 32.2^2 \times 0.1^2 \times 17{,}000 = 4.9$

$H_D =$ 408 ft

= 150 psi

Note: Closing pressure rise is greater than opening pressure drop in this example. Use 6000 + 875 = 6875 psi for pipe-stress calculations.

ing should have a loop radius equal to three times the tubing outside diameter.

- A muffler can be simply an enlarged section of pipe.
- Spring-type or bladder-type accumulators are good surge absorbers, but they slow down response of the system to a change in pressure.
- Each pulsation damper adds some resistance to flow—negligible in most cases.
- Dampers containing elastomer parts are limited to moderate temperatures.
- Involute and toroidal dampers depend on fluid velocity for the damping effect.

The involute damper has an interesting principle of operation: the velocity of a fluid particle increases towards the center and is inversely proportional to the radius. The result: centrifugal forces smooth out pulses.

EXAMPLE 2.. Penstock for water turbine

Fluid = water		Bulk modulus β = 29.4×10^4 psi		Density ρ = 62.4	lb/ft³
Flow = 7200 gpm		Velocity V_F = 9.2	ft/sec	System head H = 1300	ft
Pipe dia, d_i = 15 in.		Length L = 3280	ft	= 562	psi
Pipe material = steel		Modulus E = 30×10^6	psi	Wall thickness e = 0.5	in.

GENERAL EQUATIONS: Answers:

(2) V_P = (read approximate value from Fig 4) — V_P = 3860 ft/sec

(3) $T_C = 2L/V_P = 2 \times 3280/3860$ — T_C = 1.7 sec

FAST CLOSING ($\leqslant T_C$):

(5) $H_{R\,(\max)} = V_P V_F/g = 3860 \times 9.2/32.2$ — $H_{R\,(\max)}$ = ~~1100~~ ft = ~~476~~ psi It's too high

SLOW CLOSING: (For example, 20 sec)

(8) $H_R - M - H - \sqrt{M^2 - m_1^2} = 2793 - 1300 - \sqrt{2793^2 - 2400^2}$ — H_R = 73 ft = 31.6 psi

$m_1 = H + H_{R\,(\max)} = 1300 + 1100 = 2400$

$m_2 = V_F^2\,(V_P - 2L/T)^2/2g^2H = \dfrac{9.2^2}{2600 \times 32.2^2}\left(3860 - \dfrac{6560}{20}\right)^2 = 393$

$M = m_1 + m_2 = 2400 + 393 = 2793$

FAST OPENING ($\leqslant T_C$): Does not apply

(9) $H_D = \sqrt{n_F(2H + n_F)} - n_F =$ — H_D = ft = psi

$n_F = H_{R\,(\max)}^2/2H =$

SLOW OPENING: (For example, 8 sec)

(10) $H_D = \sqrt{n_S\,(2H + n_S)} - n_S = \sqrt{21(2600 + 21)} - 21$ — H_D = 213 ft = 92.3 psi

$n_S = 2V_F^2L^2/g^2T^2H = 2 \times 9.2^2 \times 3280^2/32.2^2 \times 8^2 \times 1300 = 21$

Note: Opening pressure drop (and the corresponding reflected rise) is greater than the closing drop because the valve is opened faster than it is closed. Use 562 + 92.3 = 654 psi for pipe-stress calculations

Selection procedure for

Hydraulic accumulators

If system flow is intermittent, an accumulator will permit smaller pumps and fewer controls. Here's a technique for picking the right one.

LOUIS DODGE, Consultant, New Richmond, Ohio

ACCUMULATOR means *energy-storer*. In hydraulics it is a tank or chamber built to hold reserve fluid at system pressure which is immediately available to provide or absorb momentary flow. The major types are described below. Oil is assumed to be the hydraulic fluid, but any liquid can be used if the seals are made compatible.

When you have an accumulator in the system and the flow requirements are intermittent, you can get away with a smaller pump and fewer controls. The accumulator discharges its stored fluid at high or low flow rates on demand and also absorbs line shocks that would otherwise damage system components. To replenish the accumulator during lulls, only a small pump is needed. the tables on page 289 describe most of the common applications, including some that are not related to

Six accumulator designs

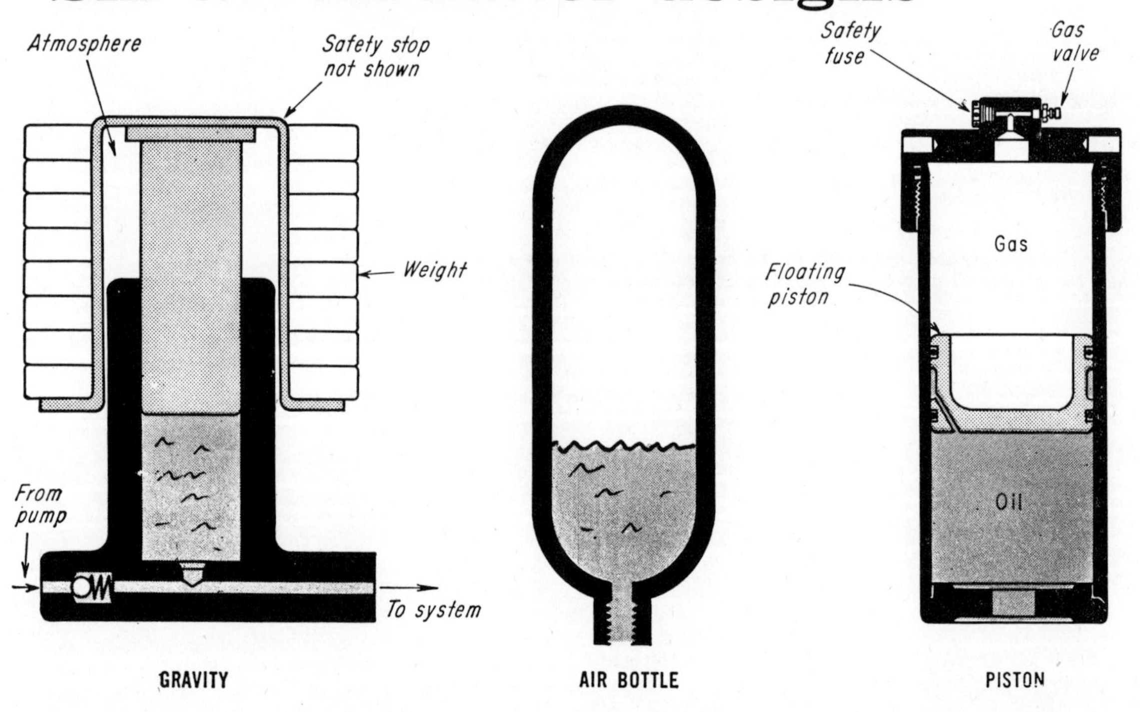

Performance guide for preliminary selection

(special designs can exceed these values)

TYPICAL PERFORMANCE	GRAVITY (machine applic.)	AIR BOTTLE	GAS PISTON	SPRING PISTON	BLADDER
Capacity (oil volume)	**To 250 in.^{3}**	**10 to 1500 gal**	**3 in.^{3} to 20 gal**	**To 100 in.^{3}**	**To 20 gal**
Pressure range, psi	**To 500**	**50 to 500**	**To 5000**	**50 to 500**	**To 3000**
Temperature range, F	**To 250**	**Limited by fluid only**	**–40 to 250**	**To 175**	**To 175**
Response	**Sluggish**	**Fast**	**Fairly fast**	**Fairly fast**	**Fast**
Remarks	**Requires full-pressure seals**	**Has air-oil interface**	**Needs good piston seals**	**Needs safety stop if spring fails**	**Depends on bladder and discharge valve**

energy storage.

You can choose an accumulator by following three simple steps:

- Quickly eliminate unlikely types with the preliminary selection table shown above. This table gives basic performance limits.
- Calculate approximate size, basing your estimate on the system demand flow and allowable pressure drop. Equations for each major type are given in this article.
- Contact accumulator manufacturers for a final check on your selection.

Gravity accumulator

Force F available for acceleration is the differential force between the weighted piston and the back pressure of the fluid. Assuming that the piston weight is equiv-

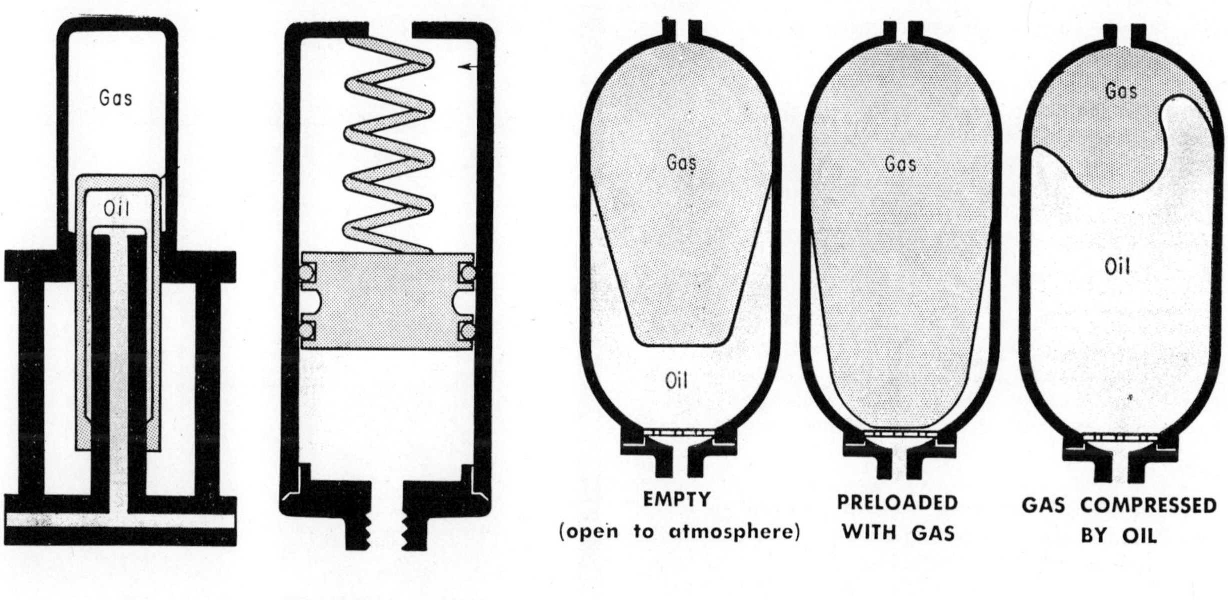

DUAL-DIAMETER PISTON SPRING-LOADED PISTON BLADDER

TYPICAL RESPONSE OF BLADDER ACCUMULATOR

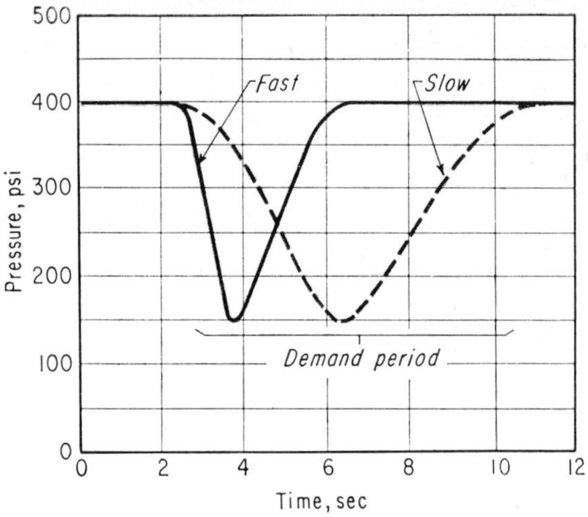

alent to 300 psi and that the fluid pressure drops to 270 psi during discharge, only 30 psi (10% of the nominal pressure) is available to accelerate the piston mass. Therefore piston acceleration cannot exceed 1/10 g, or about 3.2 ft/sec^2.

Air bottle

Air loss from the chamber into the system oil by leakage or absorption can occur, because there is no liquid-air barrier. The opposite—release of entrained or dissolved air from the oil to the gas—is also possible. High charging pressures and large pressure changes aggravate the condition. Avoid overdischarge and never tilt the chamber, because the air might escape directly into the system. And do not place the chamber at a high point in the system, because it will become a trap for rising bubbles of air, particularly at idling.

Piston

The single-diameter piston accumulator is similar in function to an air bottle, except that it has a floating piston as a barrier to keep the system oil from contacting the gas (usually air or nitrogen).

A dual-diameter piston accumulator doubles as an intensifier. Operation is the same as that of the single-diameter accumulator, except that the oil pressure is greater than the gas pressure by the ratio of gas-side piston area to oil-side piston area. The volume of oil flow is less by the same ratio. One version is shown in the sketch on p 285; another on p 289.

Spring-loaded piston

The stored energy is in the spring, and the spring chamber is open to the atmosphere. Since force is limited by spring strength, this accumulator is recommended for low pressures only. Oil pressure equals spring force (rate × displacement) divided by effective piston area.

Bladder or bag

The gas—usually compressed air or nitrogen—is contained in a synthetic-rubber bag inside a steel chamber. This accumulator has low inertia and can respond very fast—it is good for absorbing line shocks. Special supports or valving (one version shown in sketches) prevent the bladder from extruding into the oil port.

SYMBOLS

Gas side

V_p =precharge volume, in.^{3}

V_{LP} =volume at lowest operating pressure

V_{HP} =volume at highest operating pressure

Liquid side

x =pressure drop, expressed as ratio $(P_{HP}—P_{LP})/P_{HP}$

ΔV =demand volume, in.^{3}

Q =pump capacity, in.^{3}/accumulator cycle

s =piston (cylinder) stroke, in.

d =piston diameter, in.

N =piston strokes/accumulator cycle

Both sides

P_p =precharge pressure, psi

P_{LP} =lowest system operating pressure

P_{HP} =highest system operating pressure

V_{ACC} =total accumulator volume, in.^{3}

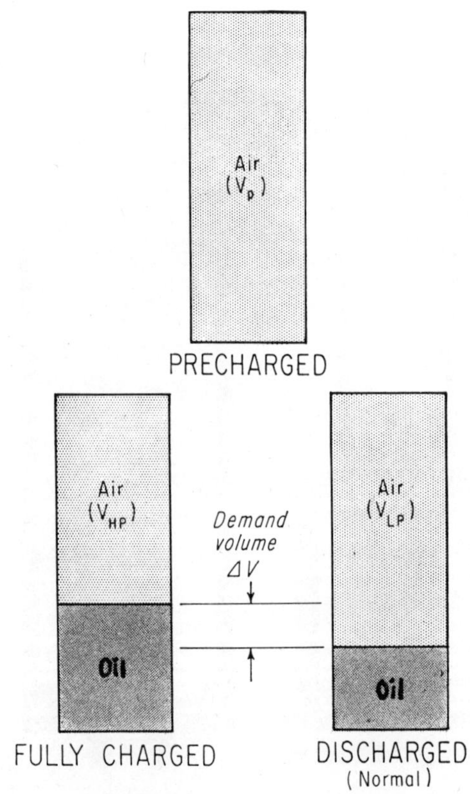

TYPICAL AIR-BOTTLE CHARGE
(oil volume $= V_{HP}/3$)

A variant of the bladder type is the spring-loaded diaphragm. The spring and its end plate guide the rolling diaphragm smoothly along the cylinder walls. Another variant is a simple synthetic hose supported inside a perforated tube. The hose is pressurized with gas, sealed at both ends, and inserted directly into a large pipeline. The purpose is to absorb shocks.

GAS-LIQUID ACCUMULATOR PERFORMANCE

Air-bottle, piston, and bladder-type accumulators are governed by the same principles. The following discussion pertains to any gas-liquid (hydropneumatic) accumulator for industrial service.

Precharging

The gas side must be precharged, and most accumulators have a gas valve for this purpose. Exceptions are accumulators for low-pressure service—a few hundred psi—because atmospheric-pressure gas or air can be compressed by the system fluid directly.

If the maximum system pressure is 3000 psi, a typical precharge for an aircraft accumulator is 1000 psi. Then the gas will be compressed to about one third of its precharged volume (assuming isothermal compression) when system pressure rises to 3000 psi. (Air volume to oil volume ratio is small.) A 300-psi precharge will result in a 10:1 compression when system pressure rises to 3000 psi. An extreme case is atmospheric precharge: here the compression ratio would be $3000/14.7 = 204$, and the accumulator over-all volume would have to be at least 204 times the volume of the gas at 3000 psi, which is not practical for any conventional accumulator.

A lot depends on the application. If the system must maintain constant pressure within 5 or 10%, the accumulator gas volume at system pressure should be large in proportion to the oil volume so that the accumulator can discharge considerable liquid without excessive pressure drop. But if all that's needed is a large volume of oil at a rapidly decreasing pressure, gas volume at highest operating pressure can be small.

Demand volume and pressure

Demand volume ΔV of an accumulator is the oil that must be discharged into the system during the period of need. The energy comes from expanding the stored gas. The volume relationship is:

$$\Delta V = V_{LP} - V_{HP} \qquad (1)$$

Computing the initial V_{HP} and final V_{LP} gas volumes is simple, using Boyle's gas law:

$$P_{HP}V_{HP}^n = P_{LP}V_{LP}^n = PV^n \text{ at any point} \qquad (2)$$

where exponent $n = 1$ for isothermal expansion and 1.4 for adiabatic expansion. Actual expansion is never wholly one or the other. So for the purpose of this discussion, polytropic expansion is assumed, with $n = 1.25$. The error is small for ordinary applications using air or nitrogen where pressure changes are less than 25% during operation.

If greater accuracy is desired, assume isothermal compression for the charging cycle, which is gradual. Assume close to adiabatic expansion during discharge, which usually is rapid.

Pressure requirements are set by the system. To calculate accumulator size you must know the maximum pressure P_{HP} and the allowable pressure drop expressed as the ratio x:

$$x = (P_{HP} - P_{LP})/P_{HP} \text{ or}$$
$$P_{LP}/P_{HP} = 1 - x \qquad (3)$$

Combining Eq 1, 2, and 3:

$$V_{HP}/V_{LP} = V_{HP}/(V_{HP} + \Delta V) = (1 - x)^{1/n} \qquad (4)$$

which is the basic equation relating demand volume and pressure drop. Expressed another way:

$$V_{HP} = \left[\frac{(P_{LP}/P_{HP})^{1/n}}{1 - (P_{LP}/P_{HP})^{1/n}}\right] \times \Delta V = m\,\Delta V \qquad (5)$$

Values of m are plotted on the next page.

Accumulator size = gas volume + oil volume. Desired volume ratio of gas to oil must be specified before combined volume of gas and oil is calculated. As a rule the gas volume is specified as a multiple of the oil volume at the fully charged condition. A typical value for industrial service is 3:1. Then total accumulator volume $V_{ACC} = 1.33\ V_{HP}$.

Predicting demand volume

Before you can specify a demand volume for the accumulator, you have to estimate peak demands of the system. Over-all demand volume is the summation of all the component demands in the system. It can best be expressed graphically, as shown in the volume-cycle-time graph on page 288. The summation, which includes all working cylinders and the system pump, can be expressed:

$$V = (\pi/4)(d_1^2 s_1 N_1 + d_2^2 s_2 N_2 + d_3^2 s_3 N_3 \text{ etc}) - Q \qquad (6)$$

where d is piston diameter, s is stroke, N is strokes per cycle, and Q is pump delivery/cycle.

SYSTEM APPLICATIONS

The Editor's Note on the last page lists recent articles and books that describe many different accumulator systems. For this article, a few examples will suffice.

High-speed piston: Schematic on page 289 shows a basic job for an accumulator: generating high volume flow for a short-time cyclic operation.

Pressure compensator: Bottom schematic is of a holding circuit; the pump can be idled or stopped, while line pressure is maintained by the accumulator despite leakage from the piston. Pressure switch starts pump if pressure drops below a predetermined value.

Pump pulsation damper for slow-speed pumps: Sinusoids in graph depict typical piston or vane flow pulsations. The degree of irregularity of pulses, K = (maximum flow − minimum flow)/(average flow), %. Smoothness of delivery depends on K, number of pistons or vanes, system characteristics, accumulator volume, and accumulator response.

For example, the factor K for a single-acting plunger pump is 400%, for a double-acting pump 200%. Value of K decreases with an increasing number of pistons. It decreases more rapidly with a growing uneven number of pistons than with an even number. Here are test values arranged in order of increasing smoothness:

Number of pistons	K, %
4	40
6	15
5	8.5
8	7.8
10	3.6
7	2.2
12	2.1
9	1.2
11	0.7

Assuming a 30% safety factor, the accumulator demand volume $\Delta V = 1.3\ K \times$ gpm/rpm. If the accumulator can respond rapidly, pump speed can be as high as 150 rpm. Above 150 rpm an accumulator is not very effective, but at high speeds pulsation is not as much of a problem anyway.

FOUR EXAMPLES

1) Computing chamber volume

Conditions: Bladder-type accumulator with gas-to-oil volume ratio = 3:1 at fully charged condition. Exponent $n = 1.25$. System demand volume $\Delta V = 6$ in.^{3} Allowable pressure drop 5% ($x = 0.05$).

Calculation: Using Eq 4:

$$V_{HP}/(V_{HP} + 6) = (1 - 0.05)^{1/1.25} = 0.96$$

$$V_{HP} = 144 \text{ in.}^3$$

$$V_{ACC} = 144 + 144/3 = 192 \text{ in.}^3$$

2) Accumulator for fast-response valve

Conditions: A large valve has to be opened rapidly by hydraulic power within a few seconds, but it is infrequently operated. Recharge time approximately ½ hr. Maximum pressure 120 psi. Demand volume 50 gal. Temperature level and air contamination are not problems.

Selection: First choice (from performance table on page 285) is the air bottle for fast response.

If reliability must be improved, a battery of gas-piston accumulators is an alternative. For both choices, the fluid-power system requires pump capacity of 2 gpm and an electric motor of 1 hp at 1200 rpm.

3) Reducing fluid hammer

Conditions: Hammer is caused by rapid closing of valves. Small demand volume. System pressure = 250 psi.

Selection: First check maximum pressure intensity (peak pressure on gage). Compute oil volume in pressure lines to offending valves. Approximately 3% of this quantity is a good initial estimate of accumulator demand volume. Calculate or estimate frequency of "hammer."

Small demand volume and low pressure suggest use of spring-type accumulator. Selection of spring is important. The effective spring energy (force × distance) should be equivalent to the fluid pressure range from 250 psi to slightly above the recorded peak pressure. The natural frequency of spring should be asynchronous to periodicity of fluid hammer. Install accumulator close to source of hammer. Provide for slight leakage. Spring chamber must be open to atmosphere.

If spring size becomes too large and leakage cannot be tolerated, a small accumulator of the bladder or piston type can be substituted.

A brief theory of fluid hammer, including a shortcut method of application, is given in a recent *Your Turn* exercise (*PE*—April 2 '62, p 105). It relates speed of sound, fluid compressibility, mechanical strength of the pipe, pipe dimensions, and the modulus of elasticity. The pressure pulse is caused by the valve's closing faster than a pressure wave can travel a distance equal to twice the length of the pipe conveying the fluid.

4) Precision circuit

Conditions: Select an accumulator for a circuit demanding high operating precision. Pressure variation

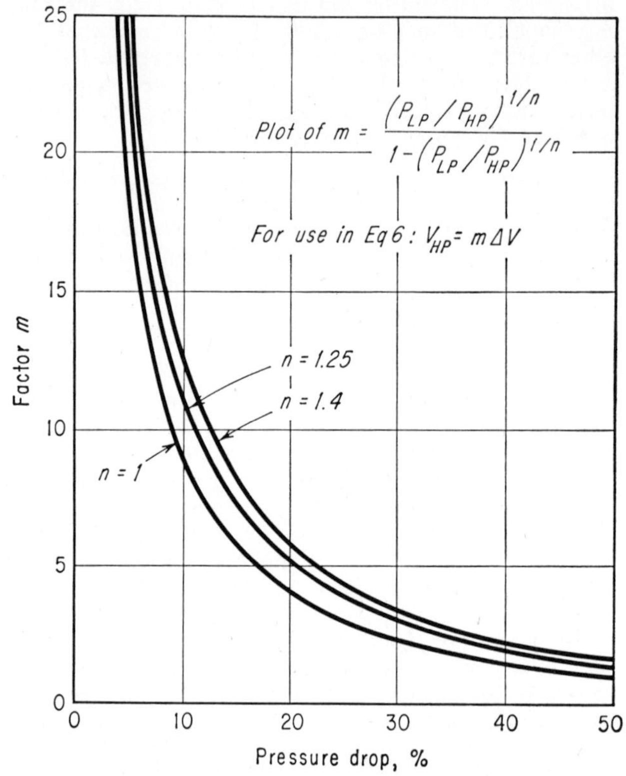

CALCULATION OF CHAMBER AIR VOLUME is simple when the math is eliminated by this graph. The curves can be read accurately from about 10% to 50% pressure drop; below that it is better to calculate m factor directly.

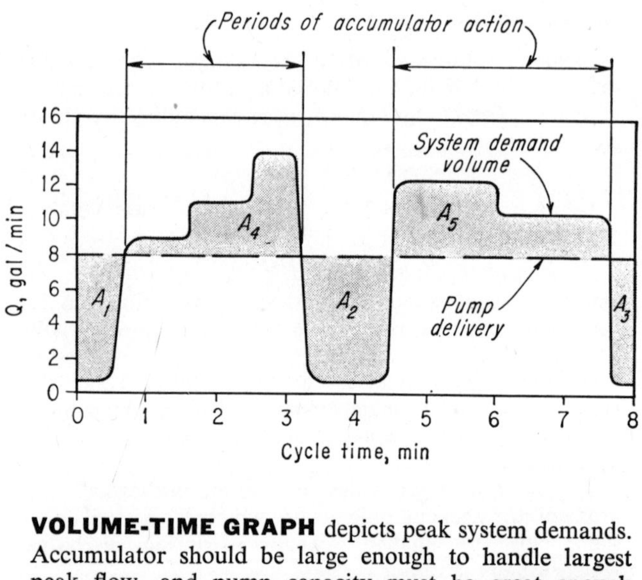

VOLUME-TIME GRAPH depicts peak system demands. Accumulator should be large enough to handle largest peak flow, and pump capacity must be great enough to replenish accumulator during lulls. Graphically, area under pump delivery line at least must equal the area under system demand line. $A_1 + A_2 + A_3$ must be $\geqq A_4 + A_5$. If $A_4 > A_2$, demand volume $\Delta V = A_4 + A_5 - A_2$. If $A_4 < A_2$ and $A_5 > A_4$, $\Delta V = A_5$. Increase calculated ΔV 25% to provide a safety factor and to cover losses.

Eight common uses for accumulators

IF A SYSTEM REQUIREMENT IS:	AND PUMP HAS THIS LIMITATION:	THEN YOU CAN USE AN ACCUMULATOR AS:
Large intermittent flow, always available	**Small capacity**	**Power-storage chamber, to be pumped up slowly, released fast**
Large flow in emergency	**Might fail**	**Emergency storage**
Leakage make-up flow only	**Sometimes shut down or unloaded**	**Pressure sustainer**
Extremely small metered flows (lubricator)	**No pump at all**	**Fluid dispenser, to be replenished periodically**
Short-time operation (remote location)	**No pump at all**	**Power-storage chamber, to be charged initially then discharged until depleted**
Limited peak pressure	**Excessive discharge pressure if flow is blocked**	**Safety surge tank to absorb pressure rise**
Nonpulsing flow	**Slow speed, pulsing**	**Pulse damper (100-lb pulses are not unusual)**
Small changes in system volume (thermal or mechanical reasons)	**Constant-volume system**	**Expansion chamber**

Three special uses for accumulators

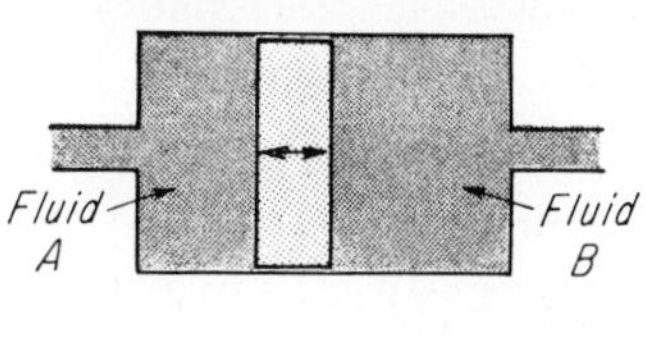

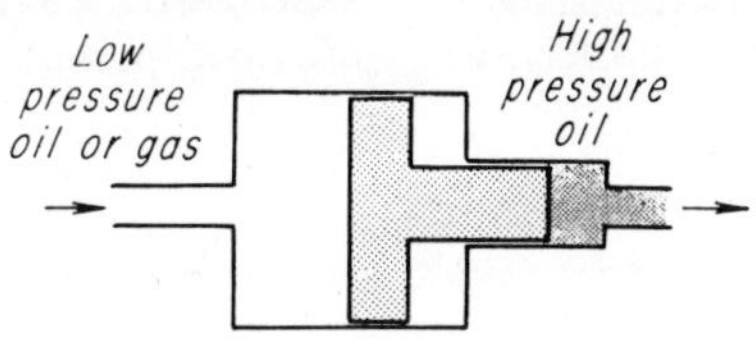

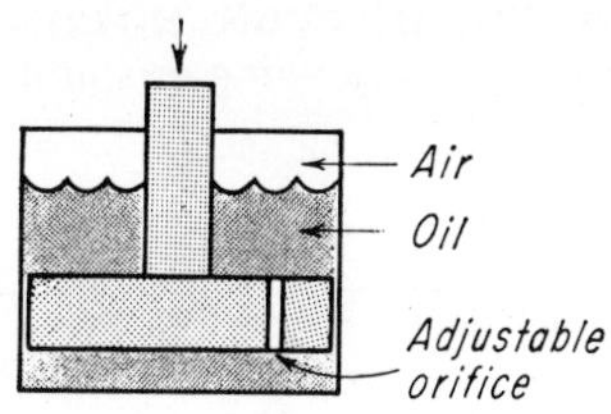

TYPE	**Piston or bladder**	**Piston, with unequal diameters**	**Piston, air-oil damper**
APPLICATION	**Barrier to separate two fluids in a system**	**Pressure intensifier**	**Mechanical shock absorber**

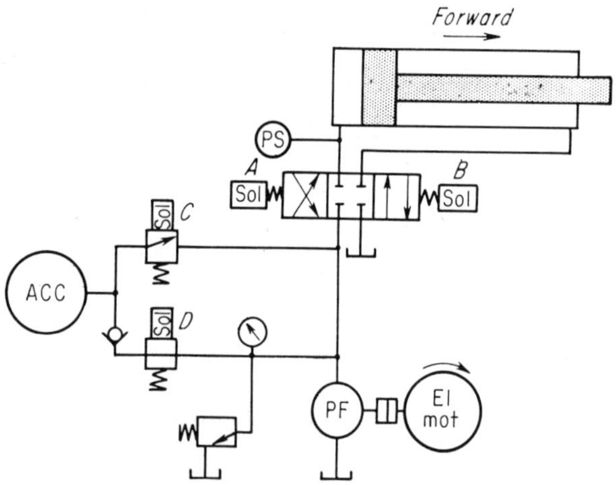

ACCUMULATOR ASSISTS PUMP during high-speed cylinder cycling; therefore pump can be smaller. Key: *PF* = fixed displacement pump; *PS* = pressure switch; + = energized; − = de-energized.

Using key and schematic diagram cycle is: − (A, B,C,D) = prestart condition (shown). Start *PF* = charge *ACC*. + B = slow forward. + B + C = fast forward. + B − C + D + *PS* = slow forward under load and high pressure (*PS* de-energizes C when high pressure is reached). + A − B − *PS* = slow return. + A + C − D = fast return. − A − C = cycle completed.

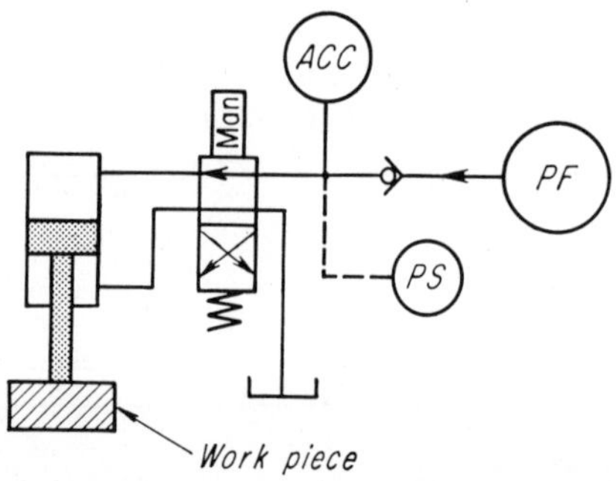

ACCUMULATOR HOLDS PRESSURE constant during pump idling or shutdown despite piston leakage.

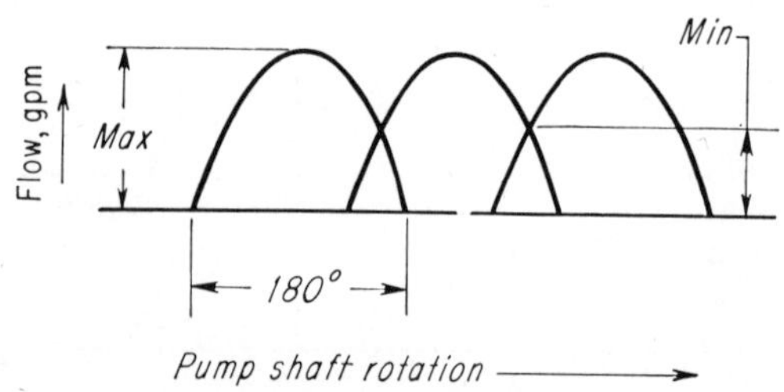

ACCUMULATOR CAN DAMP PULSES of pump provided pump speed is less than 150 rpm.

must be held within 5% of system pressure. Air contamination of oil cannot be tolerated. Temperature variation will not exceed 10 F. Maximum temperature 115 F. System pressure 350 psi. Estimated cyclical oil requirements are less than 2 gal.

Selection: A bladder type of accumulator is the best choice. Calculate demand volume from cyclical oil volume requirements according to procedure shown on previous page. For accumulator size proceed according to sample calculation in example 1. Make sure bladder material is compatible at all temperatures with the gas and oil used.

EDITOR'S NOTE:

System calculations involving accumulator response time are covered by chapters in two recent books published by McGraw-Hill: *Handbook of Fluid Dynamics* by Streeter describes surge tanks for pipelines and derives theoretical relationships for water hammer and other transients; *Design of Hydraulic Control Systems* by Lewis and Stern develops graphs of accumulator response times for several hypothetical systems. —*Franklin D. Yeaple*

11

SEALS AND GASKETS

Dynamic and static seals for liquids and gases

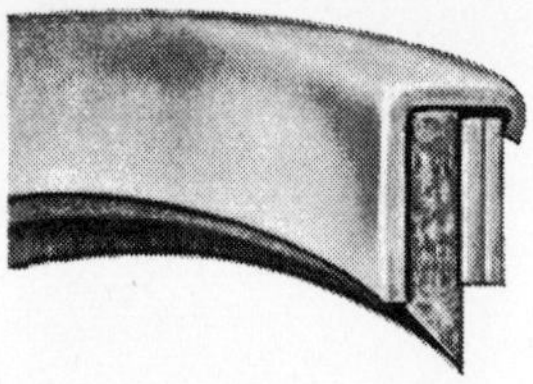

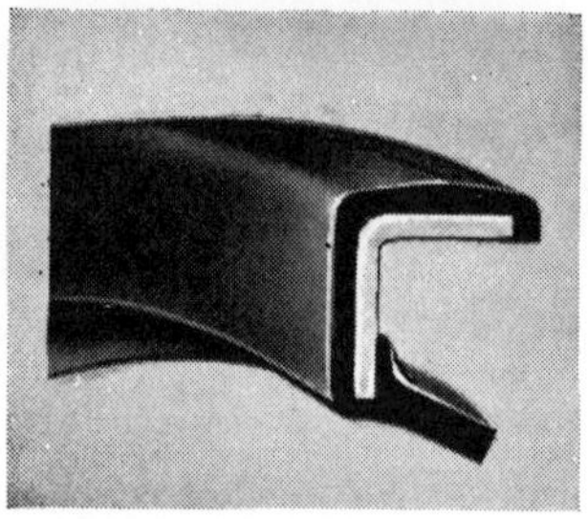

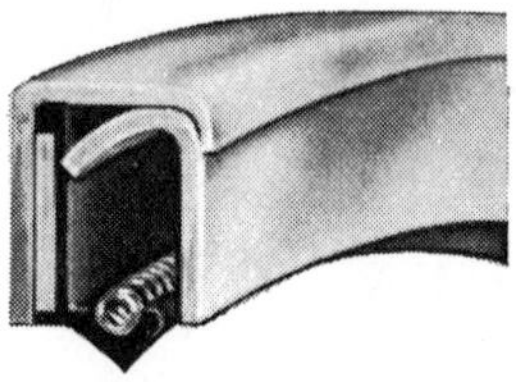

HOW TO CHOOSE A DYNAMIC SEAL

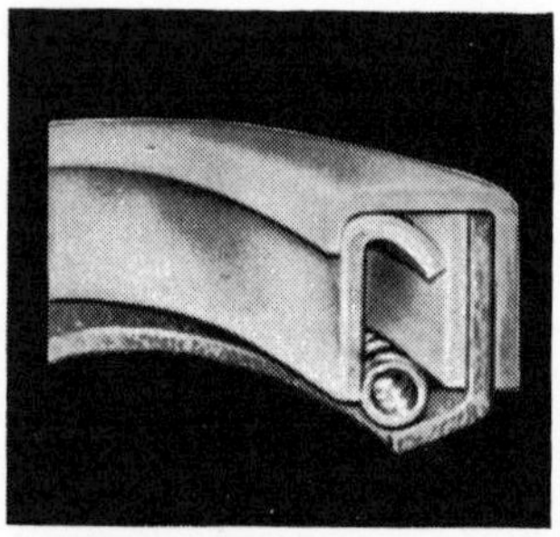

V L PEICKII, *director of research & engineering*
DAN A CHRISTENSEN, *research engineer*
National Seal Division
Federal-Mogul-Bower Bearings Inc,
Redwood City, Calif

What is the true function of a seal? It may be used to:

1. Seal-in oil, grease or other fluid
2. Seal-out dirt, fluids and contaminants
3. Maintain applied pressure or vacuum
4. Or perform a combination of these jobs

But a seal will have to do more. Each application imposes its own special problem.

SEALING IN

- What about oil surge pressures?
- What are the leakage tolerances (in and out)?
- Do adjacent bearings, gears, etc., keep the lubricant away from the seal; or do they create local high-pressure areas or jet streams against the seal?
- Is the housing vented to relieve pressure caused by raised temperature?

SEALING OUT

- Does the housing allow dirt to pack against the seal?
- When the assembly cools, is it vented to prevent contaminants being sucked in by the created vacuum?
- Is the seal protected against direct splash?

PRESSURE OR VACUUM

- Are static conditions better, same, or worse than operating conditions?
- Is there alternating vacuum and pressure caused by operating conditions or between static and dynamic conditions?

The designer has 5 basic types of seals to serve him. He has 17 materials to choose among. He has hundreds of design configurations. Beginning on next page is a complete review of all the types and an exhaustive analysis of each in terms of basic selection factors, materials, and design recommendations.

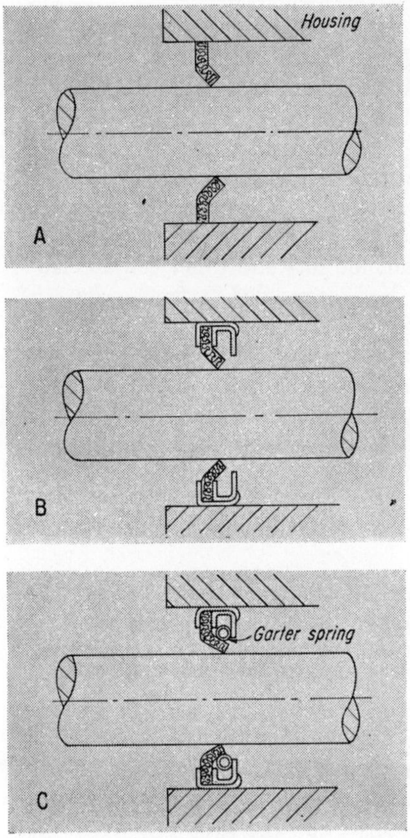

1 . . . BASIC SHAFT SEAL (A) has leather or rubber lip attached to housing and held against shaft. Steel case (B) facilitates mounting, and garter spring (C) holds lip against the shaft.

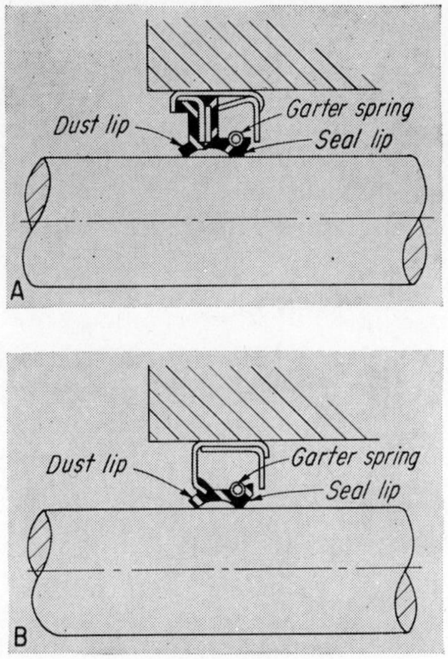

2 . . . TO EXCLUDE DUST and other contaminants from shaft seal, a dust lip is added. It can be a separate element (A), or integral (B).

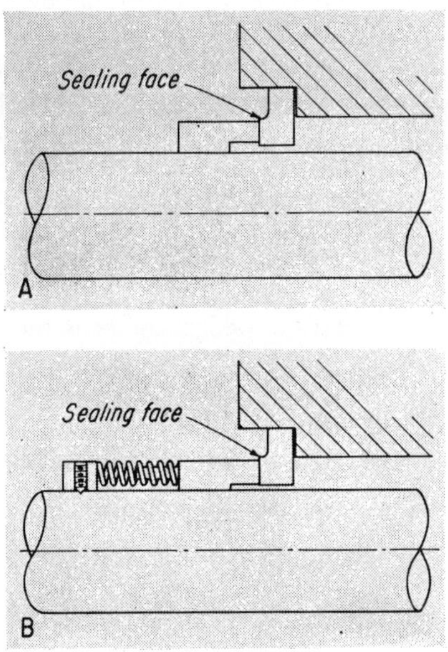

3 . . . BASIC FACE SEAL (A) has two flat radial faces—one connected to the housing, the other to the shaft. Spring force, as in (B), is usual means of holding the faces together to take up wear and axial movement.

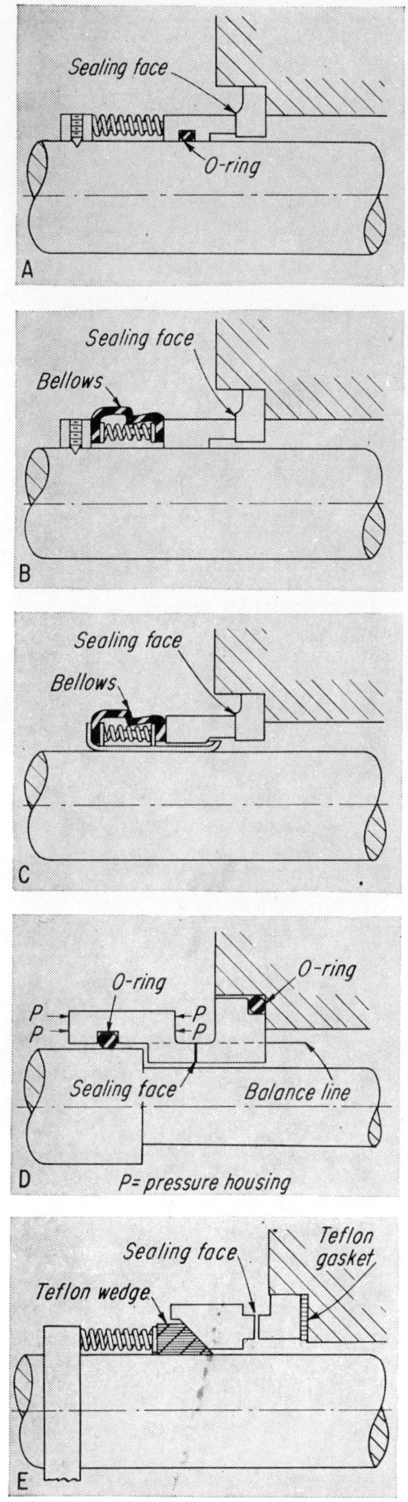

4 . . . AUXILIARY SEALING with face seal is performed by an O-ring (A) or bellows (B). To counteract shaft torque, bellows type can be mounted in a metal case with torque locks (C). Where internal pressures are high, pressure-balanced faces must be provided as in (D). For high temperatures a Teflon wedge (E) compensates for expansion.

THE TYPES OF SEALS

There are five basic types to choose among. The choice depends on many functional, environmental and economic factors. Each type of seal has its own configurations and principles of operation which make it particularly suitable for specific applications.

Shaft seals

A leather or synthetic-rubber flange or lip attached to the housing and held against the shaft, Fig 1(A), is the basic sealing element. A metal case usually holds this element in position Fig 1(B).

Misalignment, shaft runout, seal shrinkage, surge pressures and other cyclical factors tend to lift the lip off the shaft either intermittently or continuously. Consequently, a finger or garter spring is added to hold it against the shaft, Fig 1(C).

A dust lip is often added to exclude dirt and other contaminants. It can be a separate element, Fig 2(A), or molded integrally, Fig 2(B), with the primary rubber sealing lip—with either or both lips bonded to a washer or to the case. This produces rigidity, permits accurate positioning, and represents the basic shaft seal design.

Primary characteristics of this seal are: small space requirements; wide variety of standard sizes, materials and lip combinations; good sealing with many types of fluids; and good exclusion of dirt or contaminants.

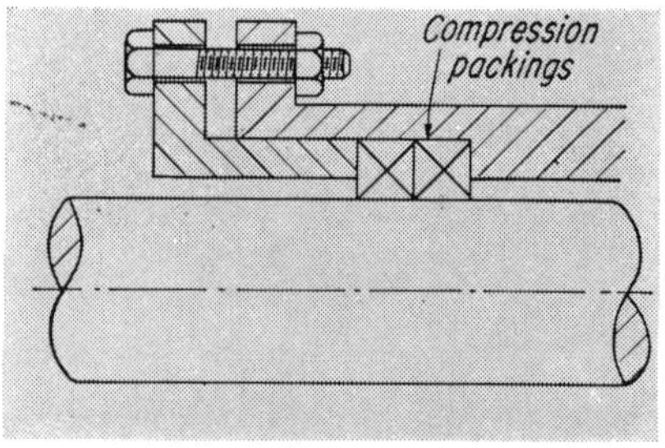

5 . . . COMPRESSION PACKINGS consist of twisted, woven or braided strands of vegetable fibers, asbestos or soft metals.

Face seals

Sealing is performed by continuous contact between two smooth, essentially flat radial faces—one attached to the shaft and the other to the housing, Fig 3(A). One face is designed to move axially to take up wear or play, usually by means of a spring, Fig 3(B). The rotating face must seal to the shaft as with an O-ring, Fig 4(A), or a rubber bellows, 4(B).

To counteract torque between shaft and sealing face, the seal assembly is often mounted in a metal case with integral torque locks and with gasketing, or press-fit between the shaft and the case, Fig 4(C). The seal assembly can be made either internal or external, and rotating or nonrotating. Other variations of face-type seals are the pressure-balanced type, Fig 4(D), where face pressure is independent of housing pressure; and the high-temperature seal, Fig 4(E), in which a Teflon wedge compensates for axial play and runout. O-rings or similar sliding seals can serve instead of Teflon. In high temperatures or other rubber-degrading atmospheres, metal bellows are sometimes used.

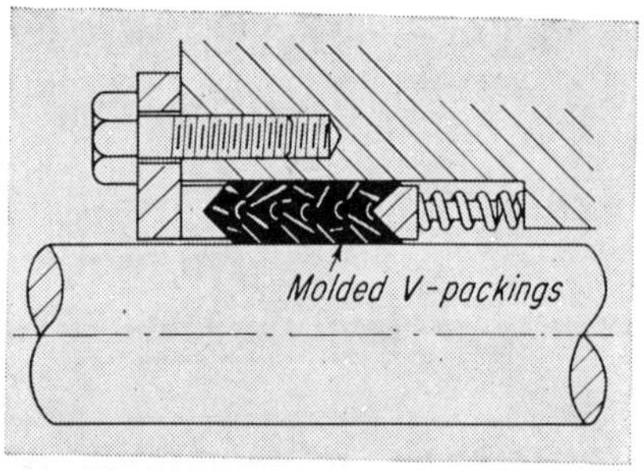

6 . . . MOLDED PACKINGS are usually made in U or V shapes and assembled in multiples with a spring-loaded take-up for wear.

Face seals are most useful where high pressure, temperature, speed or contaminating conditions exist. They are usually specially designed for pumps, process machinery, high-speed and high-temperature motors, earthmoving equipment and military vehicles. However, simple, inexpensive designs are commonly used on automotive water pumps and similar applications.

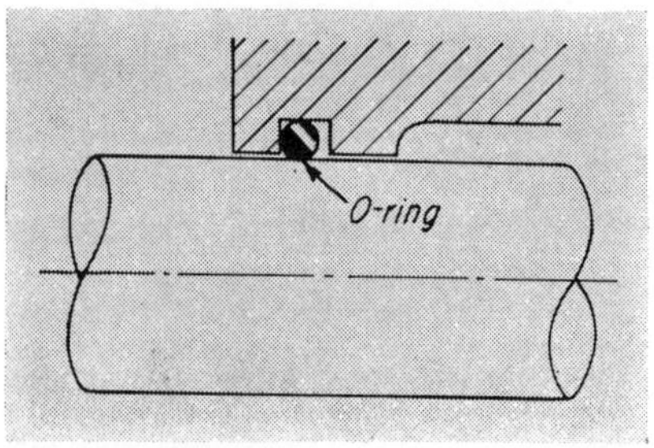

7 . . . SPECIAL FORM of molded packings is the O-ring (also made in cross-sectional variations of D or square shapes).

Compression packing

The oldest type of seal still widely used consists of twisted, woven or braided strands of vegetable fibers, asbestos, or soft metals packed between shaft and housing. These are often combined with a binder and a lubricant. In practice, the packing is usually placed in one or more coils in a bore and compressed with a flanged unit (Fig 5).

Compression packing is used at relatively low speed, and in high-pressure and temperature applications where occasional leakage is

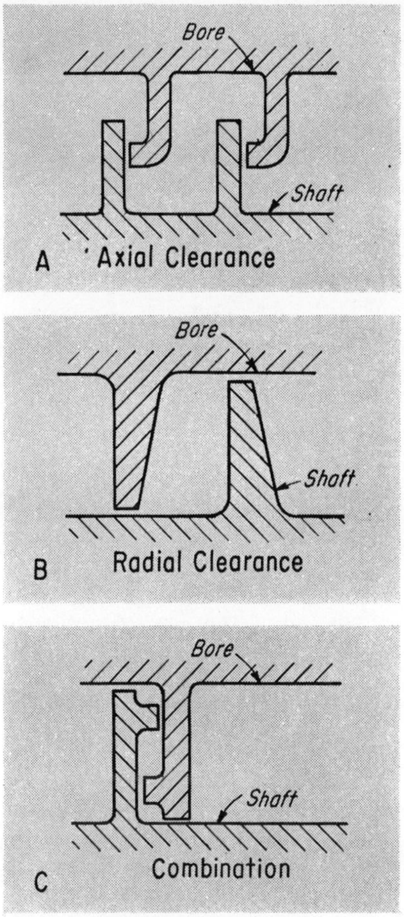

8 . . . LABYRINTH or positive clearance seals limit leakage by means of abrupt cross-sectional changes with axial clearance (A), radial clearance (B) or a combination (C).

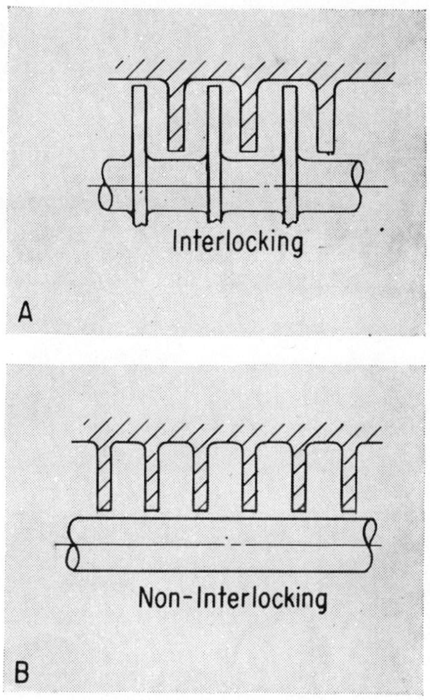

9 . . . LABYRINTH SEALS are also classified as interlocking (A) or non-interlocking (B).

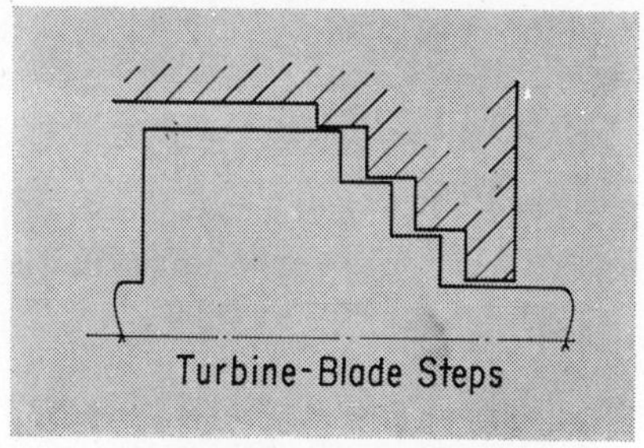

10 . . . STEPPED LABYRINTH is a variation of the interlocking type.

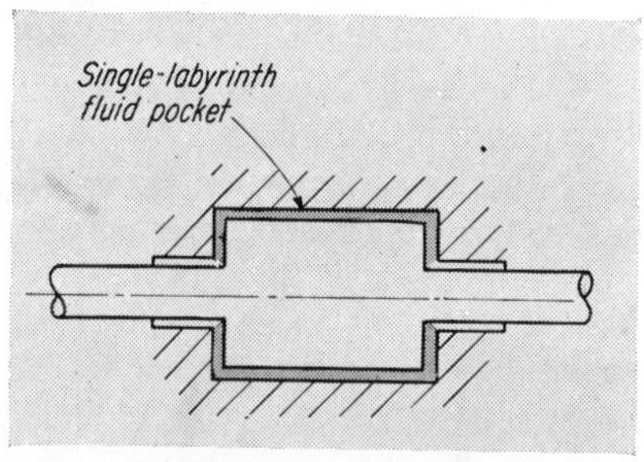

11 . . . SINGLE LABYRINTH uses a dense fluid to seal gas or light fluids.

tolerable. This type has been supplanted in many applications, but is still used on pumps, valves, and pistons to seal hot water, steam and other very hot fluids.

Molded packings

Here the loose packing is molded so that it assumes a precise cross-sectional form. The elements can be used individually, or in multiples with a spring-loaded takeup for wear. The usual cross-sectional designs are of the U- or V-type (Fig 6). A special form of the molded packing is the O-ring (Fig 7)—sometimes made with cross-sectional variations in the form of a D or square.

The inherent characteristic of most molded packings is that increasing pressure tends to increase sealing pressure. This is one reason for their popularity in high-pressure and high-temperature piston applications. Another reason: moderate degradation of the material does not affect performance as much as in some other types of seals. They are also often used in combination rotating and sliding applications of relatively slow speed, and the O-ring type frequently serves as a static seal.

Labyrinth or positive-clearance seals

Where no other type will perform satisfactorily, a labyrinth seal can be designed to limit the leakage to a predetermined amount. Such custom-designed seals are called on where conditions of temperature, surface speed and pressure prevent the use of positive-contact seals. They usually have built-in provision to bleed off the leakage.

All labyrinth seals limit the leakage flow by means of

abrupt changes in cross-section. The positive clearance in these seals is of two basic types Fig 8(A and B) or a combination (C). Axial clearance is used where close radial clearance is impossible because of excessive shaft runout or radial expansion. Radial clearance is preferred for applications that require excessive end-play or axial movement. In certain applications a combination of axial and radial clearance labyrinths is possible—this makes them twice as effective.

Another way of classifying labyrinth seals (Fig 9) is according to whether they are interlocking (A), or non-interlocking (B). The interlocking type is more effective, but is also more difficult to assemble.

A variation of the interlocking labyrinth is the step labyrinth (Fig 10) used in single-flow-stream turbines. Here, the reactive end-thrust counteracts the turbine blade end-thrust, and ordinary labyrinth sealing is compounded with centrifugal sealing at each of the steps.

Sometimes, when a sealing fluid denser than the sealed fluid can be used, a single or multiple labyrinth, as in Fig 11, is effective and economical. The dense sealant in such design performs its function by centrifugal action.

SELECTION FACTORS

For a good perspective on each of the basic types, it is useful to summarize the most important selection factors, Table I. Most of the limitations here can be exceeded with special designs or under carefully controlled conditions. But, of course, extremes of all factors cannot be achieved simultaneously by any one type of seal. Evaluations in the table are general and therefore approximate; their usefulness is based on the following definitions and qualifications.

General adaptability indicates how readily the basic type of seal can be adapted to particular applications.

Availability refers to off-the-shelf availability as well as the degree to which standard sizes of each type can be modified to suit special applications.

Positive action evaluations indicate how well each basic type will seal.

Minimum space requirements listed are approximate values for standard small-size seals. Special designs can often be built with smaller cross-sections.

Life estimates are based on average operating conditions of temperature, speed, surface finish and other factors.

Temperature limits are approximate maximum oil-sump temperature, provided other temperature generating conditions are not severe. Life and performance factors can be improved if the design provides for dissipating heat from seal friction. Face seals are often cooled by auxiliary means, particularly in hot-oil pumps; this usually extends their range of usefulness by several hundred degrees. However, there is no advantage in picking a seal to operate, for example, at 400 F if the oil breaks down rapidly at 300 F, because the carbon buildup on the shaft will cause leakage before the seal fails.

Pressure limits are approximate maximum values that will not shorten life of a standard seal excessively. Special designs can often be developed to exceed the listed pressures, but are expensive and must be operated under closely controlled conditions. Applications with high vacuums and minimum pressure-drops are usually sealed best with face seals or molded packings whose relatively large contact area allows good lubrication and minimum leakage. A problem with most shaft seals and some face seals and packings is to guard against reversal of pressure or vacuum—these tend to lift the seal lip or packing, and suck in dirt or other contaminant. Usually, this can be avoided by properly venting the housing to prevent pressure buildup or vacuum caused by temperature changes.

Speed values listed can seldom be exceeded without shortening seal life, unless the design provides for cooling.

Chemical resistance is a general evaluation that indicates what is possible in each basic type. Resistance to specific types of chemicals depends on choice of seal materials (see Table II). It should be noted that trace elements present in oils and fluids can be very destructive to seal materials.

Dirt exclusion varies widely. But in general, shaft seals with auxiliary lips do an excellent job in moderately severe conditions, while face seals are best for extremely dirty conditions, such as earthmoving vehicles. Flushing the grease or oil during periodic lubrication may permit a less costly or otherwise more suitable seal. Protective housing or shroud design is also helpful.

Runout tolerance refers to shaft movement perpendicular to the axis which can be caused by shaft misalignment or eccentricity. When runout tolerance exceeds the limits shown, special seal designs must be developed. In general, all seals perform better and last longer when shaft runout is low—runout at high speed gives dynamic forces that can destroy the seal rapidly. Excessive runout can also cause heavy leakage, even in a newly installed seal. Since packings have little flexibility, they will not tolerate much runout. In labyrinth seals, clearances are too low to allow much runout.

Axial shaft movement is generally a limiting factor only in face seals where axial movement is limited by spring travel. However, high axial acceleration can make any seal leak. In the face seal it causes the sealing faces to separate, and leakage in other contact seals results from the high pressure built up ahead of the contact area, and ineffective wiping. Packings are generally best suited to this condition.

Lubrication in correct amounts is beneficial to any seal: it minimizes friction and acts as a cooling medium. Shaft seals operate more effectively with intermittent or splash lubrication, because of their low contact area and pressure, particularly at low speed and pressures. Leather shaft seals and some leather and fiber packings absorb enough lubricant to operate well with intermittent lubrication.

Cost varies widely and the relative evaluations are intended to indicate which type might be suitable for detailed study of actual costs in any particular application. The simplest seals, such as O-rings and loose packings,

I—COMPARISON OF BASIC TYPES

Selection Factors	Shaft Seal	Face Seal	Compression Packings	Molded Packings	Positive Clearance
General adaptability	Excellent	Good	Limited	Limited	Specialized
Availability	Excellent	Good	Good	Good	Partly or fully integral with machine
Positive action	Excellent	Excellent	Good under limited conditions	Good under limited conditions	No
Space (radial x axial), in.	¼ x ¼	⅜ x ⅝	¼ x ½	¼ x ½ except O-rings: ⅛ x ⅜ long	Variable but usually large except for single loose rings
Life (hours)	2000	3000	1000	1000	Unlimited in labyrinth type, 1000 hr in others
Temperature limits, °F	−65 to +300 (to 350 intermittent)	−65 to +850	No practical lower limit, to +1500	−65 to +500	No practical lower limit; to +2000
Pressure limits, psi (and vacuum)	15—std. design 100—special	1000	10,000 with rotation and reciprocation	3000 with backup rings and rotation and reciprocation	Depends on number of elements and tolerable leakage; usually 1500
Speed (fpm)	3000	15,000	1000 with reciprocation and rotation	1000 with reciprocation and rotation	10,000 with rings, unlimited with labyrinth
Resistance to attack by oils and chemicals	Good	Excellent	Excellent	Good	Excellent
Dirt and exclusion	Excellent under moderate conditions	Excellent under severe conditions	Good	Good	Fair (leakage usually excludes dirt during operation)
Runout tolerance, in.	0.025	0.025	0.0005	0.002	Depends on design; usually .0005
Axial shaft movement	Unlimited	0.075 in. (depends on design)	Unlimited	Unlimited	Wide variation, depending on design
Lubrication	Intermittent	Continuous	Continuous, may be self-lubricated	Continuous	Continuous with loose rings, none with labyrinth
Cost	Moderate	High, with a few exceptions	Low	Low	Moderate to very high

may cost only a few cents, some face and shaft seals go up to $100. Specially designed positive-clearance seals usually cost even more.

SEAL MATERIALS

Materials for the sealing elements of the first four basic types of seals are evaluated in Table II; these are the materials found in the most popular commercial seals now available. Their listed characteristics are generally applicable, but they will necessarily vary from one seal manufacturer to another. For example, silicone material varies widely and its use in shaft seals or molded packings for specific operating conditions should be checked by the seal manufacturer. Similarly, the performance of carbon face-seals is likely to vary widely, because carbon materials are numerous and can be identified only by a proprietary grade designation. Particle sizes, compositions and processing procedures all affect the performance characteristics of carbon face-seals.

Materials for positive-clearance seals are not listed in the table because the choice is not critical for performance by these seals. Generally, their materials are metals compatible with the shaft or the bore to which they are fastened.

EFFECT OF DESIGN VARIATIONS

Shaft seals

These are used in more dynamic applications than all others combined. This is because of their positive sealing ability; versatility in meeting requirements of temperature, shaft speed and pressure; low cost; relatively high insensi-

II—APPLICATION CHARACTERISTICS OF SEAL MATERIALS

	Temperature, °F	Abrasion Resistance	Acceptable Fluids
SHAFT-SEAL MATERIALS			
Nitrile	−30 to 225, continuous. To 300, intermittent	Excellent; R_c 28 or harder and approximately 15-rms shaft recommended	Acid solutions—except strong oxidizing acids; alkali solutions except 10–20% sodium hydroxide; coolants, antifreeze solutions; EP gear oils, all forms; fuels—petroleum, gasoline, diesel fuel; grease—petroleum-base; hydraulic fluids, glycol; lubricating oils; sal solutions; synthetic fluids—polyglycol, silicone and alkyl silicates
Polyacrylic	−30 to 300 Continuous 350 Intermittent	Fair; R_c 28 or harder and approximately 10 rms shaft finish recommended; lubrication must be continuous	Same as for Nitrile
Leather	−50 to 200	Excellent; will operate satisfactorily on 30 rms shaft at slow speeds; best choice for intermittent lubrication	Same as for Nitrile except no aqueous solutions or glycols
Silicone (consult seal manufacturer)	−80 to 500 in air or suitable fluids	Good; R_c 28 or harder and 10 rms shaft finish recommended	Commercial silicone up to 250 F in lube oils. Special silicones in lubricating oils up to 400 F; most industrial chemicals up to 500 F
FACE-SEAL MATERIALS			
Resin (filled, molded, or reinforced)	−60 to 212; lower limit is brittle point of resins; higher limit possible when face-heat generation is low	Excellent in absence of foreign solids when materials are chosen for specific applications	Mineral oils with resins. Hydraulic fluids; synthetic oils, water
Carbon	Up to 500 (usually limited by other seal components)	Excellent with clean fluids	Practically all industrial chemicals and corrosives as well as oils, solvents, water and steam
Ceramic (sometimes bonded to metal)	Up to 500 (usually limited by other seal components)	Excellent; will tolerate high solid contaminants but life will be shortened proportionately	Most industrial chemicals, oils, fuels, and solvents
Metal (steel, stainless, bronze, ni-resist., satellite) Most face seals are available with metallic cases; corrosion and differential thermal expansions must be considered, and corrosion protection applied when necessary	Up to 500 (usually limited by other seal components)	Excellent; governed by material of mating face; when 2 face materials are properly matched, will tolerate contaminating solids and large amounts of external dirt	Oils, fuels, and fluids compatible with corrosion resistance of the metal
Face-seal rubber bellows or boots	32 to 212		Synthetic and mineral oil and hydraulic fluids; most industrial chemicals, water and vegetable acids
Face-seal O-rings and static sealing rings	32 to 212; Kel-F or Teflon up to 500; auxiliary cooling of face seal extends the range		Same as for nitrile above; Teflon and Kel-F are inert to almost all industrial chemicals, corrosives and solvents
MOLDED PACKING MATERIALS			
Nitrile (including fiber-reinforced)	−30 to 225, continuous; to 300 intermittent	Excellent; R_c 45 and 10 rms shaft finish recommended; continuous lubrication required	Same as for Nitrile in shaft seals
Polyacrylic	−30 to 350; consult mftr. on fluids and speeds when nearing these temperatures	Good; must have continuous lubrication	Same as for Nitrile
Silicone	Poor with hot lubricants	Poor	Seal manufacturer should be consulted before using
Neoprene	−65 to 225	Excellent	Used primarily with water, salt water, Freon; poor against most lubricants, solvents, hydraulic fluids and industrial chemicals
Metallic Packings (aluminum copper, lead)	−70 to 1000 fiber-filled; packing good to 212 only	Must be matched to shaft material; shaft finish and size tolerance must be held closely	Solvents, hydraulic fluids, most industrial chemicals, steam, etc. limited only by metal corrosion
COMPRESSION PACKING MATERIAL			
Asbestos	To 600	Good on smooth shafts only	Hot water, superheated steam, hot oils, air and other hot gases
Fibers (jute, flax, linen, duck)	To 212	Good; R_c 28 or harder and 10 rms or better shaft finish recommended	Cold and hot water, low pressure steam, oils, ammonia

III—OPERATING CONDITIONS FOR BASIC SHAFT SEALS

	Spring-Loaded Types		Non-Spring-Loaded Types	
	 Leather	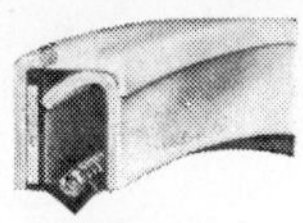 Rubber	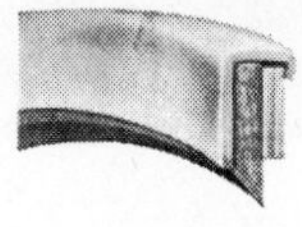 Leather	 Rubber

SHAFT SPEED, fpm — Shaft speed is the prime factor in selection of the proper seal. It affects the limitations of all other operating conditions, thus it is shown as slow, moderate or high expressed in feet per minute.

	Spring-Loaded Leather	Spring-Loaded Rubber	Non-Spring-Loaded Leather	Non-Spring-Loaded Rubber
Slow	0 to 800	0 to 1000	0 to 300	0 to 500
Moderate	800 to 1500	1000 to 2000	300 to 600	500 to 1000
High	1500 to 2000	2000 to 3000	600 to 1000	1000 to 2000

TEMPERATURE LIMITS, °F — Temperature limits are the fluid temperatures of the medium to be sealed. They represent the point beyond which the sealing material or medium sealed become ineffective. Operation at higher or lower temperatures can be accomplished by the use of special compounds under controlled conditions.

	Spring-Loaded Leather	Spring-Loaded Rubber	Non-Spring-Loaded Leather	Non-Spring-Loaded Rubber
Continuous	−65° +200°	−65° +225°	−65° +200°	−60° +225°
Intermittent	−65° +225°	−60° +250°	−65° +225°	−60° +250°

PRESSURE, max psi — Standard design fluid seals are not intended to withstand pressure. Any internal pressure materially reduces seal life, and means should be provided to relieve internal pressure build-up. Special seal design is required for high-pressure sealing.

	Spring-Loaded Leather	Spring-Loaded Rubber	Non-Spring-Loaded Leather	Non-Spring-Loaded Rubber
Slow	15	10	0	25
Moderate	10	7	0	10
High	5	5	0	5

SHAFT FINISH, max microin — Fine shaft finishes assure maximum sealing efficiency. The coarser the finish, the greater the risk of leakage and wear on the seal. Surface finish value (RMS) is not, however, the only criterion for shaft finishes. Type of finish, direction of finishing marks, and spiral lead are all factors. Polished or ground finishes with concentric finish marks are preferred. When a finish lead is present, its direction should lead the fluid inward.

	Spring-Loaded Leather	Spring-Loaded Rubber	Non-Spring-Loaded Leather	Non-Spring-Loaded Rubber
Slow	25	25	25	25
Moderate	20	20	20	20
High	20	20	20	20

SHAFT HARDNESS, suggested Rockwell — Successful sealing is possible on shafts as soft as cold rolled steel. Better performance and longer seal life are obtained, however, on shafts above C20 Rockwell. Fluid starvation, abrasives, and high surface speeds require hard shafts.

	Spring-Loaded Leather	Spring-Loaded Rubber	Non-Spring-Loaded Leather	Non-Spring-Loaded Rubber
Abrasives	above C-45	above C-45	above C-45	above C-45
No abrasives	above B-80	above B-80	above B-80	above B-80

SHAFT TO BORE MISALIGNMENT, Total indicator — This is misalignment of the center of shaft rotation with bore center, which causes wear on one side of the seal. This condition becomes more severe as surface speeds increase. (See "Shaft Runout" below.)

	Spring-Loaded Leather	Spring-Loaded Rubber	Non-Spring-Loaded Leather	Non-Spring-Loaded Rubber
Slow	0.010	0.015	0.005	0.005
Moderate	0.005	0.010	0.005	0.005
High	0.005	0.010	0.005	0.005

SHAFT RUNOUT, Total indicator rpm — This is misalignment of the sealing surface on the shaft with the center of shaft rotation, or movement of the center of shaft rotation through bearing looseness or shaft whip. This type of eccentric wear is more evenly distributed about the circumference of the sealing lip. Runout should be kept to a minimum. Combination of shaft runout with shaft-to-bore misalignment imposes very difficult sealing conditions.

	Spring-Loaded Leather	Spring-Loaded Rubber	Non-Spring-Loaded Leather	Non-Spring-Loaded Rubber
0–800	0.010	0.025	0.003	0.005
800–2200	0.005	0.020	0.003	0.005
2200–4200	0.003	0.015	0.003	0.005

tivity to dimensional variations; and tolerance of a variety of lubricants and lubricating conditions. Basically, there are four variations of shaft seals, or a combination of these four. Recommended operating conditions for these are given in Table III, and the most useful combinations are shown in Table IV.

A special version is the dual-purpose seal, Fig 12, which incorporates the liquid and dust lips in a single unit bonded to the seal ring.

External-type synthetic shaft-seal, Fig 13, is used where the bore rotates—but seldom where the shaft rotates, because centrifugal force has a large effect on the sealing member and results in performance charactertistics inferior to conventional types.

Suitable shaft finishes are important for performance of shaft seals. One way to minimize the expense and bother of shaft finishes is to use unitized seals, Fig. 14, where the wear ring is an integral part of the seal. Sometimes they also solve difficult installation problems, and usually can be supplied in all the basic shaft-seal types.

(text continued, p. 302)

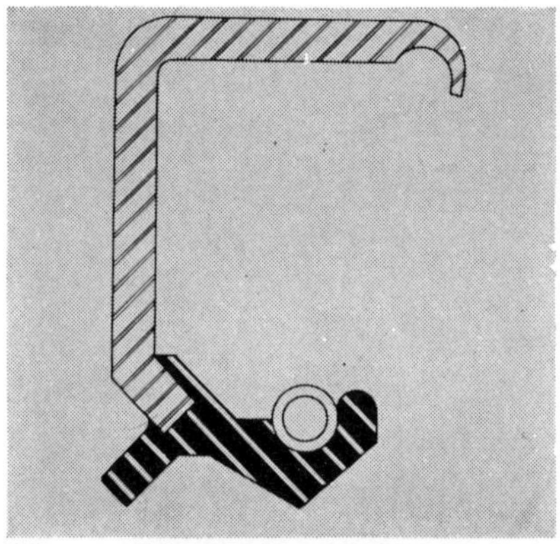

12 . . . INTEGRAL DUST LIP is incorporated in this special version of shaft seal.

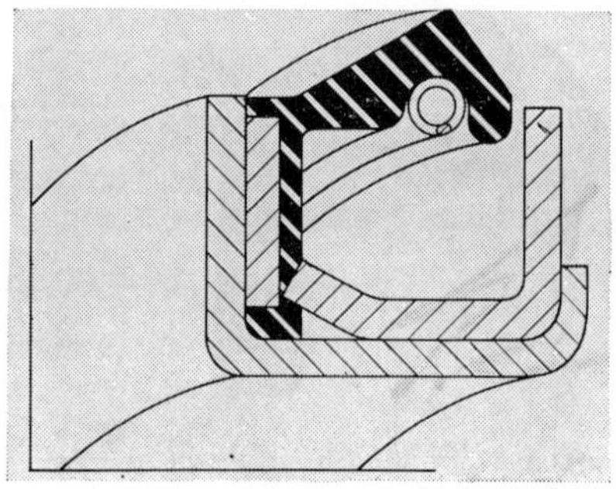

13 . . . EXTERNAL-TYPE SHAFT SEAL is one where the bore rotates.

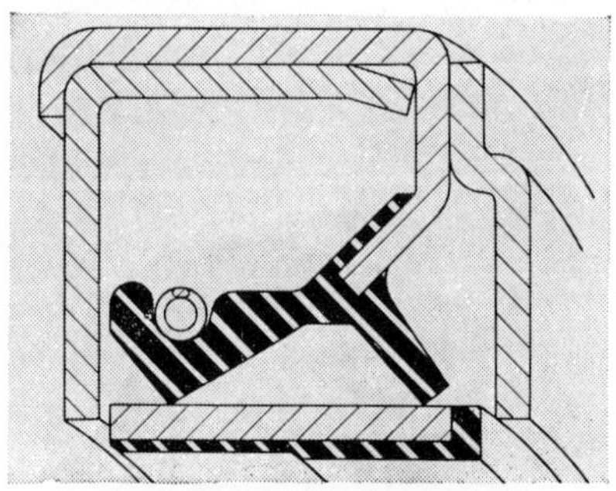

14 . . . SHAFT FINISHES A PROBLEM? One way to minimize it is to use a shaft seal with an integral wear ring—usually pressed on the shaft.

IV—COMBINATION TYPES OF SHAFT SEALS

LEATHER / SYNTHETIC	
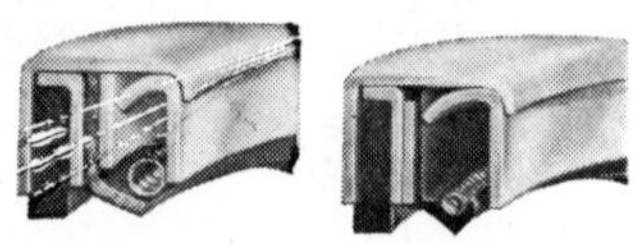	**A felt auxiliary** adjacent to the leather or synthetic lip is recommended where external conditions require the added protection.
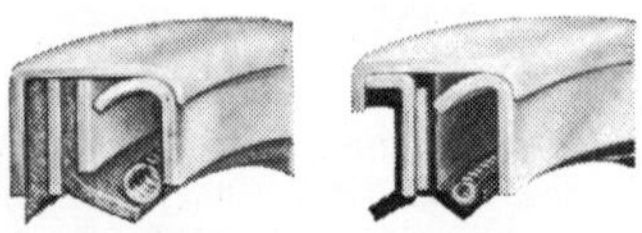	Designed for applications where shaft enters seal from the spring-loaded side; the leather or synthetic auxiliary lip acts as a dust baffle.
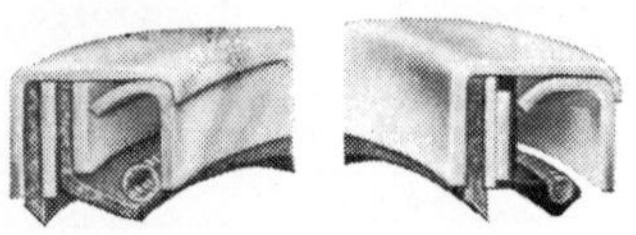	For applications where the shaft entry is from the auxiliary side. The leather auxiliary is intended to prevent the entry of dirt or other foreign matter.
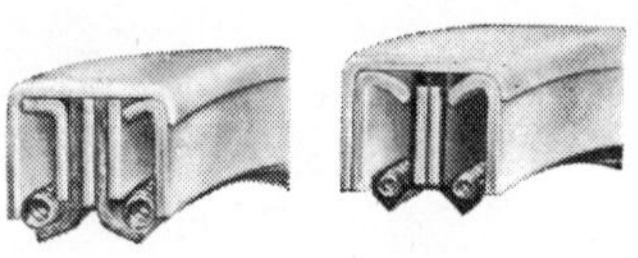	This design basically consists of two sealing members facing opposite directions for separating two fluids. The synthetic seal is designed for separating low-viscosity fluids where positive sealing is required, especially at higher speeds and temperatures.
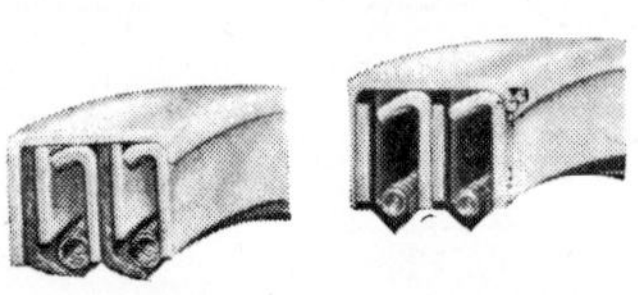	This design has two sealing members facing in the same direction for special operating conditions where inner lip is needed to stem a heavy flood of lubricant or surge pressures, and second lip is free to function under normal conditions. Synthetic type is for very unusual conditions where the hydraulic action is so severe that a single lip seal is incapable of providing zero leakage.

V—APPLICATION RANGE FOR BASIC FACE SEALS

	Description	Temperature, °F	Pressure, psi	Speed, fpm	Comments
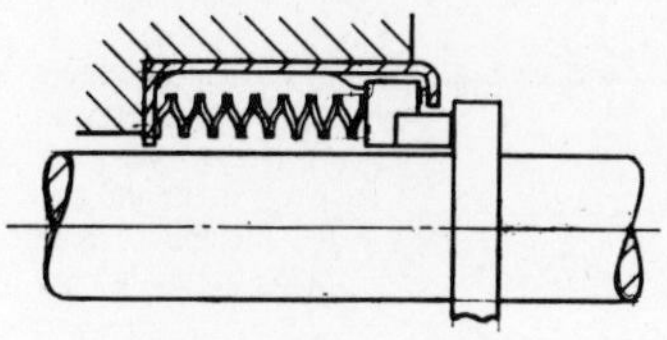	Metal bellows, carbon face	−65 to +650	200	15,000	Inert to many chemicals which degrade elastomers. Bellows and face materials available in a wide range of materials.
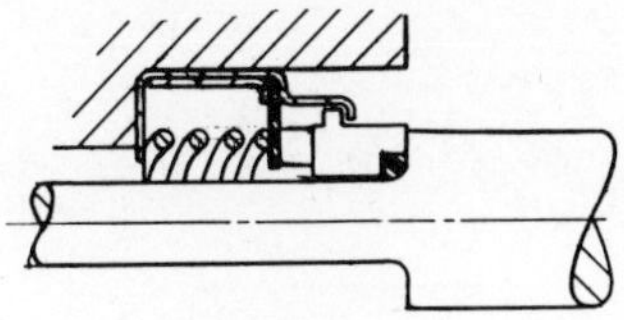	Unitized with O-ring sealing of mating face	−30 to +300	15	5000	Readily available in stock sizes for pumps, etc. Pressure limit is low because of diaphragm type of bellows. Unitized principle is often applied to other designs.
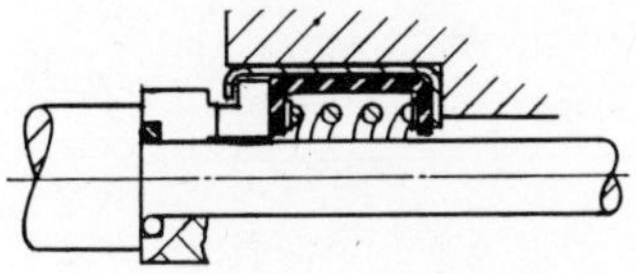	General-purpose bellows or boot seal	−30 to +250	250	15,000	Allows large axial movement because of spring type and bellows construction
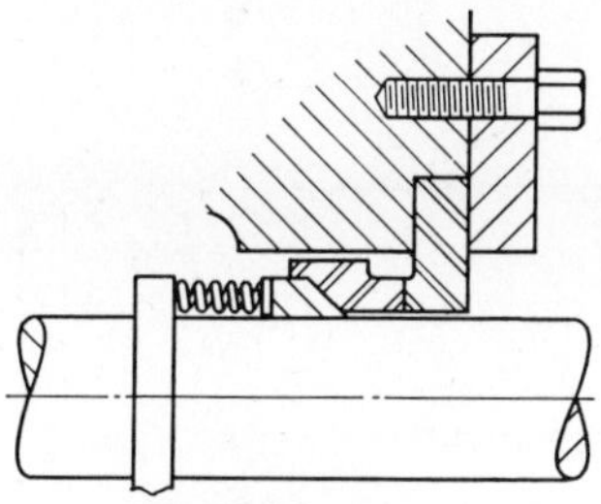	Teflon-wedge packing	−65 to +500	2000	15,000	Especially suitable for corrosive liquids and high temperature, as well as pressure.
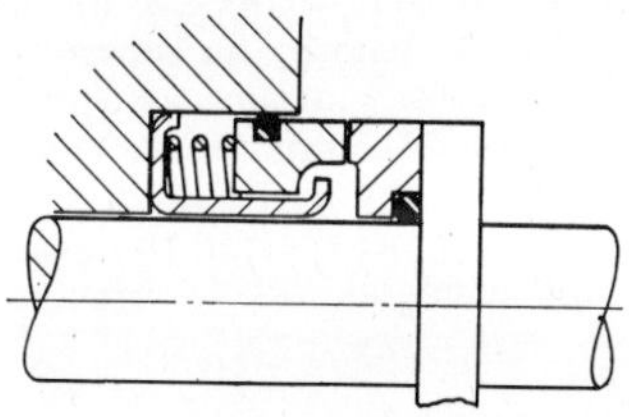	O-ring packing	−30 to +300	750	15,000	Moderate-pressure, high-speed seal
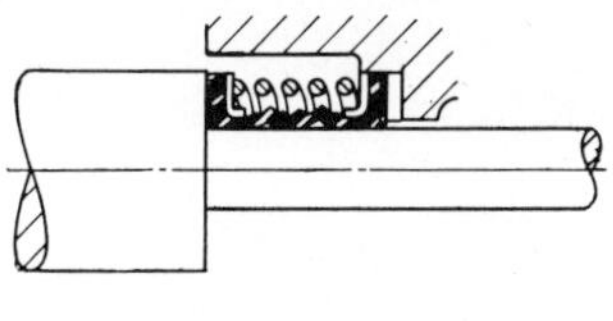	Inexpensive bellows type	−30 to +250	0–15	5000	Inexpensive low-pressure design for mild service. Bellows construction limits pressure. Face often seals on housing or shaft shoulder, depending on particular seal design.
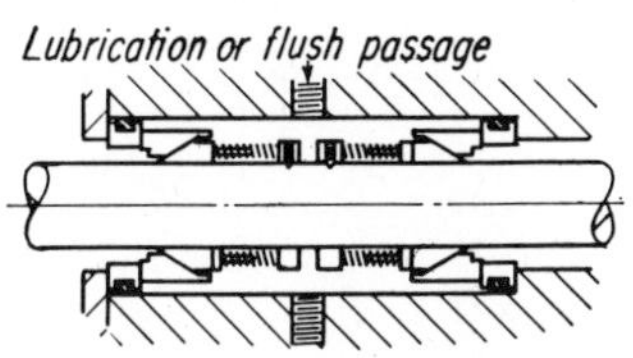	Double installation with flushing	−65 to +500	2000	15,000	Flushing permits cooling, prevents escape of toxic fumes, serves where auxiliary seal lubrication is necessary.
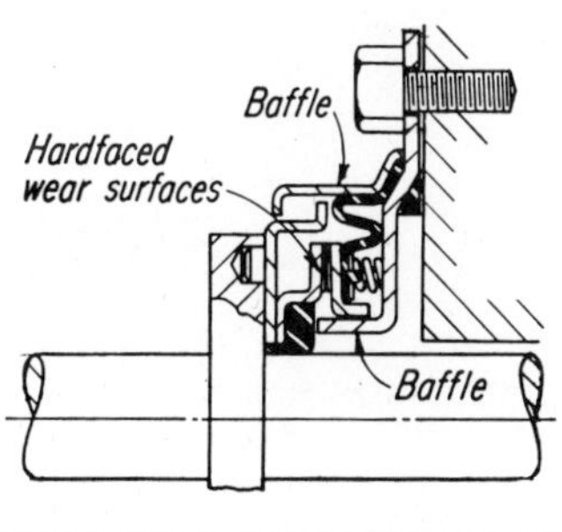	Track-roller type	−65 to +300	0–15	5000	Design especially suitable for low-speed, very abrasive conditions. Hard facing on both faces resists wear. Unitized design also includes dirt baffle.

Two factors that can limit the use of shaft seals are pressure (or vacuum), and temperature. High temperatures cause hardening and degradation of rubber or leather elements in the shaft seal. Adequate lubrication and auxiliary cooling may be used to keep temperatures low. Another method uses sealing elements with low compressive force on the shaft. Reducing rubbing speed, shaft runout and misalignments is also helpful. For applications with inherently high- or low-temperature problems, special compounds such as the polyacrylics are required.

Pressure sensitivity of shaft seals is inherent in their design; although some have been used successfully up to 100 psi. Typical action of forces on a shaft seal are shown in Fig. 15. In (A), in order to have equilibrium at the lip, the spring force S and the flexing moment M must counterbalance three forces: F_R, the force caused by shaft runout; I, the inertia force; and the opposing components of the unit pressure P. Generally, all these various forces can be resolved into two forces, P_H and P_V acting horizontally and vertically as in (B). P_H tends to displace the sealing element along the shaft and with high enough pressure may blow the lip out. In the usual sealing element configuration, however, the vertical force P_V is larger and tends to flatten the sealing element on the shaft as in (C) well before such blowout occurs. The high contact pressure and resultant high contact width causes excessive heat generation which hardens and degrades the sealing element material. This, in turn, causes cracks in the contact surface which become leak channels. Using a harder compound, shortening and thickening the seal element, and using an abrasion resistant compound can improve the performance. But such changes would also cause lower runout tolerance, higher initial heat generation, and in general shorter life. For this reason shaft seals are best for low pressures.

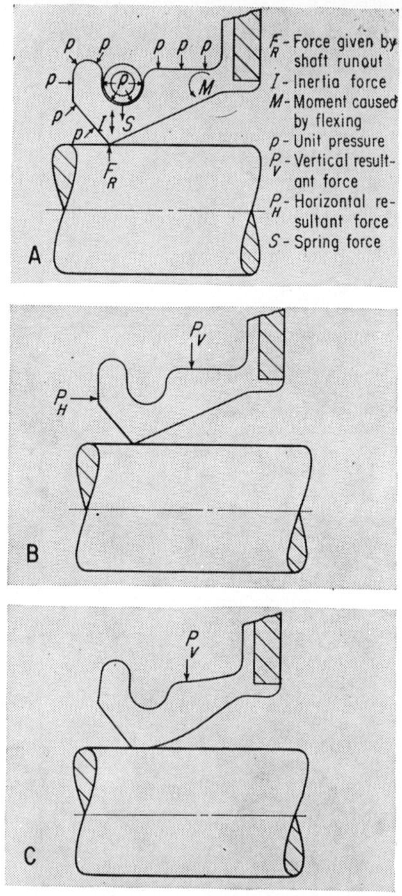

15 . . . ACTION OF FORCES on a shaft seal (A) can usually be summarized into two forces acting at right angles, shown in (B). For proper sealing, vertical force should be greater—but if excessive it will flatten the seal as in (C) and cause excessive wear.

Face Seals

These complement shaft seals, and the following three features are inherent in their design:

- High speeds and pressures are readily handled with appropriate designs.
- They can be made of metals, elastomers, plastics, leather, cork, or other materials to suit a wide range of applications.
- The faces are usually lapped to within a few light-bands of flatness (0.000,030 in.) to provide a fine seal.

Although face seals are not as standardized as shaft seals, they fall into approximately eight types (Table V). Variations of these types can usually be made for internal or external sealing, and either rotating or nonrotating.

In the table all high-pressure seals except the metal-bellows type have a sliding packing (sometimes called static seal) instead of bellows, or diaphragm. This is because the projected areas of a bellows or diaphragm are so great that the resulting force is sufficient to overcome the spring force. This may cause the spring to bind or may overcome the spring Fig 16(A) or it may overload the face if the pressure is from the inside as in Fig 16(B). The effect is similar to that of vacuum acting on the bellows or diaphragm.

Face seals for high pressure incorporate a pressure-balanced design with various degrees of balance (Fig 17). The unbalanced type (A) is generally for low pressures

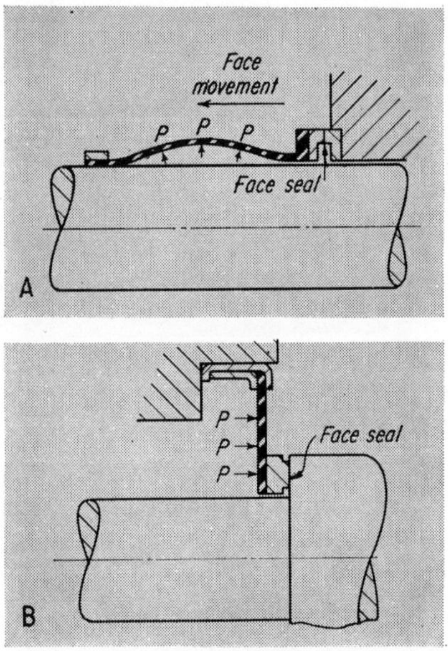

16 . . . IN FACE SEALS, use of bellows or diaphragm is impractical because pressure causes bellows to bind and face to lift as in (A), or give excessive face pressures as in (B).

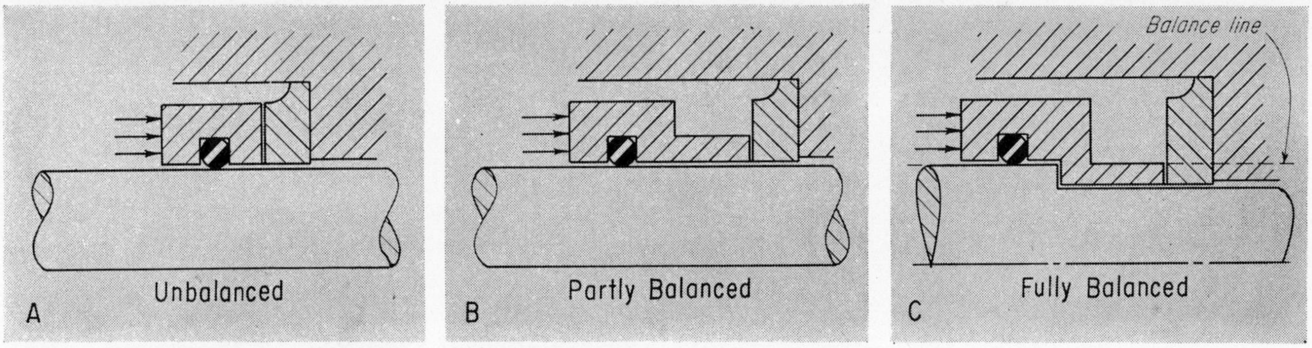

17 . . . FOR HIGH PRESSURES, face seals use static O-rings as secondary seals with various degrees of pressure balancing.

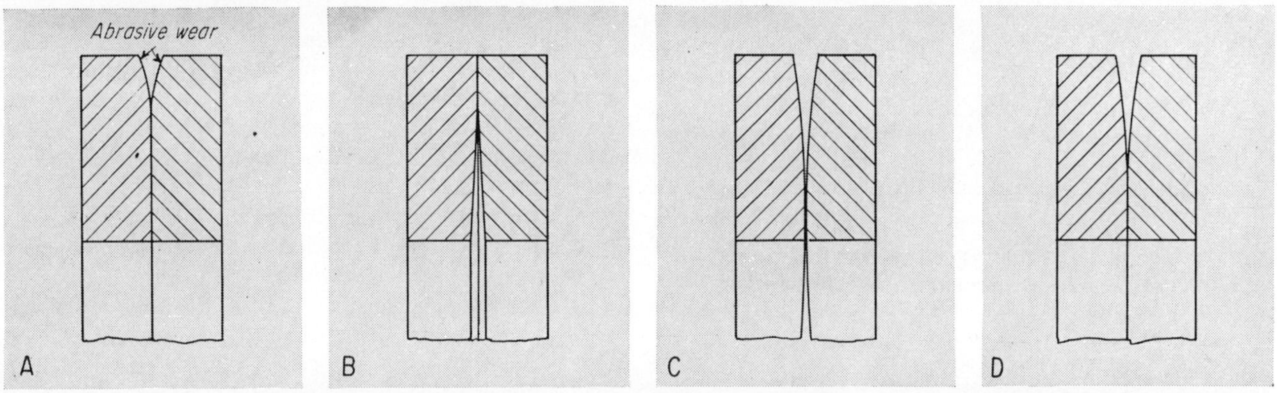

18 . . . To prolong life of face seals in abrasive applications faces are sometimes made concave. Dirt entry and wear on a flat face as in (A) promotes early failure. Wear on a concave face as in (B) results in progressively exposing a new contact surface as in (C) and (D).

that will not affect adversely the lubricating film between the sliding faces. The partly balanced (B) is used where one wants a portion of the fluid pressure to complement the predetermined spring force acting on the sealing faces. Generally, this is desirable in medium-pressure applications where surge pressures could separate the sealing faces and cause leakage. The partly unbalanced pressure acts as a safety factor to keep the sealing faces together. For high pressures it is sometimes desirable to use the fully balanced type (C). Here, the spring force must be designed to take care of any pressure surges. Most face seals of moderate to high pressure are partly balanced. Special constructions using this principle are good for pressures up to 5000 psi.

One of the most exacting specifications for face seals is quality of the mating surfaces. There are many theories about how smooth these surfaces should be for minimum friction on the one hand, and effective lubrication on the other. For minimum friction the surfaces are sometimes specified smooth and flat to within 3 microin rms, but for effective lubrication micropools of lubricants are required, and this suggests that the finish should be coarser. Many finishing techniques have been designed to compromise these two opposing requirements, but they are still subject to experimental verification. In general, where abrasion resistance must be high and leakage requirements low the faces must be flat to within three light bands (0.000,030 in.), surface finish as good as possible (3 rms desirable), hard (66 R_c), and structurally dense as possible.

There are, however, some helpful precautions—particularly in dirty applications. One of the most effective is shrouding or adding a dirt baffle. Abrasive wear usually occurs first as a rounding-off of the outer edges of the face seals, Fig 18(A). Where a dirt baffle is insufficient to counteract this type of wear, face seals can be lapped to a spherical or conical shape; Fig 18(B) gives an exaggerated view. This has a tendency to extend seal life so that, as the outer edges wear, the contact area moves toward the center, Fig 18 (C and D).

Compression packings

The design variations here are usually limited to selection of materials to suit the operating conditions, and these will be found in Table I. Typical design, (Fig 19), is sometimes used in rotating applications. However, compression packings in general are best suited for high-pressure reciprocating, or combination of reciprocating and rotating motion. Some designs will hold pressures up to 10,000 psi, but lubrication and surface speed are usually limiting factors. Since friction is so high in this type of seal, speed is usually limited to 1000 fpm, except in centrifugal pumps where water seepage through the seal can be used to lubricate and cool it.

The temperature limit for fiber and combination fiber-rubber packings is about 250 F; for asbestos, about 700 F; and for metallic or metallic-covered packings, to 1500 F.

Packings have a tendency to score the shaft more than do most other types of contact seals. This is because foreign particles tend to enter and become imbedded in the packing. Cleanliness is necessary and the shaft should be as hard and smooth as practical. Separate wear rings can be installed where scoring is likely to damage an expensive shaft.

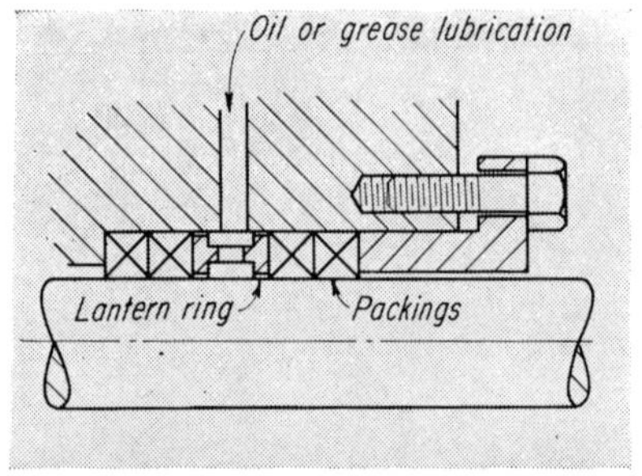

19 . . . COMPRESSION PACKING in this typical design for low rotating speeds uses oil or grease pocket.

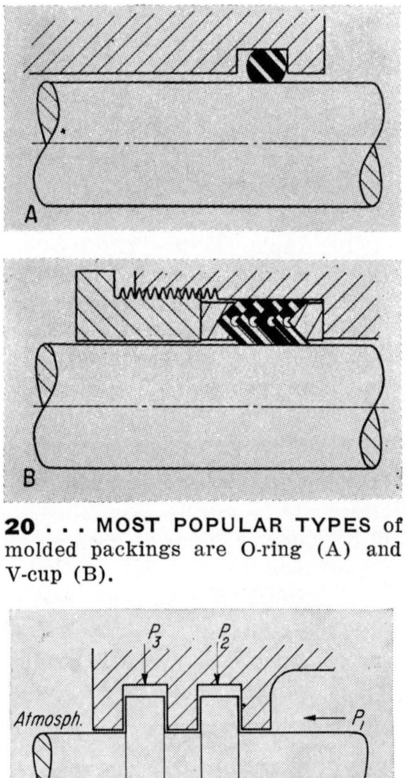

20 . . . MOST POPULAR TYPES of molded packings are O-ring (A) and V-cup (B).

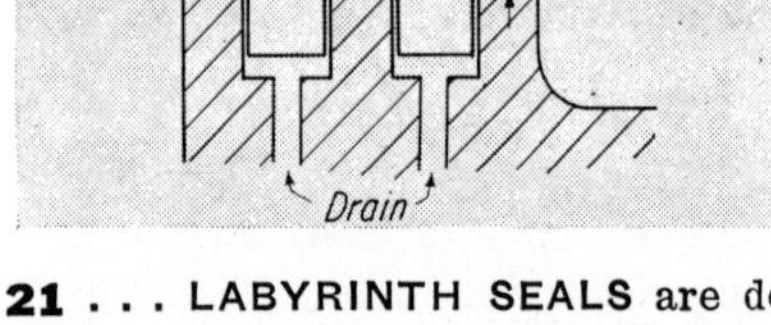

21 . . . LABYRINTH SEALS are designed to limit the leakage in accordance with various empirical formulas which have been derived from the well-known Martin formula.

Molded Synthetic Packings

Designs here are available in a variety of geometric shapes, such as O-ring, V-cup or U-cup. They may also be found in the shape of an X, a D, a square or even a triangle. Each variety is supposed to minimize extrusion under pressure when it is installed according to instructions. The O-ring and V-cup (Fig 20) are the most popular, but not usually recommended for rotating service. They are most suitable for high-pressure static and reciprocating service. O-rings, however, can be used at speeds up to 100 rpm at low pressure. V- or U-cups will work up to 1500 psi in applications that combine reciprocating and low-rotating motion.

Back-up or end rings of leather are often used with O-rings and packings, primarily to prevent extrusion but also to help maintain lubrication. If the sealed medium does not provide adequate lubrication, grease or oil fittings are needed for auxiliary lubrication, as shown with compression packings.

Labyrinth or positive clearance

These seals are so specialized that no standard types or designs have evolved. Design is usually controlled by the tolerable leak-rate, from which one can calculate gap clearances and number of elements required.

1. **Number of rings** to limit leakage to a given flow can be found as follows:

$$N = (40\,P - 2600)\,(W/A)/540\,(W/A) - P$$

where N = Number of rings
W = Permissible leakage, lb/sec
A = $C\pi D$, cross-sectional area, in.^{2}
C = Clearance, in.
D = Diameter, in.
P = Absolute pressure, psia

In this equation, to find N_{12} for a leakage from pressure level P_1 to a lower pressure level P_2, the N for each must be found, then $N_{12} = N_1 - N_2$.

2. **Leakage flow rate** can be found as follows:

$$W = 25\,KA\sqrt{\frac{\left(\frac{P_1}{V_1}\right) - \left[1 - \left(\frac{P_2}{P_1}\right)^2\right]}{\left[N - \mathrm{Log}_n\left(\frac{P_2}{P_1}\right)\right]}}$$

where W = Flow rate, lb/hr
V_1 = Initial specific volume, ft^3/lb
K = Experimental coefficient

For interlocking labyrinths, $K = 55$ approximately, if the velocity is effectively throttled between labyrinths. It is independent of clearance in the usual range. For non-interlocking labyrinths, K varies with ratio of labyrinth spacing divided by radial clearance. For a ratio of 5, $K = 100$; for a ratio of 50, $K = 60$.

Noncontacting dynamic packing

Labyrinth shaft seals

Flow friction is all that keeps the fluid in when you have positive clearance seals. Here's your guide to how much leaks out.

LOUIS DODGE, Hydraulics Consultant, New Richmond, Ohio

HOLDING fluid behind a labyrinth shaft seal is a little like stopping the wind with a door ajar. You won't stop all of it. But a labyrinth or some other clearance seal may be your only choice when the rubbing friction and wear of a conventional mechanical seal can't be tolerated.

Shaft seals are primarily to reduce leakage of fluid from the system, and the greater the obstruction to flow the more efficient the seal. Although only ten basic designs are sketched, there are hundreds of variants.

Some companies, including Kentucky Metals Inc and Clark Bros Co, have built special labyrinth shaft seals made of honeycomb metal (see cover) to seal low-pressure gases. The advantage is that the rotating shaft can easily machine away any interference, assuring minimum operating clearance.

The majority of labyrinth seals are machined of bronze or a similar alloy. They can be made accurately, they are mechanically strong, and they withstand high temperature. But the metal selected must not gall or melt during wear-in. There must be no residue or distortion to lessen the effectiveness of the seal. For high-speed turbine shaft seals, the stationary labyrinth teeth often are machined of an alloy that effectively sublimates (evaporates without melting) when rubbed by the shaft, which is usually of a different material. After the wear-in period, the seals are noncontacting, and the final operating clearance depends on thermal expansion and eccentricity of the shaft relative to the seal teeth.

There is a limit to the complexity of

TURBINE WITH LABYRINTH SEALS (*GE gas turbine*)

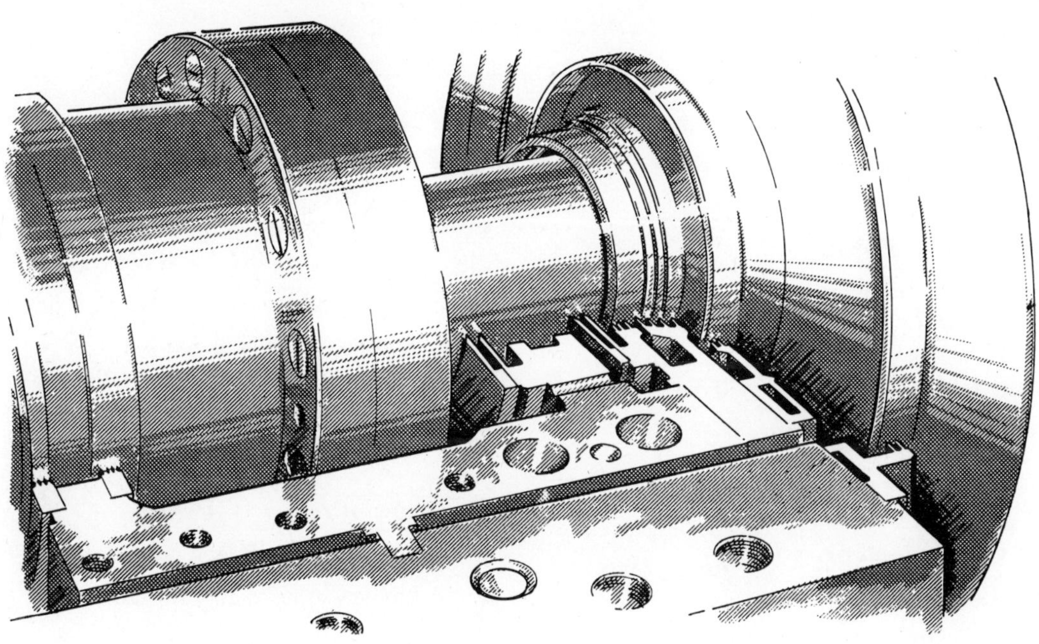

1 . . Sleeve

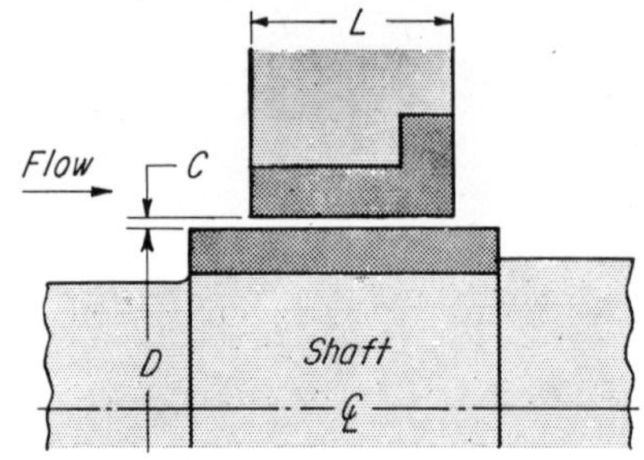

$$H_T = \frac{V^2}{2g}\left[1.5 + \frac{fL}{2C}\right]$$

$$V = \sqrt{\frac{2g\ H_T}{1.5 + \frac{fL}{2C}}}$$

For centrifugal pumps, and other liquid-handling applications. Has lowest cost.

2 . . Interlocking labyrinths

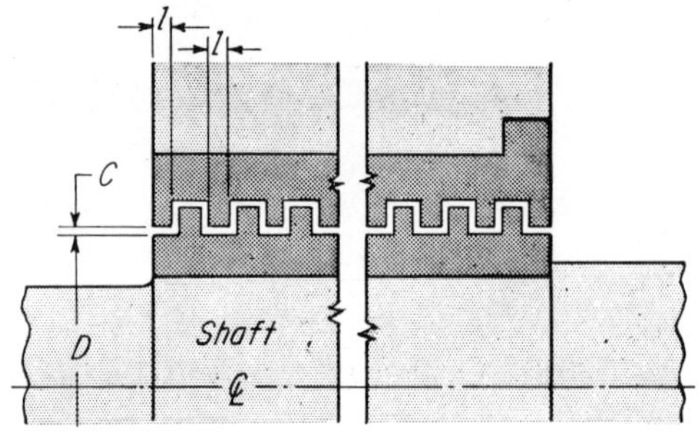

$$H_T = \frac{V^2}{2g}\left[1.5 + \frac{fL}{2C} + 2.0\ (n\text{-}1)\right]$$

$$V = \sqrt{\frac{2g\ H_T}{1.5 + \frac{fL}{2C} + 2.0\ (n\text{-}1)}}$$

For sealing liquid and gas. Requires split assembly, but has lower leakage.

---

the seal. Tests indicate a diminishing advantage beyond a certain number of stages, particularly where the shaft doesn't stay centered according to calculations. In such cases, the simpler labyrinths are as good as the more serpentine, and much less expensive.

The tapered teeth in Fig 4 and 6 are considered the most efficient for sealing gas. Usually the teeth are rings inserted into the seal sleeve and staked. They are tapered to an edge of about 0.010 in. at the shaft diameter, and no clearance is allowed. The tips are quickly ground off when the machine begins to operate. Water lubrication can be used to prevent excessive heating during wear-in.

Seal calculations

In the hydraulic field, it is conventional practice to set minimum and maximum limits for the clearance C

EQUATIONS AND SYMBOLS

A = area of annular clearance, ft² = $D\pi C$

C = radial clearance, in.

D = diameter of shaft, in.

f = friction factor (dimensionless)
$f = 64/N_R$ for laminar flow ($N_R < 2320$)
$f = 0.316/N_R^{0.25}$ for turbulent flow

H_E = entrance head loss, ft = $0.5\ V^2/2g$

H_S = head loss for one stage, ft
$H_S = 2V^2/2g$ for Fig 2 (4 edges)
$H_S = V^2/2g$ for Figs 3 and 4 (2 edges)
$\Sigma H_S = H_S\ (n-1)$

H_V = velocity head, ft = $V^2/2g$

H_F = friction head loss, ft = $f\ \frac{L}{2C}\ \frac{V^2}{2g}$

H_T = total head loss of all stages, ft
$H_T = H_E + \Sigma H_S + H_V + H_F$

L = total axial length of shaft seal, in.

n = number of seals or teeth
$n - 1$ = number of stages or spaces

N_R = Reynolds number, dimensionless
$N_R = 2CV/12\ \upsilon = 2 \times 10^5\ CV$ for 80 F water

Q = flow, ft³/sec = VA

V = fluid velocity, ft/sec

υ = kinematic viscosity, ft²/sec

3 . . Non-interlocking

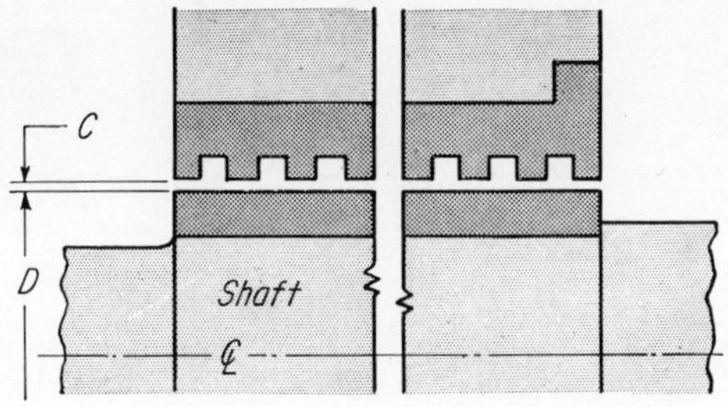

$$H_T = \frac{V^2}{2g}\left[1.5 + \frac{fL}{2C} + 1.5\,(n\text{-}1)\right]$$

$$V = \sqrt{\frac{2g\ H_T}{1.5 + \frac{fL}{2C} + 1.0\,(n\text{-}1)}}$$

Compromise between Fig 1 and Fig 2. Seals liquid and gas. No split necessary.

4 . . Tapered teeth

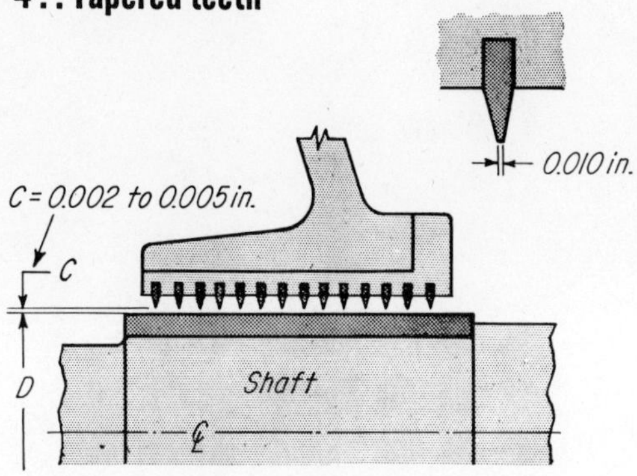

$$H_T = \frac{V^2}{2g}\left[1.5 + 1.5\,(n\text{-}1)\right]$$

$$V = \sqrt{\frac{2g\ H_T}{1.5 + 1.5\,(n\text{-}1)}}$$

For low-leakage seals. Commonly used on steam turbines and gas turbines.

---

between shaft and seal. The values depend on the strength of the rotating shaft and its deflection. As proved by tests, speed of the shaft has no effect on performance of the seal. Typical recommended values of clearance are plotted at the right.

The head loss and flow equations on these two pages are approximate and cannot take the place of tests. They show trends accurately, however, and are workable once you've applied correction factors for your own designs.

Briefly, the effective total resistance to flow, measured in feet of lost head, is the sum of three types of unit resistance: turbulent conversion of initial static pressure to velocity; wall friction; and turbulence caused by abrupt changes of section in the flow path.

The first two, H_V and H_F, are explained sufficiently by the equations and symbols table at the left. The third type of resistance includes both H_E and H_S and is explained as follows: The first abrupt change of section is at the entrance, and head loss according to the conventional entrance equation is $H_E = 0.5V^2/2g$. The succeeding changes in section depend on the number of edges in each labyrinth. For Fig 2, there are four edges per

For Figs 1, 2 and 3 . . RECOMMENDED CLEARANCES

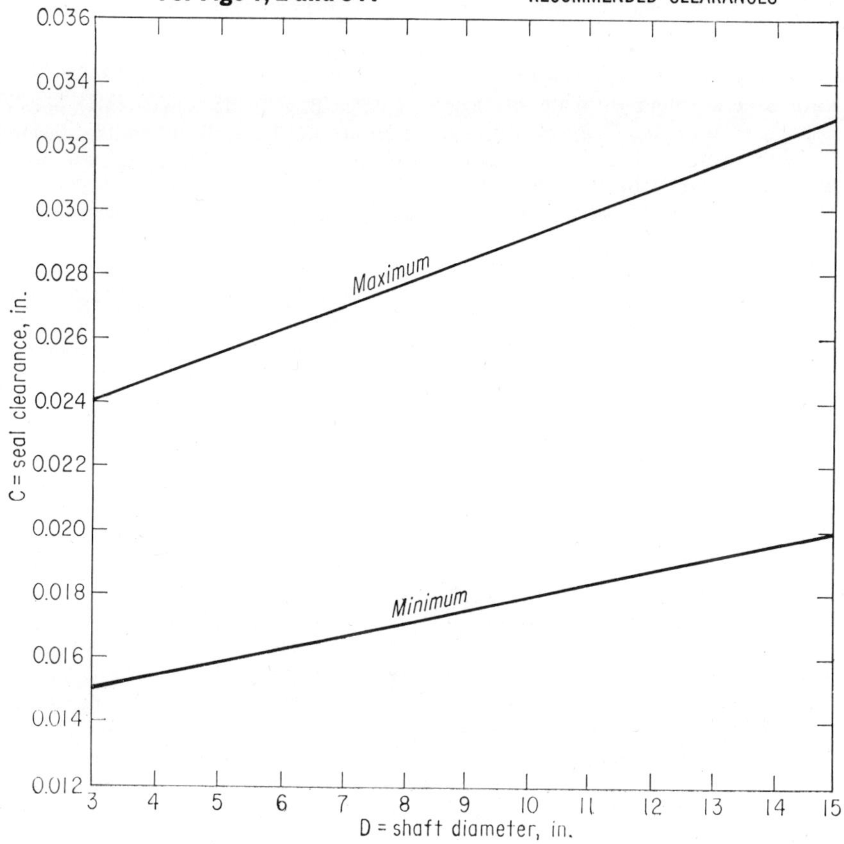

5.. Sleeve seals

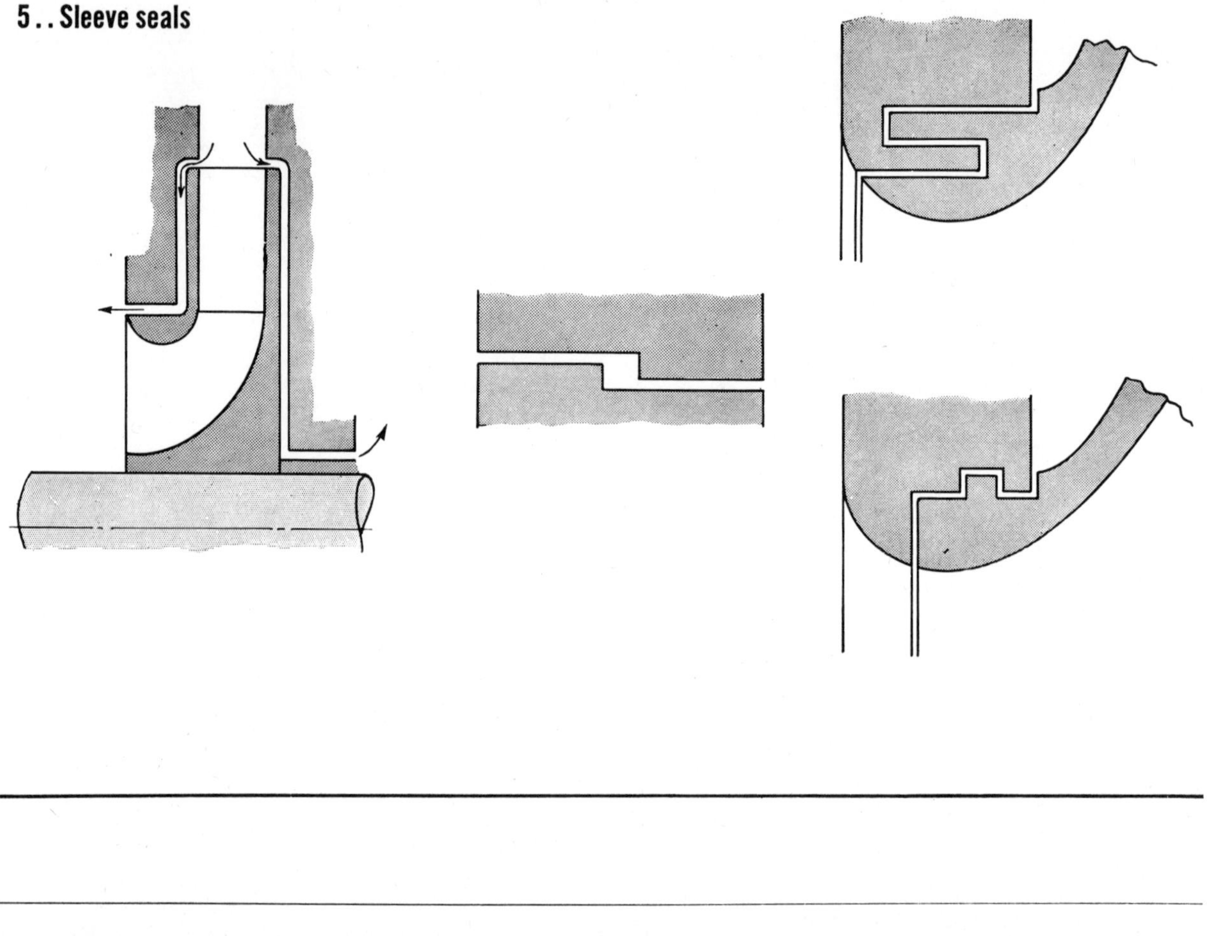

stage and a rough estimate of head loss $H_S = 4 \times 0.5V^2/2g = 2V^2/2g$. For Fig 3 and 4, there are two edges per stage, and the relationship is $H_S = 1.0V^2/2g$. The summation is

$$H_T = H_V + H_F + H_E + \Sigma H_S \quad (1)$$

where the value of ΣH_s is explained in the equations and symbols table.

To express Eq 1 in terms of velocity, insert the known expressions for head loss given in the equations and symbols table. This useful relationship results:

$$H_T = [V^2/2g]\,[1 + fL/2C + 0.5 + a\,(n-1)] \quad (2)$$

where $a = 2$ for the seal of Fig 2 and $a = 1$ for the seal of Fig 3 and 4.

Eq 2 applies to gases and liquids. For gases the accuracy is problematical because velocity and density change from state to stage and you have to assume average values.

Leakage flow at any given point, for gas or liquid, is simply the average velocity V at that point times the cross-sectional flow area A. The relationship is Q

For gases, accurate predictions of leakage flow using a particular equation are not possible unless the design of the seal happens to match the conditions of the tests upon which the flow equation is based. There are many more published equations than are discussed here, and variations in calculated leakage flow might exceed 2:1.

A good analysis of the literature including discussions of the different equations, and a thorough bibliography, may be found in papers 58-A-118 and 60-GTP-12, ASME (American Society of Mechanical Engineers). However, for a first approximation, the equations in this article will suffice. Once you've built and tested a prototype, you can modify the equation constants and get closer results.

One way to improve the accuracy of calculation is to use trial and error. Assume a Reynolds Number, calculate friction factor, compute leakage flow. Using the computed flow, check out the assumed Reynolds Number, correct it if necessary, and try again. The graph on the following page gives viscosity values for calculating Reynolds Number.

Calculation Examples (water)

1—Sleeve seal (Fig 1)

Known pressure drop $H_T = 300$ psi = 693 ft head. $D = 10$ in., $L = 2$ in., $C = 0.020$ in., and Reynolds number $N_R = 2 \times 10^5\, CV$. Velocity is assumed to be 180 ft/sec through the clearance, which is 85% of the free discharge velocity $\sqrt{2g \times 693}$.

Then $N_R = 2 \times 10^5 \times 0.020/12 \times 180 = 6 \times 10^4$; $f = 0.316/(6 \times 10^4)^{1/4}$; $V = \sqrt{62.4 \times 693/(1.5 + 0.020 \times 2/2 \times 0.02)} = 134$ ft/sec; $A = 10\pi 0.020/144 = 0.00436$ ft^2; and $Q = 134 \times 0.00436 = 0.58$ ft^3/sec.

2—Labyrinth seal (Fig 3)

Known pressure drop $H_T = 300$ psi = 693 ft. $D = 10$ in., $L = 0.125$ in., $C = 0.02$ in., $n = 8$ stages. Assume velocity is 35% of free discharge.

Velocity $V = 0.35\sqrt{2g\ 963} = 74$ ft/sec. $N_R = 2 \times 10^5\ 0.02/12 \times 74 = 2.48 \times 10^4$. $f = 0.316/(248 \times 10^4)^{1/4} = 0.025$. $V = \sqrt{62.4 \times 693/(1.5 + 0.025 \times 0.125/2 \times 0.02 + 1.0 \times 7)} = 72$ ft/sec. $Q = 72 \times 0.00436 = 0.314$ ft^3/sec.

6 . . Labyrinths

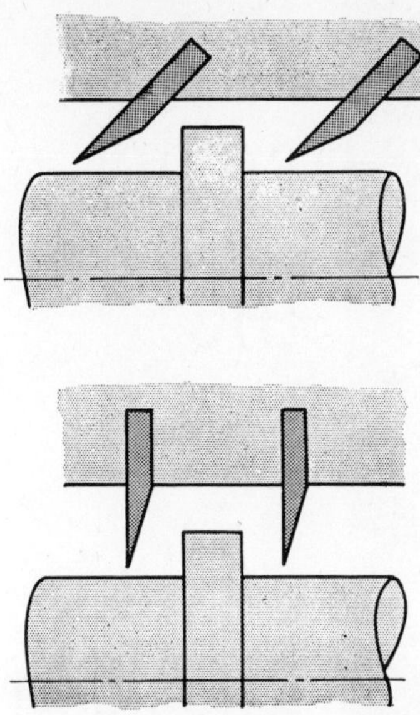

7. . Honeycomb design

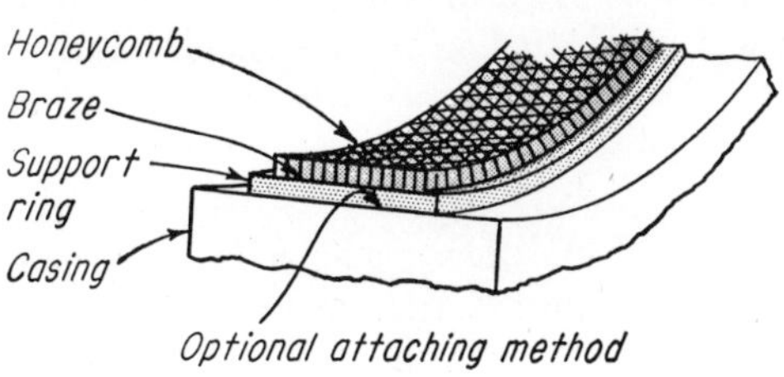

The approximate equation is:

$$H_T = \frac{V^2}{2g}[1.5 + 1.5(n - 1)]$$

where H_T = total head loss, ft, V = fluid velocity, ft/sec, and n = number of cell walls intercepted by the camber of the seal edge (length in direction of flow).

KINEMATIC VISCOSITY OF FLUIDS

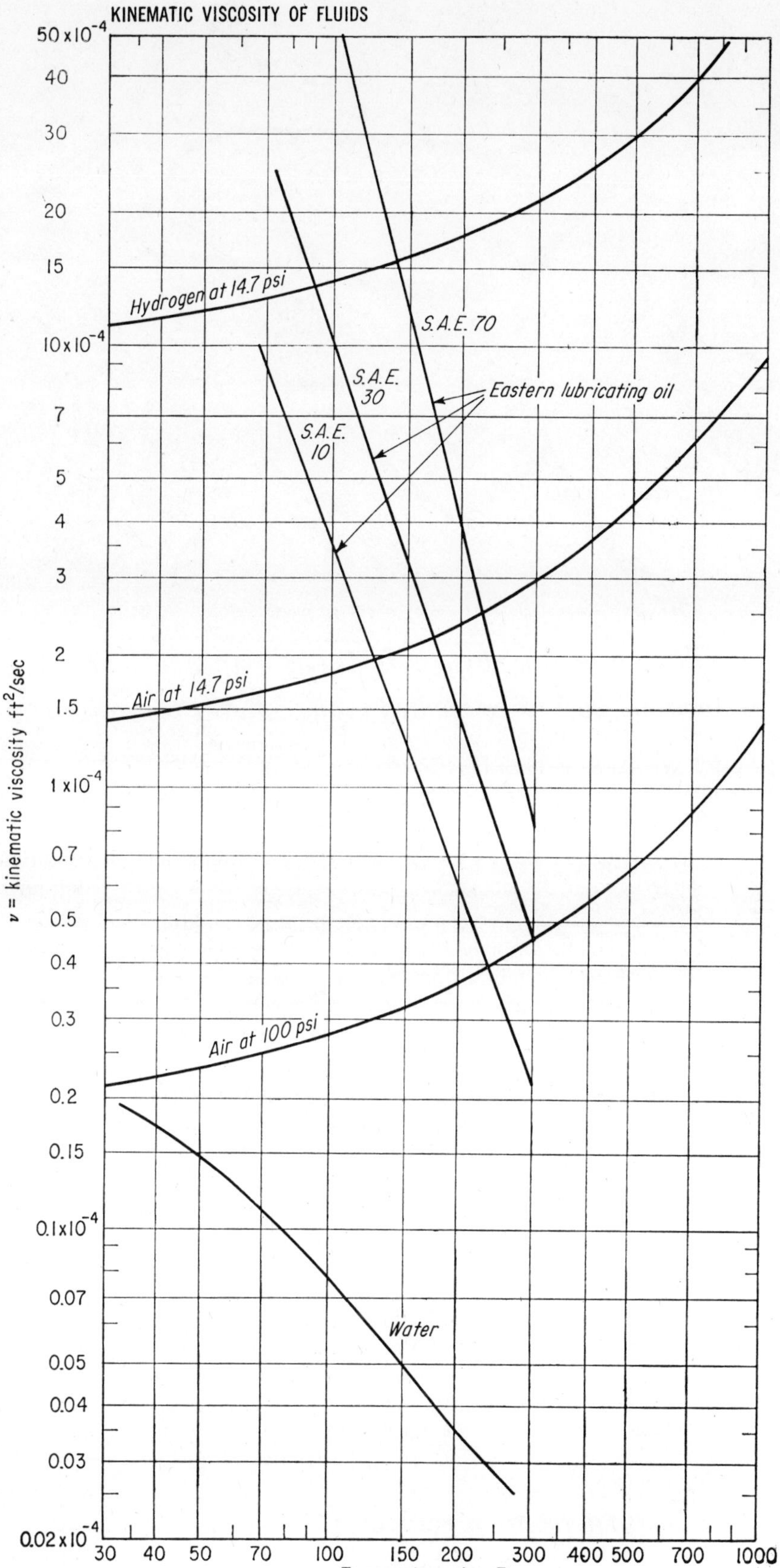

how to select the right gasket material

This 4-step procedure puts design emphasis where it belongs—on the total bolt force applied to the joint at time of installation.

J J WHALEN, *staff engineer, Johns-Manville Corp, NYC*

Three main design factors govern the selection of a gasket material—whether sheet packing, metal, or a combination of materials. The factors are:

Fluid compatibility at the pressure-temperature condition being designed for must be checked first. A recent survey of the technical literature from reliable packing manufacturers indicates a sufficiency of such information. You should collect these publications, file them, and keep them up-to-date.

Pressure-temperature combination determines whether the gasket material is inherently strong enough to resist blow-out. One rule-of-thumb criterion is the product of operating pressure P_i and operating temperature T. Table I lists maximum values of $P_i \times T$ for several basic types of gasket material. These figures are based on experience, test data, and analysis of current technical literature.

Total bolt force at installation must be sufficient to: (1) flow the gasket surface into the flange surface to make an effective seal; (2) prevent the internal pressure from opening the flanges. This demands carefully juggling gasket material, seating area, bolt selection, and flange-surface finish. Here is a logical way to achieve the right balance among these factors for the majority of gasket joint problems.

1. CALCULATE THE TOTAL BOLT FORCE

Fig 1 shows forces and stresses at work in a typical gasketed joint. Total force F_b required to generate the proper gasket-seating stress is not a design value; it is determined at installation. The problem at design stage is to calculate it by the method that will come closest to conditions at installation. There are three such calculation methods, each with certain advantages and limitations:

Bolt-torque method is based on using a torque wrench or impact wrench at installation. Knowing the installation bolt-torque T, total bolt force is calculated from the following empirical equation:

$$F_b = N_b T/0.2D_b \qquad (1)$$

The number of bolts N_b and their diameter D_b are either known or are assumed, subject to later adjustment for adequate sealing stress or resistance to internal pressure force. Remember that torque-wrench readings are only accurate to ± 25% because of variations in bolt lubrication, fit, and surface finish. Impact wrenches are more accurate, but manufacturers should be consulted as to how much torque is applied to a given bolt at a given setting.

Estimation method is used if the flanged joint is to be made by standard wrenches and with little control over qualifications of the mechanic. The following is an old rule-of-thumb relation based on investigations of bolt tension in steamtight fittings,

$$F_b = 16{,}000 D_b N_b \qquad (2)$$

and is limited largely to flange systems using alloy steel bolts.

Bolt-stress method is the most precise; it takes into account the bolt material and the thread series specified. The equation is:

$$F_b = N_b S_b A_b \qquad (3)$$

MECHANICS OF A GASKETED JOINT

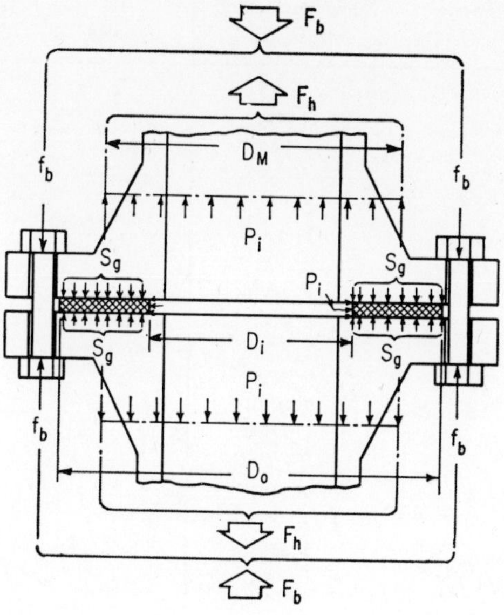

LIST OF SYMBOLS

N_b	Number of bolts
P_i	Operating pressure, psi
P_t	Test pressure, psi
T	Bolt torque, in.-lb
f_b	Force per bolt, lb
F_b	Total bolt force, lb
F_h	Hydrostatic end force, lb
D_i	Gasket ID, in.
D_o	Gasket OD, in.
D_b	Nominal bolt diameter, in.
D_m	Gasket mid-diameter, in.
A_g	Gasket contact area, sq in.
A_m	Area subject to pressure P_i
A_b	Stress area per bolt, sq in.
S_g	Gasket seating stress, psi
S_b	Bolt stress, psi
K	Safety factor, dimensionless

Here, you decide what maximum bolt stress S_b is to be tolerated for the bolt material selected, then pick out the bolt-stress area A_b from Table II for the thread series used. Note that stress area (and hence bolt force) can differ as much as 10% between thread series. This stress area, the latest ASA standard (B1.1-1960), is somewhat larger than the conventional root-diameter area. For convenience in multiple calculations, Table II can be converted to a table of bolt forces simply by multiplying the stress areas, in turn, by a series of stress values—say 10, 20, 30, 45, and 60 kips per sq in.

After the assumption of bolt stress S_b is made, a bolt torque must be specified to ensure that this stress is approximated at installation. Correct torque is calculated by transposing Eq (1) in terms of stress:

$$T = 0.2D_bS_bA_b \tag{4}$$

The stress S_b used in Eq (3) and (4) is a matter of judgment. Most common figure for carbon- and stainless-steel bolts is a stress just below yield strength of the bolt material. For alloy-steel bolts, the yield strength is difficult to approach at installation. For these, practical S_b values may be estimated from Eq (2) and the stress area from Table II.

Hydraulically operated bolt tensioners provide an accurate way of ensuring proper bolt loading. Essentially a jack, the tensioner has a pulling rod equipped with a threaded adapter for the bolt at one end, and a piston at the other. Oil pressure is admitted to the tensioner housing, which is designed to carry the compressive reaction to the flange. With the bolt preloaded to proper tension, the nut is spun down and oil pressure released. Commercial accuracy is ± 10%, much better than preloading with torque wrenches or bolt heaters. Main disadvantage is extra cost of tensioners—3 required for uniform closure.

TABLE I—PRESSURE-TEMPERATURE VALUE FOR GASKET MATERIALS

Material	Max $P_i \times T$	Max Temp, °F
Rubber	15,000	300
Vegetable fiber	40,000	250
Rubberized cloth	125,000	400
Compressed asbestos	250,000	850
Metal types	>250,000	depends on metal

Table II—STRESS AREAS FOR FLANGE BOLTS

Nominal dia, in	Coarse Threads: Threads per in	Coarse Threads: Stress area, sq in	Fine Threads: Threads per in	Fine Threads: Stress area, sq in
0.125 (No. 5)	40	0.0079	44	0.0082
0.138 (No. 6)	32	0.0090	40	0.0101
0.164 (No. 8)	32	0.0139	36	0.0146
0.190 (No. 10)	24	0.0174	32	0.0199
0.216 (No. 12)	24	0.0240	28	0.0257
1/4	20	0.0317	28	0.0362
5/16	18	0.0522	24	0.0579
3/8	16	0.0773	24	0 0876
7/16	14	0.1060	20	0.1185
1/2	13	0.1416	20	0.1597
1/2	12	0.1374		
9/16	12	0.1816	18	0.2026
5/8	11	0 2256	18	0.2555
3/4	10	0.3340	16	0.3724
7/8	9	0.4612	14	0.5088
1	8	0.6051	12	0.6624
1 1/8	7	0.7627	12	0.8549
1 1/4	7	0.9684	12	1.0721
1 3/8	6	1.1538	12	1.3137
1 1/2	6	1.4041	12	1.5799

Table III—MINIMUM SEATING STRESSES FOR TYPICAL GASKET MATERIALS

	Material	Type (see Table IV)	Thickness, in.	Minimum seating stress, psi	Flange-surface finish 1st choice	Flange-surface finish 2nd choice
NONMETAL	**Rubber sheet**					
	SBR (75 Durometer)	Flat	1/32 and up	200		
	CR (60 Durometer)	Flat	1/32 and up	175		
	Compressed asbestos					
	SBR binder	Flat	1/64	3,000		
			1/32	2,000		
			1/16	1,600		
			1/8	1,200	Concentric-serrated	All other types
	CR binder	Flat	1/64	3,750		
			1/32	2,500		
			1/16	2,000		
			1/8	1,500		
	Rubberized cloth	Folded	2-ply	2,500		
			3-ply	2,100		
			4-ply	1,800		
	Vegetable-fiber sheet	Flat	all	750		
	Fluorocarbon (TFE)					
	Virgin	Flat	1/64	14,000		
			1/32	6,500		
			1/16	3,700	80 rms	
			1/8	1,600	(1/64 in. only)	All other types
	Glass-filled	Flat	1/64	14,000	Concentric-serrated	
			1/32	11,000	(all other thicknesses)	
			1/16	6,000		
			1/8	3,000		
	Asbestos cloth (impreg.)	Flat	3/32	1,600		
METAL	**Flat metals**					
	Aluminum	Flat	1/32 and 1/16	20,000	Concentric-serrated	80 rms or less
	Copper			45,000		
	Carbon steel			68,750		
	Monel			81,250		
	Stainless			93,750		
	Stamped metals					
	Lead			500		
	Aluminum			1,000		
	Copper	Corrugated jacket-metal filler	9/64 only	2,500		
	Carbon steel			3,500		
	Monel			4,500		
	Stainless			6,000	80 rms or less	150 to 200 rms
	Machined metals					
	Aluminum	Profile (1/8-in. pitch)	All thicknesses	25,000		
	Copper			35,000		
	Carbon steel			55,000		
	Monel			65,000		
	Stainless			75,000		

2. SELECT A MATERIAL THAT WILL SEAT

The total bolt force F_b calculated in step 1 must be enough to seat the gasket into the flange. In most cases, gasket geometry is fixed by the design, and gasket ID and OD are known. If not, they can be assumed and a tentative gasket area A_g computed—to be adjusted later as balanced design dictates. The actual compressive stress available to seat the gasket is then:

$$S_g = F_b/A_g \tag{5}$$

Next, select a gasket material that will seat satisfactorily under this stress. Check through Table III for materials that have minimum seating stresses equal to or less than the available stress computed in Eq (5). The available stress can be as much as four times higher than the Table III value without overcompressing that gasket material.

Table III—*continued*

	Material	Type (see Table IV)	Thickness, in.	Minimum seating stress, psi	Flange-surface finish 1st choice	2nd choice
METAL-ASBESTOS	**Metal (asbestos-filler)** Aluminum Copper Carbon steel Stainless Monel	Corrugated and corded	1/8 only	 2,000 2,500 3,000 3,500 4,000	150 to 200 rms	Concentric-serrated
	Lead Aluminum Copper Carbon steel Nickel Monel Stainless	Plain metal jacket	1/8 only	500 2,500 4,000 6,000 6,000 7,500 10,000	80 rms or less	150 to 200 rms
	Same as stamped metals	Corrugated jacket	Same as for stamped metals			
	Stainless	Spiral-wound	0.125 and 0.175	3,000 to 30,000		

Table IV—TYPICAL GASKET CROSS-SECTIONS

Type (see Table I)	Cross-section	Description
Flat	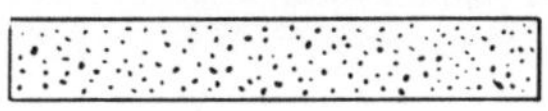	Simplest gasket form; available in wide variety of materials for different conditions of fluid media, temperature and pressure; most easily manufactured for nonsymmetrical shapes.
Folded	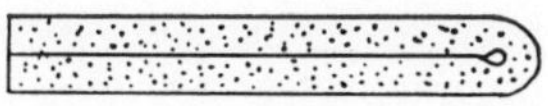	Fabric-based gaskets especially designed for extremely rough and wavy flanges with low bolting loads.
Plain metal jacket	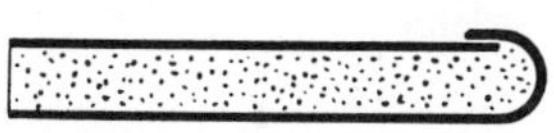	Asbestos filler enclosed partially or totally by metal jacket; combines easy compression with resistance to high pressure. Temperature resistance depends on jacket material chosen.
Corrugated jacket: **metal-filled** **asbestos-filled**		Corrugated metal or asbestos core enclosed by a corrugated-metal jacket and top washer. Better than flat gaskets for rough flanges—corrugations give greater resilience.
Corrugated and corded		Deeply corrugated metal with twisted asbestos cord cemented into the corrugations on both faces; aluminized for non-sticking. Well-suited to rough or warped flanges.
Profile		Heavy solid metal with concentric V-shape contact ridges that provide multiple seating surfaces. Each ridge is 0.010 in. wide. For N ridges, effective area is then: $\pi \times$ mean contact dia $\times 0.010 \times N$. Seating stress increases with decrease in pitch.
Spiral-wound		Interlocked plies of preformed metal strip are spiral-wound with an interleaving cushion of asbestos or fluorocarbon plastic strip. Ratio of metal to filler can be closely controlled to vary seating stress over wide range. Well suited to fluctuating-temperature applications.

However, if the available stress is less than the tabulated minimum, the total bolt force must be increased or the gasket area reduced. Or a different type of gasket may be required. For best seating, select the thinnest gasket consistent with flange flatness and parallelism.

Questions of fluid compatibility and $P_t \times T$ combination will have been checked at the outset. Table III does not show all available gasket types, so search technical literature for minimum seating stresses for other materials, as well as new ones that appear.

3. CHECK THE HYDROSTATIC END FORCE

Total bolt force must be more than enough to overcome hydrostatic end force—defined here as the product of the internal pressure and a mean internal area A_m. Extensive testing shows that, just prior to leakage, internal pressure acts up to the mid-diameter of the gasket. The internal pressure used for this calculation must always be the test pressure P_t, not the operating pressure P_i. Thus, for balance,

$$F_b \geqq K P_t A_m \qquad (6)$$

Where K is a safety factor selected to fit particular design conditions, as indicated in Table V.

4. SPECIFY THE FLANGE-SURFACE FINISH

Once the statics of the gasket joint have been resolved and an appropriate gasket material selected, pay close attention to flange-surface finish. Table II shows the preferred finish for each material, plus a second choice. A concentric-serrated surface is one machined with 90° V-grooves $\frac{1}{64}$ in. deep on a $\frac{1}{32}$-in. pitch. Machining concentrically requires special tools and techniques, but gives multiple sealing lands as compared to the continuous exit path provided by conventional spiral machining. A third choice: spiral-serrated flanges will usually work satisfactorily on spiral-wound gaskets or sheet packing—materials that flow easily.

Table V—SAFETY FACTORS FOR GASKETED JOINTS

K-factor	When to Apply
1.2 to 1.4	For minimum-weight applications where all installation factors (bolt lubrication, tension, parallel seating, etc.) are carefully controlled; low-temperature applications; where adequate proof pressure is applied.
1.5 to 2.5	For most normal designs where weight is not a major factor, vibration is moderate, and temperatures do not exceed 700 to 800 F. Use high end of range where bolts are not lubricated.
2.6 to 4.0	For cases of extreme fluctuations in pressure, temperature or vibration; where no test pressure is applied, or where uniform bolt tension is difficult to ensure.

SAMPLE PROBLEM

A gasket is needed to seal industrial hydraulic fluid at 1200 psi and 180 F. Flanges are $1\frac{1}{2}$-in., raised-face, 600-lb weld-neck type made from Type 304 stainless steel. There are four bolts, $\frac{3}{4}$ in.–10 NC, made from ASTM-A193 grade B7 alloy steel. Hydrotest pressure is specified as 2.5 times operating pressure.

1. **Total bolt force.** Assuming torque wrenches will check installation, select bolt-stress method to calculate total bolt force. From Table II, stress area for $\frac{3}{4}$–10NC bolt is 0.3340 sq. in. Bolt material specified can easily take 30,000 psi without yielding (Eq 2 gives: 16,000 × 0.75/0.3340 = 36,000) thus from Eq (3):

$$F_b = 4 \times 30{,}000 \times 0.3340$$
$$F_b = 40{,}080 \text{ lb}$$

Torque required to produce this stress level at installation comes from Eq (4):

$$T = 0.2 \times 0.75 \times 30{,}000 \times 0.3340/12$$
$$T = 125 \text{ ft-lb (to be specified on print)}$$

2. **Gasket-material selection:** The $P_t \times T$ relation is 1200 × 180 = 216,000 which, from Table I, suggests choosing compressed asbestos type. This would be compatible with hydraulic fluid. Gasket area is 4.73 sq in., calculated from OD of $2\frac{7}{8}$ in. (same as OD of raised flange, per ASA–B16.5) and ID of $1\frac{1}{2}$ in. Thus seating stress available is calculated from Eq (5):

$$S_g = 40{,}080/4.73$$
$$S_g = 8470 \text{ psi}$$

Table III shows this stress will easily seat selected compressed asbestos gasket. Tentatively choose asbestos with CR (neoprene) binder for oil resistance; thickness, $\frac{1}{32}$ in.

3. **Hydrostatic end force:** Mean area acted upon by pressure is defined by a diameter of $(2\frac{7}{8} + 1\frac{1}{2})/2$ and computes to 3.74 sq in. Selecting a safety factor of 1.5 from Table V, the end force is calculated and balanced against total bolt force by Eq (6):

$$40{,}080 \geqq 1.5 \times 2.5 \times 1200 \times 3.74$$
$$40{,}080 \text{ lb} \geqq 16{,}830 \text{ lb}$$

Thus there is no end-force balance problem with bolts stressed to 30,000 psi.

4. **Surface finish:** Table III shows a concentric-serrated surface finish on the flange is best. Economy may dictate a conventional spirel-serrated surface, which Table III shows will work in this case.

5. **Final specification:** Include not only material type and dimensions, but also the bolt-torque value calculated in step 1. For greater torque-wrench accuracy, specify uniform fit on all bolts and lubrication before installation.

12

SYSTEM PERFORMANCE

Flow theory, optimization, reliability

THEORY OF HYDRAULIC FLOW CONTROL

NILS M. SVERDRUP
Aerojet-General Corp., Azusa, Calif.

In the design of any hydraulic circuit, the rate of flow is determined by the resistance of the various components and by the pressure differential across the circuit. Because, in spite of its importance, there is a serious lack of published data on hydraulic resistance, equations are derived herein for the flow rate of any circuit — including both parallel and series types. This is followed by a flow analysis of such components as valves, venturis, orifices and other flow limiting devices. Comprehensive examples illustrate the application of the analyses to practical design problems.

SYMBOLS

H = loss of head, ft
V = maximum velocity, ft/sec
g = gravitational constant, = 32.2 ft/sec²
ϵ = coefficient of energy loss of a passage
P = pressure loss, psi
S = specific gravity of fluid
G = flow rate, US gal/min
W = flow rate, lb/sec
h = potential head, ft
ν = kinematic viscosity, centistokes
μ = absolute viscosity, centipoises
p = fluid pressure, psia
$\Sigma = \sum_{i=1}^{i=n} \epsilon_i D_i^{-4}$ = sum of hydraulic resistances
$\sum_{i=1}^{i=n} P_i$ = sum of pressure losses, psi
$\psi = \epsilon_0 D_0^{-4}$ = hydraulic resistance
C_D = coefficient of discharge
R = Reynolds number
f = coefficient of friction
D = dia of fluid passage, in.
Δp = pressure differential, psi

When fluid flows through a passage such as a pipe, venturi tube, flow control valve, etc., hydraulic resistance will cause a dissipation of energy in the form of heat with a resulting loss in pressure. This loss of energy can be expressed as a function of the velocity head:

$$H = \epsilon V^2/2g \tag{1}$$

The loss of head H and the pressure loss P are related by:

$$P = 0.4333\, SH \tag{2}$$

or

$$H = 2.3077\, P/S \tag{3}$$

Combining Eqs (1) and (3),

$$P = 6.729 \times 10^{-3} S \epsilon V^2 \tag{4}$$

In most applications, the velocity V is not known and varies throughout the system. The rate of flow, however, is generally given, either in US gpm (G) or lb per sec (W). These quantities, V, W and G are related as follows:

$$V = 0.4085\, GD^{-2} = 2.938\, WS^{-1}D^{-2} \tag{5}$$

Substituting the preceding flow relations in Eq (4), the pressure loss in any fluid passage is expressed by:

$$P = 1.123 \times 10^{-3}\, G^2 S \epsilon D^{-4} \tag{6}$$

and

$$P = 5.809 \times 10^{-2}\, W^2 S^{-1} \epsilon D^{-4} \tag{7}$$

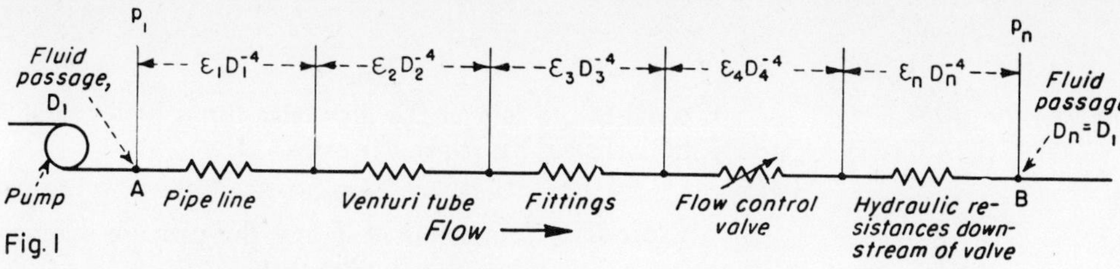

Fig. 1

Fig. 1—Basic hydraulic resistance diagram for series circuit in which pipe diameter and elevation are constant between points A and B. Eqs. (9), (10) and (11) give the pressure drop and resistance for this fundamental circuit; Eqs (18) and (19) give the flow rate.

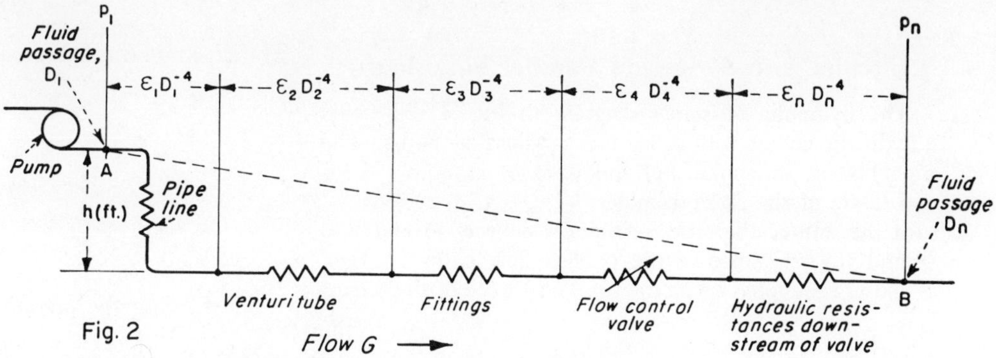

Fig. 2

Fig. 2—Resistance diagram for series circuit in which elevation and pipe diameter vary between points A and B. Eqs (14) and (15) give the pressure drop; Eqs (16) and (17) give the flow rate.

where the product ϵD^{-4} is called the hydraulic resistance, and ϵ is the coefficient of energy loss.

Because most passages are circular in cross-section, all succeeding equations will be derived in terms of diameters. In the preceding relationships D is the diameter of the smallest section of the fluid passage, where the velocity is a maximum. If the flow area A is noncircular, it can be converted into an equivalent diameter as follows:

$$D = 1.128\sqrt{A} \tag{8}$$

If in Fig. 1, the fluid passage is horizontal and has the same diameter at points A and B ($D_1 = D_n$), Eq (6), which involves volume flow-rates can be written as:

$$p_1 - p_n = \sum_{i=1}^{i=n} P_i = 1.123 \times 10^{-3} G^2 S \sum_{i=1}^{i=n} \epsilon_i D_i^{-4} \tag{9}$$

and, for mass flow-rate units, Eq 7 is

$$p_1 - p_n = \sum_{i=1}^{i=n} P_i = 5.809 \times 10^{-2} W^2 S^{-1} \sum_{i=1}^{i=n} \epsilon_i D_i^{-4} \tag{10}$$

where p_1 and p_n are the absolute pressures at A and B, respectively, and the summation of P_i is the total pressure loss, or the sum of the pressure losses in the various passages. The last term represents the sum of the various hydraulic resistances, and will be represented by the symbol Σ in all succeeding equations where:

$$\Sigma = \sum_{i=1}^{i=n} \epsilon_i D_i^{-4} = \epsilon_1 D_1^{-4} + \epsilon_2 D_2^{-4} + \epsilon_3 D_3^{-4} + \cdots \epsilon_n D_n^{-4} \tag{11}$$

If point A is located at a different elevation than B, as shown in Fig. 2, and the fluid passage at points A and B are of different diameter (D_1 and D_n), Eqs (9) and (10) must be modified to include the effect of the potential head h and the change in pressure caused by variations in velocities. The expression for the pressure differential, $p_1 - p_n$, can be obtained from the Bernoulli equation by expressing the velocities at the two points in terms of diameters and flow rates as given by Eq (5). The equations for the pressure differential then become:

$$p_1 - p_n = \sum_{i=1}^{i=n} P_i + 1.123 \times 10^{-3} G^2 S (D_n^{-4} - D_1^{-4}) - 0.4333\, S\, h \tag{12}$$

and

$$p_1 - p_n = \sum_{i=1}^{i=n} P_i + 5.809 \times 10^{-2} W^2 S^{-1} (D_n^{-4} - D_1^{-4}) - 0.4333\, S\, h \tag{13}$$

In these equations, the second term on the right hand side represents the change in pressure resulting from a change in velocity. For example, an increase in the diameter D_n at B results in a decrease in velocity with consequent increase in pressure p_n. A decrease in D_n would increase the velocity and decrease the pressure.

The last term represents the intensity of pressure resulting from the potential head h. This can be either positive or negative; if A is at a higher elevation than B as shown in Fig. 2, it is negative since h is positive. If A is lower than B, it is positive since h is negative.

Substituting the value of the total pressure loss from Eqs (9) and (10) in Eqs (12) and (13):

$$p_1 - p_n = 1.123 \times 10^{-3} G^2 S\, [\Sigma + (D_n^{-4} - D_1^{-4})] - 0.4333\, S\, h \tag{14}$$

and

$$p_1 - p_n = 5.809 \times 10^{-2} W^2 S^{-1}\, [\Sigma + (D_n^{-4} - D_1^{-4})] - 0.4333\, S\, h \tag{15}$$

Solving the preceding equations for G and W, the rate of flow through the circuit is:

$$G = 29.843 \sqrt{\frac{(p_1 - p_n) + 0.4333\, S\, h}{[\Sigma + (D_n^{-4} - D_1^{-4})]\, S}} \tag{16}$$

and

$$W = 4.149 \sqrt{\frac{[(p_1 - p_n) + 0.4333\, S\, h]\, S}{\Sigma + (D_n^{-4} - D_1^{-4})}} \tag{17}$$

For the conditions of Fig. 1 ($h = O$; $D_n = D_1$), Eqs

(16) and (17) simplify as follows:

$$G = 29.843\sqrt{\frac{p_1 - p_n}{S\Sigma}} \quad (18)$$

and

$$W = 4.149\sqrt{\frac{(p_1 - p_n)S}{\Sigma}} \quad (19)$$

Circuits with Series and Parallel Flow Paths

The hydraulic resistance diagram in Fig. 3 represents a hydraulic circuit with series and parallel flow paths. The pipe line is horizontal and the passages at points A, B and C are of the same diameter.

If the sum of the series resistances between A and B is Σ and the resistance of the parallel flow paths between B and C is denoted by Z, the total resistance of the circuit is $\Sigma + Z$. (20)

Then, in accordance with Eqs (18) and (19), the rate of flow through the series circuit is:

$$G = 29.843\sqrt{\frac{p_1 - p_n}{S(\Sigma + Z)}} \quad (21)$$

and

$$W = 4.149\sqrt{\frac{(p_1 - p_n)S}{\Sigma + Z}} \quad (22)$$

The expression for resistance Z of the parallel flow paths can be derived as follows:

The rate of flow G in the main circuit between A and B is equal to the sum of the flow rates through the parallel paths. Hence, for n parallel paths,

$$G = G_1 + G_2 + G_3 + \cdots + G_n \quad (23)$$

Expressing G in terms of Z and the pressure differential $p_2 - p_n$ across the parallel paths,

$$G = 29.843\sqrt{\frac{p_2 - p_n}{SZ}} \quad (24)$$

The flow rate through each of the parallel paths is:

$$G_1 = 29.843\sqrt{\frac{p_2 - p_n}{S\Sigma_1}} \quad (25)$$

$$G_2 = 29.843\sqrt{\frac{p_2 - p_n}{S\Sigma_2}} \quad (26)$$

$$G_n = 29.843\sqrt{\frac{p_2 - p_n}{S\Sigma_n}} \quad (27)$$

Substituting the preceding values in Eq (23),

$$29.843\sqrt{\frac{p_2 - p_n}{SZ}} = 29.843\sqrt{\frac{p_2 - p_n}{S\Sigma_1}} + 29.843\sqrt{\frac{p_2 - p_n}{S\Sigma_2}} + \cdots + 29.843\sqrt{\frac{p_2 - p_n}{S\Sigma_n}} \quad (28)$$

This resolves into

$$\frac{1}{\sqrt{Z}} = \frac{1}{\sqrt{\Sigma_1}} + \frac{1}{\sqrt{\Sigma_2}} + \cdots + \frac{1}{\sqrt{\Sigma_n}} \quad (29)$$

Solving for Z, the general equation for n parallel flow path is:

$$Z = \frac{1}{[(\Sigma_1)^{-1/2} + (\Sigma_2)^{-1/2} + \cdots + (\Sigma_n)^{-1/2}]^2} \quad (30)$$

EXAMPLE 1 Assume that in Fig. 3, the sum of the series hydraulic resistances between points A and B including that of the flow control valve, is $\Sigma = 4.68$, and that the resistances of the parallel flow path are $\Sigma_1 = 8.9$, $\Sigma_2 = 15.4$, and $\Sigma_3 = 28.6$. If the pump discharge pressure at point A is $p_1 = 580$ psi, abs, the pressure at point C is $p_n = 400$ psi, abs and the specific gravity of the fluid is $S = 0.785$, what is the rate of flow through the main circuit between points A and B and through the parallel paths in U. S. gpm?

SOLUTION: In accordance with Eq (21), the rate of flow through the hydraulic circuit is:

$$G = 29.843\sqrt{\frac{580 - 400}{0.785(4.68 + Z)}}$$

The value of the resistance Z of the parallel flow paths can be computed from Eq (30). Hence:

$$Z = \frac{1}{(8.9^{-1/2} + 15.4^{-1/2} + 28.6^{-1/2})^2}$$

$$= \frac{1}{(0.3352 + 0.2548 + 0.1870)^2} = 1.656$$

Thus, the flow through the main circuit between points A and B is:

$$G = 29.843\sqrt{\frac{580 - 400}{0.785(4.680 + 1.656)}}$$

$$= 29.843\sqrt{\frac{180}{0.785 \times 6.336}} = 179.53 \text{ gpm}$$

From Eq (9), the pressure loss between points A and B resulting from the series resistances is:

$$p_1 - p_2 = 1.123 \times 10^{-3} \times 179.53^2 \times 0.785 \times 4.68 = 132.95 \text{ psi}$$

The pressure at point B is:

$$p_2 = 580 - 132.95 = 447.05 \text{ psi, abs}$$

The rate of flow through the parallel flow paths can now be computed with Eqs (25), (26) and (27). Thus:

$$G_1 = 29.843\sqrt{\frac{447.05 - 400}{0.785 \times 8.9}} = 77.45$$

$$G_2 = 29.843\sqrt{\frac{447.05 - 400}{0.785 \times 15.4}} = 58.88$$

$$G_3 = 29.843\sqrt{\frac{447.05 - 400}{0.785 \times 28.6}} = 43.20$$

As a check on the analyses, the sum of the rate of flow through the parallel flow paths should equal the rate of flow in the main circuit between points A and B. Thus:

$$G = 77.45 + 58.88 + 43.20 = 179.53 \text{ U. S. gpm}$$

This is in agreement with the value previously computed.

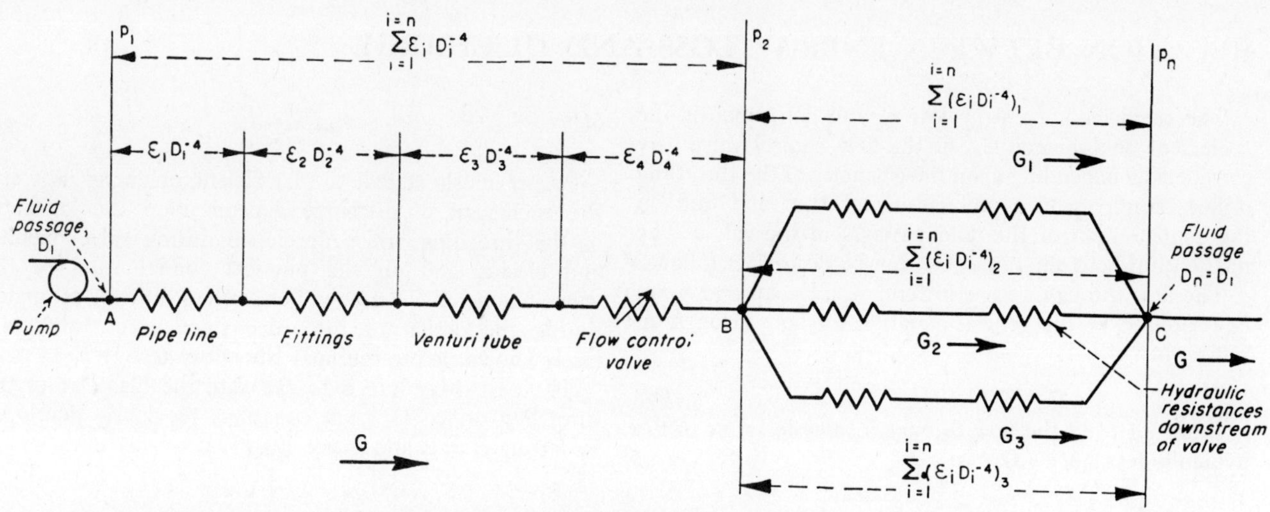

Fig. 3—Resistance diagram with series and parallel flow paths. Expressions for the flow rate in each of the parallel paths are given by Eqs (25)—(28); Eq (30) gives the hydraulic resistance for *n* parallel paths.

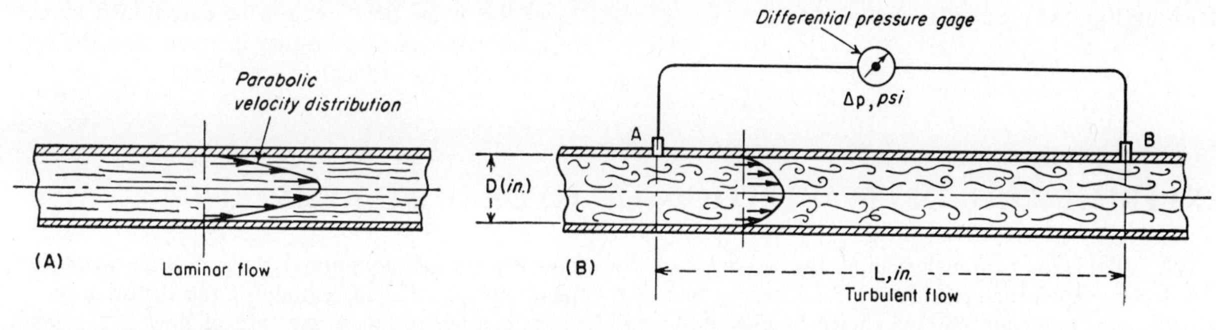

Fig. 4—Velocity distribution for laminar flow (A), and turbulent flow (B). Differential gage in (B) measures the pressure drop.

EQUATIONS FOR THROTTLING ORIFICE OF FLOW CONTROL VALVE

The diameter of the throttling orifice of a flow control valve can be obtained from Eqs (14) and (15) by letting

$$\Sigma = \sum^{1} \epsilon_i D_i^{-4} + \epsilon_4 D_4^{-4} \tag{31}$$

Substituting the value of the resistance given by Eq (31),

$$D_4 = \sqrt[4]{\frac{\epsilon_4}{\dfrac{890.608[(p_1 - p_n) + 0.4333\, S\, h]}{G^2 S} - \sum^{1} \epsilon_i D_i^{-4} - (D_n^{-4} - D_1^{-4})}} \tag{32}$$

or

$$D_4 = \sqrt[4]{\frac{\epsilon_4}{\dfrac{17.214[(p_1 - p_n) + 0.4333\, S\, h]\, S}{W^2} - \sum^{1} \epsilon_i D_i^{-4} - (D_n^{-4} - D_1^{-4})}} \tag{33}$$

The resistance term for the hydraulic circuit, in Eqs (32) and (33) then becomes:

$$\sum^{1} \epsilon_i D_i^{-4} = \epsilon_1 D_1^{-4} + \epsilon_2 D_2^{-4} + \epsilon_3 D_3^{-4} + \cdots + \epsilon_n D_n^{-4} \tag{34}$$

where the resistance for the valve $\epsilon_4 D_4^{-4}$ is omitted.

Note that in Eqs (16), (17), (32) and (33), the potential head h is positive when point A is at a higher elevation than B, and becomes negative when A is lower than B.

For the fundamental circuit, Fig. 1, Eqs (32) and (33) become:

$$D_4 = \sqrt[4]{\frac{\epsilon_4}{\dfrac{890.608(p_1 - p_n)}{G^2 S} - \sum^{1} \epsilon_i D_i^{-4}}} \tag{35}$$

and

$$D_4 = \sqrt[4]{\frac{\epsilon_4}{\dfrac{17.214(p_1 - p_n) S}{W^2} - \sum^{1} \epsilon_i D_i^{-4}}} \tag{36}$$

Similarly, the diameter of the throttling orifice of the flow control valve in circuits with series and parallel flow paths, Fig. 3, can be derived:

RELATION BETWEEN ENERGY LOSS AND DISCHARGE

The coefficient of energy loss ε_4, and consequently the coefficient of discharge C_{D4} of the flow control valve vary considerably depending upon the geometry of the throttling orifice during opening and closing of the valve and on the configuration of the fluid passages in the valve. The relationship between ε_4 and C_{D4} can be derived as follows:

The flow through a valve in terms C_{D4},D_4, Δp and S can be expressed by the general equation for flow through an orifice, or

$$G = 29.843 C_{D4} D_4^2 \sqrt{\Delta p/S} \tag{37}$$

From Eq (18) the flow through a valve in terms of the hydraulic resistance $\varepsilon_4 D_4^{-4}$ is:

$$G = 29.843 \sqrt{\frac{\Delta p}{S \varepsilon_4 D_4^{-4}}} \tag{37a}$$

In Eqs (37) and (37a) it is assumed that the inlet and outlet ports are of the same diameter, which usually applies. Equating Eqs (37) and 37a) and sloving for ϵ_4:

$$\epsilon_4 = \frac{1}{C_{D4}^2} \tag{38}$$

or

$$C_{D4} = \frac{1}{\sqrt{\epsilon_4}} \tag{39}$$

As previously stated, the coefficient of energy loss and the coefficient of discharge depend upon the geometry of the throttling orifice developed during valve opening and closing and on the physical configuration of the valve in general. Therefore, these values must be estimated during the design of the valve since exact values are rarely known before the final laboratory test.

The best procedure is to calculate the diameter of the throttling orifice D_4 from equations previously developed by letting $\varepsilon_4 = 1$, and hence Eq (39),

$$C_{D4} = 1/\sqrt{1} = 1$$

In practice, the geometry of the throttling orifice varies and often is not circular. Under these conditions, D_4 can be converted into an equivalent flow area. A curve representing orifice area versus valve lift can then be constructed. Later the actual flow area can be determined by choosing realistic values of C_{D4}, bearing in mind that the flow area is inversely proportional to this factor.

EXPERIMENTAL DETERMINATION OF HYDRAULIC RESISTANCE

When fluid flows through a pipe, the character of the flow depends upon such factors, as specific gravity, viscosity, velocity, pipe diameter, etc. As shown in Figs. 4(A) and (B) the flow may be either laminar or turbulent.

The criterion applicable to all fluids for determining the character of the flow is the dimensionless Reynolds number:

$$R = 7{,}741.920 VD/\nu = 3{,}162.556 G/\nu D$$

or

$$R = 22{,}747.606 W/\nu SD = 22{,}747.606 W/\mu D \tag{40}$$

When the flow is increased slowly in a smooth straight pipe through a carefully rounded entrance, laminar flow has been maintained for values of Reynolds number as large as 50,000 depending upon the care taken to prevent initial disturbing effects such ast those caused by fittings or valves. Under normal conditions laminar flow does not prevail for values of Reynolds numbers above 2,400 to 3,000. When the velocity is gradually reduced the flow pattern reverts back to laminar. The lower critical Reynolds number below which turbulent flow cannot persist is 1,500 to 2,000. Most of the flow dealt with in hydraulic engineering, however, is turbulent. In installations having a large number of bends, fittings, and valves, the flow may be turbulent for values of R as low as 1,000.

The differential gage in Fig. 4(B) measures the pressure drop Δp in the length of pipe L between points A and B If a valve or fitting is installed in the pipe line as shown in Fig. 5(A), considerable disturbance of the flow pattern will occur.

This increase in turbulence may be present in a considerable length of straight pipe downstream of the valve or fitting with consequent increase in energy loss. The disturbance usually persists from 40 or more diameters downstream before normal pattern is re-established, depending upon the magnitude of the disturbance.

If, in each instance, the rate of flow is the same, the pressure loss $(\Delta p)_1$ between points A and B in Fig. 5(A) will be greater than the pressure loss Δp in Fig. 4, because of the resistance of the valve or fitting and the dissipation of energy resulting from the increased turbulence in the line. This later excess loss obviously should be charged to the valve or fitting.

If the length of straight pipe between A and B is $L = L_1 + L_2$, then the pressure loss caused by the valve or fitting is $p_{loss} = (\Delta p)_1 - \Delta p$.

In accordance with Eqs (18) and (19):

$$G = 29.843 \sqrt{\frac{(\Delta p)_1 - \Delta p}{S \epsilon_o D_o^{-4}}} \tag{41}$$

and

$$W = 4.149 \sqrt{\frac{[(\Delta p)_1 - \Delta p] S}{\epsilon_o D_o^{-4}}}$$

Solving for $\varepsilon_o D_o^{-4}$, the resistance of the valve or the fitting is

$$\epsilon_o D_o^{-4} = 890.608 \frac{(\Delta p)_1 - \Delta p}{G^2 S} = \Psi \tag{42}$$

and

$$\epsilon_o D_o^{-4} = 17.214 \frac{[(\Delta p)_1 - \Delta p] S}{W^2} = \Psi \tag{43}$$

From the preceding equations the value of the coefficient of energy loss ε_o is readily obtained:

$$\epsilon_o = \Psi D_o^4 \tag{44}$$

In Fig. 5(A) the pipe diameter on the upstream and the

downstream side of the device to be tested is the same. In Fig. 5(B), the pipes are of different size, so that the velocity heads at A and B are of different magnitude. The expression for the flow through the device can be obtained from Eqs (16) and (17) by letting the potential head h equal zero.

Hence:

$$G = 29.843\sqrt{\frac{(\Delta p)_1 - \Delta p}{[\epsilon_o D_o^{-4} + (D_2^{-4} - D_1^{-4})]S}}$$

and

$$W = 4.149\sqrt{\frac{[(\Delta p)_1 - \Delta p)]S}{\epsilon_o D_o^{-4} + (D_2^{-4} - D_1^{-4})}}$$

where $(\Delta p)_1$ is the pressure differential between A and B in Fig. 5(B); Δp is the pressure loss in the pipes of length L_1, L_2 and diameter D_1, D_2, respectively, under normal (undisturbed) flow and at the same flow rate.

Solving for $\epsilon_o D_o^{-4}$, the resistance of the device is

$$\epsilon_o D_o^{-4} = 890.608\frac{(\Delta p)_1 - \Delta p}{G^2 S} - (D_2^{-4} - D_1^{-4}) = \Psi \quad (42a)$$

and

$$\epsilon_o D_o^{-4} = 17.214\left[\frac{(\Delta p)_1 - \Delta p}{W^2}\right]S - (D_2^{-4} - D_1^{-4}) = \Psi \quad (43a)$$

In the case of a flow control valve, its resistance is comparatively high and the term Δp usually is negligible in practical analyses. For a fitting or check valve with a relatively low hydraulic resistance, however, the value of Δp becomes significant.

A study of laboratory test data indicates that for fittings, such as tees, elbows, unions, etc., and valves for a given throttling orifice area, the quantity ϵD^{-4}, and consequently the coefficients ϵ and C_D remain essentially constant with varying Reynolds number when the flow is well in the turbulent region.

This, however, does not apply to the pipe line where ϵ is a function of the coefficient of friction (and consequently also of the Reynolds number R), and the ratio of the pipe length to the pipe diameter, or

$$\epsilon = fL/D \quad (45)$$

When the flow is laminar, the coefficient of friction f is independent of the roughness of the pipe wall and has the value

$$f = 64/R \quad (46)$$

However, experiments show that when the flow is turbulent, the value of f varies considerably with the degree of wall roughness.

Because of the complexity of turbulent flow along with the difficulty in establishing a numerical scale of roughness, a satisfactory theoretical equation for computing the friction coefficient so far has not been developed. However, by analyzing experimental data, Blasius found that for smooth pipes the friction coefficient f can be expressed by:

$$f = 0.316/\sqrt[4]{R} \quad (47)$$

Eq (47), which is valid for Reynolds numbers up to about 100,000 is represented graphically in Fig. 6, as is Eq (46).

When the Reynolds number exceeds 100,000 the relationship f and R for flow in smooth pipes is given by

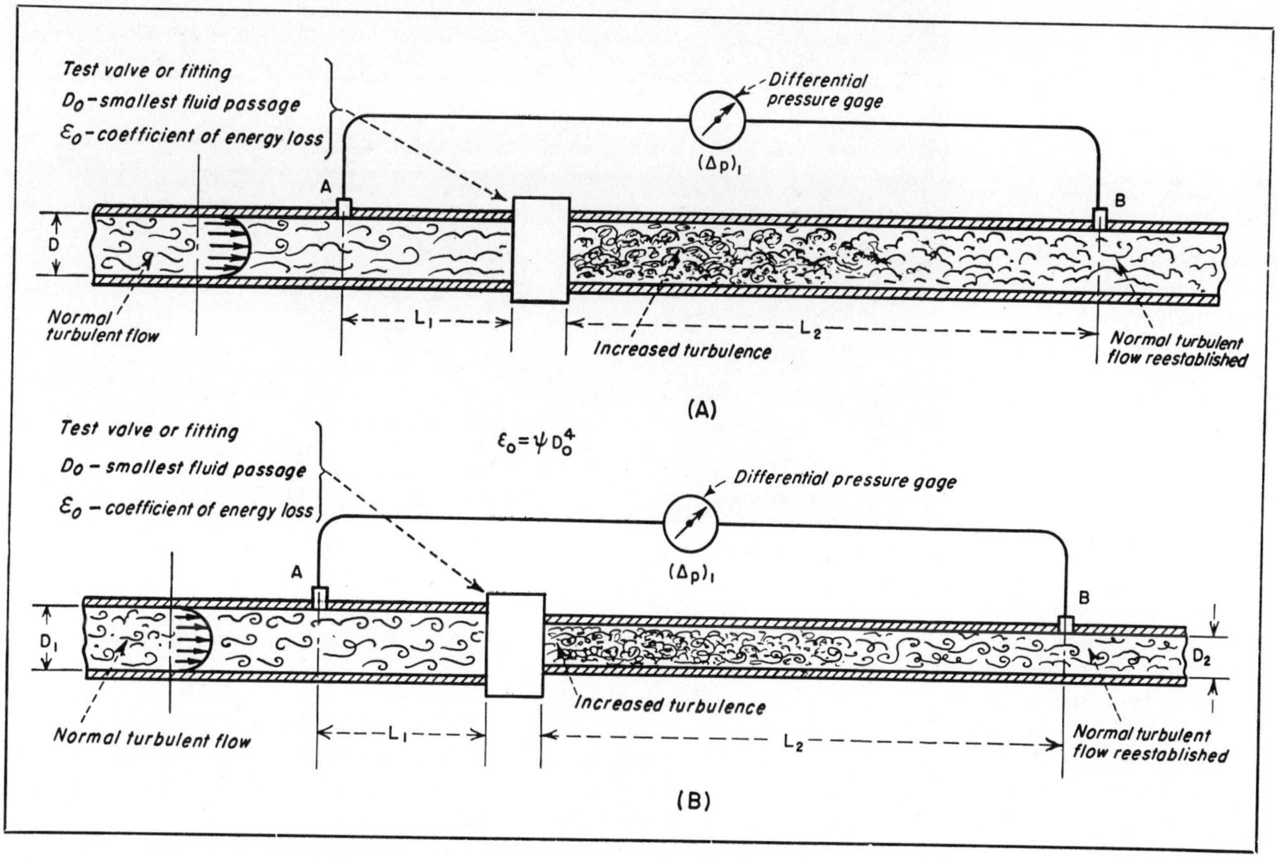

Fig. 5—Test set-ups for checking the flow characteristics of valves or fittings and typical flow patterns that will result. In (A) pipe diameter is constant; in (B) diameters at inlet and outlet valve or fitting are different.

the Karman-Prandtl friction equation:

$$1/\sqrt{f} = -0.8 + 2\log_{10}(R\sqrt{f}) \quad (48)$$

It is not practicable to solve this equation directly for f. The best procedure is to solve for R for various values of f and plot a curve of R versus f as shown in Fig. 7.

In measuring the pressure loss and the rate of flow through hydraulic devices, a test fluid is frequently used with different physical properties than that of the actual service fluid. It will then be necessary to modify the values thus obtained to correspond with actual operating conditions as follows:

If the rate of flow is G (gpm), the pressure loss, specific gravity, and the coefficient of energy loss when using the test fluid and the actual service fluid are P_1, S_1, ϵ_1, and P_2, S_2, ϵ_2, respectively, in accordance with Eq (6):

$$P_2 = P_1 \frac{S_2\epsilon_2}{S_1\epsilon_1} \quad (49)$$

If the flow is measured in W (lb per sec), then from Eq (7):

$$P_2 = P_1 \frac{S_1\epsilon_2}{S_2\epsilon_1} \quad (50)$$

As previously stated the coefficients of energy loss for fittings, such as tees, elbows, unions, etc., and valves for a given throttlings orifice area, remain essentially constant with varying Reynolds number when the flow is well in the turbulent region, in which case $\epsilon_1 = \epsilon_2$ in Eqs (49) and (50). This however, does not apply to pipe lines where ϵ varies considerably with the Reynolds number.

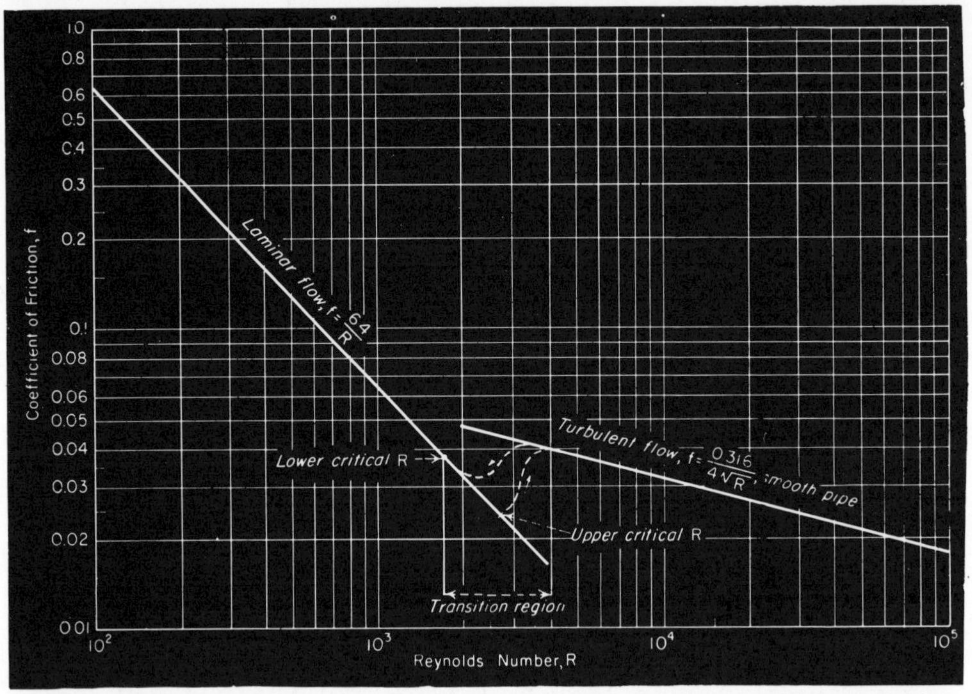

Fig. 6—Coefficient of friction, *f*, versus Reynolds number, *R*, for laminar and turbulent flow in smooth pipes.

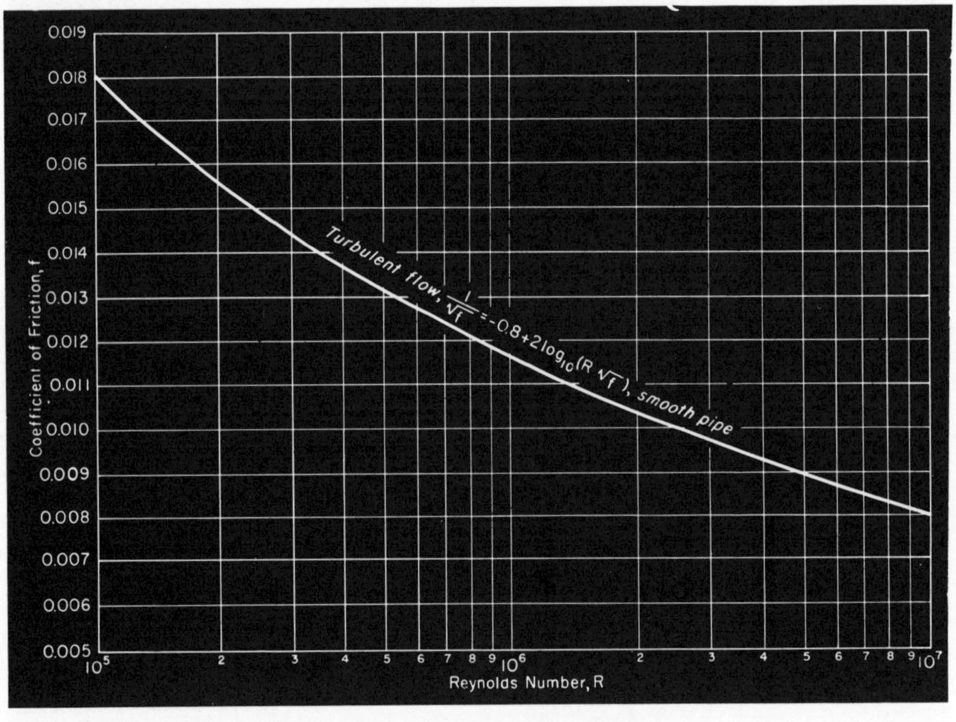

Fig. 7—Extended plot of turbulent flow curve shown in Fig. 6.

Since, in accordance with Eq (45) the coefficient of energy loss for a straight pipe is defined as $\epsilon = fL/D$ it follows that Eqs (49) and (50) can be written in the following forms for a straight pipe:

$$P_2 = P_1 \frac{S_2 f_2}{S_1 f_1} \text{ (flow measured in gpm)} \quad (51)$$

and

$$P_2 = P_1 \frac{S_1 f_2}{S_2 f_1} \text{ (flow measured in lb per sec)} \quad (52)$$

After the Reynolds number has been determined, the values of the coefficients of friction f_1 and f_2 can be obtained from charts Figs. 6 and 7.

It is readily seen that the Reynolds number may vary considerably between the laboratory test fluid and the actual service fluid. The test fluid usually is a low viscosity fluid. For example, if the kinematic viscosity of the test fluid is $\nu_1 = 1.1$ centistokes and the Reynolds number is $R_1 = 20{,}000$ then if the service fluid is MIL-O-5606 hydraulic oil at 60 deg F with a kinematic viscosity of $\nu_2 = 25$ centistokes, the Reynolds number when using the service fluid becomes for the same flow rate

$$R_2 = R_1 \frac{\nu_1}{\nu_2} = 20{,}000 \times \frac{1.1}{25} = 880 \text{ (for flow in gpm)} \quad (53)$$

$$R_2 = R_1 \frac{\nu_1 S_1}{\nu_2 S_2} = R_1 \frac{\mu_1}{\mu_2} \text{ (for flow in lb per sec)} \quad (53a)$$

In this case, the flow is turbulent using the test fluid but becomes laminar under actual operating conditions when the service fluid is used. The value of the coefficients of friction in Eqs (51) and (52) therefore would vary considerably, from $f_1 = 0.0266$ when using the test fluid to $f_2 = 0.072$ when using the service fluid.

It is also seen from Eq (6) that if the pressure loss P is constant and the flow rate is the variable quantity, the following relationship exists:

$$\frac{P}{P} = \frac{1.123 \times 10^{-3} G_2^2 S_2 \epsilon_2 D^{-4}}{1.123 \times 10^{-3} G_1^2 S_1 \epsilon_1 D^{-4}}$$

From this,

$$G_2 = G_1 \sqrt{\frac{S_1 \epsilon_1}{S_2 \epsilon_2}} \quad (54)$$

and similarly from Eq (7),

$$W_2 = W_1 \sqrt{\frac{S_2 \epsilon_1}{S_1 \epsilon_2}} \quad (55)$$

where G_1, W_1 and G_2, W_2 refer to the flow rates using the laboratory test fluid and the actual service fluid, respectively.

For flow through a straight pipe, the preceding expressions assume the following form:

$$G_2 = G_1 \sqrt{\frac{S_1 f_1}{S_2 f_2}} \quad (54a)$$

and

$$W_2 = W_1 \sqrt{\frac{S_2 f_1}{S_1 f_2}} \quad (55a)$$

EXAMPLE 2

Assume that fuel with a specific gravity $S = 0.785$ and kinematic viscosity $\nu = 1.06$ centistokes flows through the circuit represented by the resistance diagram in Fig. 8. The discharge pressure, p_1 of the centrifugal pump at A is represented graphically in Fig. 9. Also assume that the fuel flows through a fuel injector at B. The coefficient of energy loss of the injector assembly is $\epsilon_5 = 5.3$. The pressure, p_n, immediately downstream of the injector due to combustion is represented graphically in Fig. 10. The centrifugal pump is located 10 ft above the injector. The hydraulic line consists of a 180 in. long, 1.00 in. OD, 0.049 in. thick smooth pipe, three 90 deg elbows and one 45 deg elbow. A venturi meter is installed in the pipe line. The pressure loss through the venturi is 6.1 psi at a flow rate of 4.75 lb per sec. The required flow rate versus valve lift is shown in Fig. 10. It is desired to determine the required area of the throttling orifice of the flow control valve.

SOLUTION: The resistance of the various components of the circuit is calculated as follows:

Valve—The graph, Fig. 10, indicates that the required flow rate with the valve fully open at 0.800 in. lift is 4.75 lb per sec.

Pipe Line—From Eq (40a), the Reynolds number is:

$$R = 22{,}747.606 \times \frac{4.75}{1.06 \times 0.785 \times 0.902} = 143{,}962$$

and from Fig. 7 the coefficient of friction is $f_1 = 0.0167$:

From Eq (45), the coefficient of energy loss is:

$$\epsilon_1 = 0.0167 \times \frac{180}{0.902} = 3.33$$

and the hydraulic resistance of the pipe line is:

$$\epsilon_1 D_1^{-4} = 3.33 \times 0.902^{-4} = 5.04.$$

Venturi Meter—Since the pressure loss through the venturi meter measures 6.1 psi, at 4.75 lb per sec flow, its hydraulic resistance, Eq (43), is:

$$\epsilon_2 D_2^{-4} = 17.214 \times \frac{6.1 \times 0.785}{4.75^2} = 3.65$$

Fittings—The coefficients of energy loss for the 90 and 45 deg elbows (0.844 in. ID) are about 1.3 and 0.43 respectively. The hydraulic resistance for four fittings is therefore:

$$\epsilon_3 D_3^{-4} = (3 \times 1.3 + 0.43)\ 0.844^{-4} = 8.54$$

Fuel Injector—The fuel issues through n parallel bellmouth injector orifices of diameter D (coefficient of contraction = 1). These parallel orifices then can be converted into one equivalent which has the value:

$$D_n = \sqrt{\sum_{i=1}^{i=n} D_i^2}$$

It is also assumed that the equivalent orifice for this specific injector is $D_n = 0.627$ in., and that the coefficient of energy loss for the injector assembly is $\epsilon_5 = 5.3$. The hydraulic resistance of the injector assembly then is:

$$\epsilon_5 D_5^{-4} = 5.3 \times 0.627^{-4} = 34.29$$

Flow Control Valve—The coefficient of discharge for the type of flow control valve shown in Fig. 12 is about $C_{D4} = 0.82$, and from Eq (38), the coefficient of energy loss is:

$$\epsilon_4 = \frac{1}{0.82^2} = 1.49$$

The coefficient of discharge tends to increase slightly (to about $C_{D4} = 0.90$) as the lift decreases, because of the more favorable entrance condition at the lower lifts.

The required diameter D_4 of the throttling orifice can now be computed from Eqs (33) and (34).

Thus, from Eq (34), the resistance term is:

$$\sum^{1} \epsilon_i D_i^{-4} = 5.04 + 3.65 + 8.54 + 34.29 = 51.52$$

In accordance with Fig. 10, the rate of flow at 0.8 in. valve lift is $W = 4.75$ lb per sec, and the pressure immediately downstream of the fuel injector is $p_n = 350$ psi, abs.

From Fig. 9, the pump discharge pressure at this flow rate is $p_1 = 445$ psi, abs. Then, substituting in Eq (33),

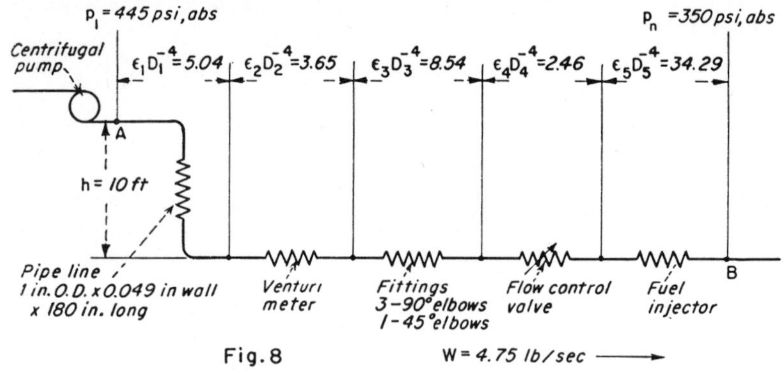

Fig. 8—Hydraulic resistance diagram for the fuel system of Example 2.

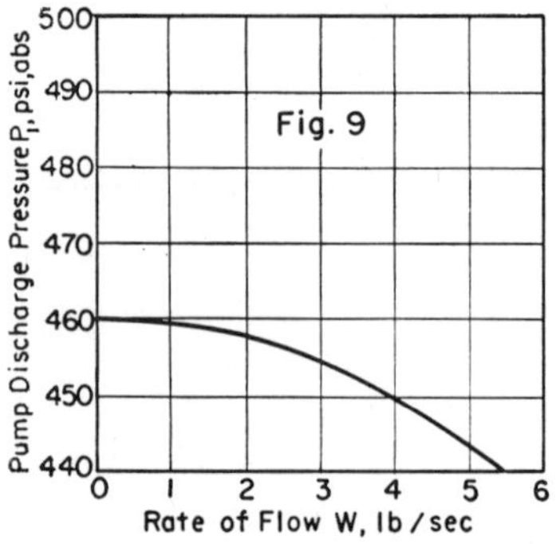

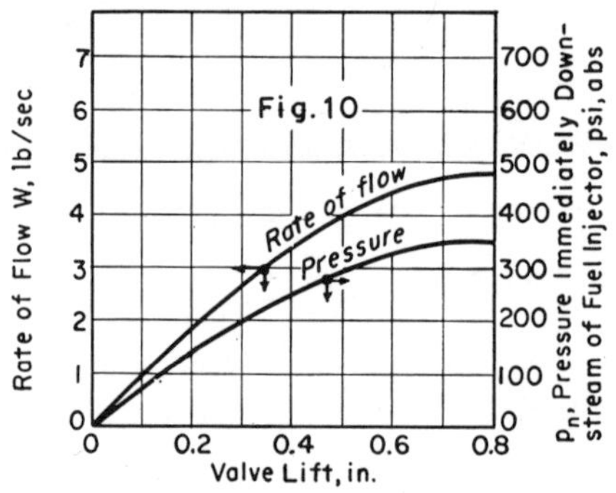

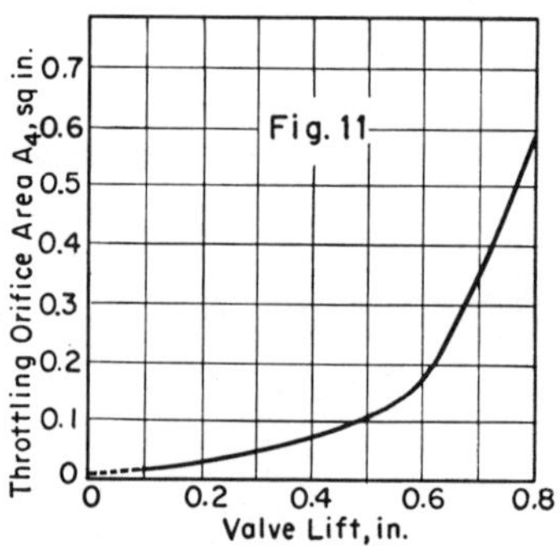

Fig. 9—Pump discharge pressure versus rate of flow.

Fig. 10—Rate of flow, and pressure immediately downstream of fuel injector, versus valve lift.

Fig. 11—Throttling orifice area versus valve lift.

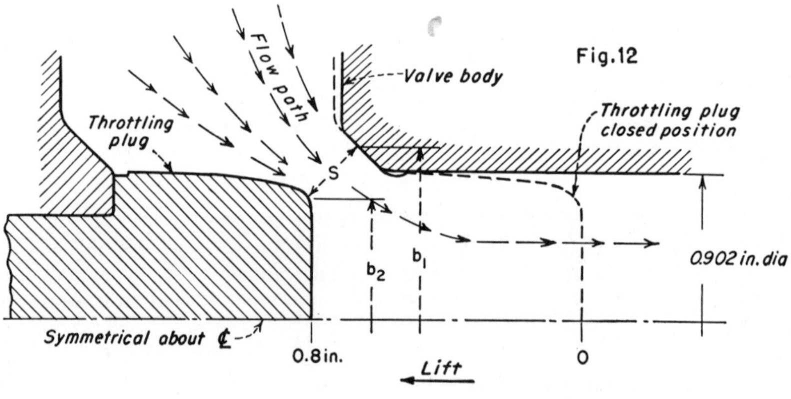

Fig. 12—Flow control valve in which flow rate is determined by an annular throttling device.

the diameter of the throttling orifice is:

$$D_4 = \sqrt[4]{\frac{1.49}{\frac{17.214[(445 - 350) + 0.4333 \times 0.785 \times 10]0.785}{4.75^2} - 51.52 - (0.627^{-4} - 0.902^{-4})}}$$

$$= \sqrt[4]{\frac{1.49}{58.94 - 51.52 - 4.96}} = 0.882 \text{ in.}$$

which converted into an equivalent flow area gives:

$$A_4 = \frac{\pi 0.882^2}{4} = 0.611 \text{ sq. in.}$$

As a check, the rate of flow can now be computed from Eqs (11) and (17).

The coefficient of energy loss of the flow control valve has previously been determined as $\epsilon_4 = 1.49$. The hydraulic resistance of valve is:

$\epsilon_4 D_4^{-4} = 1.49 \times 0.882^{-4} = 2.46$

By Eq. (11), the hydraulic resistance of the circuit including flow control valve is:

$$\sum_{i=1}^{i=n} \epsilon_i D_i^{-4} = 5.04 + 3.65 + 8.54 + 2.46 + 34.29 = 53.98$$

and substituting in Eq (17):

$$W = 4.149\sqrt{\frac{[(445 - 350) + 0.4333 \times 0.785 \times 10]0.785}{53.98 + (0.627^{-4} - 0.902^{-4})}}$$

$$= 4.149\sqrt{\frac{(95 + 3.40)0.785}{53.98 + 4.96}} = 4.75 \text{ lb per sec}$$

This is the required flow rate with the valve fully open at

0.800 in. lift.

In a similar manner, the required throttling orifice area can be computed for different valve lifts and flow rates. These data are represented graphically in Fig. 11. In the type of valve shown in Fig. 12, the flow is controlled by means of an annular throttling orifice. The rate of flow at any lift is governed by the smallest flow area developed. The flow area A (sq. in.) is given by the equation:

$$A = 1.5708 \times (d_1 + d_2)\, s$$

where s (*in.*) is the distance between the boundary surfaces; d_1 and d_2 the respective diameters (in.).

EXAMPLE 3 The pressure loss P_1 through a valve measures 23.5 psi at a flow rate W_1 of 3.5 lb per sec. The test fluid is water at 68 *F* (S_1 = 0.998, ν_1 = 1.011 centistokes). The diameter of the throttling orifice is 0.500 in. What is the pressure loss under actual operating conditions if the service fluid is JP-4 fuel at 120 *F* (S_2 = 0.762, ν_2 = 0.715 centistokes) and the flow rate is the same?

SOLUTION: In accordance with Eq (40a), the Reynolds number using water as test fluid is:

$$R_1 = 22747.606 \times \frac{3.50}{1.011 \times 0.998 \times 0.500} = 157{,}816$$

and using *JP*-4 fuel

$$R_2 = 22747.606 \times \frac{3.50}{0.715 \times 0.762 \times 0.500} = 292{,}262$$

Because of the high value of the Reynolds numbers, which is well in the turbulent region, the coefficient of energy loss for the valve is essentially the same when water or the actual service fluid is used, so that $\varepsilon_1 = \varepsilon_2$.

Then from Eq (50), the pressure loss through the valve under actual operating conditions using *JP*-4 fuel is:

$$P_2 = 23.5 \times \frac{0.998}{0.762} = 30.8 \text{ psi}$$

ESTABLISHMENT OF FLOW

When the velocity of the fluid is increased by opening a valve, pressure energy is dissipated in accelerating the mass of fluid in the pipe line. If the line is long and the valve opens rapidly, the energy dissipated in accelerating the fluid may be of considerable magnitude, and the flow rate will vary with time. For example, let Fig. 13 represent a pipe line in which flow is established by opening a valve. Then, from Newton's second law:

$$\text{Force} = \Delta P \frac{\pi D^2}{4} = m \frac{\Delta V}{\Delta T} \text{ (lb)} \tag{56}$$

Substituting for m the mass of the column of fluid of length L and expressing the velocity in terms of the rate of flow G (gpm) or W (lb per sec) as given by Eq (5):

$$\Delta P \frac{\pi D^2}{4} = \left(\frac{62.4\pi D^2 LS}{4 \times 12^3 g}\right) \frac{0.4085 D^{-2} \Delta G}{\Delta T}$$

$$= \left(\frac{62.4\pi D^2 LS}{4 \times 12^3 g}\right) \frac{2.938 S^{-1} D^{-2} \Delta W}{\Delta T}$$

The preceding expression resolves into:

$$\Delta P = 4.581 \times 10^{-4} \frac{SL}{D^2} \frac{\Delta G}{\Delta T} = 3.295 \times 10^{-3} \frac{L}{D^2} \frac{\Delta W}{\Delta T} \tag{57}$$

It is seen from Eq (57), that the pressure ΔP required to accelerate the fluid is directly proportional to the length of the pipe and to the rate of change in flow and is inversely proportional to the square of the inside diameter of the pipe.

Introducing the preceding acceleration terms in Eqs (16) and (17), the rate of flow at any instant during the opening transient of the valve is:

$$G = 29.843 \sqrt{\frac{(p_1 - p_n) + 0.4333 Sh - 4.581 \times 10^{-4} \frac{SL}{D^2} \frac{\Delta G}{\Delta T}}{[\Sigma + (D_n^{-4} - D_1^{-4})] S}} \tag{58}$$

and

$$W = 4.149 \sqrt{\frac{\left[(p_1 - p_n) + 0.4333 Sh - 3.295 \times 10^{-3} \frac{L}{D^2} \frac{\Delta W}{\Delta T}\right] S}{\Sigma + (D_n^{-4} - D_1^{-4})}} \tag{59}$$

Separation

The phenomenon of separation is responsible for many of the difficulties encountered in the design of hydraulic devices and is a frequent cause of differences between theory and experiment. When the pressure loss in a hydraulic device is unexpectedly high, the presence of separation may be suspected. Separation is, in addition, a source of instability because of the backflow along the boundary, the formation of eddies and consequent increase in turbulence. Separation in a fluid passage depends upon the shape of the solid boundaries, the velocity of the fluid and the pressure gradient.

Figs. 14 and 15 illustrate some typical cases of separation. In Fig. 14(A) the fluid approaches the sharp

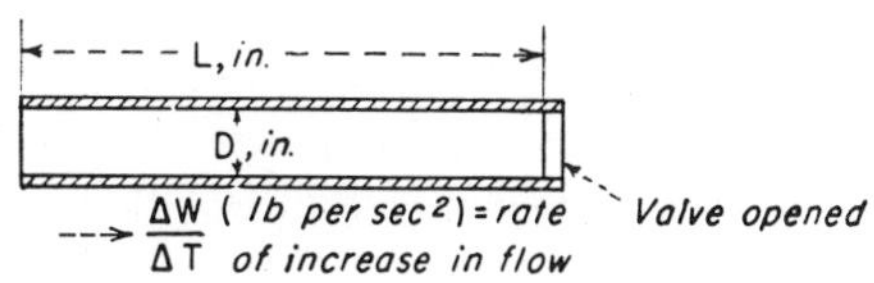

ΔT = *increment of time in operating valve, sec*

ΔV = *increase in velocity in time T, ft/sec*

Δp = *pressure dissipated in accelerating the mass of fluid, psi.*

m = *mass of fluid the velocity of which is changed, slugs*

D = *inside diameter of pipe, in.*

Fig. 13—Sketch and symbols for illustrating transient flow in pipe after opening of valve. Eqs (58) and (59) give flow rate.

entrance edge of the throttling orifice with high velocity. On the downstream side of the edge, the fluid separates from the boundary and formation of eddies develops. The formation of eddies and increased turbulence result in dissipation of energy. Energy thus dissipated into thermal energy cannot be recovered in useful form and manifests itself as an excessive pressure loss through the hydraulic device.

Separation can be minimized and often eliminated by streamlining the solid boundaries and making all edges around which the fluid is flowing well rounded as shown in Fig. 14(B).

Fig. 15 shows separation in the throat and in the diffuser cone of a venturi tube. In Fig. 15(A), separation is caused by a too large entrance cone angle. In the diverging section, Fig. 15(B), the pressure is increased and the velocity is decreased in the direction of flow. The pressure gradient is therefore positive; often referred to as an adverse gradient.

The layer of fluid near the wall moves with a relatively slow motion. At some downstream point A, the pressure becomes sufficiently high to overcome the momentum of this layer and separation occurs. Because of the adverse pressure gradient backward flow is established and eddies of considerable size may be formed.

Separation in the diffuser cone can usually be eliminated by decreasing the angle of divergence to about 6 to 10 degrees.

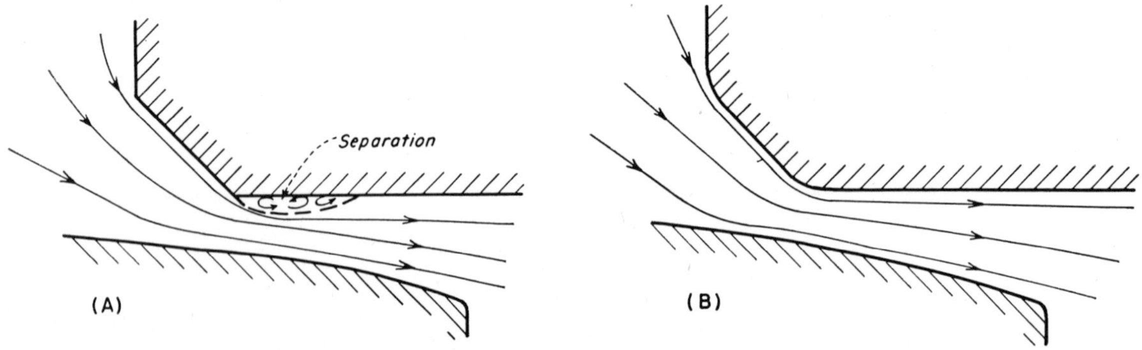

Fig. 14—Separation in throttling orifice is illustrated in (A) when sharp edges are present; (B) shows how rounded boundary eliminates separation.

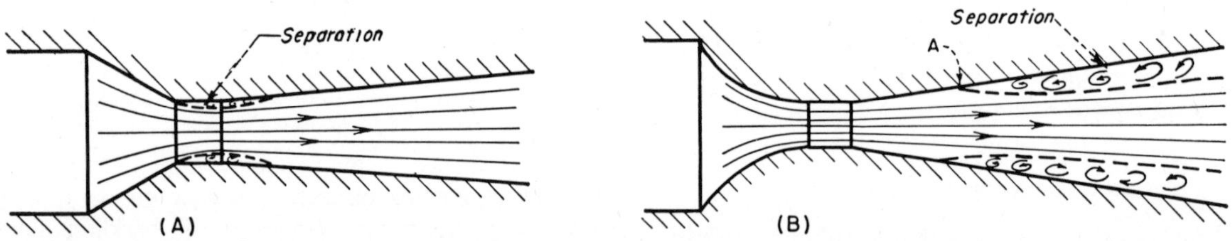

Fig. 15—Flow separation in venturi tubes. In (A) separation (see arrow) in the throat caused by too large entrance cone angle. In (B) separation (see arrow) in the diffuser cone caused by too large angle of divergence.

EXAMPLE 4 The rate of flow G_1 through a smooth straight pipe is 120 U. S. gpm and the pressure loss P_1 is 24.5 psi. The test fluid is water at 68 *F* ($S_1 = 0.998$, $\nu_1 = 1.011$ centistokes). The *ID* of the pipe is 0.902 in. What is the rate of flow under actual operating conditions if the pressure loss is the same and the service fluid is MIL-0-5606 hydraulic oil at 80 *F* ($S_2 = 0.838$, $\nu_2 = 18.5$ centistokes)?

SOLUTION: In accordance with Eq (40), the Reynolds number, using water as test fluid, is:

$$R_1 = 3162.556 \times \frac{120}{1.011 \times 0.902} = 416{,}161$$

and from Fig. 7, the coefficient of friction is $f_1 = 0.0136$.

However, the coefficient of friction f_2 when the service fluid is used is a function of the Reynolds number R_2 which in turn depends upon the rate of flow G_2, which is an unknown quantity. A direct solution to the problem is possible, although somewhat cumbersome. The problem, however, is readily solved by means of approximations:

As a first approximation in determining the coefficient of friction f_2, it can be assumed that the Reynolds number using the service fluid is approximately:

$$R_2' = R_1 \frac{\nu_1}{\nu_2} = 416{,}161 \times \frac{1.011}{18.5} = 22{,}742$$

and in accordance with Fig. 6, $f_2' = 0.0257$.

Then from Eq (54a):

$$G_2' = 120\sqrt{\frac{0.998 \times 0.0136}{0.838 \times 0.0257}} = 95.3 \text{ U.S. gpm}$$

The Reynolds number for this flow rate becomes:

$$R_2'' = 3162.556 \times \frac{95.3}{18.5 \times 0.902} = 18{,}061$$

and from Fig. 6, $f_2'' = 0.0273$.

It is therefore concluded that $f_2 > 0.0273 > 0.0257$. As

a final approximation it is assumed that $f_2 = 0.0275$. The rate of flow using the service fluid then becomes:

$$G_2 = 120\sqrt{\frac{0.998 \times 0.0136}{0.838 \times 0.0275}} = 92.1 \text{ U.S. gpm}$$

The Reynolds number for this flow rate is:

$$R_2 = 3162.556 \times \frac{92.1}{18.5 \times 0.902} = 17{,}455$$

and from Fig. 6, $f_2 = 0.0275$, which is in agreement with the previous assumption.

The following direct solution to the problem is also of interest:

Inspection of Eq (53), which gives the relationship between the Reynolds number and the kinematic viscosity, indicates that the Reynolds number when using the service fluid will be less than 100,000 because of its relatively high viscosity.

In accordance with Eq (47), the value of the coefficient of friction for Reynolds numbers up to about 100,000 is:

$$f_2 = \frac{0.316}{\sqrt[4]{R_2}}$$

Substituting the value of R_2 as given by Eq (40), the preceding expression resolves to:

$$f_2 = \frac{0.316}{\sqrt[4]{3162.556}}\left(\frac{\nu_2 D}{G_2}\right)^{0.25} = 4.214 \times 10^{-2}\left(\frac{\nu_2 D}{G_2}\right)^{0.25}$$

Combining Eq (54a) and the above equation:

$$G_2 = G_1\left[\frac{S_1 f_1}{4.214 \times 10^{-2}\left(\frac{\nu_2 D}{G_2}\right)^{0.25} S_2}\right]^{0.5}$$

The preceding equation can now be solved for G_2; hence:

$$G_2 = 6.108 G_1^{1.143}\left(\frac{S_1 f_1}{S_2}\right)^{0.571}\left(\frac{1}{\nu_2 D}\right)^{0.143}$$

This equation provides a direct means for computing the rate of flow of the service fluid. It is, however, to be noted that this equation is valid only when the flow of the service fluid is turbulent, and when the Reynolds number is less than 100,000. Eqs (49a) to (55a), inclusive, are valid for either laminar or turbulent flow and for any value of Reynolds number.

Substituting numerical values in the above equation:

$$G_2 = 6.108 \times 120^{1.143}\left(\frac{0.998 \times 0.0136}{0.838}\right)^{0.571}\left(\frac{1}{18.5 \times 0.902}\right)^{0.143} = 92.1 \text{ U.S. gpm}$$

This checks with the value previously computed by the approximation method.

Cavitation

Cavitation occurs in a hydraulic device wherever the static pressure is reduced to the vapor pressure of the liquid. The liquid then will vaporize and a cavity filled with vapor will form. Cavitation results in vibration, inefficiency, and/or loss in performance. In addition, cavitation is extremely damaging to the material of the hydraulic device. As the cavities are carried by the fluid into regions of higher pressure, the cavities abruptly collapse and the vapor changes back into the liquid phase. This abrupt collapse of the cavities is accompanied by sudden shock forces of exceedingly high magnitude, causing erosion or pitting of the metal surfaces and eventual failure by fatigue.

If the downstream pressure of a venturi tube is gradually reduced while the inlet pressure is maintained constant, the rate of flow and consequently the velocity of the fluid will increase. According to Bernoulli, the pressure in a system decreases with increasing velocity. Therefore, as the velocity at the throat increases, the throat pressure will decrease until it equals the vapor pressure of the liquid. The liquid then will vaporize and cavitation occurs. After cavitation has been established, the rate of flow through the venturi will remain constant regardless of further decrease in the downstream pressure.

This characteristic of the venturi tube is used in the design of hydraulic control circuits to limit the flow to a certain maximum value if the downstream pressure suddenly should diminish; for example, a result of combustion failure in a rocket circuit.

Fig. 16 shows a cavitating venturi and the pressure distribution along the flow path. The curves of pressure gradient represent conditions of constant inlet pressure, constant flow rate and diminishing downstream pressure. As the downstream pressure is gradually reduced, the zone of cavitation will lengthen until the vapor pocket surrounding the high velocity jet extends from the throat for some distance downstream into the diffuser cone. The intensity of cavitation therefore depends upon the magnitude of the downstream pressure. The end of the visible vapor pocket marks the point where cavitation is suppressed. At this point the pressure abruptly begins to rise; for example, as shown at point A.

The pressure loss, $p_1 - p_3$, through the venturi in Fig. 16 is approximately 15 per cent of the pressure differential, $p_1 - p_2$, between the inlet and throat, depending upon the surface finish of the boundary surface. A rough finish will increase the pressure loss considerably. The total pressure loss through the diffuser is determined by the friction, which depends upon the finish of the boundary surface and the length of the diffuser cone; and the loss from vortex motion which depends upon the angle of divergence. A certain degree of vortex motion is always present in a diffuser without separation. However, the intensity of the vortex motion and the dissipation of energy increase rapidly with increasing angle of divergence.

The friction loss in the diffuser can be reduced by increasing the cone angle α, thus shortening the cone. However, the increase in cone angle results in additional dissipation of energy, because of increased vortex motion. It is therefore to be expected that the sum of the friction loss and the loss from vortex motion becomes a minimum for a certain angle of divergence. Experiments indicate

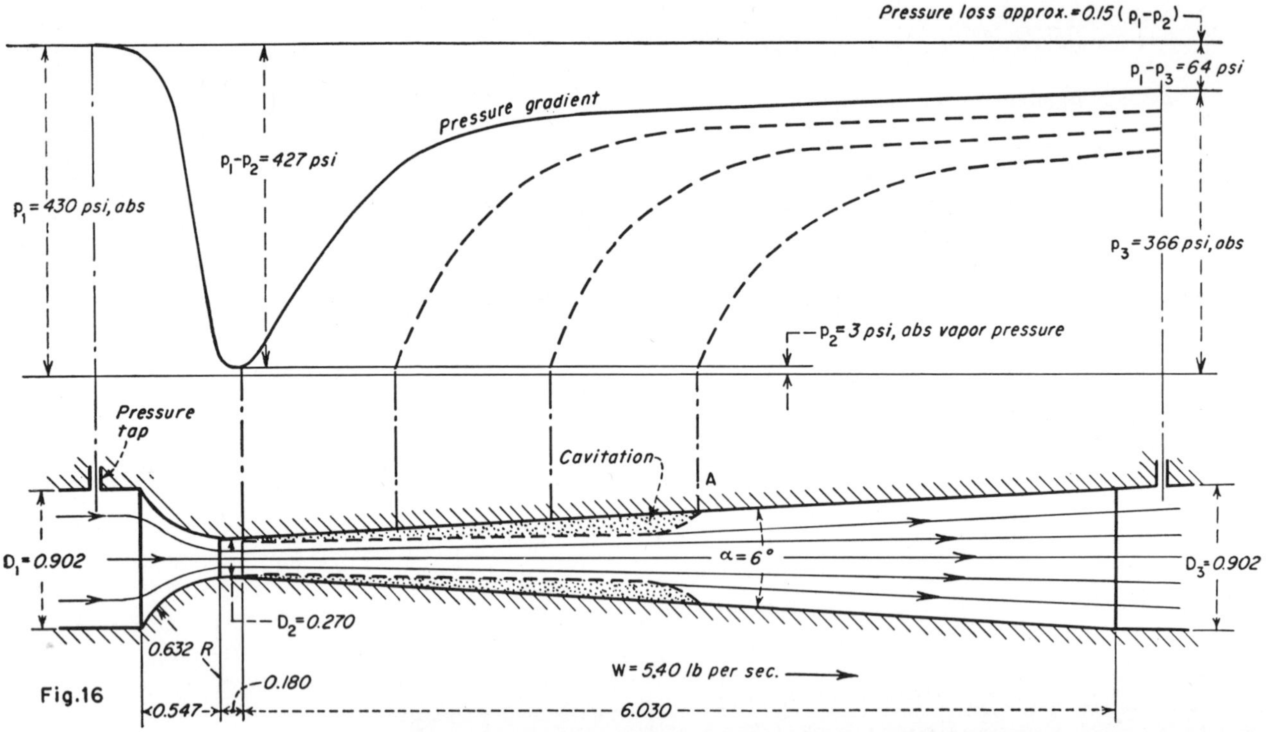

Fig. 16—Cavitating venturi and pressure distribution along the flow path.

Fig. 17—Coefficient of energy loss for diffuser cone.

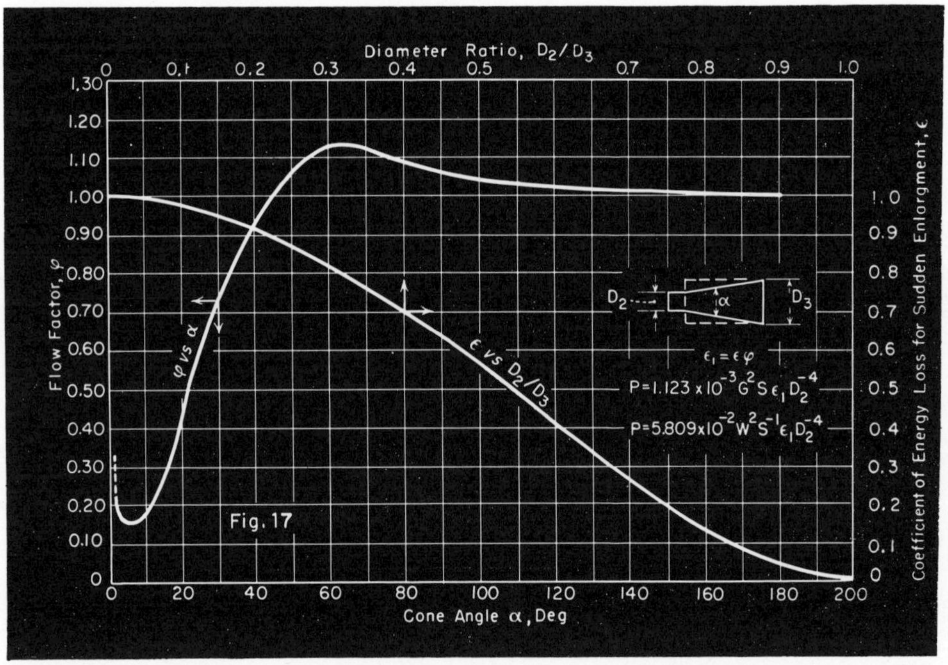

that in a diffuser with smoothly finished boundary surface the pressure loss becomes a minimum for an angle of divergence α of 6 degrees. If the finish is rougher, the total loss will be greater. The minimum loss then would occur at a slightly larger angle from 7 to 8 degrees.

The coefficient of energy loss ε_1 for a smoothly finished diffuser may be determined from Fig. 17. The coefficient ε_1 is the product of the coefficient of energy loss ε for abrupt expansion and a factor ϕ which is a function of the angle of divergence α. For example, if the diameter ratio D_2/D_3 is 0.45 and the angle of divergence α is 15 deg, then the coefficient of energy loss is $\varepsilon_1 = \varepsilon\phi = 0.63 \times 0.26 = 0.16$. With this coefficient, the pressure loss through the diffuser can be computed by Eqs (6) and (7). The pressure loss between the inlet and throat of a venturi is about 3 to 6 per cent of the velocity head at the throat, depending on shape and finish of the entrance.

Derivation of Design Equations

The rate of flow at which cavitation occurs in a venturi can be derived from Eqs (16) and (17).

If in Fig. 16, p_1, D_1 and p_2, D_2 denote pressures and diameters at the inlet and throat, and the potential head $h = O$ (i.e., the venturi is horizontal), then in accordance with Eq (16), the rate of flow through the venturi is:

$$G = 29.843\sqrt{\frac{p_1 - p_2}{[\varepsilon D_2^{-4} + (D_2^{-4} - D_1^{-4})]S}} \qquad (60)$$

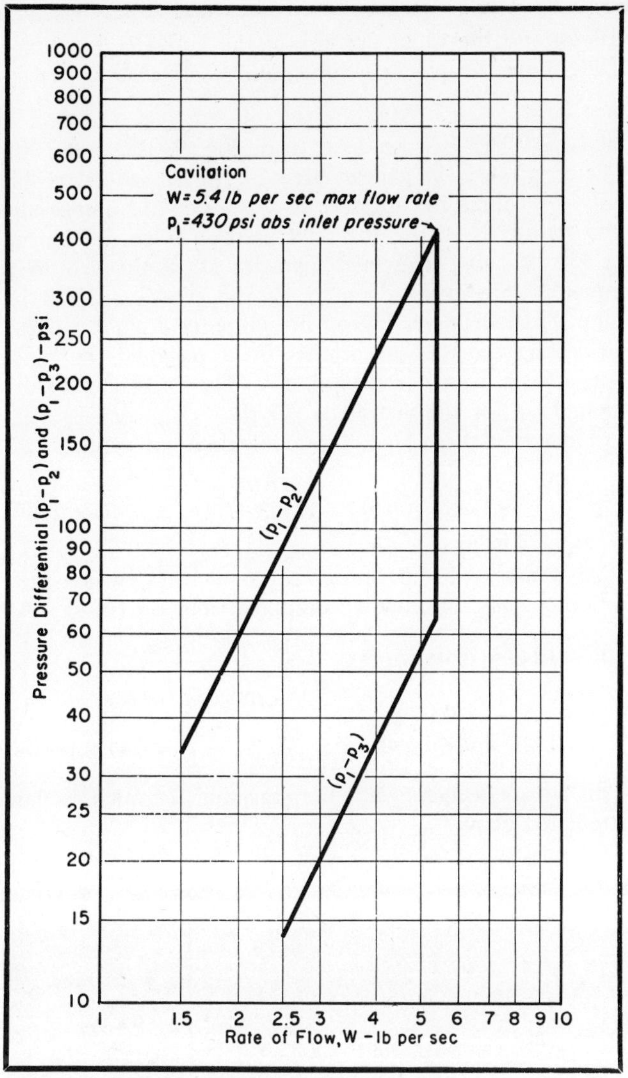

Fig. 18—Hydraulic characteristics of venturi shown in Fig. 16.

In this equation, the resistance term εD_2^{-4} considers the energy dissipated between the inlet and throat. However, in venturi analyses it is convenient to eliminate this term by introducing a coefficient of discharge C_D which then corrects for the dissipation of energy. Eq (60) then takes the form:

$$G = 29.843 C_D \sqrt{\frac{p_1 - p_2}{(D_2^{-4} - D_1^{-4})S}}$$

which can be written

$$G = 29.843 C_D D^2_1 \sqrt{\frac{p_1 - p_2}{\left[\left(\frac{D_1}{D_2}\right)^4 - 1\right]S}} \tag{60a}$$

From Eq (17):

$$W = 4.149 C_D D_1^2 \sqrt{\frac{(p_1 - p_2)S}{(D_1/D_2)^4 - 1}} \tag{60b}$$

The rate of flow G or W at which cavitation occurs can now be computed from Eqs (60a) or (60b) by letting p_2 equal the vapor pressure of the liquid.

The equations for the throat diameter of the venturi can be obtained by solving Eqs (60a) and (60b) tor D_2; hence:

$$D_2 = D_1 \sqrt[4]{\frac{G^2S}{890.608 C_D^2 D_1^4 (p_1 - p_2) + G^2S}} \tag{62}$$

and

$$D_2 = D_1 \sqrt[4]{\frac{W^2}{17.214 C_D^2 D_1^4 (p_1 - p_2)S + W^2}} \tag{62a}$$

The equations for the pressure differential between the inlet and the throat become:

$$p_1 - p_2 = 1.123 \times 10^{-3} \frac{G^2S}{D_1^4}\left[\left(\frac{D_1}{D_2}\right)^4 - 1\right]\frac{1}{C_D^2} \tag{63}$$

and

$$p_1 - p_2 = 5.809 \times 10^{-2} \frac{W^2}{D_1^4 S}\left[\left(\frac{D_1}{D_2}\right)^4 - 1\right]\frac{1}{C_D^2} \tag{63a}$$

In Fig. 16 is shown the physical configuration of the venturi of Example 6 and the pressure gradient. The entrance is well rounded to prevent possible separation in the throat under normal flow conditions due to the high fluid velocity. For example, the velocity in the throat is 244 ft per sec at 4.75 lb per sec normal flow and increases to a maximum of 277 ft per sec at a flow rate of 5.40 lb per sec when cavitation occurs. To reduce the pressure loss through the venturi to a minimum, the angle of divergence α of the diffuser cone should be 6 deg as indicated by Fig. 17, and the boundary surface should be smoothly finished. Because of the small cone angle, separation will not occur in the diffuser under normal operating conditions. A diffuser cone angle in excess of 9 to 10 deg is not advisable for this type of venturi as separation may develop because of an adverse pressure gradient. The graph, Fig. 18, represents the pressure differential, $p_1 - p_2$, between the inlet and the throat and the pressure loss $p_1 - p_3$, versus rate of flow. The pressure loss through the venturi is about 15 per cent of the pressure differential between inlet and throat. A pressure loss of 50 psi must therefore be tolerated under the normal flow condition of 4.75 lb per sec.

When the venturi tube is used as a flow rate meter and not as a flow limiting device, the pressure loss can be reduced by increasing the throat diameter, thus reducing the fluid velocity in the throat. The pressure differential between the inlet and throat as well as the pressure loss through the venturi then will be greatly reduced. The pressure loss through the venturi meter in Example 2, for example, is only 6.1 psi at 4.75 lb per sec rate of flow as compared to 50 psi pressure loss through the flow limiting venturi in this example.

Venturi meters often are designed with a conically shaped entrance which blends smoothly into the throat. The included angle of the entrance cone is about 20 degrees. The coefficient of discharge of the venturi meter is a function of the Reynolds number at the throat and the ratio of the throat to inlet diameter. Results of flow tests conducted with cast iron venturi tubes have been published by ASME. Fig. 19 is based on such data and

EXAMPLE 5 If in Fig. 10, of Example 2, the increase in flow rate ΔW of $4.40 - 3.95 = 0.45$ lb per sec from 0.5 in. to 0.6 in. lift occurs over a time ΔT of 0.05 sec, the pressure required to accelerate the column of fluid in the pipe line is in accordance with Eq (57):

$$\Delta P = 3.295 \times 10^{-3} \times \frac{180}{0.902^2} \times \frac{0.45}{0.05} = 6.57 \text{ psi}$$

The flow rate at 0.6 in. lift therefore will be slightly less than the required 4.40 lb per sec. The pressure of 6.57 psi required to accelerate the fluid, however, is small compared to the total available differential pressure, or about 4 per cent. Since the rate of flow is proportional to the square root of the pressure differential across the circuit, this discrepancy becomes of negligible value. However, it is readily seen from Eqs (58) and (59) that if the valve opens rapidly, (the rate of change in flow rate is large), the required acceleration pressure becomes significant. This will also be the case with long pipe lines.

EXAMPLE 6 Assume that in the hydraulic circuit, Fig. 8, the downstream pressure abruptly diminishes due to combustion failure, thus causing a sudden increase in flow. It is desired to limit the flow to a maximum value of 5.40 lb per sec by installing a cavitating venturi in the circuit. The pressure loss between the pump and the venturi is 10 psi and the vapor pressure of the liquid is 3 psi, abs. Determine the physical dimensions of the venturi.

SOLUTION: In accordance with Fig. 9, the pump discharge pressure at a flow rate of 5.40 lb per sec is 440 psi, abs. Since the pressure loss between the pump and the venturi is 10 psi, the inlet pressure p_1 to the venturi is 430 psi, abs. The inlet diameter D_1 of the venturi is equal to the inside diameter of the pipe, or 0.902 in. The throat diameter D_2 can be computed from Eq (62a) by letting the pressure at the throat p_2 equal the vapor pressure of the fuel, or 3 psi, abs. The coefficient of discharge C_D for the venturi is 0.970.

Then from Eq (62a), the required throat diameter is:

$$D_2 = 0.902 \sqrt[4]{\frac{5.40^2}{17.214 \times 0.970^2 \times 0.902^4 (430 - 3) 0.785 + 5.40^2}}$$

$$= 0.270 \text{ in.}$$

As a check, the maximum rate of flow (at which cavitation occurs) can now be computed from Eq (60b):

$$W = 4.149 \times 0.970 \times 0.902^2 \sqrt{\frac{(430 - 3) 0.785}{\left(\frac{0.902}{0.270}\right)^4 - 1}}$$

$$= 4.149 \times 0.970 \times 0.902^2 \sqrt{\frac{427 \times 0.785}{124.558 - 1}} = 5.40 \text{ lb per sec}$$

This is in agreement with the maximum permissible flow stipulated above.

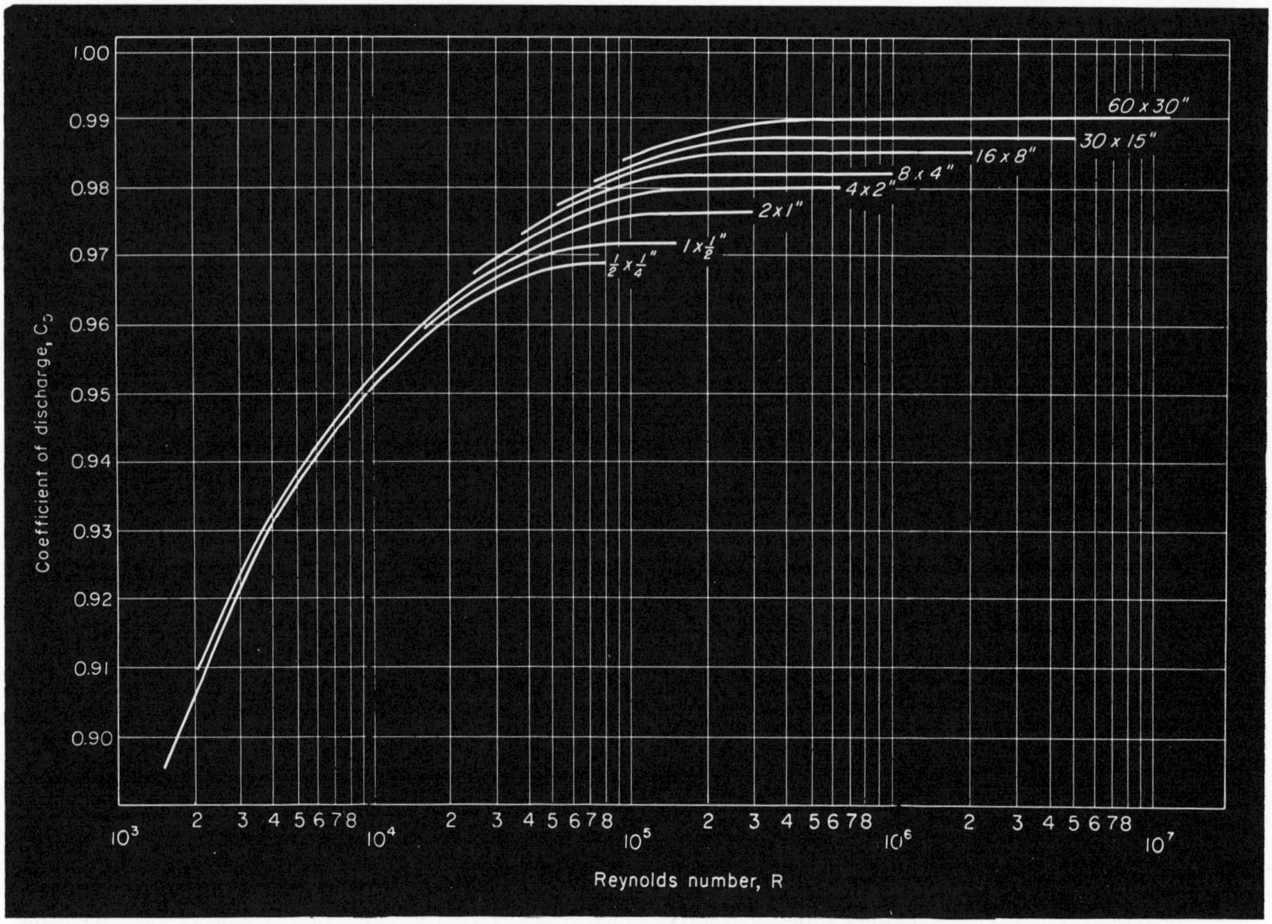

Fig. 19—Coefficients of discharge versus Reynolds number for venturi tubes with throat to inlet diameter ratios of 0.50.

gives the coefficient of discharge C_D, versus Reynolds number, R, at the throat for venturi tubes with a throat to inlet diameter ratio of 0.5. The coefficient of discharge C_β, for other diameter ratios may be computed from Professor Pardoe's equation:

$$C_\beta = \left[\frac{1 - \beta^4}{(1 - \beta^4) + \left(\frac{0.9375}{C_D^2} - 0.9375 \right)} \right]^{1/2} \quad (64)$$

where β is the ratio of the throat to the inlet diameter, and C_D is the coefficient of discharge for a venturi tube of throat to inlet diameter ratio of 0.5 and having the same diameter of throat as obtained from Fig. 19. For example, if from Fig. 19, the coefficient of discharge, C_D, is 0.962 then from chart, Fig. 20, the coefficient of discharge, C_β, is 0.959 for a venturi tube with the same throat diameter, but with a throat to inlet diameter ratio, β, of 0.6.

The coefficient of discharge is also affected by boundary roughness. Since the values of C_D given in Fig. 19 are for venturi tubes with machined throat but with unfinished entrance cone and diffuser, C_D for a venturi tube with smoothly finished boundary surface will be slightly higher because of the smaller frictional resistance between inlet and throat. However, the values given in the chart probably will be accurate within one percent. When the venturi has been in service some time, the boundary surface may become rough and the coefficient of discharge will decrease somewhat.

The character of the flow entering the venturi tube also will affect its hydraulic characteristics. Considerable disturbance may be caused to the flow pattern by hydraulic devices, bends, fittings, etc. on the upstream side of the venturi. It is therefore of importance that the venturi be preceded by a sufficient length of straight pipe to permit normal flow pattern to be re-established before the flow reaches the upstream pressure tap. In some instances 100 or more diameters of straight pipe may be required, depending on the magnitude of the disturbance caused to the flow. A spiral flow, for example, will increase the pressure at the wall due to the centrifugal effect. The measured pressure differential between the inlet and the throat therefore will be too low and calibration of the venturi will show a too high value of C_D, in some instances exceeding unity. Spiral flow may be eliminated by installing straightening vanes upstream of the venturi meter.

When great accuracy is required it is always advisable to calibrate the venturi meter under laboratory conditions identical to, or simulating, the actual installation.

Fig. 20—Variation of venturi tube coefficients with diameter ratio; the throat diameter and Reynolds number being constant.

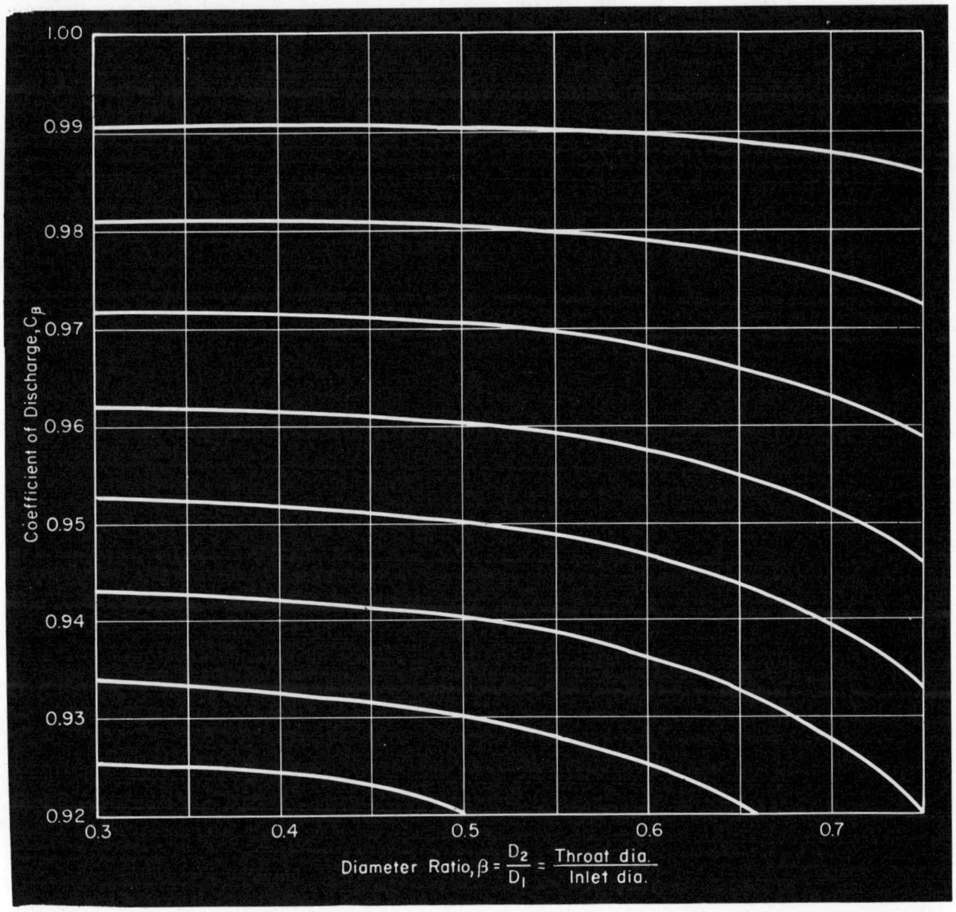

One-pump hydraulic systems may save money

Split the flow not the pump

A ten-step technique for selecting and sizing a multiple-output pump to energize several circuits of different functions

WES MASTER, Vice President, Engineering, The Dynex Company, Pewaukee, Wis;
FRANK YEAPLE, Associate Editor

FIRST, overcome the tradition that you must provide a separate pump for every function—or if you pick a single central pump that it must drive everything all at once. The concept of split flow, where a single pump has multiple outlets serving different functions, can reduce the pump size and the sizes of the driven motors and cylinders.

Then, follow good design practice. The key to successful multiple-circuit design is strict adherence to some very simple rules. Each rule in itself is obvious, but the sequence is critical, and you must give full attention to each step. These are the rules:

1. Determine all **output requirements** (speed, torque, force, and power, for each function, through the full range).
2. Categorize them according to **type of motion or force** (linear, rotary, vibratory, and holding).
3. Determine the total number of **simultaneous functions** (maximum flow and power to be expected, in each branch and for the system).
4. Determine the number of **series non-simultaneous functions** or groups of functions (taking hydraulic power off the same branch but not all at the same time).
5. Determine the number of **parallel simultaneous functions** (functions or groups of functions occurring at the same time but in different branches).
6. **Establish priority** (pick the best locations in the circuit for each function, so that the lowest priority functions are at the tail ends of the branch and the critical functions are at the head of the branch, preferably not in the same branch as another critical function).
7. **Size the piping and specify the valving** for the system and for each branch.
8. Add up the **simultaneous power requirements** to help determine the needed power for the prime mover.
9. Finish the **circuit layout** and specify the central hydraulic system.
10. **Select a prime mover.**

The method might be obvious, but its execution is difficult and demands a fair amount of intuition and compromise on your part. For example, how do you predict a range of speed, torque, force, and power for the shovel drive on a bulldozer? The fact is you'll have to guess, or consult with someone who has done it. But you cannot proceed to steps 2-9 until you've put down numbers for every function in step 1.

Step 2 is the easiest—unless you can't decide whether to use a linear actuator or a fluid motor. But this step is only a check on step 1, and doesn't add much new information.

Steps 3 through 6 are pure LOGIC and you can even use the AND-OR-NOT concepts of transistorized circuits to find the optimum combinations. (See *PE* reprint "Binary Logic," $2 from Reader Service Manager. See "Switching Circuits." *PE*—Aug 2 '65 p 65.)

For example, a road machine might use 8 functions in moving forward during a work cycle, and a different 4 functions during standstill. The control would call for the 8, OR the 4, not both. And each of the 8 is a part of an AND: 1 AND 2 AND so forth.

Experience (yours or borrowed) is an important ingredient. In the example following (a 4-wheel bulldozer) the wheels will have outputs in 4 basic modes: forward straight, forward turn, backward straight, and backward turn. But what about turning on a bank, while pushing rocks ahead of the dozer? And what about turning with both wheels at full power in opposite directions in heavy clay soil? Or driving straight ahead at full speed on concrete? It is in these areas the compromises must be made.

You should set the approximate power levels; then you can calculate the actual speeds and torques for any combination of problems. One method is to set the power level to the wheels on one side at 60% or more of the available engine power, assuming that the other wheels are idle. A more systematic approach, if wheel powers are known, is to work from the wheels (output) back to the prime mover.

The non-simultaneous functions can be added conveniently to any of the main branches. Because they will not operate at the same time, there is no extra drain on the available flow. Series-parallel valving can meter the flow to each auxiliary function separately or in parallel. If a split-flow pump is used, outlets can be assigned for any given auxiliary function, thus isolating the function.

An example

A rubber-tired tractor bulldozer, 4-wheel drive, will demonstrate the method nicely. (Allan Heinrich, proj-

SYMBOLS AND EQUATIONS

Basic circuit arrangements

SERIES FUNCTIONS

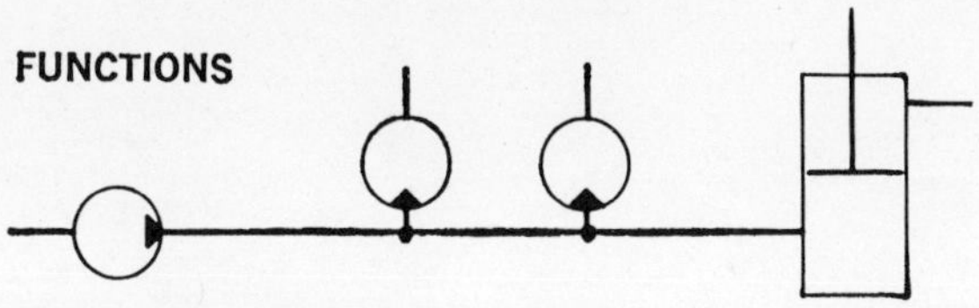

PARALLEL FUNCTIONS

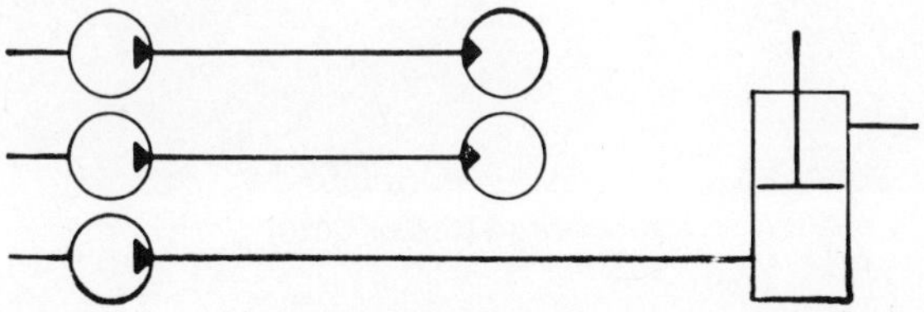

Symbols for schematics

SIMPLE PUMP

SIMPLE HYDRAULIC MOTOR

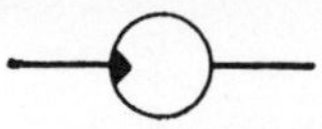

SPLIT-FLOW, fixed-displacement pump (inlet not shown)

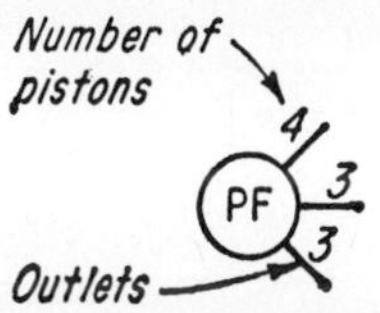

3-POSITION, 4-way, tandem center, manually operated directional valve

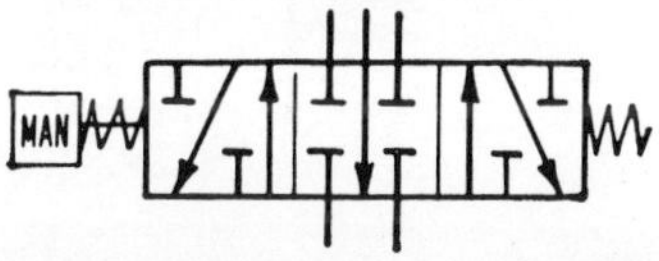

CHECK VALVE

TO TANK

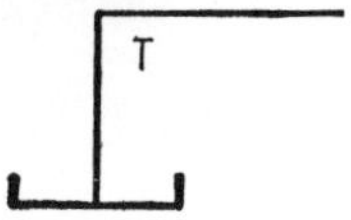

RELIEF VALVE

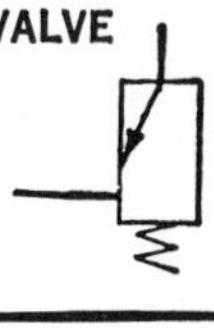

SPRING RETURN

Power and torque relationships

Power: $hp = TN/63{,}000$

$hp = QP/1714$

Where: T = torque, in.-lb

N = speed, rpm

P = pressure, psi

Q = flow, gpm

ect engineer at Dynex, prepared this example from typical data. Photo is of typical design.) Let's assume that it has these functions:

1. Propel vehicle
2. Operate dozer
3. Drive winch

The *propel* function is to be handled by an hydraulic motor on each wheel. Performance of an individual wheel must be: maximum speed, 59.2 rpm; torque at maximum speed, 30,000 lb-in.; speed at maximum torque, 29.6 rpm; and maximum torque at low speed, 74,500 lb-in. Tractor speed must be adjustable in two ways: for overall forward or reverse motion; and for turning, where the outside wheels turn at a faster rate than the inside ones.

The *dozer* blade will be operated with two cylinders. Each must produce a maximum force of 10,000 lb, at a maximum speed of 10 in./sec.

The winch will be turned with a single hydraulic motor. Maximum line pull must be 20,000 lb, maximum lin speed 280 ft/min at 20,000 lb, maximum drum torque 200,000 lb-in., and maximum drum speed 53.5 rpm.

Normally, the *winch* will not have to operate at full speed while the vehicle is moving. Also, the winch motor and dozer cylinders will not have to operate together. The dozer and propel functions will be simultaneous.

STEP 1—Output requirements

Propel: Condition I (high speed) is where wheel speed = 59.2 rpm and torque = 30,000 lb-in.; condition II (full torque) is where wheel speed = 29.6 rpm and torque = 74,500 lb-in.; and condition III (full torque and full speed), although it doesn't occur in practice, is where speed = 59.2 rpm and torque = 74,500 lb-in. The motor must be able to handle all three.

Propel power can be calculated for the three conditions using the conventional torque-speed formula:

$$\text{hp} = \frac{TN}{63{,}000}$$

For condition I, hp = 30,000 × 59.2/63,000 = 28.2; for II, hp = 74,500 × 29.6/63,000 = 35; for III, hp = 74,500 × 59.2/63,000 = 70.

The wheel motors will be geared down to wheel speed. For this example, 3000-rpm motors of the step-variable type are selected. Each motor operates at either of two displacements. At maximum vehicle loads, the higher displacement is chosen to provide maximum torque (low speed); at light loads, where higher speed is desired, the lower displacement is chosen (reduced torque).

Motor specifications are: maximum displacement, 2.1 in.^{3} per revolution; rated pressure, 6000 psi; rated speed, 3000 rpm; power output at rated speed and pressure, 90.5 hp; and torque at rated pressure, 1900 lb-in.

Gear reduction between motor and wheel = (output torque required)/(input torque × gear reduction efficiency). Assume efficiency = 92%, and calculate gear reduction, *GR,* for maximum torque:

$$GR = \frac{74{,}500}{1900 \times 0.92} = 42.6:1$$

Therefore, maximum motor speed = 59.2 × 42.6 = 2520 rpm, and motor speed at full torque = 29.6 × 42.6 = 1260 rpm. Required oil flow for four motors (each with 2.1 in.^{3} displacement) at 1260 rpm = about 50 gpm (12.5 gpm/motor), including leakage allowance. Power output is about 35 hp for one motor, 140 hp for four.

Dozer: The cylinder specifications are: Maximum force = 10,000 lb per cylinder; and maximum speed = 10 in./sec. Assume 3500 psi maximum system operating pressure. Area A = 10,000/3500 = 2.86 in.^{2} (about a 2-in. bore). Based on a 2-in. bore, at 3200 psi, required flow to each cylinder = 3.141 × 10 × 60/231 = 8.15 gpm, or 16.3 gpm for both. Power to cylinders is computed with a conventional flow-pressure formula:

$$\begin{aligned} \text{hp} &= \text{gpm} \times \text{psi}/1714 \\ &= 16.3 \times 3200/1714 = 30.4 \text{ hp} \end{aligned}$$

Winch: The drum horsepower = TN/63,000 = 200,000 × 53.5/63,000 = 170 hp. Motor horsepower must be that much or greater.

The motor selected has these specifications: displacement = 6 in.^{3} per revolution; rated pressure = 6000 psi; rated speed = 2500 rpm; torque at rated pressure = 5500 lb-in.; and output at rated speed and pressure = 218 hp.

Gear reduction, *GR,* between the hydraulic motor and the winch drum, based on maximum motor torque, is:

$$GR = \frac{200{,}000}{5500 \times 0.92} = 39.5:1$$

Therefore, maximum motor speed = 53.5 × 39.5 = 2110 rpm, and approximately 57.2 gpm will be required.

STEP 2—Categorize outputs

Propel = rotary
Dozer = linear
Winch = rotary

Rubber-tired tractor-dozer

WHAT ARE SPLIT-FLOW AND STEP-VARIABLE?

A — SPLIT-FLOW PUMP

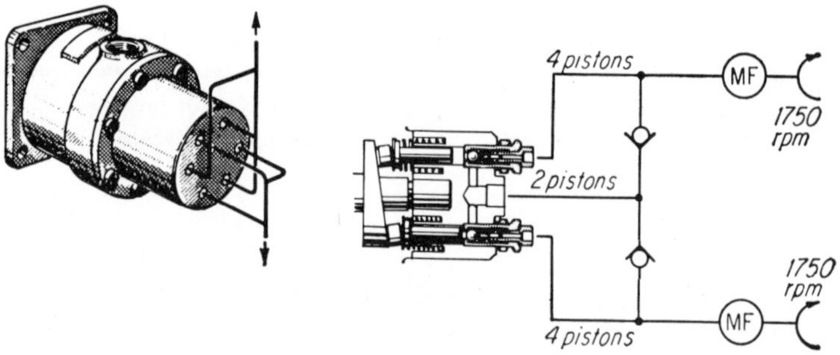

B — STEP-VARIABLE MOTOR

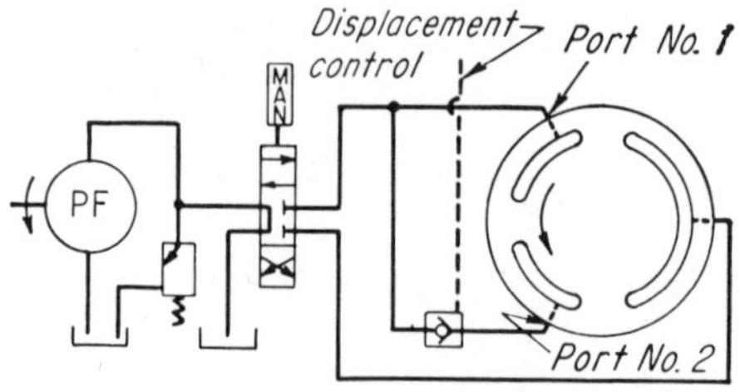

Split flow means that the individual pistons or groups of pistons are made to discharge into independent lines, free to develop different pressures depending on the loads. **Step variable** means that at least one pressure port of the motor has been divided in two, and by selection of valving the incoming fluid can enter one or both halves. If both halves are pressurized, fluid enters during full displacement of the motor (piston type) as it turns, and normal speed results. If only one half is pressurized, fluid enters during only half of the displacement and the motor has to run twice as fast to handle the fixed flow from the pump. Efficiency is less at partial displacement because the intake is blocked before the end of the stroke, but many applications can justify this to save a second pump or a speed changer.

STEP 3—Determine the total number of simultaneous functions

A- Propel motors and dozer cylinder; also B- Propel motors (at slow speed) and winch

For A- Maximum flow = 50 + 16.3 = 66.3 gpm

Maximum propel motor pressure = 6000 psi

Maximum dozer cylinder pressure = 3200 psi

STEP 4—Determine the number of series, non-simultaneous functions

Dozer, propel, and winch circuits

STEP 5—Determine the number of parallel simultaneous functions

Propel and dozer circuits

STEP 6—Establish priority

Propel and dozer have priority over the winch

STEP 7—Size the piping and valving

See the table for normal functions. Each of the valves incorporates additional functions: The step variable selector valve has a built-in check valve; the propel directional valve and winch directional valve have built-in relief valves and motor overload valves; the dozer directional valve has a built-in relief valve and a fourth position called "float." In float position, all ports (pressure, cylinder Port 1, cylinder Port 2, and tank) are interconnected, thus allowing the dozer blade to move up or down as the ground contour varies.

STEP 8—Add up the simultaneous power requirements

Horsepower for propel and dozer = 205.4 hp

Horsepower for winch = 200 hp

Because the propel-dozer functions do not operate at the same time as the winch, the prime mover power needs to provide only 205.4 hp.

STEP 9—Finish the circuit layouts, specify the central hydraulic system.

To provide the independent simultaneous flow to each of the four propel motors plus the dozer cylinder, choose two split-flow pumps (independent outlet ports). Each pump will be split into three independent flows. Two pumps rated for 33.15 gpm at 6000 psi will provide the oil.

When steering the vehicle, additional flow is required by the outside wheels. The circuit is designed to insure flow from three pump pistons to each motor. Four pistons from one split flow pump are connected through check valves to all four motors. In this way flow will go to the motors with the least resistance.

One of the objectives is to make use of all or part of the oil from the propel-dozer circuits for the winch circuit. So if the outlet series ports of the propel and dozer valves are connected into the winch circuit, the winch circuit is inoperative only when both the propel *and* the dozer are operating. But when only the propel function is in operation, the winch is able to operate slowly but at full torque.

At this point, an adjustment of the winch gear ratio, *GR*, might be desirable. In the example, we initially based the winch gear ratio on torque. Now, because we have a known gpm available for the winch motor (from propel and dozer circuits when not used), we can base the gear ratio on motor speed resulting from the available gpm. Flow from propel and dozer circuit = 66.3 gpm, winch motor speed = 2450 rpm, and required winch drum speed = 53.5 rpm. Thus, $GR = 2450/53.5 = 45.8 : 1$.

With the proposed circuit, the winch gear reduction should be increased from 39.5:1 to 45.8:1. Winch circuit pressure becomes 5100 instead of 6000 psi for the required drum torque. Winch tubing can be reduced to 0.219 in.

STEP 10—Select a prime mover.

Based on a maximum pump output horsepower of 205.4, an engine of approximately 230 hp is required.

PIPING AND VALVING (STEP 7)

A — Piping

BRANCH OF CIRCUIT	PROPEL MOTOR	DOZER CYLINDER	WINCH MOTOR
Max flow, gpm	12.5	16.3	57.2
Max pressure psi	6000	3200	6000
Tube size, in.	¾	¾	1½
Tube material, ASTM	4130	4130	4130
Tube wall in.	0.120	0.109	0.250

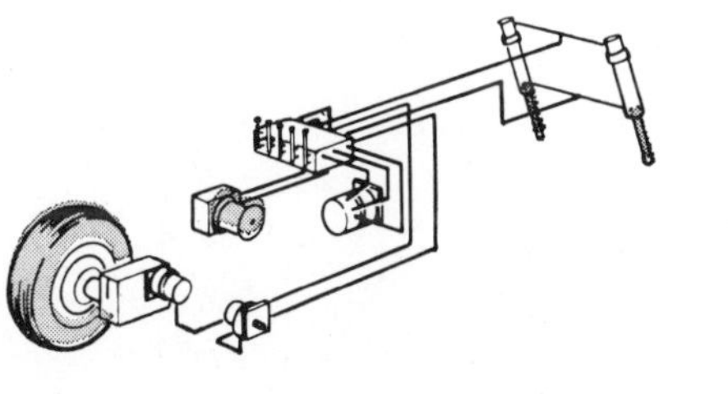

B — Valving

FUNCTION	TYPE OF VALVE
Step variable selector	3-way, 2 position
Propel directional	4-way, 3 position, tandem center
Winch directional	4-way, 3 position, tandem center
Dozer directional	4-way, 4 position

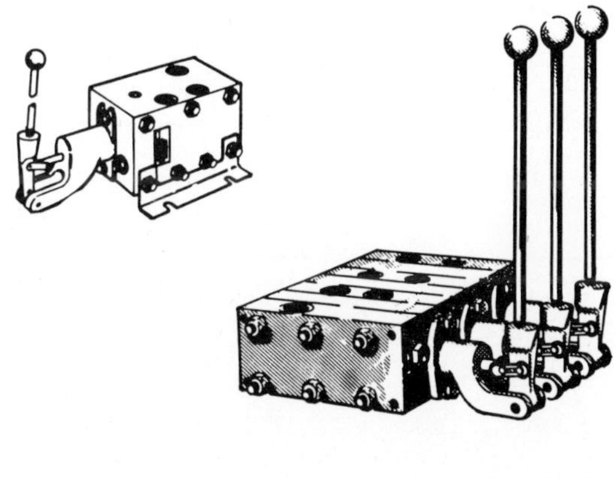

FINAL TRACTOR-DOZER CIRCUITS

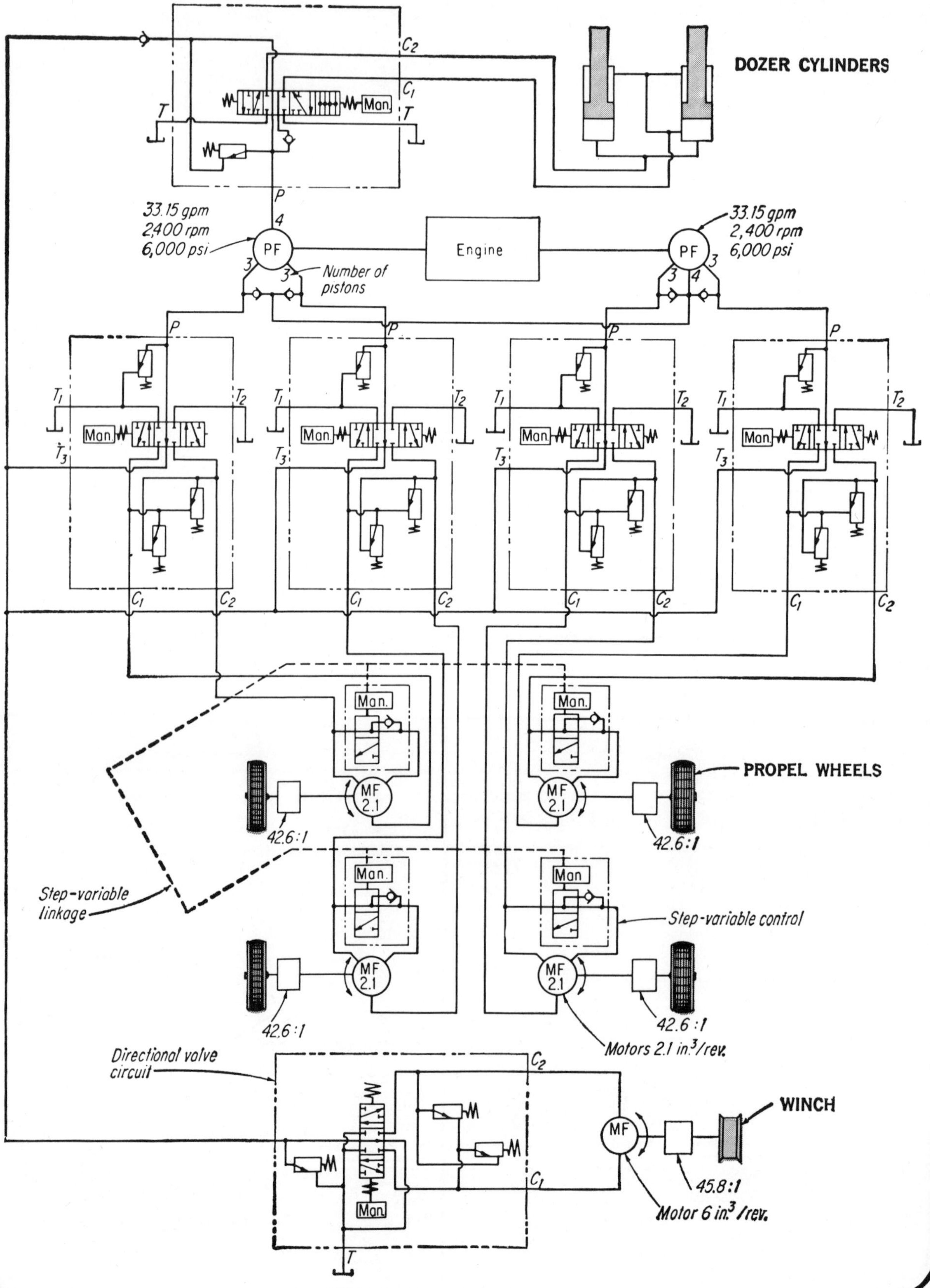

INDEX